Vineet Bafna   S. Cenk Sahinalp (Eds.)

Research in Computational Molecular Biology

15th Annual International Conference, RECOMB 2011
Vancouver, BC, Canada, March 28-31, 2011
Proceedings

Springer

Volume Editors

Vineet Bafna
University of California San Diego
EBU3b, #4218
9500 Gilman Drive
La Jolla, CA 92093-0404, USA
E-mail: vbafna@cs.ucsd.edu

S. Cenk Sahinalp
Simon Fraser University
School of Computing Science
8888 University Drive
Burnaby, BC, V5A 1S6, Canada
E-mail: cenk@cs.sfu.ca

ISSN 0302-9743   e-ISSN 1611-3349
ISBN 978-3-642-20035-9   ISBN 978-3-642-20036-6 (eBook)
DOI 10.1007/978-3-642-20036-6
Springer Heidelberg Dordrecht London New York

Library of Congress Control Number: 2011923333

CR Subject Classification (1998): J.3, H.2.8, I.2, F.2.2, I.6, F.2

LNCS Sublibrary: SL 8 – Bioinformatics

Typesetting: Camera-ready by author, data conversion by Scientific Publishing Services, Chennai, India

Printed on acid-free paper

Springer is part of Springer Science+Business Media (www.springer.com)

Lecture Notes in Bioinformatics 6577

Edited by S. Istrail, P. Pevzner, and M. Waterman

Subseries of Lecture Notes in Computer Science

Preface

This volume contains the papers presented at RECOMB 2011: the 15th Annual International Conference on Research in Computational Molecular Biology held in Vancouver, Canada, during March 28–31, 2011. The RECOMB conference series was started in 1997 by Sorin Istrail, Pavel Pevzner, and Michael Waterman. RECOMB 2011 was hosted by the Lab for Computational Biology, Simon Fraser University, and took place at the Fairmont Hotel Vancouver. This year, 43 papers were accepted for presentation out of 153 submissions. The papers presented were selected by the Program Committee (PC) assisted by a number of external reviewers. Each paper was reviewed by at least three members of the PC, or by external reviewers, and there was an extensive Web-based discussion over a period of two weeks, leading to the final decisions. Accepted papers were also invited for submission to a special issue of the *Journal of Computational Biology*. The Highlights track, first introduced during RECOMB 2010, was continued, and resulted in 53 submissions, of which 5 were selected for oral presentation.

In addition to the contributed talks, RECOMB 2011 featured keynote addresses on the broad theme of Next-Generation Sequencing, and Genomics. The keynote speakers were Evan Eichler (University of Washington), Daphne Koller (Stanford University), Elaine Mardis (Washington University, St. Louis), Marco Marra (BC Genome Sciences Centre), Karen Nelson (J. Craig Venter Institute), and Jun Wang (Beijing Genomics Institute). Finally, RECOMB 2011 featured a special industry panel on Next-Generation Sequencing and Applications, chaired by Inanc Birol.

RECOMB 2011 was made possible by the dedication and hard work of many individuals and organizations. We thank the PC and external reviewers who helped form a high-quality conference program, the Organizing Committee, coordinated by Cedric Chauve, the Organization and Finance Chair and Amanda Casorso, the event coordinator, for hosting the conference and providing the administrative, logistic, and financial support. We also thank our sponsors, including Genome BC, CIHR, SFU, NSERC, BC Cancer Foundation, PIMS, MITACS, Ion Torrent and Pacific Biosystems, as well as our Industrial Relations Chair, Inanc Birol and Sponsorship Chair Martin Ester. Without them the conference would not have been financially viable. We thank the RECOMB Steering Committee, chaired by Martin Vingron, for accepting the challenge of organizing this meeting in Vancouver. Finally, we thank all the authors who contributed papers and posters, as well as the attendees of the conference for their enthusiastic participation.

February 2011

Vineet Bafna
S. Cenk Sahinalp

Conference Organization

Program Committee

Tatsuya Akutsu	Kyoto University, Japan
Can Alkan	University of Washington, USA
Nancy Amato	Texas A&M University, USA
Rolf Backofen	Albert Ludwigs University Freiburg, Germany
Joel Bader	Johns Hopkins University, USA
Vineet Bafna (Chair)	University of California, San Diego, USA
Nuno Bandeira	University of California, San Diego, USA
Vikas Bansal	Scripps Translational Science Institute, USA
Ziv Bar-Joseph	Carnegie Mellon University, USA
Bonnie Berger	Massachusetts Institute of Technology, USA
Jadwiga Bienkowska	BiogenIdec, USA
Mathieu Blanchette	McGill University, Canada
Phil Bradley	Fred Hutchinson Cancer Research Center, USA
Michael Brent	Washington University in St. Louis, USA
Michael Brudno	University of Toronto, Canada
Lenore Cowen	Tufts University, USA
Colin Dewey	University of Wisconsin, Madison, USA
Dannie Durand	Carnegie Mellon University, USA
Eleazar Eskin	University of California, Los Angeles, USA
David Gifford	Massachusetts Institute of Technology, USA
Bjarni Halldorsson	Reykjavik University, Iceland
Eran Halperin	International Computer Science Institute, Berkeley, USA
Trey Ideker	University of California, San Diego, USA
Sorin Istrail	Brown University, USA
Tao Jiang	University of California, Riverside, USA
Simon Kasif	Boston University, USA
Mehmet Koyuturk	Case Western Reserve University, USA
Jens Lagergren	Royal Institute of Technology, Sweden
Thomas Lengauer	Max Planck Institute for Informatics, Germany
Michal Linial	The Hebrew University of Jerusalem, Israel
Satoru Miyano	University of Tokyo, Japan
Bernard Moret	EPFL, Switzerland
William Noble	University of Washington, USA
Lior Pachter	University of California, Berkeley, USA
Ron Pinter	Technion, Israel
Teresa Przytycka	National Institutes of Health, USA
Predrag Radivojac	Indiana University, USA
Ben Raphael	Brown University, USA

Knut Reinert	Freie Universität Berlin, Germany
Marie-France Sagot	INRIA, France
S. Cenk Sahinalp	Simon Fraser University, Canada
David Sankoff	University of Ottawa, Canada
Russell Schwartz	Carnegie Mellon University, USA
Eran Segal	Weizmann Institute of Science, Israel
Roded Sharan	Tel Aviv University, Israel
Adam Siepel	Cornell University, USA
Mona Singh	Princeton University, USA
Donna Slonim	Tufts University, USA
Terry Speed	Walter+Eliza Hall Institute, Australia
Haixu Tang	Indiana University, USA
Anna Tramontano	University of Rome, Italy
Alfonso Valencia	Spanish National Cancer Research Centre, Spain
Martin Vingron	Max Planck Institute for Molecular Genetics, Germany
Tandy Warnow	University of Texas, Austin, USA
Eric Xing	Carnegie Mellon University, USA
Jinbo Xu	Toyota Technology Institute, Chicago, USA
Michal Ziv-Ukelson	Ben Gurion University of the Negev, Israel

Steering Committee

Serafim Batzoglou	Stanford University, USA
Bonnie Berger	Massachusetts Institute of Technology, USA
Sorin Istrail	Brown University, USA
Michal Linial	The Hebrew University of Jerusalem, Israel
Martin Vingron (Chair)	Max Planck Institute for Molecular Genetics, Germany

Organizing Committee

Cedric Chauve (Finance Chair)	Simon Fraser University, Canada
Martin Ester (Sponsorship Chair)	Simon Fraser University, Canada
Inanc Birol (Industry Relations and Panels Chair)	BCGSC, Canada
Fiona Brinkman	Simon Fraser University, Canada
Jack Chen	Simon Fraser University, Canada
Artem Cherkasov	Prostate Centre, VGH, Canada
Colin Collins	Prostate Centre, VGH, Canada
Anne Condon	University of British Columbia, Canada
Steven Jones	BCGSC, Canada
Tamon Stephen	Simon Fraser University, Canada
Stas Volik	Prostate Centre, VGH, Canada
Wyeth Wasserman	University of British Columbia, Canada
Kay Wiese	Simon Fraser University, Canada

Previous RECOMB Meetings

Dates	Hosting Institution	Program Chair	Conference Chair
January 20-23, 1997 Santa Fe, NM, USA	Sandia National Lab	Michael Waterman	Sorin Istrail
March 22-25, 1998 New York, NY, USA	Mt. Sinai School of Medicine	Pavel Pevzner	Gary Benson
April 22-25, 1999 Lyon, France	INRIA	Sorin Istrail	Mireille Regnier
April 8-11, 2000 Tokyo, Japan	University of Tokyo	Ron Shamir	Satoru Miyano
April 22-25, 2001 Montreal, Canada	Université de Montreal	Thomas Lengauer	David Sankoff
April 18-21, 2002 Washington, DC, USA	Celera	Gene Myers	Sridhar Hannenhalli
April 10-13, 2003 Berlin, Germany	German Federal Ministry for Education and Research	Webb Miller	Martin Vingron
March 27-31, 2004 San Diego, USA	University of California, San Diego	Dan Gusfield	Philip E. Bourne
May 14-18, 2005 Boston, MA, USA	Broad Institute of MIT and Harvard	Satoru Miyano	Jill P. Mesirov and Simon Kasif
April 2-5, 2006 Venice, Italy	University of Padova	Alberto Apostolico	Concettina Guerra
April 21-25, 2007 San Francisco, CA, USA	QB3	Terry Speed	Sandrine Dudoit
March 30-April 2, 2008 Singapore, Singapore	National University of Singapore	Martin Vingron	Limsoon Wong
May 18-21, 2009 Tucson, AZ, USA	University of Arizona	Serafim Batzoglou	John Kececioglu
August 12-15, 2010 Lisbon, Portugal	INESC-ID and Instituto Superior Técnico	Bonnie Berger	Arlindo Oliveira

External Reviewers

Andreotti, Sandro
Arndt, Peter
Arvestad, Lars
Atias, Nir
Bebek, Gurkan
Benos, Panayiotis
Bercovici, Sivan
Bielow, Chris
Borgwardt, Karsten
Caglar, Mehmet Umut
Cai, Yizhi
Cakmak, Ali
Chalkidis, Georgios
Chen, Ken
Chitsaz, Hamid
Choi, Jeong-Hyeon
Choi, Sang Chul
Clark, Wyatt
Condon, Anne
Conrad, Tim

Costa, Fabrizio
Danko, Charles
Dao, Phuong
DasGupta, Bhaskar
DeConde, Robert
Del Pozo, Angela
Do, Chuong
Donmez, Nilgun
Dutkowski, Janusz
Edwards, Matthew
Ekenna, Chinwe
Elberfeld, Michael
Emde, Anne-Katrin
Erten, Sinan
Frank, Ari
Fujita, Andre
Furlanello, Cesare
Furlotte, Nicholas
Gitter, Anthony
Goldenberg, Anna

Gomez, Gonzalo
Gonzalez-Izarzugaza, Jose Maria
Gottlieb, Assaf
Gronau, Ilan
Haas, Stefan
Hajirasouliha, Iman
Han, Buhm
Hannenhalli, Sridhar
Hannum, Gregory
Hansen, Niels
He, Dan
He, Xin
Hildebrandt, Andreas
Hofree, Matan
Holtgrewe, Manuel
Hormozdiari, Fereydoun
Hosur, Raghavendra
Hower, Valerie
Huson, Daniel
Iqbal, Zam
Jacob, Arpith
Jiang, Rui
Jiao, Dazhi
Ji, Chao
Kamisetty, Hetunandan
Kang, Eun Yong
Kang, Hyun Min
Karakoc, Emre
Kayano, Mitsunori
Kelleher, Neil
Kim, Sangtae
Kim, Sun
Koller, Daphne
Krause, Roland
Kundaje, Anshul
Kuo, Dwight
Lacroix, Vincent
Lan, Alex
Lange, Sita
Lasserre, Julia
Lee, Byoungkoo
Lee, Seunghak
Le, Hai-Son
Lengauer, Thomas
Lenhof, Hans-Peter

Lin, Tien-ho
Li, Sujun
Liu, Xuejun
Liu, Yu
Li, Yong
Loh, Po-Ru
Lopez, Gonzalo
Lozano, Jose A.
Lugo-Martinez, Jose
Magger, Oded
Maier, Ezekial
Manavi, Kasra
Martins, Andre
Maticzka, Daniel
Mayampurath, Anoop
McPherson, Andrew
Medvedev, Paul
Melsted, Pall
Menconi, Giulia
Milo, Nimrod
Milo, Ron
Möhl, Mathias
Mongiovì, Misael
Murali, T.M.
Nath, Shuvra
Nibbe, Rod K.
Niida, Atsushi
Novembre, John
O'Donnell, Charles
Ofran, Yanay
Owen, Megan
Patel, Vishal
Pejavar, Vikas
Peterlongo, Pierre
Pevzner, Pavel
Pinhas, Tamar
Pisanti, Nadia
Platts, Adrian
Pop, Mihai
Rahman, Atif
Rausell, Antonio
Reeder, Christopher
Reynolds, Sheila
Ritz, Anna
Robert, Adam

Roch, Sebastien
Rolfe, Alex
Saglam, Mert
Salari, Raheleh
Schlicker, Andreas
Schulz, Marcel
Schulz, Marcel
Schweiger, Regev
Scornavacca, Celine
Sennblad, Bengt
Shibuya, Tetsuo
Shiraishi, Yuichi
Shlomi, Tomer
Shringarpure, Suyash
Sindi, Suzanne
Singer, Meromit
Singh, Ajit
Singh, Rohit
Siragusa, Enrico
Sjoestrand, Joel
Snir, Sagi
Sommer, Ingolf
Spencer, Sarah
Spivak, Marina
Srivas, Rohith
Steinhoff, Christine
Sul, Jae Hoon
Sutto, Ludovico
Szczurek, Ewa
Tamada, Yoshinori
Tanaseichuk, Olga

Tesler, Glenn
Thomas, Shawna
Thompson, James
Torda, Andrew
Ulitsky, Igor
Veksler-Lublinsky, Isana
Viksna, Juris
Volouev, Anton
Waldispühl, Jérôme
Wang, Jian
Whitney, Joseph
Will, Sebastian
Wu, Wei
Wu, Yufeng
Xie, Minzhu
Xin, Fuxiao
Yanover, Chen
Ye, Chun
Yeger-Lotem, Esti
Yeh, Hsin-Yi
Yooseph, Shibu
Yosef, Nir
Yu, Chuan-Yih
Zaitlen, Noah
Zakov, Shay
Zakov, Shay
Zhang, Shaojie
Zhang, Xiuwei
Zinmann, Guy
Zotenko, Elena

Table of Contents

Bacterial Community Reconstruction Using Compressed Sensing

Amnon Amir[1,*] and Or Zuk[2,*]

[1] Department of Physics of Complex Systems,
Weizmann Institute of Science, Rehovot, Israel
[2] Broad Institute of MIT and Harvard, Cambridge, Massachusetts, USA
amnon.amir@weizmann.ac.il, orzuk@broadinstitute.org

Abstract. Bacteria are the unseen majority on our planet, with millions of species and comprising most of the living protoplasm. We propose a novel approach for reconstruction of the composition of an unknown mixture of bacteria using a single Sanger-sequencing reaction of the mixture. Our method is based on compressive sensing theory, which deals with reconstruction of a sparse signal using a small number of measurements. Utilizing the fact that in many cases each bacterial community is comprised of a small subset of all known bacterial species, we show the feasibility of this approach for determining the composition of a bacterial mixture. Using simulations, we show that sequencing a few hundred base-pairs of the 16S rRNA gene sequence may provide enough information for reconstruction of mixtures containing tens of species, out of tens of thousands, even in the presence of realistic measurement noise. Finally, we show initial promising results when applying our method for the reconstruction of a toy experimental mixture with five species. Our approach may have a potential for a simple and efficient way for identifying bacterial species compositions in biological samples.

Availability: supplementary information, data and MATLAB code are available at: http://www.broadinstitute.org/~orzuk/publications/BCS/

1 Introduction

Microorganisms are present almost everywhere on earth. The population of bacteria found in most natural environments consists of multiple species, mutually affecting each other, and creating complex ecological systems [28]. In the human body, the number of bacterial cells is over an order of magnitude larger than the number of human cells [37], with typically several hundred species identified in a given sample taken from humans (for example, over 400 species were characterized in the human gut [17], while [38] estimates a higher number of 500-1000, and 500 to 600 species were found in the oral cavity [36,13]). Changes in the human bacterial community composition are associated with physical condition,

---

* These authors contributed equally to this work.

V. Bafna and S.C. Sahinalp (Eds.): RECOMB 2011, LNBI 6577, pp. 1–15, 2011.

and may indicate [33] as well as cause or prevent various microbial diseases [22]. In a broader aspect, studies of bacterial communities range from understanding the plant-microbe interactions [40], to temporal and meteorological effects on the composition of urban aerosols [4], and is a highly active field of research [35].

Identification of the bacteria present in a given sample is not a simple task, and technical limitations impede large scale quantitative surveys of bacterial community compositions. Since the vast majority of bacterial species are non-amenable to standard laboratory cultivation procedures [1], much attention has been given to culture-independent methods. The golden standard of microbial population analysis has been direct Sanger sequencing of the ribosomal 16S subunit gene (16S rRNA) [25]. However, the sensitivity of this method is determined by the number of sequencing reactions, and therefore requires hundreds of sequences for each sample analyzed. A modification of this method for identification of small mixtures of bacteria using a single Sanger sequence has been suggested [29] and showed promising results when reconstructing mixtures of 2-3 bacteria from a given database of ~ 260 human pathogen sequences.

Recently, DNA microarray-based methods [21] and identification via next generation sequencing (reviewed in [23]) have been used for bacterial community reconstruction. In microarray based methods, such as the Affymetrix PhyloChip platform [4], the sample 16S rRNA is hybridized with short probes aimed at identification of known microbes at various taxonomy levels. While being more sensitive and cheaper than standard cloning and sequencing techniques, each bacterial mixture sample still needs to be hybridized against a microarray, thus the cost of such methods limit their use for wide scale studies. Methods based on next generation sequencing obtain a very large number of reads of a short hyper-variable region of the 16S rRNA gene [2, 12, 24]. Usage of such methods, combined with DNA barcoding, enables high throughput identification of bacterial communities, and can potentially detect species present at very low frequencies. However, since such sequencing methods are limited to relatively short read lengths (typically a few dozens and at most a few hundred bases in each sequence), the identification is non unique and limited in resolution, with reliable identification typically up to the genus level [26]. Improving resolution depends on obtaining longer read lengths, which is currently technologically challenging, and/or developing novel analytical methods which utilize the (possibly limited) information from each read to allow in aggregate a better separation between the species.

In this work we suggest a novel experimental and computational approach for sequencing-based profiling of bacterial communities (see Figure 1). We demonstrate our method using a single Sanger sequencing reaction for a bacterial mixture, which results in a linear combination of the constituent sequences. Using this mixed chromatogram as linear constraints, the sequences which constitute the original mixture are selected using a Compressed Sensing (CS) framework.

Compressed Sensing (CS) [5, 14] is an emerging field of research, based on statistics and optimization with a wide variety of applications. The goal of CS is recovery of a signal from a small number of measurements, by exploiting the fact that many natural signals are in fact sparse when represented at a certain appropriate basis. Compressed Sensing designs sampling techniques that condense the information of a compressible signal into a small amount of data. This offers the possibility of performing fewer measurements than were thought to be needed before, thus lowering costs and simplifying data-acquisition methods for various types of signals in many distantly related fields such as magnetic resonance imaging [32], single pixel camera [16], geophysics [30] and astronomy [3]. Recently, **CS** has been applied to various problems in computational biology, e.g. for pooling designs for re-sequencing experiments [18,39], for drug-screenings [27] and for designing multiplexed DNA microarrays [10], where each spot is a combination of several different probes.

The classical **CS** problem is solving the under-determined linear system,

$$\mathcal{A}\mathbf{v} = \mathbf{b} \tag{1}$$

where $\mathbf{v} = (v_1, ..., v_N)$ is the vector of unknown variables, $\mathcal{A}$ is the *sensing* matrix, often called also the *mixing* matrix and $\mathbf{b} = (b_1, ..., b_k)$ are the measured values of the k equations. The number of variables N, is far greater than the number of equations k. Without further information, $\mathbf{v}$ cannot be reconstructed uniquely since the system is under-determined. Here one uses an additional sparsity assumption on the solution - by assuming that we are interested only in solution vectors $\mathbf{v}$ with only at most s non-zero entries, for some $s \ll N$. According to the **CS** theory, when the matrix $\mathcal{A}$ satisfy certain conditions, most notably the Restricted Isometry Property(RIP) [6,7], one can find the sparsest solution uniquely by using only a logarithmic number of equations, $k = O(s \log(N/s))$, instead of a linear number (N) needed for general solution of a linear system. Briefly, RIP for a matrix $\mathcal{A}$ means that any subset of $2s$ columns of $\mathcal{A}$ is almost orthogonal (although since $k < N$ the columns cannot be perfectly orthogonal). This property makes the matrix $\mathcal{A}$ invertible for sparse vectors v with sparsity s, and allows accurate recovery of $\mathbf{v}$ from eq. (1) - for more details on the RIP condition and the reconstruction guarantees see [6,7].

In this paper, we show an efficient application of pooled Sanger-sequencing for bacterial communities reconstruction using **CS**. The sparsity assumption is fulfilled by noting that although numerous species of bacteria have been characterized and are present on earth, at a given sample typically only a small fraction of them are present at significant levels. The proposed Bacterial Compressed Sensing (**BCS**) algorithm uses as inputs a database of known 16S rRNA sequences and a single Sanger-sequence of the unknown mixture, and returns the sparse set of bacteria present in the mixture and their predicted frequencies. We show a successful reconstruction of simulated mixtures containing dozens of bacterial species out of a database of tens of thousands, using realistic biological parameters. In addition, we demonstrate the applicability of our method for a real sequencing experiment using a toy mixture of five bacterial species.

2   The BCS Algorithm

In the Bacterial Community Reconstruction Problem we are given a bacterial mixture of unknown composition. In addition, we have at hand a database of the orthologous genomic sequences for a specific known gene, which is assumed to be present in a large number of bacterial species (in our case, the gene used was the 16S rRNA gene). Our purpose is to reconstruct the identity of species present in the mixture, as well as their frequencies, where the assumption is that the sequences for the gene in all or the vast majority of species present in the mixture are available in the database. The input to the reconstruction algorithm is the measured Sanger sequence of the gene in the mixture (see Figure 1). Since Sanger sequencing proceeds independently for each DNA strand present in the sample, the sequence chromatogram of the mixture corresponds to the linear combination of the constituent sequences, where the linear coefficients are proportional to the abundance of each species in the mixture.

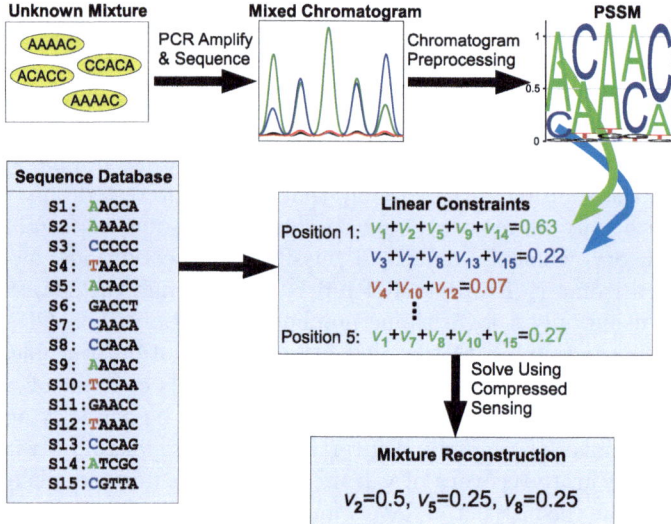

Fig. 1. Schematics of the proposed BCS reconstruction method. The 16S rRNA gene is PCR-amplified from the mixture and then subjected to Sanger sequencing. The resulting chromatogram is preprocessed to create the Position Specific Score Matrix (PSSM). For each sequence position, four linear mixture equations are derived from the 16S rRNA sequence database, with v_i denoting the frequency of sequence i in the mixture, and the frequency sum taken from the experimental PSSM. These linear constraints are used as input to the **CS** algorithm, which returns the sparsest set of bacteria recreating the observed PSSM.

Let N be the number of known bacterial species present in our database. Each bacterial population is characterized by a vector $\mathbf{v} = (v_1, ..., v_N)$ of frequencies of the different species. Denote by $s = \|\mathbf{v}\|_{\ell_0}$ the number of species present in the sample, where $\|.\|_{\ell_0}$ is the ℓ_0 norm which simply counts the number of non-zero elements of a vector $\|\mathbf{v}\|_{\ell_0} = \sum_i 1_{\{v_i \neq 0\}}$. While the total number of known species N is usually very large (in our case on the order of tens to hundreds of thousands), a typical bacterial community consists of a small subset of the species, and therefore in a given sample, $s \ll N$, and $\mathbf{v}$ is a sparse vector. The database sequences are denoted by a matrix S, where S_{ij} is the j'th nucleotide in the orthologous sequence of the i'th species ($i = 1, .., N, j = 1, .., k$).

We represent the results of the mixture Sanger sequencing as a $4 \times k$ Position-specific-Score-Matrix (PSSM) comprised of the four vectors $\mathbf{a}, \mathbf{c}, \mathbf{g}, \mathbf{t}$, representing the measured frequencies of the four nucleotides in sequence positions $1..k$. The frequency of each nucleotide at a given position j gives a linear constraint on the mixture:

$$\sum_{i=1}^{N} v_i 1_{\{S_{ij} = 'A'\}} = a_j \tag{2}$$

and similarly for the nucleotides 'C','G' and 'T'.

Define the $k \times N$ mixture matrix A for the nucleotide 'A':

$$A_{ij} = \begin{cases} 1 & S_{ij} = 'A' \\ 0 & \text{otherwise} \end{cases} \tag{3}$$

and similarly for the nucleotides 'C', 'G', 'T'. The constraints given by the sequencing reaction can therefore be expressed in matrix form as:

$$A\mathbf{v} = \mathbf{a}, C\mathbf{v} = \mathbf{c}, G\mathbf{v} = \mathbf{g}, T\mathbf{v} = \mathbf{t} \tag{4}$$

The crucial assumption we make in order to cope with the insufficiency of information is the sparsity of the vector $\mathbf{v}$, which reflects the fact that only a small number of species are present in the mixture. We therefore seek a sparse solution for the set of equations (4). **CS** theory shows that under certain conditions on the mixture matrix and the number of measurements (see below), the sparse solution can be recovered uniquely by solving the following minimization problem [8, 15, 41]:

$$\mathbf{v}^* = argmin_{\mathbf{v}} \|\mathbf{v}\|_{\ell 1} = argmin_{\mathbf{v}} \sum_{i=1}^{N} |v_i| \quad s.t. \quad A\mathbf{v} = \mathbf{a}, C\mathbf{v} = \mathbf{c}, G\mathbf{v} = \mathbf{g}, T\mathbf{v} = \mathbf{t} \tag{5}$$

which is a convex optimization problem whose solution can be obtained in polynomial time. The above formulation requires our measurements to be precisely equal to their expected value based on the species frequency and the linearity assumption for the measured chromatogram. This description ignores the effects of noise, which is typically encountered in practice, on the reconstruction. Clearly, measurements of the signal mixtures suffer from various types of noise and biases. Fortunately, the **CS** paradigm is known to be robust to measurement noise [6,9]. One can cope with noise by enabling a trade-off between sparsity and accuracy in the reconstruction merit function, which in our case is formulated as:

$$\mathbf{v}^* = argmin_{\mathbf{v}} \frac{1}{2} \left(\|\mathbf{a} - A\mathbf{v}\|_{\ell 2}^2 + \|\mathbf{c} - C\mathbf{v}\|_{\ell 2}^2 + \|\mathbf{g} - G\mathbf{v}\|_{\ell 2}^2 + \|\mathbf{t} - T\mathbf{v}\|_{\ell 2}^2 \right) + \tau \|\mathbf{v}\|_{\ell 1} \tag{6}$$

This problem represents a more general form of eq. (5), and accounts for noise in the measurement process. This is utilized by insertion of an $\ell 2$ quadratic error term. The parameter τ determines the relative weight of the error term vs. the sparsity promoting term. Many algorithms which enable an efficient solution of problem (6) are available, and we have chosen the widely used GPSR algorithm described in [19]. The error tolerance parameter was set to $\tau = 10$ for the simulated mixture reconstruction, and $\tau = 100$ for the reconstruction of the experimental mixture. These values achieved a rather sparse solution in most cases (a few species reconstructed with frequencies above zero), while still giving a good sensitivity. The performance of the algorithm was quite robust to the specific value of τ used, and therefore further optimization of the results by fine tuning τ was not followed in this study.

3   Results

3.1   Simulation Results

In order to asses the performance of the proposed **BCS** reconstruction algorithm, random subsets of species from the greengene database [11] were selected. Within these subsets, the relative frequencies of each species were drawn at random from a uniform frequency distribution normalized to sum to one (results for a different, power-law frequency distribution, are shown later), and the mixture Sanger-sequence PSSM was calculated . This PSSM was then used as the input for the **BCS** algorithm, which returned the frequencies of database sequences predicted to participate in the mixture (see Figure 1 and online Supplementary Methods).

A sample of a random mixture of 10 sequences, and a part of the corresponding mixed sequence PSSM, are shown in Figure 2A,B respectively. Results of the **BCS** reconstruction using a 500 bp long sequence are shown in Figure 2C. The **BCS** algorithm successfully identified all of the species present in the original mixture, as well as several false positives (species not present in the original mixture). The largest false positive frequency was 0.01, with a total fraction of 0.04 false positives. In order to quantify the performance of the **BCS** algorithm, we used two main measures: RMSE and recall/precision. RMSE is the Root-Mean Squared-Error between the original mixture vector and the reconstructed vector, defined as $RMSE(\mathbf{v}, \mathbf{v}^*) = \|\mathbf{v} - \mathbf{v}^*\|_{\ell 2} = \left(\sum_{i=1}^{N} (v_i - v_i^*)^2 \right)^{1/2}$. This measure accounts both for the presence or absence of species in the mixture, as well as their frequencies. In the example shown in Figure 2 the RMSE score of the reconstruction was 0.03. As another measure, we have recorded the *recall*, defined as the fraction of species present in the original vector $\mathbf{v}$ which were also present in the reconstructed vector $\mathbf{v}^*$ (this is also known as sensitivity), and the *precision*, defined as the fraction of species present in the reconstructed vector $\mathbf{v}^*$ which were also present in the original mixture vector $\mathbf{v}$. Since the predicted frequency is a continuous variable, whereas the recall/precision relies on a binary categorization, a minimal threshold for calling a species present in the reconstructed mixture was used before calculating the recall/precision scores.

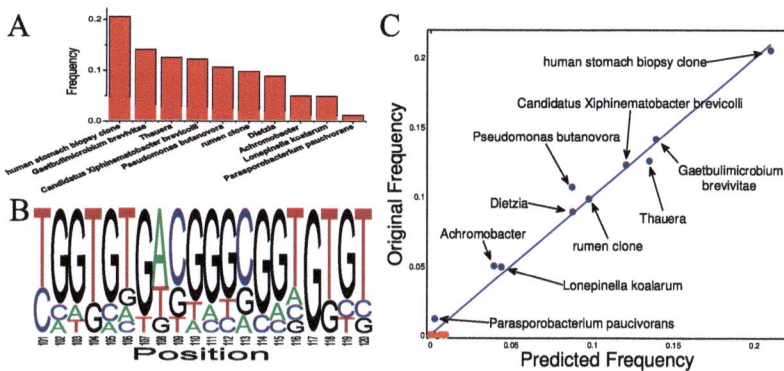

Fig. 2. Sample reconstruction of a simulated mixture. A. Frequencies and species for a simulated random mixture of $s = 10$ sequences. Species were randomly selected from the 16S rRNA database, with frequencies generated from a uniform distribution. **B.** A 20 nucleotide sample region of the PSSM for the mixture in (A). **C.** True vs. predicted frequencies for a sample **BCS** reconstruction for the mixture in (A) using $k = 500$ bases of the simulated mixture. Red circles denote species returned by the **BCS** algorithm which are not present in the original mixture.

Effect of Sequence Length. To determine the typical sequence length required for reconstruction, we tested the **BCS** algorithm performance using different sequence lengths. In Figure 3A (black line) we plot the reconstruction RMSE for random mixtures of 10 species. To enable faster running times, each simulation used a random subset of $N = 5000$ sequences from the sequence database for mixture generation and reconstruction. It is shown in Figure 3A that using longer sequence lengths results in a larger number of linear constraints and therefore higher accuracy, with ~300 nucleotides sufficing for accurate reconstruction of a mixture of 10 sequences. The large standard deviation is due to a small probability of selection of a similar but incorrect sequence in the reconstruction, which leads to a high RMSE. Due to a cumulative drift in the chromatogram peak position prediction, typical usable experimental chromatogram lengths are in the order of $k \sim 500$ bases rather than the ~1000 bases usually obtained when sequencing a single species (see online Supplementary Methods for details).

In order to asses the effect of similarites between the database sequences (which leads to high coherence of the mixing matrix columns) on the performance of the **BCS** algorithm, a similar mixture simulation was performed using a database of random nucleotide sequences (i.e. each sequence was composed of i.i.d. nucleotides with 0.25 probability for 'A','C','G' or 'T'). Using a mixing matrix derived from these random sequences, the **BCS** algorithm showed better performance (green line in Figure 3A), with ~ 100 nucleotides sufficing for a similar RMSE as that obtained for the 16S rRNA database using 300 nucleotides.

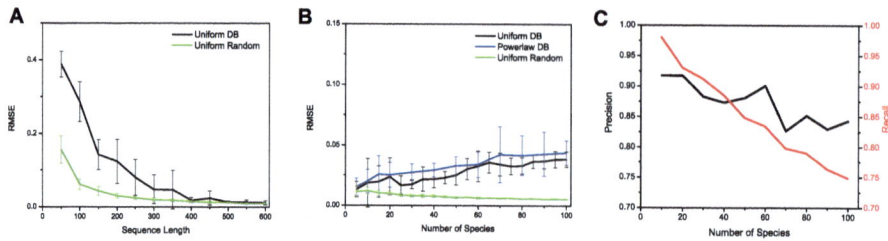

Fig. 3. Reconstruction of simulated mixtures. A. Effect of sequence length on reconstruction performance. RMSE between the original and reconstructed frequency vectors for uniformly distributed random mixtures of $s = 10$ species from the 16S rRNA database (black) or randomly generated sequences (green). Error bars denote the standard deviation derived from 20 simulations. **B.** Dependence of reconstruction performance on number of species in the mixture. Simulation is similar to (A) but using a fixed sequence length ($k = 500$) and varying the number of species in the mixture. Blue line shows reconstruction performance on a mixture with power-law distributed species frequencies ($v_i \sim i^{-1}$). **C.** Recall (fraction of sequences in the mixtures identified, shown in red) and precision (fraction of incorrect sequences identified, shown in black) of the **BCS** reconstruction of uniformly distributed database mixtures shown as black line in (B). The minimal reconstructed frequency for a species to be declared as present in the mixture was set to 0.25%.

Effect of Number of Species. For a fixed value of $k = 500$ nucleotides per sequencing run, the effect of the number of species present in the mixture on reconstruction performance is shown in Figure 3B,C. Even on a mixture of 100 species, the reconstruction showed an average RMSE less than 0.04, with the highest false positive reconstructed frequency (i.e. frequency for species not present in the original mixture) being less than 0.01. Using a minimal frequency threshold of 0.0025 for calling a species present in the reconstruction, the **BCS** algorithm shows an average recall of 0.75 and a precision of 0.85. Therefore, while the sequence database did not perform as well as random sequences, the 16S rRNA sequences exhibit enough variation to enable a successful reconstruction of mixtures of tens of species with a small percent of errors.

The frequencies of species in a biologically relevant mixture need not be uniformly distributed. For example, the frequency of species found on the human skin [20] were shown to resemble a power-law distribution. We therefore tested the performance of the **BCS** reconstruction on a similar power-law distribution of species frequencies with with $v_i \sim i^{-1}$. Performance on such a power-law mixture is similar to the uniformly distibuted mixture (blue and green lines in Figure 3B respectively) in terms of the RMSE. A sample power-law mixture and reconstruction are shown in Figure S4A,B. The recall/precision of the **BCS** algorithm on such mixtures (Figure S4C) is similar to the uniform distribution for mixtures containing up to 50 species, with degrading performance on larger mixtures, due to the long tail of low frequency species.

Effect of Noise on BCS Solution. Experimental Sanger sequencing chromatograms contain inherent noise, and we cannot expect to obtain exact measurements in practice. We therefore turned to study the effect of noise on the accuracy of the **BCS** reconstruction algorithm. Measurement noise was modeled as additive i.i.d. Gaussian noise $z_{ij} \sim N(0, \sigma^2)$ applied to each nucleotide read at each position. Noise is compensated for by the insertion of the $\ell2$ norm into the minimization problem (see eq. (6)), where the factor τ determines the balance between sparsity and error-tolerance of the solution. The effect of added random i.i.d. Gaussian noise to each nucleotide measurement is shown in Figure 4. The reconstruction performance slowly degrades with added noise both for the real 16S rRNA and the random sequence database.

Using a noise standard deviation of $\sigma = 0.15$ (which is the approximate experimental noise level - see later) and sequencing 500 nucleotides, the reconstruction

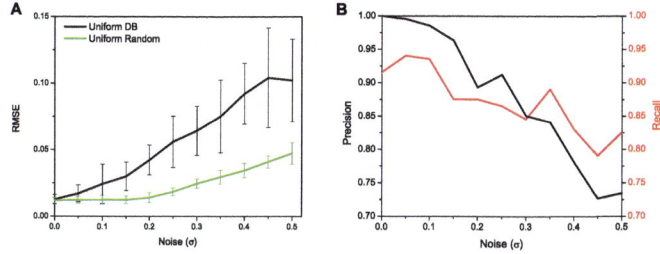

Fig. 4. Effect of noise on reconstruction. A. Reconstruction RMSE of mixtures of $s = 10$ sequences of length $k = 500$ from the 16S rRNA sequence database (black) or random sequences (green), with Gaussian noise added to the chromatogram. **B.** Recall (red) and precision (black) of the 16S rRNA database mixture reconstruction shown in (A).

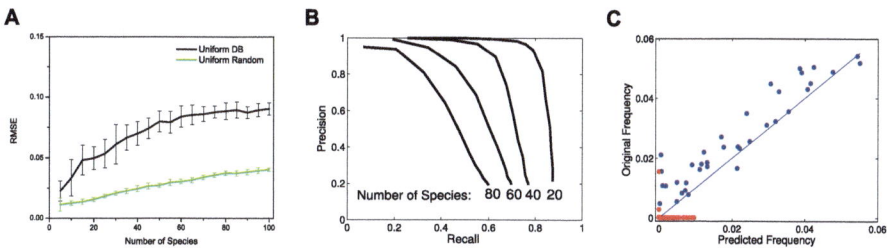

Fig. 5. Reconstruction with experimental noise level. A. Reconstruction RMSE as a function of number of species present in the mixture. Frequencies were sampled from a uniform distribution. Noise is set to $\sigma = 0.15$. Sequence length is set to $k = 500$. Black and green lines represent 16S rRNA and random sequences respectively. **B.** Recall vs. precision curves for different number of 16S rRNA sequences as in (A) obtained by varying the minimal inclusion frequency threshold. **C.** Sample reconstruction of $s = 40$ 16S rRNA sequences from (A).

performance as a function of the number of species in the mixture is shown in Figure 5. Under this noise level, the **BCS** algorithm reconstructed a mixture of 40 sequences with an average RMSE of 0.07 (Figure 5B), compared to ~ 0.02 when no noise is present (Figure 3B). By using a minimal frequency threshold of 0.006 for the predicted mixture, **BCS** showed a recall (sensitivity) of ~ 0.7, with a precision of ~ 0.7 (see Figure 5B), attained under realistic noise levels. To conclude, we have observed that the addition of noise leads to a graceful degradation in the reconstruction performance, and one can still achieve accurate reconstruction with realistic noise levels.

3.2   Reconstruction of an Experimental Mixture

While these simulations show promising results, they are based on correctly converting the experimentally measured chromatogram to the PSSM used as input to the **BCS** algorithm (see Figure 1). A major problem in this conversion is the large variability in the peak heights and positions observed in Sanger sequencing chromatograms (see Figure S2). It has been previously shown that a large part of this variability stems from local sequence effects on the polymerase activity [31]. In order to overcome this problem, we utilize the fact that both peak position and height are local sequence dependent, in order to accurately predict the chromatograms of the sequences present in the 16S rRNA database. The **CS** problem is then stated in terms of reconstruction of the measured chromatogram using a sparse subset of predicted chromatograms for the 16S rRNA database. This is achieved by binning both the predicted chromatograms and the measured mixture chromatogram into constant sized bins, and applying the **BCS** algorithm on these bins (see online Supplementary Methods and Figure S1).

We tested the feasibility of the **BCS** algorithm on experimental data by reconstructing a simple bacterial population using a single Sanger sequencing chromatogram. We used a mixture of five different bacteria: (*Escherichia coli W3110, Vibrio fischeri, Staphylococcus epidermidis, Enterococcus faecalis* and *Photobacterium leiognathi*). A sample of the measured chromatogram is shown in Figure 6A (solid lines). The **BCS** algorithm relies on accurate prediction of the chromatograms of each known database 16S rRNA sequence. In order to asses the accuracy of these predictions, Figure 6A shows a part of the predicted chromatogram of the mixture (dotted lines) which shows similar peak positions and heights to the ones experimentally measured (solid lines). The sequence position dependency of the prediction error is shown in Figure 6B. On the region of bins 125-700 the prediction shows high accuracy, with an average root square error of 0.08. The loss of accuracy at longer sequence positions stems from a cumulative drift in predicted peak positions, as well as reduced measurement accuracy. We therefore used the region of bins 125-700 for the **BCS** reconstruction.

Results of the reconstruction are shown in Figure 6C. The algorithm successfully identifies three of the five bacteria (*Vibrio fischeri, Enterococcus faecalis* and *Photobacterium leiognathi*). Out of the two remaining strains, one (*Staphylococcus epidermidis*) is identified at the genus level, and the other (*Escherichia coli*) is mistakenly identified as *Salmonella enterica*. While *Escherichia coli* and

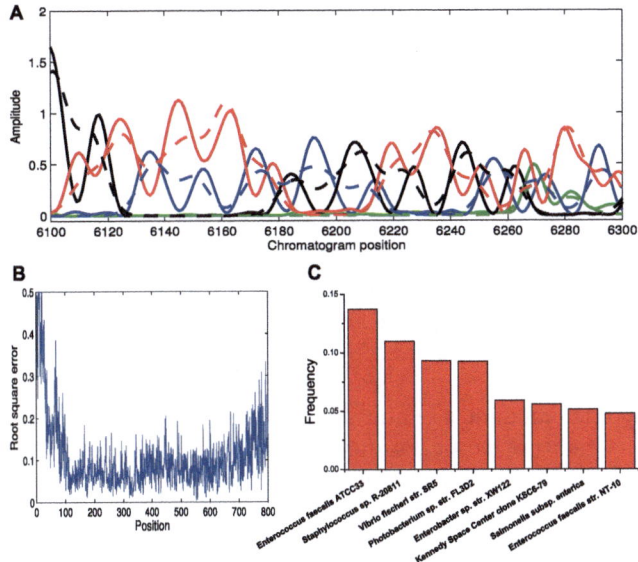

Fig. 6. Reconstruction of an experimental mixture. A. Sample region of the mixed chromatogram (solid lines). 16S rRNA from five bacteria was extracted and mixed at equal proportions. Dotted lines show the local-sequence corrected prediction of the chromatogram using the known mixture sequences. **B.** Square root distance between the predicted and measured chromatograms shown in (A) as a function of bin position, representing nucleotide position in the sequence. Prediction error was low for sequence positions $\sim 100 - 700$. **C.** Reconstruction results using the **BCS** algorithm. Runtime was ~ 20 minutes on a standard PC. Shown are the 8 most frequent species. Original strains were : *Escherichia coli, Vibrio fischeri, Staphylococcus epidermidis, Enterococcus faecalis* and *Photobacterium leiognathi* (each with 20% frequency).

Salmonella enterica show a sequence difference in 33 bases over the PCR amplified region, only two bases are different in the region used for the **BCS** reconstruction, and thus the *Escherichia coli* sequence was removed in the database preprocessing stage (see online Supplementary Methods). When this sequence is manually added to the database (in addition to the *Salmonella enterica* sequence), the **BCS** algorithm correctly identifies the presence of *Escherichia coli* rather than *Salmonella enterica* in the mixture. Another strain identified in the reconstruction - the Kennedy Space Center clone KSC6-79 - is highly similar in sequence (differs in five bases over the region tested) to the sequence of *Staphylococcus epidermidis* used in the mixture.

4   Discussion

In this work we have proposed a framework for identifying and quantifying the presence of bacterial species in a given population using information from a single Sanger sequencing reaction. Simulation results with noise levels comparable to the measured noise in real chromatograms indicate that our method can reconstruct mixtures of tens of species. When not enough information is present in the sequence (for example when a large number of sequences is present in the mixture), performance of the reconstruction algorithm decays gracefully, and still retains detection of the prominent species.

In order to test the applicability of the **BCS** algorithm to real experimental data, we performed a reconstruction of a toy mixture containing five bacterial species. Results of the sample reconstruction (identification of 3 out of 5 species at the strain level, and the additional 2 at the genus level, when E. coli is not omitted from the database) indicate that with appropriate chromatogram preprocessing, **BCS** can be applied to experimental mixtures. However, further optimization of the sequencing and preprocessing is required in order to obtain more accurate results.

Essentially, the amount of information needed for identifying the species present in the mixture is logarithmic in the database size [5, 14], as long as the number of the species present in the mixture is kept constant. Therefore, a single sequencing reaction with hundreds of bases contains in principle a very large amount of information and should suffice for unique reconstruction even when the database contains millions of different sequences. Compressed Sensing enables the use of such information redundancy through the use of linear mixtures of the sample. However, the mixtures need to be RIP in order to enable an optimal extraction of the information. In our case, the mixtures are dictated by the sequences in the database, which are clearly dependent. While two sequences which differ in a few nucleotides have high coherence and clearly do not contribute to RIP, even a single insertion or deletion completely brings the two sequences to being 'out of phase', thus making it easier to distinguish between them using **CS**(provided that the insertion/deletion did not occur to close to the end of the sequenced region). Since the mixing matrix is built using each sequence in the database separately, we do not rely on correct alignment of the database sequences, and, moreover, while a species actually present in the mixture is likely to appear in the solution with high frequency, sequences of similar species which are different by one or a few insertion or deletion events, will violate the linear constrains present in our optimization criteria, and are not likely to 'fool' the reconstruction algorithm.

While limited to the identification of species with known 16S rRNA sequences, the **BCS** approach may enable low cost simple comparative studies of bacterial population composition in a large number of samples. Our method, like any other method, can perform only as well as is allowed by the inherent interspecies variation in the sequenced region. For example, if two species are completely identical at the 16S rRNA locus, no method will be able to distinguish between them based on this locus alone. In the simulations presented, we defined a species reconstruction to be accurate having up to 1 nucleotide difference from the original sequence. Since sequence lengths used were typically around 500bp, the reconstruction sequence accuracy was approx. 0.2%. Average sequence difference between genus was measured as approx. 3%, whereas between species is approx. 2% [42], and therefore simulation performance was measured at sub-species resolution. However, there are cases of species with identical or nearly identical 16S rRNA sequences, and therefore these species can not be discriminated based on 16S rRNA alone. Sequencing of additional loci (such as in the MLST database [34]) are likely to be required in order to achieve higher reconstruction resolution. Our proposed method can easily be extended to more than one sequencing reaction per mixture, whether they come from the same region or distinct regions, by simply joining all sequencing results as linear constraints. Such an extension can lead to a larger number of linear constraints. This increases the amount of information available for our reconstruction algorithm, which will enable us to both overcome experimental noise present in each sequencing, and distinguish between species more accurately and at a higher resolution.

Acknowledgments

We thank Amit Singer, Yonina Eldar, Gidi Lazovski and Noam Shental for useful discussions, Eytan Domany for critical reading of the manuscript, Joel Stavans for supporting this research and Chaime Priluski for assistance with chromatogram peak prediction data.

References

1. Amann, R., Ludwig, W., Schleifer, K.: Phylogenetic identification and in situ detection of individual microbial cells without cultivation. Microbiological Reviews 59(1), 143–169 (1995)
2. Armougom, F., Raoult, D.: Use of pyrosequencing and DNA barcodes to monitor variations in firmicutes and bacteroidetes communities in the gut microbiota of obese humans. BMC Genomics 9(1), 576 (2008)
3. Bobin, J., Starck, J., Ottensamer, R.: Compressed sensing in astronomy. Journal of Selected Topics in Signal Processing 2, 718–726 (2008)
4. Brodie, E., DeSantis, T., Parker, J., Zubietta, I., Piceno, Y.M., Andersen, G.L.: Urban aerosols harbor diverse and dynamic bacterial populations. Proceedings of the National Academy of Sciences 104(1), 299–304 (2007)
5. Candes, E.: Compressive sampling. In: Int. Congress of Mathematics, Madrid, Spain, pp. 1433–1452 (2006)
6. Candes, E., Romberg, J., Tao, T.: Stable signal recovery from incomplete and inaccurate measurements. Arxiv preprint math/0503066 (2005)
7. Candes, E., Tao, T.: Decoding by linear programming. IEEE Transactions on Information Theory 51(12), 4203–4215 (2005)
8. Candes, E., Tao, T.: Near-optimal signal recovery from random projections: Universal encoding strategies? IEEE Transactions on Information Theory 52(12), 5406–5425 (2006)
9. Candes, E., Tao, T.: The Dantzig selector: statistical estimation when p is much larger than n. Annals of Statistics 35(6), 2313–2351 (2007)
10. Dai, W., Sheikh, M., Milenkovic, O., Baraniuk, R.: Compressive sensing dna microarrays. EURASIP Journal on Bioinformatics and Systems Biology (2009), doi:10.1155/2009/162824

11. DeSantis, T., Hugenholtz, P., Larsen, N., Rojas, M., Brodie, E., Keller, K., Huber, T., Dalevi, D., Hu, P., Andersen, G.: Greengenes, a chimera-checked 16S rRNA gene database and workbench compatible with ARB. Applied and Environmental Microbiology 72(7), 5069 (2006)

12. Dethlefsen, L., Huse, S., Sogin, M., Relman, D.: The pervasive effects of an antibiotic on the human gut microbiota, as revealed by deep 16S rRNA sequencing. PLoS Biology 6(11), e280 (2008)

13. Dewhirst, F., Izard, J., Paster, B., et al.: The human oral microbiome database (2008)

14. Donoho, D.: Compressed sensing. IEEE Transaction on Information Theory 52(4), 1289–1306 (2006)

15. Donoho, D.: For most large underdetermined systems of linear equations the minimal l1-norm solution is also the sparsest solution. Communications on Pure and Applied Mathematics 59(6), 797–829 (2006)

16. Duarte, M., Davenport, M., Takhar, D., Laska, J., Sun, T., Kelly, K., Baraniuk, R.: Single-pixel imaging via compressive sampling. IEEE Signal Processing Magazine 25(2), 83–91 (2008)

17. Eckburg, P., Bik, E., Bernstein, C., Purdom, E., Dethlefsen, L., Sargent, M., Gill, S., Nelson, K., Relman, D.: Diversity of the human intestinal microbial flora. Science 308(5728), 1635–1638 (2005)

18. Erlich, Y., Gordon, A., Brand, M., Hannon, G., Mitra, P.: Compressed Genotyping. IEEE Transactions on Information Theory 56(2), 706–723 (2010)

19. Figueiredo, M., Nowak, R., Wright, S.: Gradient projection for sparse reconstruction: Application to compressed sensing and other inverse problems. IEEE Journal of Selected Topics in Signal Processing 1(4), 586–597 (2007)

20. Gao, Z., Tseng, C., Pei, Z., Blaser, M.: Molecular analysis of human forearm superficial skin bacterial biota. Proceedings of the National Academy of Sciences 104(8), 2927 (2007)

21. Gentry, T., Wickham, G., Schadt, C., He, Z., Zhou, J.: Microarray applications in microbial ecology research. Microbial Ecology 52(2), 159–175 (2006)

22. Guarner, F., Malagelada, J.: Gut flora in health and disease. Lancet 361(9356), 512–519 (2003)

23. Hamady, M., Knight, R.: Microbial community profiling for human microbiome projects: Tools, techniques, and challenges. Genome Research 19(7), 1141–1152 (2009), PMID: 19383763

24. Hamady, M., Walker, J., Harris, J., Gold, N., Knight, R.: Error-correcting barcoded primers for pyrosequencing hundreds of samples in multiplex. Nature Methods 5(3), 235–237 (2008)

25. Hugenholtz, P.: Exploring prokaryotic diversity in the genomic era. Genome Biology 3(2), reviews0003.1–reviews0003.8 (2002)

26. Huse, S., Dethlefsen, L., Huber, J., Welch, D., Relman, D., Sogin, M.: Exploring microbial diversity and taxonomy using SSU rRNA hypervariable tag sequencing. PLoS Genetics 4(11), e1000255 (2008)

27. Kainkaryam, R., Woolf, P.: Pooling in high-throughput drug screening. Current Opinion in Drug Discovery & Development 12(3), 339 (2009)

28. Keller, M., Zengler, K.: Tapping into microbial diversity. Nature Reviews Microbiology 2(2), 141–150 (2004)

29. Kommedal, O., Karlsen, B., Sabo, O.: Analysis of mixed sequencing chromatograms and its application in direct 16S rDNA sequencing of poly-microbial samples. Journal of Clinical Microbiology (2008)

30. Lin, T., Herrmann, F.: Compressed wavefield extrapolation. Geophysics 72 (2007)
31. Lipshutz, R., Taverner, F., Hennessy, K., Hartzell, G., Davis, R.: DNA sequence confidence estimation. Genomics 19(3), 417–424 (1994)
32. Lustig, M., Donoho, D., Pauly, J.: Sparse mri: The application of compressed sensing for rapid MR imaging. Magnetic Resonance in Medicine 58, 1182–1195 (2007)
33. Mager, D., Haffajee, A., Devlin, P., Norris, C., Posner, M., Goodson, J.: The salivary microbiota as a diagnostic indicator of oral cancer: A descriptive, non-randomized study of cancer-free and oral squamous cell carcinoma subjects. J. Transl. Med. 3(1), 27 (2005)
34. Maiden, M., Bygraves, J., Feil, E., Morelli, G., Russell, J., Urwin, R., Zhang, Q., Zhou, J., Zurth, K., Caugant, D., et al.: Multilocus sequence typing: a portable approach to the identification of clones within populations of pathogenic microorganisms. Proceedings of the National Academy of Sciences 95(6), 3140–3145 (1998)
35. Medini, D., Serruto, D., Parkhill, J., Relman, D., Donati, C., Moxon, R., Falkow, S., Rappuoli, R.: Microbiology in the post-genomic era. Nat. Rev. Micro. 6(6), 419–430 (2008)
36. Paster, B., Boches, S., Galvin, J., Ericson, R., Lau, C., Levanos, V., Sahasrabudhe, A., Dewhirst, F.: Bacterial diversity in human subgingival plaque. J. of Bacteriology 183(12), 3770–3783 (2001)
37. Savage, D.: Microbial ecology of the gastrointestinal tract. Annual Reviews of Microbiology 31, 107–133 (1977)
38. Sears, C.: A dynamic partnership: Celebrating our gut flora. Anaerobe 11(5), 247–251 (2005)
39. Shental, N., Amir, A., Zuk, O.: Identification of rare alleles and their carriers using compressed se(que)nsing. Nucleic Acid Research 38(19), e179 (2010)
40. Singh, B., Millard, P., Whiteley, A., Murrell, J.: Unravelling rhizosphere-microbial interactions: opportunities and limitations. Trends Microbiol. 12(8), 386–393 (2004)
41. Tropp, J.A.: Just relax: Convex programming methods for identifying sparse signals in noise. IEEE Transactions on Information Theory 52(3), 1030–1051 (2006)
42. Yarza, P., Richter, M., Peplies, J., Euzeby, J., Amann, R., Schleifer, K.H., Ludwig, W., Glckner, F.O., Rossell-Mra, R.: The all-species living tree project: A 16s rrna-based phylogenetic tree of all sequenced type strains. Systematic and Applied Microbiology 31(4), 241–250 (2008)

Constrained De Novo Sequencing of Peptides with Application to Conotoxins

Swapnil Bhatia[1,2], Yong J. Kil[1], Beatrix Ueberheide[3], Brian Chait[3],
Lemmuel L. Tayo[4,5], Lourdes J. Cruz[5], Bingwen Lu[6,7],
John R. Yates III[6], and Marshall Bern[1]

[1] Palo Alto Research Center
[2] Department of Electrical and Computer Engineering, Boston University
[3] Rockefeller University
[4] Mapua Institute of Technology, The Philippines
[5] Marine Science Institute, University of the Philippines
[6] The Scripps Research Institute
[7] Pfizer Inc.

Abstract. We describe algorithms for incorporating prior sequence knowledge into the candidate generation stage of de novo peptide sequencing by tandem mass spectrometry. We focus on two types of prior knowledge: homology to known sequences encoded by a regular expression or position-specific score matrix, and amino acid content encoded by a multiset of required residues. We show an application to de novo sequencing of cone snail toxins, which are molecules of special interest as pharmaceutical leads and as probes to study ion channels. Cone snail toxins usually contain 2, 4, 6, or 8 cysteine residues, and the number of residues can be determined by a relatively simple mass spectrometry experiment. We show here that the prior knowledge of the number of cysteines in a precursor ion is highly advantageous for de novo sequencing.

1 Introduction

There are two basic approaches to peptide sequencing by tandem mass spectrometry (MS/MS): *database search* [16], which identifies the sequence by finding the closest match in a protein database, and *de novo sequencing* [5], which attempts to compute the sequence from the spectrum alone. Database search is the dominant method in shotgun proteomics because it can make identifications from lower quality spectra with less complete fragmentation. *De novo* sequencing finds use in special applications for which protein databases are difficult to obtain. These applications include unsequenced organisms [31], biotech products such as monoclonal antibodies [3,28], phosphopeptide epitopes [13], endogenous antibodies [29], and peptide toxins [2,27,37].

In many de novo sequencing applications, partial knowledge of the sequence is relatively easy to obtain. For example, antibodies contain long conserved segments and 10- to 13-residue hypervariable segments (complementarity determining regions), and a peptide from a digest may overlap both types of regions. A fairly simple database-search program can recognize MS/MS spectra of peptides with N- or C-terminus in a known conserved segment, and these spectra can then be de novo sequenced to determine the variable segment. As another example, nerve toxins from arthropods and

V. Bafna and S.C. Sahinalp (Eds.): RECOMB 2011, LNBI 6577, pp. 16–30, 2011.

mollusks contain highly conserved cysteine scaffolds with the number and positions of the cysteines well-conserved but the other residues variable. The numbers of cysteines in various precursors can be determined by a relatively simple mass spectrometry experiment: derivatize cysteines and measure the mass shifts. In the absence of such an experiment, the researcher can simply try each guess at the number of cysteines. These two examples are by no means exhaustive. Partial knowledge may also be obtained from previous experiments or computations, sequenceable overlapping peptides, "split" isotope envelopes from certain post-translational modifications, residue-specific derivatizations, amino acid analysis, Edman sequencing, manual inspection, comparative genomics, and so forth.

In this paper we explore the possibility of using partial knowledge to guide de novo sequencing. Related previous work has used close homology to a database protein to help assemble de novo peptide sequences into a protein sequence [3] and also to correct sequencing errors [25]. We apply partial knowledge to the candidate generation stage rather than the later stages (scoring, protein assembly, and correction of mistakes) for several reasons. First, we aim to use much weaker partial knowledge, for example, the number of cysteines rather than close homology (say 90% identity) to a known sequence. Second, our partial knowledge is often exact rather than probabilistic. Third, it is logically cleaner to maintain the scorer as a function of only the candidate sequence and the mass spectrum, independent of any protein database or biological knowledge. Fourth, we find it convenient to use a single scorer (ByOnic) for all peptide identification tasks, so that we can freely combine database search and de novo sequencing results, even within a single run of the program. Almost all MS/MS data sets contain numerous spectra identifiable by database-search, from keratin and trypsin if nothing else, and leaving these spectra to be identified *de novo* reduces the number of true identifications and falsely increases the number of "interesting" de novo sequences.

The rest of the paper is organized into the following sections: problem formulation, algorithms, validation of the approach on known conotoxins, and announcement of novel conotoxins. At this point, we believe we have completely sequenced about 15 novel mature conotoxins from two species (*Conus stercusmuscarum* and *Conus textile*), but in this bioinformatics paper we report only two new toxins, one from each species, while we wait for peptide synthesis to validate our sequences. Currently only about 130 mature conotoxins (meaning exact termini and modifications) are known after 40 years of study [23], so 15 novel conotoxins represents a substantial contribution to the field. (We have also observed about 35 mature conotoxins that match database sequences in *C. textile*, slightly exceeding the original analysis of the same data sets [36,37].) Most studies add only one or two de novo sequences at a time. For example, Nair et al. [27] and Ueberheide et al. [37] each manually sequenced one novel toxin.

2 Problem Formulation

In a tandem mass spectrometer, charged peptides break into a variety of charged and neutral fragments. The mass spectrometer measures the mass over charge (m/z) of these fragments and outputs a tandem mass spectrum, a histogram of ion counts (intensities) over an m/z range from zero to the total mass of the peptide. Given a mass spectrum, the goal of *de novo* peptide sequencing is to generate a sequence of possibly modified amino acid residues whose fragmentation would best explain the given spectrum.

Formally, a *spectrum* S is a triple (S, M, c) where S is a set of pairs of positive real numbers $\{(m_1, s_1), \ldots, (m_n, s_n)\}$, M is a positive real number, and c is an integer. M denotes the total mass of the peptide whose fragmentation produced S and is the sum of the masses of the amino acids in its sequence. The peptide charge is c, which is typically in the range +1 to +4 for the spectra we consider. Each pair (m_i, s_i) in S denotes a peak in the spectrum at m/z of m_i of intensity s_i. Let $\mathcal{A}$ be a set of symbols representing amino acid residues and modifications. We define a *peptide* p as a nonempty string over the alphabet $\mathcal{A}$.

We assume that we have access to a *peptide scoring function* h which, given a peptide p, spectrum S, and a set of allowable modifications, returns the probability that S is produced by p. Let A be a set of distinct positive numbers representing the fixed masses of the symbols in $\mathcal{A}$. The problem of de novo *candidate generation* is this: Given an integer k (say $k = 100,000$), tandem mass spectrum S, a set of symbols $\mathcal{A}$ and their masses A, find a set C of k candidate peptides p over the alphabet $\mathcal{A}$ such that $\max_{p \in C} h(S, \mathcal{A}, A, p)$ is maximized.

The parameter k above sets a limit on the number of candidate sequences we can afford to score. We cannot afford to score all possible sequences, because the number of possible peptides of a given mass M is exponential in the length of the peptide. Prior work [2,12,30] has shown the advantage of considering sets of spectra, but in this paper we generally focus attention on the de novo sequencing of single spectra.

In accord with almost all de novo sequencing programs, such as Lutefisk [35],PEAKS [26], EigenMS [7], NovoHMM [17] and PepNovo [19], we have factored the problem into candidate generation and scoring phases. Candidate generation typically uses a dynamic programming best-path algorithm [10,11,26], to compute thousands of possible sequences. The scoring phase then scores each of these candidates, using more detailed global information such as proton mobility, fragmentation propensies, and mass measurement recalibration [7], that does not conform to the separability requirement (the "principle of optimality") of dynamic programming. Here we describe how to incorporate partial knowledge into the candidate-generation phase. For scoring, we use the scorer in ByOnic [6], which is primarily a database-search program.

De novo sequencing is well known to be a difficult problem, due to incomplete fragmentation, noise peaks, mixture spectra, and the large numbers of peptides and fragments within error tolerance for any given mass. The best de novo sequencing programs rarely give a completely correct answer on a peptide of mass 2000 Da. High-accuracy instruments [20] and CID/ETD pairs [12,30] help, yet conotoxins remain especially challenging targets due to prevalent modifications and high proline content, which tends to suppress fragmentation.

3   Constraints and Algorithms for Constrained De Novo Search

Sequence constraints restrict the search from the space of all possible peptides of the given precursor mass to a proper subset of the space, in which all peptides satisfy certain *a priori* criteria. For example, we might assume that the peptide contains 4 cysteines, as do all α-conotoxins.

To demonstrate the feasibility and utility of such a constrained search approach to de novo sequencing, and to explore its role in a de novo sequencing protocol, we implemented two types of constraints in our peptide candidate generator: a multiset constraint and a regular expression constraint. We describe these constraints and our algorithm to generate candidates satisfying them below. We also implemented a simple search algorithm, similar to SALSA [24], for searching for spectra that satisfy mass and regular expression constraints. We describe this below.

3.1   Multiset Constraint

Let $\mathcal{A}$ be the set of amino acid symbols (including modifications). A *multiset constraint* is a vector $c : \mathcal{A} \to \mathbb{N}$ describing a subset of $\mathcal{A}^*$—the set of all strings over $\mathcal{A}$. We denote this subset by $S(c)$. Thus, the vector

$$c(\mathsf{G}) = 1; \ c(\mathsf{V}) = 2; \ c(\mathsf{C}) = 4; \ \text{and } c(x) = 0, \ \forall x \in \mathcal{A} \setminus \{\mathsf{G}, \mathsf{V}, \mathsf{C}\};$$

is an example of a multiset constraint. A multiset constraint defines $S(c)$ in the following way: if $c(x) = n$, then x must appear at least n times in every string in $S(c)$. All x such that $c(x) = 0$ impose no constraints on $S(c)$. Thus, in the above example,

$$S(c) = \{w : w \in \mathcal{A}^* \text{ and } w \text{ contains at least one } \mathsf{G}, \text{ at least two } \mathsf{V}, \text{ and at least four } \mathsf{C}\}.$$

For example, VGCCQCPARCKCCV satisfies the constraint in the above example, but CCPARCCVR does not.

3.2   An Algorithm for Generating Multiset-Constrained Candidates

Let $\mathcal{A}$ be the set of amino acid symbols (including modifications). Let $\mathcal{S} = (T, M)$ be a given (deisotoped and decharged) spectrum, let c be a multiset constraint, and let N be a positive integer. The objective of the de novo candidate generation algorithm is to output a set of N peptides, all satisfying the multiset constraint c, containing a peptide that best explains the spectrum $\mathcal{S}$. Our algorithm proceeds in two stages. In the first stage, we construct a directed multigraph G, in which each vertex is a tuple containing an integer mass in the interval $[0, M]$ and a count of the number of each of the symbols in c consumed by a prefix ending at the vertex. Arcs are added between vertices whose mass differs by that of an amino acid and have a compatible count. An arc of G is assigned a cost obtained as a function of the best peaks in T supporting the vertices of the arc. In the second stage, we obtain the N shortest paths in G. Each path must start at the vertex representing mass zero with no symbols from the multiset constraint consumed, and must reach a vertex representing the mass M in which all the symbols appearing in the multiset constraint are consumed.

Intuitively, our dynamic programming algorithm generates a graph with multiple stages where a stage represents a partial set of constraints satisfied so far. More formally, let $V(G)$ denote the vertex set of the directed multigraph G. Let A denote the set of masses of the amino acids represented by the symbols in $\mathcal{A}$. By $span(A)$ we mean the union of the set of numbers that can be written as a sum of elements of A, and the set $\{0\}$. We denote by $\mathcal{A}_c$ the set of symbols $\{a_1, \ldots, a_n\}$ in the constraint c—i.e.,

$c(a_i) > 0$—and by A_c the corresponding masses of the amino acids they represent. Then,

$$V(G) = \left\{ (m, v) : m \in span(A) \text{ and } m \leq M; \quad v \in \prod_{i=1}^{n} \{0, \dots, c(a_i)\} \right\}, \quad (1)$$

where the product is the usual cartesian product of sets. Thus, each vertex (m, v) represents the mass of a prefix weighing m, and n bounded counters, which we denote by $v_1, \dots, v_n$. The i-th counter keeps a count of the number of a_i symbols consumed by the prefix—ending at that vertex—of any peptide passing through that vertex.

Vertices $x = (m_1, u)$ and $y = (m_2, v)$ in $V(G)$ are related by an arc from x to y if and only if either of the following conditions is satisfied:

1. $m_2 - m_1 \in A \setminus A_c$, and $u = v$, or

2. $m_2 - m_1$ is the mass of $a_i \in A_c$, and $v_k = \begin{cases} u_k + 1 \text{ if } k = i, \text{ and} \\ u_k, \text{ otherwise.} \end{cases}$

Figure 1 (a) shows a visual representation of the directed multigraph constructed from a small multiset constraint.

We annotate each vertex of the multigraph G with supporting peaks, if any, from the given spectrum. For example, consider the directed multigraph constructed under a constraint $c(C) = 4$, and consider the vertex $(320, (2))$. This vertex represents a mass of 320 Da, and represents a prefix containing two C out of the minimum of four required, assuming carbamidomethylated Cysteine. We then search the peak list in the spectrum for b-ions (e.g., peaks in the interval $321.00728 \pm \epsilon$ Da) and y-ions (e.g., peaks in the interval $M - 300.98 \pm \epsilon$) supporting this vertex, for a given fragment mass error tolerance of ϵ. After annotating all vertices in this way, we assign costs to each arc in G. In determining the cost of each arc, we use this information about the presence of supporting peaks, their intensity, and the agreement of the mass difference of peaks across an arc with an amino acid mass. Vertices with no support contribute to a penalty for all their arcs. Finally, we attempt to obtain the K least cost paths between the starting vertex of mass zero and a final vertex of mass M and with its prefix symbol counts matching or exceeding the multiset constraint.

More formal details of the algorithm are listed in pseudocode form in Algorithm 1. When A_c is empty, the algorithm guarantees that every peptide is considered, as is clear from lines 13-19. Line 5 guarantees that no peptide of a mass larger than that reported by the spectrum is considered. Line 26 guarantees that the list of peptides, implied by the list of paths considered, must be of mass M. This argument also holds for unconstrained symbols when A_c is not empty. When A_c is not empty, consider any prefix of any peptide $w \in S(c)$. If the prefix contains no constrained letters, then by the arguments above, it is guaranteed to be present as a path in the directed multigraph. If it contains some constrained letters, then their counts and the prefix's mass together must be represented by some vertex in $V(G)$, because of lines 3-12. Finally, only paths ending in a vertex who counts match the multiset constraint and whose mass matches the mass M reported in the spectrum are used for generating peptides. The converse argument proceeds along a similar path. (Note that our algorithm does not generate

---

Algorithm 1. GENERATING MULTISET-CONSTRAINED DE NOVO CANDIDATES

Require: Constraint $c : \mathcal{A} \to \mathbb{N}, \mathcal{A}_c, A_c$; Spectrum $\mathcal{S} = (T, M)$; Number of candidates K

1. $V(G) \leftarrow (0, (0, \dots, 0))$
2. **while** more vertices in $V(G)$ remain to be expanded **do**
3. $(m, (v_1, \dots, v_n)) \leftarrow$ next unexpanded vertex from $V(G)$
4.   **for** every $a \in \mathcal{A}$ **do**
5.   **if** $m + mass(a_i) \leq M$ **then**
6.   **if** $a \in \mathcal{A}_c$ **then**
7.   Let a be the i-th symbol in $\mathcal{A}_c$, denoted by a_i
8.   **if** $(m + mass(a_i), (v_1, \dots, v_i + 1, \dots, v_n)) \notin V(G)$ **then**
9. $(m', v') \leftarrow (m + mass(a_i), (v_1, \dots, v_i + 1, \dots, v_n))$
10. $V(G) \leftarrow V(G) \cup \{(m', v')\}$
11.   Mark (m', v') as unexpanded
12.   **end if**
13.   **else**
14.   **if** $(m + mass(a_i), (v_1, \dots, v_n)) \notin V(G)$ **then**
15. $(m', v') \leftarrow (m + mass(a_i), (v_1, \dots, v_n))$
16. $V(G) \leftarrow V(G) \cup \{(m', v')\}$
17.   Mark (m', v') as unexpanded
18.   **end if**
19.   **end if**
20.   Add arc from (m, v) to (m', v')
21.   **end if**
22.   **end for**
23. **end while**
24. Annotate each vertex with peaks in T corresponding to its mass
25. Assign weights to each arc
26. Obtain K shortest paths between $(0, (0, \dots, 0))$ and $(M, (c(a_1), \dots, c(a_n)))$
27. **if** no such path exists **then**
28.   Stop and report an unsatisfiable constraint error
29. **else**
30.   Translate each path of vertices into a string over $\mathcal{A}$
31.   Stop and return this set of peptides
32. **end if**

---

unreachable vertices—for example, $(0, (1))$ in the example above—though we choose to ignore this detail in equation 1 above.)

We have omitted several details about some of the steps of our algorithm, such as the arc weighting computation, presence of duplicate and conflated paths, incorporation of terminal modifications, and speed and memory optimizations. While these details may be necessary in an implementation of the algorithm—and our own implementation includes them—they are largely independent of the focus of this paper: demonstrating the feasibility and utility of constrained de novo search. We comment on these details where necessary. A complexity analysis of similar constrained shortest paths problems was carried out by Barrett et al. [4].

3.3 Acyclic Regular Expression Constraint

Let $\mathcal{A}$ be the set of amino acid symbols (including modifications) and let n be a positive integer. An *n-letter acyclic regular expression constraint* is a string $c \in (\mathcal{A} \cup \{\mathcal{A}\})^n$ describing a subset of $\mathcal{A}^*$, which we denote by $S(c)$. Thus, the string $\mathcal{A}CC\mathcal{A}\mathcal{A}\mathcal{A}K\mathcal{A}CC$ is an example of a 10-letter acyclic regular expression (or regex) constraint. An n-letter regex constraint c has the following interpretation. Every string in $S(c)$ must belong to $\mathcal{A}^n$, and must agree with c at every position, except those containing an $\mathcal{A}$. In the above example of a regex constraint,

$$S(c) = \{w : w \in \mathcal{A}^n \text{ and } w \text{ has C in positions 2,3,9, and 10, and K in position 7}\}$$

For example, GCCPTCKPCC satisfies the regex constraint but CCPCKPCC and AGC-CPTCKCC do not.

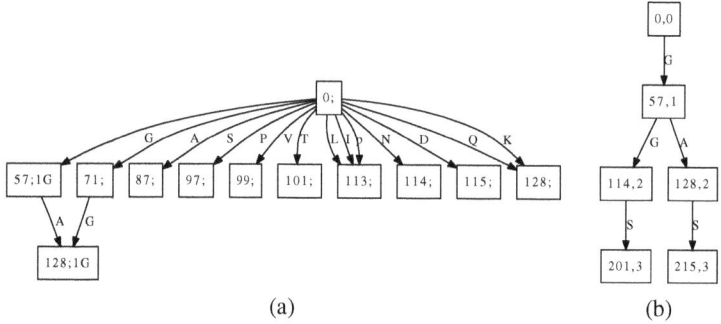

(a)   (b)

Fig. 1. (a) The directed multigraph resulting from the multiset constraint "$c(\mathsf{G}) = 1$" given a spectrum of 128.06 Da. Vertices are labeled with the integer mass of and a count of the constrained symbols in the prefix they represent. In this case, only two paths—GA and AG—satisfy the multiset and mass constraint. (b) The directed multigraph resulting from the regex constraint "G$\mathcal{A}$S" given a spectrum of total mass 215.09 Da. Vertices are labeled with the integer mass and the length of the prefix they represent. In this case, the path that satisfies the regex and mass constraint—GAS—is unique.

3.4 An Algorithm for Generating Regex-Constrained Candidates

In its graph and flow, our algorithm for generating regex-constrained peptides is similar to the algorithm for multiset-constrained peptides given above. The main difference is in the information represented in each vertex. Let c be an n-letter regex constraint. In this case,

$$V(G) = \{(m, v) : m \in span(A) \text{ and } m \le M; \quad v \in \{0, \ldots, n\}\}, \qquad (2)$$

Each vertex represents the mass of a prefix of every path passing through it, and a count of the number of letters in the prefix. Vertices $x = (m_1, v)$ and $y = (m_2, v+1)$ in $V(G)$ are related by an arc from x to y if and only if $m_2 - m_1 \in A$. Other details are similar to the multiset algorithm described above; the differences are formally presented in Algorithm 2 below. Figure 1 (b) shows a visual representation of the directed multigraph constructed for a small regex constraint.

3.5   Constrained Spectral Search and Clustering

Given a spectrum, in many cases, *de novo* sequencing of the complete peptide may be difficult. Typically, this is due to the quality of the spectrum, unavailability of the complete ladder of peaks in any single spectrum due to digestion, or low mass accuracy of the fragments or the precursor. In such instances, it is desirable to have tool that can quickly search for other spectra that describe the unknown peptide under consideration. We implemented a simple spectral search tool for this purpose, similar to SALSA [24]. Given a spectrum, we consider the set of its peaks as vertices and construct a directed multigraph G in which we add an arc between any two peaks separated by the mass of some amino acid, including modified amino acids. Then, we enumerate all maximal distinct paths in G. This results in a list of short peptide fragments, not necessarily of the mass reported in the spectrum, all of which are supported by peaks in the given spectrum. This "spectral fingerprint" can be used to search for spectra containing peaks supporting a particular peptide fragment. In the context of conotoxin spectra, we have found this tool to be useful for filtering out spectra that contain a "CC fingerprint" and thus, are likely to be sequenceable conotoxins. It can also be used for clustering spectra.

Algorithm 2. GENERATING REGEX-CONSTRAINED DE NOVO CANDIDATES

Require: Constraint $c : \{1, \ldots, n\} \rightarrow \mathcal{A}$; Spectrum $\mathcal{S} = (T, M)$; Number of candidates K

1. $V(G) \leftarrow (0, 0)$
2. **while** more vertices in $V(G)$ remain to be expanded **do**
3. $(m, i) \leftarrow$ next unexpanded vertex from $V(G)$
4.   **if** $i = n$ **then**
5.   Go to line 23
6.   **end if**
7.   **if** $c(i + 1) = $ "$\mathcal{A}$" **then**
8. $\mathcal{B} \leftarrow \mathcal{A}$
9.   **else**
10. $\mathcal{B} \leftarrow \{c(i + 1)\}$
11.   **end if**
12.   **for** every $a \in \mathcal{B}$ **do**
13.   **if** $m + mass(a) \leq M$ **then**
14.   **if** $(m + mass(a), i + 1) \notin V(G)$ **then**
15. $(m', i') \leftarrow (m + mass(a), i + 1)$
16. $V(G) \leftarrow V(G) \cup \{(m', i')\}$
17.   Mark (m', i') as unexpanded
18.   **end if**
19.   Add arc from (m, i) to (m', i')
20.   **end if**
21.   **end for**
22. **end while**
23. (Same as lines 24-25 in Algorithm 1 above)
24. Obtain K shortest paths between $(0, 0)$ and (M, n)
25. (Same as lines 27-32 in Algorithm 1 above)

In addition to the above algorithms, we have also implemented algorithms that allow ordered multiset-constraints (e.g., two C followed, not necessarily immediately, by a W), combine multiset and regex constraints (e.g. GCCKP followed by two C), and impose mass intervals in which a constraint must be satisfied. We postpone discussion of these algorithms to future work.

4   Application to Conotoxins

We obtained MS/MS data of *Conus textile* venom from Brian Chait's laboratory at Rockefeller University and of *C. textile* and *C. stercusmuscarum* venom from John Yates's laboratory at the Scripps Research Institute. The Rockefeller data [37] were LTQ MS/MS spectra, both CID and ETD, with low-accuracy precursor and fragment masses. We did not obtain Rockefeller's charge-enhanced precursor data [37], only the standard carbamidomethylated cysteine. Sample preparations and data acquisition strategies were as described previously [37,36]. The Scripps data [36] were HCD and CID Orbitrap MS/MS spectra with high-accuracy precursor and fragment masses.

Both *C. textile* data sets had been analyzed previously by database search and, in the case of the Rockefeller data, a limited amount of manual de novo sequencing. *C. textile* is one of the better studied cone snails, with a large amount of venom, and GenBank contains about 100 (redundant) *C. textile* entries, more than half of which are putative toxins. One of the goals in proteomic analysis is to observe the toxins in their mature forms, meaning with the post-translational modifications and exact termini. Conotoxins are heavily modified peptides of lengths about 10–40 residues, and known modifications include bromotryptophan, hydroxyproline, hydroxyvaline, oxidized methionine, and amidated C-terminus. Both the Rockefeller and Scripps studies claimed 31 *C. textile* toxins observed in their final form. The venom contains about 90 toxins, as estimated by the number of disulfide-bonded precursors [37]. For *C. stercusmuscarum*, there is very little sequence data available, only seven GenBank entries, none of which are annotated as toxins, so this data was essentially unanalyzed when we received it.

4.1   Validation on Known Sequences

We implemented the algorithms for constrained *de novo* search listed above into a single command-based interactive tool which we call CONOVO. The tool is capable of reading in a set of CID or ETD spectra and accepting a sequence of commands to operate on them. These include commands for de-isotoping and de-charging spectra, adding, deleting and ignoring peaks in spectra, normalizing peak intensities in a spectrum, generating a spectral fingerprint, collecting the top peaks from several spectra into a single spectrum, constructing, examining, and modifying directed multigraphs for *de novo* search, and generating candidate peptides. We wrote scripts for processing all the spectra that we received from the Yates and Chait laboratories. Our scripts processed each spectrum by issuing commands to CONOVO to load, deisotope and decharge the spectrum, and then generate candidates under various constraints, or without any constraints. After candidate generation in each case, the candidates were scored by the ByOnic scorer [6] and the highest scoring candidate was logged along with a detailed report explaining the score. We pointed our scripts to the spectra and executed the scripts without any subsequent human intervention.

We first report results from 79 *C. textile* CID spectra from both laboratories. These spectra describe cysteine-rich conotoxins whose complete and correct sequences are known. Our scripts sequenced these spectra using purely multiset-constrained and purely regex-constrained *de novo* search. We also ran an unconstrained *de novo* search under the same conditions as the multiset-constrained *de novo* search. The answer found under all three conditions on these spectra agreed with the correct answer, modulo **K-Q**, **I-L**, **M-F**, and **GG-N** substitutions, if any. Our scripts executed all of the following multiset-constraints on each spectrum: $c(\mathsf{C}) = 2$, $c(\mathsf{C}) = 3$, ..., $c(\mathsf{C}) = 6$. We obtained regex constraints directly from the correct answer by retaining all C and substituting all other letters with $\mathcal{A}$ in the correct answer.

Figure 2 shows the decimal logarithm of the position of the correct answer in the generated candidate list in the constrained case (X-axis) and the unconstrained (Y-axis) case. The left plot shows the multiset-constrained case and the right plot shows the regex-constrained case. Points near or on the diagonal $x = y$ line result from spectra on which both the constrained search and the unconstrained search produced the right answer at the same or similar position in their respective candidates list. Points below the diagonal result from spectra on which the constrained search produced the correct candidate at a position in its candidates list that was an order of magnitude lower than the position of the correct candidate on the unconstrained candidates list. For example, the point $(4.47, 2.03)$ in the multiset case (left) corresponds to the correct *C. textile* conotoxin scaffold precursor **SCCNAGFCRFGCTPCCY**, which was generated at position 29,610 by the unconstrained search and at positions 109 and 1 under the $c(\mathsf{C}) = 6$ and the $\mathcal{ACCAAAACAAACAACCA}$ constraints. The plot confirms our hypothesis that constraints can be extremely effective in improving the efficiency of the *de novo* search by reducing the search space to a subset where all the candidates satisfy *a priori* knowledge. We note that a regex constraint is more effective than a multiset constraint, but it requires much stronger *a priori* knowledge: one must supply the exact position of every letter in the constraint.

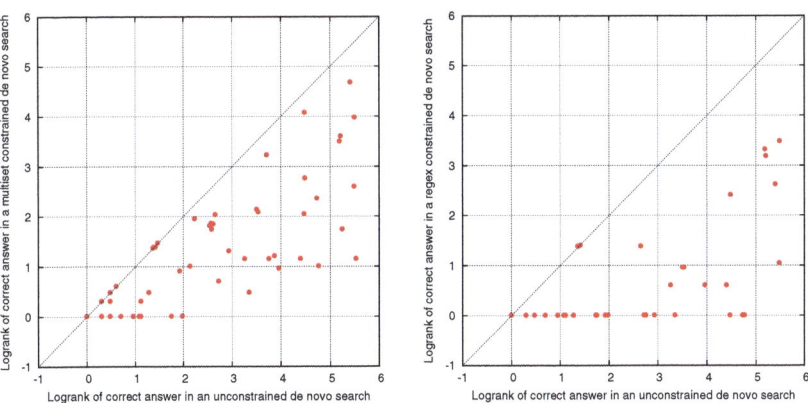

Fig. 2. A comparison of a multiset-constrained (left) or regex-constrained (right) *de novo* search with an unconstrained *de novo* search for peptide candidates for 79 cysteine-rich conotoxin spectra of the venom of *C. textile*

4.2   Discovery of Novel Conotoxins

In addition to known conotoxins, CONOVO also found peptides that appear to be new. In this paper, we report two sequences that were, to the best of our knowledge, unknown; we will report all other sequences once they have been verified by synthesis.

Figures 3(a) and 3(b) show the spectra describing the novel conotoxin found in the *C. textile* venom and the *C. stercusmuscarum* venom respectively. Figures 3(c) and 3(d) show spectra describing a prefix and a suffix of the novel *C. stercusmuscarum* toxin. In the *C. Textile* data, CONOVO found the sequence

C[+57]C[+57]GP[+16]TAC[+57]LAGC[+57]KPC[+57]C[+57][-1]

in at least 16 spectra of mass 1786 Da and 1802 Da. The 1802 Da spectra indicate a PTM on the second proline. The identifications were obtained from the multiset-constrained search described in Section 4.1. Figure 3(a) shows one of the spectra describing this novel conotoxin.

In the *C. stercusmuscarum* data, CONOVO found the following sequence:

APAC[+57]C[+57]GPGASC[+57]PRYFKDNFLC[+57]GC[+57]C[+57]

The prefix APACCGPGASCPR, but with a few incorrect letters, was found in the multiset-constrained search described above. After the constrained search completed, we collected potentially related spectra with a spectral fingerprint search described above.

We sequenced the spectrum in Figure 3(c) and obtained APACCGPGASCPRYF, which, due to its odd number of C, we guessed was a prefix of the complete sequence of the toxin. We then found a spectrum for the complete toxin (Figure 3(b)) using a wild-card with mass up to 2000 Da. (Notice that this spectrum would be hard to find by spectrum similarity to Figure 3(a).) This spectrum is not sequenceable on its own. We then found the spectrum shown in Figure 3(d) using spectral search.

5   Discussion

Constrained de novo sequencing is a new peptide identification approach that is especially well suited to studies focused on diverse but homologous protein families such as conotoxins or antibodies. We found the approach advantageous in the conotoxin studies described here, and by the end of the project, our data-analysis approach was fully automated, with human intervention required only to inspect ByOnic's scoring reports, and reconcile spectra that clearly contained closely related or identical peptides, but had incompatible top-scoring sequences.

Constrained sequencing is advantageous in different ways. First, in the case of sequenceable spectra with complete fragmentation, constraints reduce the space of plausible candidates while incorporating specific expert knowledge, thus boosting the efficacy of the scorer by eliminating spurious candidates. Such a reduction in candidate space could in principle be achieved by running an unconstrained search followed by an elimination step imposing the constraints, but for long peptides, this is not feasible. If the size of the generated candidates list is limited—as is the case in practice for long peptides—then the correct answer may not even be generated by an unconstrained search, rendering the elimination step ineffective. In this case, a constrained search is

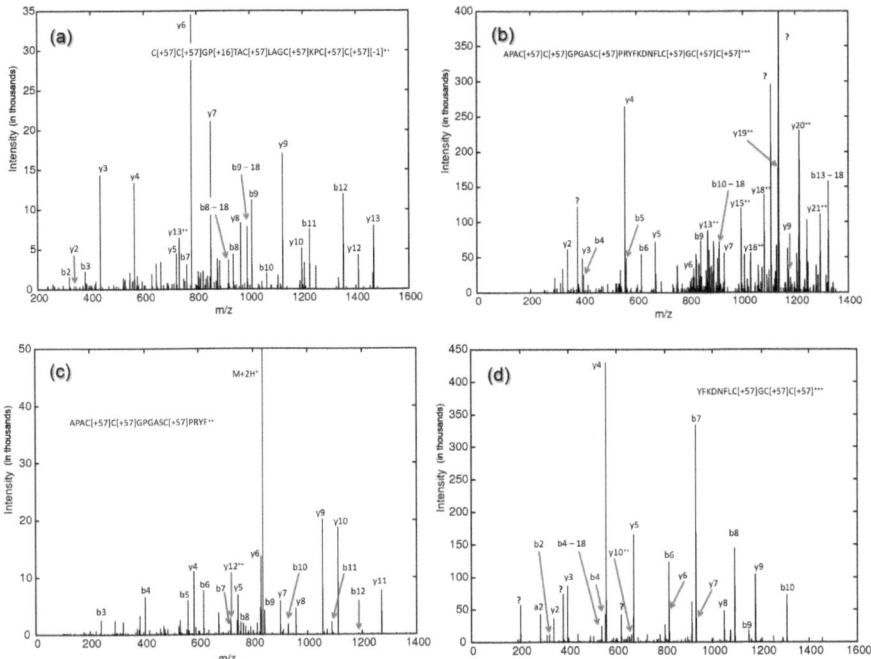

Fig. 3. (a) A novel *C. textile* toxin. All mass errors are less than 4 ppm. Despite the high mass accuracy, this spectrum would be quite challenging to sequence without some prior knowledge, because of the two PTMs (hydroxyproline and amidated C-terminus) and the missing cleavages at b1/y14 and b4/y11 (after hydroxyproline). The closest known conotoxin is **CCGPTACMAGCR-PCC**, two substitutions away. (b) A novel *C. stercusmuscarum* toxin with no BLAST hits in GenBank with E-value below 1.0. This spectrum gives what we believe is the complete toxin, observed in the undigested venom. Mass errors are less than 10 ppm, but with software recalibration of the m/z readings, the errors can be reduced to less than 4 ppm. (c) The N-terminal half of the novel *C. Stercus muscarum* toxin, also observed in the undigested venom. All errors are less than 4 ppm. (d) The C-terminal half of the novel *C. Stercus muscarum* toxin, observed in a tryptic digest of the venom. With software recalibration, all errors are less than 4 ppm.

a natural and effective solution. The reduction in the size of the candidate space may span orders of magnitude as revealed by a simple counting argument (see Figure 4) and illustrated by our experiments.

Second, constraints can actually bridge missing cleavages and sequence otherwise unsequenceable spectra. Consider a spectrum containing all the peaks supporting any candidate of the form PEPTIDE$\mathcal{A}'\mathcal{A}'\mathcal{A}'\mathcal{A}'$ where $\mathcal{A}'$ does not contain C, and all the peaks supporting PEPTIDECCCC. It is plausible that a scorer may rank candidates from both sets equally or even prefer the former candidates over the latter as a result of peak position noise. Yet, even a simple regex constraint like PEPTIDEC$\mathcal{A}\mathcal{A}\mathcal{A}$ would be sufficient for the scorer to rule out the former set in favor of the latter candidate. In such a case, the gain from a single-letter regex constraint is more significant than the reduction in space provided by fixing a single letter.

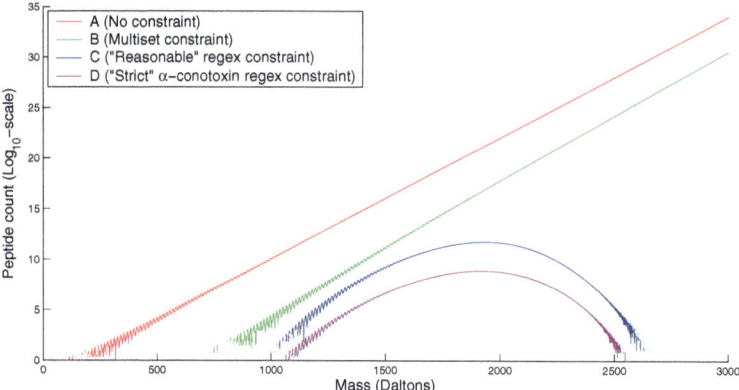

Fig. 4. Size of the candidate space: A without constraints; B with multiset constraints; C with regex constraints like $CC\mathcal{A}^{3\,or\,4}C\mathcal{A}^{3\,to\,7}C$; and D with an α-conotoxin regex constraint [39]

Nevertheless, constraints are no panacea for *de novo* sequencing. While the multiset-constrained candidate generation succeeded in sequencing a majority of the known sequences and discovering some unknown sequences, it was not successful on all spectra. In most such cases, we were able to obtain a candidate within an edit distance of three or four of the correct answer. We were then able to complete the sequence either manually, or by running a search on a database of newly discovered sequences, or by using the letters found so far, as a regex constraint. We used ByOnic's wild-card feature [8,9] to be useful in computing missing masses in incomplete *de novo* sequences, which we were later able to fill using constrained or unconstrained de novo search.

For lower accuracy spectra, there were instances where our *de novo* search produced several plausible candidates. We discovered a heuristic for separating false positives in such instances. We checked whether the mass errors of the b- and y-ions matched in magnitude. We rejected candidates, for example, in which y-ions had mass errors of 50 ppm which b-ions had mass errors of 5 ppm.

We also found that constraints were not very useful on spectra of very high or low quality, since the former were readily sequenceable without constraints while the latter were mostly unsequenceable. We also note that incorrect constraints—e.g., requiring a letter that is absent in the correct sequence—can ruin a *de novo* search. The user should start with a weak constraint and gradually strengthen it as more of the sequence becomes known. Thus, several iterations of a combination of multiset and regex constraints and ByOnic's wild card may prove to be an effective *de novo* protocol.

We found spectral search to be a handy tool for gathering spectral evidence for low confidence letters and for ruling out competing candidates. While spectral clustering would have been helpful, we did not use any clustering in this project.

Acknowledgments

This work was supported by NIGMS grant R21 GM094557, an ARRA supplemental to grant R21 GM085718, and a NSF Computing Innovations Fellowship.

References

1. Bandeira, N., Tsur, D., Frank, A., Pevzner, P.A.: Protein identification by spectral networks analysis. Proc. Natl. Acad. Sci. USA 104, 6140–6145 (2007)
2. Bandeira, N., Clauser, K.R., Pevzner, P.A.: Assembly of peptide tandem mass spectra from mixtures of modified proteins. Molecular Cell. Proteomics 6, 1123–1134 (2007)
3. Bandeira, N., Pham, V., Pevzner, P., Arnott, D., Lill, J.R.: Automated de novo protein sequencing of monoclonal antibodies. Nature Biotechnology 26, 1336–1338 (2008)
4. Barrett, C., Jacob, R., Marathe, M.: Formal language constrained path problems. SIAM J. on Computing 30, 809–837 (2000)
5. Bartels, C.: Fast algorithm for peptide sequencing by mass spectrometry. Biomedical and Environmental Mass Spectrometry 19, 363–368 (1990)
6. Bern, M., Cai, Y., Goldberg, D.: Lookup peaks: a hybrid of de novo sequencing and database search for protein identification by tandem mass spectrometry. Anal. Chem. 79, 1393–1400 (2007)
7. Bern, M., Goldberg, D.: De novo analysis of peptide tandem mass spectra by spectral graph partitioning. J. Computational Biology 13, 364–378 (2006)
8. Bern, M., Phinney, B.S., Goldberg, D.: Reanalysis of Tyrannosaurus rex Mass Spectra. J. Proteome Res. 8, 4328–4332 (2009)
9. Bern, M., Saladino, J., Sharp, J.S.: Conversion of methionine into homocysteic acid in heavily oxidized proteomics samples. Rapid Commun. Mass Spectrom. 24, 768–772 (2010)
10. Chen, T., Kao, M.-Y., Tepel, M., Rush, J., Church, G.M.: A dynamic programming approach to de novo peptide sequencing by mass spectrometry. J. Computational Biology 8, 325–337 (2001)
11. Dančik, V., Addona, T.A., Clauser, K.R., Vath, J.E., Pevzner, P.A.: De novo peptide sequencing via tandem mass spectrometry. J. Computational Biology 6, 327–342 (1999)
12. Datta, R., Bern, M.: Spectrum fusion: using multiple mass spectra for de novo peptide sequencing. J. Comput. Biol. 16, 1169–1182 (2009)
13. Depontieu, F.R., Qian, J., Zarling, A.L., McMiller, T.L., Salay, T.M., Norris, A., English, A.M., Shabanowitz, J., Engelhard, V.H., Hunt, D.F., Topalian, S.L.: Identification of tumor-associated, MHC class II-restricted phosphopeptides as targets for immunotherapy. Proc. Natl. Acad. Sci. USA 106, 12073–12078 (2009)
14. Duda, R.O., Hart, P.E., Stork, D.G.: Pattern Classification. Wiley-Interscience, Hoboken (2000)
15. Elias, J.E., Gibbons, F.D., King, O.D., Roth, F.P., Gygi, S.P.: Intensity-based protein identification by machine learning from a library of tandem mass spectra. Nature Biotechnology 22, 214–219 (2004)
16. Eng, J.K., McCormack, A.L., Yates III, J.R.: An approach to correlate tandem mass spectral data of peptides with amino acid sequences in a protein database. J. Am. Soc. Mass Spectrom. 5, 976–989 (1994)
17. Fischer, B., Roth, V., Roos, F., Grossmann, J., Baginsky, S., Widmayer, P., Gruissem, W., Buhmann, J.M.: NovoHMM: A hidden Markov model for de novo peptide sequencing. Anal. Chem. 77, 7265–7273 (2005)
18. Eppstein, D.: Finding the k shortest paths. SIAM J. Computing 28, 652–673 (1998)
19. Frank, A., Pevzner, P.: PepNovo: De Novo Peptide Sequencing via Probabilistic Network Modeling. Anal. Chem. 77, 964–973 (2005)
20. Frank, A.M., Savitski, M.M., Nielsen, M.L., Zubarev, R.A., Pevzner, P.A.: De Novo Peptide Sequencing and Identification with Precision Mass Spectrometry. J. Proteome Research 6, 114–123 (2007)

21. Graehl, J.: Implementation of David Eppstein's k Shortest Paths Algorithm, `http://www.ics.uci.edu/~eppstein/`
22. Havilio, M., Haddad, Y., Smilansky, Z.: Intensity-based statistical scorer for tandem mass spectrometry. Anal. Chem. 75, 435–444 (2003)
23. Kaas, Q., Westermann, J.C., Halai, R., Wang, C.K., Crak, D.J.: ConoServer, a database for conopeptide sequences and structures. Bioinformatics 445, 445–446 (2008)
24. Liebler, D.C., Hansen, B.T., Davey, S.W., Tiscareno, L., Mason, D.E.: Peptide sequence motif analysis of tandem MS data with the SALSA algorithm. Anal. Chem. 74, 203–210 (2002)
25. Liu, X., Han, Y., Yuen, D., Ma, B.: Automated protein (re)sequencing with MS/MS and a homologous database yields almost full coverage and accuracy. Bioinformatics 25, 2174–2180 (2009)
26. Ma, B., Zhang, K., Hendrie, C., Liang, C., Li, M., Doherty-Kirby, A., Lajoie, G.: PEAKS: powerful software for peptide de novo sequencing by tandem mass spectrometry. Rapid Comm. in Mass Spectrometry 17, 2337–2342 (2003), `http://www.bioinformaticssolutions.com`
27. Nair, S.S., Nilsson, C.L., Emmett, M.R., Schaub, T.M., Gowd, K.H., Thakur, S.S., Krishnan, K.S., Balaram, P., Marshall, A.G.: De novo sequencing and disulfide mapping of a bromotryptophan-containing conotoxin by Fourier transform ion cyclotron resonance mass spectrometry. Anal. Chem. 78, 8082–8088 (2006)
28. Pham, V., Henzel, W.J., Arnott, D., Hymowitz, S., Sandoval, W.N., Truong, B.-T., Lowman, H., Lill, J.R.: De novo proteomic sequencing of a monoclonal antibody raised against OX40 ligand. Analytical Biochemistry 352, 77–86 (2006)
29. Resemann, A., Wunderlich, D., Rothbauer, U., Warscheid, B., Leonhardt, H., Fuschser, J., Kuhlmann, K., Suckau, D.: Top-Down de Novo Protein Sequencing of a 13.6 kDa Camelid Single Heavy Chain Antibody by Matrix-Assisted Laser Desorption Ionization-Time-of-Flight/Time-of-Flight Mass Spectrometry. Anal. Chem. 82, 3283–3292 (2010)
30. Savitski, M.M., Nielsen, M.L., Kjeldsen, F., Zubarev, R.A.: Proteomics-Grade de Novo Sequencing Approach. J. Proteome Research, 2348–2354 (2005)
31. Shevchenko, A., et al.: Charting the proteomes of organisms with unsequenced genomes by MALDI-quadrupole time-of-flight mass spectrometry and BLAST homology searching. Anal. Chem. 73, 1917–1926 (2001)
32. Syka, J.E., Coon, J.J., Schroeder, M.J., Shabanowitz, J., Hunt, D.F.: Peptide and protein sequence analysis by electron transfer dissociation mass spectrometry. Proc. Natl. Acad. Sci. USA 101, 9528–9533 (2004)
33. Tabb, D.L., Smith, L.L., Breci, L.A., Wysocki, V.H., Lin, D., Yates III., J.R.: Statistical characterization of ion trap tandem mass spectra from doubly charged tryptic digests. Anal. Chem. 75, 1155–1163 (2003)
34. Tabb, D.L., MacCoss, M.J., Wu, C.C., Anderson, S.D., Yates III., J.R.: Similarity among tandem mass spectra from proteomic experiments: detection, significance, and utility. Anal. Chem. 75, 2470–2477 (2003)
35. Taylor, J.A., Johnson, R.S.: Implementation and uses of automated de novo peptide sequencing by tandem mass spectrometry. Anal. Chem. 73, 2594–2604 (2001)
36. Tayo, L.L., Lu, B., Cruz, L.J., Yates III., J.R.: Proteomic analysis provides insights on venom processing in *Conus textile*. J. Proteome Research 9, 2292–2301 (2010)
37. Ueberheide, B.M., Fenyö, D., Alewood, P.F., Chait, B.T.: Rapid sensitive analysis of cysteine rich peptide venom components. Proc. Natl. Acad. Sci. USA 106, 6910–6915 (2009)
38. Zhang, Z., McElvain, J.S.: De novo peptide sequencing by two-dimensional fragment correlation mass spectrometry. Anal. Chem. 72, 2337–2350 (2000)
39. Alpha-conotoxin family signature. Accession number PS60014, ProSite ExPASy Proteomics Server (March 2005)

Metabolic Network Analysis Demystified

Leonid Chindelevitch[1,2], Aviv Regev[1,2], and Bonnie Berger[1,2,*]

[1] Mathematics Department,
Computer Science and Artificial Intelligence Laboratory, MIT,
Cambridge, MA 02139
[2] Broad Institute, 7 Cambridge Center, Cambridge, MA 02142
bab@mit.edu

Motivation. Metabolic networks are a representation of current knowledge about the metabolic reactions available to a given organism. These networks can be placed into various mathematical frameworks, of which the constraint-based framework [1] has received the most attention over the past 15 years. This results in a predictive model of metabolism. Metabolic models can yield predictions of two types: quantitative, such as the growth rate of an organism under given experimental conditions [2], and qualitative, such as the viability of a mutant [3] or minimal media required for growth [4]. Qualitative predictions, on which we focus, tend to be more robust and reliable than quantitative ones, while remaining experimentally testable and biologically relevant.

Here, we summarize new theoretical results related to metabolic models. These results are transformed into an algorithmic pipeline that reveals key structural properties of metabolic networks, such as blocked reactions, enzyme subsets, elementary modes, essential reactions and synthetic lethality. While the constraint-based approach to metabolic network analysis is over 15 years old, this work is, to our knowledge, the first time the theory of linear programming is used to reveal structural elements of metabolic models, rather than just predict a growth phenotype. We believe that a deeper understanding of these models will ultimately result in their wider applicability to biological questions.

Methods. Theorems 1 and 2 state that cut sets and modes are closely related in both fully reversible as well as fully irreversible networks. This relationship is based on the duality between the rowspace and the nullspace of a matrix, which is different from the Boolean duality described by Klamt and Gilles [5]. The same relationship holds between minimal cut sets and elementary modes. Theorems 1 and 2 provide a characterization of cut sets, which yields both an efficient method for identifying such sets as well as several important structural insights.

Theorem 3, based on a special case of Theorem 2, provides an efficient method for identifying all blocked reactions in a metabolic network. It is helpful to identify these both for simplifying subsequent analysis (they can be deleted) and for pinpointing the areas in which our knowledge of metabolism may currently

---

* Corresponding author.

V. Bafna and S.C. Sahinalp (Eds.): RECOMB 2011, LNBI 6577, pp. 31–33, 2011.
© Springer-Verlag Berlin Heidelberg 2011

be incomplete. Theorem 4 is another application of Theorem 2 and provides an efficient method for identifying all enzyme subsets in a metabolic network. This method was used previously by Gagneur and Klamt [6], but the fact that it actually identifies all enzyme subsets had not been established to our knowledge.

Theorem 5 is a result about the reduction of a stoichiometric matrix to canonical form. We say that S is in *canonical form* if it contains no blocked reactions, no effectively unidirectional reactions (reversible reactions which can only proceed in the forward or only in the reverse direction), no enzyme subsets, and no linearly dependent rows. We propose a 4-step reduction process that eliminates each of these undesirable structural elements in turn. The highly technical theorem 5 states that this 4-step reduction process we propose is guaranteed to converge after a single iteration, unlike the one proposed by Gagneur and Klamt [6]. Finally, Theorem 6 is an auxiliary result about the numerical stability of blocked reactions. It says that if a reaction is blocked, then any reaction obtained from it by a small perturbation will be blocked as well.

Results. We have applied the algorithmic pipeline based on the methods above to each of the 52 genome-scale metabolic networks representing 37 different species, downloaded from the UCSD Systems Biology group website [7] and parsed by a script we developed. The most significant result is our finding that, of the 45 networks containing a well-defined biomass reaction, 20 are certifiably unable to exhibit growth. Another remarkable result for these networks is that their canonical form (obtained by the process outlined in the discussion of Theorem 5) tends to be significantly smaller than their original stoichiometric matrix, providing an average 23-fold reduction in size (the average is computed only over networks that contain a biomass reaction). The fraction of blocked reactions in these networks was 35.3% on average. This could mean one of the following: either the state of our knowledge of the metabolism in these organisms is incomplete, and the reactions are indications of systematic gaps in our knowledge, or the constraint-based formalism is too stringent for these reactions (in our opinion, the former is much more likely to be the case). In addition, a significant fraction of the remaining (unblocked) reactions in each network are part of an enzyme subset. The average fraction of unblocked reactions in an enzyme subset is 54%. It is also interesting that the biomass reaction is always in an enzyme subset when it is not blocked, likely because of the large number of metabolites it typically involves.

References

1. Price, N., Reed, J., Palsson, B.: Genome-scale models of microbial cells: evaluating the consequences of constraints. Nature Reviews Microbiology 2, 886–897 (2004)
2. Varma, A., Palsson, B.: Stoichiometric Flux Balance Models Quantitatively Predict Growth and Metabolic By-Product Secretion in Wild-Type Escherichia coli W3110. Applied and Environmental Microbiology, 3724–3731 (1994)
3. Wunderlich, Z., Mirny, L.: Using the topology of metabolic networks to predict viability of mutant strains. Biophysics Journal 91(6), 2304–2311 (2006)

4. Suthers, P., Dasika, M., Kumar, V., Denisov, G., Glass, J., Maranas, C.: A genome-scale metabolic reconstruction of Mycoplasma genitalium, iPS189. PLoS Computational Biology 5(2), e1000285 (2009)
5. Klamt, S., Gilles, E.: Minimal cut sets in biochemical reaction networks. Bioinformatics 20, 226–234 (2004)
6. Gagneur, J., Klamt, S.: Computation of elementary modes: a unifying framework and the new binary approach. BMC Bioinformatics 5(175) (2004), doi:10.1186/1471-2105-5-175
7. In Silico Organisms,
    http://gcrg.ucsd.edu/In_Silico_Organisms/Other_Organisms

Causal Reasoning on Biological Networks: Interpreting Transcriptional Changes
(Extended Abstract)

Leonid Chindelevitch[1], Daniel Ziemek[1,*], Ahmed Enayetallah[2],
Ranjit Randhawa[1], Ben Sidders[3], Christoph Brockel[4], and Enoch Huang[1]

[1] Computational Sciences Center of Emphasis,
Pfizer Worldwide Research and Development, Cambridge, MA, USA
[2] Compound Safety Prediction,
Pfizer Worldwide Medicinal Chemistry, Groton, CT, USA
[3] eBiology, Pfizer Worldwide Research and Development, Sandwich, Kent, UK
[4] Translational and Bioinformatics,
Pfizer Business Technologies, Cambridge, MA, USA

1 Introduction

Over the past decade gene expression data sets have been generated at an increasing pace. In addition to ever increasing data generation, the biomedical literature is growing exponentially. The PubMed database (Sayers et al., 2010) comprises more than 20 million citations as of October 2010. The goal of our method is the prediction of putative upstream regulators of observed expression changes based on a set of over 400,000 causal relationships. The resulting putative regulators constitute directly testable hypotheses for follow-up.

2 Methods

In order to find those regulators, we first construct a *causal graph* G_C whose nodes are transcript levels, compound concentrations, and states of biological processes. To represent causality, each node appears twice, once with a + sign (upregulation) and once with a − sign (downregulation). A directed edge from node a to node b means that the abundance or activity of b is regulated by the abundance or activity of a. Each node is annotated with various identifiers, and each edge is annotated with the article it is based on and the specific excerpt that gave rise to it, to facilitate hypothesis validation for the scientists. We licensed the substrate for our method from two vendors: Ingenuity Inc. and Genstruct Inc. This yields 250,000 unique relationships covering 65,000 full-text articles indexed by PubMed. Pollard et al., 2005 presented a similar approach, but did not provide any details on implementation.

The gene expression data determines the subset G^+ of all genes that are significantly overexpressed and the subset G^- of all genes that are significantly

---

* Joint first author, corresponding author: daniel.ziemek@pfizer.com

V. Bafna and S.C. Sahinalp (Eds.): RECOMB 2011, LNBI 6577, pp. 34–37, 2011.

underexpressed. We define $G^{\pm} := G^{+} \cup G^{-}$. We choose a distance threshold Δ which determines the maximum length of the paths we consider. Given a hypothesis $h \in V(G_C)$, we classify each node of G_C into one of three possible sets: $S_h^{+} := \{v \in V(G_C) | d(h, v) \leq \Delta, d(h, v) < d(h, -v)\}$, $S_h^{-} := \{v \in V(G_C) | d(h, -v) \leq \Delta, d(h, -v) < d(h, v)\}$, $S_h^{0} := \{v \in V(G_C) | d(h, v) > \Delta \text{ or } d(h, v) = d(h, -v)\}$, where $d(\cdot, \cdot)$ is the distance between two nodes in the graph G_C. In order to evaluate the goodness-of-fit of a hypothesis h to the observed expression data, we score 1 for each *correct* prediction, -1 for each *incorrect* prediction and 0 for each *ambiguous* prediction made by h about $G^{\pm}$. We define $n_{\sigma, \tau} := |S_h^{\sigma} \cap G^{\tau}|$ for $\sigma, \tau \in \{+, -\}$. That is, the score of hypothesis h is $s(h, G^{\pm}) = n_{++} + n_{--} - n_{+-} - n_{-+}$.

However, a good score does not necessarily mean good explanatory power, because of possible connectivity differences between the nodes of G_C. Therefore we also look at statistical significance. For a given hypothesis h and a given score $s_0 := s(h, G^{\pm})$, we would like to compute the probability of h scoring s_0 or better with a *random* set of genes $G_R^{\pm} := G_R^{+} \cup G_R^{-}$, chosen with $|G_R^{+}| = |G^{+}|$ and $|G_R^{-}| = |G^{-}|$. We have developed a method for computing this probability in time cubic in $|G^{\pm}|$.

When processing a particular data set, our algorithm begins by computing the scores for each hypothesis and ranks the set of all hypotheses by their score. The correctness p-value p of a hypothesis is typically required to be below a certain threshold. The enrichment p-value p_E of a hypothesis is also required to pass a certain threshold. p_E is the probability of finding $n_{++} + n_{--} + n_{+-} + n_{-+}$ differentially expressed transcripts for a putative hypothesis h under the null model and represents a standard measure in gene set overrepresentation methods (e.g. Draghici et al., 2003) . Finally, we may also filter out those hypotheses whose number of correct predictions, $C := n_{++} + n_{--}$, is below a certain user-defined threshold.

Table 1. The top five causal hypotheses from the three oncogene expression signatures (Bild et al., 2006) are shown in the table, where C is the number of transcript changes correctly explained by the hypothesis; I incorrectly & A ambiguously. A +/- indicates the inferred directionality of the hypothesis.

Myc							E2F3							H-Ras						
Gene	Rank	Score	C	I	A	p	Gene	Rank	Score	C	I	A	p	Gene	Rank	Score	C	I	A	p
MYC+	1	22	23	1	1	$2 \cdot 10^{-14}$	CDKN2A -	1	12	13	1	1	$3 \cdot 10^{-9}$	TNF +	1	36	47	11	6	$1 \cdot 10^{-15}$
ZBTB16 -	2	10	10	0	0	$4 \cdot 10^{-11}$	E2F1 +	2	10	11	1	0	$8 \cdot 10^{-6}$	IL1B +	2	28	32	4	1	$5 \cdot 10^{-15}$
ALK +	3	9	9	0	0	$3 \cdot 10^{-12}$	E2F family +	3	5	5	0	0	$7 \cdot 10^{-5}$	F2 +	3	23	27	4	0	$4 \cdot 10^{-16}$
TP53 -	4	8	12	4	0	$2 \cdot 10^{-3}$	PROX1 +	4	4	4	0	0	$4 \cdot 10^{-6}$	EGF +	4	21	26	5	0	$1 \cdot 10^{-12}$
HDAC6 -	5	3	3	0	0	$6 \cdot 10^{-5}$	ITGB1 -	5	3	3	0	0	$6 \cdot 10^{-5}$	TGFB1 +	5	21	31	10	2	$5 \cdot 10^{-8}$
...							...							HRAS +	10	15	19	4	0	$5 \cdot 10^{-9}$

3   Validation and Results

Using simulations we established that our method is able to recover embedded regulators given our causal graphs with high-accuracy in the presence of noise. In order to test the performance of the causal reasoning algorithm on a biological data set we sought out experimental data which had a single, well defined perturbation that should be identified by the algorithm. Bild et al., 2006 used recombinant adenoviruses to infect non-cancerous human mammary epithelial cells with a construct to overexpress one of five oncogenes; c-Myc, H-Ras, c-Src, E2F3 and β-catenin. The data from this paper was not present in the causal interaction knowledge base when we applied our causal reasoning algorithm to these published signatures. For three signatures (c-Myc, H-Ras, E2F3) either the overexpressed protein or a protein immediately downstream from it, is correctly identified by our algorithm as the top-ranked predicted hypothesis (Table 1). c-SRC and β-catenin had very few matching genes. Our method did not return highly significant results in those cases, meaning that no confident predictions were possible.

We also used our algorithm to compare myocardial gene expression changes associated with isoprenaline-induced (pathological) hypertrophy with exercise-induced (adaptive) hypertrophy in mice, obtained from the public domain (Galindo et al., 2009). In the isoprenaline group, the analysis supports biological networks of several hallmarks of cardiac disease and cardiomyocyte stress (e.g. Aragno et al., 2008). These include hypotheses indicative of increased hypoxia, increased NOS production, oxidative stress, inflammatory response and endoplasmic reticulum stress. In contrast, the exercise-induced hypertrophy demonstrates perturbation of the same biological networks as in the isoprenaline group but with reversed direction of regulation, e.g. decreased hypoxia.

The outlined results provide evidence that method based on the outlined score and statistical measures can accurately detect the underlying cause of a biological gene expression signature and identify regulatory modules from within a larger, more complex data set. In our experience the output of our method was easy to interpret for biologists, and several hypotheses have already been selected for follow-up. It is our hope that the interplay between experimental work based on our method, the discovery of novel biology and the subsequent enrichment of the causal graph will lead to a virtuous cycle allowing for the continued expansion of the boundaries of biological knowledge.

References

[Aragno et al.,2008]Aragno, M., Mastrocola, R., Alloatti, G., Vercellinatto, I., Bardini, P., Geuna, S., Catalano, M.G., Danni, O., Boccuzzi, G.: Oxidative stress triggers cardiac brosis in the heart of diabetic rats. Endocrinology 149(1), 380–388 (2008)

[Bild et al., 2006]Bild, A.H., Yao, G., Chang, J.T., Wang, Q., Potti, A., Chasse, D., Joshi, M.-B., Harpole, D., Lancaster, J.M., Berchuck, A., Olson, J.A., Marks, J.R., Dressman, H.K., West, M., Nevins, J.R.: Oncogenic pathway signatures in human cancers as a guide to targeted therapies. Nature 439(7074), 353–357 (2006)

[Draghici et al., 2003]Draghici, S., Khatri, P., Martins, R.P., Ostermeier, G.C., Krawetz, S.A.: Global functional profiling of gene expression. Genomics 81(2), 98–104 (2003)

[Galindo et al., 2009]Galindo, C.L., Skinner, M.A., Errami, M., Olson, L.D., Watson, D.A., Li, J., McCormick, J.F., McIver, L.J., Kumar, N.M., Pham, T.Q., Garner, H.R.: Transcriptional profile of isoproterenol-induced car- diomyopathy and comparison to exercise-induced cardiac hypertrophy and human cardiac failure. BMC Physiol. 9, 23 (2009)

[Pollard et al., 2005]Pollard, J., Butte, A.J., Hoberman, S., Joshi, M., Levy, J., Pappo, J.: A computational model to define the molecular causes of type 2 diabetes mellitus. Diabetes Technol. Ther. 7(2), 323–336 (2005)

[Sayers et al., 2010]Sayers, E.W., Barrett, T., Benson, D.A., Bolton, E., Bryant, S.H., Canese, K., Chetvernin, V., Church, D.M., Dicuccio, M., Federhen, S., Feolo, M., Geer, L.Y., Helmberg, W., Kapustin, Y., Landsman, D., Lipman, D.J., Lu, Z., Madden, T.L., Madej, T., Maglott, D.R., Marchler-Bauer, A., Miller, V., Mizrachi, I., Ostell, J., Panchenko, A., Pruitt, K.D., Schuler, G.D., Sequeira, E., Sherry, S.T., Shumway, M., Sirotkin, K., Slotta, D., Souvorov, A., Starchenko, G., Tatusova, T.A., Wagner, L., Wang, Y., Wilbur, W.J., Yaschenko, E., Ye, J.: Database resources of the national center for biotechnology information. Nucleic Acids Res. 38(Database issue), D5–D16 (2010)

Hapsembler: An Assembler for Highly Polymorphic Genomes

Nilgun Donmez and Michael Brudno

Department of Computer Science, The Donnelly Centre and Banting & Best
Department of Medical Research, University of Toronto
{nild,brudno}@cs.toronto.edu

Abstract. As whole genome sequencing has become a routine biological experiment, algorithms for assembly of whole genome shotgun data has become a topic of extensive research, with a plethora of off-the-shelf methods that can reconstruct the genomes of many organisms. Simultaneously, several recently sequenced genomes exhibit very high polymorphism rates. For these organisms genome assembly remains a challenge as most assemblers are unable to handle highly divergent haplotypes in a single individual. In this paper we describe Hapsembler, an assembler for highly polymorphic genomes, which makes use of paired reads. Our experiments show that Hapsembler produces accurate and contiguous assemblies of highly polymorphic genomes, while performing on par with the leading tools on haploid genomes. Hapsembler is available for download at http://compbio.cs.toronto.edu/hapsembler.

Keywords: Genome assembly, polymorphism.

1 Introduction

In the last decade the sequencing and assembly of genomes have become routine biological experiments. Almost all genome projects today use the whole-genome shotgun sequencing approach: many copies of the genome are randomly sheared, and each part is sequenced to generate a read. Based on the overlaps between these reads an assembly algorithm then reconstructs contigs (contiguous regions) of the original, sequenced genome. Genome assembly has been the topic of extensive research, with a plethora of off-the-shelf methods that can reconstruct the genomes of many species. Simultaneously, de novo assembly of some organisms remains a challenge. Assembly tools currently available are optimized for the assembly of large mammalian genomes [1], [8], or smaller, bacterial genomes from High Throughput Sequencing (HTS) data [14], [3], [2]. In both of these settings the difficulty of genome assembly is due to the relative sizes of the reads and the genome: short reads cannot span longer repeats, and make the assembly of the genome difficult. While many of these assembly tools also consider the possibility of polymorphisms in the sequenced individuals, the low frequency of SNPs and other variants in these genomes makes addressing these polymorphisms a relatively tractable problem.

V. Bafna and S.C. Sahinalp (Eds.): RECOMB 2011, LNBI 6577, pp. 38–52, 2011.

Simultaneously, there are now several known organisms with extremely high polymorphism rates [11],[13]: for example the C. savignyi genome has a SNP rate of 5% (50-fold higher than human). Because this variability is also present between the sequenced individual's paternal and maternal chromosomes, assembling these genomes is very difficult with current methods. Previous efforts to assemble these highly polymorphic genomes have used one of two approaches: either they allow for promiscuous overlaps, to connect the reads from different haplotypes[20], or they enforce strict overlapping requirements, thus separating the haploid genomes (haplomes) [10]. The first approach has the drawback that the spurious overlaps make it difficult to reconstruct long segments of the genome. The second approach will unnecessarily subdivide the genome, as many real overlaps will be removed. Furthermore, higher coverage will be needed, as the effective total size of the genome is doubled. In the C. savignyi genome assembly the two haplomes were aligned to each other to better reconstruct the genome [12], however only 60% of the genome could be assembled automatically, and the rest was finished via manual analysis of the contigs [10].

In this paper we present Hapsembler, a haplotype-aware genome assembly algorithm. Hapsembler combines the classical overlap-layout-consensus approach for genome assembly with a haplotype-aware, Bayesian approach for error correction and a novel structure we term the matepair graph, that helps to reconstruct the genome via a thorough analysis of the paired sequencing reads.

2   Methods

Hapsembler consists of three main modules: the alignment module, the error correction module, and the graph module. The alignment module is used to compute pairwise sequence alignments between reads that obey certain criteria. This procedure employs an efficient kmer hashing technique to detect overlaps longer than a user defined length. This initial set of overlaps is used to correct sequencing errors by a probabilistic error correction procedure based on Naive Bayes [15]. The corrected reads are then passed through another overlapping stage to compute the final set of overlaps.

The graph module builds an overlap graph (a.k.a string graph [7]) using the overlaps reported by the alignment module. This graph is used to construct "path sets" representing possible tilings between paired reads. A mate pair graph is then constructed in which nodes represent mate pairs, and edges represent possible overlaps between mate pairs as constrained by the path sets. Finally, contigs are determined by finding maximal paths in the mate pair graph.

In the following subsections we give a detailed explanation of these steps.

2.1   Overlap Computation

To compute overlaps between the reads, we employ a kmer hashing technique similar to [9]. After masking frequent kmers, reads that share a sufficient number of kmers (see Appendix 1 for these and other parameters) are aligned using a modified Needleman-Wunsch algorithm. This modified algorithm uses the positions of matching kmers to estimate the overlapping portion of the reads. For instance, if a kmer is at index 60 in read R and at index 45 in read R', the start of the reads are assumed to be 15 base pairs apart. Consequently, only the corresponding diagonal and diagonals within a distance of d are computed in the dynamic programming matrix. The parameter d is dynamically set to $e * l$, where e is a user defined error tolerance rate and l is the length of the read. In general, different kmers that are shared between two reads may suggest different starting indices. To prevent redundant calls to the Needleman-Wunsch procedure, we bundle such indices based on their proximity. Overlaps greater than a minimum length l_{min} with identity $(1 - e)$ or more are reported.

2.2   Error Correction

Real sequence reads often have errors even after aggressive quality trimming. To correct these errors at an early stage, we employ a double pass overlap computation approach. In this approach, the initial set of overlaps are used to correct the reads. For each read, the set of all pairwise alignments initially found are used to decide, for each base of the read, whether the position is correct or is a sequencing error. Note that, in the case of diploid organisms, error correction is further complicated, as disagreements between reads may be due to either errors, or polymorphisms (SNPs or small indels). To avoid overcorrecting reads with SNPs, we employ a Naive Bayesian method at this stage.

Briefly, this method works by scoring each pairwise alignment using base quality values and using this score to derive the probability that two reads are sampled from the same haploid genome (haplome). The consensus sequence for the read is then computed by weighting each alignment's contribution with this probability.

Formally, let $C_x \in \{A, C, G, T\}$ denote the xth base of a read R. Ambiguous bases (eg. N) are not used to correct other errors; they can, however, themselves be corrected. Suppose that R has pairwise alignments with the reads $\{S^1, S^2, ..., S^n\}$. Let F_i denote the base aligned with x in the pairwise alignment between R and S^i. Then the probability of C_x assuming a particular nucleotide is given by:

$$p(C_x | F_{i=1,2,...,n}) = p(C_x) \prod_{i=1}^{n} p(F_i | C_x) \tag{1}$$

Above $p(C_x)$ is the prior probability of C_x and it is equal to $1 - 10^{-q/10}$ if it has the same value as the nucleotide present in the read, where q is the Phred style [16] quality score associated with the base. Otherwise it is equal to $(10^{-q/10})/3$. $p(F_i | C_x)$ is given by the equation:

$$p(F_i | C_x) = p(F_i | C_x, B_i)p(B_i) + p(F_i | C_x, \neg B_i)p(\neg B_i) \tag{2}$$
$$= p(F_i | C_x, B_i)p(B_i) + p(F_i)p(\neg B_i) \tag{3}$$

$p(F_i|C_x, B_i)$ is the conditional probability of F_i given C_x and B_i where B_i is a binary variable indicating whether the two reads belong to the same locus of the genome/haplome or not. When two reads belong to different loci, the bases are independent of each other hence $p(F_i|C_x, \neg B_i) = p(F_i)$ (i.e. we have conditional independence). Here, we abuse the notation slightly and use $p(B_i)$ to denote the posterior probability of S^i and R belonging to the same locus. To estimate $p(B_i)$, we first have to compute the following probabilities:

$$p(B_{R,S^i}|R, S^i) = p(B) \prod_{j=1}^{k} p(R_j, S_j^i|B) \tag{4}$$

$$p(\neg B_{R,S^i}|R, S^i) = p(\neg B) \prod_{j=1}^{k} p(R_j, S_j^i|\neg B) \tag{5}$$

$$= p(\neg B) \prod_{j=1}^{k} p(R_j)p(S_j^i) \tag{6}$$

Above, k is the number of bases in the pairwise alignment between R and S^i. $p(R_j, S_j^i|B)$ denotes the conditional probability of the jth position in the alignment. If the reads R and S^i belong to the same region, this means the disagreements between the two sequences should be due to sequencing errors alone. In other words, if the two bases differ, at least one of them must be an error. Let q denote the quality value of the jth base in read R and q^i denote the quality value of the corresponding base in read S^i. If $R_j \neq S_j^i$:

$$p(R_j, S_j^i|B) = (1 - 10^{-q/10})((10^{-q^i/10})/3) \tag{7}$$
$$+(1 - 10^{-q^i/10})((10^{-q/10})/3) \tag{8}$$
$$+2((10^{-q/10})/3)((10^{-q^i/10})/3) \tag{9}$$

If $R_j = S_j^i$, on the other hand, either both reads are correct or they both have a sequencing error:

$$p(R_j, S_j^i|B) = (1 - 10^{-q/10})(1 - 10^{-q^i/10}) \tag{10}$$
$$+((10^{-q/10}))((10^{-q^i/10})/3) \tag{11}$$

The prior probability $p(B)$, of two reads belonging to the same location is set to a value near 1.0 since we only compare reads that have a sufficient overlap. The posterior probability $p(B_i)$ is then estimated as:

$$p(B_i) = \frac{1}{1 + exp(\log p(\neg B_{R,S^i}|R, S^i) - \log p(B_{R,S^i}|R, S^i))} \tag{12}$$

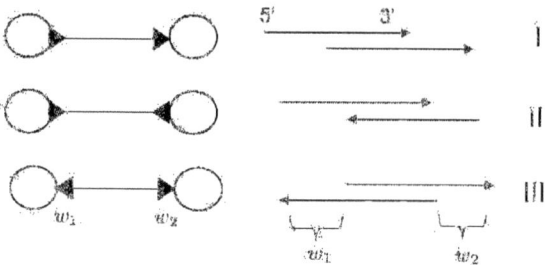

Fig. 1. *Possible edges in a bidirected overlap graph.* The directed lines to the right represent the reads where the arrowed end is the 3' end while the flat end is the 5' end. w_1 and w_2 are the lengths of the reads that are not covered by the overlap.

The consensus nucleotide of read R for position x is chosen to be the nucleotide that gives the highest probability $p(C_x|F_{i=1,2,...,n})$. In practice, we use the log probabilities to avoid multiplication. We also compute a look-up table in advance to avoid doing the same computations many times.

Note that the equations above do not account for indel (insertion/deletion) errors. Although indels could be handled similarly, there are no associated quality values with missing bases. Consequently, we handle indels separately. If a significant fraction of reads are calling for a deletion (at least 3 votes for deletion and at most 1 vote for no deletion; these parameters are conservative for the relatively low coverage levels used in this study, to avoid over-correction) the base is deleted. A similar rule is applied for insertion and the insertion base is selected using the same procedure as above where $\log p(C_x)$ is taken to be $\log(1/4)$. For the computation of $p(B_i)$, we use a default gap quality value. After all the reads are corrected as above, we do another pass of the overlapping phase, now with the corrected reads.

2.3   Building the Overlap Graph

Once the overlaps between the reads are finalized we build a bidirected overlap graph where the nodes are reads and the edges represent overlaps between the reads. Formally, a bidirected graph G is a graph with node and edge sets V and E, where each edge can acquire either of the two types of arrows at each vertex; *in-arrow* and *out-arrow* [6], [19]. As a result, there are 3 possible types of edges in a bidirected graph; *in-out*, *out-out* and *in-in* (Figure 1). A walk in a bidirected graph must obey the following rule: If we come to a node using an in-arrow we must leave the node using an out-arrow. Similarly, if we come to a node using an out-arrow we must leave using an in-arrow. In the former case, the node is said to be *in-visited* and in the latter case it is said to be *out-visited*.

The Figure 1 omits non-proper overlaps that are caused by reads that are entirely contained by other reads. These reads are excluded when building the overlap graph.

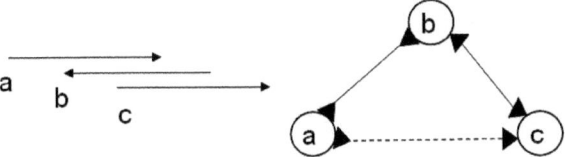

Fig. 2. *Transitive edge reduction.* Since we can reach the node **c** from **a** via **b**, we do not need the edge between **a** and **c**. Note that during this procedure, we check whether the two paths have the same overall length within a permissible difference f, where f is defined as the total number of indels that are present in the pairwise alignments associated with the overlaps. Otherwise, the edge is not deleted.

Many edges in the initial overlap graph will be redundant since they can be inferred by other edges. These edges are removed from the graph using an operation called *transitive edge reduction* [7] as illustrated in Figure 2.

2.4 Finding Paths between Mate Pairs

At this stage, we have a simplified overlap graph in which most nodes have 1 incoming and 1 outgoing edge. However, some nodes will have degrees of 3 or more due to repeats or polymorphic regions. In this section, we will show how to use the mate pairs to resolve such nodes to generate longer paths.

For each mate pair in our dataset, we assume there is a given approximate insert size mean μ_l and standard deviation σ_l. Ideally, we would like to find a single path between each pair with a length in the range $\mu_l \pm (k\sigma_l)$, where k is a real number controlling the largest deviation from the mean we are willing to allow. In general, there may be an exponential number of such paths in the graph. However, we can identify the subgraph containing all paths between two nodes *shorter than* a given length in polynomial time.

This idea can be summarized as follows. Let nodes a and a' be a mate pair. First, we perform Dijkstra's [4] shortest path finding algorithm starting from a (and leaving the node using an out-arrow). While Dijkstra's algorithm is originally invented for directed or undirected graphs, the generalization to bidirected graphs is straightforward. The only difference is that instead of a single distance from the source, we have to keep track of the shortest in-distance and out-distance separately for each node. We also modify the algorithm so that only the nodes that are within a distance of $\mu^* = \mu_l + (k\sigma_l)$ are enqueued. During this search, if we do not encounter a' it means there is no path in the graph between a and a' less than the given length. This situation can arise for several reasons: (1) the insert size deviation might be higher than the expected for that mate pair, (2) there might be a region with no coverage between the mates, or (3) we might be missing overlaps (due to sequencing errors, short overlaps, etc). If we encounter a' during the search, we do another pass of Dijkstra's, this time starting from a'. During this second pass, we enqueue a node if and only if the sum of its shortest distance from a, its current distance from a' and the read's

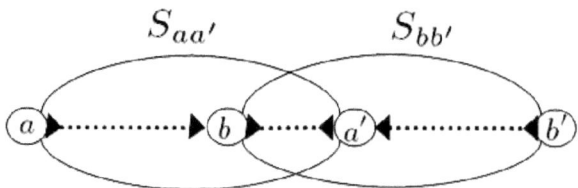

Fig. 3. *Overlapping mate pairs.* By comparing the previously computed sets of $S_{aa'}$ and $S_{bb'}$, we can determine whether the two mate pair nodes should be connected or not.

length is less than μ^*. Furthermore, we put such nodes into a set which we shall call $S_{aa'}$ together with their shortest in/out-distances from a and a'.

After this second pass, we end up with a set of nodes, $S_{aa'}$, that are guaranteed to lie on at least one path that has length less than μ^* between a and a'. It is easy to show that this set is also exhaustive; that is all vertices v that satisfy $indist(a, v) + outdist(a', v) + length(v) < \mu^*$ or $outdist(a, v) + indist(a', v) + length(v) < \mu^*$, where $in/outdist(x, y)$ denotes the in/out distance of node y from node x, are included in $S_{aa'}$.

Note that we find this set of nodes in polynomial time even though there might be an exponential number of paths between a and a'.

2.5   Building the Mate Pair Graph

The process described above is repeated for each mate pair, yielding a set of sets $\bar{S}$. We then use these sets to build the mate pair graph.

Consider two mate pair sets $S_{aa'}$ and $S_{bb'}$. To decide if these mate pairs have paths that overlap with each other, we first check whether each of the following conditions hold:

$$a \in S_{bb'} \tag{13}$$
$$a' \in S_{bb'} \tag{14}$$
$$b \in S_{aa'} \tag{15}$$
$$b' \in S_{aa'} \tag{16}$$

Whenever there is less than two positive answers to the checks, an overlap of paths is not possible. Otherwise, we check if the length of the paths and orientations are compatible. For example, if we find that $a' \in S_{bb'}$ and $b \in S_{aa'}$, we check whether the following inequalities hold (Figure 3):

$$indist(a, b) + outdist(a', b) + length(b) < \mu^* \tag{17}$$
$$outdist(b, a') + indist(b', a') + length(a') < \mu^* \tag{18}$$

where μ^* is defined as above. Recall that we store these distances together with the nodes, hence these checks are done in constant time. However, this algorithm may become problematic in terms of memory for large insert sizes since we store a set proportional to the size of the insert for each mate pair (for fixed coverage and read length). In practice, we use a slightly different version of this algorithm which can be implemented to give linear space complexity independent of the insert size. In this version, we perform two extra Dijkstra's starting from each end of a mate pair, this time in opposite directions (i.e. leaving the node using an in-arrow). As before, we only enqueue nodes that are within the distance cutoff. This gives us two additional sets $S_{\hat{a}}$ and $S_{\hat{a}'}$. Then for each node b in $S_{aa'}$ we check if its mate pair b' is in $S_{\hat{a}}$ or $S_{\hat{a}'}$ depending on the orientation of the node. If the path lengths are compatible with the insert sizes (see above) then we put an edge between aa' and bb' in the mate pair graph. Since we can immediately determine which edges aa' should be incident to, we do not have to store the set of aa' after we process aa'. As a result, this alternative algorithm takes only linear space in the number of nodes.

2.6   Processing the Mate Pair Graph

The mate pair graph is structurally similar to the overlap graph and can undergo similar simplifying procedures, namely nested mate pair removal and transitive edge reduction.

When a mate pair lies entirely within an other mate pair, it creates an unnecessary bubble or dead end in the mate pair graph . Such nested mate pairs are detected and marked while building the mate pair graph. For a mate pair aa', this is done by checking the sets $S_{\hat{a}}$ and $S_{\hat{a}'}$ to see if there is any mate pair bb' such that $b \in S_{\hat{a}}$, $b' \in S_{\hat{a}'}$ and:

$$outdist(a, b) + dist(a, a') + outdist(a', b') < \mu^* \tag{19}$$

where $dist(a, a')$ is the shortest distance between the pair aa' as computed during Dijkstra's algorithm. If there is any such mate pair, aa' is removed from the mate pair graph.

Finally, we perform transitive edge reduction on the mate pair graph. When creating the mate pair graph, three values are stored with each edge. For example, for the mate pair edge between aa' and bb' of Figure 3, we would store the distances $indist(a, b)$, $outdist(b, a')$ and $indist(a', b')$. Given three mate pair nodes where each node is connected to the other two, we decide whether one of the edges can be inferred from the other two using these distances.

2.7   Polymorphism and Repeat Resolution with the Mate Pair Graph

In essence, polymorphism resolution is very similar to repeat resolution and Hapsembler is designed to exploit paired reads to handle both problems at once. Figure 4 illustrates how the graph module works on a toy example. In this

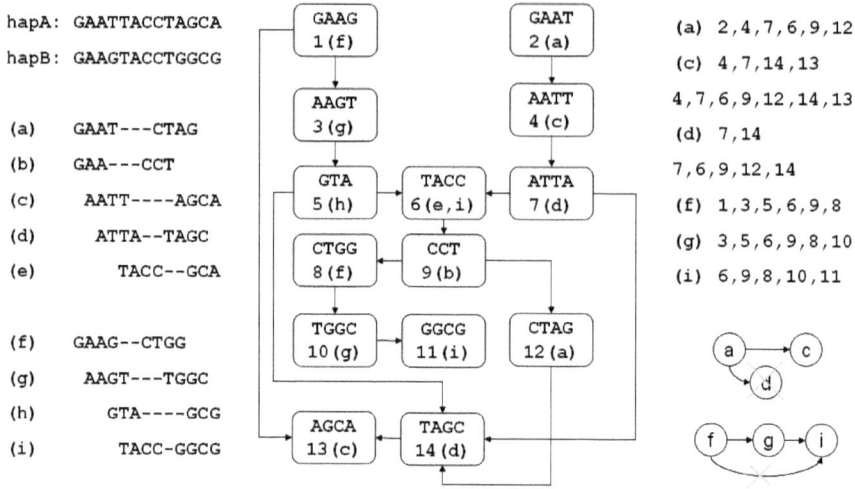

Fig. 4. *Polymorphism and repeat resolution.* **Left:** The diploid genome and mate pairs sampled from it. We do not know the exact distances between the pairs but we assume that we are given an upper bound (in this case 13bp). **Middle:** The overlap graph after removal of contained reads (i.e. GAA, GCA and GCG) and transitive edge reduction. The nodes are labelled with the mate pairs they belong to and arbitrarily numbered. The minimum overlap size is set to 2bp. **Right:** The paths between the mate pairs shorter than the given upper bound and the resulting mate pair graph. In practice, we do not need the exact paths and we only compute the set of nodes that lie on at least one path. The node d is removed since it is contained by node c. In addition, the edge between f and i is removed during transitive edge reduction. The resulting paths correspond to the two haplomes.

example, we have a diploid genome[1] which has several SNPs. After the overlap graph is built and simplified, we still have several ambiguous nodes (nodes with degree 3 or more). Some of these ambiguities are due to short repeats and some due to regions that are identical in both haplomes. Nevertheless, the mate pair graph built upon this overlap graph is less tangled. Indeed, the simplified mate pair graph has exactly two disjoint paths, each spelling the sequence of one haplome.

Although the toy example given in Figure 4 is completely resolved by our algorithm, in general the mate pair graph might still have ambiguous nodes. In particular, as with repeats, haplotype resolution is limited to the size of the inserts. After simplification, we report each uninterrupted path in the mate pair graph as a contig. If there are long chains in the simplified overlap graph that have not been visited during the mate pair graph traversal (or if all the reads are unpaired), these are also reported as contigs. The consensus sequence for each contig is generated using a greedy multiple sequence alignment.

---

[1] For simplicity, we assume the genome is single stranded.

3   Results

We analyzed the performance of Hapsembler in two categories: read correction and assembly. For each category, we performed experiments with simulated and real reads. Since Hapsembler is built for polymorphic data, we compiled a reference sequence using the C. savignyi reference assembly as described in [11]. This draft assembly is organized in 374 "hypercontigs", where each hypercontig is a pairwise alignment of two sequences, each representing a single haplotype. To use in our experiments, we picked the largest three hypercontigs, totalling roughly 33mbp (haploid size = 16.5mbp).

Since C. savignyi has a very complex genome with many long repeats and a very high polymorphism rate, we only used Sanger reads in our experiments with this genome. However, we also give results on a bacterial genome using Roche/454 reads. For these experiments we downloaded 454 reads from an ongoing sequencing project on evolution of antibiotic resistant E. coli (NCBI Short Read Archive, accession code: SRR024126). This dataset includes 110,388 reads amounting to 10x sequence coverage. The reads were downloaded using the clipping option and barcode trimmed. As our reference we used the NCBI reference E. coli sequence (NC_000913.2).

3.1   Error Correction

To test the effect of our error correction we first simulated Sanger reads at different coverage levels from the C. savignyi reference. For these simulations, we downloaded the raw Sanger reads from the C. savignyi sequencing project [18] to use as templates. Using the length of these reads as a distribution, we uniformly sampled Sanger reads generating 13x, 10x, and 7x haploid coverage levels. The templates are also used to model errors by converting the quality scores to error probabilities. For instance, if a base has quality 20, we introduce an error with probability 0.01.

We compare the performance of our error correction method to the method of [17] (referred to as H-SHREC throughout this text). We choose this implementation of SHREC for its ability to handle indel type errors. The results on simulated Sanger reads is summarized in table 1. Even though the reads are sampled from a highly polymorphic reference, Hapsembler is able to correct a large fraction of errors while introducing relatively few errors. H-SHREC fails to reduce the total numbers of errors for all three datasets.

Next we assessed the effect of our error correction on two real datasets. As our first dataset we used a subset of the real Sanger C. savignyi reads as follows. After vector and quality trimming, we mapped the reads to the reference sequence we described above using MUMmer (version 3.22)[5]. A read is considered eligible if at least 90% of its sequence maps to a location with a minimum of 95% identity. For each eligible read, we also included its mate pair. This procedure yielded 558,936 reads totalling 358mbp. As our second dataset we used the 454 E. coli reads described above.

Table 1. *The effect of error correction on simulated Sanger reads.* The reads are first quality trimmed and then subjected to error correction. Total bases is calculated as the sum of all of the reads after trimming. Miscorrections denote errors introduced by the error correction procedure. Number of errors and miscorrections are calculated via alignments to the original error-free reads.

Coverage	Total (mbp)	No. of errors after trimming	Method	No. of errors after correction	No. of miscorrections
13x	194	3,924,331	Hapsembler	598,111	45,911
			H-SHREC	4,750,235	2,221,266
10x	148	3,056,019	Hapsembler	631,957	48,997
			H-SHREC	4,370,374	2,443,059
7x	102	2,199,861	Hapsembler	741,785	51,746
			H-SHREC	2,893,788	1,453,007

Since in this case we do not have the ground truth, we assessed the performance of both methods by mapping the reads to the reference sequences before and after error correction. Results are summarized in table 2. In the C. savignyi dataset, after correcting with Hapsembler, the number of reads mapping perfectly increases by more than 6-folds. H-SHREC moderately improves the number of reads mapping perfectly, however the number of reads mapping at the 95% threshold decreases, suggesting that the overall number of errors might have increased. On the E. coli dataset, Hapsembler and H-SHREC perform similarly.

3.2   Assembly

We first evaluated the performance of Hapsembler on the 454 E. coli dataset. We compare the results achieved by Hapsembler with Velvet [14] and Euler [3], which have support for 454 reads. Since Hapsembler is designed to work with paired reads we also simulated an artificial pairing of these reads as follows. Reads are mapped to the reference sequence and sorted by their mapping positions allowing duplicate mappings. For each read mapping to the forward strand, an unpaired read mapping on the opposite strand that has distance closest to 8000bp is taken. If there is no such read with distance 8000 ± 2400, then the read is left unpaired. This mapping yielded 33,160 pairs with insert size mean and standard deviation of 8534.67 and 713.35 respectively. The rest of the reads are left as single reads. Table 3 show the results for contigs of size 500bp or greater.

To evaluate Hapsembler on a polymorphic genome, we simulated 13x coverage of paired Sanger reads from the C. savignyi reference haplomes with errors added as described above. The start of the first reads are chosen uniformly and the start the paired reads are selected using a normal distribution with mean and standard deviation 10kbp and 1kbp respectively. The results are summarized in table 4. Of particular emphasis for polymorphic assembly is the ability of an algorithm to report haplotype-specific contigs, rather than mozaics from the two

Table 2. *The effect of error correction on real Roche/454 and Sanger reads.* C. savignyi dataset consists of 558,936 Sanger reads and E. coli dataset consists of 110388 Roche/454 reads. Reads are mapped to the reference sequences using MUMmer. Number of reads mapped is calculated by counting the reads that map with at least 95% identity and 95% coverage. A read is considered to be perfect if the entire read maps with 100% identity. The numbers in paranthesis denote the number of discarded reads (by H-SHREC) that map in each category. H-SHREC discards a total of 13600 and 1132 reads from the C. savignyi and E. coli datasets respectively.

Data	Error correction	No. of reads mapped	No. of reads mapped perfectly	Total size of perfect reads (mbp)
C. savignyi	Hapsembler	421,819	126,306	83.9
	H-SHREC	(11,401) 391,016	(761) 48,994	32.6
	None	411,626	20,689	13.6
E. coli	Hapsembler	89,817	10,292	3.9
	H-SHREC	(738) 88,814	(10) 9,573	3.7
	None	88,624	4,154	1.6

Table 3. *Assembly of E.coli with real and artificially paired 454 reads.* Coverage and accuracy are computed by mapping contigs to E. coli reference sequence (4639kbp) using MUMmer. N50 is defined as the largest contig size such that the sum of contigs at least as long is greater than half the genome size.

Reads	Tool	N50 (kbp)	No. of Contigs	Total Size (kbp)	Coverage (%)	Accuracy (%)
	Hapsembler	72.4	128	4600	90.3	98.6
unpaired	Velvet	41.2	199	4585	89.5	98.4
	Euler	8.1	913	4731	88.6	98.6
	Hapsembler	103.7	111	4693	90.9	98.6
paired	Velvet	41.9	189	4607	89.5	98.2
	Euler	9.4	765	4593	88.4	98.6

haplotypes. However, conserved sequences and low coverage regions make this difficult, and the longer contigs may have several "jumps" between the haplomes. To estimate the long-range linkage information in the contigs, we computed maximal haplotype blocks in the assembly, for which all of the SNPs match one haplome. In Figure 5 we plot the fraction of genome covered by haplotype specific blocks of a certain size or greater, measured in the number of SNPs. The figure shows that half of the genome is covered by haplotype-specific blocks containing 300 adjacent SNPs or more.

Table 4. *Assembly of three hypercontigs of C. savignyi with simulated paired reads.* Contigs are mapped to the reference haplomes using MUMmer. Coverage is calculated by taking only the best hit of each location of the contigs. Accuracy is calculated as percent identity excluding SNPs. N50 is defined as the largest contig size such that such that the sum of contigs at least as long is greater than half the genome size, where the genome size is taken as the sum of the two haplomes.

Tool	N50 (kbp)	No. of Contigs	Total Size (mbp)	Coverage (%)	Accuracy (%)
Hapsembler	23.4	2886	34	87.4	99.4

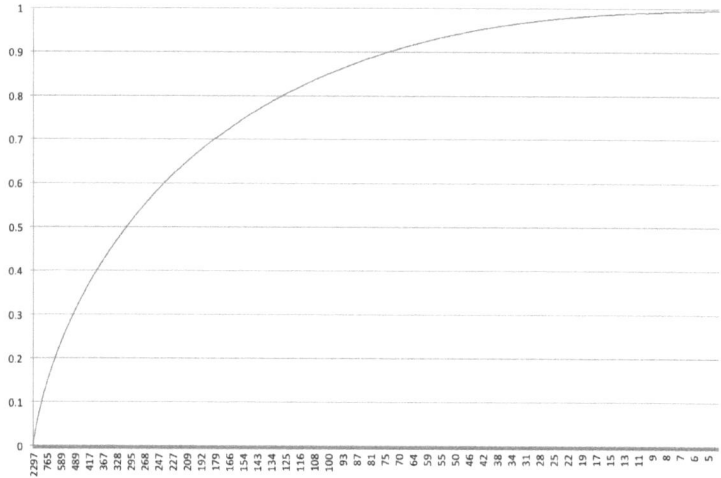

Fig. 5. *Fraction of genome in haplotype blocks.* X axis denote the number of adjacent SNPs covered by a contiguous region of a contig. Y axis shows the fraction of genome covered by the haplotype blocks.

4   Discussion

In this paper we presented Hapsembler, an assembly algorithm for whole genome shotgun data that is optimized for highly polymorphic genomes. Due to the large number of differences between the maternal and paternal copies of the chromosomes, these genomes have classically been difficult to assemble, with custom algorithms [12] and extensive manual intervention [10] required to achieve a high quality assembly. Hapsembler, to our knowledge, is the first tool that specifically targets this problem.

Hapsembler combines the use of mate pairs with a sophisticated error correction procedure to achieve a better assembly. Nevertheless, the methods required for this improvement are computationally expensive. Currently the most time consuming steps are read overlapping and mate pair graph building stages. While Hapsembler takes 36 minutes to assemble the E. coli dataset, Euler and Velvet take only a few minutes each. Fortunately, both of these bottlenecks are suitable for parallel computation. For example, the overlapping stage of 558k reads from the C. savignyi dataset takes less than 40 minutes using four quad-core Intel 3 GHz Xeon compute nodes. Nevertheless, further improvements are necessary to make Hapsembler work on large scale whole-genome datasets. Similarly, additional future work is necessary to take advantage of high-coverage High Throughput Sequencing reads (Solexa/Illumina or AB/SOLiD) in combination with lower-coverage Sanger reads. Finally we believe that exploring additional representation methods for polymorphic genome assemblies is a fruitful area for future research. We believe that representing haplotypes as paths on a genome graph (similar to Allpaths [2]) may allow for representation of the inherent complexity of polymorphic genomes.

Acknowledgements

This research was supported by an NSERC Discovery Grant to MB.

References

1. Batzoglou, S., Jaffe, D.B., Stanley, K., Butler, J., Gnerre, S., Mauceli, E., Berger, B., Mesirov, J.P., Lander, E.S.: ARACHNE: A Whole-Genome Shotgun Assembler. Genome Research 12, 177–189 (2002)
2. Butler, J., et al.: ALLPATHS: De novo assembly of whole-genome shotgun microreads. Genome Research 18, 810–820 (2008)
3. Chaisson, M.J., Pevzner, P.A.: Short read fragment assembly of bacterial genomes. Genome Research 18, 324–330 (2008)
4. Dijkstra, E.W.: A note on two problems in connexion with graphs. Numerische Mathematik 1, 269–271 (1959)
5. Kurtz, S.: Versatile and open software for comparing large genomes. Genome Biology 5, R12 (2004)
6. Medvedev, P., Georgiou, K., Myers, E.W., Brudno, M.: Computability of Models for Sequence Assembly. In: Giancarlo, R., Hannenhalli, S. (eds.) WABI 2007. LNCS (LNBI), vol. 4645, pp. 289–301. Springer, Heidelberg (2007)
7. Myers, E.W.: The fragment assembly string graph. Bioinformatics 21(2), 79–85 (2005)
8. Myers, E.W., et al.: A Whole-Genome Assembly of Drosophila. Science 287(5461), 2196–2204 (2000)
9. Rasmussen, K., Stoye, J., Myers, E.W.: Efficient q-Gram Filters for Finding All e-matches Over a Given Length. J. of Computational Biology 13, 296–308 (2005)
10. Small, K.S., Brudno, M., Hill, M.M., Sidow, A.: A haplome alignment and reference sequence of the highly polymorphic Ciona savignyi genome. Genome Biology 8(1) (2007)
11. Small, K.S., Brudno, M., Hill, M.M., Sidow, A.: Extreme genomic variation in a natural population. PNAS 104(13), 5698–5703 (2007)

12. Sundararajan, M., Brudno, M., Small, K., Sidow, A., Batzoglou, S.: Chaining Algorithms for Alignment of Draft Sequence. In: Jonassen, I., Kim, J. (eds.) WABI 2004. LNCS (LNBI), vol. 3240, pp. 326–337. Springer, Heidelberg (2004)
13. Weinstock, G.M., et al.: The Genome of the Sea Urchin Strongylocentrotus purpuratus. Science 314, 941–952 (2006)
14. Zerbino, D.R., Birney, E.: Velvet: Algorithms for de novo short read assembly using de Bruijn graphs. Genome Research 18, 821–829 (2008)
15. Domingos, P., Pazzani, M.: On the Optimality of the Simple Bayesian Classifier under Zero-One Loss. Machine Learning 29(2-3), 103–130 (1997)
16. Ewing, B., Hillier, L., Wendl, M.C., Green, P.: Base-calling of automated sequencer traces using phred. II. Error probabilities. Genome Research 8, 175–185 (1998)
17. Salmela, L.: Correction of sequencing errors in a mixed set of reads. Bioinformatics 26, 1284–1290 (2010)
18. Ciona savignyi database at Broad Institute,
    `http://www.broadinstitute.org/annotation/ciona/`
19. Kececioglu, J.: Exact and Approximation Algorithms for DNA Sequence Reconstruction. PhD dissertation, Technical Report 91-26, Department of Computer Science, University of Arizona (December 1991)
20. Dehal, P., et al.: The Draft Genome of Ciona intestinalis: Insights into Chordate and Vertebrate Origins. Science 298(5601), 2157–2167 (2002)

Appendix 1: Parameters

To run H-SHREC (version 1.0) we use the largest strictness value the program accepts for each dataset. For the C. savignyi datasets, we set the number of iterations to 1 (more iterations introduced more errors). For E. coli, 3 iterations are used. The other parameters are left at defaults. Velvet (version 1.0.13) is tested with all odd kmer sizes between 17 and 27 inclusive. The results are reported with the kmer size (19) that achieved the highest N50 value. The expected coverage is set to 10. For Euler (version 1.1.2), we test all odd kmer sizes between 21 and 27 and choose the size (23) that maximized the N50 value.

Hapsembler is run with an error threshold of 0.07 and minimum overlap size of 30bp. The kmer size is set to 13. Kmers that appear more than 100 times the expected coverage in the data are masked. The minimum number of kmers required to perform Needleman-Wunsch is set to 1. These parameters are kept constant for all the datasets reported.

Discovery and Characterization of Chromatin States for Systematic Annotation of the Human Genome

Jason Ernst and Manolis Kellis

MIT

A plethora of epigenetic modifications have been described in the human genome and shown to play diverse roles in gene regulation, cellular differentiation and the onset of disease. Although individual modifications have been linked to the activity levels of various genetic functional elements, their combinatorial patterns are still unresolved and their potential for systematic de novo genome annotation remains untapped. Here, we use a multivariate Hidden Markov Model to reveal chromatin states in human T cells, based on recurrent and spatially coherent combinations of chromatin marks. We define 51 distinct chromatin states, including promoter-associated, transcription-associated, active intergenic, large-scale repressed and repeat-associated states. Each chromatin state shows specific enrichments in functional annotations, sequence motifs and specific experimentally observed characteristics, suggesting distinct biological roles. This approach provides a complementary functional annotation of the human genome that reveals the genome-wide locations of diverse classes of epigenetic function.

V. Bafna and S.C. Sahinalp (Eds.): RECOMB 2011, LNBI 6577, p. 53, 2011.
© Springer-Verlag Berlin Heidelberg 2011

Disease Gene Prioritization Based on Topological Similarity in Protein-Protein Interaction Networks

Sinan Erten[1], Gurkan Bebek[2,3], and Mehmet Koyutürk[1,2,*]

[1] Dept. of Electrical Engineering & Computer Science
[2] Center for Proteomics & Bioinformatics
Case Western Reserve University, Cleveland, OH 44106, USA
[3] Genomic Medicine Institute, Cleveland Clinic, Cleveland, OH 44195, USA
`{mse10,gurkan,mxk331}@case.edu`

Abstract. In recent years, many algorithms have been developed to narrow down the set of candidate disease genes implicated by genome wide association studies (GWAS), using knowledge on protein-protein interactions (PPIs). All of these algorithms are based on a common principle; functional association between proteins is correlated with their connectivity/proximity in the PPI network. However, recent research also reveals that networks are organized into recurrent network schemes that underlie the mechanisms of cooperation among proteins with different function, as well as the crosstalk between different cellular processes. In this paper, we hypothesize that proteins that are associated with similar diseases may exhibit patterns of "topological similarity" in PPI networks. Motivated by these observations, we introduce the notion of "topological profile", which represents the location of a protein in the network with respect to other proteins. Based on this notion, we develop a novel measure to assess the topological similarity of proteins in a PPI network. We then use this measure to develop algorithms that prioritize candidate disease genes based on the topological similarity of their products and the products of known disease genes. Systematic experimental studies using an integrated human PPI network and the Online Mendelian Inheritance (OMIM) database show that the proposed algorithm, VAVIEN, clearly outperforms state-of-the-art network based prioritization algorithms. VAVIEN is available as a web service at `http://www.diseasegenes.org`.

1 Introduction

Characterization of disease-associated variations in human genome is an important step toward enhancing our understanding of the cellular mechanisms that drive complex diseases, with profound applications in modeling, diagnosis, prognosis, and therapeutic intervention [1]. Genome-wide linkage and association studies in healthy and affected populations provide chromosomal regions containing hundreds of polymorphisms that are potentially associated with certain genetic diseases [2]. These polymorphisms often implicate up to 300 genes, only a few of which may have a role in the manifestation of the disease. Investigation of that many candidates via sequencing is clearly an expensive task, thus not always a feasible option. Consequently, computational methods

---

[*] Corresponding author.

V. Bafna and S.C. Sahinalp (Eds.): RECOMB 2011, LNBI 6577, pp. 54–68, 2011.

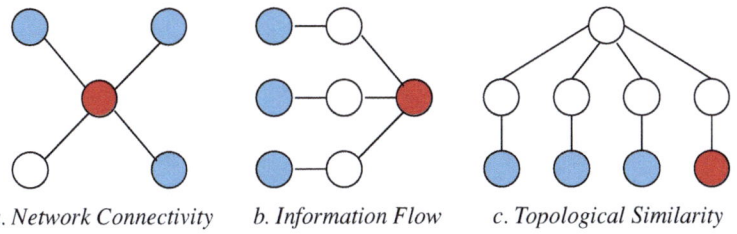

a. Network Connectivity   b. Information Flow   c. Topological Similarity

Fig. 1. Key principles in network-based disease gene prioritization. Nodes and edges respectively represent proteins and interactions. Seed proteins (proteins known to be associated with the disease of interest) are shown in light blue, proteins that are implicated to be associated with the same disease by the respective principle are shown in dark red, other proteins are shown in white. (a) *Network Connectivity* [3, 10, 11, 12, 13] infers association of the red protein with the seed proteins because it interacts heavily with them. (b) *Information Flow* [14, 15, 16, 17] infers association of the red protein with seed proteins because it exhibits crosstalk to them via indirect interactions through other proteins. (c) *Topological Similarity*, proposed in this paper, infers association of the red protein with the seed proteins because it (indirectly) interacts with a hub protein in a way topologically similar to them.

are primarily used to prioritize and identify the most likely disease-associated genes by utilizing a variety of data sources such as gene expression [3, 4] and functional annotations [5, 6, 7]. Protein-protein interactions provide an invaluable resource in this regard, since they provide functional information in a network context and can be obtained at a large scale via high-throughput screening [8].

In the last few years, many algorithms have been developed to utilize protein-protein interaction (PPI) networks in disease gene prioritization [9, 10, 11, 12, 13, 14, 15, 16, 17, 18, 19]. These algorithms take as input a set of *seed proteins* (coded by genes known to be associated with the disease of interest or similar diseases), *candidate proteins* (coded by genes in the linkage interval for the disease of interest), and a network of interactions among human proteins. Subsequently, they use protein-protein interactions to infer the relationship between seed and candidate proteins and rank the candidate proteins according to these inferred relationships. The key ideas in network-based prioritization of disease genes are illustrated in Figure 1.

Network connectivity is useful in disease gene prioritization. Network-based analyses of diverse phenotypes demonstrate that products of genes that are implicated in similar diseases are clustered together into highly connected subnetworks in PPI networks [20, 21]. Here, the similarity between diseases refers to the similarity in clinical classification of diseases. Motivated by these observations, many studies search the PPI networks for interacting partners of known disease genes to narrow down the set of candidate genes implicated by GWAS [10, 11, 12, 13] (Figure 1(a)). These algorithms are also extended to take into account the information provided by the genes implicated in diseases similar to the disease of interest [3].

Information flow based methods take into account indirect interactions. Methods that consider direct interactions between seed and candidate proteins do not utilize knowledge of PPIs to their full potential. In particular, they do not consider interactions among proteins that are not among the seed or candidate proteins, which might also

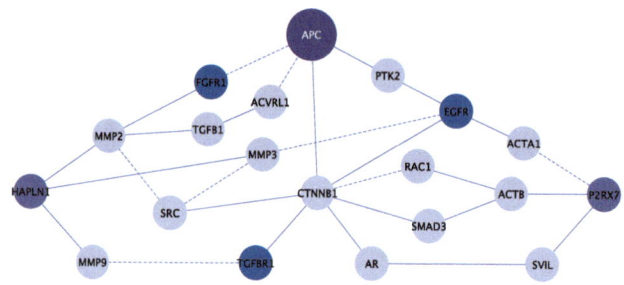

Fig. 2. Motivating example for using topological similarity to prioritize candidate disease genes. Two PPI subnetworks connecting key cancer driver genes, *APC-HAPLN1* ($p < 0.0068$) and *APC-P2RX7* ($p < 0.0212$), were found significant when bimodality of coexpression with proteomic targets were calculated [23]. Darker nodes represent proteins coded by genes that carry "driver mutations". Blue nodes represent growth factor receptors (GFRs). Although *APC-HAPLN1* and *APC-P2RX7* do not directly interact or exhibit significant crosstalk with growth factors and products of driver genes, their relative locations with respect to these proteins exhibit similarities.

indicate indirect functional relationships between candidate and seed proteins. For this reason, connectivity-based ("local") methods are vulnerable to false negative and positive interactions [22]. *Information flow* based ("global") methods ground themselves on the notion that products of genes that have an important role in a disease are expected to exhibit significant network crosstalk to each other in terms of the aggregate strength of paths that connect the corresponding proteins (Figure 1(b)). These methods include random walk with restarts [14,15] and network propagation [16,17], which significantly outperform connectivity based methods [9].

Topological similarity indicates functional association. Despite their differences, all network-based disease gene prioritization algorithms are based on a unique principle: the association between proteins is correlated with their connectivity/proximity in the PPI network. However, recent research also reveals that networks are organized into recurrent network schemes that underlie the interaction patterns among proteins with different function [24, 25]. A well-known network schema, for example, is a chain of membrane-bound receptors, protein kinases, and transcription factors, which serves as a high-level description of the backbone of cellular signaling. Dedicated mining algorithms identify more specific network schemes at a higher resolution, indicating that similar principles are used recurrently in interaction networks [26,27]. Inspired by these results, in this paper, we develop a network-based disease gene prioritization algorithm that uses topological similarity to infer the association between seed and candidate proteins (Figure 1(c)). Below, we further motivate this approach with an example from the systems biology of cancer.

Motivating example. While the *APC* gene has been identified to be one of the most important genes that plays a role in the development of colorectal cancer, there are multiple proteins that work in parallel with Apc to create these cancers [28, 29]. Although the actual mechanisms of selection are not clear, it is known that, proteins that are not directly interacting with *APC*, and have similar functions in a cell, such as tumor suppressor genes *PTEN* [30], *TRP53* [31] and *p21* [32] when mutated with *APC* increase the tumor burden. In a recent study, Bebek *et al.* [23] present a pipeline where bimodality of coexpresssion is used to prioritize proteomics targets identified in a mouse model of colorectal cancer. Some of the significant proteins identified are shown in Figure 2 in a PPI network. The identified targets *HAPLN1, P2RX7* (colored purple in the figure) are linked to growth factor receptors (GFRs) (*EGFR, TGFR1, FGFR1*, colored blue in the figure), but not connected to each other. As seen in the figure, similarities of these two proteomic targets in their function and role in disease are also reflected in their relative topology with respect to *APC* and growth factors.

Contributions of this study. We propose a topological similarity based disease gene prioritization scheme in this paper. For this purpose, we develop a measure of topological similarity among pairs of proteins in a PPI network and use the network similarity between seed and candidate proteins to infer the likelihood of disease association for the candidates. We present the proposed methods in Section 2. Systematic experimental studies using an integrated human PPI network and the Online Mendelian Inheritance (OMIM) database are presented in Section 3. These results show that the proposed algorithm, VAVIEN[1], clearly outperforms state-of-the-art network based prioritization algorithms. We conclude our discussion in Section 4.

2   Methods

In this section, we first describe the disease gene prioritization problem within a formal framework. Subsequently, we formulate the concept of topological similarity of pairs of proteins in terms of their proximity to other proteins in the network. Finally, we discuss how topological similarity of proteins is used to prioritize candidate disease genes.

2.1   Disease Gene Prioritization Problem

Let D denote a disease of interest, which is potentially associated with various genetic factors (*e.g.*, sleep apnea, Alzhemier's disease, autism). Assume that a genome-wide association study (GWAS) using samples from control and affected populations is conducted, revealing a linkage interval that is significantly associated with D. Potentially, such a linkage interval will contain multiple genes, which are all candidates for being mechanistically associated with D (*i.e.*, the mutation in a gene in the linkage interval might have a role in the manifestation of disease). This set of candidate genes, denoted $\mathcal{C}$, forms the input to the disease gene prioritization problem.

The aim of disease gene prioritization is to rank the genes in $\mathcal{C}$ based on their potential mechanistic association with D. For this purpose, a set of genes that are already known to be associated with D or diseases similar to D is used (where similarity between diseases is defined phenotypically, *e.g.*, based on the clinical description of diseases). The idea here is that genes in $\mathcal{C}$ that are mechanistically associated with D are likely to exhibit patterns of association with such genes in a network of PPIs. This set

---

[1] From va-et-vient (*Fr.*); an electrical circuit in which multiple switches in different locations perform identical tasks (*e.g.*, control lighting in a stairwell from either end).

of genes is referred to as the *seed* set and denoted $\mathcal{S}$. Each gene $v \in \mathcal{S}$ is assigned a disease-association score $\sigma(v, D) \in (0, 1]$, representing the known level of association between v and D. The association score for v and D is set to 1 if it is a known association listed in OMIM database. Otherwise, it is computed as the maximum clinical similarity between D and any other disease associated with v [33] (a detailed discussion on computation of similarity scores can be found in [34]).

In order to capture the association of the genes in $\mathcal{C}$ with those in $\mathcal{S}$, network-based prioritization algorithms utilize a network of known interactions among human proteins. The human protein-protein interaction (PPI) network $\mathcal{G} = (\mathcal{V}, \mathcal{E}, w)$ consists of a set of proteins $\mathcal{V}$ and a set of undirected interactions $\mathcal{E}$ between these proteins, where $uv \in \mathcal{E}$ represents an interaction between $u \in \mathcal{V}$ and $v \in \mathcal{V}$. Since PPI networks are noisy and incomplete [35], each interaction $uv \in \mathcal{E}$ is also assigned a confidence score representing the reliability of the interaction between u and v [36, 37, 25]. Formally, there exists a function $w : \mathcal{E} \rightarrow (0, 1]$, where $w(uv)$ indicates the reliability of interaction $uv \in \mathcal{E}$.

In this paper, the reliability score is derived through a logistic regression model where a positive interaction dataset (MIPS Golden PPI interactions [38]) and a negative interaction dataset (Negatome [39]) are used to train a model with three variables: (i) co-expression measurements for the corresponding genes derived from multiple sets of tissue microarray experiments (normal human tissues measured in the Human Body Index Transcriptional Profiling (GEO Accession: GSE7307) [40]), (ii) the proteins' small world clustering coefficient, and (iii) the protein subcellular localization data of interacting partners [41]. Co-expression values are used since co-regulated genes are more likely to interact with each other than others [36, 25]. On the other hand, the network feature that we are extracting, the small world clustering coefficient, is a measure of connectedness. This coefficient shows how likely the neighbors (interacting peers) of a protein are neighbors of each other [42]. We also incorporate the protein subcellular localization data into the logistic model, since this would eliminate interactions among proteins that are not biologically significant [25]. The logistic regression model is trained on randomly selected 1000 positive and negative training data sets for 100 times and regression constants are determined to score each PPI.

Given $\mathcal{S}$ and $\mathcal{G}$, network-based disease gene prioritization aims to compute a score $\alpha(v, D)$ for each $v \in \mathcal{C}$, representing the potential association of v with disease D. For this purpose, we develop a novel method, VAVIEN, to rank candidate genes based on their topological similarity to the seed genes in $\mathcal{G}$.

2.2   Topological Similarity of Proteins in a PPI Network

Recent research shows that molecular networks are organized into functional interaction patterns that are used recurrently in different cellular processes [24, 26]. In other words, proteins with similar function often interact with proteins that are also functionally similar to each other [27]. Motivated by this observation, VAVIEN aims to assess the functional similarity between seed and candidate proteins based on their *topological similarity*, that is the similarity of their relative location with respect to other proteins in the network. For this purpose, we first define the topological profile of a protein in a PPI network.

Topological profile of a protein. For a given protein $v \in V$ and a PPI network $\mathcal{G}$, the *topological profile* β_v of v is defined as a $|V|$-dimensional vector such that for each $u \in V$, $\beta_v(u)$ represents the proximity of protein v to protein u in $\mathcal{G}$. Clearly, the proximity between two proteins can be computed in various ways. A well-known measure of proximity is the shortest path (here, the most reliable path) between the two proteins, however, this method is vulnerable to missing data and noise in PPI networks [22]. A reliable measure of network proximity is effective conductance, which is based on a model that represents the network as an electrical-circuit. In this model, each edge is represented as a capacitor with capacitance proportional to its reliability score. Effective conductance can be computed using the inverse of the Laplacian matrix of the network, however, this computation is quite costly since it requires computation of the inverse of a sparse matrix [43]. Fortunately, however, computation of effective conductance and random walks in a network are known to be related [44] and proximity scores based on random walks can be computed efficiently using iterative methods.

VAVIEN computes the proximity between pairs of proteins using random walk with restarts [45, 46]. This method is used in a wide range of applications, including identification of functional modules [47] and modeling the evolution of social networks [48]. It is also the first information flow based method to be applied to disease gene prioritization [15, 14] and is shown to clearly outperform connectivity based methods.

Random walk with restarts computes the proximity between a protein v and all other proteins in the network as follows: A random walk starts at v. At each step, if the random walk is at protein u, it either moves to an interacting partner t of u (i.e., $ut \in \mathcal{E}$) or it restarts the walk at v. The probability $P(u, t)$ of moving to a specific interacting partner t of u is proportional to the reliability of the interaction between u and t, i.e., $P(u, t) = w(ut)/W(u)$ where $W(u) = \sum_{t':t'u \in \mathcal{E}} w(ut')$ is the weighted degree of u in the network. The probability of restarting at a given time step is a fixed parameter denoted r. After a sufficiently long time, the probability of being at node u at a random time step provides a measure of the proximity between v and u, which can be computed iteratively as follows:

$$x_v^{(k)} = (1 - r)P x_v^{(k-1)} + r e_v. \tag{1}$$

Here $x_v^{(k)}$ denotes a probability vector such that $x_v^{(k)}(u)$ equals the probability of being at protein u at the kth iteration of the random walk, $x_v^{(0)} = e_v$, and e_v is the restart vector such that $e_v(u) = 1$ if $u = v$ and 0 otherwise. For a given value of r, the topological profile of protein v is defined as $\beta_v = \lim_{k \to \infty} x_v^{(k)}$.

Note that the concept of topological profile introduced here is not to be confused by the *gene closeness profile* used by the CIPHER algorithm for disease gene prioritization [18]. Here, topological profile is constructed using the proximity of a protein of interest to every other protein in the network. It is therefore a global signature of the location of the protein in the PPI network. In contrast, gene closeness profile is based only on the proximity of a protein of interest to proteins coded by known disease genes. Furthermore, the proposed algorithm is different from random walk based prioritization algorithms in that these algorithms score candidate proteins directly based on random walk proximity to seed proteins [15]. In contrast, VAVIEN uses random walk proximity as a feature to assess the topological similarity between seed and candidate proteins, which in turn is used to score candidate proteins. We now describe this approach in detail.

Topological similarity of two proteins. Let u and $v \in \mathcal{V}$ denote two proteins in the network. The topological similarity of u and v is defined as

$$\rho(\beta_u, \beta_v) = corr(\beta_u, \beta_v) = \frac{\sum_{t \in \mathcal{V}} (\beta_u(t) - \frac{1}{|\mathcal{V}|})(\beta_v(t) - \frac{1}{|\mathcal{V}|})}{\sqrt{\sum_{t \in \mathcal{V}} (\beta_u(t) - \frac{1}{|\mathcal{V}|})^2} \sqrt{\sum_{t \in \mathcal{V}} (\beta_v(t) - \frac{1}{|\mathcal{V}|})^2}}, \quad (2)$$

where $corr(X, Y)$ denotes the Pearson correlation coefficient of random variables X and Y. The idea behind this approach is that, if two proteins interact with similar proteins, or lay on similar locations with respect to hub proteins in the network, then their topological profiles will be correlated, which will be captured by $\rho(\beta_u, \beta_v)$.

2.3   Using Topological Similarity to Prioritize Candidate Genes

The core idea behind the proposed algorithm is that candidate genes whose products are topologically similar to the products of seed genes are likely to be associated with D. Based on this idea, we propose three schemes to prioritize candidate genes based on their topological similarity with seed genes. All of these schemes are implemented in VAVIEN.

Proritization based on average topological similarity with seed genes (ATS). For each $u \in \mathcal{C}$, the topological profile vector β_u is computed using random walk with restarts. Similarly, topological profile vectors β_v of all genes $v \in \mathcal{S}$ are computed separately. Subsequently, for each $u \in \mathcal{C}$, the association score of u with D is computed as the weighted average of the topological similarity of u with the genes in $\mathcal{S}$, where the contribution of each seed gene is weighted by its association with D, i.e.:

$$\alpha_{\text{ATS}}(u, D) = \frac{\sum_{v \in \mathcal{S}} \sigma(v, D)\rho(u, v)}{\sum_{v \in \mathcal{S}} \sigma(v, D)}. \quad (3)$$

Prioritization based on topological similarity with average profile of seed genes (TSA). Instead of computing the topological similarity for each seed gene separately, this approach first computes an average topological profile that is representative of the seed genes and computes the topological similarity of the candidate gene and this average topological profile. More precisely, the association score of $u \in \mathcal{C}$ with D is computed as:

$$\alpha_{\text{TSA}}(u, D) = \rho(\beta_u, \bar{\beta}_{\mathcal{S}}), \quad (4)$$

where

$$\bar{\beta}_{\mathcal{S}} = \frac{\sum_{v \in \mathcal{S}} \sigma(v, D)\beta_v}{\sum_{v \in \mathcal{S}} \sigma(v, D)}. \quad (5)$$

Prioritization based on topological similarity with representative profile of seed genes (TSR). The random walk with restarts model can be easily extended to compute the proximity between a group of proteins and each protein in the network. This can be done by generalizing the random walk to one that makes frequent restarts at any of the proteins in the group. This is indeed the idea of disease gene prioritization using random walk with restarts [15]. This method is also useful for directly computing a representative topological profile for $\mathcal{S}$, instead of taking the average of the topological profiles of the genes in $\mathcal{S}$. More precisely, for given seed set $\mathcal{S}$ and association scores σ for all genes in $\mathcal{S}$, the proximity of the products of genes in $\mathcal{S}$ to each protein in the network is computed by replacing the restart vector in Equation 1 with vector $e_{\mathcal{S}}$ where

$$e_{\mathcal{S}}(t) = \frac{\sigma(t, D)}{\sum_{v \in \mathcal{S}} \sigma(v, D)}, \tag{6}$$

if $t \in \mathcal{S}$ and $e_{\mathcal{S}}(t) = 0$ otherwise. Then, the topological profile $\beta_{\mathcal{S}}$ of $\mathcal{S}$ is computed as $\beta_{\mathcal{S}} = \lim_{k \to \infty} x^{(k)}$. The random walk based approach to disease gene prioritization estimates the association of each candidate gene with the disease as the proximity between the product of the candidate gene and $\mathcal{S}$ under this model, *i.e.*, it directly sets $\alpha = \beta_{\mathcal{S}}$. In contrast, we compute the association of $u \in \mathcal{C}$ with D as

$$\alpha_{\text{TSR}}(u) = \rho(\beta_u, \beta_{\mathcal{S}}). \tag{7}$$

Once α is computed using one of (3), (4), or (7), VAVIEN ranks the candidate genes in decreasing order of α.

3   Results

In this section, we systematically evaluate the performance of VAVIEN in capturing true disease-gene associations using a comprehensive database of known disease-gene associations. We start by describing the datasets and experimental settings. Next, we analyze the performance of different schemes implemented in VAVIEN and the effect of parameters. Subsequently, we compare the performance of VAVIEN with three state-of-the-art network based prioritization algoritms.

3.1   Datasets

Disease association data. The Online Mendelian Inheritance in Man (OMIM) database provides a publicly accessible and comprehensive database of genotype-phenotype relationship in humans. We acquire disease-gene associations from OMIM and map the gene products known to be associated with disease to our PPI network. The dataset contains 1931 diseases with number of gene associations ranging from 1 to 25, average being only 1.31. Each gene v in the seed set $\mathcal{S}$ is associated with the similarity score $\sigma(v, D)$, indicating the known degree of association between v and D as mentioned before.

Human protein-protein interaction (PPI) network. In our experiments, we use the human PPI data obtained from NCBI Entrez Gene Database [49]. This database integrates interaction data from several other databases available, such as HPRD, BioGrid, and BIND. After the removal of nodes with no interactions, the final PPI network contains 8959 proteins and 33528 interactions among these proteins. We assign reliability scores to these interactions using the methodology described in Section 2.1.

3.2   Experimental Setting

In order to evaluate the performance of different methods in prioritizing disease-associated genes, we use leave-one-out cross-validation. For each gene u that is known to be associated with a disease D in our dataset, we conduct the following experiment:

- We remove u from the set of genes known to be associated with D. We call u the *target gene* for that experiment. The remaining set of genes associated with D becomes the seed set S.
- We generate an artificial linkage interval, containing the target gene u with other 99 genes located nearest in terms of genomic distance. The genes in this artificial linkage interval (including u) compose the candidate set C.
- We apply each prioritization algorithm to obtain a ranking of the genes in C.
- We assess the quality of the ranking provided by each algorithm using the evaluation criteria described below.

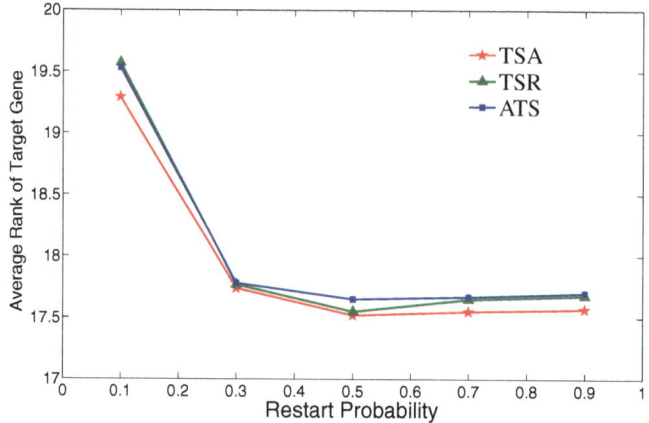

Fig. 3. The performance of the three prioritization algorithms implemented in VAVIEN as a function of the restart probability used in computing proximity via random walk with restarts. The performance here is measured in terms of the average rank of the target gene among 100 candidate genes, a lower value indicating better performance.

Evaluation criteria. We first plot ROC (precision *vs.* recall) curves, by varying the threshold on the rank of a gene to be considered a "predicted disease gene". *Precision* is defined as the fraction of true disease genes among all genes ranked above the particular threshold, whereas *recall* is defined as the fraction of true disease genes identified (ranked above the threshold) among all known disease genes. Note that, this is a conservative measure for this experimental set-up since there exists only one true positive (the target gene) for each experiment. For this reason, we also compare these methods in terms of the *average rank* of the target gene among 100 candidates, computed across all disease-gene pairs in our experiments. Clearly, lower average rank indicates better performance. Finally, we report the percentage of true disease genes that are ranked as one of the genes in the *top* 1% (practically, the top gene) and also in the *top* 5% among all candidates.

3.3   Performance Evaluation

Performance of methods implemented in VAVIEN **and the effect of restart parameter.** We compare the three different algorithms (ATS, TSA and TSR) implemented in VAVIEN in Figure 3. Since the topological profile of a protein depends on the restart probability (the parameter r) in the random walk with restarts, we also investigate the effect of this parameter on the performance of algorithms. In the figure, the average rank of the target gene among 100 candidate genes is shown for each algorithm as a function of restart probability. As seen in the figure, the three algorithms deliver comparable performance. However, TSA, which makes use of the average profile of seed genes to compute the topological similarity of the candidate gene to seed genes achieves the best performance. Furthermore, the performance of all algorithms implemented in VAVIEN appears to be robust to the selection of parameter r, as long as it is in the range $[0.3 - 0.9]$. In our experiments, we set $r = 0.5$ and use TSA as the representative algorithm since this combination provides the best performance.

Performance of VAVIEN **compared to existing algorithms.** We also evaluate the performance of VAVIEN in comparison to state-of-the-art algorithms for network-based disease gene prioritization. These algorithms are the following:

– *Random walk with restarts*: This algorithm prioritizes candidate genes based on their proximity to seed genes, using a random walk with restarts model, i.e., α is set to β_S [15].
– *Network prioritization*: This algorithm is very similar to random walk with restarts, with one key difference. In network prioritization, the stochastic matrix in (1) is replaced with a flow matrix in which both the incoming and outgoing flow to a protein is normalized (*i.e.*, $P(u, t) = w(ut)/\sqrt{W(u)W(t)}$ in network propagation) [16].
– *Information flow with statistical correction*: Based on the observation that the performance of information flow based algorithms (including random walk with restarts and network propagation) depend on network degree, this algorithm applies statistical correction to the random walk based association scores based on a reference model that takes into account the degree distribution of the PPI network [33].

While software implementing these algorithms are available (*e.g.*, PRINCE [16] implements network propagation, DADA [50] implements statistical correction), we here report results based on our implementation of each algorithm. We implement all algorithms using identical settings for data integration and incorporation of disease similarity scores, differing from each other only in how network information is utilized in computing disease association scores. The objective of this approach to provide a setting in which the algorithmic ideas can be directly compared, by removing the influence of implementation details and datasets used. It should be noted, however, that the performance of these algorithms could be better than the performance reported here if available software and/or different PPI datasets are used.

The ROC curves for the three existing methods and VAVIEN are shown in Figure 4, demonstrating the relationship between precision and recall for each algorithm. Other performance measures for all methods are listed in Table 1. As seen in both the figure and the table, VAVIEN clearly outperforms all of the existing algorithms in ranking candidate disease genes. In particular, it is able to rank 40% of true disease genes as the

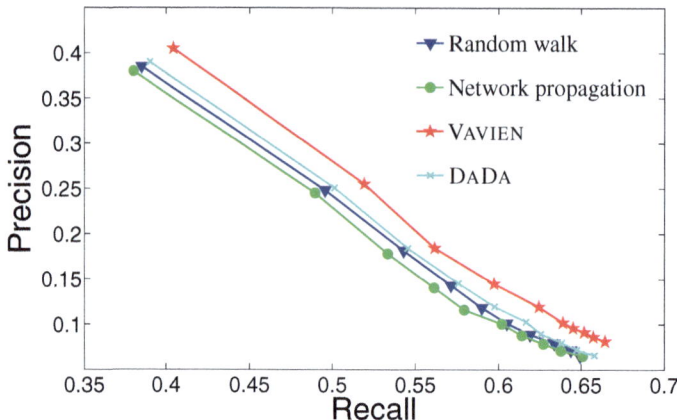

Fig. 4. ROC curves comparing the performance of the proposed method with existing information-flow based algorithms

top candidate among 100 candidates and it ranks 62% of true disease genes in the top 5% of all candidates.

Information flow based algorithms are previously shown to be biased with respect to the degree of the target genes [33]. In other words, these methods work poorly in identifying loosely connected disease genes. Previous efforts reduce this bias to a certain extent by introducing several statistical correction schemes [33]. Motivated by these observations, we here investigate the effect of the bias introduced by degree distribution on the performance of different algorithms. The results of these experiments are shown in Figure 5. In this figure, the change on the performance (average rank of the target gene) of different methods is plotted with respect to the degree of the target gene. As clearly seen, VAVIEN is the algorithm that is affected least by this bias and it outperforms other methods in identifying loosely connected disease genes. It is particularly impressive that VAVIEN's performance is less affected by degree distribution as compared to DADA, since DADA is designed explicitly with the purpose of removing the effect of network degree.

As argued in the previous sections, information flow based proximity and topological similarity capture different aspects of the relationship between functional association

Table 1. Comparison of VAVIEN with existing algorithms for network-based disease gene prioritization. VAVIEN outperforms state-of-the-art information flow based algorithms with respect to all performance criteria.

METHOD	Avg. Rank	Ranked in top 1%	Ranked in top 5%
VAVIEN	17.52	40.48	62.46
Random walk	18.58	38.42	59.01
Network propagation	18.28	37.97	57.96
Random walk with statistical correction	17.86	39.41	59.76

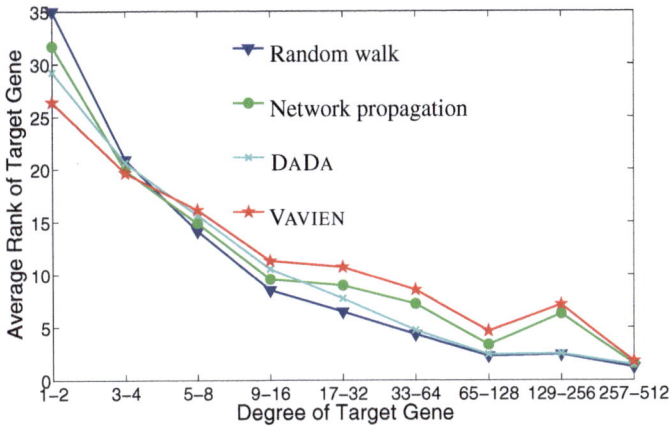

Fig. 5. Relation between the degree of target disease gene and its corresponding rank among 100 candidates for VAVIEN and existing algorithms

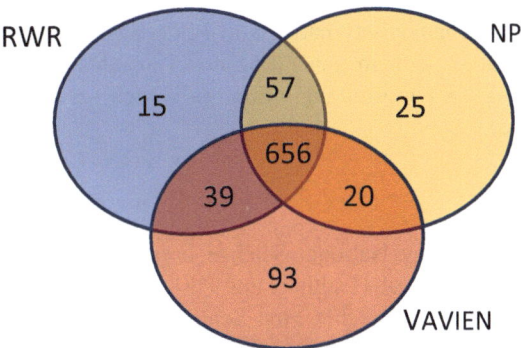

Fig. 6. Venn Diagram comparing the true disease genes ranked by each method as the most likely candidate. The sets labeled RWR, NP, and VAVIEN represent the set of true disease genes that are ranked top by random walk with restarts, network propagation, and topological similarity, respectively. Each number in an area shows the number of true candidates in that set (*e.g.*, 20 true disease genes were ranked top by network propagation and VAVIEN, but not random walk with restarts).

and network topology. Consequently, we expect that the proposed topological similarity and information flow based algorithms will be successful in identifying different disease associated genes. In order to investigate whether this is the case, we compare target genes that are correctly identified as the true disease gene by each algorithm. These results are shown by a Venn diagram in Figure 6. In this figure, each value represents the number of true disease genes that are ranked 1st among 100 candidates by the corresponding algorithm(s). Among 1996 disease-gene associations, VAVIEN is able to rank the true candidate first in 808 of the cases. 93 of these genes are not ranked as the top candidate by neither random walk with restarts nor network propagation. On the other hand, the number of true candidates that are uniquely identified by each of the other two algorithms is lower (15 for random walk with restarts, 25 for network propagation), demonstrating that VAVIEN is quite distinct in its approach and it is more powerful in extracting information that is missed by other algorithms. Furthermore, the 93 candidates uniquely identified by VAVIEN mostly code for loosely connected proteins (with 67 of them having $<= 5$ known interactions). This observation supports our claim that VAVIEN is indeed less effected by the bias introduced by degree distribution, as compared to information flow based network proximity.

4   Conclusion

In this paper, we present an algorithm, called VAVIEN, for harnessing the topological similarity of proteins in a network of interactions to prioritize candidate disease-associated genes. After investigating the performance of the three schemes implemented in VAVIEN with respect to the restart parameter, we conduct a comprehensive set of experiments on OMIM data and show that VAVIEN outperforms existing information flow based models, as well as their statistically adjusted version, in terms of ranking the true disease gene highest among other candidate genes. These results demonstrate that in addition to the connectivity patterns in PPI networks, topological patterns in these networks are also useful in generating novel insights into systems biology of complex diseases. VAVIEN is available online as a web service at http://www.diseasegenes.org.

Acknowledgments

This work was supported by National Science Foundation CAREER Award CCF-0953195, National Institutes of Health grants P30-CA043703, UL1-RR024989, and R01- HL106798, and a Choose Ohio First Scholarship from the State of Ohio. We would like to thank Vishal Patel, Rob Ewing, and Mark R. Chance (Case Western Reserve University) for many useful discussions. We would also like to note the contribution of anonymous reviewers whose queries and suggestions have helped improve this paper significantly.

References

1. Brunner, H.G., van Driel, M.A.: From syndrome families to functional genomics. Nat. Rev. Genet. 5(7), 545–551 (2004)
2. Glazier, A.M., Nadeau, J.H., Aitman, T.J.: Finding Genes That Underlie Complex Traits. Science 298(5602), 2345–2349 (2002)
3. Lage, K., Karlberg, E., Storling, Z., Olason, P., Pedersen, A., Rigina, O., Hinsby, A., Tumer, Z., Pociot, F., Tommerup, N., Moreau, Y., Brunak, S.: A human phenome-interactome network of protein complexes implicated in genetic disorders. Nat. Bio 25(3), 309–316 (2007)
4. Nica, A.C., Dermitzakis, E.T.: Using gene expression to investigate the genetic basis of complex disorders. Human Molecular Genetics 17(R2), ddn134– ddn285 (2008)
5. Adie, E., Adams, R., Evans, K., Porteous, D., Pickard, B.: SUSPECTS: enabling fast and effective prioritization of positional candidates. Bioinformatics 22(6), 773–774 (2006)

6. Chen, J., Bardes, E.E., Aronow, B.J., Jegga, A.G.: Toppgene suite for gene list enrichment analysis and candidate gene prioritization. Nucleic Acids Research 37(Web Server issue), gkp427+ (2009)
7. Turner, F., Clutterbuck, D., Semple, C.: Pocus: mining genomic sequence annotation to predict disease genes. Genome Biology 4(11), R75 (2003)
8. Ewing, R.M., Chu, P., Elisma, F., Li, H., Figeys, D.: Large-scale mapping of human protein-protein interactions by mass spectrometry. Molecular Systems Biology 3 (2007)
9. Navlakha, S., Kingsford, C.: The power of protein interaction networks for associating genes with diseases. Bioinformatics 26(8), 1057–1063 (2010)
10. Franke, L., Bakel, H., Fokkens, L., de Jong, E.D., Egmont-Petersen, M., Wijmenga, C.: Reconstruction of a functional human gene network, with an application for prioritizing positional candidate genes. Am. J. Hum. Genet. 78(6), 1011–1025 (2006)
11. Ideker, T., Sharan, R.: Protein networks in disease. Genome Research 18(4), 644–652 (2008)
12. Karni, S., Soreq, H., Sharan, R.: A network-based method for predicting disease-causing genes. Journal of Computational Biology 16(2), 181–189 (2009)
13. Oti, M., Snel, B., Huynen, M.A., Brunner, H.G.: Predicting disease genes using protein-protein interactions. J. Med. Genet., jmg.2006.041376 (2006)
14. Chen, J., Aronow, B., Jegga, A.: Disease candidate gene identification and prioritization using protein interaction networks. BMC Bioinformatics 10(1), 73 (2009)
15. Köhler, S., Bauer, S., Horn, D., Robinson, P.N.: Walking the interactome for prioritization of candidate disease genes. Am. J. Hum. Genet. 82(4), 949–958 (2008)
16. Vanunu, O., Magger, O., Ruppin, E., Shlomi, T., Sharan, R.: Associating genes and protein complexes with disease via network propagation. PLoS Comp. Bio. 6(1) (January 2010)
17. Zhang, L., Hu, K., Tang, Y.: Predicting disease-related genes by topological similarity in human protein-protein interaction network. Central European Journal of Physics 8, 672–682 (2010), 10.2478/s11534-009-0114-9
18. Wu, X., Jiang, R., Zhang, M.Q., Li, S.: Network-based global inference of human disease genes. Molecular Systems Biology 4 (May 2008)
19. Missiuro, P.V.V., Liu, K., Zou, L., Ross, B.C., Zhao, G., Liu, J.S., Ge, H.: Information flow analysis of interactome networks. PLoS Computational Biology 5(4), e1000350+ (2009)
20. Goh, K.I., Cusick, M.E., Valle, D., Childs, B., Vidal, M., Barabási, A.L.: The human disease network. PNAS 104(21), 8685–8690 (2007)
21. Rhodes, D.R., Chinnaiyan, A.M.: Integrative analysis of the cancer transcriptome. Nat. Genet. 37 Suppl. (June 2005)
22. Pandey, J., Koyutürk, M., Grama, A.: Functional characterization and topological modularity of molecular interaction networks. BMC Bioinformatics 11(Suppl. 1), S35 (2010)
23. Bebek, G., Patel, V., Chance, M.R.: Petals: Proteomic evaluation and topological analysis of a mutated locus signaling. BMC Bioinformatics 11, 596 (2010)
24. Pandey, J., Koyutürk, M., Kim, Y., Subramaniam, S., Szpankowski, W., Grama, A.: Functional annotation of regulatory pathways. Bioinformatics Suppl. on ISMB/ECCB 2007 23(13), i377–i386 (2007)
25. Bebek, G., Yang, J.: Pathfinder: mining signal transduction pathway segments from protein-protein interaction networks. BMC Bioinformatics 8, 335 (2007)
26. Banks, E., Nabieva, E., Peterson, R., Singh, M.: Netgrep: fast network schema searches in interactomes. Genome Biology 9(9) (2008)
27. Kirac, M., Özsoyoglu, G.: Protein function prediction based on patterns in biological networks. In: Vingron, M., Wong, L. (eds.) RECOMB 2008. LNCS (LNBI), vol. 4955, pp. 197–213. Springer, Heidelberg (2008)
28. Sjöblom, T., Jones, S., Wood, L.D., Parsons, D.W., Lin, J., Barber, T.D., et al.: The consensus coding sequences of human breast and colorectal cancers. Science 314(5797), 268–274 (2006)

29. Wood, L.D., Parsons, D.W., Jones, S., Lin, J., Sjöblom, T., Leary, R.J.: The genomic land-scapes of human breast and colorectal cancers. Science 318(5853), 1108–1113 (2007)
30. Marsh, V., Winton, D.J., Williams, G.T., Dubois, N., Trumpp, A., Sansom, O.J., Clarke, A.R.: Epithelial pten is dispensable for intestinal homeostasis but suppresses adenoma de-velopment and progression after apc mutation. Nat. Genet. 40(12), 1436–1444 (2008)
31. Halberg, R.B., Chen, X., Amos-Landgraf, J.M., White, A., Rasmussen, K., Clipson, L., et al.: The pleiotropic phenotype of apc mutations in the mouse: allele specificity and effects of the genetic background. Genetics 180(1), 601–609 (2008)
32. Patel, V.N., Bebek, G., Mariadason, J.M., Wang, D., Augenlicht, L.H., Chance, M.R.: Predic-tion and testing of biological networks underlying intestinal cancer. PLoS One 5(9) (2010)
33. Erten, S., Koyutürk, M.: Role of centrality in network-based prioritization of disease genes. In: Pizzuti, C., Ritchie, M.D., Giacobini, M. (eds.) EvoBIO 2010. LNCS, vol. 6023, pp. 13–25. Springer, Heidelberg (2010)
34. van Driel, M.A., Bruggeman, J., Vriend, G., Brunner, H.G., Leunissen, J.A.: A text-mining analysis of the human phenome. EJHG 14(5), 535–542 (2006)
35. Stumpf, M.P.H., Thorne, T., de Silva, E., et al.: Estimating the size of the human interactome. Proc. Natl. Acad. Sci. USA 105(19), 6959–6964 (2008)
36. Sharan, R., Suthram, S., et al.: Conserved patterns of protein interaction in multiple species. Proc. Natl. Acad. Sci. USA 102(6), 1974–1979 (2005)
37. Suthram, S., Shlomi, T., Ruppin, E., Sharan, R., Ideker, T.: A direct comparison of protein interaction confidence assignment schemes. BMC Bioinformatics 7, 360+ (2006)
38. Mewes, H.W., Heumann, K., Kaps, et al.: Mips: a database for genomes and protein se-quences. Nuc. Ac. Res. 27(1), 44–48 (1999)
39. Smialowski, P., Pagel, P., Wong, P., Brauner, B., Dunger, I., Fobo, G., Frishman, G., Mon-trone, C., Rattei, T., Frishman, D., Ruepp, A.: The negatome database: a reference set of non-interacting protein pairs. Nucleic Acids Res. 38(Database issue), D540–D544 (2010)
40. Barrett, T., Troup, D.B., et al.: Ncbi geo: archive for high-throughput functional genomic data. Nucleic Acids Res. 37(Database issue), 885–890 (2009)
41. Sprenger, J., Lynn Fink, J., Karunaratne, S., Hanson, K., Hamilton, N.A., Teasdale, R.D.: Lo-cate: a mammalian protein subcellular localization database. Nucleic Acids Res. 36(Database issue), D230–D233 (2008)
42. Goldberg, D.S., Roth, F.P.: Assessing experimentally derived interactions in a small world. Proc. Natl. Acad. Sci. USA 100(8), 4372–4376 (2003)
43. Spielman, D.A., Srivastava, N.: Graph sparsification by effective resistances. In: STOC, pp. 563–568 (2008)
44. Tetali, P.: Random walks and the effective resistance of networks. Journal of Theoretical Probability 4(1), 101–109 (1991)
45. Lovász, L.: Random walks on graphs: A survey. Combinatorics, Paul Erdos is Eighty 2, 353–398 (1996)
46. Tong, H., Faloutsos, C., Pan, J.Y.: Random walk with restart: fast solutions and applications. Knowledge and Information Systems 14(3), 327–346 (2008)
47. Macropol, K., Can, T., Singh, A.: Rrw: repeated random walks on genome-scale protein networks for local cluster discovery. BMC Bioinformatics 10(1), 283 (2009)
48. Tong, H., Faloutsos, C.: Center-piece subgraphs: problem definition and fast solutions. In: KDD 2006: Proceedings of the 12th ACM SIGKDD, pp. 404–413. ACM, NY (2006)
49. Maglott, D., Ostell, J., Pruitt, K.D., Tatusova, T.: Entrez Gene: gene-centered information at NCBI. Nucl. Acids Res. 35(suppl-1), D26–D31 (2007)
50. Erten, S., Bebek, G., Ewing, R.M., Koyutürk, M.: Dada - degree-aware disease gene priori-tization. BioData Mining (to be published)

Understanding Gene Sequence Variation in the Context of Transcription Regulation in Yeast

Irit Gat-Viks[1], Renana Meller[2], Martin Kupiec[2], and Ron Shamir[2]

[1] Broad Institute of MIT and Harvard
[2] Tel-Aviv University

The availability of expression quantitative trait loci (eQTL) data can help understanding the genetic basis of variation in gene expression. However, it has proven difficult to accurately predict functional genetic changes due to low statistical power. To address this challenge, we developed a novel computational approach for combining eQTL data with complementary regulatory network to identify modules of genes, their underlying genetic polymorphism and their shared regulatory proteins activity. The resulting eQTL model implicates novel central protein complexes that share not only a regulatory protein but also an underlying genetic variation. Our method manifests higher sensitivity than prior computational efforts.

Computationally, we tackle the important problem of automatic prediction of eQTL-target relations. The integrated approach makes it possible to capture weaker linkage signals and to avoid groups of genes that happen by chance to be linked to the same genomic interval. In terms of biological and medical discovery, using our framework on eQTL data in yeast, we implicate a novel role of eQTLs on genes comprising protein complexes, including the aerobic cellular respiration complex, affected by genetic changes in the mitochondrial inner membrane proteins Crd1/Cat5, and the Sum1p/Rfm1p/Hst1p middle sporulation repression complex, which is influenced by genetic variation residing within the Rfm1 itself.

Our discovery of previously uncharacterized modules in the well-studied segregating yeast population underscores the utility of our integrated methods in genetic analysis. Thus, our study establishes a broadly applicable, comprehensive approach to reveal eQTL-target relationships.

V. Bafna and S.C. Sahinalp (Eds.): RECOMB 2011, LNBI 6577, p. 69, 2011.

Identifying Branched Metabolic Pathways by Merging Linear Metabolic Pathways

Allison P. Heath[1], George N. Bennett[2], and Lydia E. Kavraki[1,3,4]

[1] Department of Computer Science, Rice University, Houston, TX, USA
[2] Department of Biochemistry and Cell Biology, Rice University, Houston, TX, USA
[3] Department of Bioengineering, Rice University, Houston, TX, USA
[4] Structural and Computational Biology and Molecular Biophysics,
Baylor College of Medicine, Houston, TX, USA
{aheath,gbennett,kavraki}@rice.edu

Abstract. This paper presents a graph-based algorithm for identifying complex metabolic pathways in multi-genome scale metabolic data. These complex pathways are called *branched pathways* because they can arrive at a target compound through combinations of pathways that split compounds into smaller ones, work in parallel with many compounds, and join compounds into larger ones. While most previous work has focused on identifying linear metabolic pathways, branched metabolic pathways predominate in metabolic networks. Automatic identification of branched pathways has a number of important applications in areas that require deeper understanding of metabolism, such as metabolic engineering and drug target identification. Our algorithm utilizes explicit atom tracking to identify linear metabolic pathways and then merges them together into branched metabolic pathways. We provide results on two well-characterized metabolic pathways that demonstrate that this new merging approach can efficiently find biologically relevant branched metabolic pathways with complex structures.

1 Introduction

The quantity and quality of metabolic data has greatly increased in the last few decades, as indicated by the growth of such databases as the Kyoto Encyclopedia of Genes and Genomes (KEGG) [17] and MetaCyc [8]. Gaining understanding from these vast quantities of metabolic data requires novel computational tools that enable automatic identification and thorough analysis of biologically relevant metabolic pathways. These computational tools may reveal novel or alternative metabolic pathways, potentially spanning multiple species, that could not have been identified by manual means. Importantly, the ability to find metabolic pathways in multi-genome scale data has applications in fields such as metabolic engineering, which focuses on discovering and implementing new metabolic schemes.

The central problem in computational metabolic path finding is the following: given a start and target compound, find and return *biologically relevant* or *realistic* pathways of enzymatic reactions that produce the target compound from

V. Bafna and S.C. Sahinalp (Eds.): RECOMB 2011, LNBI 6577, pp. 70–84, 2011.

the start compound. Previous work in this area has primarily focused on finding linear sequences of reactions between start and target compounds [25]. However, more complex metabolic pathways, termed *branched pathways*, are dominant in metabolic networks and provide a more complete picture of metabolic processes [22,24]. Branched pathways consist of multiple pathways that interact biochemically. For example, a start compound may be split into smaller compounds which, in parallel, undergo several different chemical reactions. The resulting products can then combine to form the target compound. The identification of branched pathways enables the analysis of metabolic processes with a more comprehensive perspective as compared to the limited picture provided by linear pathways.

The main contribution of this paper is a novel algorithm for identifying branched metabolic pathways by using atom tracking information to merge linear pathways. The merging approach of the presented algorithm is different from previous graph-based approaches, which start from a single linear pathway and then find new linear pathways to attach as branches [24,16]. The results demonstrate that the new algorithm is able to efficiently find different network topologies in multi-genome scale data obtained from KEGG. The rest of the paper proceeds as follows: Section 2 describes the relevant previous work in the area of graph-based metabolic pathfinding; Section 3 describes how our new algorithm merges linear pathways to find branched pathways; Section 4 contains the results of our algorithms, validated on two well-characterized branched metabolic pathways; Section 5 concludes the paper.

2   Previous Work

Graph-based Models for Finding Metabolic Pathways. Graphs provide a natural, well-studied computational model for identifying biologically relevant pathways in metabolic networks [11]. Graph-based metabolic path finding algorithms complement stoichiometric approaches, as they focus on different aspects of modeling and understanding metabolism [25,12,13]. Stoichiometric models are typically utilized for modeling specific organisms or metabolic systems [15]. Most stoichiometric models are based on the steady-state assumption and therefore require explicit labeling of internal and external compounds [19]. This can be a disadvantage as there are feasible biochemical pathways that do not obey the steady-state assumption and/or a compound that labeled as internal could easily be provided as an external compound [25]. However, both types of models are important for gaining insights into metabolic networks.

Graph-based methods have suffered from the disadvantage of finding pathways with spurious connections [3]. Several approaches have been developed to try to overcome this problem, such as removing certain currency metabolites from the graph [31,14,27], adding weights based on the degree of the nodes [9,12] or using measures of structural similarity between compounds [21,26]. However, an approach more closely related to the underlying biochemistry is to use *atom mapping data* [2,3]. Atom mapping data provides a systematic way of understanding a biochemical reaction by providing a specific mapping between each atom in the input compounds of a reaction to an atom in the output compounds. In the last few years, the availability of atom mapping data has been steadily increasing, with one of the primary sources being the KEGG RPAIR database [18,20].

Previous work has mainly used atom mapping data for finding metabolic pathways by only allowing connections through reactions where at least one atom is being transfered from input to output compound [12,23] or only returning pathways that conserve at least one atom, typically carbon [2,3,4,5]. However, there are often instances where it is biochemically relevant to find pathways which conserve a high percentage of atoms from start to target compounds [6,24]. The algorithm presented in this paper is based on our earlier work that finds atom conserving pathways by explicitly tracking multiple atoms in metabolic networks [16].

Foundation for the Presented Work. The algorithm presented in this paper utilizes a graph-based structure that incorporates atom mapping data called an *atom mapping graph*, G_{am}, whose design is based on the observation that the same atom mapping pattern between two compounds often appears in multiple reactions [2]. G_{am} is a directed bipartite graph containing *compound nodes* and *mapping nodes*. The compound nodes have unique identifiers for both the compound as well as its atoms. The compound nodes are connected by directed edges to mapping nodes that explicitly specify, via the unique identifiers, what atoms from the input compound become the atoms in the output compound. Each atom mapping only maps the atoms between a pair of compounds and so the mapping nodes in G_{am} only have one input edge and one output edge connected to two different compound nodes, while the compound nodes have a degree equal to the number of mappings they participate in. Since the same atom mapping can occur in many different reactions, a correspondence is stored between the mapping nodes and the reactions in which they occur. A more detailed description of the construction of G_{am} can be found in [16].

Previously, we developed and validated an algorithm for identifying the k shortest linear pathways in G_{am} that conserve at least a given number of atoms between desired start and target compounds [16]. This problem has been shown to be PSPACE-complete and NP-complete when a compound can only be used once in a pathway [6]. Previously unnamed, we will call the linear path finding algorithm from [16] LPAT, for Linear Pathfinding with Atom Tracking. Our results demonstrated that LPAT is able to search across the thousands of reactions and compounds from multiple species contained in KEGG and return realistic metabolic pathways in a few minutes. Based on LPAT, we also developed and validated a novel graph-based algorithm for identifying branched metabolic pathways. Also previously unnamed, we will call the branched path finding algorithm from [16] BPAT-S, for Branched Pathfinding using Atom Tracking and Seed pathways. The first step of BPAT-S uses LPAT to obtain a set of linear pathways between the desired start and target compounds. BPAT-S then annotates and stores the linear pathways with information about the specific reactions and compounds through which atoms are lost or gained. These annotated linear pathways are called *seed pathways* and indexed for efficient processing and attachment of branches. The branches are identified by calling LPAT to find linear pathways between compounds through which atoms are lost to compounds through which atoms are gained. These linear branches are then attached to the seed pathway to give rise to branched pathways, which are ranked first by the number of atoms they conserve and then by the total number of reactions they contain. In our previous study, we demonstrated that BPAT-S can efficiently find and return branched pathways that correspond to known branched pathways [16]. To the best of our knowledge, the only other algorithm with similar abilities is the recently developed ReTrace algorithm. ReTrace takes a similar approach as BPAT-S, but is based on pathways that only conserve one atom [24]. In this paper, we present a novel algorithm that takes a significantly different approach from BPAT-S or ReTrace by merging linear pathways to form branched pathways and extends the topologies of pathways that can be identified automatically.

3   BPAT-M: Branched Pathfinding by Merging Linear Pathways

This section describes a new algorithm, Branched Pathfinding using Atom Tracking and Merging (BPAT-M), for finding branched pathways by merging linear pathways returned by LPAT. BPAT-M removes the division between seed and branch pathways found in BPAT-S, thus enabling BPAT-M to find pathway topologies that BPAT-S cannot. BPAT-M utilizes the observation that a significant portion of time is spent finding the branches in BPAT-S, but these branches may already be contained in the set of linear pathways found by LPAT. This redundancy is eliminated in BPAT-M by carefully inventorying the linear pathways. BPAT-M takes advantage of the fact that linear pathways can only be merged together if the pathways do not have overlapping atoms in their target compounds. The atom tracking information from the linear pathways provided by LPAT are processed by BPAT-M to construct three data structures Q, C, and M. These data structures enable the efficient merging of linear pathways to find branched pathways. The construction of Q, C and M is described in 3.1. Section 3.2 then describes Algorithm 1, which harnesses the extensive indexing of linear pathways contained in Q, M and C to find and return n branched pathways ranked first by the number of atoms conserved and second by the number of reactions.

3.1   Construction of Q, C and M from Target Atom Markings (TAMs)

A target atom marking (TAM) of a linear pathway is a set of indices corresponding to the specific atoms in the target compound that have been conserved from the start compound. Typically, the number of TAMs found is much less than the theoretical maximum number due to chemical constraints. On the right of Figure

1 there are three linear pathways from α-D-glucose 6-phosphate to stachyose and their associated TAMs. TAMs play a central role in the performance of BPAT-M because they allow a quick way to determine which linear pathways can not be merged together. Two pathways can not be merged together if the intersection of their TAMs is nonempty, that is if they contain the same atom index or indices. If two pathways have disjoint TAMs, they can not necessarily be merged because the algorithm must check whether they share a common reaction. However, if two pathways are mergeable, then the TAM of the merged pathway is the union of the TAMs of the pathways.

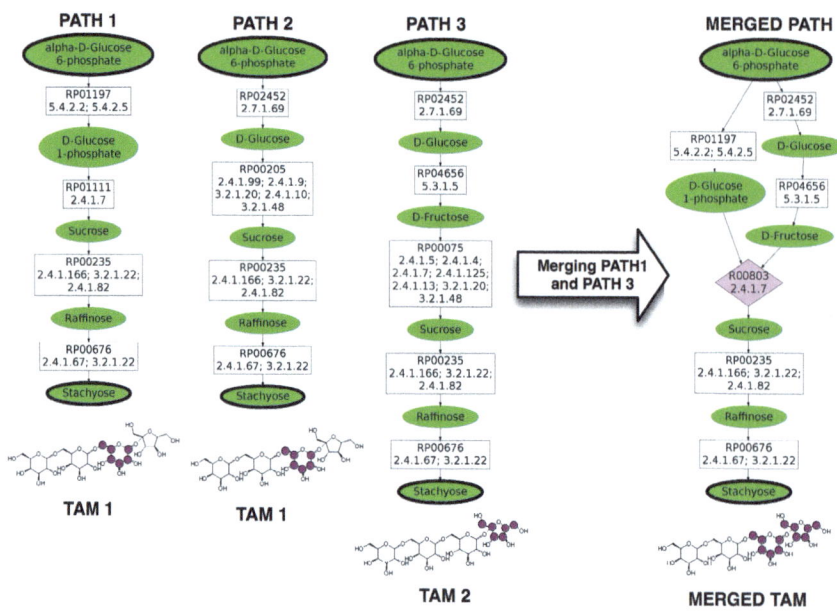

Fig. 1. Three linear pathways from α-D-glucose 6-phosphate to stachyose and their associated carbon TAMs, as indicated by the magenta circles. The two potentially mergeable pairs of paths are PATH 1 with PATH 3 and PATH 2 with PATH 3. The result of merging PATH 1 and PATH 3 is displayed on the right.

The ability to use the TAMs to quickly determine if pathways are not mergeable motivates the construction of the data structure, Q, which maps a TAM to a list of linear pathways containing that TAM. For a particular TAM, t, this means that $Q[t]$ returns all linear pathways whose TAM is equal to t, sorted by their length. For example, using the pathways depicted in Figure 1, $Q[\text{TAM 1}]$ would return the pathways labeled PATH 1 and PATH 2. After Q is constructed, all disjoint combinations of the TAMs from the linear pathways are computed and stored in a list C. For example, for the hypothetical TAMs, $t_1 = \{0,1,2\}$, $t_2 = \{0,1\}$, $t_3 = \{2,3\}$ and $t_4 = \{4,5\}$, C would contain $\{t_1,t_4\}$, $\{t_2,t_3,t_4\}$, $\{t_2,t_3\}$, $\{t_2,t_4\}$ and $\{t_3,t_4\}$ as all disjoint combinations. C is then sorted in decreasing order by the total size of the combination. In this example, $\{t_2,t_3,t_4\}$

would be first entry in C because it is of size six. Sorting C this way is important because the goal is to find pathways that conserve a larger number of atoms. For each combination $c \in C$, the TAMs are accessed by their indices, so if $c = \{t_2, t_3, t_4\}$, $c[1] = t_2$, $c[2] = t_3$ and $c[3] = t_4$. C is then used to dictate how the search proceeds to merge combinations of linear pathways to obtain branched pathways.

Once potentially mergeable linear pathways have been identified using Q and C, they must be further compared to see if they can be merged through a common reaction, r. The data structure M is constructed to store the results of comparing pairs of linear pathways for mergability, thus the comparison is only performed once. M maps all pairs of mergeable linear pathways to a tuple containing r and the number of mapping nodes from the target compound that r occurs. M is constructed by first identifying all pairs of pathways, p_1 and p_2, with disjoint TAMs. The mapping nodes of p_1 and p_2 are compared starting from the target compound. This comparison identifies the position, m, of the mapping nodes closest to the target compound that differs between the two pathways is identified. In Figure 1, the comparison between PATH 1 and PATH 3 would identify RP01111 and RP00075 as the different mapping nodes closest to the target compound and m would be 2, using zero-based indexing. The final step is to look up the reactions that are associated with the two mapping nodes at m and determine if the mapping nodes share common reaction that can be used to merge the two pathways. If there is no common reaction, then the pathways are not mergable. In the case of PATH 1 and PATH 3 in Figure 1, both RP01111 and RP00075 are found in the reaction R00803 (EC Number 2.4.1.7) in KEGG. The right side of Figure 1 depicts PATH 1 and PATH 3 merged by R00803. This information about the mergability of PATH 1 and PATH 3 in Figure 1 would then be stored as $M[PATH1, PATH3] = (R00803, 2)$. In the construction of M, only the mapping nodes that differ closest to the target compound are considered as potential merge points because if merging two paths results in a larger TAM, they must share a common reaction at this point. It is possible that two paths may interact closer to the start compound, but this is currently not considered by the algorithm because it does not impact the TAM.

3.2   Finding Branched Pathways Using Q, C and M

After processing and indexing the linear pathways to construct Q, C and M, these data structures are given as input to Algorithm 1 along with n number of branched pathways to return and a fixed beam width w, which can be used to bound the search. Algorithm 1 then returns the final result of BPAT-M, the top n branched pathways it finds ranked first by the number of atoms conserved and then by the number of reactions. Despite reducing the number of linear pathways combinations that need to be tested by using biochemical constraints, the number of such combinations sometimes remains quite large. Therefore, the algorithm performs a beam search with a fixed beam width, w, which is provided by the user. The heuristic used for the beam search is discussed in more detail later in the section, and its usage means that BPAT-M does not guarantee

---

Algorithm 1. BPAT-M Search

Input: Pathways organized by their TAMs, Q; Sorted list of all combinations of disjoint TAMs, C; Mergeable pairs of paths, M; Number of pathways to return, n; Limit on Intermediate Branched Pathways (IBPs), w;

Output: Sorted list of branched pathways $\mathcal{P}$, containing linear pathways and merge points, sorted first by number of atoms conserved, then by total number of nodes

1: $\mathcal{P} \leftarrow \{\}$
2: **for each** c in C **do**
3:     **if** $\mathcal{P}$ contains more than n pathways and the nth pathway conserves more atoms than the size of c **then**
4:         Truncate $\mathcal{P}$ to n pathways
5:         Break
6:     $\mathcal{T} \leftarrow \{\}$ //for storing the IBPs, sorted by the same criteria as $\mathcal{P}$
7:     **for each** pair of linear pathways (p_i, p_j) in $(Q[c[1]] \times Q[c[2]])$ **do**
8:         **if** $M(p_i, p_j)$ exists **then**
9:             Add IBP containing $p_i, p_j, M(p_i, p_j)$ to $\mathcal{T}$
10:     **for** $k = 3$ to size of c **do**
11:         $\mathcal{N} \leftarrow \mathcal{T}$
12:         $\mathcal{T} \leftarrow \{\}$
13:         **for each** IBP P in $\mathcal{N}$ **do**
14:             **for each** linear pathway p_q in $Q[c[k]]$ **do**
15:                 **for each** linear pathway p_l in P **do**
16:                     **if** $M(p_q, p_l)$ exists and is a valid merge point in P **then**
17:                         Add new IBP containing P merged with p_q and $M(p_q, p_l)$ to $\mathcal{T}$
18:         **if** $\mathcal{T}$ contains more than w pathways **then**
19:             Truncate $\mathcal{T}$ to w pathways
20:     Add all pathways in $\mathcal{T}$ to $\mathcal{P}$
21: Return $\mathcal{P}$

---

finding the optimal combination. However, the results demonstrate that the search performs well in practice.

Algorithm 1 works by taking each combination of TAMs $c \in C$ in turn and using them to build branched pathway combinations. The first two TAMs, $c[1]$ and $c[2]$, are used to obtain the set of associated pathways for each TAM from Q and all pairs of pathways are tested for mergeability using M (lines 7-9). If a pair of pathways are mergeable, then they are stored in the set of Intermediate Branched Pathways (IBPs), $\mathcal{T}$. The IBPs store a list of mergeable linear pathways and their merge points. Then, for each subsequent TAM in c (line 10), all of the pathways associated with the TAM $c[k]$, $p_q \in Q[c[k]]$ are retrieved (line 14). Each $p_q \in Q[c[k]]$ is then tested for mergeability with each linear pathway in each IBP (lines 13-16). If p_q is mergeable with a linear pathway, p_l in IBP, that is $M[p_q, p_l]$ contains a merge point, p_q can potentially be merged with the IBP to create a branched pathway that conserves more atoms. However, because p_l has already been merged with other pathways, it must be verified that the merge point between p_q and p_l is still valid (line 16).

A merge point is always valid if p_l has not been merged with another pathway in the IBP at the same mapping node it would use to merge with the new pathway, p_q. However, if p_l has been previously merged at the same point with another pathway in the IBP, the merge point with p_q can still be valid if the reaction in the merge point of p_q and p_l is the same reaction used previously and the substrate compound in p_q is not contained in the other pathways. Otherwise, the merge point is invalid. As an example, there could be a reaction r that takes the substrate compounds a, b and c. If two pathways, p_1 and p_2 were merged together through r, with p_l containing a and p_q containing b, there are two possibilities for a third pathway p_3, that is potentially mergable with p_1 at r. If p_3 contains c, then the merge point is still valid and the resulting branched pathway would contain p_1, p_2 and p_3 merged through r. However, if p_3 contained b, then the merge point is invalid, as p_2 has already been merged through b. By checking for validity, multiple pathways being merged through the same reaction are handled in a general way and only limited by the substrates used in the reaction.

In Algorithm 1, if the merge point is valid, p_q is merged with the IBP and the resulting branched pathway is stored as another IBP (line 17). Therefore, each IBP gives rise to a number of new IBPs equal to the number of p_q that have valid merge points with the IBP. This means that there is a theoretical combinatorial explosion of IBPs for each C_i and we have observed that very large numbers of IBPs can be generated in practice. This resulted in the introduction of the beam width, w, to limit the number of combinations generated. After adding the pathways for each TAM, only the top w IBPs, sorted by number of atoms conserved and the sum of the length of the linear pathways, are carried over for each subsequent TAM (lines 18-19). Since the pathways are first ranked by the number of atoms they conserve and C is sorted by the size of each combination, the search can terminate when n pathways have been found and the next combination to try is smaller than the TAM of the nth pathway (lines 3-5).

The final way in which the run time and/or space required by BPAT-M is reduced is by limiting the number of pathways that are kept for each TAM in Q. This is done by sorting the pathways by length and only keeping a user specified number of the shortest pathways for each TAM. Future work is needed to investigate the impact of these parameters and develop easier ways for users to understand and select the proper limitations for their application. At this point, it's recommended to perform a larger search by setting the parameter values high to utilize the computational and storage resources available. Our results demonstrate that even with the heuristic limits, BPAT-M performs well in practice.

4   Results

We present two representative, biologically interesting, test cases for branched metabolic pathways. Both begin from α-D-Glucose 6-Phosphate (G6P), a common form of intracellular glucose. The target compound of the first pathway is lycopene and the target compound of the second pathway is cephalosporin C.

4.1   Experimental Setup

All KEGG data used in the following experiments was downloaded on February 10, 2010. After processing the KEGG RPAIR to obtain a universal index for each atom in each compound, G_{am} was constructed using 11,892 RPAIR entries involving 6,002 compounds and corresponding to 7,510 reactions from more than 1,200 organisms. In the results presented in this paper the full G_{am} was used, but subgraphs of G_{am} corresponding to particular organisms, reactions or compounds of interest can easily be created and searched. Additionally, reversibility information was obtained from XML representations of the KEGG metabolic

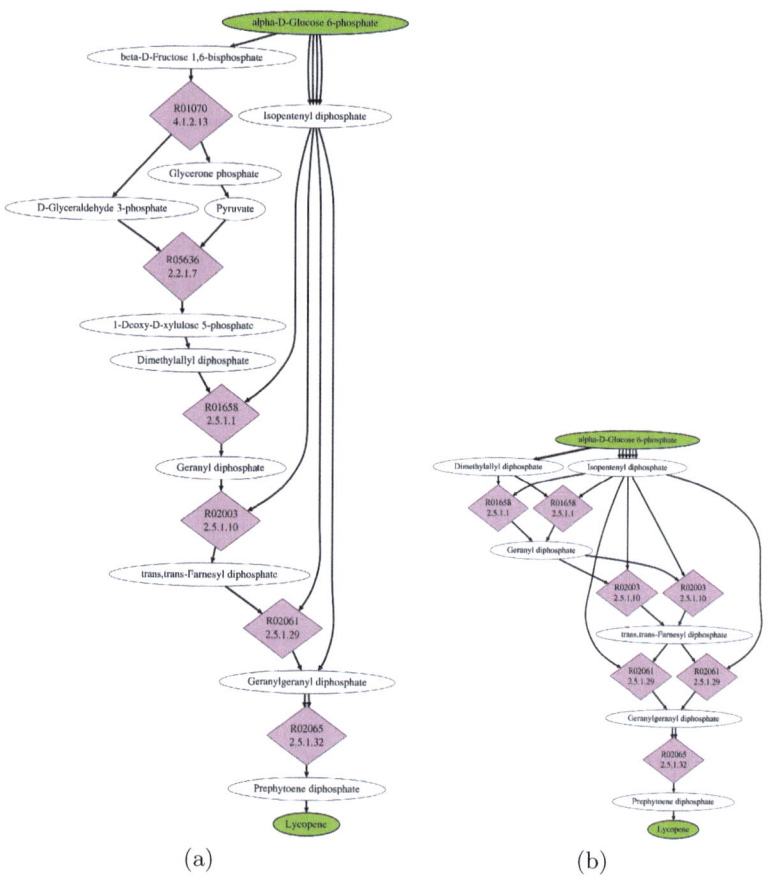

(a)   (b)

Fig. 2. Top ranked branched metabolic pathways for G6P to lycopene, with the linear pathways conserving 2 carbons, as found by (a) BPAT-S and in (b) BPAT-M. In the interest of space, in (a) 34 mapping nodes and 28 compound nodes and (b) 14 mapping nodes and 11 compound nodes along the linear paths between merge points are hidden from view. The full figures can be viewed in online supplementary material (URL at the end of the paper).

pathway maps, distributed in the the KEGG Markup Language (KGML). A reaction is considered irreversible if it is consistently labeled as such across all of the KEGG metabolic pathway maps. Otherwise, the reaction is considered reversible. The processing of the KGML pathway maps resulted in 4,360 reactions being labeled irreversible. Once the reaction direction is determined, this information is then used to label RPAIR entries reversible or irreversible, which is used in the construction of G_{am}. For each RPAIR entry, all associated reactions have to be checked for directionality. If all of the reactions are irreversible and consistent in the labeling of the compounds as substrates and products then the RPAIR entry is considered irreversible. Otherwise, the entry is labeled as reversible. This resulted in 6,386 RPAIR entries being considered irreversible.

The implementation was done in Java using the Chemical Development Kit [29] and the Java Universal Network/Graph Framework (http://jung.sourceforge. net/). All result figures are drawn using Graphviz (http://www.research.att.com/ sw/tools/graphviz/). All experiments were run on the Shared University Grid at Rice (SUG@R), using a single core from a 2.83GHz Intel Xeon E5440 with access to 16GB of RAM for each pathway. The parameters for BPAT-S and BPAT-M were generally chosen to perform the most exhaustive search, given the resources available. In the branched pathway figures the ellipses are compounds, with the start and target compounds highlighted in green, the pink diamonds contain the KEGG ID and EC numbers for the reactions that occur at the branching points of the pathway. Each edge corresponds to one molecule of each compound. In the interest of space and clarity, the intermediate mapping and compound nodes of linear pathways composing the branched pathway have been removed from the figures, leaving only the reactions that occur at the branching points and their immediate input and output compounds, as well as the start and target compounds. The figures depicting the full branched pathways can be found in the online supplementary material (URL at the end of the paper).

4.2   α-d-Glucose 6-Phosphate to Lycopene

Lycopene is a C_{40} carotenoid having a bright red color and is found in fruits and vegetables, such as tomatoes and watermelons. Lycopene's nutritional and pharmaceutical potential has resulted in a number of investigations on using metabolic engineering techniques to increase yield and/or produce lycopene in microbial hosts [1,32]. The known biosynthesis pathway of lycopene is relatively well understood and has an interesting "woven" topology; isopentenyl diphosphate (IPP) and dimethylallyl diphosphate (DMAPP) are produced by either the 2-C-methyl-D-erythritol 4-phosphate/1-deoxy-D-xylulose 5-phosphate (MEP/DOXP) or mevalonate (MVA) pathways, and then DMAPP combines with IPP to make two molecules of geranyl diphosphate which are combined with IPP in two more sequential reactions resulting in two molecules of geranylgeranyl diphosphate, which combine to make the C_{40} molecule prephytoene diphosphate that becomes lycopene [28]. BPAT-S and BPAT-M were given as the start compound G6P, the target compound lycopene and three as the number of carbons to conserve. For BPAT-M, k for LPAT was set to 1,000,000, resulting in

36,405 linear pathways without cycles, which had 16 mutually exclusive target atom markings and generated 6,301 combinations. The number of pathways in each cluster was limited to 2,500 and w was set to 500, due to memory limitations. For BPAT-S, due to run time limitations, k for LPAT was set to 500,000, resulting in 22,064 linear pathways without cycles. In both cases, LPAT required about one minute to find the linear pathways.

The top ranked results from BPAT-S and BPAT-M are depicted in Figures 2(a) and 2(b), respectively. BPAT-M performs better than BPAT-S, in that it can find find the known "woven" topology starting with IPP and DMAPP. BPAT-S cannot identify the known topology because it only allows branches off of an initial seed pathway. Additionally, BPAT-M completed in 74.1 minutes, while BPAT-S required 463.2 minutes. Due to space constraints, the intermediates to IPP and DMAPP are not displayed in Figure 2. The full figures found in the online supplementary material (URL at the end of the paper) show that the BPAT-M result correctly finds the lycopene pathway that utilizes the MEP pathway to synthesize IPP and DMAPP; the BPAT-S result utilizes both the MEP/DOXP pathway for DMAPP and MVA pathway for IPP thus revealing the variety available. The MEP/DOXP and MVA pathways demonstrate how search tools for metabolic pathways can illuminate different alternative pathways that may be found in different organisms, which can have applications in areas such as metabolic engineering. At the same time, the ability to find alternative or novel pathways makes it more difficult to judge the performance of different algorithms. Both the BPAT-M and BPAT-S results are biochemically correct in the usage of MEP/DOXP and MVA pathways, but one may be preferred over the other based on the specific application in mind.

4.3 α-d-Glucose 6-Phosphate to Cephalosporin C

Cephalosporin C is a β-lactam antibiotic, synthesized by certain bacteria and fungi, but not used clinically because of its low potency [10]. However, it is an important precursor for a number of related antibiotics and has been a target for increased production using metabolic engineering approaches [30]. The biosynthetic pathway for cephalosporin C includes an reaction that synthesizes δ-(L-α-aminoadipyl)-L-cysteinyl-D-valine (ACV) from L-valine, L-cysteine and L-2-aminoadipate. The pathway then proceeds through isopenicillin N which then undergoes a series of reactions resulting in cephalosporin C [30]. BPAT-M and BPAT-S were given as input G6P as the start compound, cephalosporin C as the target compound and three as the minimal number of carbons to conserve. For both BPAT-S and BPAT-M, k for LPAT was set to 1,000,000, resulting in 31,280 linear pathways without cycles found in less than one minute. For BPAT-M the number of pathways in each cluster was limited to 5,000 and w was set to 1,000. BPAT-M found 5 mutually exclusive target atom markings resulting in 18 combinations. BPAT-M took 1.9 minutes to complete the search and BPAT-S took 366.4 minutes.

The top ranked pathways for BPAT-S and BPAT-M are depicted in Figures 3(a) and 3(b), respectively. The simplified figures demonstrate that both BPAT-M and

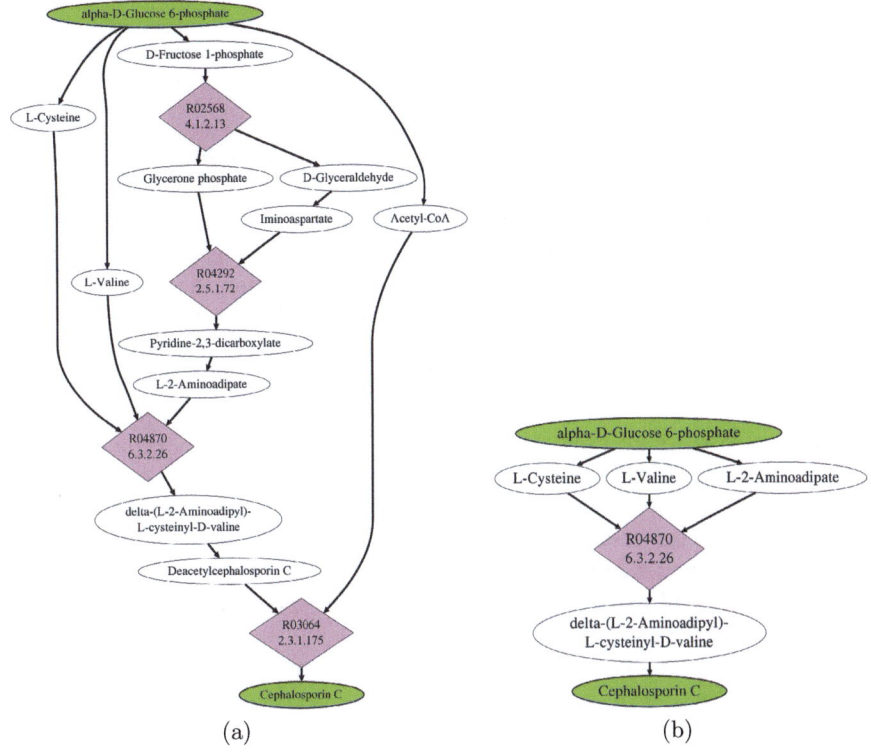

Fig. 3. Top ranked branched metabolic pathways for G6P to cephalosporin C, with the linear pathways conserving 3 carbons, as found by (a) BPAT-S and (b) BPAT-M. In the interest of space, in (a) 33 mapping nodes and 25 compound nodes and in (b) 27 mapping nodes and 21 compound nodes along the linear paths between merge points are hidden from view. The full figures can be viewed in the online supplementary material (URL at the end of the paper).

BPAT-S find the correct overall metabolic scheme, containing the crucial reaction catalyzed by ACV synthetase, which requires three different substrate compounds and produces ACV as the product [7]. Similar to the lycopene pathway, BPAT-M is able to identify this scheme much quicker than BPAT-S. BPAT-S is able to identify the branch through acetyl-CoA which contributes its acetyl group, containing two carbons, to deacetylcephalosporin C to make cephalosporin C. BPAT-M does not identify the pathway through acetyl-CoA since it is not in the original set of linear pathways that conserved three carbons. The full pathway figures, found in the online supplementary material (URL at the end of the paper), reveal that BPAT-M is able to find shorter and pathways more similar to the known pathways through L-valine and L-2-aminoadipate. While BPAT-S finds a similar branched pathway to L-2-aminoadipate, it returns a long and unlikely pathway to L-valine. The unlikely pathway to L-valine is due the approach taken by BPAT-S of attaching branches that maximize the number of atoms to a single seed pathway.

5   Discussion

We have described and tested a new algorithm, BPAT-M, for identifying branched metabolic pathways that utilizes atom tracking information to efficiently merge together biochemically interacting linear pathways. The experimental results highlight both the strengths and weakness of the approach taken by BPAT-M. These results reveal that the algorithm's performance may depend on the underlying structure of the pathway they seek to find. The merging approach used by BPAT-M will likely perform better if all of the branches conserve at least the given number of atoms and are of similar length. If this is the case, then BPAT-M can also find more complex topologies, because it is not limited to requiring all branches to start and end from the same pathway. This is typical of large compounds made from similar components, and the lycopene result highlights this ability of BPAT-M. The lycopene pathway utilizes a "woven" topology that is returned as the top result by BPAT-M, but BPAT-S is unable to find. In both cases, BPAT-M took significantly less time than BPAT-S.

Despite the promising results, it is clear that the approach would benefit from a number of improvements. Future work will allow multiple start and target compounds to be used as input, utilize other sources of metabolic data and examine ways to mine the resulting pathways to help the user understand them. The results also highlight that comparing the resulting branched metabolic pathways is nontrivial, especially if the goal is to find alternative or novel pathways, and further work is required to develop meaningful evaluation methods. Results from ReTrace, as described in [24], are not presented because ReTrace has several parameters that affect the search. We observed that the results and runtime varied widely depending on the parameters given to ReTrace. We found that ReTrace, given the same KEGG data, efficiently finds branched pathways similar to those presented in Section 4, but do not contain as many branches. This may be due to a number of factors and performing a valid and exhaustive comparison will be the subject of future work. While it can be difficult to identify *a priori* which method will perform better, it is reasonable to try different algorithms and analyze the results to gain better understanding of metabolic pathways.

Acknowledgments. APH and LEK were supported in part by NSF ABI-0960612, NHARP 01907, a fund from the John & Ann Doerr Fund for Computational Biomedicine at Rice University, and a Sloan Fellowship. APH is a fellow with the Shell Center for Sustainability at Rice University and is partially supported by an SCS grant. Computational resources were provided by the Shared University Grid at Rice, funded by NSF under Grant EIA-0216467 and partnership between Rice University, Sun Microsystems, and Sigma Solutions, Inc.

Online Supplementary Material. The full pathways of Figure 2 and Figure 3 can be found at: http://www.kavrakilab.org/metapath/recomb-2011-supp.

References

1. Alper, H., Miyaoku, K., Stephanopoulos, G.: Construction of lycopene-overproducing E. coli strains by combining systematic and combinatorial gene knockout targets. Nature Biotechnology 23(5), 612–616 (2005)
2. Arita, M.: In silico atomic tracing by substrate-product relationships in Escherichia coli intermediary metabolism. Genome Research 13(11), 2455–2466 (2003)
3. Arita, M.: The metabolic world of Escherichia coli is not small. Proceedings of the National Academy of Sciences of the United States of America 101(6), 1543–1547 (2004)
4. Blum, T., Kohlbacher, O.: MetaRoute: fast search for relevant metabolic routes for interactive network navigation and visualization. Bioinformatics 24(18), 2108–2109 (2008)
5. Blum, T., Kohlbacher, O.: Using Atom Mapping Rules for an Improved Detection of Relevant Routes in Weighted Metabolic Networks. Journal of Computational Biology 15(6), 565–576 (2008)
6. Boyer, F., Viari, A.: Ab initio reconstruction of metabolic pathways. Bioinformatics 19(90002), 26ii–34ii (2003)
7. Byford, M.F., Baldwin, J.E., Shiau, C.Y., Schofield, C.J.: The Mechanism of ACV Synthetase.. Chemical Reviews 97(7), 2631–2650 (1997)
8. Caspi, R., Altman, T., Dale, J.M., Dreher, K., Fulcher, C.A., Gilham, F., Kaipa, P., Karthikeyan, A.S., Kothari, A., Krummenacker, M., Latendresse, M., Mueller, L.A., Paley, S., Popescu, L., Pujar, A., Shearer, A.G., Zhang, P., Karp, P.D.: The MetaCyc database of metabolic pathways and enzymes and the BioCyc collection of pathway/genome databases. Nucleic Acids Research 38(Database issue), D473–D479 (2010)
9. Croes, D., Couche, F., Wodak, S.J., van Helden, J.: Inferring meaningful pathways in weighted metabolic networks. Journal of Molecular Biology 356(1), 222–236 (2006)
10. Demain, A.L., Elander, R.P.: The β-lactam antibiotics: past, present, and future. Antonie van Leeuwenhoek 75(1), 5–19 (1999)
11. Deville, Y., Gilbert, D., van Helden, J., Wodak, S.J.: An overview of data models for the analysis of biochemical pathways. Briefings in Bioinformatics 4(3), 246–259 (2003)
12. Faust, K., Croes, D., van Helden, J.: Metabolic pathfinding using RPAIR annotation. Journal of Molecular Biology 388(2), 390–414 (2009)
13. de Figueiredo, L.F., Schuster, S., Kaleta, C., Fell, D.A.: Response to comment on 'Can sugars be produced from fatty acids? A test case for pathway analysis tools. Bioinformatics 25(24), 3330–3331 (2009)
14. Gerlee, P., Lizana, L., Sneppen, K.: Pathway identification by network pruning in the metabolic network of Escherichia coli. Bioinformatics 25(24), 3282–3288 (2009)
15. Gombert, A.K., Nielsen, J.: Mathematical modelling of metabolism. Current Opinion in Biotechnology 11(2), 180–186 (2000)
16. Heath, A.P., Bennett, G.N., Kavraki, L.E.: Finding Metabolic Pathways Using Atom Tracking. Bioinformatics 26(12), 1548–1555 (2010)
17. Kanehisa, M., Araki, M., Goto, S., Hattori, M., Hirakawa, M., Itoh, M., Katayama, T., Kawashima, S., Okuda, S., Tokimatsu, T., Yamanishi, Y.: KEGG for linking genomes to life and the environment. Nucleic Acids Research 36(Database issue), D480–D484 (2008)

18. Kanehisa, M., Goto, S., Hattori, M., Aoki-Kinoshita, K.F., Itoh, M., Kawashima, S., Katayama, T., Araki, M., Hirakawa, M.: From genomics to chemical genomics: new developments in KEGG. Nucleic Acids Research 34(Database issue), D354–D357 (2006)
19. Kauffman, K.J., Prakash, P., Edwards, J.S.: Advances in flux balance analysis. Current Opinion in Biotechnology 14(5), 491–496 (2003)
20. Kotera, M., Hattori, M., Oh, M., Yamamoto, R., Komeno, T., Goto, S., Yabuzaki, J., Kanehisa, M.: RPAIR: a reactant-pair database representing chemical changes in enzymatic reactions. Genome Informatics 15, P062 (2004)
21. McShan, D.C., Rao, S., Shah, I.: PathMiner: predicting metabolic pathways by heuristic search. Bioinformatics 19(13), 1692–1698 (2003)
22. Michal, G.: Biochemical Pathways: An Atlas of Biochemistry and Molecular Biology. John Wiley & Sons, Inc., New York (1999)
23. Mithani, A., Preston, G.M., Hein, J.: Rahnuma: hypergraph-based tool for metabolic pathway prediction and network comparison. Bioinformatics 25(14), 1831–1832 (2009)
24. Pitkänen, E., Jouhten, P., Rousu, J.: Inferring branching pathways in genome-scale metabolic networks. BMC Systems Biology 3(1), 103 (2009)
25. Planes, F.J., Beasley, J.E.: A critical examination of stoichiometric and path-finding approaches to metabolic pathways. Briefings in Bioinformatics 9(5), 422–436 (2008)
26. Rahman, S.A., Advani, P., Schunk, R., Schrader, R., Schomburg, D.: Metabolic pathway analysis web service (Pathway Hunter Tool at CUBIC). Bioinformatics 21(7), 1189–1193 (2005)
27. Ranganathan, S., Maranas, C.D.: Microbial 1-butanol production: Identification of non-native production routes and in silico engineering interventions. Biotechnology Journal 5(7), 716–725 (2010)
28. Sandmann, G.: Carotenoid biosynthesis and biotechnological application. Archives of Biochemistry and Biophysics 385(1), 4–12 (2001)
29. Steinbeck, C., Hoppe, C., Kuhn, S., Floris, M., Guha, R., Willighagen, E.L.: Recent Developments of the Chemistry Development Kit (CDK) - An Open-Source Java Library for Chemo- and Bioinformatics. Current Pharmaceutical Design 12(17), 2111–2120 (2006)
30. Thykaer, J.: Metabolic engineering of β-lactam production. Metabolic Engineering 5(1), 56–69 (2003)
31. Wagner, A., Fell, D.A.: The small world inside large metabolic networks. Proceedings of the Royal Society B: Biological Sciences 268(1478), 1803–1810 (2001)
32. Yoon, S.H., Kim, J.E., Lee, S.H., Park, H.M., Choi, M.S., Kim, J.Y., Lee, S.H., Shin, Y.C., Keasling, J.D., Kim, S.W.: Engineering the lycopene synthetic pathway in E. coli by comparison of the carotenoid genes of Pantoea agglomerans and Pantoea ananatis. Applied Microbiology and Biotechnology 74(1), 131–139 (2007)

A Probabilistic Model for Sequence Alignment with Context-Sensitive Indels

Glenn Hickey and Mathieu Blanchette

McGill Centre for Bioinformatics,
McGill University, Montreal, Quebec, Canada
{hickey,blanchem}@mcb.mcgill.ca

Abstract. Probabilistic approaches for sequence alignment are usually based on pair Hidden Markov Models (HMMs) or Stochastic Context Free Grammars (SCFGs). Recent studies have shown a significant correlation between the content of short indels and their flanking regions, which by definition cannot be modelled by the above two approaches. In this work, we present a context-sensitive indel model based on a pair Tree-Adjoining Grammar (TAG), along with accompanying algorithms for efficient alignment and parameter estimation. The increased precision and statistical power of this model is shown on simulated and real genomic data. As the cost of sequencing plummets, the usefulness of comparative analysis is becoming limited by alignment accuracy rather than data availability. Our results will therefore have an impact on any type of downstream comparative genomics analyses that rely on alignments. Fine-grained studies of small functional regions or disease markers, for example, could be significantly improved by our method. The implementation is available at http://www.mcb.mcgill.ca/~blanchem/software.html

Keywords: context-sensitive indel model, statistical alignment, tree-adjoining grammar.

1 Introduction

Short insertions and deletions (indels) play a critical role in human evolution, accounting for a comparable amount of sequence divergence from chimpanzee as point mutations [32]. They are also significant factors in a variety of genetic diseases [1,6]. Short indels have a similarly large impact, albeit implicit, on any conclusions drawn from comparative sequence analysis, as their accurate detection is necessary to obtain meaningful sequence alignments. The increasing availability of sequence data from different species has prompted recent genome-wide studies of the indel process and its effects [5,18,21,30,31]. The aim of our current work is to use a key result from these studies, namely that indel rates are highly dependent on the surrounding sequence context, to revisit and improve current methods of probabilistic sequence alignment. We do not intend for our model to scale to genome-sized sequences, but rather for it to be used to refine uncertain regions previously aligned using faster and simpler models.

V. Bafna and S.C. Sahinalp (Eds.): RECOMB 2011, LNBI 6577, pp. 85–103, 2011.
© Springer-Verlag Berlin Heidelberg 2011

Score-based models of evolution used in sequence alignment are in the process of being replaced by probabilistic models [19,23]. Similar transitions have successfully occurred in other areas of bioinformatics such as phylogenetics [8], spurred by advantages of using stochastic models that apply equally to sequence alignment. These advantages include more informative parameter estimates, the ability to assess uncertainty in the results, and a neutral model under which tests for selection can be performed. Research on probabilistic indel models and their application to sequence alignment dates back to seminal work by Bishop and Thompson [4] which was refined by Thorne, Kishino and Felsenstein [33]. Most subsequent work in this field, such as allowing overlapping indels [22], or extensions to multiple alignment [10,13], builds on the observation that indel models can be formulated as pair Hidden Markov Models (pair-HMMs) [7]. The pair-HMM representation offers the benefit of its accompanying generic algorithms such as Forward, Viterbi and Baum-Welch, which can be used for alignment and training without the need for model-specific algorithms. It has also led to the use of more general formalisms from linguistics in order to create new, more powerful models. For example, Stochastic Context Free Grammars (SCFGs) have been used to develop a model of evolution where the dependence between nucleotides resulting from pair-bonds in RNA stems could be explicitly modelled [25,12]. Tree-Adjoining Grammars (TAGs) were used to generalize this work to incorporate pseudoknots in the RNA secondary structure [20,34].

One form of inter-site mutational dependence with genome-wide prevalence that is not addressed by any of the current probabilistic models is the relationship between short inserted and deleted sequences and their flanking regions. More specifically, these indels often have a tandem match nearby in the sequence (Figure 1). Two biological mechanisms, namely replication slippage and unequal crossing over during recombination, are known to cause exactly this phenomenon [17]. Recent genome-wide studies have revealed that the vast majority (roughly 90%) of short indels have at least a partial tandem match [21,31], making it reasonable to assume that these, or similar, mechanisms are the primary driving force behind the processes of short insertion and deletion and, consequently, evolution in general. Sequence alignment is far from a solved problem [23], so it is logical to seek to use this context signal to improve accuracy through better detection of indels. Some work toward this end has already been done for score-based pairwise alignment, beginning with Benson who extended the Smith-Waterman algorithm to include duplications events which insert an arbitrary number of copies of the repeat sequence [2], increasing the time complexity to $O(n^4)$. On sequences with pre-identified tandem repeats, the running time is reduced to $O(n^2)$. This problem of aligning sequences of annotated repeats was revisited under a more general model that includes tandem excisions by Berard et al., who proposed a $O(n^4)$ alignment algorithm [3]. Subsequent work allowed for additional events including multiple-excisions, but at the cost of an exponential running time [26].

Our aim in this report is to introduce a similar notion of context-sensitive indels into the more modern statistical alignment framework, noting that a

```
T C T C C A - - - - T C - C T C T
T A G C C A T C C A T A T C T - -
```

context insert   *context delete*

Fig. 1. Short indels are often flanked by a tandem match. The crossing dependencies shown in the insertion and deletion identified in this sample alignment cannot be modelled by pair-HMMs SCFGs.

similar project was successfully undertaken for context-sensitive substitutions [29]. We propose a pair-TAG framework for doing this. In Section 3, we show how certain properties of this model lead to dynamic programming algorithms that are more efficient than previous work using TAGs and even SCFGs. A pairwise aligner based on our model was implemented and tested on a variety of genome sequences, and the increased precision and statistical power of this model are shown on simulated and real genomic data.

2   TAG Indel Model

We seek to generalize the current HMM-based alignment approach to allow for an indel rate that increases in proportion to the similarity between the inserted or deleted sequence and its flanking sequence as shown in Figure 1. Unfortunately, the "crossing" dependencies between the flanking and indel sites as shown cannot, by definition, be modelled by a HMM or SCFG since languages of the type $x_1, x_2, \ldots, x_n, x_1, x_2, \ldots, x_n$, i.e. repeated strings, are neither regular nor context-free. We therefore turn to a slightly more powerful formalism called the Tree-Adjoining Grammar (TAG), a mildly context-sensitive generalization of CFGs introduced by Joshi in 1975 for application in computational linguistics [14]. The key property of TAGs that we are interested in is their ability to recognize tandem repeats of arbitrary strings [15].

A TAG is defined by a quintuple $\{\Sigma, NT, \Gamma, A, S\}$, where Σ is the set of terminal symbols, NT the non-terminal symbols, and S is the Start symbol. Γ and A are the sets of initial and auxiliary trees respectively. TAG trees are binary, with each node associated with either a terminal or non-terminal symbol from the grammar. All interior nodes of initial trees are labelled with non-terminal nodes, except the root which may be labelled with the Start symbol. All leaf nodes of an auxiliary tree are labelled with terminal nodes except for one designated foot node, which is labelled with the same non-terminal as the tree's root. A TAG combines initial and auxiliary trees to form new trees using two operations: substitution and adjunction. A substitution replaces the leaf of one tree with the root of another while an adjunction operation replaces the internal node of one tree with an auxiliary tree. Please see Figure 2 for examples of these operations. A TAG tree can be mapped to a string by reading the terminal symbols in its

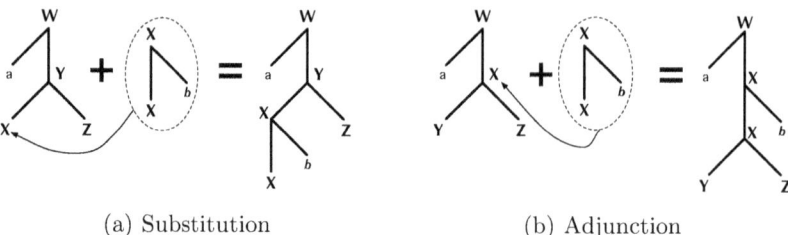

(a) Substitution   (b) Adjunction

Fig. 2. The two TAG operations. (a) An auxiliary tree is substituted onto a leaf of an initial tree. (b) An auxiliary tree is adjoined at the internal node of an initial tree.

leaves in a pre-order traversal. The set of strings generated by a TAG is therefore the set of strings associated with all trees that it can generate.

Substitution, adjunction, and emission probabilities can be added to a TAG in order to generate a stochastic model [27]. Such models have not been widely adopted in practice as TAG parsing algorithms have time complexity $O(n^6)$ [35], as opposed to the less-prohibitive $O(n)$ and $O(n^3)$ of their respective HMM and SCFG counterparts. Still, they have seen some application in bioinformatics. Uemura et $al.$ developed a model to detect RNA secondary structures including pseudoknots using a custom $O(n^5)$ algorithm [34] that was later extended by Matsui et $al.$ to align a sequence to a known secondary structure [20]. Sequence alignment requires a pair TAG which, analogous to the pair-HMM, would emit a pair of sequences with the corresponding quadratic increase in time complexity ($O(n^{10})$ for the pseudoknot model ($O(n^{12})$ in the general case). It is therefore of little surprise that our model, presented below, is, to the best of our knowledge, the first TAG-based sequence aligner. Our approach relies on creating a TAG

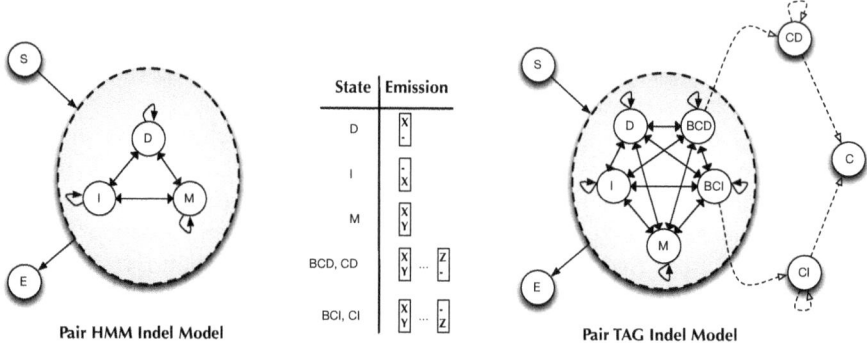

Fig. 3. Comparison of the standard pair-HMM model with a HMM-like projection of the TAG model. Substitutions are mapped to solid arrows and adjunctions to dashed arrows. The Emission format for each emitting state is provided, where X, Y, Z are nucleotides.

that is expressive enough to model repeats while still being simple enough to be parsed in an almost-left-to-right manner.

We begin the description of our model by briefly reviewing the pair HMM sequence aligner, which we wish to generalize. A detailed description can be found in Durbin *et al.* [7]. This model is illustrated in Figure 3, and has a Start state, an End state, and three emitting states: Match, Insert and Delete. The pair alignment is generated from left to right, with each state emitting a single alignment column. The transition and emission probabilities are derived from reversible indel and substitution models, respectively. The Viterbi algorithm (and traceback) can be used to compute the maximum likelihood (ML) alignment in $O(n^2)$. Our air-TAG indel model is illustrated in Figure 4. It is comprised of Start tree, S, and initial trees $\Gamma = \{M, I, D, E\}$, corresponding to Match, Insert, Delete, and End states, respectively. Context sensitive indel operations are supported through the addition of the auxiliary trees, $A \in \{BCD, BCI, CD, CI, C\}$, corresponding to Begin Context Delete, Begin Context Insert, (continue) Context Delete, (continue) Context Insert, and Close context, respectively. A new context indel is begun by substituting either a BCD or BCI tree onto a tree from the first group. The indel (and flanking region) can then be extended by adjoining CD or CI trees to nodes marked with a "*". The C tree is adjoined in order to "Close" the indel. Subsequent events are added by substituting new trees onto the BCD or BCI leaf nodes. This model is projected onto a HMM-like graph in Figure 3 for comparison purposes, where solid and dashed arrows represent substitutions and adjunctions, respectively. It is not a true HMM since multiple paths are required to generate an alignment. Similar to pair-HMM aligners, it is based on the assumption that indels cannot overlap.

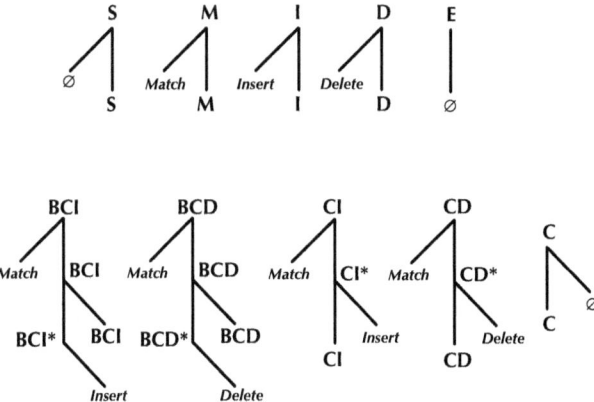

Fig. 4. The pair-TAG indel grammar. Left: Non-context initial trees, and the start tree, are combined through substitution (Start, Match, Insert, Delete, End). Right: Auxiliary trees are combined through adjunction to nodes marked with a "*" (**Begin Context Insert, Begin Context Delete, Continue Context Insert, Continue Context Delete, Close** context).

Let $sub(X,Y)$ and $adj(X,Y)$ denote the probability of substituting or adjoining, respectively, tree Y onto tree X. We determine these values using a simple reversible model defined by the following six parameters:

R_I: Exponential Indel Rate P_I: Geometric Indel Length
R_{CI}: Exponential Context Indel Rate P_{CI}: Geometric Context Indel Length
t: Total Time (2 · divergence time) P_A: Geometric Alignment Length

The substitution probabilities are listed in Table 1, where the function at row i and column j corresponds to $sub(i,j)$. Due to the symmetry of the model, the D and BCD columns need not be shown since, for example $sub(I,D) = sub(D,I)$. All other substitutions, such as $sub(I,S)$, have probability 0. The derivation of $sub(M,j)$, where $j \in \{E, M, I, D, BCI, BCD\}$, is illustrated in Figure 5(a). The probability is the product of the probabilities on each edge in the path from M to j. Similar schematics were used to derive each entry in Table 1. There are only two possible adjunctions, "continue indel" and "close indel", so the only nonzero adjunction probabilities are $adj(BCD, CD) = adj(BCI, CI) = P_{CI}$ and $adj(BCD, C) = adj(BCI, C) = 1 - P_{CI}$.

We now describe the emission model. Indel (I, D), Match (M) and Context Indel (BCI, BCD, CI, CD) trees emit one, two and three nucleotides, respectively, at their leaves. Emission probabilities are represented using the following notation: $emi(x)$ is the probability an indel emits x; $emi(x, y)$ is the probability a match emits x in sequence 1 and y in sequence 2; and $emi(x, y, z)$ is the probability a context indel emits a match of x, y followed by an indel of z to the right in the alignment. The probabilities of emissions of one and two nucleotides are described using a Jukes Cantor model with rate λ [16]. Let $\Pr[x]$ be the stationary probability of nucleotide x, and $\Pr[y \mid x, n]$ be the probability that x mutates into y, given expected number of substitutions, n, according to this model. It follows that $emi(x) = \Pr[x]$ and $emi(x, y) = \Pr[x] \cdot \Pr[y \mid x, t\lambda]$.

Non-context-dependent emission models would have $emi(x, y, z) = emi(x, y) \cdot emi(z)$. The fundamental contribution of our model is in introducing a dependency between the match and indel sites. We accomplish this by introducing a second Jukes Cantor parameter, γ, that represents the expected number of substitutions between the two sites. We can then use Equation 1 to compute the distribution of the probability of the indel having occurred via a tandem repeat event. Since the exact time of the indel event is unknown, we must integrate

Table 1. TAG Substitution probabilities where $X \in \{S, M, CI, CD\}$. Entry at row i and column j corresponds to $sub(i,j)$. In the case where $j = E$, $sub(i,j) = 1 - P_A$.

	M	I	BCI
X	$P_A \cdot e^{-2t(R_I + R_{CI})}$	$\dfrac{P_A \cdot (1 - e^{-2t(R_I + R_{CI})}) \cdot R_I}{2(R_I + R_{CI})}$	$\dfrac{P_A \cdot (1 - e^{-2t(R_I + R_{CI})}) \cdot R_{CI}}{2(R_I + R_{CI})}$
I	$P_A \cdot (1 - P_I) \cdot e^{-t(R_I + 2R_{CI})}$	$P_A \cdot P_I$	$\dfrac{P_A \cdot (1 - P_I) \cdot (1 - e^{-t(R_I + 2R_{CI})}) \cdot R_{CI}}{R_I + 2R_{CI}}$
D	$P_A \cdot (1 - P_I) \cdot e^{-t(R_I + 2R_{CI})}$	$\dfrac{P_A \cdot (1 - P_I) \cdot (1 - e^{-t(R_I + 2R_{CI})}) \cdot R_I}{R_I + 2R_{CI}}$	$\dfrac{P_A \cdot (1 - P_I) \cdot (1 - e^{-t(R_I + 2R_{CI})}) \cdot R_{CI}}{R_I + 2R_{CI}}$

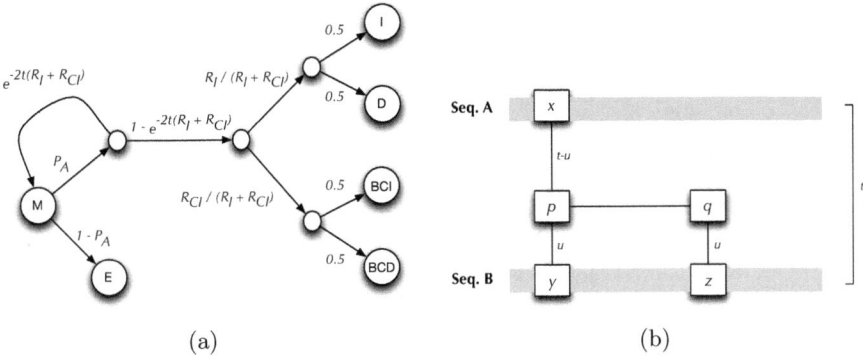

(a)   (b)

Fig. 5. (a) Example how substitution probabilities for the Match tree, as listed in Table 1, are derived. The derivation is most easily shown as a HMM-like projection, where silent states are inserted to ensure that probabilities total one. (b) The terms of Equation 1, as they relate to an illustration of a context insertion. All possible states of p and q are summed over, and all values of u are integrated over, in order to compute $emi(x, y, z)$.

over all time points, and sum over each possible state in the context site (p) and indel site (q) at each time. These variables are illustrated in Figure 5(b). The five terms in the equation correspond to branches connecting the five states in the figure. The level of dependence can be tuned using the context parameter γ, from strictly allowing tandem repeats with $\gamma = 0$ to full independence as γ approaches infinity. Equation 1 expands to a sum of exponential terms when the Jukes-Cantor probabilities are used, and can be solved exactly.

$$emi(x, y, z) = \sum_{p,q \in \{A,C,G,T\}} \int_0^t \Pr[x] \cdot \Pr[p \mid x, (t - u)\lambda] \cdot \Pr[y \mid p, u\lambda] \cdot \Pr[q \mid p, \gamma] \cdot \Pr[z \mid q, u\lambda] \, du$$

(1)

Figure 6 shows an example of a TAG derivation for a single base insertion followed by a two-base context deletion. An Insert tree is first substituted into the Start tree. A Begin Context Delete tree is then substituted into the resulting tree. The context deletion is continued then closed by adjoining the Context Delete and Close trees, respectively. The process is terminated by substituting the End tree. The resulting alignment is emitted as shown in the figure, and the probability of this alignment is a product of all $sub(\cdot), adj(\cdot)$, and $emi(\cdot)$ terms shown.

3   Pairwise Alignment and Parameter Estimation

The most basic question that can be answered using a probabilistic model of evolution is to determine the probability that two sequences are homologous.

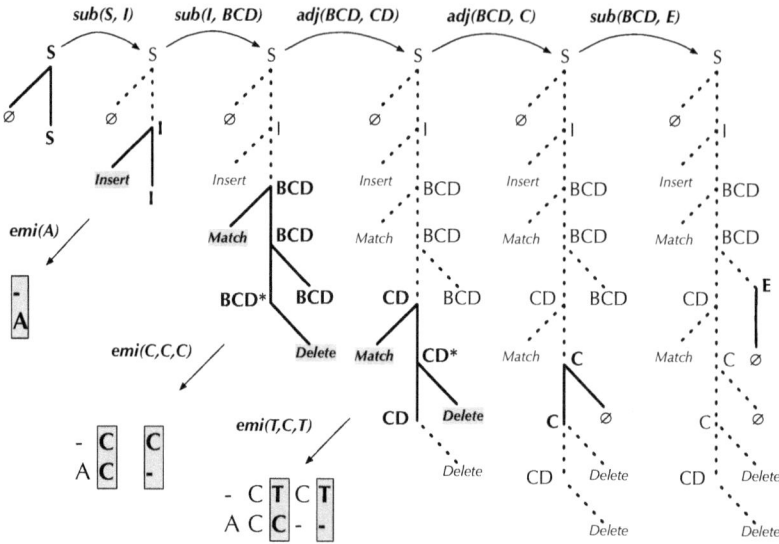

Fig. 6. An example of a TAG derivation as it emits an alignment. The derivation tree is illustrated after each of the three substitutions and two adjunctions. The emitted alignment is also displayed in the lower left. The probabiliy of this derivation can be computed as the product of the (eight total) emission, substitution, and adjunciton operations shown.

Algorithmic solutions to this problem can usually be trivially modified to produce ML alignments along with posterior probabilities for alignment columns. Two model properties are often used to speed up these algorithms. The first is reversibility. Felsenstein's pulley principle states that under a reversible model, the probability that sequences A and B are descended from common ancestor C can be computed by assuming one descendant is the ancestor of the other [9]. This allows likelihoods to be computed directly, without exploring all possible ancestral states. The second property is Markov dependence, which allows all alignments to be efficiently totalled by the Forward algorithm using dynamic programming [7]. An example of a table entry used for this algorithm is illustrated in Figure 7, where $F^1(i, p)$ stores the probability that the pair of subsequences formed by the first i characters of A and the first p characters of B are homologous. Markov dependence enables $F^1(i, p)$ to be computed from $F^1(i - 1, p)$, $F^1(i, p - 1)$, and $F^1(i - 1, p - 1)$, allowing the table to be quickly filled from left to right. The Markov property applies to pair-HMM models, but is lost when moving to more general formalisms such as pair-SCFGs or pair-TAGs, whose respective parsing algorithms, Inside/CYK [7] and Tag-Parse [35] necessitate tables of increasingly higher dimension (also illustrated in Figure 7). The cost of updating each table entry also increases from $O(1)$ for the Forward algorithm to $O(n \cdot m)$ for the Inside algorithm, and to $O(n^2 \cdot m^2)$ for TAG-Parse, for a pair of sequences of lengths m and n. Assuming $n \geq m$ and a constant model size, the space/time

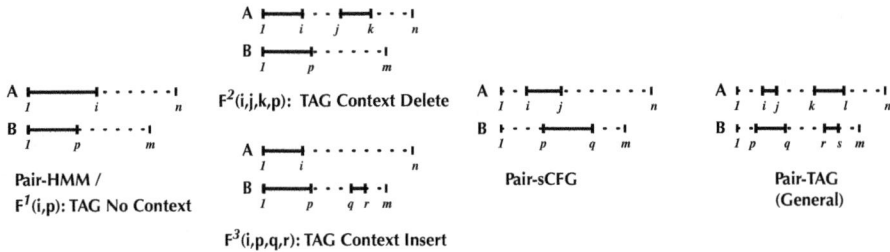

Fig. 7. Illustration of the subproblems considered in the dynamic programming algorithms used to parse our TAG Indel model (TAG No Context; TAG Context Delete; TAG Context Insert) along side those required for a pair-HMM, pair-SCFG and general pair-TAG. For each type of model, we show the input strings A and B, the substrings "covered" by each subproblem, and the indices required to denote them.

complexity of Forward is $O(n^2)/O(n^2)$, Inside is $O(n^4)/O(n^6)$, and TAG-Parse is $O(n^8)/O(n^{12})$.

The TAG Indel model we presented in Figure 4 was designed to mitigate most of the computational overhead, noted above, of moving beyond pair-HMMs. By limiting substitutions and adjunctions to a single point on each tree to the right of previously emitted columns, we ensure a unique "left-to-right" derivation for any alignment. The only exception is columns inserted or deleted by the context operations emitted to the right of adjunction points. These "context windows" are independent since indels do not overlap and can, without any loss of generality, also be derived in a left-to-right fashion. The TAG-Indel derivation can therefore be described as a combination of two different left-to-right processes, context and non-context, linked together in series. These ideas are perhaps best visualized in the HMM-like projection of the TAG-Indel model shown in Figure 3. The black arrows refer to TAG substitutions and the dashed arrows refer to context adjunctions. Alignments are generated as in the pair-HMM by beginning at the S state and then repeatedly transitioning to connected states, emitting columns as specified. When a BCD or BCI state is reached, the model then switches into a context window, transitioning on the dotted connections and simultaneously emitting two columns at a time. The model eventually moves to the C state and reverts back to a non-context mode.

Let s be a TAG Indel tree that is substituted or adjoined in the process of deriving an alignment of sequences A and B. It follows directly from the definition of our model that the range of characters during the derivation up to and including s can be described as one of the three cases illustrated in Figure 7: *TAG No Context, TAG Context Insert or TAG Context Delete*. The total probability of all possible alignments of the emitted characters given that s was the last tree used in the derivation is denoted $F^1(s, i, j), F^2(s, i, j, k, p)$, and $F^3(s, i, p, q, r)$ for the three cases respectively. The recursions to compute these values are provided in Appendix Section A.1, as is pseudocode for Context-Forward, a $O(n^4)$

context-sensitive generalization of the Forward Algorithm. The vast majority of context indels are extremely short, however [31], and by bounding this length with some small constant L, the complexity reduces to $O(n^2 L^2)$. Context-Forward can be modified into Context-Viterbi by changing the summations to max operators, and storing traceback information in addition to likelihoods. The resulting procedure can be used to obtain the ML alignment, or to sample from the ML alignments in a similar manner as is done for HMMs [7]. We define the Context-Backward tables, $B^{\{1,2,3\}}$, as storing the probability that the remaining subsequences of A and B (dashed lines in Figure 7) are emitted, *given that* state s was used in the corresponding Forward entry. For example, $B^2(CD, i, j, k, p)$ is the probability that the model emits $A_{[i+1,j-1]}, A_{[k+1,n]}$ and $B_{[p+1,m]}$ given that it already emitted $A_{[1,i]}, A_{[j,k]}$ and $B_{[1,p]}$, and s was the last tree used in the latter derivation. The details of Context-Backward are given in Appendix Section A.2. The Context-Forward and Context-Backward tables can be combined to generate ML estimators for the substitution, emission and adjunction probabilities as is done by the Baum-Welch algorithm for HMMs [7]. We use these estimators in an Expectation-Maximization (EM) loop to train the model, which is described in full in Appendix Section A.3.

4   Results and Discussion

The training and alignment algorithms described above were implemented in C++. The program succeeds at aligning pairs of 100-bp sequences in less than 5 seconds on a typical desktop computer, eventually allowing whole-genome piece-wise realignment on a large cluster in a few hours. The first results we present are from a simulation experiment designed to validate the model design and training procedure, as well as the statistical test used later to measure goodness of fit to real data. We simulated a set of 250 pairwise alignments using our model for each possible parameter assignment (144 total) from the following values, limiting context indel lengths to ten to speed up training:

$$R_I, R_{CI} \in \{0, 0.01, 0.1, 0.5\}, \ \lambda, \gamma \in \{0.01, 0.1, 0.5\}, \ P_I = 0.75, \ P_{CI} = 0.5, \ P_A = 0.9875.$$

These values were chosen in an attempt to represent those that could be encountered when aligning human to species as close as chimpanzee and distant as mouse. $t = 1$ was used for all tests, as time cannot be estimated independently from the rates. For each of the 144 parameter assignments, both the HMM and TAG indel models were trained on the generated data. The estimated parameters were accurate to within a few percent of the true parameters in most cases. There were exceptions in pathological cases, such as $\{R_I = 0.5, R_{CI} = \lambda = \gamma = 0.01\}$ where the estimate of λ was completely wrong but these cases are unrealistic, and as liable to fool other models. To show that estimating these context indel parameters is meaningful, we contrasted the fit of the TAG indel model to that of the pair-HMM using two goodness of fit tests: the Likelihood Ratio Test (LRT) and the Bayesian Information Criterion (BIC), both of which correct for the number of free parameters. After training both models on the data as described above, we computed the total log likelihoods of the Viterbi alignments for each parameter set. The LRT showed that the TAG-Indel model fits the data significantly better than the HMM ($p < 10^{-6}$) whenever the rate of context indel is non-trivial ($R_{CI} \geq 0.01$). The BIC is a stricter test that corrects for the data size in addition to assigning an increased penalty for the number free parameters. Under this criterion, the model with the lower BIC value is to be chosen. Figure 8(a) plots the relationship between $\Delta BIC = BIC[HMM] - BIC[TAG]$ and the parameters used to generate the data ($\Delta BIC > 0$ favours the TAG model). The TAG Indel model fits the data better under most conditions and as expected, the fit increases in proportion to R_{CI}, R_{CI}/R_I, and in inverse proportion to λ and γ.

Table 2. Estimated parameters obtained by training our model on batches of subsequences randomly sampled from the human genomes, along with homologus subsequences from the genomes of seven mammalian species

Species	λ	γ	R_I	R_{CI}	P_I	P_{CI}
Chimp	0.020968	0.144747	0.000111	0.000669	0.576667	0.633721
Gorilla	0.026970	0.118685	0.000193	0.000672	0.553240	0.691680
Macaque	0.094694	0.236328	0.000579	0.001951	0.587768	0.672405
Marmoset	0.158098	0.209717	0.000951	0.003485	0.450678	0.778480
Tarsier	0.341867	0.332006	0.003834	0.007663	0.453935	0.634864
Dog	0.414850	0.473482	0.004827	0.010788	0.539543	0.537534
Mouse	0.528795	0.388574	0.007306	0.013690	0.494740	0.646847

Having validated the training algorithm and goodness of fit test, we proceed to experiments on genomic sequence data. Our model in its current form is not designed to align entire genomes. Rather, we propose it as a method to resolve uncertain gaps within sequences that have at least been roughly aligned. We therefore use short subsequences from BlastZ [28] pairwise alignments of the human genome to other species as downloaded from the UCSC Genome Browser [24]. The subsequences were sampled as follows. The chromosome 22 alignment (.axt file) was divided into blocks of length 60 (including gaps), and 500 blocks were chosen at random[1]. Implementation improvements such as banding [11] and faster numeric estimation procedure could be used to vastly increase the sample lengths in the future. The reference alignments between human (hg19) and chimp (panTro2), gorilla (gorGor1), rhesus macaque (rheMac2), marmoset (calJac2), tarsier (tarSyr1), dog (canFam2), and mouse (mm9), were sampled in

---

[1] Alignment blocks with gaps near the block's boundary (gap of length $\geq k$ within k sites of the boundary) were excluded, to ensure all gaps' contexts are present. We observe that that this procedure will bias the observed indel length distribution (against longer indels), but note that 1) most indels are short enough to not be significantly affected [5,31], and 2) we are less interested in the absolute rates detected than in the relative context signal and presenting the potential benefits of a proof-of-concept version of our model.

this way. Gaps were then removed from each BlastZ alignment block, and the TAG indel model was trained on each sample. We repeated this procedure ten times and report the average results below.

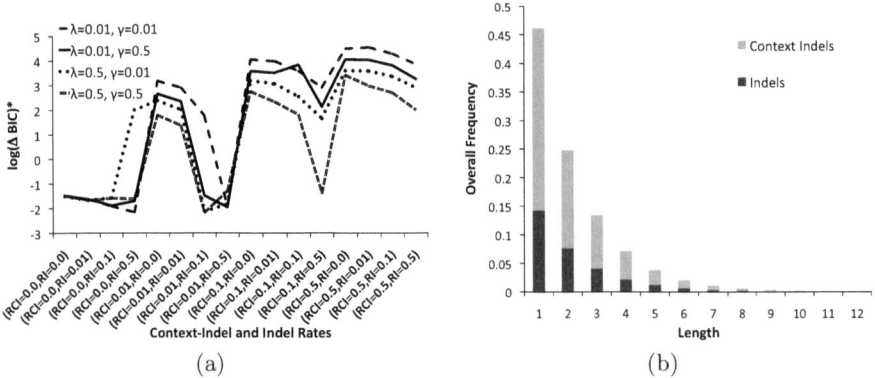

(a)   (b)

Fig. 8. (a) BIC analysis on simulated data. Lines for $\lambda = 0.1$ and $\gamma = 0.1$ not shown, but they are bound by below and above by $\lambda = 0.5, \gamma = 0.5$ and $\lambda = 0.01$ and $\gamma = 0.01$ respectively. *: $\log \Delta\text{BIC} = \log_{10}(\text{BIC(HMM)} - \text{BIC(TAG)})$ if $\text{BIC(HMM)} > \text{BIC(TAG)}$, 0 if $\text{BIC(HMM)} = \text{BIC(TAG)}$, and $-\log_{10}|\text{BIC(HMM)} - \text{BIC(TAG)}|$ if $\text{BIC(HMM)} < \text{BIC(TAG)}$. (b) Indel length distribution for Human/Dog alignment.

The estimated parameters for the alignments between human and the selected species are given in Table 2. In each case the context indel rate was much higher than the non-context rate, supporting the hypothesis that most short indels are the result of tandem events. The relative rate of context events does lower as the species become more diverged. The γ values, on the other hand, increase, reflecting higher numbers of substitutions occurring in or around the indel. It is likely that subsequent insertions and deletions, which we do not model, also disrupt the context signal for more diverged species and are the source, rather than a mechanistic difference, of the lower relative rates. The length distributions of both the context and non-context indels are similar to each other, and the numbers for human/chimp are comparable to what has previously been observed using other methods [31]. Figure 8(b) illustrates, as an example, the indel length distributions for human/dog. Figure 9(a) shows the ΔBIC score in relation to divergence from human (as measured by the substitution rate λ). Our model provides a superior fit to the data at low divergence (such as chimp and gorilla), and improves as the divergence increases. There is a point where, as mentioned above, the context signal begins to be lost due to excessive divergence (after dog in Figure 9(a)). Still, ΔBIC remained significantly above zero even for mouse.

Finally, we demonstrate how modelling context-sensitive indels improves alignment accuracy. Since the true alignments are not known, we used the parameters estimated from the genomic data to simulate "gold standard" alignments for each organism. Using the same number of trials and dimensions of data as above, we

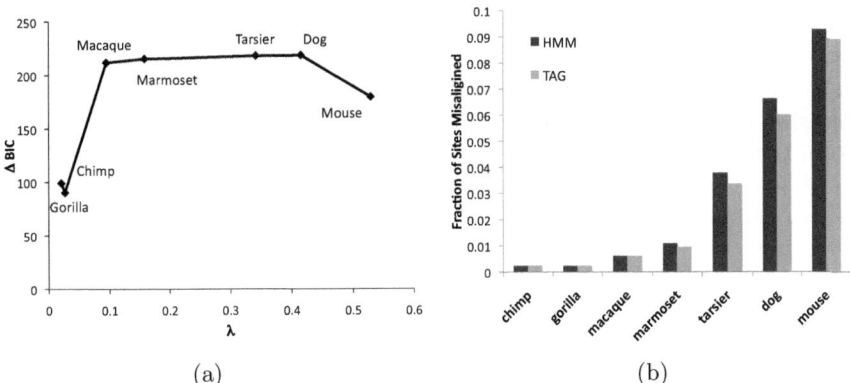

(a)   (b)

Fig. 9. (a) BIC test on pairs of genomic sequences involving human and each of the species listed. Δ BIC is the difference in BIC score between the TAG model and the HMM model. (b) Alignment accuracy results for simulated pairs of sequences involving human and each of the listed species.

again trained both the TAG and HMM models on pairs of simulated orthologous sequences, and computed the maximum-likelihood pairwise alignments with each model[2]. The fraction of alignment errors (pair of aligned nucleotides present in the gold standard and not in the computed alignment) made by each method is reported in Figure 9(b). For closely related species, both models produce roughly equally accurate alignments. This is likely due to the fact that the sequences surrounding indels are sufficiently conserved to accurately pinpoint the indel location without the need for a context-dependent model. However, for more diverged species, the TAG alignments make significantly fewer alignment errors, allowing a $\sim$ 10% reduction in alignment errors for marmoset, tarsier, dog, and mouse. As in the BIC test, we see the performance gains begin to drop off at mouse, as the distance becomes large enough to degrade the context signal.

In conclusion, we developed a probabilistic indel model that generalizes current pair-HMM approaches by adding context-indel events whose probabilities are governed by their similarity to their flanking sequences. This approach was motivated by recent studies showing the prevalence of such events across the human genome, and the fact that many gaps in alignments produced by context-free approaches are still very uncertain, especially in regions exhibiting high divergence. We therefore claim that the accuracy of current methods can be improved by taking into account this context signal. As probabilistic models based on pair-HMMs are being increasingly favoured over score-based approaches, we elected to generalize the pair-HMM using a pair-TAG to include context sensitivity. Our simulation experiments demonstrated that our TAG indel model can accurately

---

[2] To adjust for the fact that the true direction of replication is unknown, and to attempt to remove unfair bias in favour of the TAG model, we reversed the direction of half the simulated sequence pairs.

detect context-sensitive indels and their associated parameters from sequence data. We also showed that it fits pair sequence data from alignments between human and species ranging from chimpanzee to mouse significantly better than a pair-HMM. The estimated rates from these alignments were used to generate "gold-standard" alignments, which the TAG model was able to more accurately recover than the pair-HMM as divergence increased. We therefore conclude that our TAG indel model provides a significant improvement over current methods, and has demonstrated the potential to be used to refine uncertain regions in existing alignments, especially in diverged sequences. These results will have an impact on any analyses that rely on alignments, such as fine-grained studies of small functional regions or disease markers. In the future, we are interested in adding more complex events to our model such as indels within or between the context indel and flanking region, scaling the input data sizes, and applying it to multiple alignment and ancestral sequence reconstruction.

Acknowledgements

Both authors were funded in part by NSERC, and are grateful to the anonymous reviewers for their helpful feedback.

References

1. Ball, E., Stenson, P., Abeysinghe, S., Krawczak, M., Cooper, D., Chuzhanova, N.: Microdeletions and microinsertions causing human genetic disease: common mechanisms of mutagenesis and the role of local DNA sequence complexity. Human Mutation 26(3), 205–213 (2005)
2. Benson, G.: Sequence alignment with tandem duplication. Journal of Computational Biology 4(3), 351–367 (1997)
3. Bérard, S., Rivals, E.: Comparison of minisatellites. Journal of Computational biology 10(3-4), 357–372 (2003)
4. Bishop, M., Thompson, E.: Maximum likelihood alignment of DNA sequences* 1. Journal of Molecular Biology 190(2), 159–165 (1986)
5. Chen, F., Chen, C., Li, W., Chuang, T.: Human-specific insertions and deletions inferred from mammalian genome sequences. Genome Research 17(1), 16 (2007)
6. Chuzhanova, N., Anassis, E., Ball, E., Krawczak, M., Cooper, D.: Meta-analysis of indels causing human genetic disease: mechanisms of mutagenesis and the role of local DNA sequence complexity. Human Mutation 21(1), 28–44 (2003)
7. Durbin, R., Eddy, S., Krogh, A., Mitchison, G.: Biological sequence analysis south asia edition: probabilistic models of proteins and nucleic acids (2003)
8. Felenstein, J.: Inferring phylogenies. Sinauer Associates Sunderland, Mass (2003)
9. Felsenstein, J.: Evolutionary trees from DNA sequences: a maximum likelihood approach. Journal of Molecular Evolution 17(6), 368–376 (1981)
10. Hein, J., Jensen, J., Pedersen, C.: Recursions for statistical multiple alignment. Proceedings of the National Academy of Sciences of the United States of America 100(25), 14960 (2003)
11. Hein, J., Wiuf, C., Knudsen, B., Moller, M., Wibling, G.: Statistical alignment: computational properties, homology testing and goodness-of-fit. Journal of Molecular Biology 302(1), 265–280 (2000)

12. Holmes, I.: A probabilistic model for the evolution of RNA structure. BMC Bioinformatics 5(1), 166 (2004)
13. Holmes, I., Bruno, W.: Evolutionary HMMs: a Bayesian approach to multiple alignment. Bioinformatics 17(9), 803 (2001)
14. Joshi, A., Levy, L., Takahashi, M.: Tree adjunct grammars. Journal of Computer and System Sciences 10(1), 136–163 (1975)
15. Joshi, A., Schabes, Y.: Tree-adjoining grammars. Handbook of Formal Languages, Beyond Words 3, 69–123 (1997)
16. Jukes, T., Cantor, C.: Evolution of protein molecules. Mammalian Protein Metabolism 3, 21–132 (1969)
17. Levinson, G., Gutman, G.: Slipped-strand mispairing: a major mechanism for DNA sequence evolution. Molecular Biology and Evolution 4(3), 203 (1987)
18. Lunter, G.: Probabilistic whole-genome alignments reveal high indel rates in the human and mouse genomes. Bioinformatics 23(13), i289 (2007)
19. Lunter, G., Drummond, A., Miklós, I., Hein, J.: Statistical alignment: Recent progress, new applications, and challenges. Statistical Methods in Molecular Evolution, 375–405 (2005)
20. Matsui, H., Sato, K., Sakakibara, Y.: Pair stochastic tree adjoining grammars for aligning and predicting pseudoknot RNA structures. Bioinformatics 21(11), 2611–2617 (2005)
21. Messer, P., Arndt, P.: The majority of recent short DNA insertions in the human genome are tandem duplications. Molecular Biology and Evolution 24(5), 1190 (2007)
22. Miklós, I., Lunter, G., Holmes, I.: A" long indel" model for evolutionary sequence alignment. Molecular Biology and Evolution 21(3), 529 (2004)
23. Miklos, I., Novak, A., Satija, R., Lingso, R., Hein, J.: Stochastic models of sequence evolution including insertion-deletion events. Statistical Methods in Medical Research 18(5), 448–453 (2009)
24. Rhead, B., Karolchik, D., Kuhn, R., Hinrichs, A., Zweig, A., Fujita, P., Diekhans, M., Smith, K., Rosenbloom, K., Raney, B., et al.: The UCSC genome browser database: update 2010. Nucleic Acids Research (2009)
25. Rivas, E., Eddy, S.: Noncoding RNA gene detection using comparative sequence analysis. BMC Bioinformatics 2(1), 8 (2001)
26. Sammeth, M., Stoye, J.: Comparing tandem repeats with duplications and excisions of variable degree. IEEE/ACM Transactions on Computational Biology and Bioinformatics, 395–407 (2006)
27. Schabes, Y.: Stochastic lexicalized tree-adjoining grammars. In: Proceedings of the 14th Conference on Computational Linguistics, vol. 2, pp. 425–432. Association for Computational Linguistics (1992)
28. Schwartz, S., Kent, W., Smit, A., Zhang, Z., Baertsch, R., Hardison, R., Haussler, D., Miller, W.: Human–mouse alignments with BLASTZ. Genome Research 13(1), 103 (2003)
29. Siepel, A., Haussler, D.: Phylogenetic estimation of context-dependent substitution rates by maximum likelihood. Molecular Biology and Evolution 21(3), 468 (2004)
30. Sinha, S., Siggia, E.: Sequence turnover and tandem repeats in cis-regulatory modules in Drosophila. Molecular Biology and Evolution 22(4), 874 (2005)
31. Tanay, A., Siggia, E.: Sequence context affects the rate of short insertions and deletions in flies and primates. Genome Biology 9(2), R37 (2008)

32. The Chimpanzee Genome Sequencing and Analysis Consortium: Initial sequence of the chimpanzee genome and comparison with the human genome. Nature 437(7055), 69 (2005)
33. Thorne, J., Kishino, H., Felsenstein, J.: Inching toward reality: an improved likelihood model of sequence evolution. Journal of Molecular Evolution 34(1), 3–16 (1992)
34. Uemura, Y., Hasegawa, A., Kobayashi, S., Yokomori, T.: Tree adjoining grammars for RNA structure prediction* 1. Theoretical Computer Science 210(2), 277–303 (1999)
35. Vijay-Shankar, K., Joshi, A.: Some computational properties of tree adjoining grammars. In: Proceedings of the Workshop on Strategic Computing Natural Language, p. 223. Association for Computational Linguistics (1986)

A   Appendix

A.1   Context-Forward Algorithm

The dynamic programming recursions for Context-Forward are listed below.

$$F^1(D, i, p) = emi(A_i) \cdot \sum_{s \in \{S,D,I,M,BCD,BCI\}} F^1(s, i-1, p) \cdot sub(s, D)$$

$$F^1(I, i, p) = emi(B_p) \cdot \sum_{s \in \{S,D,I,M,BCD,BCI\}} F^1(s, i, p-1) \cdot sub(s, I)$$

$$F^1(M, i, p) = emi(A_i, B_p) \cdot \sum_{s \in \{S,D,I,M,BCD,BCI\}} F^1(s, i-1, p-1) \cdot sub(s, M)$$

$$F^1(E, i, p) = \sum_{s \in \{S,D,I,M,BCD,BCI\}} F^1(s, i, p) \cdot sub(s, E)$$

$$F^2(BCD, i, j, k, p) = emi(A_i, A_k, B_p) \cdot \sum_{s \in \{S,D,I,M,BCD,BCI\}} F^1(s, i-1, p-1) \cdot sub(s, BCD)$$

$$F^3(BCI, i, p, q, r) = emi(A_i, B_p, B_r) \cdot \sum_{s \in \{S,D,I,M,BCD,BCI\}} F^1(s, i-1, p-1) \cdot sub(s, BCI)$$

$$F^2(CD, i, j, k, p) = emi(A_i, A_k, B_p) \cdot \sum_{s \in \{BCD,CD\}} F^2(s, i-1, j, k-1, p-1) \cdot adj(s, CD)$$

$$F^3(CI, i, p, q, r) = emi(A_i, B_p, B_r) \cdot \sum_{s \in \{BCI,CI\}} F^3(s, i-1, p-1, q, r-1) \cdot adj(s, CI)$$

$$F^1(BCD, i, p) = \sum_{s \in \{BCD,CD\}} \sum_{0 < u < i} F^2(su, u+1, i, p) \cdot adj(CD, C)$$

$$F^1(BCI, i, p) = \sum_{s \in \{BCI,CI\}} \sum_{0 < v < p} F^3(s, i, v, v+1, r) \cdot adj(CI, C)$$

The first four equations, which update non context entries of the form $F^1(s, i, p)$ are computed exactly as in the Forward algorithm. The context entries are similar except an extra two indices are kept for the context indel, and three characters are emitted instead of two or one. Finally, context indels can be "closed" once the inserted or deleted sequence is immediately flanking the current alignment. Pseudocode for the Context-Forward algorithm is shown below.

CONTEXT-FORWARD

```
1   Initialize all entries in F to 0
2   F¹(S, 0, 0) ← 1
3   Compute F¹(D, 1, 0) and F¹(I, 0, 1)
4   for i ← 1 to n
5       for p ← 1 to m
6           for s ∈ {S, M, D, I, BCD, BCI}
7               Compute F¹(s, i, p)
8           for j ← i + 1 to n
9               for k ← j to n
10                  Compute F²(BCD, i, j, k, p) and F²(CD, i, j, k, p)
11          for q ← p + 1 to m
12              for r ← q to m
13                  Compute F³(BCI, i, p, q, r) and F³(CI, i, p, q, r)
14  Compute and Return F(E, n, m)
```

A.2   Context-Backward Algorithm

The recursions for computing $B^{\{1,2,3\}}$ are presented below. Each entry is associated with a state s and a range (ranges and indices are illustrated in 7), and stores the probability that the model emits the subsequences not in the range, given that the last emission inside the range was s.

$$
\begin{aligned}
B^1(s,i,p) & \\
s\in\{S,M,D,I,E,BCD,BCI\} & = \sum \begin{cases}
emi(A_{i+1}) \cdot B^1(D, i+1, p) \cdot sub(s, D) \\
emi(B_{p+1}) \cdot B^1(I, i, p+1) \cdot sub(s, I) \\
emi(A_{i+1}, B_{p+1}) \cdot B^1(M, i+1, p+1) \cdot sub(s, M) \\
\sum_{k>i+1} emi(A_{i+1}, A_k, B_{p+1}) \cdot B^2(BCD, i+1, k, k, p+1) \cdot sub(s, BCD) \\
\sum_{r>p+1} emi(A_{i+1}, B_{p+1}, B_r) \cdot B^3(BCI, i+1, p+1, r, r) \cdot sub(s, BCI)
\end{cases}
\end{aligned}
$$

$$
B^2(BCD, i, k, k, p) = \begin{cases}
B^2(BCD, k, p) \cdot adj(CD, C) & \text{(if } k = i+1) \\
emi(A_{i+1}, A_{k+1}, B_{p+1}) \cdot B^2(CD, i+1, k, k+1, p+1) \cdot adj(BCD, CD) & \text{(otherwise)}
\end{cases}
$$

$$
B^3(BCI, i, p, r, r) = \begin{cases}
B^3(BCI, i, r) \cdot adj(CI, C) & \text{(if } r = p+1) \\
emi(A_{i+1}, B_{p+1}, B_{r+1}) \cdot B^3(CI, i+1, p+1, r, r+1) \cdot adj(BCI, CI) & \text{(otherwise)}
\end{cases}
$$

$$
B^2(CD, i, j, k, p) = \begin{cases}
B^2(BCD, k, p) \cdot adj(CD, C) & \text{(if } j = i+1) \\
emi(A_{i+1}, A_{k+1}, B_{p+1}) \cdot B^2(CD, i+1, j, k+1, p+1) \cdot adj(CD, CD) & \text{(otherwise)}
\end{cases}
$$

$$
B^3(CI, i, p, q, r) = \begin{cases}
B^3(BCI, i, r) \cdot adj(CI, C) & \text{(if } q = p+1) \\
emi(A_{i+1}, B_{p+1}, B_{r+1}) \cdot B^3(CI, i+1, p+1, q, r+1) \cdot adj(CI, CI) & \text{(otherwise)}
\end{cases}
$$

The pseudocode for Context-Backward is provided. Its running time is identical to Context-Forward.

CONTEXT-BACKWARD

```
1   Initialize all entries in B to 0
2   B¹(E, n + 1, m + 1) ← 1
3   for i ← n to 0
4       for p ← m to 0
5           for s ∈ {E, M, D, I, BCD, BCI}
6               Compute B¹(s, i, p)
7           for j ← n to i + 1
8               for k ← j to i + 1
9                   Compute B²(BCD, i, j, k, p) and B²(CD, i, j, k, p)
10          for q ← m to p + 1
11              for r ← q to p+
12                  Compute B³(BCI, i, p, q, r) and B³(CI, i, p, q, r)
13  Compute and Return B¹(S, 0, 0)
```

A.3 Expectation-Maximization Loop for Estimating Parameters

Let $\Phi = (emi(\cdot), sub(\cdot), adj(\cdot))$, and $\Psi = (\lambda, \gamma, R_I, R_{CI}, P_i, P_{CI}, P_A)$. The parameters are learned though iterative re-estimation, beginning with a random seed. First the ML estimates of the substitution, adjunction and emission probabilities, $\widehat{\Phi}$, are computed from the Context-Forward and Context-Backward tables. We define $t(\Psi) \to \Phi$ to be the system of equations (Table 1, Equation 1), that maps the model parameters to the TAG probabilities. In general, $t^{-1}(\Phi) \to \Psi$ cannot be computed analytically so we use gradient descent to obtain an approximate solution, $\widehat{\Psi}$, that minimizes $w(t(\widehat{\Psi})) - w(\widehat{\Phi}))^2$. $w(\cdot)$ is a heuristic scaling factor that weights each probability by the likelihood of the state that produces it (obtainable directly from the dynamic programming tables). For example, $w(emi(s,x,y)) = \Pr[s] \cdot emi(s,x,y)$ and $w(adj(s_1, s_2)) = \Pr[s_1] \cdot adj(s_1, s_2)$. This ensures that probabilities associated with transitioning out of, or emitting from, unlikely states, which can be high and meaningless, do not dominate the objective function. The entire training procedure is outlined in the following pseudocode.

CONTEXT-TRAIN

1   $\widehat{\Phi} \leftarrow$ Random Seed.
2   $\Pr[A, B] \leftarrow 0$
3   converge $\leftarrow$ false
4   **while** converge = false
5            $\Pr'[A, B] \leftarrow$ Context-Forward()
6   Context-Backward()
7   Compute $\widehat{\Phi}$
8   Estimate $\widehat{\Psi} \approx t^{-1}(\widehat{\Phi})$
9   converge $\leftarrow$ $(\Pr'[A, B] = \Pr[A, B])$
10           $\Pr[A, B] \leftarrow \Pr'[A, B])$
11 **Return** $\widehat{\Psi}$

$\widehat{\Phi}$ is estimated directly from the Context-Forward and Context-Backward tables:

For $s \in \{S, E, D, I, M, BCD, BCI\}$:

$$\Pr[s] = \sum_{i,p} F^1(s, i, p) \cdot B(s, i, p)$$

$$sub(s, D) = \sum_{i,p} F^1(s, i, p) \cdot emi(A_{i+1}) \cdot sub(s, D) \cdot B^1(D, i+1, p)/\Pr[s]$$

$$sub(s, I) = \sum_{i,p} F^1(s, i, p) \cdot emi(B_{p+1}) \cdot sub(s, I) \cdot B^1(I_x, i, p+1)/\Pr[s]$$

$$sub(s, M) = \sum_{i,p} F^1(s, i, p) \cdot emi(A_{i+1}, B_{p+1}) \cdot sub(s, M) \cdot B^1(M, i+1, p+1)/\Pr[s]$$

$$sub(s, E) = \sum_{i,p} F^1(s, i, p) \cdot sub(s, E) \cdot B^1(E, i+1, p)/\Pr[s]$$

$$sub(s, BCD) = \sum_{i,p} F^1(s, i, p) \sum_{k>i+1} emi(A_{i+1}, A_k, B_{p+1}) \cdot sub(s, BCD) \cdot B^2(BCD, i+1, k, k, p+1)/\Pr[s]$$

$$sub(s, BCI) = \sum_{i,p} F^1(s, i, p) \sum_{r>p+1} emi(A_{i+1}, B_{p+1}, B_r) \cdot sub(s, BCI) \cdot B^3(BCI, i+1, p+1, r, r)/\Pr[s]$$

For $s \in \{BCD, CD\}$:

$$\Pr[s] = \sum_{i,j,k,p} F^2(s, i, j, k, p) \cdot B^2(s, i, j, k, p)$$

$$adj(s, C) = \sum_{i,k,p} F^2(s, i, i+1, k, p) \cdot adj(s, C) \cdot B^2(BCD, k, p)/\Pr[s]$$

$$adj(s, CD) = \sum_{i,j,k,p} F^2(s, i, j, k, p) \cdot emi(A_{i+1}, A_{k+1}, B_{p+1}) \cdot adj(s, CD) \cdot B^2(CD, i+1, j, k+1, p+1)/\Pr[s]]$$

$$\Pr[s] = \sum_{i,p,q,r} F^3(s, i, p, q, r) \cdot B(s, i, p, q, r)$$

$$adj(s, C) = \sum_{k,p,r} F^3(s, i, p, p+1, r) \cdot adj(s, C) \cdot B^3(BCI, k, p)$$

$$adj(s, CI) = \sum_{i,p,q,r} F^3(s, i, p, q, r) \cdot emi(A_{i+1}, B_{p+1}, B_{r+1}) \cdot adj(s, CI) \cdot B^3(CI, i+1, p+1, q, r+1)/\Pr[s]$$

$$emi(x) = \frac{\sum_{i,p|A_i=\alpha, B_p=-} F^1(D, i, p) \cdot B^1(D, i, p) + \sum_{i,p|A_i=-, B_p=x} F^1(i, i, p) \cdot B^1(I, i, p)}{\sum_{i,p} F^1(D, i, p) \cdot B^1(D, i, p) + \sum_{i,p} F^1(i, i, p) \cdot B^1(I, i, p)}$$

$$emi(x, y) = \frac{\sum_{i,p|A_i=x, B_p=y} F^1(M, i, p) B^1(M, i, p)}{\sum_{i,p} F^1(M, i, p) \cdot B^1(M, i, p)}$$

$$emi(x, y, z) = \frac{\sum_{s \in \{BCD, CD\}, i,j,k,p|A_i=x, B_p=y, A_k=z} F^2(s, i, j, k, p) \cdot B(s, i, j, k, p) + \sum_{s \in \{BCI, CI\}, i,p,q,r|A_i=x, B_p=y, B_r=z} F^3(s, i, p, q, r) \cdot B(s, i, p, q, r)}{\sum_{s \in \{BCD, CD\}} F^2(s, i, j, k, p) \cdot B(s, i, j, k, p) + \sum_{s \in \{BCI, CI\}} F^3(s, i, p, q, r) \cdot B(s, i, p, q, r)}$$

Simultaneous Structural Variation Discovery in Multiple Paired-End Sequenced Genomes

Fereydoun Hormozdiari[1,*], Iman Hajirasouliha[1,*], Andrew McPherson[1], Evan E. Eichler[2,**], and S. Cenk Sahinalp[1,**]

[1] School of Computing Science, Simon Fraser University, Burnaby, BC, Canada
[2] Department of Genome Sciences, University of Washington, and HHMI
eee@gs.washington.edu, cenk@cs.sfu.ca

Motivation. Next generation sequencing technologies have been decreasing the costs and increasing the world-wide capacity for sequence production at an unprecedented rate, making the initiation of large scale projects aiming to sequence almost 2000 genomes [1]. Structural variation detection promises to be one of the key diagnostic tools for cancer and other diseases with genomic origin. In this paper, we study the problem of detecting structural variation events in two or more sequenced genomes through high throughput sequencing . We propose to move from the current model of (1) detecting genomic variations in single next generation sequenced (NGS) donor genomes independently, and (2) checking whether two or more donor genomes indeed agree or disagree on the variations (in this paper we name this framework *Independent Structural Variation Discovery and Merging - ISV&M*), to a new model in which we detect structural variation events among multiple genomes simultaneously.

Combinatorial Modeling and Problem Definition. Here we introduce the problem of Simultaneous Structural Variation discovery in Multiple Genomes (SSV-MG). Given one reference genome and a number of paired-end sequenced genomes, our aim is to predict structural variation (SV) events in these genomes simultaneously. In [2], the MPSV problem was defined to compute a unique *assignment* of each discordant paired-end read to a maximal SV cluster such that the total number of implied SV events is minimized. The SSV-MG problem also aims to identify maximal SV clusters and assign each discordant paired-end read to one of the SV clusters under a maximum parsimonious criteria which we formally define in this paper. The goal is to *simultaneously* predict SVs in several donor genomes by means of minimizing a *weighted sum of* structural differences between the donor genomes as well as the reference genome. For each SV event identified by an SSV-MG algorithm, a weight (cost) is associated based on the set of the distinct genomes sharing the SV event (i.e. having a discordant paired-end read which is assigned to the SV cluster). If an SV event is shared among many distinct genomes, its weight is relatively small, while an SV event which is unique to only one individual has a larger weight.

---

* Joint First Authors.
** Corresponding authors.

V. Bafna and S.C. Sahinalp (Eds.): RECOMB 2011, LNBI 6577, pp. 104–105, 2011.
© Springer-Verlag Berlin Heidelberg 2011

Complexity, Approximate Algorithms and Heuristics. We show that the SSV-MG problem is NP-hard and there exists a constant c for which SSV-MG has no approximation factor within $c\frac{\omega_{\max}}{\omega_{\min}}\log n$, unless $P = NP$, where $\omega_{\max}$ and $\omega_{\min}$ are the maximum and minimum possible weights among all SV clusters. We present a tight approximation algorithm (which we name SSC throughout the manuscript) for the SSV-MG problem, based on the greedy algorithm for the set cover problem. In addition, we provide two heuristics for solving the SSV-MG problem as well as alternative algorithms for special cases of the problem. The first heuristic, SSC-W, uses the weights of the SV clusters to calculate the *cost-effectiveness* of each cluster, while the second heuristic, SSC-W-CR, deploys the concept of *conflict resolution* (introduced in [3]) to obtain more accurate results. SSC-W is a greedy method similar to the weighted set cover algorithm with one major difference. Here the weight of each subset is not fixed throughout the algorithm, but rather is dependent to the elements which are assigned to the subsets.

Results. We compared our proposed SSV-MG framework against an ISV&M framework (i.e. VariationHunter [2,3]) using two different data sets. The first dataset is a Yoruba family which constitute a father-mother-child trio (NA18507, NA18508 and NA18506), while the second dataset is a CEU trio (NA12878, NA12891 and NA12892) which was sequenced in the 1000 genomes project [1]. The total number of Alu insertion predictions which match a loci reported in dbRIP was consistently higher for SSC and SSC-W in comparison to ISV&M. This suggests that using the SSV-MG approach will improve the true positive rate for Alu insertion predictions. Note that an ISV&M analysis [3] reported that among the top 3000 predicted loci, 410 were predicted as de novo (that is, unique to the child). This number clearly is extremely high and far from being true. However, using the SSC algorithm, this number was reduced to only 20 de novo events, while the number was reduced to zero when we used SSC-W. We also predicted deletion events in both the YRI and CEU trios using the ISV&M and SSV-MG approaches. For a positive control, we used the fosmid deletion calls reported for NA18507 by Kidd et al. and deletions reported for YRI population by Conrad et al. For the CEU trio, we used the set of validated deletions reported by 1000 genome project for the same trio [1]. The SSV-MG algorithms detected more known positive predictions than the ISV&M approach for both datasets. SSC-W-CR found the best set of results.

References

1. The 1000 Genomes Project Consortium: Nature 467, 1061–1073 (2010)
2. Hormozdiari, F., Alkan, C., Eichler, E., Sahinalp, S.C.: Combinatorial Algorithms for Structural Variation Detection in High Throughput Sequenced Genomes. Genome Research 19: 1270–1278 (2009)
3. Hormozdiari, F., Hajirasouliha, I., Dao, P., Hach, F., Yorukoglu, D., Alkan, C., Eichler, E.E., Sahinalp, S.: Next-generation VariationHunter: combinatorial algorithms for transposon insertion discovery. Bioinformatics 26(12), i350–i357 (2010)

Variable Selection through Correlation Sifting

Jim C. Huang and Nebojsa Jojic

eScience Group, Microsoft Research
One Microsoft Way, Redmond, WA, 98052, USA

Abstract. Many applications of computational biology require a variable selection procedure to sift through a large number of input variables and select some smaller number that influence a target variable of interest. For example, in virology, only some small number of viral protein fragments influence the nature of the immune response during viral infection. Due to the large number of variables to be considered, a brute-force search for the subset of variables is in general intractable. To approximate this, methods based on ℓ_1-regularized linear regression have been proposed and have been found to be particularly successful. It is well understood however that such methods fail to choose the correct subset of variables if these are highly correlated with other "decoy" variables. We present a method for sifting through sets of highly correlated variables which leads to higher accuracy in selecting the correct variables. The main innovation is a filtering step that reduces correlations among variables to be selected, making the ℓ_1-regularization effective for datasets on which many methods for variable selection fail. The filtering step changes *both* the values of the predictor variables and output values by projections onto components obtained through a computationally-inexpensive principal components analysis. In this paper we demonstrate the usefulness of our method on synthetic datasets and on novel applications in virology. These include HIV viral load analysis based on patients' HIV sequences and immune types, as well as the analysis of seasonal variation in influenza death rates based on the regions of the influenza genome that undergo diversifying selection in the previous season.

1 Introduction

With the advent of high-throughput technologies for profiling biological systems, scientists are faced with the problem of sifting through large volumes of data to identify some subset of variables that explain some target variable y. Because of the asymmetries in acquisition costs, such datasets typically consist of a modest number n of $p-$dimensional observations, with $n \ll p$. Each observation typically consists of a scalar target y and a vector of input variables $\mathbf{x} \in \mathbb{R}^p$, among which it is expected that some subset of variables is useful for predicting the target. The task faced by the scientist is one of identifying the $q \ll p$ relevant variables given n observations, with little *a priori* knowledge that would allow us to differentiate among the p variables.

V. Bafna and S.C. Sahinalp (Eds.): RECOMB 2011, LNBI 6577, pp. 106–123, 2011.

In such situations, the statistics literature often prescribes the use of well-studied linear models of the form $y = \sum_{j=1}^{p} x_j \beta_j + \epsilon$ for describing the relation between input variables and the target, where the amplitudes and signs of these non-zero β_j's then determine the influence that each variable has on the output. Under such models, the problem of *variable selection* consists of estimating weights β_j under the additional assumption that the weights are *sparse*, or that only a few of the weights are non-zero. The variable selection problem can be posed as the problem of solving, given n observations indexed by i, the optimization problem of the form

$$\min_{\beta} \sum_{i=1}^{n} \left(y^i - \sum_{j=1}^{p} x_j^i \beta_j\right)^2 \quad \text{s.t} \quad \sum_{j=1}^{p} \left[|\beta_j| \neq 0\right] = q, \tag{1}$$

where q is the desired number of variables to be included in the model. In contrast to the problem of regression, where the emphasis is on estimating the sign and magnitude of the effect of each variable on the output, the problem of variable selection can here be seen as the recovery of the support of the vector of weights $\boldsymbol{\beta}$, or the set of indices j for which $\beta_j \neq 0$. It is well understood that the above problem is generally intractable for even moderate values of p, as it requires a combinatorial search through all possible subsets of input variables. An alternative to solving the above problem is to instead constrain the sum of absolute values β_j so that we obtain a convex optimization problem for which there are efficient solvers [4]. Among the most effective approaches that use this strategy is the Lasso [20], which formulates the problem of estimating the weights as

$$\min_{\beta} \sum_{i=1}^{n} \left(y^i - \sum_{j=1}^{p} x_j^i \beta_j\right)^2 + \lambda \sum_{j=1}^{p} |\beta_j|, \tag{2}$$

where $\lambda > 0$ is a regularization parameter that controls the sparsity of the resulting solution $\hat{\boldsymbol{\beta}}$. Solutions to the above problem have been widely studied in recent years under various assumptions: in particular, the behavior of solutions has been studied in terms of asymptotic signed support recovery error [17,21]. It has been shown that recovering the correct support requires that there be *few correlations* among variables [22]. Unfortunately, the presence of such correlations is the norm rather than an exception for biological data. These variables typically represent different aspects of the biological processes of interest and are often well-modeled as being generated from a smaller set of shared latent variables. For instance, gene expression levels are typically correlated, as they are a function of the unobserved states of cellular pathways. Similarly, immune assays are correlated, as they are driven by a single set of pathogens in the patient and by the unique properties of the patient's immune system. For such problems, latent variables act that influence input variables must be accounted for in order to select the correct subset of predictive variables.

2   The Problems Caused by Correlations

In terms of the effect of shared latent variables support recovery, there are two regimes to be considered:

A) The target variable y is generated from latent variables $\mathbf{z}$ and the variables in $\mathbf{x}$ correlate with the targets as proxies for these latent variables (Fig. 1(a)).

B) The target variable y is generated from a small number of input variables in $\mathbf{x}$, but other input variables may also correlate weakly with y, as they are influenced by the shared latent variables $\mathbf{z}$ (Fig. 1(b)). This can lead the estimation algorithms astray, especially when the true solution contains a large number of input variables.

Some examples of A) include survival analysis and genome-wide association studies, where both the output and the input variables to be selected (e.g.: such as genetic markers or gene expression profiles) may be influenced by confounding variables (e.g.: gender or ethnicity). In such situations, the sparsity assumption is inappropriate, as it is sufficient to obtain accurate estimates of the confounders in order to minimize the prediction error. For example, consider the case where several variables among $\mathbf{x}$ are simply multiple noisy versions of a single latent variable z which, were it directly observable, would be most predictive of the target variable y. Although any subset of $\mathbf{x}$ containing a single noisy version of z would be predictive, the best estimate of z will make use of the average of all of z's proxy variables, and so a non-sparse solution is the one that correctly minimizes prediction error.

In situations of type B), the assumption of sparsity in Equations (1) and (2) is more appropriate, but the use of the ℓ_1-norm can lead to errors in variable selection. To illustrate this, let us again assume that several variables in $\mathbf{x}$ are

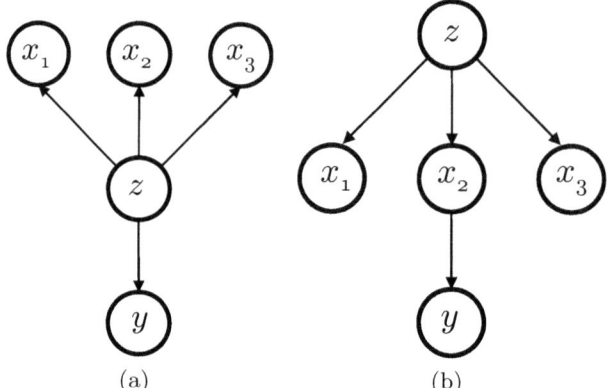

(a)   (b)

Fig. 1. Directed graphical models demonstrating two examples of situations A and B in which a single latent variable z influences the input variables $\mathbf{x}$. a) The target variable y is generated from latent variable z and the variables x_1, x_2, x_3 correlate with y as proxies for z; b) y is generated from x_2 alone, but other observed variables x_1, x_3 may also correlate weakly with y, as they share the influence of z with x_2

perturbed versions of a single latent variable z, but this time y is determined as a noisy version of one of the variables in $\mathbf{x}$. In this case, other variables in both $\mathbf{x}$ and z may be merely proxies of the true correlate and so cannot further help in predicting y. Instead, the problem of finding the single true correlate among $\mathbf{x}$ is made more difficult by these correlations, as other observed variables serve as decoys. The problem is especially difficult if one of these decoy variables has a large sample variance, as under the ℓ_1-norm penalty in Equation (2), it is possible to use a lower weight to capture its effect on y than is needed for capturing the effect of the true correlate. Thus the optimization criterion may in fact be lower for incorrect choices of the support, even though this problem would not plague the original ℓ_0-regularized problem of Equation (1), where the magnitude of the weight of the selected variable only matters in affecting the prediction error. Simple normalization of the data is not a sufficient remedy, especially when multiple variables are involved ($q > 1$), as normalized versions of correlated variables remain correlated.

For many datasets in virology and immunology, it is often the case that we encounter situation B), where we are interested in discovering correlated causes that directly explain an output variable. Consider, for example, the task of analyzing the immune responses of different patients to different viral peptides associated with a particular viral infection, which has important applications in vaccine design. Detecting which of these immune responses are effective in controlling viral infection requires that we account for the dependence between the immune response and the patients' latent immune types. Without accounting for these dependencies, the variable selection process may fail to detect the correct subset of responses that together control the infection, selecting instead proxy variables whose measurements are correlated with those of the relevant ones. In the next section we describe this problem in more detail using datasets from virology and immunology.

3   Datasets

3.1   Sifting through HIV Epitopes to Explain Viral Load

The Human Immunodeficiency Virus (HIV) is strongly affected by the cytotoxic lymphocyte (CTL) immune responses of the host. The virus tends to mutate at sites exposed to CTL surveillance through the cellular presentation of relevant HIV peptides by human leukocyte antigen (HLA) molecules. The presentation and subsequent immune recognition of particular peptides has been reported to associate with a drop in the concentration of viral particles in the patient's blood (viral load), while the recognition of other peptides seems to be less effective. The relationship between the immune response and viral replication patterns is of vital interest in immunology, but understanding it is complicated by the correlations among the immune responses, the involvement of multiple peptides in an effective immune response and the existence of many latent causes. Among the latent causes are the HLA molecules themselves, each with its own particular binding preferences, and each responsible for a number of different peptides to be presented for potential CTL recognition. Evasive mutations have different fitness costs to the virus, and these costs again may depend not on a single mutation, but on multiple correlated mutations. A number of other latent variables are also expected to exist, such as prior infections and other immune mechanisms that can affect either viral mutation or recognition in different sites.

To analyze the relationships between viral peptides that are recognized by the immune response (epitopes) and the corresponding viral load, several datasets can be used. A patient's own HLA molecules and invading HIV sequences can be sequenced, and then the known HLA binding preferences can be used to estimate the evolutionary pressure at different points in the HIV proteome. Here we use a Western Australia cohort of HIV-positive patients [6,15], where the targets $y^1, \cdots, y^n$ consist of the log of viral load measurements in $n = 140$ HIV-positive patients, and the variables $\mathbf{x}$ are derived from each patient's HIV sequence and the HLA types. In particular, each variable corresponds to one of the $p = 492$ 9-mer peptides from the HIV Gag protein, which plays an important role in immune surveillance. For each of these peptides, the patient-specific binding score is then estimated as the minimum over binding energies of the bound configurations of that peptide and each of the patient's six HLA variants [6]. This produces p binding score variables $\mathbf{x}$ for each of the n patients with observed HIV Gag sequences. The resulting set of 140×492 binding measurements x^i_j can be collected into a *binding matrix* $\mathbf{X}$ whose columns correspond to the immune responses for different peptides, from which we would like to select a subset that predicts the log-viral load for each patient.

3.2 The Analysis of the Seasonal Variation in Influenza Genomic Sites and Their Relationship with the Pneumonia and Influenza Index for the Subsequent Season

The adaptations of viruses are generally localized to a small number of sites in their genome, with correlations between viral mutations and immune responses. We assembled a dataset that relates the seasonal variation in influenza genomic sites and the pneumonia and influenza index (P&I), a measure of seasonal mortality rates. It has been previously established that only a handful of sites can be used to characterize different strains [1]. However, in each season, a certain set of sites can be found where the variation across sequenced strains will be atypically high in the presence of correlations between increases and decreases in seasonal variation across sites. Under the hypothesis that such evidence of diversifying selection indicates the likelihood of certain cluster transitions [13] in the *subsequent* season, we investigate whether seasonal variation in entropy of different sites may be predictive of P&I, whereby selected sites may also further inform both influenza research and public health policy.

We focused on $p = 2232$ sites in the influenza virus polymerase, which contains the acidic (PA) and basic 1, 2 (PB1, PB2) protein subunits, as the polymerase is known to regulate the replication and transcription of the influenza viral RNA in infected cells [18]. We obtained protein sequences collected during the flu seasons of 1968-1997 in the US for the PA,PB1,PB2 proteins [2]. For each of these flu

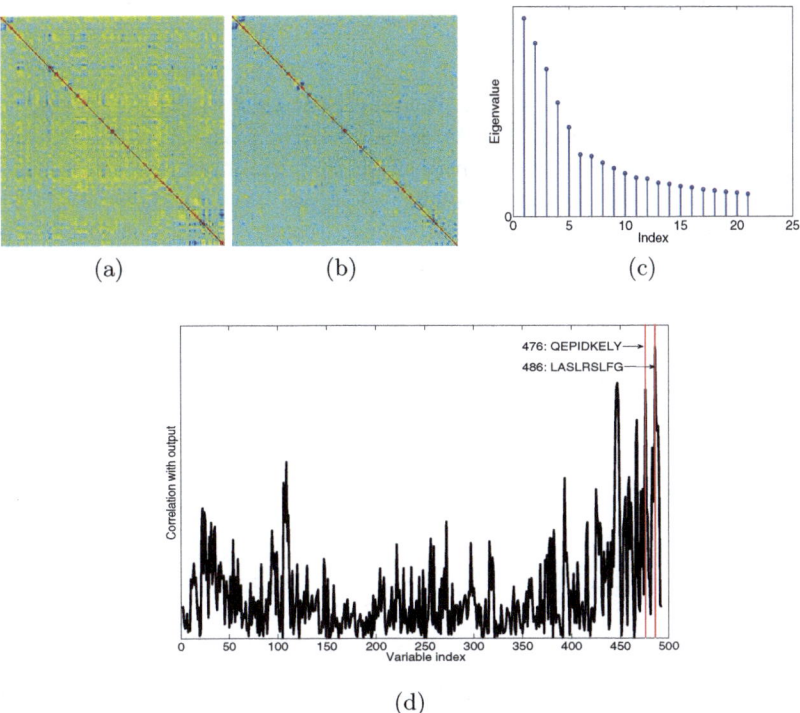

Fig. 2. An example of variable selection in the presence of highly-correlated input variables. a), b) Correlation matrices for $\mathbf{X}$ before and after removing five principal components of $\mathbf{X}$; c) The distribution over the highest 20 eigenvalues of $\mathbf{X}^T\mathbf{X}$; d) The correlation between columns of $\mathbf{X}$ and viral load $\mathbf{y}$, with epitopes found by variable selection under ℓ_0 and ℓ_1-norm constraints indicated. The epitope LASLRSLFG at $j = 486$ corresponds to the epitope whose binding scores are maximally correlated with viral load. Once the first principal component of $\mathbf{X}$ is removed, the epitope QEPIDKELY is selected by variable selection using both ℓ_0 and ℓ_1-norm constraints.

seasons, we obtained the total percentage of deaths among patients aged 65 or older that could be attributed to flu and pneumonia (the P&I index) in the next flu season [9]. For each of the polymerase protein subunits and for each of the 30 flu seasons, we then computed the entropy of the frequency of amino acid occurrence at each site for each protein subunit. We excluded the seasons of 1973 and 1984 due to an insufficient number of sequences collected during these years. The resulting set of 27×2232 site-specific entropies for the influenza polymerase subunits then form the *entropy matrix* $\mathbf{X}$ whose columns correspond to sites from which we would like to identify some subset that predict influenza mortality. As in the previous HIV example, this problem is made difficult by the presence of many pairwise correlations between the entropies at different sites in the polymerase subunits.

4   Correlation Sifting

The binding and entropy matrices $\mathbf{X}$ for the HIV and influenza datasets, like many other scientific datasets, exhibit many pairwise correlations among variables. This can be seen from the matrix $\mathbf{X}^T\mathbf{X}$ (Figure 2(a)), which exhibits a sharply decaying spectrum of eigenvalues, so that the input variables in $\mathbf{X}$ are well approximated using a small number of latent factors (Figure 2(c)). Consider a principal components analysis (PCA) given by $\mathbf{X}^T\mathbf{X} = \mathbf{VDV}^T$, where the eigenvalues of the matrix $\mathbf{X}^T\mathbf{X}$ correspond to elements along the diagonal of $p \times p$ matrix $\mathbf{D}$, and $\mathbf{V}$ is a $p \times p$ matrix whose columns are corresponding eigenvectors, or principal components of $\mathbf{X}$. We denote by $\mathbf{V}_k$ the $k < p$ columns of $\mathbf{V}$ corresponding to the k largest eigenvalues of $\mathbf{X}^T\mathbf{X}$. Similarly, define $\mathbf{V}_{\bar{k}}$ as the remaining $n - k$ columns of $\mathbf{V}$. We define the pair of matrices

$$\mathbf{Z} = \mathbf{XV}_k, \quad \mathbf{R} = \mathbf{X} - \mathbf{ZV}_k^T. \tag{3}$$

Now, we can rewrite the problems of either (1) or (2) in matrix-vector form as

$$\min_{\boldsymbol{\beta}} \left\| \mathbf{y} - \mathbf{ZV}_k^T\boldsymbol{\beta} - \mathbf{R}\boldsymbol{\beta} \right\|_2^2 + \lambda\|\boldsymbol{\beta}\|_1, \tag{4}$$

where $\mathbf{y} = [y^1 \cdots y^n]^T$ and $\boldsymbol{\beta} = [\beta_1 \cdots \beta_p]^T$. We then relax the above problem by introducing $\boldsymbol{\theta}$ as a proxy for $\mathbf{V}_k^T\boldsymbol{\beta}$, but allowing it to change independently of $\boldsymbol{\beta}$:

$$\min_{\boldsymbol{\beta},\boldsymbol{\theta}} \left\| \mathbf{y} - \mathbf{Z}\boldsymbol{\theta} - \mathbf{R}\boldsymbol{\beta} \right\|_2^2 + \lambda\|\boldsymbol{\beta}\|_1 \tag{5}$$

The above minimization over $\boldsymbol{\beta}$ and $\boldsymbol{\theta}$ is decoupled and can be performed independently. Because $\mathbf{R}$ and $\mathbf{Z}$ are orthogonal, the optimal solution $(\hat{\boldsymbol{\beta}}, \hat{\boldsymbol{\theta}})$ satisfies

$$\hat{\boldsymbol{\theta}} = \left(\mathbf{Z}^T\mathbf{Z}\right)^{-1}\mathbf{Z}^T\mathbf{y}, \tag{6}$$

and so $\hat{\boldsymbol{\theta}}$ is the solution to a least-squares problem with outputs $\mathbf{y}$ and matrix of inputs $\mathbf{Z}$. As the above equation does not depend on $\hat{\boldsymbol{\beta}}$, this allows us to solve for $\hat{\boldsymbol{\theta}}$ first and then solve for $\hat{\boldsymbol{\beta}}$ as a function of $\hat{\boldsymbol{\theta}}$.

The correlation sifting method is summarized in Table 1. We see here that for $k = 0$, our method reduces to the regular Lasso, and for $\hat{\boldsymbol{\beta}} = 0$ (or $\lambda \to \infty$), our method consists of principal components regression [7] in which we perform ordinary least-squares regression on the principal components of $\mathbf{X}$. For any fixed λ and k, the method consists of a single PCA step followed by solving a modified Lasso problem, the latter which presents a unique global optimum and can be solved efficiently [4]. The effect of regressing on k leading principal components is that the matrix $\mathbf{R}^T\mathbf{R}$ computed from the remaining components now has a lower maximum eigenvalue relative to the noise variance σ^2 (see Figures 2(a) and 2(b) for an example). If we set k to be too high, we may remove signal power necessary for correct support recovery and so we may increase the corresponding error rate. However, we have also shown that by setting k to be low, we must contend with

Table 1. The correlation sifting algorithm. The algorithm consists of computing the principal components of $\mathbf{X}$, performing a least-squares estimate and then solving a modified Lasso problem.

- For $\mathbf{X} \in \mathbb{R}^{n \times p}$, compute the principal components $\mathbf{V}$, where $\mathbf{X}^T\mathbf{X} = \mathbf{VDV}^T$.
- Let $\mathbf{V}_k$ denote the $k < p$ columns of $\mathbf{V}$ corresponding to the k largest eigen-values of $\mathbf{X}^T\mathbf{X}$.
- Compute $\mathbf{Z} = \mathbf{X}\mathbf{V}_k, \mathbf{R} = \mathbf{X} - \mathbf{Z}\mathbf{V}_k^T$
- Compute $\hat{\theta} = \left(\mathbf{Z}^T\mathbf{Z}\right)^{-1}\mathbf{Z}^T\mathbf{y}$ and let $\tilde{\mathbf{y}} = \mathbf{y} - \mathbf{Z}\hat{\theta}$.
- Compute the optimum $\hat{\beta}$ to the modified Lasso problem

$$\min_{\beta} \left\|\tilde{\mathbf{y}} - \mathbf{R}\beta\right\|_2^2 + \lambda\|\beta\|_1. \tag{7}$$

---

correlations among variables to be selected. In practice, we recommend that the value of k should be selected based on cross-validation prediction error, the usefulness of which we will demonstrate using synthetic experiments in the next section.

4.1   The Effect of Removing Correlations

To illustrate how the above algorithm fixes the problem of variable correlations for variable selection, we will use examples with both synthetic data and real HIV data. For the sake of illustration, in both problems we will assume that there is a single variable to be selected. In the first example, we have generated data for $p = 2$ correlated variables x_1, x_2, shown in Figure 3 (left). The underlying model was generated as $y = 0.01x_2 + \epsilon$, where ϵ is a normal random variable. Suppose that we were to apply the Lasso to this synthetic dataset with the regularization parameter λ set such that a single variable is to be included in the solution. The choice of which variable is to be included will be determined solely on the basis of which variable has greater correlation (up to a scaling factor) $\sum_{i=1}^{n} x_i y_i$ with the output variable. Because the variables x_1, x_2 are both highly correlated and exhibit uneven variances, the variable with the largest sample variance is the one that is selected, which in this example yields the erroneous selection of variable x_1 (Figure 3, left). Note that standardization of the variances of the variables does not remove the problem of x_1 being more correlated with y than x_2, despite the fact that x_2 is the correct variable to be selected (Figure 3, center). However, applying the correlation sifting method to yield transformed variables r_1, r_2 and transformed output $\tilde{y}$ corrects this problem (Figure 3, right) such that now r_2 is more correlated with $\tilde{y}$ than r_1 and so r_2 is selected. The effect of the correlation sifting method is to project the input variables onto the remaining PCA components such that the influence of proxy variables on the output via the principal components of $\mathbf{X}$ that have large eigenvalues is reduced.

The effect of removing correlations can also be seen in our second example, which consists of HIV-gag binding scores and HIV viral load measurements in 140 patients. Here we suppose as above that only one of the peptides influences patient viral load under a linear model $\mathbf{y} = \mathbf{X}\boldsymbol{\beta}^* + \boldsymbol{\epsilon}$ where ϵ_i is a zero-mean noise variable. Let this peptide be indexed by j and let the corresponding element of $\boldsymbol{\beta}^*$ be β_j^*. If we were to perform variable selection under an ℓ_0-norm constraint, we would solve the problem in Equation (1) by minimizing the prediction error with $q = 1$, or $(j, \hat{\beta}_j) = \arg\min_{l, \beta_l} \|\mathbf{y} - \mathbf{x}_l \beta_l\|^2$. Figure 2(d) shows the selected variable $j = 476$ (the peptide the peptide $QEPIDKELY$) from the above procedure for the 140×492 binding matrix $\mathbf{X}$. However, performing variable selection using the Lasso (Equation (2)) with the ℓ_1-norm penalty λ set sufficiently high to select a single variable yields $j' = 486$, or the peptide $LASLRSLFG$, a solution that differs from the ℓ_0-norm constrained solution. Figure 2(d) provides some insight by showing the sample correlation between the two variables $j = 476, j' = 486$ and the output. Here, the variable corresponding to the binding score for peptide $LASLRSLFG$ has significantly higher correlation with the output than peptide $QEPIDKELY$ and so the former is selected by the Lasso, despite the

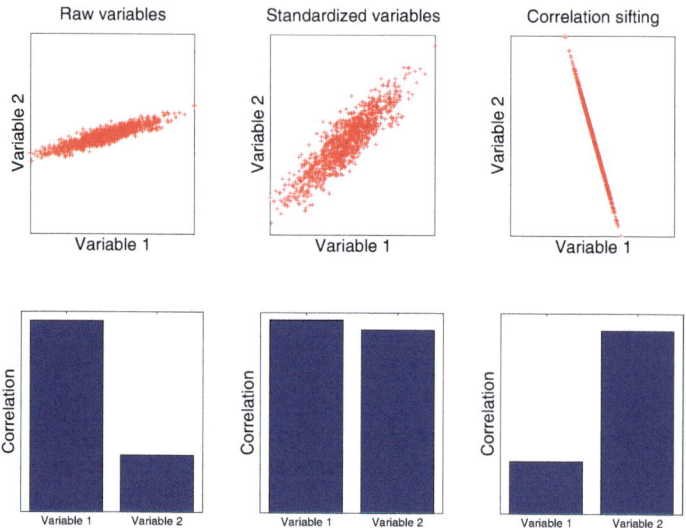

Fig. 3. The effect of removing correlations before variable selection. Suppose that we were to apply the Lasso with the regularization parameter λ set such that a single variable is to be included in the solution. The choice of which variable is to be included will be determined solely on the basis of which variable has greater correlation (up to a scaling factor) $\sum_{i=1}^{n} x_i y_i$ with the output variable. Because the variables x_1, x_2 are both highly correlated and exhibit uneven variances, the variable with the largest sample variance is the one that is selected, which in this example yields the erroneous selection of variable x_1 (left). Note that standardization of the variances of the variables does not remove the problem of x_1 being more correlated with y than x_2, despite the fact that x_2 is the correct variable to be selected (center). However, applying the correlation sifting method to yield transformed variables r_1, r_2 and transformed output $\tilde{y}$ corrects this problem (right).

fact that the latter is the one that ought to be selected according to the ℓ_0-norm constraint. As in the previous example, this is due to the both the larger variance of the binding score for peptide $LASLRSLFG$ as compared to that of $QEPIDKELY$ and the fact that both sets of binding scores are correlated. Furthermore, standardizing variables to have the same variance may remove the first problem of uneven variances, but not the second one of correlations among variables to be selected.

Having provided illustrations of the correlation sifting method, we will begin to illustrate its uses using synthetic datasets, where the set of variables that generate the output is known and so we can measure the error rate of correlation sifting in recovering the correct set of variables as compared to other methods for variable selection.

5   Results

5.1   Empirical Analysis of the Algorithm Using Synthetic Data

We generated synthetic data sets with $n = 100, p = 500$ in which many correlated and irrelevant variables were introduced in tandem with a small number of relevant variables to be selected. For any given dataset, we created a random sparse vector $\boldsymbol{\beta}^*$ with sparsity index ρ (fraction of elements of $\boldsymbol{\beta}^*$ that are non-zero) set to $\rho = 0.02, 0.05, 0.1$. We then generated an $n \times p$ matrix $\mathbf{Z}$ such that the jth column of $\mathbf{Z}$ consists of n samples drawn from a Gaussian distribution with variance $\frac{5}{1.2^j}, j = 1, \cdots, 20$ or 0.3 for $20 < j \leq p$. We then generated a $p \times p$ random non-orthogonal set of basis vectors $\mathbf{V}$ with elements drawn independently and identically from $\mathcal{N}(0, 1)$. The product $\mathbf{Z}\mathbf{V}^T$ was then standardized to create matrix $\mathbf{X}$ so that $\sum_i x_j^i = 0$ and $\sum_i (x_j^i)^2 = 1$. We then generated $\mathbf{y} = \mathbf{X}\boldsymbol{\beta}^* + \boldsymbol{\epsilon}$ where ϵ_i was drawn from $\mathcal{N}(0, \sigma^2)$ where the noise variance σ^2 is related to the signal variance by the signal-to-noise ratio $SNR = var(\mathbf{X}\boldsymbol{\beta}^*)/\sigma^2$. Finally, we generated 100 synthetic datasets in the above fashion where each dataset consists of a particular set of $\mathbf{X}, \mathbf{y}, \boldsymbol{\beta}^*, \boldsymbol{\epsilon}$ for a particular joint setting of ρ, SNR.

For each of the 100 datasets, we applied correlation sifting, where we used the Least Angle Regression (LARS) algorithm of [4] to solve the modified Lasso problem of Table 1. Once a vector $\hat{\boldsymbol{\beta}}$ with a given sparsity index has been estimated for a given dataset by solving the correlation sifting problem, we then computed the support recovery error, defined as number of entries in $\hat{\boldsymbol{\beta}}$ which disagree in sign with $\boldsymbol{\beta}^*$. Figure 4 shows the error rate over the 100 datasets as a function of the sparsity index of the solution $\hat{\boldsymbol{\beta}}$, where the sparsity index is defined as the fraction of elements of the solution that are non-zero. Note that each curve is expected to be approximately U-shaped, given that excessively sparse or dense solutions will yield a higher error. However, methods with higher accuracy should yield a lower support recovery error across sparsity index values. We show error rates obtained for $\rho = 0.02, 0.05, 0.1$ and for $SNR = 2, 4, 8$ using the correlation sifting method with $k = 1, 2, 5, 10$. The corresponding error rates obtained by regular Lasso are shown for comparison, in addition to the error rates

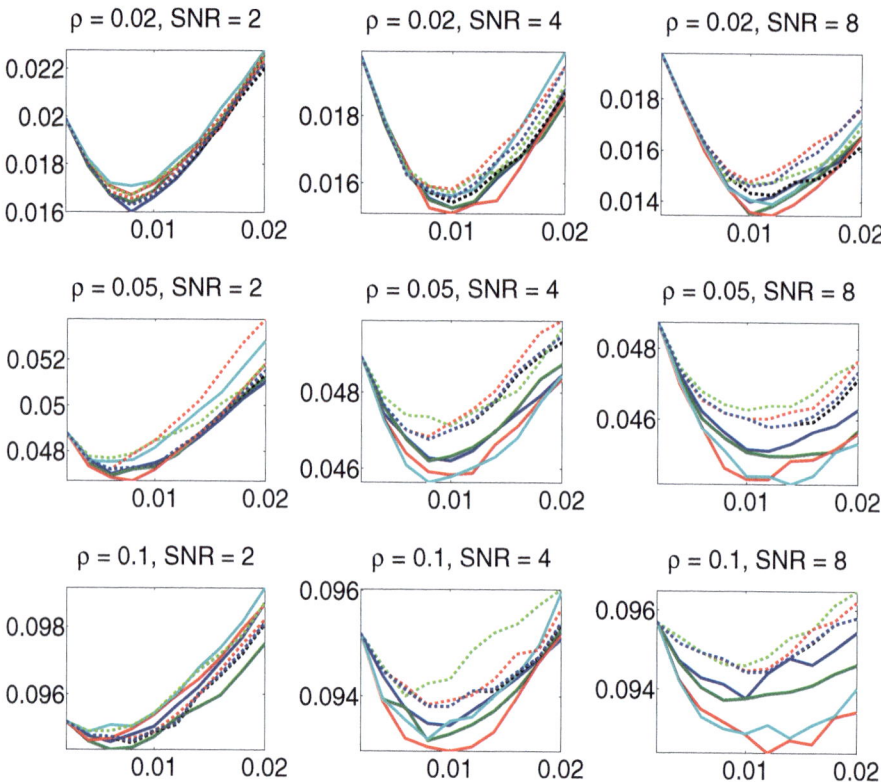

Fig. 4. Average error rates from 100 synthetically-generated data sets are shown as a function of the sparsity index of the estimated vector $\hat{\beta}$ obtained from the correlation sifting method for different values of sparsity and signal-to-noise ratios (SNRs), where an error is defined as the event $\hat{\beta}_j \neq_s \beta_j^*$ for $j = 1, \cdots, p$. Error rates are shown for the correlation sifting with $k = 1$ (blue), $k = 2$ (green), $k = 5$ (red), $k = 10$ (cyan). The dotted curves correspond to error rates obtained from regular Lasso (black dotted), the adaptive Lasso method of [24] (green dotted), the elastic net method of [23] (red dotted) and the method of [17] (blue dotted).

for the elastic net [23], adaptive Lasso [24] and method of [17] that are commonly used for variable selection. These methods consist of modified Lasso problems with a second regularization constant in addition to the ℓ_1-norm penalty of the regular Lasso. For the elastic net, the additional regularization constant penalizes the ℓ_2-norm of the solution in addition to its ℓ_1-norm. In the case of the adaptive Lasso method, the second regularization constant allows us to re-scale the ℓ_1-norm penalty for each weight individually as a function of the variance for each input variable. For the method of [17], the second regularization constant is similar to the parameter k in correlation sifting in that it consists of a threshold correlation value between inputs and output that is used to select variables with which to compute an SVD of $\mathbf{X}$.

In order to evaluate the accuracy of these methods in support recovery, we systematically explored a range of parameter values for each method. Figure 4 shows the recovery error as a function of the sparsity index of the solutions obtained from each method. In particular, for the Lasso and the methods of [17,23,24], we show the error curves corresponding to parameter values that achieved the lowest average error across sparsity index values. As can be seen, the result of correctly accounting for the correlations among variables under the correct model $y = X\beta^* + \epsilon$ is that the error rates are correspondingly lower than those obtained from solving the regular Lasso or the methods of [17,23,24]. We see that the difference in support recovery error between the correlation sifting method and other methods for variable selection increases as we increase either the amount of noise added to the outputs or the number of variables in the support. Thus, the comparative advantage between the correlation sifting method and other methods for variable selection is expected to be smaller in high SNR & sparsity index regimes and becomes more significant as the amount of noise or the size of the support is increased.

Figure 4 shows that our method achieves lower error than the other methods for a variety of the choices of k, indicating that it should also do so even if a single value for k is chosen using some standard model selection technique such as cross-validation. Furthermore, the U-shape of the curves also indicates that in terms of solution sparsity, controlled by λ, unambiguous error minima are achieved for similar values of the λ for different choices of k. This behavior suggests that this minimum would also be robust to slight variations in the dataset size, and so cross-validation for both λ and k (the only two parameters of the model) should also result in selecting the settings where our method achieves lower support recovery error relative to other methods. To show this, we repeated the above experiments in a leave-one-out cross-validation setting in which we select both the values for λ and k that minimize the total mean-squared error (MSE) on held-out data for each of the 100 synthetic datasets for each of the 9 joint settings of SNR and sparsity index ρ. We then computed the support recovery error obtained from the complete synthetic dataset for the selected value of λ and k. We also performed the same cross-validation procedure for the parameters of the regular Lasso and the methods of [17,23,24] and we computed support recovery error obtained from the complete synthetic dataset for the selected value of λ and k. We found that for the above 100 synthetic datasets with the 9 joint settings of SNR and sparsity indices, the correlation sifting method achieved a significantly lower support recovery error as compared to achieved by the Lasso and methods of [17,23,24] for 8/9 joint settings of SNR and sparsity index ($p < 0.05/9$, Bonferroni-corrected, one-sided t-test).

5.2 Identifying HIV-Gag Epitopes That Are Predictive of Viral Load

We then applied correlation sifting to the HIV data from Section 3.1. In contrast to the synthetic datasets from the previous section, here we are not provided with a known list of relevant variables and corresponding weights β^*. Instead we need to evaluate how likely it is that our method will point to HIV-Gag sites that are most likely to influence patients' viral load in clinical studies. To accomplish this, we will analyze the power of our method in two ways. First, as in the synthetic experiments, we generate β^* of varying sparsity index values and from it, the *simulated* vector $\mathbf{y}$ of log viral loads obtained by multiplying the set of HIV-Gag binding scores $\mathbf{X}$ with the simulated β^*. The second test of our method will employ cross-validation on the HIV-Gag binding matrix $\mathbf{X}$ and log viral loads whereby we evaluate the ability of different methods to predict viral load for new HIV patients. Here, it is assumed that correct support recovery will lead to models that generalize better on test outputs, so that lower prediction error on test data should indicate a low support recovery error.

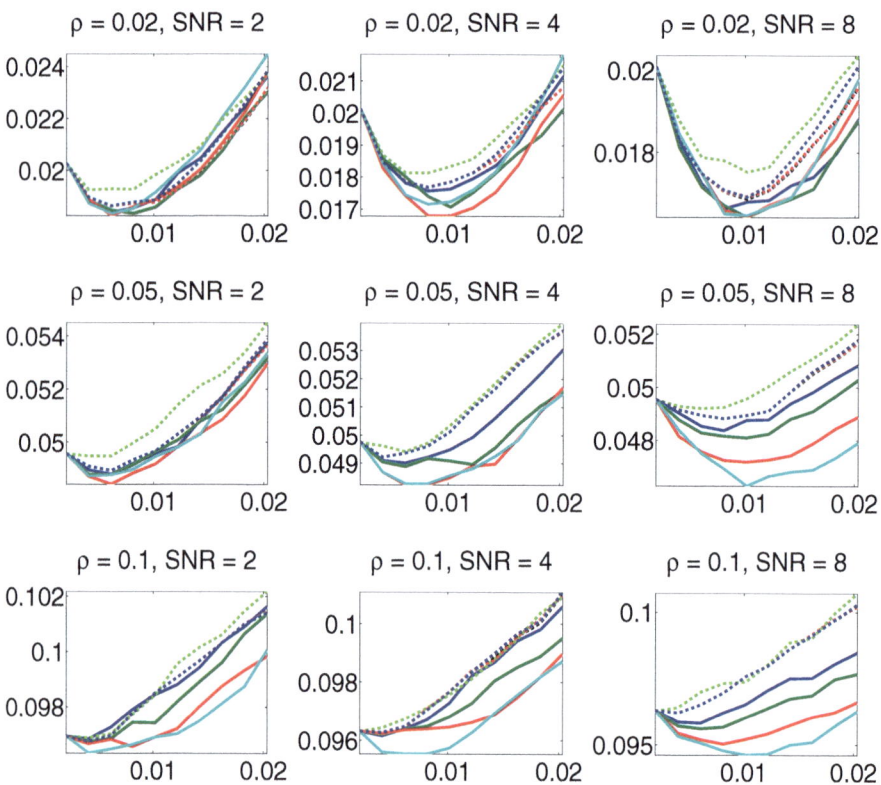

Fig. 5. Average error rates from 100 synthetically-generated HIV viral load data sets are shown as a function of the sparsity index of the estimated vector $\hat{\beta}$ obtained from the correlation sifting method for different values of sparsity and signal-to-noise ratios (SNRs). Error rates are shown for the correlation sifting with $k = 1$ (blue), $k = 2$ (green), $k = 5$ (red), $k = 10$ (cyan). The dotted curves correspond to error rates obtained from regular Lasso (black dotted), the adaptive Lasso method of [24] (green dotted), the elastic net method of [23] (red dotted) and the method of [17] (blue dotted).

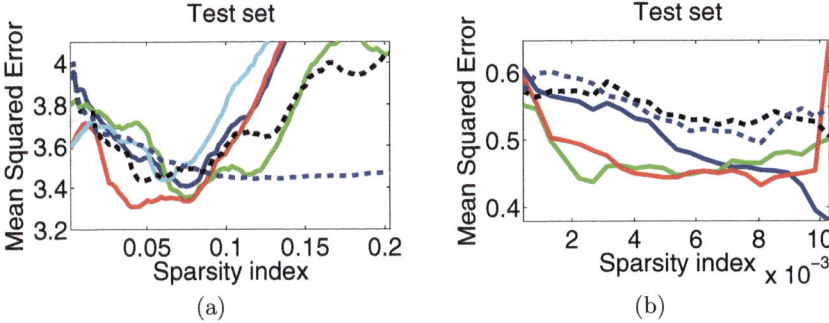

Fig. 6. a) Mean squared error (MSE) of predicting measured log viral load in HIV-positive patients using the binding scores of [6] as a function of the sparsity index of the estimated solution; b) MSE of predicting the pneumonia & influenza (P&I) index for the years 1969-1972, 1974-1983 and 1985-1997 using the entropy of amino acid variation per site in the PA protein. MSEs were computed using leave-one-out cross-validation for the regular Lasso (black dashed), the best method amongst those of [17,23,24] based on test error (blue dotted), and the method of this paper with $k = 1$ (blue), $k = 2$ (green), $k = 5$ (red), $k = 10$ (cyan).

To perform the first set of experiments for the 140×492 HIV-Gag binding matrix $\mathbf{X}$, we applied the correlation sifting method as described above for this series of datasets in tandem with the methods from the previous section. We used the same noise variances and sparsity index values as in the synthetic experiments. The resulting error rates are shown in Figure 5. As in the previous example with synthetic data, the presence of many correlated and irrelevant variables can be mitigated by correlation versus the alternative methods such as the regular Lasso, adaptive Lasso and elastic net methods. We see that concomitant with our results on synthetic data, the correlation sifting method outperforms the regular Lasso and the methods of [17,23,24] in terms of its support recovery error. Note that in contrast to the previous synthetic experiments where matrices $\mathbf{X}$ were generated from Gaussian distributions, here the variables had highly non-Gaussian distributions. The results on the synthetic HIV data then suggest that the correlation sifting method is not too sensitive to distributional assumptions about the input variables.

We then applied correlation sifting to the set of experimental log-viral load measurements $\mathbf{y}$ for the 140 patients with the matrix $\mathbf{X}$ of binding scores for each peptide in the HIV-Gag protein. Using leave-one-out cross-validation, we assessed the MSE on test data. Here we compared our method to the regular Lasso and the best method among those of [17,23,24] based on test error. The resulting MSE for test data is shown in Figure 6(a) as a function of the sparsity index of the estimated vectors $\hat{\boldsymbol{\beta}}$. As can be seen, the estimated vector $\hat{\boldsymbol{\beta}}$ obtained using our method leads to better predictions of log-viral load in unseen HIV-positive patients relative to the models learned by the regular Lasso and the methods of [17,23,24]. For the joint setting of k and λ that minimizes test error, the correlation sifting method selects 20 epitopes. From the Epitope Location Finder (ELF) database (http://www.hiv.lanl.gov/content/sequence/ELF/), we find that five of these 20 epitopes (ISPRTLNAW, LDRWEKIRL, AADTGNSSQ, EVKDTKEAL, DLNTMLNTV) have been associated with viral load and disease progression in HIV patients [3,5,8,10,12,14]. In contrast, upon examination of the epitopes discovered by the method of [17] (which achieved the best test error among the methods we compared against), we find that of the 20 epitopes discovered by this method, none were associated with viral load or disease progression in HIV patients.

5.3 Discovering the Effect of Diversifying Selection on Seasonal Influenza Epidemics

We can also use correlation sifting in order to select variables that are predictive of next year's influenza mortality from the dataset of Section 3.2, which consists of site-specific entropy scores for the current year's influenza strains. Using leave-one-out cross-validation, we assessed the MSE for test data using our method for variable selection for different values of k. For comparison, we show the test errors obtained from using the regular Lasso and the methods of [17,23,24] that achieved the minimum error on test sets. The resulting MSE for test data are shown in Figure 6(b) as a function of the sparsity index of the estimated vectors $\hat{\beta}$. As the correlation sifting method consists of solving a modified Lasso problem, the maximum sparsity index corresponds to the maximum $n - 1 = 26$ possible non-zero elements in the solution $\hat{\beta}$. As in the previous example on HIV, by removing the effect of correlations amongst variables, the estimated vector $\hat{\beta}$ obtained using our method leads to lower test error relative to the models learned by the regular Lasso and the methods of [17,23,24]. As additional validation, we performed full cross-

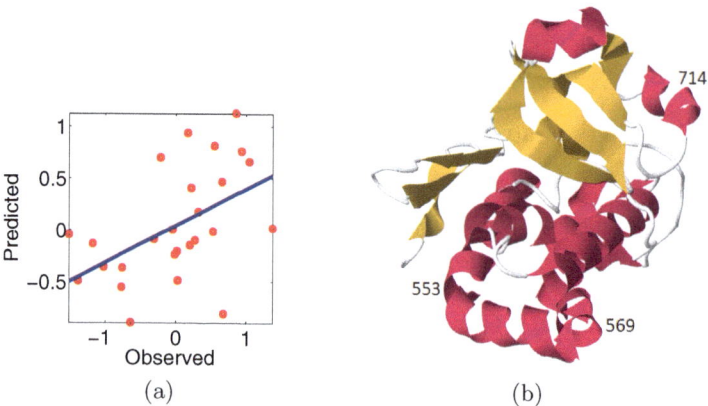

(a)   (b)

Fig. 7. a) Predicted versus observed P&I based on full leave-one-out cross-validation (Pearson's correlation $r = 0.5117, p = 0.0064$); b) The 3-D structure for the C-terminal region of the influenza A virus polymerase PB2 subunit. Sites in the region that were identified by our method are displayed.

Subunit	Site	Subunit	Site	Subunit	Site
PB2	177	PB1	12	PA	213
PB2	191	PB1	257	PA	231
PB2	231	PB1	387	PA	388
PB2	553	PB1	397	PA	394
PB2	569	PB1	464	PA	407
PB2	714	PB1	484	PA	520
PB2	756	PB1	652	PA	602
				PA	618
				PA	712

Fig. 8. Sites in the influenza PB1,PB2 and PA protein subunits identified by the correlation sifting method

validation where for each training set, we performed model selection by further splitting the training set into training and validation data and we made test predictions using the model selected based on minimum validation error. A plot of the resulting predicted versus observed P&I indices is shown in Figure 7(a): here we see that the sparse linear model provides a reasonable approximation to the relationship between the current year's sequence variation and next year's mortality rate (Pearson's correlation $r = 0.5117, p = 0.0064$). Interestingly, last year's P&I index, total protein entropy for each year and the number of sequences collected in a given year were not found to be significantly predictive of this year's P&I (data not shown). Furthermore, knowledge of the previous year's dominant influenza strain (either H1N1 or H3N2) was not found to be predictive of the next year's P&I. Figures 7(b) and 8 show the sites in the PB2, PB1 and PA subunits that led to the minimum test error in cross-validation. We find that many of these have previous experimental support: for example, sites 191 and 714 in the PB2 subunit have been implicated in PB2 binding to the importin $\alpha5$ antibody and in the significant increase in polymerase activity [19]. Site 12 in PB1 has been shown to be part of the core interface between PB1 and the PA subunit [16]. Finally, sites 553 and 569 in PB2 are thought to lie in the binding domain of PB2 (Figure 7(b)).

6   Discussion

We have developed a novel, computationally efficient method for variable selection consisting of a single PCA on the matrix of variables, an ordinary least-squares estimation, and finally solving a modified version of the Lasso problem. The most similar previous approach to mitigating the effect of variable correlations is that of [17], which used supervised principal components to filter y, but did not remove correlations among variables **X**. Other related methods include that of [11] and PCA regression [7]. The method of [11] uses an SVD to account for the presence of correlations between input variables before performing regression. However, this method differs significantly from ours, as the method of [11] is not designed for variable selection and is instead aimed at improving predictive accuracy. PCA regression, on the other hand, is a special case of the correlation sifting method with $\lambda \to \infty$ such that there is no variable selection problem being solved.

We have shown through extensive experiments on synthetic data that correlation sifting decreases the average support recovery error in comparison to this method, as well as other methods for variable selection [23,24] based on ℓ_1 regularization. We compared our method to the above methods on real-world HIV and influenza datasets in which many correlated and irrelevant variables exist. Through leave-one-out cross-validation and based on literature surveys of the biology of the HIV and influenza viruses, we found that the variables selected using our method were significantly predictive of the target variable whereas methods such as those of [17,20,23,24] were more susceptible to support recovery error due to the presence of variable correlations. We emphasize that the correlation sifting method is especially useful for situations where the output variable is best modeled as a function of observed variables that share common latent causes.

References

1. Allen, J.E., Gardner, S.N., Vitalis, E.A., Slezak, T.R.: Conserved amino acid markers from past influenza pandemic strains. BMC Microbiology 9, 77–87 (2009)
2. Bao, Y., Bolotov, P., Dernovoy, D., Kiryutin, B., Zaslavsky, L., et al.: The influenza virus resource at the National Center for Biotechnology Information. Journal of Virology 82(2), 596–601 (2008)
3. Draenert, R., Le Gall, S., Pfafferott, K.J., Leslie, A.J., Chetty, P., Brander, C., et al.: Immune selection for altered antigen processing leads to cytotoxic T lymphocyte escape in chronic HIV-1 infection. Journal of Experimental Medecine 199(7), 905–915 (2004)
4. Efron, B., Hastie, T., Tibshirani, R.: Least angle regression. Annals of Statistics 32, 407–499 (2004)
5. Frahm, N., Adams, S., Kiepiela, P., Linde, C.H., Hewitt, H.S., Lichterfield, M., et al.: HLA-B63 presents HLA-B57/B58-restricted cytotoxic T-lymphocyte epitopes and is associated with low human immunodeficiency virus load. Journal of Virology 79(16), 10218–10225 (2005)
6. Jojic, N., Reyes-Gomez, M., Heckerman, D., Kadie, C., Schueler-Fruman, O.: Learning MHC I-peptide binding. Bioinformatics 22(14), e227–e235 (2006)
7. Jolliffe, I.T.: A note on the use of principal components in regression. Journal of the Royal Statistical Society, Series C (Applied Statistics) 31(3), 300–303 (1982)
8. Goulder, P.J.R., Bunce, M., Krausa, P., McIntyre, K., Crowley, S., Morgan, B., et al.: Novel, cross-restricted, conserved and immunodominant cytotoxic T lymphocyte epitopes in slow HIV Type 1 infection. AIDS Research in Human Retroviruses 12, 1691–1698 (1996)
9. Greene, S.K., Ionides, E.L., Wilson, M.L.: Patterns of influenza-associated mortality among US elderly by geographic region and virus subtype, 19681998. American Journal of Epidemiology 163, 316–326 (2006)
10. Jones, N.A., Wei, X., Flower, D.R., Wong, M., Michor, F., Saag, M.S., et al.: Determinants of human immunodeficiency virus type 1 escape from the primary CD8+ cytotoxic T lymphocyte response. Journal of Experimental Medicine 200(10), 1243–1256 (2004)

11. Karpievitch, Y.V., Taverner, T., Adkins, J.N., Callister, S.J., Anderson, G.A., Smith, R.D., Dabney, A.R.: Normalization of peak intensities in bottom-up MS-based proteomics using singular value decomposition. Bioinformatics 25(19), 2573–2580 (2009)

12. Karlsson, A.C., Iversen, A.K.N., Chapman, J.M., de Oliviera, T., Spotts, G., McMichael, A.J., et al.: Sequential broadening of CTL responses in early HIV-1 infection is associated with viral escape. PLoS ONE 2, e225 (2007)

13. Koelle, K., Cobey, S., Grenfell, B., Pascual, M.: Epochal evolution shapes the phylodynamics of interpandemic influenza A (H3N2) in humans. Science 314, 1898–1903 (2006)

14. Mollet, L., Li, T.S., Samri, A., Tournay, C., Tubiana, R., Calvez, V., et al.: Dynamics of HIV-specific CD8+ T lymphocytes with changes in viral load. Journal of Immunology 165(3), 1692–1704 (2000)

15. Moore, C.B., John, M., James, I.R., Christiansen, F.T., Witt, C.S., Mallal, S.A.: Evidence of HIV-1 adaptation to HLA-restricted immune responses at a population level. Science 296, 1439–1443

16. Obayashi, E., Yoshida, H., Kawai, F., Shibayama, N., Kawaguchi, A., Nagata, K., et al.: The structural basis for an essential subunit interaction in influenza virus RNA polymerase. Nature 454, 1127–1131

17. Paul, D., Bair, E., Hastie, T., Tibshirani, R.: "Preconditioning" for feature selection and regression in high-dimensional problems. Annals of Statistics 36(4), 1595–1618 (2008)

18. Salomon, R., Franks, J., Govorkova, E.A., Ilyushina, N.A., Yen, H.-L., Hulse-Post, D.J., et al.: The polymerase complex genes contribute to the high virulence of the human H5N1 influenza virus isolate A/Vietnam/1203/04. Journal of Experimental Medecine 203(3), 697–698 (2006)

19. Tarendeau, F., Boudet, J., Guilligay, D., Mas, P.J., Bougault, C.M., et al.: Structure and nuclear import function of the C-terminal domain of influenza virus polymerase PB2 subunit. Nature Structural Molecular Biology 14, 229–233 (2007)

20. Tibshirani, R.: Regression shrinkage and selection via the Lasso. Journal of the Royal Statistical Society, Series B 58(1), 267–288 (1996)

21. Wainwright, M. J.: Sharp thresholds for high-dimensional and noisy sparsity recovery using ℓ_1-constrained quadratic programming (Lasso). UC Berkeley Technical Report 709, Department of Statistics (2006)

22. Zhao, P., Yu, B.: On model selection consistency of Lasso. Journal of Machine Learning Research 7, 2541–2567 (2006)

23. Zou, H., Hastie, T.: Regularization and variable selection via the elastic net. Journal of the Royal Statistical Society, Series B 67, 301–320 (2005)

24. Zou, H.: The adaptive Lasso and its oracle properties. Journal of the American Statistical Association 101(476), 1418–1429 (2006)

Weighted Genomic Distance Can Hardly Impose a Bound on the Proportion of Transpositions

Shuai Jiang and Max A. Alekseyev

Department of Computer Science and Engineering
University of South Carolina, Columbia, SC, U.S.A.

Abstract. *Genomic distance* between two genomes, i.e., the smallest number of genome rearrangements required to transform one genome into the other, is often used as a measure of evolutionary closeness of the genomes in comparative genomics studies. However, in models that include rearrangements of significantly different "power" such as *reversals* (that are "weak" and most frequent rearrangements) and *transpositions* (that are more "powerful" but rare), the genomic distance typically corresponds to a transformation with a large proportion of transpositions, which is not biologically adequate.

Weighted genomic distance is a traditional approach to bounding the proportion of transpositions by assigning them a relative weight $\alpha > 1$. A number of previous studies addressed the problem of computing weighted genomic distance with $\alpha \leq 2$.

Employing the model of multi-break rearrangements on circular genomes, that captures both reversals (modelled as *2-breaks*) and transpositions (modelled as *3-breaks*), we prove that for $\alpha \in (1, 2]$, a minimum-weight transformation may entirely consist of transpositions, implying that the corresponding weighted genomic distance does not actually achieve its purpose of bounding the proportion of transpositions. We further prove that for $\alpha \in (1, 2)$, the minimum-weight transformations do not depend on a particular choice of α from this interval. We give a complete characterization of such transformations and show that they coincide with the transformations that at the same time have the shortest length and make the smallest number of breakages in the genomes.

Our results also provide a theoretical foundation for the empirical observation that for $\alpha < 2$, transpositions are favored over reversals in the minimum-weight transformations.

1   Introduction

Genome rearrangements are evolutionary events that change genomic architectures. Most frequent rearrangements are *reversals* (also called *inversions*) that "flip" continuous segments within single chromosomes. Other common types of rearrangements are *translocations* that "exchange" segments from different chromosomes and *fission/fusion* that respectively "cut"/"glue" chromosomes.

V. Bafna and S.C. Sahinalp (Eds.): RECOMB 2011, LNBI 6577, pp. 124–133, 2011.

Since large-scale rearrangements happen rarely and have dramatic effect on the genomes, the number of rearrangements (*genomic distance*[1]) between two genomes represents a good measure for their evolutionary remoteness and often is used as such in phylogenomic studies. Depending on the model of rearrangements, there exist different types of genomic distance [10].

Particularly famous examples are the *reversal distance* between unichromosomal genomes [12] and the genomic distance between multichromosomal genomes under all aforementioned types of rearrangements [11]. Despite that both these distances can be computed in polynomial time, their analysis is somewhat complicated, thus limiting their applicability in complex setups. The situation becomes even worse when the chosen model includes more "complex" rearrangement operations such as *transpositions* that cut off a segment of a chromosome and insert it into some other place in the genome. Computational complexity of most distances involving transpositions, including the *transposition distance*, remains unknown [13,4,8]. To overcome difficulties associated with the analysis of genomic distances many researchers now use simpler models of multi-break [3], DCJ [14], block-interchange [7] rearrangements as well as *circular* instead of *linear* genomes, which give reasonable approximation to original genomic distances [1].

Another obstacle in genomic distance-based approaches arises from the fact that transposition-like rearrangements are at the same time much rare and "powerful" than reversal-like rearrangements. As a result, in models that include both reversals and transpositions, the genomic distance typically corresponds to rearrangement scenarios with a large proportion of transpositions, which is not biologically adequate. A traditional approach to bounding the proportion of transpositions is *weighted genomic distance* defined as the minimum weight of a transformation between two genomes, where transpositions are assigned a relative weight $\alpha > 1$ [10]. A number of previous studies addressed the weighted genomic distance for $\alpha \leq 2$. In particular, Bader and Ohlebusch [4] developed a 1.5-approximation algorithm for $\alpha \in [1,2]$. For $\alpha = 2$, Eriksen [9] proposed a $(1 + \epsilon)$-approximation algorithm (for any $\epsilon > 0$).

Employing the model of multi-break rearrangements [3] on circular genomes, that captures both reversals (modelled as 2-breaks) and transpositions (modelled as 3-breaks), we prove that for $\alpha \in (1, 2]$, a minimum-weight transformation may entirely consist of transpositions. Therefore, the corresponding weighted genomic distance does not actually achieve its purpose of bounding the proportion of transpositions. We further prove that for $\alpha \in (1, 2)$, the minimum-weight transformations do not depend on a particular choice of α from this interval (thus are the same, say, for $\alpha = 1.001$ and $\alpha = 1.999$), and give a complete characterization of such transformations. In particular, we show that these transformations coincide with those that at the same time have the shortest length and make the smallest number of breakages in the genomes, first introduced by Alekseyev and Pevzner [2].

---

[1] We remark that the term *genomic distance* sometimes is used to refer to a particular distance under reversals, translocations, fissions, and fusions.

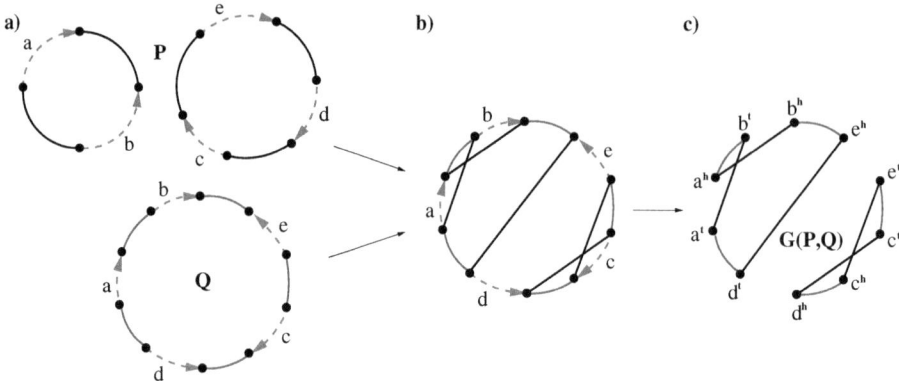

Fig. 1. a) Graph representation of a two-chromosomal genome $P = (+a-b)(+c+e+d)$ as two black-obverse cycles and a unichromosomal genome $Q = (+a+b-e+c-d)$ as a gray-obverse cycle. **b)** The superposition of the genomes P and Q. **c)** The breakpoint graph $G(P,Q)$ of the genomes P and Q (with removed obverse edges).

Our results also provide a theoretical foundation for the empirical observation of Blanchette et al. [6] that for $\alpha < 2$, transpositions are favored over reversals in the minimum-weight transformations.

2   Multi-break Rearrangements and Breakpoint Graphs

We represent a circular chromosome on n genes $x_1, x_2, \ldots, x_n$ as a cycle graph on $2n$ edges alternating between directed "obverse" edges, encoding genes and their directionality, and undirected "black" edges, connecting adjacent genes (Fig. 1a). A genome consisting of m chromosomes is then represented as m such cycles. The edges of each color form a perfect matching.

A *k-break* rearrangement [3] is defined as replacement of a set of k black edges in a genome with a different set of k black edges forming matching on the same $2k$ vertices. In the current study we consider only 2-break (representing reversals, translocations, fissions, fusions) and 3-break rearrangements (including transpositions).

For two genomes P and Q on the same set of genes,[2] represented as black-obverse cycles and gray-obverse cycles respectively, their superposition is called the *breakpoint graph* $G(P,Q)$ [5]. Hence, $G(P,Q)$ consists of edges of three colors (Fig. 1b): directed "obverse" edges representing genes, undirected black edges representing adjacencies in the genome P, and undirected gray edges representing adjacencies in the genome Q. We ignore the obverse edges in the breakpoint graph and focus on the black and gray edges forming a collection of black-gray alternating cycles (Fig. 1c).

---

[2] From now on, we assume that given genomes are always one the same set of genes.

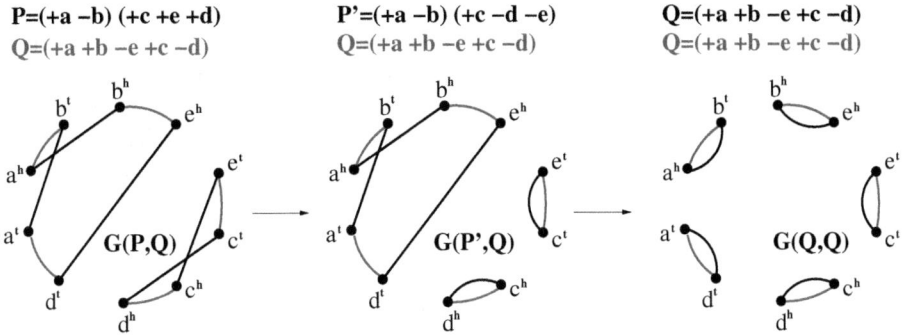

Fig. 2. A transformation between the genomes P and Q (defined in Fig. 1) and the corresponding transformation between the breakpoint graphs $G(P,Q)$ and $G(Q,Q)$ with a 2-break followed by a complete 3-break

A sequence of rearrangements transforming genome P into genome Q is called *transformation*. The length of a shortest transformation using k-breaks ($k = 2$ or 3) is called the *k-break distance* between genomes P and Q.

Any transformation of a genome P into a genome Q corresponds to a transformation of the breakpoint graph $G(P,Q)$ into the *identity breakpoint graph* $G(Q,Q)$ (Fig. 2). A close look at the increase in the number of black-gray cycles along this transformation, allows one to obtain a formula for the distance between genomes P and Q. Namely, the 2-break distance is related to the number $c(P,Q)$ of black-gray cycles in $G(P,Q)$, while the 3-break distance is related to the number $c^{odd}(P,Q)$ of *odd* black-gray cycles (i.e., black-gray cycles with an odd number of black edges):

Theorem 1 ([14]). *The 2-break distance between genomes P and Q is*

$$d_2(P,Q) = |P| - c(P,Q).$$

Theorem 2 ([3]). *The 3-break distance between genomes P and Q is*

$$d_3(P,Q) = \frac{|P| - c^{odd}(P,Q)}{2}.$$

3   Breakages and Optimal Transformations

Alekseyev and Pevzner [2] studied the number of breakages[3] in transformations. The number of breakages made by a rearrangement is defined as the actual number of edges changed by this rearrangement. A 2-break always makes 2 breakages, while a 3-break can make 2 or 3 breakages. A 3-break making 3 breakages is called *complete 3-break*. We treat non-complete 3-breaks as 2-breaks.

---

[3] In [2], the term *break* is used. We use *breakage* to avoid confusion with k-break rearrangements.

Alekseyev and Pevzner [2] proved that between any two genomes, there always exists a transformation that simultaneously has the shortest length and makes the smallest number of breakages. We call such transformations *optimal*.

For a 3-break r, we let $n_3(r) = 1$ if r makes 3 breakages (i.e., r is a complete 3-break) and $n_3(r) = 0$ otherwise. For a transformation t, we further define

$$n_2(t) = \sum_{r \in t} (1 - n_3(r)) \qquad \text{and} \qquad n_3(t) = \sum_{r \in t} n_3(r)$$

that is, $n_2(t)$ and $n_3(t)$ are correspondingly the number of 2-breaks and complete 3-breaks in t. If 2-breaks and complete 3-breaks are assigned respectively the weights 1 and α, then the weight of a transformation t is

$$W_\alpha(t) = n_2(t) + \alpha \cdot n_3(t).$$

It is easy to see that a transformation t has the length $n_2(t) + n_3(t) = W_1(t)$ and makes $2 \cdot n_2(t) + 3 \cdot n_3(t) = 2 \cdot W_{3/2}(t)$ breakages overall. Therefore, a transformation is optimal if and only if it simultaneously minimizes $W_1(t)$ and $W_{3/2}(t)$. We generalize this result in Section 4 by showing that $3/2$ can be replaced with any $\alpha \in (1, 2)$.

For a rearrangement r applied to a breakpoint graph, let $\Delta_r c^{odd}$ and $\Delta_r c^{even}$ be the resulting increase in the number of respectively odd and even black-gray cycles, respectively. Clearly, $\Delta_r c^{odd} + \Delta_r c^{even} = \Delta_r c$ gives the increase in the total number of black-gray cycles.

Lemma 1. *For any 3-break r,*

- $|\Delta_r c| \leq 1 + n_3(r)$;
- $\Delta_r c^{odd}$ *is even and* $|\Delta_r c^{odd}| \leq 2$;
- $|\Delta_r c^{even}| \leq 1 + n_3(r)$.

Proof. A 3-break r operating on black edges in the breakpoint graph $G(P, Q)$ destroys at least one and at most three black-gray cycles. On the other hand, it creates at least one and at most three new black-gray cycles. Therefore, $|\Delta_r c| \leq 3 - 1 = 2$. Similarly, if $n_3(r) = 0$, then $|\Delta_r c| \leq 2 - 1 = 1$.

By similar arguments, we also have $|\Delta_r c^{odd}| \leq 3$ and $|\Delta_r c^{even}| \leq 3$.

Since the total number of black edges in destroyed and created black-gray cycles is the same, $\Delta_r c^{odd}$ must be even. Combining this with $|\Delta_r c^{odd}| \leq 3$, we conclude that $|\Delta_r c^{odd}| \leq 2$.

If $\Delta_r c^{even} = 3$, then the destroyed cycles must be odd, implying that $\Delta_r c^{odd} = -2$. However, it is not possible for a 3-break to destroy two cycles and create three new cycles. Hence, $\Delta_r c^{even} \neq 3$. Similarly, $\Delta_r c^{even} \neq -3$, implying that $|\Delta_r c^{even}| \leq 2$. If $n_3(r) = 0$ (i.e., r is a 2-break), similar arguments imply $|\Delta_r c^{even}| \leq 1$. $\square$

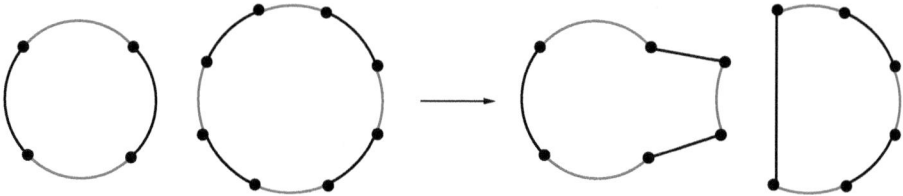

Fig. 3. A 3-break r with $\Delta_r c^{odd} = 2$ and $\Delta_r c^{even} = -2$, transforming two even black-gray cycles into two odd black-gray cycles. Such 3-breaks may appear in shortest transformations (Lemma 2) but not in optimal ones (Theorem 3).

Lemma 2. *A transformation t between two genomes is shortest if and only if $\Delta_r c^{odd} = 2$ for every $r \in t$. Furthermore, if t is a shortest transformation between two genomes, then for every $r \in t$,*

- *if $n_3(r) = 0$, then $\Delta_r c^{even} = -1$;*
- *if $n_3(r) = 1$, then $\Delta_r c^{even} = 0$ or -2.*

Proof. A transformation t of a genome P into a genome Q increases the number of odd black-gray cycles from $c^{odd}(P, Q)$ in $G(P, Q)$ to $c^{odd}(Q, Q) = |P|$ in $G(Q, Q)$ with the total increase of $|P| - c^{odd}(P, Q) = 2 \cdot d_3(P, Q)$. By Lemma 1, $\Delta_r c^{odd} \le 2$ for every $r \in t$ and thus

$$2 \cdot d_3(P, Q) = \sum_{r \in t} \Delta_r c^{odd} \le \sum_{r \in t} 2 = 2 \cdot |t|,$$

implying that $|t| = d_3(P, Q)$ (i.e., t is a shortest transformation) if and only if $\Delta_r c^{odd} = 2$ for every $r \in t$.

Now let t be a shortest transformation and thus $\Delta_r c^{odd} = 2$ for every $r \in t$. For a 2-break r to have $\Delta_r c^{odd} = 2$, it must be applied to an even black-gray cycle and split it into two odd black-gray cycles. Thus any such r also decreases the number of even black-gray cycles by 1, i.e., $\Delta_r c^{even} = -1$.

If a complete 3-break r has $\Delta_r c^{odd} = 2$, then $\Delta_r c^{even} = \Delta_r c - \Delta_r c^{odd} \le 2 - 2 = 0$. By Lemma 1, we also have $\Delta_r c^{even} \ge -2$ and $\Delta_r c^{even} \ne -1$, implying that $\Delta_r c^{even} = 0$ or -2. $\square$

By the definition, any optimal transformation is necessarily shortest. However, not every shortest transformation is optimal. The following theorem characterizes optimal transformations within the shortest transformations:

Theorem 3. *A shortest transformation t between two genomes is optimal if and only if for any $r \in t$, $\Delta_r c^{even} \ne -2$.*

Proof. Let t be a shortest transformation between two genomes. By Lemma 2, $n_3(t) = u + v$ where u is the number of complete 3-breaks with $\Delta_r c^{even} = 0$ and v is the number of complete 3-breaks with $\Delta_r c^{even} = -2$ (Fig. 3).

Hence $n_2(t)$ 2-breaks and $n_3(t) = u + v$ complete 3-breaks transform $G(P, Q)$ into $G(Q, Q)$ with $|P| = |Q|$ trivial black-gray cycles, which all are odd. By

Lemma 2, for the increase in the number of odd and even black-gray cycles in the breakpoint graph, we have:

$$\begin{cases} c^{odd}(P,Q) + 2(n_2(t) + u + v) = |P|, \\ c^{even}(P,Q) - n_2(t) - 2v = 0, \end{cases}$$

implying that

$$W_{3/2}(t) = n_2(t) + \frac{3}{2}(u+v)$$

$$= c^{even}(P,Q) - 2v + \frac{3}{2}\left(\frac{|P| - c^{odd}(P,Q)}{2} - c^{even}(P,Q) + 2v \right)$$

$$= c^{even}(P,Q) + \frac{3}{2}\left(\frac{|P| - c^{odd}(P,Q)}{2} - c^{even}(P,Q) \right) + v,$$

which is minimal if and only if $v = 0$, i.e., $\Delta_r c^{even} \neq -2$ for any $r \in t$.   □

Lemma 2 and Theorem 3 imply:

Corollary 1. *A transformation t between two genomes is optimal if and only if for any $r \in t$,*

- *if $n_3(r) = 0$, then $\Delta_r c^{odd} = 2$ and $\Delta_r c^{even} = -1$;*
- *if $n_3(r) = 1$, then $\Delta_r c^{odd} = 2$ and $\Delta_r c^{even} = 0$.*

Theorem 4. *A transformation t between genomes P and Q is optimal if and only if*

$$\begin{cases} n_2(t) = c^{even}(P,Q), \\ n_3(t) = \frac{|P| - c^{odd}(P,Q)}{2} - c^{even}(P,Q). \end{cases} \quad (1)$$

Proof. Let t be an optimal transformation between genomes P and Q. Then with $n_2(t)$ 2-breaks and $n_3(t)$ complete 3-breaks, it transforms $G(P,Q)$ into $G(Q,Q)$ with $|P| = |Q|$ trivial black-gray cycles, which are all odd. By Corollary 1, we have

$$\begin{cases} c^{odd}(P,Q) + 2(n_2(t) + n_3(t)) = |P|, \\ c^{even}(P,Q) - n_2(t) = 0, \end{cases}$$

implying formulae (1).

Vice versa, a transformation t between genomes P and Q, satisfying (1), has the length $n_2(t) + n_3(t) = \frac{|P| - c^{odd}(P,Q)}{2} = d_3(P,Q)$, implying that t is a shortest transformation. By Lemma 2, $\Delta_r c^{even} = -1$ for every 2-break $r \in t$ and $\Delta_r c^{even} = 0$ or -2 for every complete 3-break $r \in t$. Let v be the number of complete 3-breaks $r \in t$ with $\Delta_r c^{even} = -2$. Then the increase in the number of even black-gray cycles along t is

$$-c^{even}(P,Q) = -n_2(t) - 2v = -c^{even}(P,Q) - 2v,$$

implying that $v = 0$ and thus t is optimal by Theorem 3.   □

Theorem 4 implies that for some genomes, every optimal transformation consists entirely of complete 3-breaks:

Corollary 2. *For genomes P and Q with $c^{even}(P,Q) = 0$, every optimal transformation t has $n_2(t) = 0$ and thus consists entirely of complete 3-breaks.*

Corollary 3. *For an optimal transformation t between genomes P and Q,*

$$W_\alpha(t) = c^{even}(P,Q) + \alpha \cdot \left(\frac{|P| - c^{odd}(P,Q)}{2} - c^{even}(P,Q) \right).$$

4   Weighted Multi-break Distance

Let $T(P,Q)$ be the set of all transformations between genomes P and Q. For a real number $\alpha \geq 0$, we define the weighted distance $D_\alpha(P,Q)$ between genomes P and Q as

$$D_\alpha(P,Q) = \min_{t \in T(P,Q)} W_\alpha(t),$$

that is, the minimum possible weight of a transformation between P and Q.

Two important examples of the weighted distance are the "unweighted" distance $D_1(P,Q) = d_3(P,Q)$ and the distance $D_{3/2}(P,Q)$ equal the half of the minimum number of breakages in a transformation between genomes P and Q. By the definition of an optimal transformation, we have $D_{3/2}(P,Q) = W_{3/2}(t_0)$, where t_0 is an optimal transformation between genomes P and Q. Below we prove that $D_\alpha(P,Q) = W_\alpha(t_0)$ for any $\alpha \in (1,2]$.

Theorem 5. *For $\alpha \in (1,2]$,*

$$D_\alpha(P,Q) = W_\alpha(t_0),$$

where t_0 is any optimal transformation between genomes P and Q.

Furthermore, for $\alpha \in (1,2)$, if $D_\alpha(P,Q) = W_\alpha(t)$ for a transformation t between genomes P and Q, then t is an optimal transformation.

Proof. Let t be any transformation and t_0 be any optimal transformation between genomes P and Q.

We classify all possible changes in the number of even and odd black-gray cycles resulted from a single rearrangement r. By Lemma 1, $\Delta_r c^{odd}$ may take only values $-2, 0, 2$, while $|\Delta_r c| = |\Delta_r c^{odd} + \Delta_r c^{even}| \leq 1$ (if r is a 2-break) or ≤ 2 (if r is a complete 3-break). The table below lists the possible values of $\Delta_r c^{odd}$ and $\Delta_r c^{even}$, satisfying these restrictions, along with the amount of rearrangements of each particular type in t, denoted x_i for 2-breaks and y_j for complete 3-breaks.

	$n_3(r) = 0$					$n_3(r) = 1$										
$\Delta_r c^{odd}$	0	0	0	-2	2	0	0	0	0	0	2	2	2	-2	-2	-2
$\Delta_r c^{even}$	0	1	-1	1	-1	0	1	-1	2	-2	0	-1	-2	0	1	2
amount in t	x_1	x_2	x_3	x_4	x_5	y_1	y_2	y_3	y_4	y_5	y_6	y_7	y_8	y_9	y_{10}	y_{11}

For the transformation t, we have

$$\begin{cases} n_2(t) = x_1 + x_2 + x_3 + x_4 + x_5, \\ n_3(t) = y_1 + y_2 + y_3 + y_4 + y_5 + y_6 + y_7 + y_8 + y_9 + y_{10} + y_{11}. \end{cases}$$

Calculating the total increase in the number of odd and even black-gray cycles along t, we have

$$\begin{cases} -2x_4 + 2x_5 + 2y_6 + 2y_7 + 2y_8 - 2y_9 - 2y_{10} - 2y_{11} = |P| - c^{odd}(P, Q), \\ x_2 - x_3 + x_4 - x_5 + y_2 - y_3 + 2y_4 - 2y_5 - y_7 - 2y_8 + y_{10} + 2y_{11} = -c^{even}(P, Q). \end{cases}$$

Theorem 4 further implies

$$\begin{cases} n_2(t_0) = -x_2 + x_3 - x_4 + x_5 - y_2 + y_3 - 2y_4 + 2y_5 + y_7 + 2y_8 - y_{10} - 2y_{11}, \\ n_3(t_0) = x_2 - x_3 + y_2 - y_3 + 2y_4 - 2y_5 + y_6 - y_8 - y_9 + y_{11}. \end{cases}$$

Now we can evaluate the difference between the weights of t and t_0 as follows:

$$\begin{aligned} W_\alpha(t) - W_\alpha(t_0) &= n_2(t) - n_2(t_0) + \alpha \cdot (n_3(t) - n_3(t_0)) \\ &= x_1 + 2x_2 + 2x_4 + y_2 - y_3 + 2y_4 - 2y_5 - y_7 - 2y_8 + y_{10} + 2y_{11} \\ &\quad + \alpha \cdot (-x_2 + x_3 + y_1 + 2y_3 - y_4 + 3y_5 + y_7 + 2y_8 + 2y_9 + y_{10}) \\ &= x_1 + (2 - \alpha) \cdot x_2 + \alpha \cdot x_3 + 2x_4 + \alpha \cdot y_1 + y_2 + (2\alpha - 1) \cdot y_3 \\ &\quad + (2 - \alpha) \cdot y_4 + (3\alpha - 2) \cdot y_5 + (\alpha - 1) \cdot y_7 + (2\alpha - 2) \cdot y_8 \\ &\quad + 2\alpha \cdot y_9 + (\alpha + 1) \cdot y_{10} + 2 \cdot y_{11}. \end{aligned}$$

Since $\alpha \in (1, 2]$ and $x_i, y_j \geq 0$, all summands in the last expression are nonnegative and thus $W_\alpha(t) - W_\alpha(t_0) \geq 0$. Since t is an arbitrary transformation, we have

$$D_\alpha(P, Q) = W_\alpha(t_0).$$

For $\alpha \in (1, 2)$, if $D_\alpha(P, Q) = W_\alpha(t)$ then $W_\alpha(t) - W_\alpha(t_0) = 0$, implying that only x_5 and y_6 (appearing with zero coefficients in the expression for $W_\alpha(t) - W_\alpha(t_0)$) can be nonzero and thus t is optimal by Corollary 1.   □

5   Discussion

We proved that for $\alpha \in (1, 2]$, the minimum-weight transformations include the optimal transformations (Theorem 5) that may entirely consist of transposition-like operations (modelled as complete 3-breaks) (Corollary 2). Therefore, the corresponding weighted genomic distance does not actually impose any bound on the proportion of transpositions.

For $\alpha \in (1, 2)$, we proved even a stronger result that the minimum-weight transformations coincide with the optimal transformations (Theorem 5). As a consequence we have that a particular choice of $\alpha \in (1, 2)$ imposes no restrictions for the minimum-weight transformations as compared to other values of α from this interval. The value $\alpha = 3/2$ then proves that the optimal transformations coincide with those that at the same time have the shortest length and make the smallest number of breakages, studied by Alekseyev and Pevzner [2]. We further characterized the optimal transformations within the shortest transformations (i.e., the minimum-weight transformations for $\alpha = 1$) by showing that the optimal transformations avoid one particular type of rearrangements (Theorem 3, Fig. 3).

It is worth to mention that the weighted genomic distance with $\alpha \geq 2$ is useless, since it allows (for $\alpha = 2$) or even promotes (for $\alpha > 2$) replacement of every complete 3-break with two equivalent 2-breaks, thus eliminating complete 3-breaks at all.

The extension of our results to the case of linear genomes will be published elsewhere.

References

1. Alekseyev, M.A.: Multi-Break Rearrangements and Breakpoint Re-uses: from Circular to Linear Genomes. Journal of Computational Biology 15(8), 1117–1131 (2008)
2. Alekseyev, M.A., Pevzner, P.A.: Are There Rearrangement Hotspots in the Human Genome? PLoS Computational Biology 3(11), e209 (2007)
3. Alekseyev, M.A., Pevzner, P.A.: Multi-Break Rearrangements and Chromosomal Evolution. Theoretical Computer Science 395(2-3), 193–202 (2008)
4. Bader, M., Ohlebusch, E.: Sorting by weighted reversals, transpositions, and inverted transpositions. Journal of Computational Biology 14(5), 615–636 (2007)
5. Bafna, V., Pevzner, P.A.: Genome rearrangements and sorting by reversals. SIAM Journal on Computing 25, 272–289 (1996)
6. Blanchette, M., Kunisawa, T., Sankoff, D.: Parametric genome rearrangement. Gene 172(1), GC11–GC17 (1996)
7. Christie, D.A.: Sorting permutations by block-interchanges. Information Processing Letters 60(4), 165–169 (1996)
8. Elias, I., Hartman, T.: A 1.375-approximation algorithm for sorting by transpositions. IEEE/ACM Transactions on Computational Biology and Bioinformatics 3, 369–379 (2006)
9. Eriksen, N.: $(1 + \epsilon)$-Approximation of Sorting by Reversals and Transpositions. In: Gascuel, O., Moret, B.M.E. (eds.) WABI 2001. LNCS, vol. 2149, pp. 227–237. Springer, Heidelberg (2001)
10. Fertin, G., Labarre, A., Rusu, I., Tannier, E.: Combinatorics of Genome Rearrangements. The MIT Press, Cambridge (2009)
11. Hannenhalli, S., Pevzner, P.: Transforming men into mouse (polynomial algorithm for genomic distance problem). In: Proceedings of the 36th Annual Symposium on Foundations of Computer Science, pp. 581–592 (1995)
12. Hannenhalli, S., Pevzner, P.A.: Transforming Cabbage into Turnip (polynomial algorithm for sorting signed permutations by reversals). In: Proceedings of the 27th Annual ACM Symposium on the Theory of Computing, pp. 178–189 (1995); full version appeared in Journal of ACM 46, 1–27 (1995)
13. Radcliffe, A.J., Scott, A.D., Wilmer, E.L.: Reversals and Transpositions Over Finite Alphabets. SIAM J. Discrete Math. 19, 224–244 (2005)
14. Yancopoulos, S., Attie, O., Friedberg, R.: Efficient sorting of genomic permutations by translocation, inversion and block interchange. Bioinformatics 21, 3340–3346 (2005)

PSAR: Measuring Multiple Sequence Alignment Reliability by Probabilistic Sampling
(Extended Abstract)

Jaebum Kim[1] and Jian Ma[1,2]

[1] Institute for Genomic Biology, University of Illinois at Urbana-Champaign,
Urbana, IL 61801, U.S.A.
{jkim63,jianma}@illinois.edu
[2] Department of Bioengineering, University of Illinois at Urbana-Champaign,
Urbana, IL 61801, U.S.A.

1 Introduction

Multiple sequence alignment (MSA), which is of fundamental importance for comparative genomics, is a difficult problem and error-prone. Therefore, it is essential to measure the reliability of the alignments and incorporate it into downstream analyses. Many studies have been conducted to find the extent, cause and effect of the alignment errors [4], and to heuristically estimate the quality of alignments without using the true alignment, which is unknown [2]. However, it is still unclear whether the heuristically chosen measures are general enough to take into account all alignment errors. In this paper, we present a new alignment reliability score, called PSAR (Probabilistic Sampling-based Alignment Reliability) score.

2 Methods

The PSAR score is computed based on suboptimal alignments that are sampled from the posterior probability distribution of alignments, which is approximated by pairwise comparisons between each sequence and the rest of an input MSA. Specifically, given an input MSA (Figure 1A), PSAR selects one sequence at a time and makes a sub-alignment by leaving the chosen sequence out of the MSA. To re-compare the left-out sequence with the sub-alignment, all gaps in the left-out sequence and all columns in the sub-alignment that consist of only gaps are removed (Figure 1B). The pairwise comparison of the pre-processed left-out sequence and sub-alignment is based on a special type of a pair hidden Markov model (pair-HMM) that emits columns of an MSA given the left-out sequence and the sub-alignment. To sample suboptimal alignments, PSAR first constructs dynamic programming (DP) tables by using the forward algorithm [1] based on the pair-HMM (Figure 1C), and then traces back through the DP tables based on a probabilistic choice at each step [1] (Figure 1D-E). The reliability score of an input MSA is computed by measuring the consistency with the suboptimal alignments.

V. Bafna and S.C. Sahinalp (Eds.): RECOMB 2011, LNBI 6577, pp. 134–135, 2011.
© Springer-Verlag Berlin Heidelberg 2011

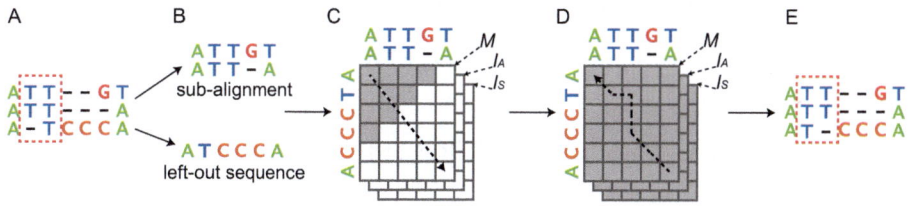

Fig. 1. Procedure of probabilistic sampling

3   Results

We evaluated the performance of our method PSAR in comparison with the GUIDANCE method [2], using the simulated data. The GUIDANCE method computes the alignment certainty score by using perturbed MSAs (alignment samples) that are generated based on perturbed phylogenetic trees. This evaluation focused on measuring how accurately each method classifies pairs of aligned characters in an input MSA into reliable or unreliable classes. By varying cut-off values, we counted the number of true positive pairs that are labeled as a reliable one as well as aligned in the true alignment, and the number of false positive pairs that are also labeled as a reliable one but not aligned in the true alignment. We found that the performance of the PSAR score is superior to the GUIDANCE score across multiple settings with different input MSAs. We also applied PSAR to compute the alignment reliability scores of the alignments of 16 amniota vertebrates for the upstream region of a gene GRIA2 downloaded from the Ensembl database. We found two regions with relatively low PSAR scores but high conservation scores by the PhastCons program [3]. We can obtain more accurate conservation scores of the above suspicious regions by analyzing the conservation scores computed from the set of suboptimal alignments as an additional information.

4   Conclusion

We propose a new probabilistic sampling-based alignment reliability score, and find that our approach is superior to existing ones. This suggests that the suboptimal alignments are highly informative source for assessing alignment reliability.

References

1. Durbin, R., Eddy, S.R., Krogh, A., Mitchison, G.: Biological sequence analysis: probabilistic models of proteins and nucleic acids. Cambridge University Press, Cambridge (1998)
2. Penn, O., et al.: An alignment confidence score capturing robustness to guide tree uncertainty. Mol. Biol. Evol. 27(8), 1759–1767 (2010)
3. Siepel, A., et al.: Evolutionarily conserved elements in vertebrate, insect, worm, and yeast genomes. Genome Res. 15(8), 1034–1050 (2005)
4. Wong, K.M., et al.: Alignment uncertainty and genomic analysis. Science 319(5862), 473–476 (2008)

Pedigree Reconstruction Using Identity by Descent

Bonnie Kirkpatrick[1], Shuai Cheng Li[2], Richard M. Karp[3], and Eran Halperin[4]

[1] Electrical Engineering and Computer Sciences, University of California, Berkeley
and International Computer Science Institute, Berkeley
[2] International Computer Science Institute, Berkeley
[3] Electrical Engineering and Computer Sciences, University of California, Berkeley
and International Computer Science Institute, Berkeley
[4] Tel Aviv University, Tel Aviv,
Israel and International Computer Science Institute, Berkeley
bbkirk@eecs.berkeley.edu, scli@icsi.berkeley.edu,
karp@cs.berkeley.edu, heran@icsi.berkeley.edu

Abstract. Can we find the family trees, or pedigrees, that relate the haplotypes of a group of individuals? Collecting the genealogical information for how individuals are related is a very time-consuming and expensive process. Methods for automating the construction of pedigrees could stream-line this process. While constructing single-generation families is relatively easy given whole genome data, reconstructing multi-generational, possibly inbred, pedigrees is much more challenging.

This paper addresses the important question of reconstructing monogamous, regular pedigrees, where pedigrees are regular when individuals mate only with other individuals at the same generation. This paper introduces two multi-generational pedigree reconstruction methods: one for inbreeding relationships and one for outbreeding relationships. In contrast to previous methods that focused on the independent estimation of relationship distances between every pair of typed individuals, here we present methods that aim at the reconstruction of the entire pedigree. We show that both our methods out-perform the state-of-the-art and that the outbreeding method is capable of reconstructing pedigrees at least six generations back in time with high accuracy.

The two programs are available at http://cop.icsi.berkeley.edu/cop/

1 Introduction

Pedigrees, or family trees, are important in computer science and in genetics. The pedigree graph encodes all the possible Mendelian inheritance options, and provides a model for computing inheritance probabilities for haplotype or genotype data. Even thirty years after the development of some of the first pedigree algorithms [19,11], pedigree graphical models continue to be a challenging graphical model to work with. Known algorithms for inheritance calculations

V. Bafna and S.C. Sahinalp (Eds.): RECOMB 2011, LNBI 6577, pp. 136–152, 2011.

are either exponential in the number of individuals or exponential in the number of loci [20]. There have been numerous and notable attempts to increase the speed of these calculations [30,1,12,7,14,22,9]. Recent work from statistics has focused on fast and efficient calculations of linkage that avoid the full inheritance calculations [5,36]. Recent contributions to genetics from pedigree calculations include fine-scale recombination maps for humans [8], discovery of regions linked to Schizophrenia [25], discovery of regions linked to rare Mendelian diseases [26], and insights into the relationship between cystic fibrosis and fertility [13].

Manual methods for constructing human pedigree graphs are very tedious. It requires careful examination of genealogical records, including marriage records, birth dates, death dates, and parental information found in birth certificates. Medical researchers then must carefully check records for consistency, for instance making sure that two married individuals were alive at the same time and making sure that children were conceived while the parents were alive. This process is very time consuming. Despite the care taken, there are sometimes mistakes [4,24,32].

For constructing non-human pedigrees, of diploid organisms, it is often impossible to know the pedigree graph since there are no genealogical records [2,6]. In this case it is particularly important to develop methods of automatically generating pedigrees from genomic data.

The problem of reconstructing pedigrees from haplotype or genotype data is not new. The oldest such method that the authors know of is due to Thompson [35]. Her approach is essentially a structured machine learning approach where the aim is to find the pedigree graph that maximizes the probability of observing the data, or likelihood. (This approach is directly analogous to maximum likelihood methods for phylogenetic reconstruction which also try to find the phylogenetic tree that maximize the likelihood.) Notice that this method reconstructs both the pedigree graph and the ancestral haplotypes which is a very time-consuming step. Thus, this approach is limited to extremely small families, perhaps 4-8 people, since the algorithms for computing the likelihood of a fixed pedigree graph are exponential [20] and there are an exponential number of pedigree graphs to consider [33].

The current state-of-the-art method is an HMM-based approximation of the number of meioses separating a pair of individuals [31]. This approach dispenses with any attempt to infer haplotypes of ancestral individuals, and instead focuses on the number of generations that separate a pair of individuals. In this approach the hidden states of the HMM represent the identity-by-descent (IBD) of a pair of individuals. Two individuals are identical-by-descent for a particular allele if they each have a copy of the same ancestral allele. The probability of the haplotype data is tested against a particular type of relationship. The main draw-back of this approach is that it may estimate a set of pair-wise relationships that are inconsistent with a single pedigree relating all the individuals.

Thatte and Steel [34] examined the problem of reconstructing arbitrary pedigree graphs from a synthetic model of the data. Their method used an HMM model for the ancestry of each individual to show that the pedigree can be reconstructed only if the sequences are sufficiently long and infinitely dense. Notice that this paper uses an unrealistic model of recombination where every individual passes on a trace of their haplotypes to all of their descendants. Kirkpatrick [17] introduced a more simple, more general version of the reconstruction algorithm introduced by Thatte and Steel.

Attempts to construct sibling relationships are known to be NP-hard, and attempts to infer pedigrees by reconstructing ancestral haplotypes are be NP-hard. Two combinatorial versions of the sibling relationship problem were proven to be NP-hard, both whole- and half-sibling problem formulations [2,29]. If ancestral haplotypes are reconstructed in the process of inferring a pedigree, as in Thompson's structured machine learning approach, then the inheritance probabilities of data must be computed on the pedigree graph. For instance, we might want to compute the likelihood, or the probability of observing the data given inheritance in the pedigree. This calculation is NP-hard for both genotype [28,21] and haplotype [16] data. This means that any efficient pedigree reconstruction method will need to find ways to avoid both these hardness problems.

Our contribution to pedigree reconstruction is two algorithms that avoid the exponential likelihood calculations. We do this by specifically *not* reconstructing ancestral haplotypes and by *not* trying to optimize sibling groups. We use estimates of the length of genomic regions that are shared identical-by-descent. In two related individuals, a region of the genome is identical-by-descent (IBD) if and only if a single ancestral haplotype sequence was the source of the sequence inherited in the two individuals. The length of IBD regions gives a statistic that accurately detects sibling relationships at multiple generations. We have two algorithms: one for constructing inbred pedigrees (CIP) and one for constructing outbred pedigrees (COP). For our outbreeding algorithm the statistic is testable in polynomial time. For our inbreeding algorithm, the statistic is computable in time dependent on the number of meioses in the predicted pedigree. Our outbreeding method works to reconstruct at least six generations back in time. Both methods are more accurate than the state-of-the-art method by Stankovich, et al [31].

The remainder of the paper is organized into sections on pair-wise IBD, practical reconstruction algorithms, and results. The section on pair-wise IBD considers the expected length of a genomic region shared between a pair of individuals. This establishes the limits of reconstruction methods that are based only on pair-wise relationships. The section on practical algorithms introduces our CIP and COP algorithms, which go beyond pair-wise relationships and actually use transitive relationship information to infer a pedigree graph. The results section considers simulation results and results running the algorithm on several HapMap Phase III populations.

2   Background

A pedigree graph has diploid individuals as nodes and edges from parents to children. The edges are typically implicitly directed down from parent to child, without drawing the actual direction arrow on the edge. Circle nodes are females, boxes are females. Let the generations be numbered backwards in time, with larger numbers being older generations. Let g be the number of generations of individuals in the graph. For example, if $g = 1$, then we are discussing only the extant individuals, whereas if $g = 2$ the graph contains the extant individuals and their parents.

In this paper, we will only consider monogamous, regular pedigrees, where a pedigree is *regular* when individuals only mate with other individuals at the same generation. Of course, a pedigree is *monogamous* if and only if every individual mates with at most one other individual, so that there are no half-siblings.

Recombination along the genome is typically modeled as a Poisson process, where the distance between recombination breakpoints is drawn from an exponential distribution. The mean of the exponential is a function of the recombination rate [10,3]. This is a model for recombination without interference, where interference means that the presence of one recombination breakpoint suppresses the occurrence of breakpoints in neighboring regions of the sequence [23]. The simulation and experimental results seem to support the use of the simplifying assumption made by using the Poisson model for recombination, however relaxing this assumption might be one way to improve on the model.

3   A Lower Bound for Pair-Wise Relationships with Out-Breeding

In order to shed light on the problem we first provide a lower bound on the best that one could do in pedigree reconstruction. Stankovich, et al [31] have been able to detect up to 3rd cousins (or relationships of 8 total meioses). We claim that this should be near optimal in the case of an infinite population size. Notice that in the infinite population size, there is no inbreeding. Therefore, the graph relating people has a path-like subgraph connecting every pair of individuals (i.e. the subgraph is a path having exactly two founders whose adjacent edges can be contracted to form a simple path). This implies that in order to estimate pedigree graphs that are more accurate than the conglomerate of a set of pairwise relationship estimates, we need to exploit features of the relationships that are not simply outbred paths between pairs of individuals. Specifically, we need to consider sets of individuals and the graphs that connect them, and we need to consider graphs, not paths, that connect pairs of individuals. This means that we need to be considering inbreeding and transitive relationships (i.e. person a is related to person c through person b).

Now, we derive a lower bound on the pair-wise outbred relationships. In an infinite population, consider two individuals i, and j, where their most recent common ancestor is g generations ago. For instance, if $g = 2$ they are siblings. Note that they have two common ancestors in this case. For general g, each individual has 2^g ancestors, where exactly two of them are shared across i and j; this is where we use the fact that the population is infinite and monogamous, since the probability of having more than two shared ancestors is zero and monogamy ensures that there are at least two shared ancestors.

Each of the ancestors of i and j has two haploids. Each of the haploids arrived from a different pedigree. Consider only the haploids that arrived from the shared pedigree (the case $g = 2$, i.e. siblings, is different since there there is IBD sharing on both haploids of i and j). These haploids of i and j are generated by a random walk over the ancestors of i and j in the gth generation. The total number of *haploid* ancestors in that generation is 2^g for each of i and j. Out of those, four are shared across i and j (two shared ancestors, each has two haploids). Let k be the number of meioses separating individuals i and j, where $k = 2(g - 1)$. For this reason, the expected number of bases shared between i and j is $\frac{4L}{2^k} = \frac{L}{2^{k-2}}$, where L is the length of the genome.

On the other hand, we can calculate the average length of a shared region between the two haploids. The number of recombinations across all generations is Poisson distributed with parameter krL, where r is the recombination rate, L is the length of the genome. Now, the length, X, of a shared region that originated from one of the four shared haploids is $X_1 + X_2$ where $X_i \sim exp(kr)$. Notice that X_i is the length of the IBD region conditioned on starting at an IBD position. Therefore from an arbitrary IBD position, we need to consider the length of the IBD region before arriving at that position, X_1, and the length after that position, X_2. So the expected length, $E[X]$, is $\frac{2}{kr}$. Since the probability to move from one shared haploid to another is negligible, we get that this is the expected length of a shared region.

Now, if t_k is the expected number of regions shared between two individuals separated by k meioses, we know that $t_k \frac{2}{kr} = \frac{L}{2^{k-2}}$, and therefore, $t_k = \frac{krL}{2^{k-1}}$, where rL is the expected number of recombinations after one generation. Therefore, $t_{10} < 1$ since $rL = 30$, and it is impossible to detect a pair-wise relationship with high probability between 4th cousins.

This is not to say that it is impossible to accurately construct a 6-generation pedigree, only that it is impossible to accurately construct a 6-generation pedigree from pair-wise relationship estimates. As noted earlier, to get accuracy on deep pedigrees, we need to consider relationships on sets of individuals, inbreeding and transitive relationships.

4   Algorithms for Constructing Pedigrees

The principle innovation of this method is to reconstruct pedigree graphs *without* reconstructing the ancestral haplotypes. This is the innovation that allows this algorithm to avoid the exponential calculation associated with inferring ancestral haplotypes, and allows the algorithm to be efficient.

The approach we employ is a *generation-by-generation* approach. We reconstruct the pedigree backwards in time, one generation at a time. Of course if we make the correct decisions at each generation, then we will construct the correct pedigree. However, since we use the predictions at previous generations to help us make decisions about how to reconstruct subsequent generations, we can accumulate errors as the algorithm proceeds backwards in time.

Given a set of extant individuals with haplotype information available, we want to reconstruct their pedigree. We construct the pedigree recursively, one generation at a time. For example, the first iteration consists of deciding which of the extant individuals are siblings. The next iteration would determine which of the parents are siblings (yielding cousin relationships on the extant individuals).

At each generation, we consider a *compatibility* graph on the individuals at generation g, where the nodes are individuals and the edges are between pairs of individuals that could be siblings. The presence or absence of edges will be determined by a statistical test, discussed later. For the moment, assume that we have such a graph.

Now, we will find sibling sets in the compatibility graph. We do this by partitioning the graph into disjoint sets of vertices with the property that each set in the partition has many edges connecting its vertices while there are few edges connecting vertices from separate sets in the partition. Of course any partitioning method can be used, and later we will introduce a partitioning heuristic. For rhetorical purposes, we will now discuss how to use a Max-Clique algorithm to partition the graph. The graph is partitioned by the following iterative procedure. Iteratively, find the Max-Clique, for all the individuals in the Max-Clique, make them siblings, by creating monogamous parents in generation $g + 1$. Remove those Max-Clique individuals from the graph. Now, we can iterate, by finding the next Max-Clique and again creating a sibling group, etc.

Next, we consider how to create the edges in the compatibility graph. Let individuals k and l be in generation g. Recall that we have an edge in the compatibility graph if k and l could be siblings. To determine this, we look at pairs i and j of descendants of k and l, respectively. Let $\hat{s}_{ij}$ be the observed average length of shared segments between haplotyped individuals i and j. This can be computed directly from the given haplotype data and need only be computed once as a preprocessing step for our algorithm. Now, for a pair of individuals k and l in the oldest reconstructed generation, $X_{i,j}$ is the random variable for the length of a shared region for individuals i, j under the pedigree model that we have constructed so far. Later, we will discuss two models for $X_{i,j}$. For now, consider the test for the edge (k, l)

$$v_{k,l} = \frac{1}{|D(k)||D(l)|} \sum_{i \in D(k)} \sum_{j \in D(l)} \frac{(\hat{s}_{ij} - \mathbb{E}[X_{ij}])^2}{var(X_{ij})} \tag{1}$$

where $D(k)$ is the set of extant individuals descended from ancestor k, and $D(k)$ is known based on the pedigree we have constructed up to this point. We compute $v_{k,l}$, making edges when $v_{k,l} < c$ for all k, l in the oldest generation, g, for some threshold c. Notice that this edge test is similar to a χ^2 test but does not have the χ^2 null distribution, because the term in the sum will not actually be normally distributed. We choose the the threshold, c, empirically by simulating many pedigrees and choosing the threshold which provides the best reconstruction accuracy.

Now, we need to calculate $\mathbb{E}[X_{i,j}]$ and $Var(X_{i,j})$. We propose two models for the random variable X_{ij}, the outbred model (COP) and the inbred model (CIP). The outbred, COP, model only allows prediction of relationships between two individuals that are unrelated at all previous generations. The inbred model, CIP, allows prediction of a relationship that relates two individuals already related in a previous generation.

4.1   IBD Model for Constructing Outbred Pedigrees (COP)

To obtain the edges in the graph, we do a test for relationship-pairs of the form shown in Figure 1. If a pair of extant individuals are related at generation g via a single ancestor at that generation, then the length of the regions they share IBD will be distributed according to the sum of two exponential variables, specifically, $exp(2(g-1)\lambda)$. This is the waiting time, where time corresponds to genome length, for a random walk to leave the state of IBD sharing. So, we have $X_{ij} = X_1 + X_2$ where $X_i \sim exp(2(g-1)\lambda)$. This means that we can quickly analytically compute $\mathbb{E}[X_{ij}]$ and $Var(X_{ij})$.

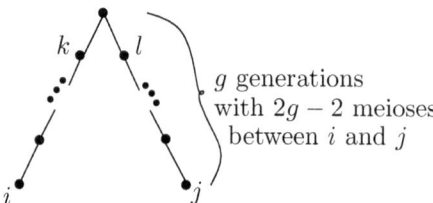

Fig. 1. Pair of Individual Related at Generation g. To test whether individuals k and l are siblings at generation g, we look at the distribution on the length of genetic regions shared IBD between all pairs of i and j descended from k and l, respectively.

4.2   IBD Model for Constructing Inbred Pedigrees (CIP)

We will do a random-walk simulation to allow for inbreeding, resulting in an algorithm with exponential running-time. The number of states in the IBD process is exponential in the number of meioses in the graph relating individuals i and j. So, the random-walk simulation is exponential in the size of the inferred pedigree.

For individuals k and l in generation g, and their respective descendants i and j, we consider the case given in Figure 2. The triangles represent the inferred sub-pedigree containing all the descendants of the individual at the point of the triangle, and individuals at the base of the triangle are extant individuals. Note that the triangles may overlap, indicating shared ancestry at an earlier generation (i.e. inbreeding).

Brief Description of the IBD Simulation. Let $X_{i,j}$ be the length of a shared region based on the pedigree structure of the model. In order to estimate this quantity, we can sample random walks in the space of inheritance possibilities. Specifically, consider the inheritance of alleles at a single position in the genome. When there are n non-founder individuals, define an inheritance vector as a vector containing $2n$ bits, where each pair of bits, $2i$ and $2i + 1$, represents the

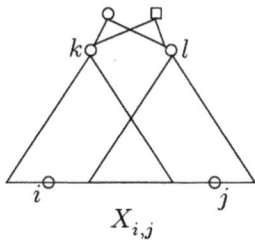

Fig. 2. Test Case. Specific individuals in the pedigree are indicated with either circles or squares. The triangle represents all the descendants of a particular individual. This represents the case where individuals i and j are cousins via the oldest generation.

grand-parental origin of individual i's two alleles. Specifically, bit $2i$ represents the maternal allele and is zero if the grand-paternal allele was inherited and is one otherwise. Similarly, bit $2i + 1$ represents the paternal allele of individual i. The set of possible inheritance vectors comprise the 2^{2n} vertices of a $2n$-dimensional hypercube, where n is the number of non-founders in the pedigree. A random walk on the hypercube represents the recombination process by choosing the inheritance vectors of neighboring regions of the genome.

Given an inheritance vector, we can model the length, in number of positions, of the genomic region that is inherited according to that inheritance vector. The end of that genomic region is marked by a recombination in some individual, and constitutes a change in the inheritance vector. The random walk on the hypercube models the random recombinations, while the length of genomic regions are modeled using an exponential distribution. This model is the standard Poisson model for recombinations. Details can be found below.

Poisson Process. Given a pedigree and individuals of interest i and j, we will compute the distribution on the length of shared regions. Here we mean sharing to be a contiguous region of the genome for which i and j have at least one IBD allele at each site.

We can model the creation of a single zygote (i.e. haplotype) as a Poisson process along the genome where the waiting time to the next recombination event is exponentially distributed with intensity $\lambda = -ln(1 - \theta)$ where θ is the probability of recombination per meiosis (i.e. per generation, per chromosome) between a pair of neighboring loci. For example, if we think of the genome as being composed of 3000 blocks with each block being 1MB in length and the recombination rate $\theta = 0.01$ between each pair of neighboring blocks, then we would expect 30 recombinations per meiosis, and the corresponding intensity for the Poisson process is $\lambda = 0.01$.

Now, we have $2n$ meioses in the pedigree, with each meiosis creating a zygote, where n is the number of non-founder individuals. Notice that at a single position in the genome, each child has two haplotypes, and each haplotype chooses one of the two parental alleles to copy. These choices are represented in an inheritance vector, a binary vector with $2n$ entries. The 2^{2n} possible inheritance vectors are the vertices of a $2n$-dimensional hypercube. We can model the recombination process as a random walk on the hypercube with a step occurring each time there is a recombination event. The waiting time to the next step is drawn from $exp(2n\lambda)$, the meiosis is drawn uniformly from the $2n$ possible meioses, and a step taken in the dimension that represents the chosen meiosis. The equilibrium distribution of this random walk is uniform over all the 2^{2n} vertices of the hypercube.

Detailed IBD Simulation. Recall that we are interested in the distribution of the length of a region that is IBD. Recall that IBD is defined as the event that a pair of alleles are inherited from the same founder allele. For individuals i and j, let D be the set of hypercube vertices that result in i and j sharing at least one allele IBD. Given x_0 a hypercube vertex drawn uniformly at random from D, we can compute the hitting time to the first non-IBD vertex by considering the random walk restricted to $D \cup \{d\}$ where d is an aggregate state of all the non-IBD vertices. The hitting time to d is the quantity of interest. In addition, we also need to consider the length of the shared region before reaching x_0, which is the time reversed version of the same process, for the same reason that we summed two exponential random variables while computing the lower bound in Section 3.

The transition matrix for this IBD process is easily obtained as $Pr[x_{i+1} = u | x_i = v] = \frac{1}{2n}$ when vertices u and v differ by exactly one coordinate, and $Pr[x_{i+1} = u | x_i = v] = 0$ otherwise. Transitions to state d are computed as $Pr[x_{i+1} = d | x_i = u] = 1 - \sum_{v \in D} Pr[x_{i+1} = v | x_i = u]$.

Now we can either analytically compute the hitting time distribution or estimate the distribution by simulating paths of this random walk. Since the number of IBD states may be exponential, it may be computationally infeasible to find eigenvectors and eigenvalues of the transition matrix [10]. We choose to simulate this random walk and estimate the distribution. This simulation is at worst exponential in the number of individuals.

4.3   Heuristic Graph Partitioning Method

The Max-Clique algorithm was used to illustrate the graph partitioning method. Max-Clique is currently used for the CIP algorithm, since the running-time there is dominated by the IBD simulation. However, the Max-Clique algorithm is exponential, thus not efficient for large input sizes. Therefore, for the COP algorithm, we consider an efficient heuristic for partitioning the vertices of the *compatibility* graph. This method is beneficial, because it looks for densely connected sets of vertices, rather than cliques, which allows for missing edges.

The algorithm is used to partition the vertices, $V(G^g)$, of graph G^g, into a partition $P = \{P_1, P_2, ..., P_C\}$, where $P_i \cap P_j = \emptyset$ for all i, j, and $V(G^g) = \cup_{i=1}^{C} P_i$. For a given partition set, let E_i be the edges of the subgraph induced by vertices P_i. We wish to find a partition such that each set in the partition is a clique or quasi-clique of vertices. The objective function is to find a partition that maximizes $\sum_{i=1}^{C}(a + 1)|E_i| - \binom{|P_i|}{2}$ where $a = 0.1$ is a parameter of the algorithm. This objective function is chosen, because it is equivalent to $\sum_{i=1}^{C} a|E_i| - \left(\binom{|P_i|}{2} - |E_i| \right)$, where the term in parentheses is the number of missing edges in the clique. Details of the partitioning method can be found in Karp and Li [15].

The running-time of this graph-partitioning heuristic largely determines the running-time of the pedigree reconstruction algorithm. The partitioning algorithm runs in polynomial time in the size of the graph, if the size of each set in the partition is constant. The step of creating the graph is polynomial in the size of the previous generation graph. Clearly it is possible, if no relationships are found, for the size of the graph at each generation to double. So, in the worst case, this algorithm is exponential. However, in practice this method performs quite well for constructing eight-generation pedigrees on large inputs.

5   Results

Pedigrees were simulated using a variant of the Wright-Fisher model with monogamy. The model has parameters for a fixed population size, n, a Poisson number of offspring λ, and a number of generations g. In each generation g, the set of n_g individuals is partitioned into $n_g/2$ pairs, and for each pair we randomly decide on a number of offspring using the Poisson distribution with expectation $\lambda = 3$.

The human genome was simulated as 3,000 regions, each of length 1MB, with recombination rate 0.01 between each region and where each founder haplotype had a unique allele for each region. The two assumptions here are 1) if two haplotypes share the same alleles in a mega-base region, then that region is identical by descent, and 2) haplotypes can be obtained for input to the pedigree reconstruction methods. Notice that Stankovich, et al also require haplotypes as input to their method [31]. The requirement that haplotypes are given is not highly restrictive since our algorithms search for haplotype regions that are shared between individuals, and since we consider regions of length 1Mb (typically > 500 SNPs), it is quite easy to determine whether two individuals have a shared haplotype across 1Mb.

In each experiment we end up having the true pedigree generated by the simulation, as well as an estimated pedigree. We evaluate the accuracy of the estimated pedigree by comparing the kinship matrices of the two pedigrees. Kinship is a model-based quantity defined as the frequency of IBD allele-sharing between a pair of individuals in a pedigree (averaged over the alleles of each individual). Since both pedigrees have the same set of haplotyped individuals, the comparison we consider is an L_1 distance between the kinship estimates of those individuals. Let K^P and K^Q be the kinship matrices of the actual pedigree P and the estimated pedigree Q, respectively. Then the evaluation method is

$$\sum_{i<j} |K_{i,j}^P - K_{i,j}^Q|$$

for haplotyped individuals i and j.

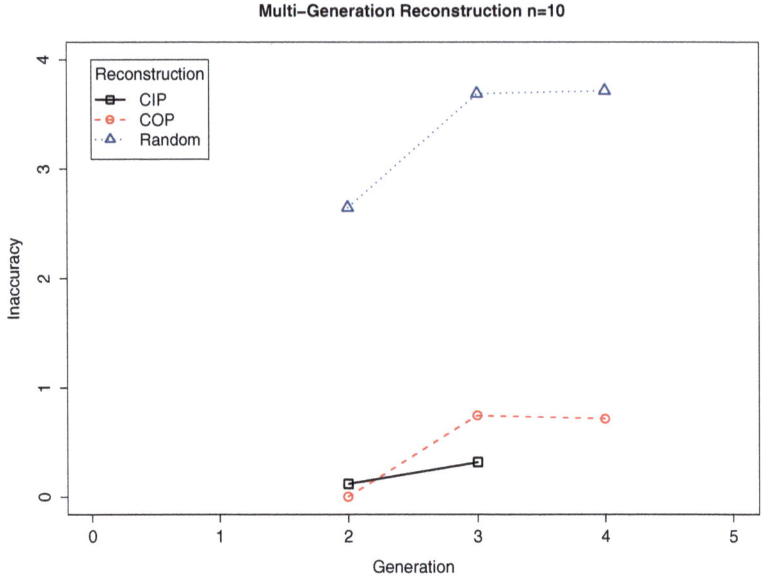

Fig. 3. Reconstruction under High Inbreeding. Here the pedigrees were simulated with a fixed population size of $n = 10$ individuals per generation. Over multiple generations, this results in a high level of inbreeding. The inaccuracy on the y-axis is measured by computing the kinship distance. (Reconstruction accuracy of 50 simulated pedigrees were averaged.)

We compare the COP and CIP methods on inbred pedigree simulations with high and moderate inbreeding, respectively $n = 10$ and $n = 50$, in Figures 3 and 4. These figures show the kinship-based inaccuracy on the y-axis and the number of generations in the reconstructed pedigree on the x-axis. As the depth of the estimated pedigree increases the error in the kinship of the estimated pedigree increases. However the accuracy is still much better than the accuracy of a randomly constructed pedigree, which is the highest, i.e. worst, line in each figure. Both methods perform better on the smaller population size.

Size of Reconstructed Pedigrees. Both the COP and CIP methods can reconstruct pedigree with four generations. The COP method for outbred pedigrees can reconstruct pedigrees going back to the most-recent common ancestor of the extant individuals. Provided with enough individuals, the method can construct pedigrees many generation deep. For example, given 400 individuals the method can construct 6 generations. As Figure 5 shows, the performance relative to a random reconstruction method is very good and so is the variance of the COP reconstruction method.

Comparison with GBIRP. We compare our two methods with the state-of-the-art method, called GBIRP, by Stankovich, et al [31]. Since GBIRP is limited to

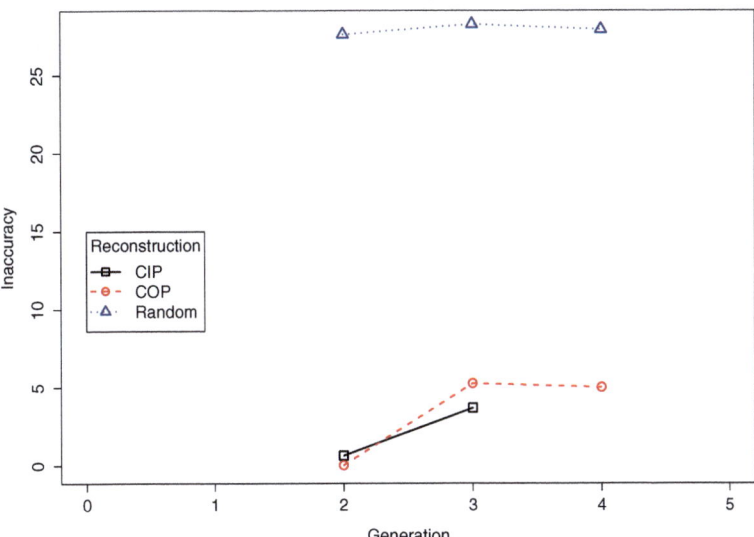

Fig. 4. Reconstruction under Less Inbreeding. Pedigrees here were simulated with a population size of $n = 50$. The y-axis show inaccuracy measured by kinship distance. (Reconstruction accuracy of 50 simulated pedigrees were averaged.)

small pedigrees, we compare the methods on three-generation simulated pedigrees with population size $n = 10$. The simulated pedigrees are connected graphs, so we can look at two accuracy measures, relationships that are mis-specified and relationships that should have been predicted but where not. GBIRP predicts meiosis distance, g_{ij}, between pairs of individuals, i, j, without inferring pedigree relationships. In order to compare GBIRP with the actual pedigree, we extract the minimum number of meiosis, a_{ij}, separating every pair of individuals i and j in the simulated pedigree. From our predicted pedigrees, we again extract a minimum meiosis distance $p_{i,j}$. Now can compute L_1 distances between the actual and predicted meiosis distances. These quantities are $\sum_{i<j:g_{i,j}\neq\infty}|a_{i,j}-g_{i,j}|$, and $\sum_{i<j:p_{i,j}\neq\infty}|a_{i,j}-p_{i,j}|$. This is the number of meioses, or edges in the pedigree graph, which are wrong on paths connecting all pairs of extant individuals. This is plotted in the left panels of Figures 6 and 7. Now, for a pair of extant individuals, there is always some relationship in the simulated pedigree, since it is a connected graph. But it is possible that one of the inference algorithms did not predict a relationship. Specifically this quantity is $\sum_{i<j:g_{i,j}=\infty} 1$, and $\sum_{i<j:p_{i,j}=\infty} 1$, and it is plotted in the right panel of both figures.

Figure 6 was done with the simulation method described above. However, in Figure 7, to obtain pedigrees with even more outbreeding, a large population size was simulated and a sub-pedigree with the desired number of extant individuals

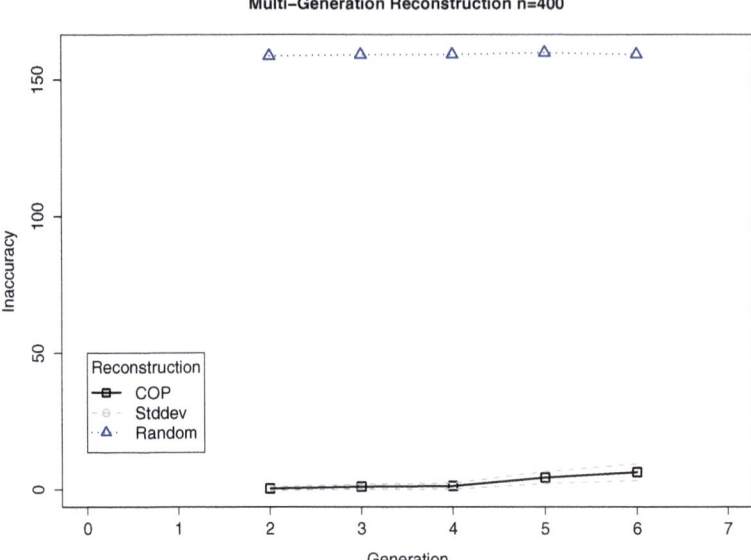

Fig. 5. Reconstruction for Deep Pedigrees. Pedigrees here were simulated with a population size of $n = 400$. (Reconstruction accuracy of 400 simulated pedigrees were averaged.)

was extracted from the large simulation. Notice that with more inbred pedigrees, under this measure of accuracy, the CIP algorithm performs superior to both the COP and the GBIRP methods. The accuracy of all of the methods improve

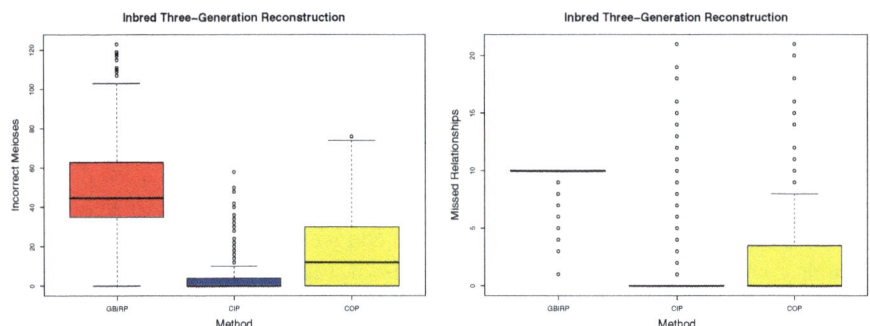

Fig. 6. Comparison with GBIRP on Inbred Simulations. The three-generation pedigrees here were simulated with $n = 10$ extant individuals, since GBIRP could not process larger pedigrees. The accuracy of 1000 simulated pedigrees were computed and plotted. Here the CIP method performs the best, i.e. closest to zero on both plots.

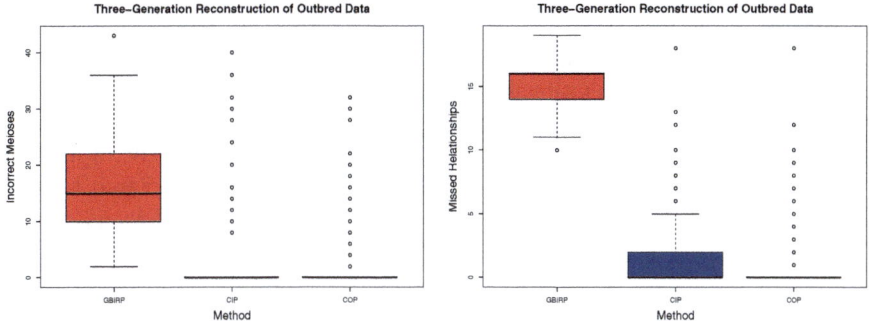

Fig. 7. Comparison with GBIRP on Outbred Simulations. The three-generation pedigrees here were simulated with $n = 10$ extant individuals, since GBIRP could not process larger pedigrees. Here, the simulated pedigree relating the extant individuals was outbred. The accuracy of 1000 simulated pedigrees were computed and plotted. All methods perform better than they did on the inbred data set. Over all, the COP method performs best on the outbred data.

when given outbred simulation data, with both CIP and COP performing very well. However, COP performs the best with outbred input data, as expected by the modeling assumptions of the method.

Relationships in the HapMap and Wellcome Trust Data. Taking haplotype data from HapMap, we ran the COP algorithm on unrelated individuals. Given the parents of the CEU and YRI trios, the algorithm discovers no relationships for eight generations. The CIP algorithm, on a subset of the CEU and YRI individuals (due to running time constraints), similarly finds no relationships for three generations. The results of on the CEU and YRI populations that we examined match the results of Pemberton, et al [27].

Taking the individuals from the Wellcome Trust data that have at least 85% IBS with some other individual, we ran COP to construct an eight-generation pedigree. There were no relationships inferred for the first seven generations, and there were several relationships inferred at the 8th generation (i.e. seventh cousin relationships).

6   Discussion

The reconstruction of pedigrees from haplotype data is undoubtedly a natural question of interest to the scientific community. Reconstructing very small families, or first generation relationships is a relatively easy task, but reconstructing a full inbred pedigree involving a few generations is inherently difficult since the traces left in our genomes by an ancestor drops exponentially with the distance to the ancestor. Here, we proposed a reconstruction method for pedigrees given haplotype data from the most recent generation. We use a generation-by-generation pedigree reconstruction approach that takes haplotype data as input and finds the pedigree(s) that relate the individuals. Notably, our methods are the first to reconstruct multi-generational pedigrees, rather than a set of pair-wise relationships which may not be consistent with each other.

We present two methods of inferring the pedigrees that relate the input haplotypes. Both our methods proceed from the bottom of the pedigree towards the top. The main difference between our methods is that in CIP we assume an inbreeding model, and in COP we assume an outbreeding model. We show that our methods perform considerably better than the state of the art.

One of the basic questions that we ask is how many generations back would it be possible to reconstruct a pedigree. By simulations, we show that one can reconstruct at least fifth cousins with some accuracy. Furthermore, we obtain a lower bound showing that given two individuals with the most-recent-common ancestor being five generations back there is a constant probability for the two not to share any genomic region inherited from the common ancestor. This bound obviously does not apply to inbred pedigrees or to multi-way relationships (i.e. rather than pair-wise relationships, consider relationships on a set of individuals). One of the open problems naturally arising from this is whether our lower bound can be extended to the case of inbreeding and to multi-way relationships. More generally, a major challenge would be to understand what are the limitations of pedigree reconstruction and under which conditions.

We note that our methods and analysis are limited to a restricted scenario in which there is monogamy and the generations are synchronous. If monogamy is broken then our approach will not work since the sibling relationships in the compatibility graph at each level will not be a simple partition. It is plausible that a different graph formulation may still provide an accurate solution to more complex pedigrees, however the exact formulation that will resolve such pedigrees is currently unknown and is left as an open challenge.

There are significant open challenges with pedigree reconstruction. For example, it would be nice to obtain confidence values on the inferred pedigree edges. However this seems very difficult, even if we can draw pedigrees from the posterior distribution of pedigree structures given the data. Since edges in a pedigree are not labeled, obtaining confidence values for a pedigree P would translate to: drawing pedigree samples, Q, from the distribution, identifying the edges in P and Q that provide the same relationships, and scoring the edges of P according to the probability of pedigree Q. As discussed in Kirkpatrick, et al [18], the second step, identifying the edges in P and Q that provide the same relationships, is a hard problem.

Acknowledgments

B.K. was supported by the NSF Graduate Research Fellowship. E.H. is a faculty fellow of the Edmond J. Safra Bioinformatics program at Tel-Aviv University. E.H was supported by the Israel Science Foundation grant no. 04514831.

References

1. Abecasis, G.R., Cherny, S.S., Cookson, W.O., Cardon, L.R.: Merlin-rapid analysis of dense genetic maps using sparse gene flow trees. Nature Genetics 30, 97–101 (2002)
2. Berger-Wolf, T.Y., Sheikh, S.I., DasGupta, B., Ashley, M.V., Caballero, I.C., Chao-valitwongse, W., Putrevu, S.L.: Reconstructing sibling relationships in wild populations. Bioinformatics 23(13), i49–i56 (2007)
3. Bickeboller, H., Thompson, E.A.: Distribution of genome shared ibd by half-sibs: Approximation by the poisson clumping heuristic. Theoretical Population Biology 50(1), 66–90 (1996)
4. Boehnke, M., Cox, N.J.: Accurate inference of relationships in sib-pair linkage studies. American Journal of Human Genetics 61, 423–429 (1997)
5. Bourgain, C., Hoffjan, S., Nicolae, R., Newman, D., Steiner, L., Walker, K., Reynolds, R., Ober, C., McPeek, M.S.: Novel case-control test in a founder population identifies p-selectin as an atopy-susceptibility locus. American Journal of Human Genetics 73(3), 612–626 (2003)
6. Brown, D., Berger-Wolf, T.: Discovering kinship through small subsets. In: Moulton, V., Singh, M. (eds.) WABI 2010. LNCS, vol. 6293, pp. 111–123. Springer, Heidelberg (2010)
7. Browning, S.R., Briley, J.D., Briley, L.P., Chandra, G., Charnecki, J.H., Ehm, M.G., Johansson, K.A., Jones, B.J., Karter, A.J., Yarnall, D.P., Wagner, M.J.: Case-control single-marker and haplotypic association analysis of pedigree data. Genetic Epidemiology 28(2), 110–122 (2005)
8. Coop, G., Wen, X., Ober, C., Pritchard, J.K., Przeworski, M.: High-Resolution Mapping of Crossovers Reveals Extensive Variation in Fine-Scale Recombination Patterns Among Humans. Science 319(5868), 1395–1398 (2008)
9. Doan, D., Evans, P.: Fixed-parameter algorithm for haplotype inferences on general pedigrees with small number of sites. In: Moulton, V., Singh, M. (eds.) WABI 2010. LNCS, vol. 6293, pp. 124–135. Springer, Heidelberg (2010)
10. Donnelly, K.P.: The probability that related individuals share some section of genome identical by descent. Theoretical Population Biology 23(1), 34–63 (1983)
11. Elston, R.C., Stewart, J.: A general model for the analysis of pedigree data. Human Heredity 21, 523–542 (1971)
12. Fishelson, M., Dovgolevsky, N., Geiger, D.: Maximum likelihood haplotyping for general pedigrees. Human Heredity 59, 41–60 (2005)
13. Gallego Romero, I., Ober, C.: CFTR mutations and reproductive outcomes in a population isolate. Human Genet. 122, 583–588 (2008)
14. Geiger, D., Meek, C., Wexler, Y.: Speeding up HMM algorithms for genetic linkage analysis via chain reductions of the state space. Bioinformatics 25(12), i196 (2009)
15. Karp, R.M., Li, S.C.: An efficient method for quasi-cliques partition (2011) (manuscript in preparation)
16. Kirkpatrick, B.: Haplotypes versus genotypes on pedigrees. In: Moulton, V., Singh, M. (eds.) WABI 2010. LNCS, vol. 6293, pp. 136–147. Springer, Heidelberg (2010)
17. Kirkpatrick, B.: Pedigree reconstruction using identity by descent. Class project, Prof. Yun Song, 2008. Technical Report No. UCB/EECS-2010-43 (2010)
18. Kirkpatrick, B., Reshef, Y., Finucane, H., Jiang, H., Zhu, B., Karp, R.M.: Algorithms for comparing pedigree graphs. CoRR, abs/1009.0909 (2010)
19. Lander, E.S., Green, P.: Construction of multilocus genetic linkage maps in humans. Proceedings of the National Academy of Science 84(5), 2363–2367 (1987)

20. Lauritzen, S.L., Sheehan, N.A.: Graphical models for genetic analysis. Statistical Science 18(4), 489–514 (2003)
21. Li, J., Jiang, T.: An exact solution for finding minimum recombinant haplotype configurations on pedigrees with missing data by integer linear programming. In: Proceedings of the 7th Annual International Conference on Research in Computational Molecular Biology, pp. 101–110 (2003)
22. Li, X., Yin, X.-L., Li, J.: Efficient identification of identical-by-descent status in pedigrees with many untyped individuals. Bioinformatics 26(12), i191–i198 (2010)
23. McPeek, M.S., Speed, T.P.: Modeling interference in genetic recombination. Genetics 139(2), 1031–1044 (1995)
24. McPeek, M.S., Sun, L.: Statistical tests for detection of misspecified relationships by use of genome-screen data. Amer. J. Human Genetics 66, 1076–1094 (2000)
25. Ng, M.Y., Levinson, D.F.: et al. Meta-analysis of 32 genome-wide linkage studies of schizophrenia. Mol. Psychiatry 14, 774–785 (2009)
26. Ng, S.B., Buckingham, K.J., Lee, C., Bigham, A.W., Tabor, H.K., Dent, K.M., Huff, C.D., Shannon, P.T., Jabs, E.W., Nickerson, D.A., Shendure, J., Bamshad, M.J.: Exome sequencing identifies the cause of a mendelian disorder. Nature Genetics 42(1), 30–35 (2010)
27. Pemberton, T.J., Wang, C., Li, J.Z., Rosenberg, N.A.: Inference of unexpected genetic relatedness among individuals in hapmap phase iii. Am. J. Hum. Genet. 87(4), 457–464 (2010)
28. Piccolboni, A., Gusfield, D.: On the complexity of fundamental computational problems in pedigree analysis. Journal of Computational Biology 10(5), 763–773 (2003)
29. Sheikh, S.I., Berger-wolf, T.Y., Khokhar, A.A., Caballero, I.C., Ashley, M.V., Chaovalitwongse, W., Chou, C., Dasgupta, B.: Combinatorial reconstruction of half-sibling groups from microsatellite data. In: 8th International Conference on Computational Systems Bioinformatics (CSB) (2009)
30. Sobel, E., Lange, K.: Descent graphs in pedigree analysis: Applications to haplotyping, location scores, and marker-sharing statistics. American Journal of Human Genetics 58(6), 1323–1337 (1996)
31. Stankovich, J., Bahlo, M., Rubio, J.P., Wilkinson, C.R., Thomson, R., Banks, A., Ring, M., Foote, S.J., Speed, T.P.: Identifying nineteenth century genealogical links from genotypes. Human Genetics 117(2-3), 188–199 (2005)
32. Sun, L., Wilder, K., McPeek, M.S.: Enhanced pedigree error detection. Hum. Hered. 54(2), 99–110 (2002)
33. Thatte, B.D.: Combinatorics of pedigrees (2006)
34. Thatte, B.D., Steel, M.: Reconstructing pedigrees: A stochastic perspective. Journal of Theoretical Biology 251(3), 440–449 (2008)
35. Thompson, E.A.: Pedigree Analysis in Human Genetics. Johns Hopkins University Press, Baltimore (1985)
36. Thornton, T., McPeek, M.S.: Case-control association testing with related individuals: A more powerful quasi-likelihood score test. American Journal of Human Genetics 81, 321–337 (2007)

A Quantitative Model of Glucose Signaling in Yeast Reveals an Incoherent Feed Forward Loop Leading to a Specific, Transient Pulse of Transcription

Sooraj KuttyKrishnan[1], Jeffrey Sabina[2], Laura Langton[3],
Mark Johnston[4], and Michael R. Brent[3]

[1] University of Washington
[2] Ion Torrent
[3] Washington University
[4] University of Colorado Denver

The ability to design and engineer organisms demands the ability to predict kinetic responses of novel regulatory networks built from well-characterized biological components. Surprisingly, few validated kinetic models of complex regulatory networks have been derived by combining models of the network components. A major bottleneck in producing such models is the difficulty of measuring in vivo rate constants for components of complex networks. We demonstrate that a simple, genetic approach to measuring rate constants in vivo produces an accurate kinetic model of the complex network that Saccharomyces cerevisiae employs to regulate the expression of genes encoding glucose transporters. The model predicts a transient pulse of transcription of HXT4 (but not HXT2 or HXT3) in response to addition of a small amount of glucose to cells, an outcome we observed experimentally. Our model also provides a mechanistic explanation for this result: HXT24 are governed by a type 2, incoherent feed forward regulatory loop involving the Rgt1 and Mig2 transcriptional repressors. The efficiency with which Rgt1 and Mig2 repress expression of each HXT gene determines which of them have a pulse of transcription in response to glucose. Finally, the model correctly predicts how lesions in the feed forward loop change the kinetics of induction of HXT4 expression.

V. Bafna and S.C. Sahinalp (Eds.): RECOMB 2011, LNBI 6577, p. 153, 2011.
© Springer-Verlag Berlin Heidelberg 2011

Inferring Mechanisms of Compensation from E-MAP and SGA Data Using Local Search Algorithms for Max Cut

Mark D.M. Leiserson, Diana Tatar, Lenore J. Cowen, and Benjamin J. Hescott

Tufts University
Department of Computer Science
161 College Ave, Medford, MA 02155

Abstract. A new method based on a mathematically natural local search framework for max cut is developed to uncover functionally coherent module and BPM motifs in high-throughput genetic interaction data. Unlike previous methods which also consider physical protein-protein interaction data, our method utilizes genetic interaction data only; this becomes increasingly important as high-throughput genetic interaction data is becoming available in settings where less is known about physical interaction data. We compare modules and BPMs obtained to previous methods and across different datasets. Despite needing no physical interaction information, the BPMs produced by our method are competitive with previous methods. Biological findings include a suggested global role for the prefoldin complex and a SWR subcomplex in pathway buffering in the budding yeast interactome.

1 Introduction

When two genes are mutated together, sometimes a surprising phenotype emerges compared to the phenotype of the individual gene mutants. When studying the yeast genome, often this can be quantified as the growth rate of the double mutant, compared with the expected growth rate of the double deletion mutant based on the growth rate of the single deletion mutants, termed *epistasis*. One of the most exciting developments in experimental data for large-scale function prediction has been technology, such as SGA [29], dSLAM [21] and E-MAP [26], that can produce high-throughput screens of massive numbers of pairwise mutant combinations. Complete E-MAPs have been published for a set of *S. cerevisiae* (budding yeast) genes involved in chromosome function [8], for a set involved in signaling pathways [11], as well as a set of genes in *S. pombe* (fission yeast) [24]. The most complete to date SGA study of *S. cerevisiae* genes was recently done by Costanzo et al. [9]. For computational biologists, pairwise genetic interaction data from E-MAP and other sources can be modeled as a complete, weighted, signed graph, that can be mined by itself, or together with other sources of interaction data, to produce functional predictions.

One of the most well-studied and useful network motifs found in genetic interaction data is the between-pathway model (BPM), introduced first by Kelley and Ideker [17] and Ulitsky and Shamir [31]. This is a network motif consisting of a particular pattern

V. Bafna and S.C. Sahinalp (Eds.): RECOMB 2011, LNBI 6577, pp. 154–167, 2011.

of genetic and physical interactions that is thought to signify two coherent sets of genes that may be compensatory or adaptive. In particular, each BPM subgraph consists of two subsets of genes, where physical interactions tend to occur between pairs of genes in the same subset, and synthetic lethal interactions tend to occur between pairs of genes in different subsets. The two subsets are called *pathways* in earlier papers [17,31,20,5], but now the term *modules* is becoming more standard (to emphasize that it is only a gene *set* that is being predicted, not directional, or temporal information in a pathway). We will also use the term *module* in this work to refer to each of the two subsets of genes in a BPM.

It was shown by Kelley and Ideker, Ulitsky and Shamir, and in subsequent work [20,5], that BPM modules in the interactome of *S. cerevisiae* (budding yeast) show significant biological enrichment for functional coherence, based on known ontological annotation. Recently, functional coherence of predicted BPM modules based on gene expression data has also been demonstrated [13]. All early methods were based on binary genetic interaction data, that is, a pair of proteins are in a synthetic lethality relationship, or they are not. For example, Brady et al. [5] showed that a search for maximal graph cuts can be used to help find BPMs based on this binary genetic interaction data.

In a 2010 Recomb paper, Kelley and Kingsford [16] considered whether the BPM paradigm could be adapted to make use of the more expressive non-binary quantitative genetic interaction data available from an E-MAP or SGA. Their approach interprets the E-MAP weights on the edges as probabilities, and they introduce a new method for clustering E-MAP data they call Expected Graph Compression based on the probabilistic graph that results. They compare the functional coherence of the modules that they found with those found by earlier papers of Bandyopadhyay et al. [2] and Ulitsky et al. [32].

In this work, we show that a new method based on local search for maximal cuts can improve the discovery of validated modules and BPMs in E-MAP data. The strength of our approach includes:

1. The method is mathematically natural, algorithmically simple, and fast in practice (though there are some open questions about theoretical convergence times, see below).
2. We achieve improved GO enrichment of BPM modules compared to previous studies.
3. Unlike all previous studies based on E-MAP data, our method makes use of the graph-theoretic structure of the genetic interaction data *only* when constructing the BPMs, allowing the location of known physical interactions to statistically validate modules. Thus, our method can find novel BPMs in network neighborhoods where less is known about physical interactions between genes.

Finally, there have been enough different studies on finding BPMs in yeast genetic interaction data that in addition to looking at the differences between what these methods can uncover, it becomes interesting to look at which modules are found again and again by all the different algorithms. Looking for these strong signals, we uncover some possible *global* mechanisms of fault-tolerance within the yeast interactome involving chaperones and chromatin remodeling.

1.1 Related work

As mentioned above, the primary studies on uncovering BPMs in binary yeast interaction data come from Kelley and Ideker [17], Ulitsky and Shamir [31], Ma et al. [20] and Brady et al. [5]. The corresponding computational problem involves finding appropriate subgraphs in an *unweighted* graph.

Papers by Bandyopadhyay et al [2] and Ulitsky et al. [32] look for modules in yeast E-MAP data; a recent paper of Kelley and Kingsford [16] explicitly tries to generalize the notion of BPMs to E-MAP data. We directly compare our BPMs and modules to all three previous methods that analyzed the Collins et al. E-MAP data. In addition, we use our LocalCut method to also generate BPMs based on the Boone lab's recent SGA map of budding yeast genetic interactions, one based on an E-MAP for budding yeast genes involved in cell signaling pathways [11], as well as an E-MAP dataset of *S. Pombe* [24]. We discuss what is similar and different on a systems scale about BPMs across different methods and different datasets.

2 Data

The Collins et al. [8], Fiedler et al. [11], and Roguev et al. [24] scalar genetic interaction datasets were downloaded from The Krogan Lab (http://interactome-cmp.ucsf.edu/). The Boone Lab [9] reports three variants (lenient, intermediate, stringent), of their SGA data. We report here results from the intermediate set (interaction values with an absolute value greater than 0.08 with a p-value < .05), though we ran the LocalCut algorithm on all three variants and saw similar results. These genetic interaction datasets span four different sets of genes: the Collins gene sets relate to chromosome function, the Fiedler gene set to signaling, the Roguev gene set to a genetic cross-section of *S. Pombe*, and the Boone to nearly 75% of the *S. cerevisiae* genome. To validate the BPMs generated by the LocalCut algorithm on each of the *S. cerevisiae* datasets, we also obtained a set of physical interactions by considering interactions between genes in these datasets in the BioGRID 3.0.66 release where the experiment type was 'physical' [27].

3 Results

Table 1 compares the results of our LocalCut algorithm to those of previous work of Bandyopadhyay et al. [2], Kelley-Kingsford [16], and Ulitsky et al. [32] on the Collins et al. [8] chromosome function E-MAP data. In order to make the results comparable across methods, we restrict to considering BPMs where each of its two modules contain between 3 and 25 genes. (This removes many BPMs from the Kelley-Kingford set, where 1 module contained only 1 or 2 genes, making it more comparable with other results). Such modules we call "accepted" in Table 1. A module is then declared *enriched* in Table 1 if, according to the program FuncAssociate [3], it is enriched for any term that describes a set of less than 500 proteins in the GO hierarchy, with a p-value $\leq$ .01. (Note that all FuncAssociate results are based on GO terms from a version of GO downloaded on 6/28/2010 except for the results of Ulitsky *et al.* which come from a slightly more recent set of GO terms updated on 1/11/2011). All methods excel at

Table 1. Comparison of our LocalCut algorithm to previous analyses of the Collins et al. E-MAP data [8]. We achieve a greater *number* of enriched modules as well as a higher *percentage* of BPMs with both modules enriched for either the same or related function.

Dataset	Modules		BPMs					
	Accepted	Enriched	Accepted	Enriched for same Function	Enriched for same or related Function	Enriched for Different Functions	One Mod Enriched	No Mods Enriched
Bandyopadhyay *et al.*	37	35	96	41 (43%)	53 (55%)	36 (38%)	7 (7%)	0 (0%)
Ulitsky *et al.*	43	43	111	43 (39%)	71 (64%)	40 (36%)	0 (0%)	0 (0%)
EGC (Kelley-Kingsford)	40	40	98	35 (36%)	52 (53%)	45 (46%)	1 (1%)	0 (0%)
Our Results (LocalCut)	**112**	**103**	**58**	**39 (67%)**	**43 (74%)**	**6 (10%)**	**9 (16%)**	**0 (0%)**

producing enriched modules, though it is perhaps impressive that we do so *looking at genetic interactions only,* whereas other methods also use physical interaction information to construct modules. However, the real strength of our method becomes apparent when looking at how the modules are combined into BPMs. A BPM is declared *enriched for the same function* if at least one enrichment term is in common between both modules. A BPM is declared *enriched for related functions* using the same definition as Kelley and Kingsford [16], i.e. if both modules are enriched, and each has an enrichment term whose most recent common ancestor describes fewer than 500 proteins. A BPM is declared *enriched for different functions* if both modules are enriched, but it doesn't meet the criteria above. LocalCut gives a much higher percentage of BPMs enriched for the same function or related functions than previous methods. We remark that LocalCut tends to produce more modules but fewer BPMs than the other methods. That's because the other methods tend to reuse modules several times as part of different BPMs, whereas LocalCut's modules tend to be unique to a particular BPM. Coupled with the enrichment results, it thus seems that other methods are reusing a smaller set of modules, and combining them into BPMs that are not necessarily functionally coherent as a module pair.

In Table 2, we seek to determine how sources of genetic interaction data affects the network of BPMs as discovered by LocalCut on the same gene set. In particular, we looked at LocalCut run on the original Collins et al. E-MAP data, as compared to the SGA data of Boone et al. *restricted to the same gene set* as the Collins et al. E-MAP data, as well as the full Boone network. On both the original data and on the restricted Boone dataset, the performance of LocalCut seems comparable as approximately the same percentage of modules are enriched, and a similar percentage of BPMs result that are enriched for the same or related functions. However, fewer total modules and BPMs are found in the Boone data restricted to this gene set, which is not surprising because there are many more 0-weight edges due to missing or corrupted data in the Boone data on the restricted set of genes (218,386 nonzero edges in the Collins et al. E-MAP data versus 15,467 nonzero edges reported by the Boone Lab on the same set of genes; the full Boone dataset has 145,805 nonzero edges). And in fact, the modules found by LocalCut based on Collins et al. interaction data are quite different than those generated from the restricted Boone data when measuring the Jaccard index (see [14,23] for definition of Jaccard index). Only 2 modules have a Jaccard index greater than .8.

Table 2. A comparison of the results of our algorithm on different datasets. Boone (restricted) refers to the Boone dataset restricted to only contain those genes also in the Collins et al. data. LocalCut finds fewer BPMs on the restricted Boone dataset than on the Collins et al. E-MAP; not surprising since there are more missing or zero-weight edges in the restricted Boone dataset. However, the proportion that are enriched for same or related function is nearly identical across both these data sources. On the full Boone dataset, while many more modules and BPMs are found by LocalCut, many more are not known to be functionally enriched, perhaps because this is a less-understood set of yeast genes with fewer annotations.

	Modules		BPMs					
Dataset	Accepted	Enriched	Accepted	Enriched for same Function	Enriched for same or related Function	Enriched for Different Functions	One Mod Enriched	No Mods Enriched
Collins et al.	112	103	58	39 (67%)	43 (74%)	6 (10%)	9 (16%)	0 (0%)
Boone (restricted)	55	52	29	18 (62%)	23 (79%)	2 (7%)	4 (14%)	0 (0%)
Boone (full)	285	104	149	8 (5%)	17 (11%)	9 (6%)	56 (38%)	67 (45%)

However, more than 50% of the modules have Jaccard indices greater than .25, meaning the modules uncovered by LocalCut are roughly as similar across datasets (E-MAP and SGA) as the modules uncovered across methods (i.e. LocalCut modules as compared to Kelley-Kingsford, Ulitsky et al. or Bandyopadhyay et al.)

In Table 3, we look at how making perturbations in the edge weights can affect the results of the LocalCut algorithm. In Variant 1 we set all positive weights to 0. In Variant 2, every weight whose absolute value is above 2.5 (the threshold for synthetic lethality and synthetic rescue as defined by [8]) is set to 2.5 or -2.5, consistent with its sign. In Variant 3, we run LocalCut on the binary version of the data, where every edge weight above 2.5 is set to +1, and any weight below -2.5 is set to -1, and all other weights are 0. Variant 4 is the negative half of Variant 3; any weight whose value is below 2.5 is set to -1 and all other weights are 0 (representing synthetic lethality or not). For Variant 5, we

Table 3. We examine how perturbing edge weights affects the performance of our LocalCut algorithm. The results validate our supposition that the nuances of scalar data are more informative than binary weights, and that positive-weight interaction edges matter as well as negative-weight edges.

	Modules		BPMs					
Dataset	Accepted	Enriched	Accepted	Enriched for same Function	Enriched for same or related Function	Enriched for Different Functions	One Mod Enriched	No Mods Enriched
LocalCut	112	103 (92%)	58	39 (67%)	43 (74%)	6 (10%)	9 (16%)	0 (0%)
LocalCut – Variant 1	50	46 (92%)	26	17 (65%)	19 (73%)	2 (8%)	5 (19%)	0 (0%)
LocalCut – Variant 2	133	61 (46%)	68	4 (6%)	6 (9%)	9 (13%)	33 (49%)	20 (29%)
LocalCut – Variant 3	54	37 (69%)	30	3 (10%)	7 (23%)	6 (20%)	17 (57%)	0 (0%)
LocalCut – Variant 4	21	14 (67%)	12	1 (8%)	2 (17%)	3 (25%)	7 (58%)	0 (0%)
LocalCut – Variant 5	98	82 (84%)	49	21 (43%)	30 (61%)	5 (10%)	12 (24%)	2 (4%)
LocalCut – *Control*	0	0 (0%)	0	0 (0%)	0 (0%)	0 (0%)	0 (0%)	0 (0%)

use the E-MAP weights augmented with interpolation: that is, we download the E-MAP weights as filled in by Ulitsky, Krogan and Shamir [30] using their method to interpolate for missing data in the original E-MAP.. The control variant aims to produce nonsense by exchanging the signs of all the weights on the edges. Of these variants, clearly all of them produce degraded performance, except possibly Variant 1, which produces fewer modules and BPMs, but has nearly the same percentage of the modules enriched for same or related function. This is an interesting result in light of the discussion in the work of Kelley and Kingsford about whether considering positive edge weights helps or hurts in the construction of BPMs [16]; LocalCut finds more BPMs with no degradation in functional enrichment if positive edge weights are included. The other variants show that, to some extent, the full range of interaction data is helpful in constructing modules and BPMs, and less is discovered if only the binary synthetic lethality or rescue data is used. We are also pleased that the control variant (exchanging the role of positive and negative edge weights in the data) yields only noise– no consistent BPMs with both modules of size between 3 and 25. This proves that the existence of meaningful negative-weight bipartite subgraphs is a true biological property of the yeast genetic interaction network, not a computational artifact; another way to say this is that there are no small bipartite subgraphs in the Collins et al. E-MAP data with positive weight between the two modules and negative weight within each module.

While previous work focused on the Collins et al. E-MAP, in this paper we look at the results of LocalCut on other high-throughput genetic interaction datasets. In particular, we look at two other genetic interaction network datasets for Baker's yeast, an E-MAP for cell signaling genes generated by Fiedler et al. [11] and the full set of genetic interactions generated using SGA by Boone et al. [9] We also ran LocalCut on the first E-MAP dataset generated for *S. pombe,* fission yeast [24]. We find the structure of the negative weight bipartite subgraphs is very different between the chromosome function network and the signaling network. In particular very few BPMs are found by LocalCut in the signaling dataset. In both the *S. pombe* and the full Boone network, a much smaller proportion of the BPMs we find are enriched for the same function. We

Table 4. Results of applying LocalCut to different datasets. Notice that the BPM network motif is very rare among the Fiedler et al. cell signaling genes, as compared to the others, implying perhaps that this network is organized at a different level of complexity or with different mechanisms of fault-tolerance. On the other hand, the fact that fewer modules and BPMs are enriched in the full Boone dataset across nearly all *S. cerevisiae* genes and the *S. pombe* network is more likely caused by the fact that our knowledge and therefore annotation of basic *S. cervisiae* cell cycle genes far exceeds our knowledge of the other networks, rather than intrinsic differences in network organization.

| Dataset | Modules | | BPMs | | | | | |
	Accepted	Enriched	Accepted	Enriched for same Function	Enriched for same or related Function	Enriched for Different Functions	One Mod Enriched	No Mods Enriched
Collins et al.	112	103	58	39 (67%)	43 (74%)	6 (10%)	9 (16%)	0 (0%)
Fiedler et al.	10	8	5	0 (0%)	4 (80%)	0 (0%)	0 (0%)	1 (20%)
Boone (Full)	285	104	149	8 (5%)	17 (11%)	9 (6%)	56 (38%)	67 (45%)
S. pombe	31	18	16	1 (6%)	1 (6%)	4 (25%)	9 (56%)	2 (13%)

Table 5. Location of physical interaction edges in LocalCut BPMs. In each dataset, many more physical interactions occur within modules than between modules; as one would expect if the modules were compensatory.

	Physical Interactions in LocalCut BPMs			
Dataset	Within Modules	Expected Within	Between Modules	Expected Between
Collins et al.	172 (8.6%)	20	18 (0.9%)	20
Fiedler et al.	13 (12.7%)	1	1 (0.9%)	1
Boone (restricted)	138 (14.4%)	27	10 (1.1%)	26
Boone (full)	147 (3.1%)	41	17 (0.3%)	39

suspect that this is because because GO annotation of function is probably *weaker* for these gene sets than for the well-studied chromosome cell machinery. Thus, LocalCut's BPMs are more likely to discover novel function in these datasets.

Finally, as remarked above, unlike competing methods, LocalCut constructs its modules and BPMs looking at genetic interactions only; thus the location of physical interaction edges can be used to *validate* the quality of the BPMs produced by LocalCut (whereas other methods take location of physical interaction edges into account when constructing BPMs). We considered 84,785 known physical interactions between pairs of genes in *S. cerevisiae*. These interactions are from the BioGRID 3.0.66 release of BioGRID where the experiment type was 'physical', excluding physical interaction hubs, i.e. genes that have more than 300 physical interactions with other genes. Of these remaining physical interactions, 2235 intersect with gene pairs in Collins et al. [8], 1900 with Fiedler, et al. [11], 441 with Boone (restricted), and 1274 with Boone (full). Not all of these physical interactions participate in BPMs, and some participate multiple times. However, if LocalCut produces meaningful instances of the BPM motif, we would expect that of the physical interaction edges that do appear, many more would appear *within* BPM modules than *between* BPM modules. Table 5 shows that this is indeed the case over all of the *S. cerevisiae* datasets.

All BPMs produced by LocalCut on all datasets are available in full at http://bcb.cs.tufts.edu/localcut

4   Discussion

We have introduced LocalCut, a method that uses maximal weighted graph cuts to find modules and BPM motifs in high-throughput genetic interaction data. We have shown that it is competitive in functional enrichment measures to other methods despite not needing to consider physical interaction data as other methods do. We ran LocalCut on different high-throughput genetic interaction datasets involving different sets of *S. cerevisiae* genes, some generated with different technologies (E-MAP or SGA) and one E-MAP dataset for *S. pombe*, and compared the resulting networks.

A recent paper of Jaimovich *et al.* tried to add directionality prediction to methods to determine fault tolerance in genetic interaction networks [15]. More specifically, while BPMs are often motivated by discussing two equally important, alternative, compensatory modules (termed *bi-directional compensation*), an alternative explanation could be that one module is crucial for functions that compensate for the abnormal cellular state resulting from the loss of the other module (termed *unidirectional compensation*) [21,4]. The work in [15] used a novel method of exploring phenotype responses to different conditions to attempt to discriminate between unidirectional and bi-directional compensation; we do not duplicate their methods here. However, looking across the set of BPMs produced by our methods and previous methods on the yeast chromosome function genes as a group, we do find several complexes that appear again and again in modules opposite different sets of genes. Could these particular complexes be agents of such unidirectional compensation, i.e. possible *global* mechanisms of fault tolerance of the cell? There are several intriguing clues that suggest that they might be. For example, two of the most popular complexes that shows up in GO enrichment in multiple BPMs, not just for us, but also in the E-MAP BPM sets and modules of Bandyopadhyay et al. [2], and Kelley-Kingsford [16] are the Prefoldin complex, and a subunit of the SWR1 complex consisting of genes ARP6, SWC3, SWC5, SWR1, VPS71, VPS72, and YAF9. The Prefoldin complex is particularly intriguing as a global mechanism of fault-tolerance because it is a chaperone; the effects of an alternative chaperone (HSP90) and its ability to buffer difficult conditions in the cell has been recently described [28]. The SWR1 complex is also intriguing because it is involved in chromatin remodeling and silencing near telomeres [18], perhaps another way for the cell to compensate when other modules go awry. For the Prefoldin complex, the genes GIM3, GIM4, GIM5, PAC10, PFD1, and YKE2 appear in multiple BPMs for us, Bandyopadhyay et al. and Kelley-Kingsford; an additional gene BUD27 annotated as being part of the Prefoldin complex appears also in the BPMs generated by Bandyopadhyay et al., but not for LocalCut or Kelley-Kingsford. The Prefoldin complex is completely missing from the BPMs when we run LocalCut on the Boone et al. SGA dataset [9], but this appears to be a missing/corrupted data problem; in particular, they report no genetic interaction data at all on two of the genes in the Prefoldin complex.

We looked at other complexes of size 3 or greater that appear in their entirety in our and in other BPMs; for example the CAF-1 complex consisting of genes CAC2, MSI1 and RLF2 appears in everyone's set of BPMs. Indeed, it has been suggested that CAF-1 participates in one of multiple redundant modules for chromatin assembly [1,12]. Opposite the module containing CAF-1, we almost always find parts of the HIR complex, where the double mutants missing both CAF-1 and HIR1 result in a synergistic reduction in silencing at telomeres. Another complex that appears in our BPMs as well as the BPMs of Boone et al. and Kelley-Kingsford (but not Bandyopadhyay et al.) is the MRE11 complex, involved in DNA damage repair [10].

Finally, we consider an interesting example BPM that LocalCut finds in *S. pombe*. Figure 1 shows the between-pathway interactions between the two modules; where the stronger the edge, the more negative the interaction. In module 1 (on the left), all 14 genes and 6 of the 7 genes in module 2 are enriched for "response to DNA damage stimulus" (GO:0006974). Some of the strongest negative edges come, in module 1, from the 3 genes HUS1, RAD1, RAD9 that make up the checkpoint clamp complex; a conserved heterotrimeric complex involved in DNA damage response [7] and from CRB2, thought to be related to the important BRCAI breast-cancer gene in humans [6].

5   Methods

5.1   The Graph Model

We model the E-MAP and SGA interactions as a weighted complete graph G, where *vertices* represent genes participating in the E-MAP or SGA, and if i and j are vertices in G, the edge e_{ij} is assigned a weight as follows:

1. If the E-MAP or SGA value for the genetic interaction of genes i and j is a negative value (i.e. of the form $-z$, for some positive $z \in \mathbb{R}$), then e_{ij} is assigned the weight $-z^2$.
2. If the E-MAP or SGA value for the genetic interaction of genes i and j is a positive value (i.e. of the form z, for some positive $z \in \mathbb{R}$), then e_{ij} is assigned the weight z^2.
3. If the E-MAP or SGA value for the genetic interaction of genes i and j is zero or missing, then e_{ij} is assigned the weight 0.

5.2   The Algorithm

Consider an arbitrary bipartition (A, B) of the vertex set of G. For such a bipartition, let $Same(v) = \sum_i e_{vi}$, for all vertices i that appear in the same subset of the partition as v, and $Opposite(v) = \sum_j e_{vj}$, for all vertices j that appear in the opposite partition to v. Call a vertex v *unhappy* if $Same(v) < Opposite(v)$, otherwise call v *happy*. Let $flip(v)$ denote the operation that, starting with a bipartition (A, B), creates a bipartition

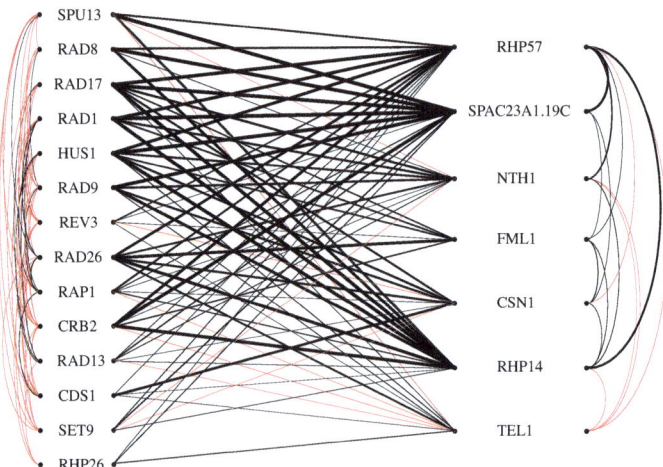

Fig. 1. A sample BPM generated by LocalCut from E-MAP data in *S. pombe*. Black lines represent negative weights (aggravating interactions) in the E-MAP data and red lines represent positive weights (alleviating interactions) in the data. The width of the line directly corresponds to how aggravating or alleviating the interaction is between the pair of genes.

that is identical in all ways, except v switches sides; that is, if v was in A, it is now placed in B, and vice versa.

Consider the following subroutine Weighted-Flip:

foreach *vertex u* **do**
  | Assign u uniformly at random with equal probability to set A or set B
end
while *there exists at least one unhappy vertex* **do**
  | Choose v at random from the set of unhappy vertices
  | `flip(v)`
end
`output bipartition(A, B)`

Theorem 1. *The subroutine terminates, and results in a bipartition (A,B) where all vertices are happy. Furthermore, if E is the set of edges with endpoints either both in A or both in B, and F is the set of edges with one endpoint in each of A and B, then when the subroutine terminates, $\sum_{e \in E} w(e) \geq \sum_{f \in F} w(f)$.*

Proof. It is first shown that there is a minimum positive amount, ϵ, dependent on the set of edge weights W, by which the weight going across the partition must decrease in any flip of an unhappy vertex. This is because for all partitions of the vertex set into two sets, we can look at the weight going across the partition, and since this is a (albeit large) finite set, there is a positive δ which is the minimum nonzero difference between the weights going across any two of these sets, and ϵ is clearly bigger than or equal to δ which is greater than 0. As the total sum of the absolute values of all the weights on all the edges is certainly an upper bound on the maximum negative weight that can cross the partition, and any flip decreases the amount of weight crossing the partition by at least a positive amount ϵ, the algorithm terminates. When the algorithm terminates, if $N(v)$ denotes the edges adjacent to v we have for every vertex v that, $\sum_{e \in \{E \cup N(v)\}} w(e) \geq \sum_{f \in \{F \cup N(v)\}} w(f)$; otherwise we could flip v. Thus summing over each edge twice (once for each endpoint) we get $\sum_{e \in E} 2w(e) \geq \sum_{f \in F} 2w(f)$, and thus $\sum_{e \in E} w(e) \geq \sum_{f \in F} w(f)$.   ◇

We note that this reduces exactly to the procedure Flip in the work of Brady et al. [5] when the weights are 0 or 1 and the graph is the unweighted graph H instead of the graph G (but all the inequality signs are reversed because all edges are given weight -1 instead of 1). Thus we have replaced a local search for maximal cuts in an unweighted graph with a version for weighted graphs, with both positive and negative edge weights.

While the above theorem proves convergence, it does not show convergence in polynomial time. In fact, the time complexity to convergence time for this algorithm is equivalent to a well-known open problem in combinatorial optimization. In particular, a partition of the vertices of a weighted graph such that all vertices are happy is called a *local max cut* of the graph. In the special case that the graph is cubic (i.e. all vertices have degree 3), Loebl [19] showed a local max cut can be found in polynomial time if all weights are nonnegative (using a more complicated algorithm than the one we describe above); a polynomial time algorithm for cubic graphs allowing negative edge weights was later found by Poljak [22]. For general graphs, convergence in the worst case is conjectured to be exponential time, because the problem is PLS-Complete, as shown by Shäeffer and Yannakakis [25].

By squaring the weights in step 1 of our algorithm, however, the absolute values of all nonzero weights greater than 1 are pulled away from 0, while weights with an absolute value less than 1 are pulled closer to zero. In practice, this speeds convergence, because any vertex that changes sides, will result in a larger gain in weight than with many weights close to 0. In practice, we found that this squaring step sufficed to allow us to run on the E-MAP and SGA data and reach convergence in a reasonable amount of time. To generate a bipartition it takes less than 20 minutes per gene. Since this is highly parallelizable to generate the potential modules for an entire dataset is very fast.

Note that the algorithmic procedure Weighted-Flip used to generate the local max cut is randomized, and it will typically generate many different local max cuts. However, if there is a large bipartite subgraph with favorable weights, it will tend to show up as a bipartite subgraph in many if not most of the local cuts, whereas subgraphs that are not naturally weighted bipartite are likely not to be conserved in all the different local cuts. We exploit this to identify such subgraphs and generate candidate BPMs.

Definition 1. *Given a gene v in G, run Weighted-Flip M times on G. Label each gene with the number of times it appears in the* same *side as v in one of the M sets (A,B) generated this way, as well as with the number of times it appears on the* opposite *side from v. If gene w appears consistently (at least $C\%$ of the time) in the same partition as v, or consistently in the opposite partition from v, then w is included in the* stable bipartite subgraph of v; *otherwise w is not included. The stable bipartite subgraph of v in G, then, is the subgraph induced by all included vertices, where v along with the vertices appearing consistently on the same side as v form one partition, and the rest of the included vertices form the other.*

For each v, we output v's stable bipartite subgraph as one of our putative between-pathway models. Note, however, that different runs of the Weighted-Flip procedure may generate different sets of results because of the random choices in the initial partition configuration and the choice of the unhappy node that needs to be flipped at each iteration step. However, we set M and C large enough so that the set of BPMs reported will be fairly consistent, regardless of the random choices made by our algorithm. In particular, we determined empirically that setting $M = 250$ and $C = 0.90$ gave relatively stable subgraphs of genes. We varied the values of C from from 0.70 to 0.95. Not surprisingly, we found that the greater values produced more consistent results. We chose 0.90 to try to maximize stability without deleting potential BPMs. We chose M high enough to ensure consistent results.

Using our algorithm, different genes v may generate the same or highly similar putative BPMs. Thus we then prune our set in order to report a collection of non-redundant BPMs as follows.

5.3   Removing BPM Redundancies

As described in the previous section, each gene's stable bipartite subgraph becomes a putative BPM. If the BPM generated is a true instance of compensatory modules, we would expect our algorithm to produce the same or similar BPM when run on another gene in the BPM. Because of the natural noise in the dataset, these BPMs are not exact matches. To fairly compare our results with alternate studies, we must remove the redundant, highly-overlapping BPMs from our set. We do this as follows. We first sort all the BPMs created by LocalCut according to their *interaction weight I*, where I is calculated by summing the edge weights of all genetic interactions within the two modules of the BPM, minus the sum of the weights of all interactions appearing between the two modules of the BPM, all divided by the number of genes in the entire BPM.

$$I = \frac{\Sigma(\text{ interactions within each module }) - \Sigma(\text{ interactions between two modules})}{\text{number of genes in BPM}}$$

Starting with the BPM with the largest interaction weight, we add a BPM to our final output set if its Jaccard index is less than the fixed threshold (set at .66 for these results) from every previously added BPM.

Acknowledgement

Thanks to Tom Zazlabsky for a useful discussion on the theory of signed graphs, to the BCB group at Tufts for helpful feedback, and to David Kelley and Carl Kingsford for access to their BPM data and discussions about its interpretation. This work was supported in part by NIH grant R01 GM080330 to LC.

References

1. Adams, C., Kamakaka, R.: Chromatin assembly: biochemical identities and genetic redundancy. Current Opinion in Genetics and Development 9, 185–190 (1999)
2. Bandyopadhyay, S., Kelley, R., Krogan, N.: Functional maps of protein complexes from quantitative genetic interaction data. PLoS Computational Biology (January 2008)
3. Berriz, G.F., King, O.D., Bryant, B., Sander, C., Roth, F.P.: Characterizing gene sets with FuncAssociate. Bioinformatics 19(18), 2502–2504 (2003)
4. Boone, C., Bussey, H., Andrews, B.J.: Exploring genetic interactions and networks with yeast. Nature Reviews Genetics 8, 437–449 (2007)
5. Brady, A., Maxwell, K., Daniels, N., Cowen, L.: Fault tolerance in protein interaction networks: Stable bipartite subgraphs and redundant pathways. PLoS ONE 4(4), e5364 (2009)
6. Callebaut, I., Mornon, J.-P.: From BRCA1 to RAP1: A widespread BRCT module closely associated with DNA repair. FEBS Letters 400, 25–30 (1997)
7. Carr, A.: DNA structure dependent checkpoints as regulators of DNA repair. DNA Repair 1, 983–994 (2002)
8. Collins, S., Miller, K., Maas, N., Roguev, A.: Functional dissection of protein complexes involved in yeast chromosome biology using a genetic interaction map. Nature (January 2007)

9. Costanzo, M., Baryshnikova, A., Bellay, J., Kim, Y., Spear, E.D., Sevier, C.S., Ding, H., Koh, J.L.Y., Toufighi, K., Mostafavi, S., Prinz, J., Onge, R.P.S., VanderSluis, B., Makhnevych, T., Vizeacoumar, F.J., Alizadeh, S., Bahr, S., Brost, R.L., Chen, Y., Cokol, M., Deshpande, R., Li, Z., Lin, Z., Liang, W., Marback, M., Paw, J., Luis, B.S., Shuteriqi, E., Tong, A.H.Y., van Dyk, N., Wallace, I.M., Whitney, J.A., Weirauch, M.T., Zhong, G., Zhu, H., Houry, W.A., Brudno, M., Ragibizadeh, S., Papp, B., Pál, C., Roth, F.P., Giaever, G., Nislow, C., Troyanskaya, O.G., Bussey, H., Bader, G.D., Gingras, A., Morris, Q.D., Kim, P.M., Kaiser, C.A., Myers, C.L., Andrews, B.J., Boone, C.: The genetic landscape of a cell. Science 327(5964), 425–431 (2010)
10. D'Amours, D., Jackson, S.: The MRE11 complex: at the crossroads of DNA repair and checkpoint signalling. Nature Reviews Molecular Cell Biology 3, 317–327 (2002)
11. Fiedler, D., Braberg, H., Mehta, M., Chechik, G., Cagney, G.: Functional organization of the *S. cerevisiae* phosphorylation network. Cell (January 2009)
12. Green, E., Antcsak, A., Bailey, A., Franco, A., Wu, K., Yates, J., Kaufman, P.: Replication-independent histone deposition by the HIR complex and asf1. Current Biology 15, 2044–2049 (2005)
13. Hescott, B.J., Leiserson, M.D.M., Slonim, D.K., Cowen, L.J.: Evaluating between-pathway models with expression data. Journal of Computational Biology 17(3), 477–487 (2010)
14. Jaccard, P.: Nouvelles recherches sur la distribution florale. Bull. Soc. Vaudoise Sci. Nat. 44, 223–270 (1908)
15. Jaimovich, A., Rinott, R., Schuldiner, M., Margalit, H., Friedman, N.: Modularity and directionality in genetic interaction maps. Bioinformatics 26(12), i228–i236 (2010)
16. Kelley, D., Kingsford, C.: Extracting between-pathway models from E-MAP interactions using expected graph compression. In: Berger, B. (ed.) RECOMB 2010. LNCS, vol. 6044, pp. 248–262. Springer, Heidelberg (2010)
17. Kelley, R., Ideker, T.: Systematic interpretation of genetic interactions using protein networks. Nature Biotechnology 23(5), 561–566 (2005), doi:10.1038/nbt1096 PMID:15877074
18. Krogan, N., Keogh, M.-C., Datta, N., Sawa, C., Ryan, O., Ding, H., Haw, R., Pootoolal, J., Tong, A., Canadien, V., Richards, D., Wu, X., Emili, A., Hughes, T., Buratowski, S., Greenblatt, J.: A Snf2 family ATPase complex required for the recruitment of the histone H2A variant Htz1. Molecular Cell 12, 1565–1576 (2003)
19. Loebl, M.: Efficient maximal cubic graph cuts. In: Leach Albert, J., Monien, B., Rodríguez-Artalejo, M. (eds.) ICALP 1991. LNCS, vol. 510, pp. 351–362. Springer, Heidelberg (1991)
20. Ma, X., Tarone, A., Li, W.: Mapping genetically compensatory pathways from synthetic lethal interactions in yeast. PLoS One 3(4), e1922 (2008), doi:10.1371/journal.pone.0001922 PMCID: PMC2275788
21. Pan, X., Ye, P., Tuan, D., Wang, X., Bader, J., Boeke, J.: A DNA integrity network in the yeast Saccharomyces cerevisiae. Cell 124, 1069–1081 (2006)
22. Poljak, S.: Integer linear programs and local search for max-cut. SIAM J. Comput. 24(4), 822–839 (1995)
23. Real, R., Vargas, J.: The probabilistic basis of Jaccard's index of similarity. Syst. Biol. 45(3), 380–385 (1996)
24. Roguev, A., Bandyopadhyay, S., Zofall, M., Zhang, K., Fischer, T., Collins, S.R., Qu, H., Shales, M., Park, H., Hayles, J., Hoe, K., Kim, D., Ideker, T., Grewal, S.I., Weissman, J.S., Krogan, N.J.: Conservation and rewiring of functional modules revealed by an epistasis map in fission yeast. Science 322(5900), 405–410 (2008)
25. Schäffer, A., Yannakakis, M.: Simple local search problems that are hard to solve. SIAM J. Comput. 20, 56–87 (1991)

26. Schuldiner, M., Collins, S.R., Thompson, N.J., Denic, V., Bhamidipati, A., Punna, T., Ihmels, J., Andrews, B., Boone, C., Greenblatt, J.F., Weissman, J.S., Krogan, N.J.: Exploration of the function and organization of the yeast early secretory pathway through an epistatic miniarray profile. Cell 123(3), 507–519 (2005)
27. Stark, C., Breitkreutz, B.-J., Reguly, T., Boucher, L., Breitkreutz, A., Tyers, M.: BioGRID: a general repository for interaction datasets. Nucleic Acids Research 34(suppl 1), D535–D539 (2005)
28. Taipale, M., Jarosz, D., Lindquist, S.: HSP90 at the hub of protein homeostasis: emerging mechanistic insights. Nature Reviews Molecular Cell Biology 11, 515–528 (2010)
29. Tong, A.H.Y., Lesage, G., Bader, G.D., Ding, H., Xu, H., Xin, X., Young, J., Berriz, G.F., Brost, R.L., Chang, M., Chen, Y., Cheng, X., Chua, G., Friesen, H., Goldberg, D.S., Haynes, J., Humphries, C., He, G., Hussein, S., Ke, L., Krogan, N., Li, Z., Levinson, J.N., Lu, H., Menard, P., Munyana, C., Parsons, A.B., Ryan, O., Tonikian, R., Roberts, T., Sdicu, A.-M., Shapiro, J., Sheikh, B., Suter, B., Wong, S.L., Zhang, L.V., Zhu, H., Burd, C.G., Munro, S., Sander, C., Rine, J., Greenblatt, J., Peter, M., Bretscher, A., Bell, G., Roth, F.P., Brown, G.W., Andrews, B., Bussey, H., Boone, C.: Global mapping of the yeast genetic interaction network. Science 303(5659), 808–813 (2004)
30. Ulitsky, I., Krogan, N., Shamir, R.: Towards accurate imputation of quantitative genetic interactions. Genome Biology (January 2009)
31. Ulitsky, I., Shamir, R.: Pathway redundancy and protein essentiality revealed in the S. cerevisiae interaction networks. Molecular Systems Biology 3(104) (2007), PMCID: PMC1865586
32. Ulitsky, I., Shlomi, T., Kupiec, M., Shamir, R.: From E-MAPs to module maps: dissecting quantitative genetic interactions using physical interactions. Molecular Systems Biology (January 2008)

IsoLasso: A LASSO Regression Approach to RNA-Seq Based Transcriptome Assembly
(Extended Abstract)

Wei Li[1], Jianxing Feng[2], and Tao Jiang[1,3]

[1] Department of Computer Science and Engineering,
University of California, Riverside, CA
[2] College of Life Science and Biotechnology, Tongji University, Shanghai, China
[3] School of Information Science and Technology, Tsinghua University, Beijing, China
{liw,jiang}@cs.ucr.edu, feng@tongji.edu.cn

Abstract. The new second generation sequencing technology revolutionizes many biology related research fields, and posts various computational biology challenges. One of them is transcriptome assembly based on RNA-Seq data, which aims at reconstructing all full-length mRNA transcripts simultaneously from millions of short reads. In this paper, we consider three objectives in transcriptome assembly: the maximization of *prediction accuracy*, minimization of *interpretation*, and maximization of *completeness*. The first objective, the maximization of prediction accuracy, requires that the estimated expression levels based on assembled transcripts should be as close as possible to the observed ones for every expressed region of the genome. The minimization of interpretation follows the parsimony principle to seek as few transcripts in the prediction as possible. The third objective, the maximization of completeness, requires that the maximum number of mapped reads (or "expressed segments" in gene models) be explained by (*i.e.*, contained in) the predicted transcripts in the solution. Based on the above three objectives, we present IsoLasso, a new RNA-Seq based transcriptome assembly tool. IsoLasso is based on the well-known LASSO algorithm, a multivariate regression method designated to seek a balance between the maximization of prediction accuracy and the minimization of interpretation. By including some additional constraints in the quadratic program involved in LASSO, IsoLasso is able to make the set of assembled transcripts as complete as possible. Experiments on simulated and real RNA-Seq datasets show that IsoLasso achieves higher sensitivity and precision simultaneously than the state-of-art transcript assembly tools.

1 Introduction

The second generation sequencing technology has become an increasingly important tool in biological and biomedical research areas, such as individual genome sequencing [1], gene expression level estimation [2], comparative genomics [3], *etc.* RNA-Seq, a technology to study transcriptome via the second generation

V. Bafna and S.C. Sahinalp (Eds.): RECOMB 2011, LNBI 6577, pp. 168–188, 2011.

sequencing, was first introduced in a series of studies in 2008 [2,4,5,6,7,8,9], and has quickly become widely accepted as a fundamental tool for transcriptome research [10,11,12,13]. The revolutionary new sequencing technology allows RNA-Seq to lower the sequencing cost and increase the data throughput substantially, but it also posts many challenging computational biology problems, one of which is transcriptome assembly and abundance estimation from RNA-Seq reads. A variety of new algorithms and tools have been developed for this problem [14,15,16,17,18,19]. Some splicing site discovery tools, for example TopHat [19] and SpliceMap [20], identify new alternative splicing events by exploring RNA-Seq reads that span different parts of the reference genome under study. Some *de novo* assembly tools, such as AbySS [14], try to assemble new transcripts solely from RNA-Seq reads. Other assembly tools (including Cufflinks [16], Scripture [17] and IsoInfer [18]) map reads to the reference genome and build transcript models (or isoforms) from these mapped reads.

Among these tools, IsoInfer [18] enumerates all possible "valid" isoforms and uses a quadratic program (QP) to estimate the expression levels of a given set of isoforms. IsoInfer then chooses the best subset of valid isoforms such that the estimated abundance of every "expressed segment" of the reference genome (*e.g.*, an exon) is proportional to the observed reads falling into the segment. On the other hand, Cufflinks [16] assembles isoforms using a parsimony strategy, *i.e.*, it attempts to identify the minimum number of isoforms to cover all the reads. To do this, Cufflinks decomposes the "overlap graph" of compatible reads into a smallest path cover, and then calculates the expression levels of the isoforms (*i.e.*, paths in the cover) using the probabilistic model proposed in [21].

The strategies that IsoInfer and Cufflinks adopted correspond to two different model selection principles: *prediction accuracy* and *interpretation* [22]. IsoInfer selects isoforms to maximize the prediction accuracy, *i.e.*, to minimize the error or discrepancy between the predicted and observed expression levels in all expressed segments. IsoInfer employs a search algorithm similar to the "best subset variable selection" algorithm [23] to find the best subset of isoforms. However, the huge search space prevents the algorithm from doing a thorough search, and many heuristic restrictions must be applied to make the search tractable. On the other hand, Cufflinks minimizes interpretation, *i.e.*, the number of variables (or isoforms) that are required to explain all the mapped reads. Here, the prediction

Table 1. Transcriptome assembly objectives of each algorithm. Although Cufflinks has a transcript abundance estimation step, the prediction accuracy is not considered explicitly during the assembly process. Also, theoretically both Cufflinks and IsoLasso take completeness into consideration, but in practice they may not fully guarantee it and thus are marked "partially" in the table.

Algorithm	Prediction accuracy	Interpretation	Completeness
IsoInfer	Yes	Partially	Yes
Cufflinks	No	Yes	Partially
Scripture	No	No	Yes
IsoLasso	Yes	Yes	Partially

accuracy is not considered explicitly during the transcriptome assembly process. By defining a "partial order" between reads, Cufflinks filters out "uncertain" paired-end reads which may result in a sub-optimal path cover in the solution, or miss some alternative splicing events. Finally, Scripture [17] reconstructs all possible isoforms by enumerating all possible paths in the "connectivity graph". This approach may lead to many incorrect isoforms for complex genes with a large number of exons, since the number of paths may be huge for such gene models.

Another important objective in transcriptome assembly is *completeness*, which requires that all exons (and exon junctions) appear in at least one isoform in the solution (as done in IsoInfer [18]), or all mapped reads be contained in at least one isoform (as done in Cufflinks [16]). In IsoInfer, the completeness is achieved by solving a set cover instance that covers all expressed segments and exon junctions. Since all the reads represented in the overlap graph are partitioned into disjoint paths in Cufflinks, they are guaranteed to be supported by at least one isoform (*i.e.*, path). However, some "uncertain" paired-end reads (*i.e.*, reads that cannot be included in partial order and thus absent in the overlap graph) may not be covered by the solution. Scripture adopts a conservative approach to enumerate all possible paths in its connectivity graph, which is guaranteed to cover all expressed segments and exon junctions. Like Cufflinks, the prediction accuracy is not considered explicitly during the transcript assembly process of Scripture. Moreover, retaining all possible isoforms clearly leads to a bad interpretation. Table 1 lists all the principles (or objectives) that IsoInfer, Cufflinks and Scripture abide by in the transcript assembly process.

In this paper, we present a new isoform assembly algorithm, IsoLasso, which balances prediction accuracy, interpretation and completeness. IsoLasso uses the LASSO algorithm, or Least Absolute Shrinkage and Selection Operator [24], which is a shrinkage least squares method in statistical machine learning. By adding an L1 norm penalty term to the least squares objective function, LASSO achieves sparsity by setting the expression levels of unrelated isoforms to zero, thus balancing both prediction accuracy and interpretation. The LASSO algorithm is widely applied in many computational biology areas, such as genome-wide association analysis [25,26], gene regulatory network [27], microarray data analysis [28], *etc.* In IsoLasso, we expand the quadratic programming problem of LASSO to take completeness into consideration. Our experiments demonstrate that IsoLasso runs efficiently and achieves overall higher sensitivity and precision than IsoInfer, Cufflinks and Scripture.

The rest of this paper is organized as follows. Section 2.1 presents our algorithm for generating (or enumerating) candidate isoforms and its relationship to minimum path covers used in Cufflinks [16]. These candidate isoforms will be fed to our LASSO algorithm described in Section 2.2 for estimating isoform expression levels (or, equivalently, for inferring expressed isoforms). Section 2.3 expands the basic LASSO approach to take completeness into consideration. Experimental results are presented in Section 3, which include comparisons between IsoLasso, IsoInfer, Cufflinks, and Scripture on simulated and real datasets.

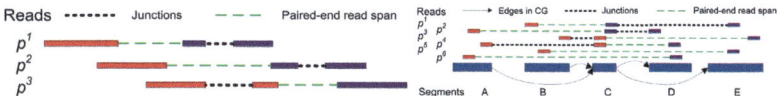

Fig. 1. (Left) Removal of "uncertain" reads may cause splicing junctions undetected in Cufflinks. Three paired-end reads, p^1, p^2 and p^3, concern different splicing junctions. Both pairs (p^1, p^2) and (p^2, p^3) are compatible, but the pair (p^1, p^3) is not. Removing any of these reads will cause one or more junctions undetected. (Right) "Infeasible" paths in the connectivity graph. In the example above, there are four possible combinations of segments: ACD, ACE, BCD, and BCE. However, ACE and BCD are infeasible since they cannot be assembled from the mapped paired-end reads.

Section 4 concludes the paper. For the convenience of the reader, we defer some mathematical definitions and the proofs of theorems to the Appendix.

2   Methods

2.1   Enumerating Candidate Isoforms

IsoInfer [18], Scripture [17] and Cufflinks [16] enumerate candidate isoforms in different ways. IsoInfer, assuming that expressed segment (or exon) boundaries in a gene are given, enumerates all possible combinations of segments. Note that it is possible that some lowly expressed segment are not hit by short reads and thus many of the isoforms enumerated by IsoInfer might have very low expression levels. Scripture enumerates all possible maximal paths in a *connectivity graph*; but some of these isoforms may be "infeasible" because they cannot be assembled from the mapped reads (Figure 1 (right) shows such an example). Cufflinks tries to build an *overlap graph* from partially ordered reads, and assembles putative transcripts by decomposing the overlap graph into a parsimonious path cover. However, a strict partial order between reads is required here. Since the actual sequence between the ends of each paired-end read is unknown, Cufflinks has to exclude some paired-end reads (called *uncertain reads*) to maintain the partial order. Removing uncertain reads may lead to two potential problems: (1) the path cover solution is actually sub-optimal and (2) some alternative splicing events are missed, if the reads including these events are removed. For instance, Figure 1 (left) provides an example that removing such "uncertain" reads leaves some splicing junctions undetected. Note that uncertain reads should be treated separately from repeat sequences or incorrectly mapped reads.

Here, we describe our method of enumerating isoforms based on the connectivity graph ([17]) in Algorithm 1, from which the enumerated isoforms will be the set of candidate isoforms to be considered in the LASSO algorithm. The algorithm first enumerates isoforms from the connectivity graph as in [17], and then uses two additional steps to remove isoforms that are impossible to assemble. We will prove some important properties of Algorithm 1: if there are no "uncertain" reads, then every isoform output by Algorithm 1 can be assembled from a maximal path in the overlap graph given in [16]. Moreover, the isoforms enumerated by Algorithm 1 form a superset of all possible maximal paths in the overlap graph. In other words, our LASSO algorithm in general considers more isoforms than Cufflinks in the transcript assembly process. Before giving a detailed description of this algorithm and proofs of these properties, we first briefly review some necessary notations first introduced in [16] and [17].

A gene sequence S of length n is an ordered character sequence $S = S_1 S_2 \cdots S_n$, $S_i \in \{A, T, G, C\}$. Define $B(n)$ as the set of binary vectors of length n. For a vector $b \in B(n)$, b_i indicates the ith element of vector b. For a subset $U \subset B(n)$, define $OR(U) = b \in B(n)$ with $b_i = 1$ iff there is an element $c \in U$ such that $c_i = 1$. For a binary vector $b \in B(n)$, define the start (or end) of b as the first (or last) non-zero index of b, and is denoted as $l(b)$ (or $u(b)$). Hence, each isoform on gene S could be represented as a binary vector $b \in B(n)$ with $b_i = 1$ iff the nucleotide S_i is included in this isoform. A single-end or paired-end read mapped to S could also be represented as an element $b \in B(n)$ with $b_i = 1$ iff this read contains S_i. A paired-end read is denoted as $p = (b^1, b^2)$, where b^1 and b^2 are the two mapped single-end reads, and $l(b^1) < l(b^2)$. Given a set of single-end or paired-end reads R, the coverage of S_i, or $cvg(S_i)$, is the number of reads b with $b_i = 1$.

A single-end read b is *compatible* with an isoform t, denoted as $b \sim t$, iff $b_i = t_i$ for $l(b) \leq i \leq u(b)$. Similarly, a paired-end read $p = (b^1, b^2)$ is compatible with isoform t, denoted as $p \sim t$, iff $b^1 \sim t$ and $b^2 \sim t$. Given a set of single-end (or paired-end) reads R mapped to gene S, the *connectivity graph (CG)* [17] is a directed acyclic graph (DAG) $G = (V, E)$, where $V = \{v_1, v_2, \ldots, v_n\}$ and $e = (v_i, v_j) \in E$ iff one of the following conditions is true:

Condition 1.   There exists a single-end read or an end of some paired-
               end read $b \in R$ such that $b_i = 1$, $b_j = 1$, and $b_k = 0$,
               $\forall i < k < j$;
Condition 2. $cvg(S_i) > 0$, $cvg(S_j) > 0$, and $cvg(S_k) = 0$, $\forall i < k < j$.

Note that Condition 2 is designed to connect two mapped reads separated by a coverage gap. Based on the definition of CG, a path h in the CG could be readily treated as an isoform by defining the isoform t as $t_i = 1$ iff $v_i \in h$. Therefore, a read b is compatible with h (denoted as $b \sim h$) iff $b \sim t$. The isoform enumeration algorithm depicted in Algorithm 1 takes the connectivity graph as the input, and outputs a set of maximal candidate isoforms T. The algorithm consists of three phases, Enumeration, Filtration and Condensation. In the Enumeration phase, all maximal paths in the connectivity graph are enumerated. However, some of these isoforms are "infeasible" in the sense that they cannot be assembled from the mapped reads (see Figure 1 (right) for an example). In this case, the second phase (*i.e.*, the Filtration phase) is required to remove such isoforms. For each isoform t generated in the Enumeration phase, the Filtration phase first finds all reads that are compatible with t, and then checks if t can be assembled from these compatible reads (it replaces t otherwise). Finally, the Condensation phase removes all the isoforms that are not maximal candidates.

Cufflinks assembles transcripts based on the *overlap graph (OG)*, which is is constructed from a set of mapped single-end or paired-end reads after remov-ing *uncertain* reads and extending reads to include their *nested* reads [16]. It

input : A CG $G = (V, E)$, and a set of mapped single-end or paired-end reads
            R
output: A set of isoforms T
begin
> **Enumeration:**
> $T \leftarrow \emptyset$
> **for** $v_j \in V$ *with* $indeg(v_j) = 0$ **do**
> > Enumerate all possible maximal paths P that begin at v_j and end at
> > some v_k with $outdeg(v_k) = 0$
> > $T \leftarrow T \cup P$
>
> **Filtration:**
> **for** $t \in T$ **do**
> > Let $t' = OR(\{b \in R | b \sim t\})$
> > $T \leftarrow (T \backslash \{t\}) \cup \{t'\}$
>
> **Condensation:**
> **for** $t \in T$ **do**
> > Let $R_t = \{b \in R |, b \sim t\}$
> > **for** $t' \in T \backslash \{t\}$ **do**
> > > Let $R_{t'} = \{b \in R |, b \sim t'\}$
> > > **if** $R_t \subset R_{t'}$ **then**
> > > > $T \leftarrow (T \backslash \{t\})$

end

Algorithm 1. Isoform Enumeration

generates transcripts by partitioning the overlap graph into a *minimum path cover*, where a path cover is a set of disjoint paths in the overlap graph such that every read appears in one and only one path. A minimum path cover is a path cover with the minimum number of paths. The following theorems and corollary state the relationship between the set of isoforms generated by Algorithm 1 and the set of transcripts that could be constructed from the overlap graph. Formal definitions of uncertain reads, nested reads and the overlap graph, and complete proofs of these theorems are given in the Appendix. Let us consider a fixed gene.

Theorem 1. *Suppose that R contains no uncertain or nested reads. If we denote the set of isoforms constructed by Algorithm 1 as T and the set of the isoforms formed by enumerating maximal paths on the OG (constructed from R) as T_{OG}, then $T = T_{OG}$.*

Corollary 1. *If R contains no uncertain or nested reads, then for every minimum path cover H of the OG, there exists a set of maximal isoforms $T' = \{t^1, \dots t^m\} \subset T$, such that $m = |H|$ and for every read b on a path $h \in H$, $b \sim t^i$, $1 \leq i \leq m$.*

Note that each nested read r in R is removed in [16] by extending the reads that r is nested in. On the other hand, if there are uncertain reads in R, Algorithm 1 may generate some isoforms that do not correspond to any paths on the OG when these uncertain reads cover some unique splicing junctions as shown in

Figure 1 (left). The following theorem states the relationship between maximal paths on the OG and the isoforms generated by Algorithm 1 when uncertain reads are present in R.

Theorem 2. *Suppose that no reads in R are nested and denote the set of isoforms constructed by Algorithm 1 as T. For every maximal path h on the OG constructed by removing uncertain reads in R, T contains an isoform which is compatible with every read on the path h.*

2.2   The LASSO Approach of Estimating Isoform Expression Levels

The Mathematical Model of RNA-Seq. Typical *alternative splicing (AS)* events include alternative 5' (or 3') splice sites, exon skipping, intron retention, mutually exclusive exons, *etc.*, but all these events can be dealt with in a unified mathematical model where a gene is partitioned into a sequence of *expressed segments* (or simply *segments*) based on exon-intron boundaries [18]. More precisely, a gene is divided into a set of segments such that every segment is a continuous region in the reference genome uninterrupted by exon-intron boundaries. Then, a given set of candidate isoforms $T = \{t^1, t^2, \ldots, t^N\}$ for a gene can be represented as a binary matrix $A = (a_{ij})_{N \times M}$, where M is the number of segments of the gene. Each isoform corresponds to a row in this matrix such that $a_{ij} = 1$ if isoform t^i includes the jth segment, and 0 otherwise.

If we assume that a read is uniformly sampled from expressed isoforms, then the number of reads falling into each segment follows a binomial distribution, which can be approximated by a Poisson distribution [21] or Gaussian distribution [18] if the number of sequenced reads is large and the length of segments is small compared with the length of the reference genome. As a result, the expected number of reads falling into the ith segment, r_i, is proportional to both the segment length l_i and the sum of the expression levels of all isoforms containing the ith segment [21,18]:

$$r_i = l_i \sum_{j=1}^{N} a_{ji} x_j \tag{1}$$

where x_j, the expected number of reads per base in isoform t^j, represents the expression level of t^j. Note that the expression level of an isoform can also be measured as RPKM (*i.e.*, Reads Per Kilobase of exon model per Million mapped reads, [2]). If there are totally E mapped reads, then an isoform t^j with expression level x_j has an expression level (in RPKM) $10^9 x_j / E$.

Notice that compared with the traditional multivariate regression model, the intercept is zero since we expect no read falling into the ith segment, if none of the isoforms contain the segment, or if the expression levels of these isoforms are all zero.

We observe that the above model simplifies the real situation. Because of the sequencing errors and repeat sequences in the reference genome, it is sometimes hard to decide whether a read really comes from a certain gene or exon (*i.e.*, the so called multi-read problem, which has been studied recently in [29]). Recent studies on RNA-Seq data also show that the above binomial model of read distribution may be an over-simplification [30,31]. Some more complicated approaches have been proposed instead, such as using generalized Poisson distribution [32], considering the locality of bases [30], applying "effective length normalization" [31,33], *etc.* In particular, the "effective length normalization" model can be easily incorporated in our model, by replacing the segment length l_i in Equation (1) with the "effective" segment length l_i', where the length is calibrated by considering repeat sequences in the reference genome [33].

The LASSO Approach. Given all mapped short reads and candidate isoforms of a gene, the expression levels $X = \{x_1, \ldots x_N\}$ of the candidate isoforms can be estimated by minimizing the following residual sum of squares:

$$X^* = \underset{X}{\arg\min} f(X) = \sum_{i=1}^{M} (\frac{r_i}{l_i} - \sum_{j=1}^{N} a_{ji} x_j)^2 \qquad (2)$$

with respect to the restrictions that $x_j \geq 0$ for all $1 \leq j \leq N$. However, such an approach may have several potential problems. For example, for a large value of N and a small value of M, the solution is not unique. It is also possible that a large number of estimated expression levels are small non-zero values which damage the interpretability. To address this latter problem, IsoInfer enumerates combinations of isoforms and chooses a minimum set of isoforms such that the error $\sum_{i=1}^{M} (\frac{r_i}{l_i} - \sum_{j=1}^{N} a_{ji} x_j)^2$ is in a specified range. To deal with an exponential number of subsets of candidate isoforms, IsoInfer has to adopt several heuristics to make the algorithm practical. Also, some "shrinkage" methods which restrict the scale of X can be used, like ridge regression [34], LASSO (or its variations like LARS [35], elastic-net [36], *etc*).

To achieve the minimization of interpretation without going through the exhaustive enumeration step in IsoInfer, we propose a new algorithm, called Iso-Lasso, based on LASSO. The LASSO approach minimizes the following objective function which seeks a balance between minimizing the overall error and minimizing the number of expressed isoforms:

$$f(X) = \sum_{i=1}^{M} (\frac{r_i}{l_i} - \sum_{j=1}^{N} a_{ji} x_j)^2 + \lambda \sum_{j=1}^{N} |x_j| \qquad (3)$$

The sparsity of variables, *i.e.*, minimizing the number of isoforms with non-zero expression levels, is obtained through the addition of an L1 normalization term, $\lambda \sum_{j=1}^{N} |x_j|$, to the original sum of squares. Since the expression level of each isoform should be non-negative, the above objective function leads to the following quadratic programming (QP) problem:

$$\min f(X) = \sum_{i=1}^{M} (\frac{r_i}{l_i} - \sum_{j=1}^{N} a_{ji} x_j)^2 + \lambda \sum_{j=1}^{N} x_j \qquad (4)$$

$$s.t. \quad x_j \geq 0, \ 1 \leq j \leq N$$

which is equivalent to the following "constrained form" [24]:

$$\min f(X) = \sum_{i=1}^{M} (\frac{r_i}{l_i} - \sum_{j=1}^{N} a_{ji}x_j)^2 \tag{5}$$

$$s.t. \quad x_j \geq 0, \ 1 \leq j \leq N$$

$$\sum_{j=1}^{N} x_j \leq \gamma$$

The parameter λ (or γ) controls the number of isoforms with non-zero expression levels in the solution. In the constrained form of LASSO (Equation (5)), a larger value of γ will exert less restriction on the values of X, which prefer a smaller sum of squares but more non-zero expression levels. In practice, a proper value of γ is selected via the "regularization path" [37], where several values of γ, $\gamma_1, \ldots \gamma_k$, are examined. If the values of the objective function in Equation (5) and the number of non-zero variables are $e_1, \ldots e_k$ and $L_1, \ldots L_k$, respectively, in these trials, then we define

$$i^* = \underset{1 \leq i \leq k}{\operatorname{argmin}} \{L_i : e_i \leq \beta * \min \{e_1, \ldots e_k\}\} \tag{6}$$

and select $\gamma = \gamma_{i^*}$, where β is a user-controlled parameter.

2.3   Completeness Requirement

To ensure completeness, *i.e.*, each segments (or junction) with mapped reads covered by at least one isoform, the sum of expression levels of all isoforms that contain this segment (or junction) should be strictly positive. Formally, we add additional constraints to the above QP:

$$\min f(X) = \sum_{i=1}^{M} (\frac{r_i}{l_i} - \sum_{j=1}^{N} a_{ji}x_j)^2 \tag{7}$$

$$s.t. \quad x_j \geq 0, \ 1 \leq j \leq N$$

$$\sum_{j=1}^{N} x_j \leq \lambda$$

$$\sum_{j=1}^{N} x_j a_{ji} \geq p, \text{ if segment } i \text{ has mapped reads} \tag{8}$$

$$\sum_{j=1}^{N} x_j a_{ji} a_{jk} \prod_{h=i+1}^{k-1} (1 - a_{jh}) \geq p, \text{ if the junction between segments}$$

i and k contains mapped reads

$$\tag{9}$$

where p is a small positive threshold value to be decided empirically. The constraints (Equation (8) and Equation (9)) will ensure that all segments and junctions with mapped reads be covered by isoforms with positive expression levels in the solution of this QP.

The above QP problem can be solved by any standard QP solver, such as the "quadprog" function in Matlab [38]. In practice, however, if a gene contains too many segments and junctions, then there will be a large number of constraints involved, which make the above QP impractical to solve. As a compromise, we introduce the above constraints only for segments (or junctions) with expression levels above a certain threshold.

3   Experimental Results

3.1   Simulated Mouse RNA-Seq Data

We use UCSC mm9 gene annotation to generate simulated single-end and paired-end reads. An *in silico* RNA-Seq data generator, Flux Simulator [39], is used to generate simulated reads. Flux Simulator first randomly assigns an expression level to every isoform in the annotation, and then simulates the library preparation process in a typical RNA-Seq experiment (including reverse transcription, fragmentation, size selection, *etc*). After that, reads are generated in the sequencing step. Various error models can be incorporated in these steps; but in our simulations, only error-free reads are simulated to compare the performance of different algorithms in the ideal situation.

The distribution of the expression levels of all 49409 isoforms in the UCSC mm9 gene annotation is plotted in Figure 2 (A).

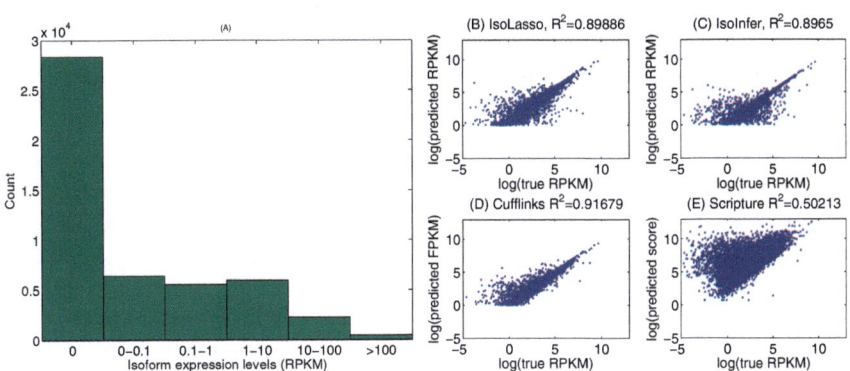

Fig. 2. The distribution of simulated isoform expression levels (A), and the expression level estimation accuracies of IsoLasso (B), IsoInfer without TSS/PAS (C), Cufflinks (D), and Scripture (E). Note that Scripture computes a "weighted score" instead of RPKM value for each predicted isoform.

Matching Criteria. All assembled isoforms (referred to as "candidate iso-forms") are matched against all known isoforms in the annotation (referred to as "benchmark isoforms"). Two isoforms match iff:

1. They include the same set of exons; and
2. All internal boundary coordinates (*i.e.*, all the exon coordinates ex-cept the beginning of the first exon and the end of the last exon) are identical.

Two single-exon isoforms match iff the overlapping area occupies at least 50% the length of each isoform.

Following [18], we use *sensitivity, precision* and *effective sensitivity* to evalu-ate the performance of different programs. Sensitivity and precision are defined as follows: if K out of M benchmark isoforms match K' out of N candidate isoforms, then

$$\text{sensitivity} = K/M \qquad (10)$$
$$\text{precision} = K'/N \qquad (11)$$

Note that several candidate isoforms may match the same benchmark isoform.

Effective sensitivity is calculated based on the isoforms satisfying *Condition I* defined in [18]. Isoforms satisfying Condition I are those with all segment junctions covered by at least one short read. If there are S benchmark isoforms satisfying Condition I and K of them are matched, then

$$\text{effective sensitivity} = K/S \qquad (12)$$

Intuitively, isoforms satisfying Condition I are those that are relatively easy to predict, since all their segment junctions are covered by short reads. It is shown in [18] that an isoform with a higher expression level is more likely to satisfy this condition.

3.2   Comparisons between IsoLasso, IsoInfer, Cufflinks, and Scripture

Sensitivity, precision and effective sensitivity. In this section, we use the sensitivity, precision and effective sensitivity defined above to compare IsoLasso with the most recent versions of IsoInfer (version V0.9.1, downloaded from web-site `http://www.cs.ucr.edu/~jianxing/IsoInfer.html`), Cufflinks (version 0.9.1, downloaded from website `http://cufflinks.cbcb.umd.edu`), and Scrip-ture (beta version, downloaded from website `http://www.broadinstitute.org/software/scripture/home`). We use TopHat [19] to map all simulated short reads with multi-reads discarded. Then, the read mapping information serves as the input for all four programs. Since IsoInfer is based on the assumption that the boundaries of all genes and exons are known, we infer exon bound-aries from mapped junction reads using TopHat and infer gene boundaries by clustering overlapping mapped reads. Note that IsoInfer is actually designed to take advantage of any known transcription start site and poly-A site (TSS/PAS)

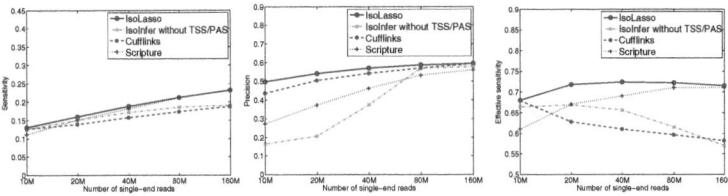

Fig. 3. Sensitivity (left), precision (middle) and effective sensitivity (right) on single-end reads

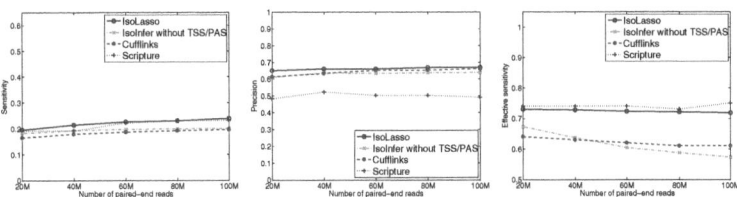

Fig. 4. Sensitivity (left), precision (middle) and effective sensitivity (right) on paired-end reads

information, although it also works without such information. Since the other three programs do not use the TSS/PAS information, neither does IsoInfer use such information in the comparison.

Figure 3 and Figure 4 plot the sensitivity, precision and effective sensitivity using various numbers of single-end and paired-end reads, respectively. On single-end reads, all transcriptome assembly tools achieve a higher sensitivity and precision as more reads are used for the assembly. Among them, IsoLasso outperforms all other programs with respect to all three criteria. This is perhaps because IsoLasso is able to maintain a good interpretation by filtering out many lowly expressed false predictions (which leads to a high precision), while keeping highly expressed isoforms and a high effective sensitivity. Scripture seems to benefit the most when more reads are available. Also, IsoInfer exhibits a sharp increase in precision from less than 20% to more than 50%, at the cost of decreased effective sensitivity (by about 10%).

On paired-end reads, IsoLasso also achieves the best precision and sensitivity as well as a good balance between precision and effective sensitivity. However, it is surprising to see that when the number of paired-end reads increases from 20M to 100M, a less than 10% increase in sensitivity and precision is observed for all the algorithms. Also, none of the algorithms have a significant increase in effective sensitivity. In fact, both Cufflinks and IsoInfer see their effective sensitivities decreased a bit when more single-end and paired-end reads are used. This is because more benchmark isoforms would satisfy Condition I of [18] as the sequencing depth increases. In this case, more isoforms are expected to be expressed for each gene, which result in a more complicated overlap graph for Cufflinks and a larger search space for IsoInfer.

Cufflinks reaches a high precision by filtering out many lowly expressed isoforms, but this sacrifices the effective sensitivity. On the other hand, Scripture achieves the highest effective sensitivity by enumerating all possible paths in the connectivity graph, but its precision is low since many of the paths are false positives.

Expression Level Estimation. All programs estimate the expression levels of predicted isoforms using different measures. Both IsoLasso and IsoInfer estimate expression levels in RPKM [2], while Cufflinks uses the term FPKM (expected number of Fragments Per Kilobase of transcript sequence per Millions base pairs sequenced) [16]. Scripture does not predict expression levels directly; instead, it computes a "weighted score" for each isoform to indicate how likely the isoform is expressed.

Fig. 2 (B) $\sim$ (E) plot the predicted and true expression levels for all predicted isoforms which are matched to the benchmark isoforms and have expression levels > 1 RPKM, using the 80M paired-end read dataset. The plots show that IsoLasso, IsoInfer and Cufflinks estimate expression levels quite accurately (the squared correlation coefficient between the predicted and true expression levels is $R^2 > 0.89$), while the "weighted score" of Scripture does not directly reflect the true expression level of isoforms ($R^2 = 0.50$). Cufflinks shows the highest prediction accuracy in expression level estimation ($R^2 = 0.91$) partly because it uses an accurate iterative statistical model to estimate the expression levels [16], which could potentially be incorporated into our method as a refinement step.

More Isoforms, More Difficult to Predict. Intuitively, genes with more isoforms are more difficult to predict. We group all the genes by their numbers of isoforms, and calculate the sensitivity and effective sensitivity of the algorithms on genes with a certain number of isoforms as shown in Figure 5 (middle) and (right). Figure 5 (left) shows the total number of isoforms and isoforms satisfying Condition I ([18]) grouped by the number of isoforms per gene.

Figure 5 shows that genes with more isoforms are more difficult to predict correctly, as both sensitivity and effective sensitivity decrease for genes with more isoforms. IsoLasso and Scripture outperform IsoInfer and Cufflinks in general. IsoLasso has a higher sensitivity and effective sensitivity on genes with at most 5 isoforms, but Scripture catches up with IsoLasso on genes containing more than 5 isoforms.

Running Time. Figure 6 plots the running time of all four transcript assembly programs using various numbers of paired-end reads. The time for data preparation is excluded, including mapping reads to the reference genome and preparing required input files for both IsoLasso and IsoInfer. Surprisingly, although employing a search algorithm, IsoInfer runs much faster than that of any other algorithm. This is partly due to the heuristic restrictions that IsoInfer adopts to

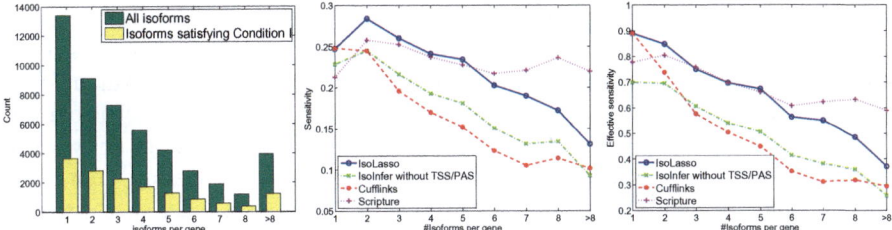

Fig. 5. The total number of isoforms and isoforms satisfying Condition I (left), and the sensitivity (middle) and effective sensitivity (right) of the algorithms grouped by the number of isoforms per gene. Here, 100M paired-end reads are simulated.

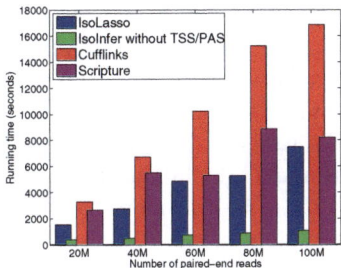

Fig. 6. The running time for all the algorithms

reduce the search space (*e.g.*, requiring the candidate isoforms to satisfy Condition I and some other conditions), and the programming languages used in each tool (IsoInfer, IsoLasso, Scripture and Cufflinks use C++, Matlab, Java, and Boost C++, respectively). All programs are run on a single 2.6 GHz CPU, but Cufflinks allows the user to run on multiple threads, which may substantially speed up the assembly process.

3.3   Real RNA-Seq Data

Reads from two real RNA-Seq experiments are used to evaluate the performance of IsoLasso, Cufflinks and Scripture. We exclude IsoInfer from the comparison because its algorithm is similar to (and improved by, as seen from the simulation results) the algorithm of IsoLasso. One RNA-Seq read dataset is generated from the C2C12 mouse myoblast cell line ([16], NCBI SRA accession number SRR037947), and the other from human embryonic stem cells (Caltech RNA-Seq track from the ENCODE project [40], NCBI SRA accession number SRR065504). Both RNA-Seq datasets include 70 million and 50 million 75 bp paired-end reads which are mapped to the UCSC *mus musculus* (mm9) and *homo sapiens* (hg19) reference genomes using Tophat [19], respectively.

Isoforms inferred by programs IsoLasso, Cufflinks and Scripture are first matched against the known isoforms from mm9 and hg19 reference genomes. There are a total of 11484 and 12193 known mouse and human isoforms recovered by at least one program, respectively (Figure 7 (A) and (B)). Among these isoforms, 4485 (39%) and 4274 (35%) isoforms are detected by all programs, while 8204 (71%) and 8084 (66%) isoforms are detected by at least two programs. These numbers show that, although there is a large overlap (more than 60%) among the known isoforms recovered by these programs, each program also identifies a substantially large number of "unique" isoforms. Such "uniqueness" of each program is shown more clearly if we compute the overlap between their predicted isoforms directly (see Figure 7 (C) and (D)). Each of the three programs predicts more than 40,000 isoforms on both dataset, but only shares 2% to 20% isoforms with other programs. About 49.5% of the mouse isoforms (46% in human) inferred by IsoLasso are also predicted by at least one of other two programs, which is substantially higher than Cufflinks (27.7% in mouse and 38.4% in human) and Scripture (4.6% in mouse and 7.4% in human). This may indicate that IsoLasso's prediction is more reliable than those of Cufflinks and Scripture since it receives more support from other (independent) programs.

Note that among all the isoforms inferred by IsoLasso, Cufflinks and Scripture, 9741 mouse isoforms and 11381 human isoforms are predicted by all three programs. These isoforms could be considered as "high-quality" ones. However, fewer than a half of these "high-quality" isoforms (4485 in mouse and 4274 in human) could be matched to the known mouse and human isoforms (see Figure 7 (A) and (B)). This suggests that the current genome annotations of both mouse and human are still incomplete. An example of the "high-quality" isoforms is shown in Figure 7 (E). Here, an isoform with an alternative 5′ end of gene Tmem70 in mouse is predicted by all three programs but cannot be found in the mm9 RefSeq annotation or GenBank mRNAs (track not shown in the figure).

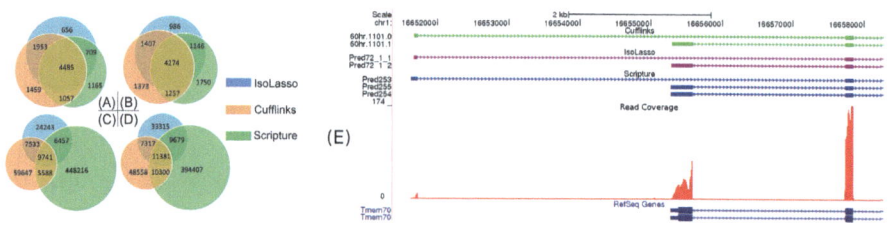

Fig. 7. The numbers of matched known isoforms of mouse (A) and human (B), and the numbers of predicted isoforms of mouse (C) and human (D), assembled by IsoLasso, Cufflinks and Scripture. (E) shows an alternative 5" start isoform of gene Tmem70 in mouse C2C12 myoblast RNA-Seq data [16]. This isoform does not appear among the known isoforms, but is detected by IsoLasso, Cufflinks and Scripture. Tracks from top to bottom: Cufflinks predictions, IsoLasso predictions, Scripture predictions, the read coverage, and the Tmem70 gene in the mm9 RefSeq annotation.

4   Conclusion

RNA-Seq transcriptome assembly is a challenging computational biology problem that arises from the development of second generation sequencing. In this paper, we proposed three fundamental objectives/principles in the transcriptome assembly: prediction accuracy, interpretation, and completeness. We also presented IsoLasso, an algorithm based on the LASSO approach that seeks a balance between these objectives. Experiments on simulated and real RNA-Seq datasets show that, compared with the existing transcript assembly tools (IsoInfer, Cufflinks and Scripture), IsoLasso is efficient and achieves the best overall performances in terms of sensitivity, precision and effective sensitivity.

Acknowledgments

IsoLasso is available at `http://www.cs.ucr.edu/~liw/isolasso.html`. We thank the anonymous referees for many constructive comments. The research is supported in part by NSF grant IIS-0711129 and NIH grant AI078885.

References

1. Wheeler, D.A., et al.: The complete genome of an individual by massively parallel dna sequencing. Nature 452, 872–876 (2008)
2. Mortazavi, A., et al.: Mapping and quantifying mammalian transcriptomes by rna-seq. Nature Methods 5, 621–628 (2008)
3. Holt, K.E., et al.: High-throughput sequencing provides insights into genome variation and evolution in salmonella typhi. Nature Genetics 40, 987–993 (2008)
4. Wilhelm, B.T., et al.: Dynamic repertoire of a eukaryotic transcriptome surveyed at single-nucleotide resolution. Nature 453, 1239–1243 (2008)
5. Lister, R., et al.: Highly integrated Single-Base resolution maps of the epigenome in arabidopsis. Cell 133(3), 523–536 (2008)
6. Morin, R., et al.: Profiling the HeLa s3 transcriptome using randomly primed cDNA and massively parallel short-read sequencing. BioTechniques 45, 81–94 (2008), PMID: 18611170
7. Marioni, J.C., et al.: RNA-seq: an assessment of technical reproducibility and comparison with gene expression arrays. Genome Research 18(9), 1509–1517 (2008)
8. Cloonan, N., et al.: Stem cell transcriptome profiling via massive-scale mRNA sequencing. Nat. Meth. 5, 613–619 (2008)
9. Nagalakshmi, U., et al.: The transcriptional landscape of the yeast genome defined by RNA sequencing. Science 320, 1344–1349 (2008)
10. Haas, B.J., Zody, M.C.: Advancing RNA-Seq analysis. Nat. Biotech. 28, 421–423 (2010)
11. Morozova, O., et al.: Applications of new sequencing technologies for transcriptome analysis. Annual Review of Genomics and Human Genetics 10(1), 135–151 (2009), PMID: 19715439
12. Wall, P.K., et al.: Comparison of next generation sequencing technologies for transcriptome characterization. BMC Genomics 10(1), 347 (2009)

13. Wang, Z., et al.: RNA-Seq: a revolutionary tool for transcriptomics. Nat. Rev. Genet. 10, 57–63 (2009)
14. Birol, I., et al.: De novo transcriptome assembly with abyss. Bioinformatics 25, 2872–2877 (2009)
15. Yassour, M., et al.: Ab initio construction of a eukaryotic transcriptome by massively parallel mrna sequencing. Proceedings of the National Academy of Sciences of the United States of America 106, 3264–3269 (2009)
16. Trapnell, C., et al.: Transcript assembly and quantification by rna-seq reveals unannotated transcripts and isoform switching during cell differentiation. Nature Biotechnology 28, 511–515 (2010)
17. Guttman, M., et al.: Ab initio reconstruction of cell type-specific transcriptomes in mouse reveals the conserved multi-exonic structure of lincrnas. Nature Biotechnology 28, 503–510 (2010)
18. Feng, J., et al.: Inference of isoforms from short sequence reads. In: Berger, B. (ed.) RECOMB 2010. LNCS, vol. 6044, pp. 138–157. Springer, Heidelberg (2010)
19. Trapnell, C., et al.: Tophat: discovering splice junctions with rna-seq. Bioinformatics 25, 1105–1111 (2009)
20. Au, K.F., et al.: Detection of splice junctions from paired-end rna-seq data by splicemap. Nucl. Acids Res., gkq211+ (April 2010)
21. Jiang, H., Wong, W.H.: Statistical inferences for isoform expression in rna-seq. Bioinformatics 25, 1026–1032 (2009)
22. Hastie, T., et al.: The Elements of Statistical Learning: Data Mining, Inference, and Prediction, ch. 3, p. 57. Springer, Heidelberg (2009)
23. Hocking, R.R., Leslie, R.N.: Selection of the best subset in regression analysis. Technometrics 9(4), 531–540 (1967)
24. Tibshirani, R.: Regression shrinkage and selection via the lasso. Journal of the Royal Statistical Society. Series B (Methodological) 58(1), 267–288 (1996)
25. Wu, T.T., et al.: Genome-wide association analysis by lasso penalized logistic regression. Bioinformatics 25, 714–721 (2009)
26. Kim, S., et al.: A multivariate regression approach to association analysis of a quantitative trait network. Bioinformatics 25, i204–i212 (2009)
27. Gustafsson, M., et al.: Constructing and analyzing a large-scale gene-to-gene regulatory network-lasso-constrained inference and biological validation. IEEE/ACM Trans. Comput. Biol. Bioinformatics 2(3), 254–261 (2005)
28. Ma, S., et al.: Supervised group lasso with applications to microarray data analysis. BMC Bioinformatics 8, 60+ (2007)
29. Paaniuc, B., et al.: Accurate estimation of expression levels of homologous genes in RNA-seq experiments. In: Berger, B. (ed.) RECOMB 2010. LNCS, vol. 6044, pp. 397–409. Springer, Heidelberg (2010)
30. Li, J., et al.: Modeling non-uniformity in short-read rates in RNA-Seq data. Genome Biology 11(5), R50+ (2010)
31. Richard, H., et al.: Prediction of alternative isoforms from exon expression levels in RNA-Seq experiments. Nucleic Acids Research 38, e112 (2010)
32. Srivastava, S., Chen, L.: A two-parameter generalized Poisson model to improve the analysis of RNA-seq data. Nucleic Acids Research 38, e170 (2010)
33. Lee, S., et al.: Accurate quantification of transcriptome from RNA-Seq data by effective length normalization. Nucleic Acids Research (November 2010)
34. Hoerl, A.E., Kennard, R.W.: Ridge regression: Biased estimation for nonorthogonal problems. Technometrics 12(1), 55–67 (1970)
35. Efron, B., et al.: Least angle regression. Annals of Statistics 32, 407–499 (2004)

36. Zou, H., Hastie, T.: Regularization and variable selection via the elastic net. Journal of the Royal Statistical Society Series B 67, 301–320 (2005)
37. Park, M.Y., Hastie, T.: L1-regularization path algorithm for generalized linear models. Journal of the Royal Statistical Society: Series B (Statistical Methodology) 69, 659–677 (2007)
38. Optimization Toolbox User's Guide. The Mathworks, Inc., Natrik (2004)
39. Sammeth, M., et al.: The flux simulator (2010), http://flux.sammeth.net
40. The ENCODE Project Consortium: Identification and analysis of functional elements in 1% of the human genome by the ENCODE pilot project. Nature 447, 799–816 (2007)

Appendix: Mathematical Definitions, Notations and Proofs of the Theorems

Definitions

The formal definitions of uncertain reads, nested reads and the overlap graph are given in [16], and are reviewed below for the reader's convenience.

A single-end read b is *nested* in another single-end read b' iff $b_i = b'_i$, $l(b) \leq i \leq u(b)$, and at least one of the following two conditions is true: (1) $l(b) \neq l(b')$ and (2) $u(b) \neq u(b')$. A paired-end read p is *nested* in another paired-end read p' iff $l(p) \geq l(p')$, $u(p) \leq u(p')$ and at least one of the following conditions is true: (1) $l(p) \neq l(p')$ and (2) $u(p) \neq u(p')$. If a single-end read b is nested in b', b can always be removed safely without losing any information.

Two single-end reads b and b' are *compatible*, denoted as $b \sim b'$, iff there exists one isoform t such that $b \sim t$, $b' \sim t$, and b and b' are not *nested* to each other. If b and b' are not compatible, we denote $b \nsim b'$. Two paired-end reads p and p' are *compatible*, denoted as $p \sim p'$, iff there exists an isoform t such that $p \sim t$, $p' \sim t$ and p is not nested in p' or *vice versa*. If p and p' are not compatible, we denote $p \nsim p'$.

Define a *partial order* $\leq$ between two single-end reads b and b': $b \leq b'$ iff $b \sim b'$ and $l(b) \leq l(b')$. It is impossible to extend the partial order to paired-end reads, since the sequence within a paired-end read is not completely known. Alternatively, for two paired-end reads p and p', define $p \leq p'$ *with respect to a given read set R* iff the following conditions are true: (1) $p \sim p'$, (2) $l(p) \leq l(p')$, $u(p) \leq u(p')$, and (3) there is no paired-end read $p'' \in R$ such that $p \sim p'$, $p \sim p''$ but $p \nsim p''$. Write $p \leq p''|R$ if $p \leq p'$ with respect to a given read set R, or write simply $p \leq p'$ if there is no ambiguity. If reads p, p' and p'' exist such that $p \sim p'$, $p' \sim p''$ and $p \nsim p''$, then p, p' and p'' are said to be *uncertain* since no partial order can be given to these reads.

Given a set of mapped single-end or paired-end reads $R = \{b^1, b^2, \ldots\}$, the overlap graph (OG) [16] is a DAG $G = (V, E)$, where $V = \{v_1, v_2, \ldots, v_{|R|}\}$ and $e = (v_i, v_j) \in E$ iff $b^i \leq b^j$. A *maximal path* of length k on the OG is a path $h = \{v_{i_1} \leq v_{i_2} \leq \cdots \leq v_{i_k}\}$ on the OG, such that there exists no path

$h' = \{v_{j_1} \leq v_{j_2} \leq \cdots \leq v_{j_{k'}}\}$ with $h \subset h'$. Because the vertices in the OG have a one-to-one relationship with the mapped reads, we also treat vertices in the OG as binary vectors to simplify notations below. For example, if a path $h = \{v_{i_1} \leq v_{i_2} \leq \cdots \leq v_{i_k}\}$, we will use $OR(h)$ to denote $OR(\{b^{i_1} \leq b^{i_2} \leq \cdots \leq b^{i_k}\})$.

Proofs of the Theorems

The following lemmas are necessary. Suppose that R is the set of reads mapped to gene S.

Lemma 1. *Denote the vertex set of the CG as $V = \{v_1, v_2, \ldots, v_n\}$. For $1 \leq i < j \leq n$, there is a path from v_i to v_j if $cvg(S_i) > 0$ and $cvg(S_j) > 0$.*

Proof. We use an induction on $n = j - i$ to prove this lemma. If $j - i = 1$, then there is an edge between v_i and v_j by Condition 2 of the CG's edge construction. Assume that $\forall k < n$, there is a path from v_i to v_j if $cvg(S_i) > 0$ and $cvg(S_j) > 0$, $j - i = k$. For $k = n$, if $cvg(S_l) = 0$ for every $i < l < j$, then there is an edge between v_i and v_j by Condition 2 of the CG's edge construction. Otherwise, if there exists $i < l' < j$ such that $cvg(S_{l'}) > 0$, then $l' - i < n$ and $j - l' < n$. Using the assumption above, there is a path from v_i to $v_{l'}$ and a path from $v_{l'}$ to v_j. Therefore, there is a path from v_i to v_j. □

Lemma 2. *For any read set $Q \subseteq R$, if every two reads in Q are compatible, then there is a maximal path h in the CG such that $\forall b \in Q, b \sim h$.*

Proof. Let $t = OR(Q)$. We construct h by defining its vertex set $V(h)$ and edge set $E(h)$ separately. For every $1 \leq i < m, t_i = 1$, if the set $\{k > i | t_k = 1\}$ is not empty, denote $j = min_k\{k > i, t_k = 1\}$. If there is a read $b \in Q$ such that $b_i = b_j = 1$ and $b_k = 0, i < k < j$, then there must be an edge e in CG from v_i to v_j by Condition 2 of CG's edge construction, and we put e in $E(h)$. Otherwise, there must be a path h' from v_i to v_j by Lemma 1, because $cvg(S_i) > 0$ and $cvg(S_j) > 0$. We put edges in h' in $E(h)$. Define $V(h)$ as the set of vertices induced by $E(h)$. A trivial case is that $|\{1 \leq i < m, t_i = 1\}| = 1$. In this case, let $V(h) = v_i, t_i = 1$ for completeness.

We claim that all reads in Q are compatible with h. This is because for a single-end read (or an end of some paired-end read) b in Q, if $b_i = 1$ then $v_i \in V(h)$. If $b_i = b_j = 1$ and $b_k = 0, i < k < j$, v_i and v_j are directly connected by edge (v_i, v_j) in h, which means that $\{v_k | i < k < j\} \cap V(h) = \emptyset$. Therefore $b \sim h$.

Once h is obtained, it is easily extended to a maximal path without violating its compatibility with every read in Q. □

Lemma 3. *Suppose that R has no uncertain or nested reads. For every maximal path h on the OG constructed based on R, $OR(h) \in T$.*

Proof. Let $t = OR(h)$ and R_t be the set of reads corresponding to path h. By Lemma 2, there is a maximal path h' on the CG such that every read $b \in R_t$ is compatible with h'. Denote the isoform corresponding to h' as t'. Then, $t' \in T$ after the Enumeration phase of Algorithm 1 and $b \sim t'$.

Let $R_{t'} = \{b \in R | b \sim t'\}$. For any $b \in R_t$, $b \sim t'$ so $b \in R_{t'}$, then we have $R_t \subseteq R_{t'}$. Furthermore, for any $b' \in R_{t'}$, $b' \sim t'$, and thus we have $b \sim b', \forall b \in R_t, \forall b' \in R_{t'}$. If there is a read $b \in R_{t'}$ but $b \notin R_t$, the vertex corresponding to b in the OG could be added to path h, because b is compatible with all the reads in R_t and b is not a nested or uncertain read. However, this contradicts the assumption that h is maximal. Therefore, $R_t = R_{t'}$ and $t \in T$ after the Filtration phase of Algorithm 1. Note that t would not be removed in the Condensation phase Algorithm 1 because t is maximal. □

Lemma 4. *Suppose that R has no uncertain or nested reads. For every isoform t output by Algorithm 1, there exists a maximal path h on the OG such that $OR(h) = t$.*

Proof. Let t be an isoform enumerated by Algorithm 1 and $R_t = \{b \in R | b \sim t\}$. Since R contains no uncertain or nested reads, the vertices corresponding to R_t in the OG form a path h. If h is not maximal, it can be "expanded" to a maximal path h' by adding some vertices not in h. According to Lemma 3, there is an isoform $t' \in T$ such that $t' = OR(h')$. Denoting $R_{t'} = \{b \in R | b \sim t'\}$, then we have $R_t \subset R_{t'}$. Therefore, t would be removed in the Condensation phase of Algorithm 1, which contradicts the fact that t is output by Algorithm 1. □

Lemmas 3 and 4 immediately lead to Theorem 1 and its corollary, Corollary 1.

Theorem 1. *Suppose that R contains no uncertain or nested reads. If we denote the set of isoforms constructed by Algorithm 1 as T and the set of the isoforms formed by enumerating maximal paths on the OG (constructed from R) as T_{OG}, then $T = T_{OG}$.*

Corollary 1. *If R contains no uncertain or nested reads, then for every minimum path cover H of the OG, there exists a set of maximal isoforms $T' = \{t^1, \ldots t^m\} \subset T$, such that $m = |H|$ and for every read b on a path $h \in H$, $b \sim t^i$, $1 \le i \le m$.*

The following theorem holds when uncertain reads are present in R.

Theorem 2. *Suppose that no reads in R are nested and denote the set of isoforms constructed by Algorithm 1 as T. For every maximal path h on the OG constructed by removing uncertain reads in R, T contains an isoform which is compatible with every read on the path h.*

Proof. The proof is similar to the proof of Lemma 3. Let $t = OR(h)$ and $1 \le l_1 < l_2 < \cdots < l_m \le n$ be indices in t such that $t_i = 1$ iff and only if $i \in \{l_1, l_2, \ldots, l_m\}$. Let R_t be the set of reads corresponding to path h. By Lemma 2, there is a maximal path h' on the CG such that every read $b \in R_t$ is compatible with h'. Denote the isoform corresponding to h' as t'. Therefore, $t' \in T$ after the Enumeration phase of Algorithm 1 and $b \sim t'$.

Let $R_{t'} = \{b \in R | b \sim t'\}$. For any $b \in R_t$, $b \sim t$ and thus we have $b \sim t'$ and $R_t \subseteq R_{t'}$. Furthermore, $t'' = OR(R_{t'})$ would be in T after the Filtration phase of Algorithm 1 and t'' is compatible with every read in R_t.

During the Condensation phase of Algorithm 1, if t'' is not removed, the theorem holds. Otherwise, there must be another $t''' \in T$ such that all reads compatible with t'' are also compatible with t'''. In other words, all reads in R_t would be compatible with t'''. □

Haplotype Reconstruction in Large Pedigrees with Many Untyped Individuals

Xin Li and Jing Li

Department of Electrical Engineering and Computer Science
Case Western Reserve University, Cleveland OH 44106, USA
`jingli@cwru.edu`

Abstract. Haplotypes, as they specify the linkage patterns between dispersed genetic variations, provide important information for understanding the genetics of human traits. However haplotypes are not directly available from current genotyping platforms, and hence there are extensive investigations of computational methods to recover such information. Two major computational challenges arising in current family-based disease studies are large family sizes and many ungenotyped family members. Traditional haplotyping methods can neither handle large families nor families with missing members. In this paper, we propose a method which addresses these issues by integrating multiple novel techniques. The method consists of three major components: pairwise identical-by-descent (IBD) inference, global IBD reconstruction and haplotype restoring. By reconstructing the global IBD of a family from pairwise IBD and then restoring the haplotypes based on the inferred IBD, this method can scale to large pedigrees, and more importantly it can handle families with missing members. Compared with existing methods, this method demonstrates much higher power to recover haplotype information, especially in families with many untyped individuals.

Availability: the program will be freely available upon request.

Keywords: haplotype inference, identical-by-descent (IBD), inheritance, linkage disequilibrium.

1 Introduction

Humans are diploid, with two homologous chromosomes each from one parent. When inherited from a parent to a child, SNPs on one chromosome tend to stay together unless meiotic recombination breaks such linkage. Haplotypes, as they represent such linkage information between SNPs, are critical for understanding the genetics of human diseases [3][2][10]. Haplotype information cannot be directly obtained in wet labs based on current genotyping technologies; therefore haplotypes need to be recovered using computational methods. Current disease studies in families pose two major challenges to haplotyping methods. The first is family size. In order to assay enough recombination breakpoints to narrow

V. Bafna and S.C. Sahinalp (Eds.): RECOMB 2011, LNBI 6577, pp. 189–203, 2011.

down the disease loci, it is desirable to recruit as many family members as possible to a study. However, current haplotyping technologies are not applicable to these large families because most methods take time exponential to family sizes. This exponential time complexity is mainly due to the fact that most of these methods are designed under the backbone of the Lander-Green algorithm [6]. Recent improvements to the Lander-Green scheme [11][5][1][4] reduce the absolute processing time but does not alter its inherent exponential nature. There exists another type of methods, which exploits the Mendelian law of inheritance instead of enumerating the inheritance patterns and works much more efficiently [7][12][8]. However, this type of methods requires direct parent-child relationships in order for the Mendelian constraints to be applied. This limitation gives rise to the second computational challenge—ungenotyped family members. As is typical in a family-based study, many individuals in a family are not available for genotyping because they are deceased or otherwise not participating. Once there are untyped individuals in a family, the Mendelian constraints cannot be applied effectively. Most rule-based methods approach this issue by enumerating the genotypes of these untyped individuals which turns out to be computationally infeasible if many family members are missing.

We have recently developed an algorithm to efficiently infer identical-by-descent (IBD) status among family members [9]. The approach overcomes these difficulties by first constructing hidden Markov models (HMMs) for all relative pairs with genotypes, and then constructing the global (pedigree-wise) IBD relationship from the inferred pairwise IBD relationships from these HMMs. We bypass the enumeration of all possible genotypes of these untyped family members using the "inheritance-generating function", which summarizes the inheritance relationship between two individuals. The inheritance-generating function can be efficiently calculated using a recursive formula similar to the calculation of kinship coefficients. Therefore, the method essentially solves the computational problem of large pedigrees. However, at the final step, the approach uses an enumerative procedure to restore the inheritance from pairwise IBD, which again is exponential to the family size. In this paper, we replace it with a much more efficient algorithm based on graph partitioning. On top of that, we integrate our previous linear system based haplotyping method [8] into the framework to recover allelic phases of each individual. The haplotyping method exhausts all available constraints imposed by inheritance and genotypes, which maximizes the usage of information in a family. All together, this work constitutes a new haplotyping scheme which can efficiently reconstruct haplotypes at a genome-wide level in a large family with many untyped individuals. The two-stage IBD inference and the subsequent haplotype reconstruction significantly alleviate the computational burden complicated by large families. We evaluate the effectiveness of our method on both real and synthetic datasets. On families with many untyped individuals, our method exhibits significantly higher power in recovering haplotypes as compared with other state-of-art haplotyping methods. The proposed method also demonstrates good scalability on large pedigrees which other methods cannot handle.

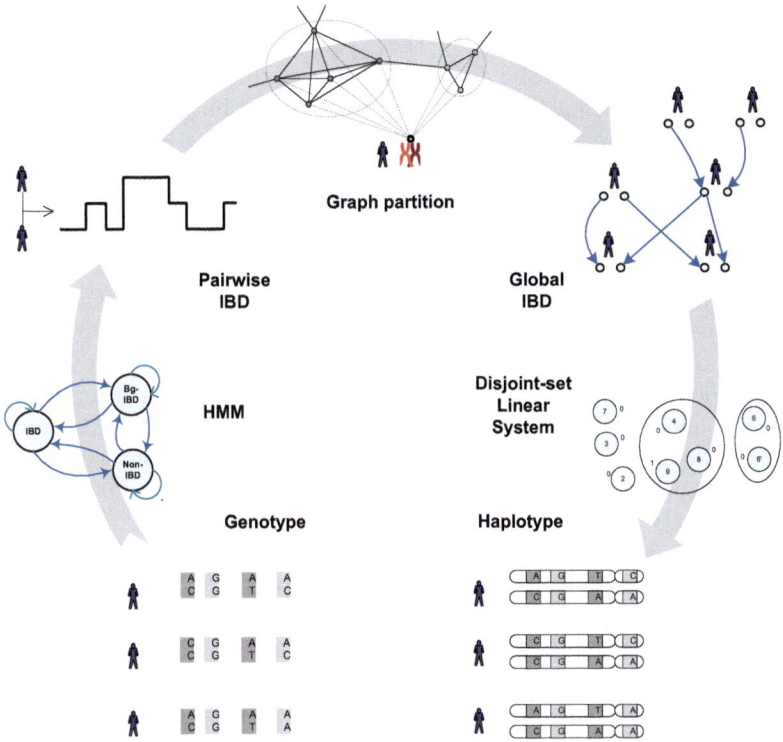

Fig. 1. The general framework of the proposed method

To give a clear picture of the method, we summarize its workflow in Fig. 1. The method consists of three steps: pairwise IBD inference from genotypes, global IBD reconstruction from pairwise IBD, and haplotype reconstruction from global IBD. Within each step, a computational technique is employed, namely, the hidden Markov model, graph partition and disjoint-set data structures. We will present the details of each of the three steps in the following sections 2.1, 2.2 and 2.3. We will show the performance of the method in section 3.

2   Methods

2.1   Inference of Pairwise IBD

The first step involves how to infer pairwise IBD sharing between relatives. This method is introduced in [9]. We briefly reiterate its essential elements here for the completeness of the paper. The method to infer pairwise IBD involves two key components: a) construct an HMM for pairs of relatives, and b) incorporate population level linkage disequilibrium into the model.

a) HMM for a pair of relatives. The IBD status between a pair of alleles can be modeled using a 2-state hidden Markov model, with transition probabilities between IBD and non-IBD states settled by profiling the degree of relatedness between the two individuals carrying these two alleles. In order to quantify such a relationship, we propose the "inheritance-generating function" to summarize all possible inheritance paths between two individuals. Intuitively, the longer the inheritance path between the two alleles, the less the probability that they descend from the same ancestral allele, because a longer inheritance path involves more segregations. In addition to transition probabilities, to fully parameterize the proposed hidden Markov models we also specify the emission probabilities. Given that two alleles are IBD, they must be the same genotype if assuming no genotyping errors. On the other hand, if two alleles are not IBD, it is purely out of chance for them to be the same genotype and such a probability can actually be determined by the allele frequencies at this locus. We further extend the model to incorporate more complex situations of missing genotypes and genotyping errors by refining the emission probabilities. HMMs between a pair of individuals can then be derived based on HMMs between pairs of alleles. The decoding process is basically the Viterbi algorithm for maximum likelihood inference or the forward-backward algorithm for point-by-point posterior probabilities. Both approaches take time linear to the number of markers. Results from both decoding algorithms will be utilized later.

b) Incorporating LD. Linkage disequilibrium at the population level, which largely reflects distant ancestral sharing among individuals, may create short identical haplotype segments among seemly unrelated individuals. We quantify such allelic dependence by adding an additional state (called the LD state) to the hidden Markov model which explicitly tags short stretches of IBD not originating from family relatedness. By doing so, one can make full use of the information embedded in the whole range of all available markers. Furthermore, by directly modeling linkage disequilibrium as distant ancestral sharing we end up with a unified hidden Markov framework, with more versatile power to fit the data because it allows synergistic interaction between IBD and LD state. The parameters related to the LD state in the model are learned from the targeted data, where we estimate the closeness of two unrelated individuals based on their genome-wide allelic sharing. The transition probabilities from the IBD state to the non-IBD and LD state are proportionally distributed according to their prior probabilities since non-IBD and LD both refer to alleles of distinct founder origins and are thus indistinguishable on a single family basis. With the help of the LD state, we can delimit the effects arising from relatedness or linkage disequilibrium.

2.2   From Pairwise IBD to Global IBD

Given that two individuals share one allele IBD, there is still ambiguity which one of the two homologous alleles of an individual is shared with the other individual. However, we want to further recover this information, or more specifically, we want to reconstruct the global IBD which explicitly labels each of the two homologous alleles of an individual with their ancestral alleles. In this section, we will first introduce a method for an ideal situation where all pairwise IBD relationships are consistent. In the second part of this subsection we will present an alternative backup approach for inconsistent data by utilizing posterior decoding.

To construct the global IBD in a family, a simple approach is to enumerate all possible inheritance patterns and check their consistency with pairwise IBD, which was implemented in [9]. However, this algorithm is not efficient. Here, we introduce a new approach in a graph theory setting. We first define an IBD graph to organize the relationships among all individuals in a pedigree. In brief, all individuals sharing two alleles will be merged into one node. Individuals sharing one allele will be connected by an edge. The groups of people who share the maternal or paternal allele of an individual can be recognized by finding two distinct cliques in her neighbors. By starting at one individual and iteratively propagating the paternal and maternal partition onto the neighboring individuals, we can settle the global IBD sharing. We will discuss the details of this approach in the following order: first we will formally define how to construct an IBD graph. Second, we will describe how to partition the neighbors of an individual into paternal sharing and maternal sharing groups and how to propagate such information further onto neighbors' neighbors. Third, we will give a proof of the correctness of this procedure. Last, we will discuss some special cases not covered in the algorithm.

We construct an IBD graph based on the pairwise IBD sharing between family members. Individuals sharing two alleles IBD with each other are identical, thus we use a single node to represent them. We use an edge to indicate the relationship of sharing one allele IBD. Formally speaking, let $G = (V, E)$, where $V = \{v_i | v_i = \{i\} \cup \{j | IBD(i, j) = 2\}\}$, $(v_a, v_b) \in E$ if $IBD(i, j) = 1, i \in v_a, j \in v_b$. Here, we assume the pairwise IBD relationships are consistent, therefore picking whichever two individuals respectively from two nodes, their relationships should be coherent.

Before getting into details of the algorithm, we first introduce some basic notations. We use $A = \{a_1, ..., a_n\}$ to indicate n different ancestral alleles in a family.[1] We define $A_i = \{x_i^1, x_i^2\}$ to be the ancestral configuration of individual i, where $x_i^1, x_i^2 \in A$ specifies the ancestral sources of each of the two homologous alleles. First, we assume no inbreeding, i.e., $x_i^1 \neq x_i^2$. We define an operation $a_k \rightarrow A_i$ to indicate assigning ancestral allele a_k to whichever x_i^1 or x_i^2 that is not yet assigned.

The algorithm starts by picking one individual from the family. Assume that this individual has two homologous alleles of distinct ancestral sources, which we denote as $\{a_1, a_2\}$, $a_1 \neq a_2$. Consider all of its neighbors in graph G, they must either carry a_1 or a_2, and based on this we can partition them into two groups, which we denote as N_1 and N_2. It is not hard to notice that both N_1 and N_2 are fully connected cliques, whereas between N_1 and N_2 there are a restricted number of edges, or more specifically, any node in N_1 can has at most one edge connected to N_2 and vice versa. Figure 2 gives an example showing

---

[1] Notice that the ancestral alleles are just labels to be assigned to different individual alleles, they are not in any particular order, nor explicitly associated with any particular founders in a family.

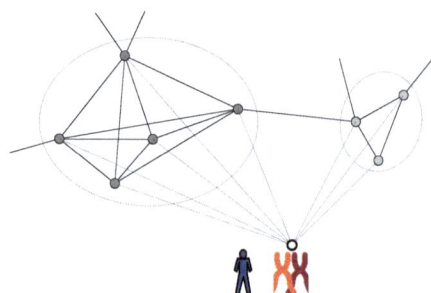

Fig. 2. Partition of the neighbors of an individual into two ancestral groups, where each group forms a clique

the neighbors of a node and how they form two cliques. This feature can help us perform such a partition quite efficiently. Once we obtain this partition, we can subsequently assign ancestral alleles to each of the two groups and further to their neighbors. These two procedures actually constitutes the two basic steps of the algorithm: we call the former one an initial "seeding" step and the latter one an iterative "propagation" step. We formally define the procedures of "seeding" and "propagation" in Algorithm 1 and 2.

In the propagation step, we make a queue to store all the newly assigned yet not fully assigned individuals. In this way all individuals will be visited at most twice in this process, hence the propagation step can be finished in linear time with respect to the number of individuals. We can prove by induction that after the seeding and the propagation step all ancestral alleles are correctly assigned to each individual.

Algorithm 1. Seeding

Find an individual s with more than 4 neighbors, $|N(s)| > 4$.
Partition $N(s)$ into two cliques: $N_1(s)$ and $N_2(s)$.
for $k = 1, 2$ **do**
  $a_k \rightarrow A_s$
  for each individual $A_j \in N_k(i)$ **do**
    $a_k \rightarrow A_j$
  end for
end for

Lemma 1. *After the seeding and propagation step, for any ancestral allele a_i, it is assigned to all of the individuals who carry this allele and none of the individuals who do not carry this allele.*

Proof. **Basis:** in the seeding step, $A_s = \{a_1, a_2\}$, any individual containing ancestral alleles a_1 or a_2 must be a neighbor of s, and on the other hand, any neighbor of s must either share a_1 or a_2 with s and we have partitioned them accordingly into two groups, therefore they will all receive proper assignment.

Algorithm 2. Propogation

```
k = 2
while there is any individual that is partially assigned do
    Find the next partially assigned individual A_i = {a_m, x}.
    k = k + 1
    a_k → A_i
    for each neighbor j of individual i do
        if a_m ∉ A_j then
            a_k → A_j
        end if
    end for
end while
```

Induction hypothesis: assume that at step k of the propagation, ancestral alleles $\{a_1, ..., a_k\}$ are all correctly assigned.

Induction: at step k+1, assume $A_j = \{a_m, x\}, m \leq k$ is an individual not yet fully assigned, applying the induction hypothesis, x must be an ancestral allele not in $\{a_1, ..., a_k\}$. Thus, we let $x = a_{k+1}$. Any neighbor of j which does not contain a_m must contain a_{k+1}, we can safely assign a_{k+1} to them. On the other hand, any individual which contains a_{k+1} must be a neighbor of j, therefore at step $k + 1$, any individual containing ancestral allele a_{k+1} receives proper assignment. ♮

The iterative propagation step will not stop until all nodes in a connected component are fully assigned. Different connected components of the IBD graph can be handled independently by applying the "seeding" and "propagation" procedures individually on each of them. Notice that it is not necessary to distinguish paternal alleles from maternal alleles when assigning ancestral allele types.

There are two situations we have not yet addressed, the first is when we cannot find a node with more than 4 neighbors in the "seeding" step. If a node has 4 or fewer neighbors, the partition of these neighbors can be ambiguous. Figure 3 shows such an example, where either way of partition can be a possible configuration. In this case, we should consider both of these two partitions and we will eventually have two alternative IBD assignments in the end. The second situation is when an individual is inbred, i.e., having two homologous alleles from the same ancestral source. Since the paternal and maternal alleles are of the same ancestry, this individual can have only one group of relatives, i.e., one

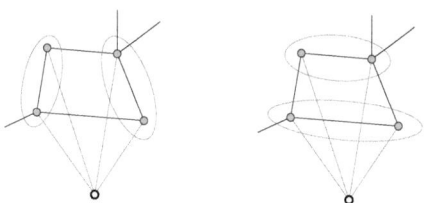

Fig. 3. The partition can be ambiguous when there are fewer than 5 neighbors

clique in the IBD graph. In this situation, we only need to propagate her allele to this single group of neighbors and beyond that everything stays the same in the "seeding" or "propagation" processes. One situation that can be confused with the inbreeding is when an individual has actually two distinct ancestral alleles but one of them is not shared with any other individual. In this case, the inbreeding and non-inbreeding situations are not distinguishable. Therefore, in the case that one person has only one clique of neighbors, the assignment of one of its two alleles is always ambiguous. Both assignments in such a case will be considered.

Finding optimal inheritance for error-prone data

The graph partition procedure introduced above takes the pairwise IBD relationships for sure (*e.g.*, using Viterbi decoding) and assumes all pairwise relationships are consistent. However when there are errors, we may not be able to find a global IBD configuration which satisfies all pairwise relationships. In these situations, we should have a backup plan which can tolerate possible inconsistencies. Here, we introduce an alternative enumerative approach utilizing results from posterior decoding. The problem is essentially formulated as an optimization problem where the search space consists of all possible inheritance patterns and the optimization criterion is defined below. Intuitively, we try to accommodate as many high probability pairwise IBD relationships as possible. We define a target function to aggregate the information over all pairwise relationships. In a straightforward way, we can use the following pseudo-likelihood function which encapsulates all pairwise relationships by multiplying their posterior probabilities.

$$L(\{A_1, ..., A_n\}) = \prod_{i,j} L(IBD(A_i, A_j)),$$

where the product is over all unordered pairs of individuals in a family and A_i is a specific ancestral allele assignment of individual i. By maximizing the target function, we are essentially trying to accommodate as many pairwise relationships of higher confidence levels as possible while sacrificing a few of those of lower confidence levels.

Since we need to maximize the target function over all possible inheritance patterns of a family, the search space could be rather huge for large families. A straightforward search involves the enumeration of 2^{2k} transmissions, where k is the number of non-founders in a family. We develop a branch-and-bound searching strategy to significantly speed up the procedure, by taking advantage of the property that a partial assignment always has a larger likelihood than that of a full assignment.

$$L(\{A_{i_1}, ..., A_{i_m}\}) \geq L(\{A_1, ..., A_n\}), \{A_{i_1}, ..., A_{i_m}\} \subset \{A_1, ..., A_n\}$$

Therefore, once the value of a partial assignment drops below the value of the current optimal solution, we can safely skip further enumerations and backtrack.

2.3   Haplotype Reconstruction

Genotypes can be phased at each marker according to the corresponding global IBD. Intuitively, we can first focus on individuals carrying homozygous alleles because they are naturally phased and we can thus resolve the genotypic values of the ancestral alleles inherited by them. This information is subsequently used to phase the other individuals sharing the same ancestral alleles and so on so forth. However, from a strict mathematical perspective, both homozygous and heterozygous loci carry some information to resolve these uncertainties. To be more specific, the constraints imposed by the global IBD and genotypes actually form a binary linear system.

Given the global IBD, the alleles of family members form two basic types of relationships and both of them can be explicitly expressed using binary linear equations. The first type of relationship is imposed by shared ancestry, which enforces that descendant alleles originating from the same ancestral allele should be the same. This type of relationship can be expressed as equivalence in a linear equation (Fig. 4, Type 1). The second type of relationship is imposed by heterozygous alleles, which dictates that the paternal allele and maternal allele of a heterozygous individual must be complementary to each other (Fig. 4, Type 2). This relationship can be expressed as $+1$ equivalence in a binary equation. A binary system naturally embeds the property that double complements should lead back to equivalence. The entire constraint system will appear as illustrated in Fig. 4, where two types of constraints and the constants are enforced. To summarize the whole process of building the system, first we treat each pair of heterozygous alleles and each missing allele as variables and each homozygous allele as a constant, second we build the binary linear system by enforcing both types of constraints and finally we solve the system and resolve the allele assignments.

Instead of using conventional techniques like Gaussian elimination, we can actually solve this linear system in a more efficient manner using the disjoint-set data structure. The general idea behind this is that we use disjoint-sets to represent independent variable sets and manipulate these sets (using

Type 1	Type 2	Constants
individuals $x_1, x_2, \ldots x_d$ sharing an ancestral allele	all heterozygous individuals: a	all homozygous individuals: b

$$\begin{cases} x_1(i_1) = x_2(i_2) \\ \quad = x_3(i_3) \\ \quad \cdots \\ \quad = x_d(i_d) \end{cases} \quad \cdots \quad i \in \{1,2\}$$

$$\begin{cases} a_1(1) = a_1(2)+1 \\ a_2(1) = a_2(2)+1 \\ \cdots \\ a_n(1) = a_n(2)+1 \end{cases}$$

$$\begin{cases} b_1(1) = b_1(2) = \text{constant} \\ b_2(1) = b_2(2) = \text{constant} \\ \cdots \\ b_k(1) = b_k(2) = \text{constant} \end{cases}$$

Fig. 4. A binary linear system of alleles. $x(1)$, $x(2)$ refer to the paternal and maternal alleles of an individual x

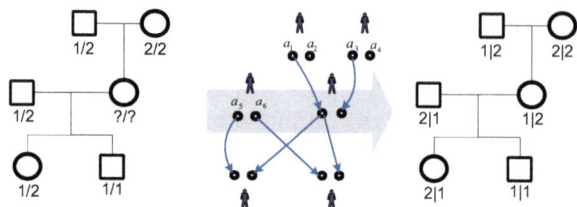

Fig. 5. Phasing the genotypes according to the global IBD

"union-find" algorithm) to encode different constraints. By doing so, we can quickly generate a solution or detect inconsistency of the system in linear time. We omit further technical details here, the algorithmic procedures for efficiently solving such a system are described in [8]. Figure 5 shows an example how the parental sources (phases) of each pair of alleles are resolved in a family by enforcing the constraints imposed by the global IBD.

3   Experiments

We examine the performance of our method on both synthetic and real datasets. To examine the power of the method to recover haplotypes, we run the method on a family with simulated genotypes. The family is drawn from a real data study assayed on the Affymetrix 6.0 SNP chip. This study has a total of 24 families and we use the available allele frequencies and haplotype segments from all these families to generate the appropriate founder haplotypes, which mimic the actual linkage disequilibrium in the data. Recombination rate is modeled at 1cM per Mb. The maps of SNP loci, the missing rate and the typing error rate used in the simulation are exactly the same as those of the real data, which assumes to be typical of the Affymetrix 6.0 SNP chip.

Here, we examine the contribution of different relatives in determining the haplotypes of an individual. We start with a family with only three typed individuals and gradually increase the number of typed individuals one by one in each subsequent experiment. We compare the efficacy of our method (named PED-IBD) with that of MERLIN. MERLIN [1] is a popular linkage analysis software package implementing the Lander-Green algorithm. As far as we know, MERLIN is probably the fastest program among all implementations of the Lander-Green algorithm and most of these programs are actually not feasible on large families. Fig. 6 shows the family structures and the ratio of phased loci. In the first setting, three of the family members are genotyped however none of them form direct parent-child relationships. In this situation, our method can correctly phase approximately 22% of the heterozygous loci. In comparison, MERLIN cannot recover any loci. The purpose of this first setting is to examine the powers of the methods when no parent-child Mendelian constraints can be obtained. PED-IBD has obviously higher phased ratio than MERLIN in the beginning three settings. The performance of MERLIN catches up only after

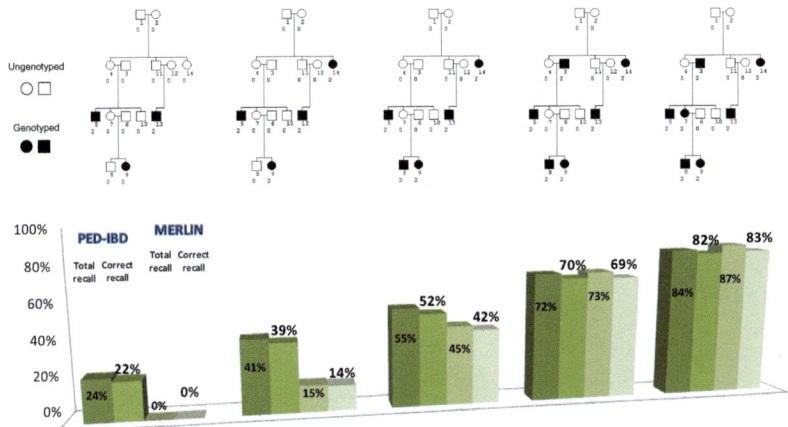

Fig. 6. Proportion of total phased and correctly phased loci in families with different numbers of genotyped individuals. From left to right, four grouped bars represent total phased loci and correctly phased loci of PED-IBD, total phased loci and correctly phased loci of MERLIN, out of all heterozygous loci. In the pedigree diagrams, shaded nodes indicate genotyped individuals

direct parent-child pairs are added as in the last two settings. This phenomenon suggests that MERLIN relies heavily on close relationships to resolve the uncertainty, but our method can make better use of all available information in the pedigree. Comparing the precision, which is the correctly phased loci out of all phased loci, two methods are similar, with the precision of our method at 94.04%, 93.69%, 96.40%, 97.81%, and MERLIN at 94.21%, 94.41%, 95.38%, 95.90% for the last four settings. We also simulate a big family of 21 members with 11 genotyped individuals, MERLIN quitted halfway in running this family presumably because of the exponential memory requirement or time complexity involved. Fig. 7 shows the ratio of phased loci yielded by our method on different members of the family. The leftmost two bars show the overall ratio of total phased loci and correctly phased loci of this family. The other bars are results from individual family members. The result agrees with common sense that individuals with more close relatives generally get higher ratios of their allelic phases resolved. Direct parent-child relationships also offer a major contribution here, where individuals having a genotyped parent or child have significantly more phased loci than others. The running time of the method scales quadratically with the number of genotyped individuals and linearly with the number of markers. On this specific family of 11 genotyped individuals, the program takes around 5 minutes to finish 10K markers on a regular PC.

The second data set we use is from a real data study of hypertension. These families and their members are collected according to familial aggregations of the disease therefore the family sizes and their structures should presumably reflect one typical pattern in many family based studies. Here we want to evaluate the power of our method under a realistic distribution of family structures,

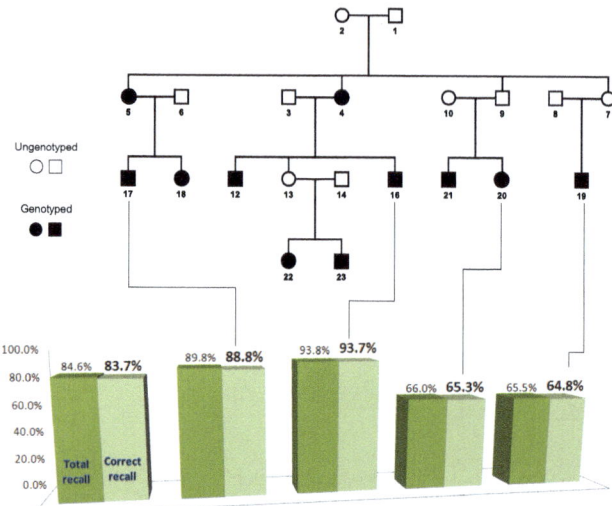

Fig. 7. Proportion of total phased and correctly phased loci for different members of a family. Shaded nodes are genotyped individuals

as this may provide some empirical basis in assessing the effectiveness of the method for other real datasets. We have a total of 196 families, among them 141 families have more than 1 typed members and there are an average of 4 typed individuals in each of these families. All families are genotyped on Affymetrix 6.0 SNP array. We want to examine the impacts of four important factors on the efficacy of phasing: family size (number of typed members), relationship between family members, missing genotypes and genotyping errors. Statistics (Fig. 8, 9) of different families are binned according to the number of typed individuals in each family. The line indicates the averaged values of all families in each bin. We exclude singleton individuals because there is no available information to phase them.

Fig. 8 (a) shows the proportion of loci that are phased out of all heterozygous loci given different numbers of typed individuals in a family. In general more typed individuals add more information to the family and lead to a higher phased ratio. However, the relationships between individuals also make a difference. Breaking down the phased ratios in families of two typed members (Fig. 8 (b)), we can observe that parent-child relationships are much more powerful than others in resolving the phase uncertainty. It may seem counterintuitive that full-sibship does not offer any gain than half-sibship, but we can understand this result by considering the fact that full siblings can actually share both paternal and maternal haplotypes at certain chromosomal regions and in these regions they are like identical twins hence mutually non-informative. The influence of missing genotypes is minor, as shown in Fig. 9(a), most of the missing genotypes can be imputed. However this could also be due to the relative low missing rate of the data which is just 0.3%. The disturbance caused by genotyping errors turns

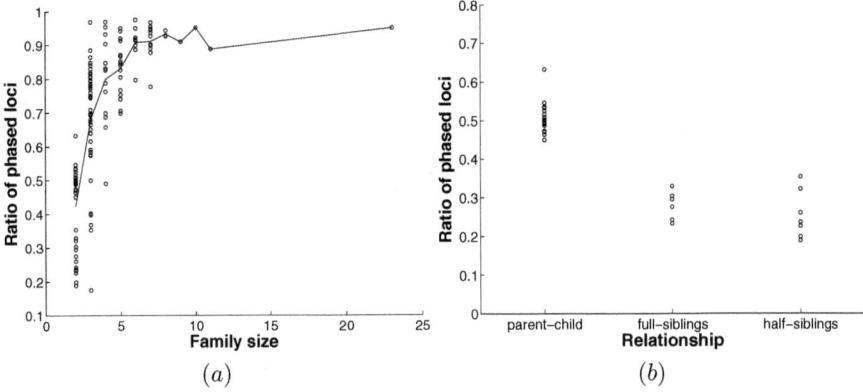

Fig. 8. (a) Proportion of heterozygous loci that are phased in different families. (b) Proportion of phased loci for different relationships.

out to be the final bottleneck on the haplotyping effectiveness. As demonstrated in Fig. 8 (a), most loci can be unambiguously phased given a large enough family size, however that proportion quickly approaches an upper bound. The total phased ratio fluctuates around 96% for families above the size of 5. To see how this major drawback is caused by genotyping errors, examining genotypes against the inheritance patterns of these families (Fig. 9 (b)), we can observe that around 3% of loci are not consistent with inheritance. Large families are generally more sensitive to typing errors because one such error in a single individual affects the entire locus of the family. In a summary, these four major factors: family size, family structure, missing genotypes and typing errors are exhibiting intertwined effects on haplotyping effectiveness, therefore in assessing the phasing capacity for a real study, all these factors should be taken into account.

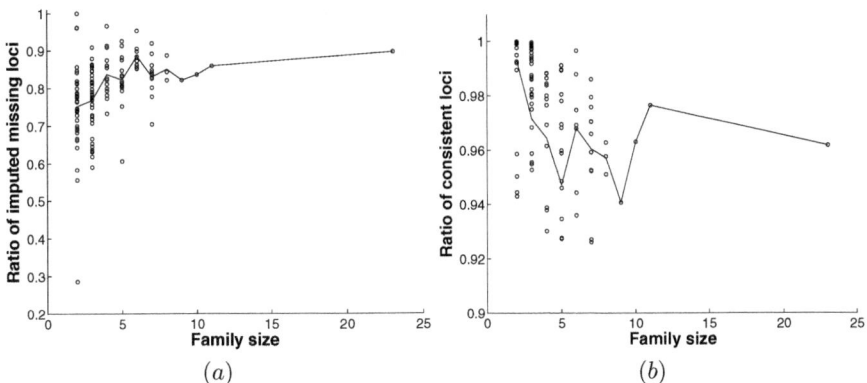

Fig. 9. (a) Proportion of missing loci that are imputed. (b) Proportion of consistent loci.

4   Discussions

We introduce a new method to efficiently reconstruct haplotypes in large families with many ungenotyped individuals. Our approach stands on three major components: pairwise IBD inference, global IBD reconstruction and haplotype restoring. By taking a two-step—genotype to pairwise IBD, pairwise IBD to global IBD—approach, we can significantly reduce the time complexity for resolving the IBD sharing pattern among family members. This makes our method scale well to large families which traditional methods cannot handle. The subsequent haplotyping algorithm is based on linear systems, it exhausts all available constraints imposed by global IBD and genotypes, thus maximizes the usage of information in a family. Compared with other popular methods, our method has much higher power to recover allelic phases in families with many missing members. On a real dataset of 196 families, the method yield more than 90% phased loci on families with more than five typed individuals. The proposed method constitutes an important advance in haplotyping technology which bridges the technical gap in existing methods on large families with missing members.

Acknowledgement

We would like to thank Dr. Xiaofeng Zhu for helpful discussions. This research is supported by National Institutes of Health/National Library of Medicine [grant LM008991], and in part by National Institutes of Health/National Center for Research Resources [grant RR03655].

References

1. Abecasis, G.R., Cherny, S.S., Cookson, W.O., Cardon, L.R.: Merlin–rapid analysis of dense genetic maps using sparse gene flow trees. Nat. Genet. 30(1), 97–101 (2002)
2. Bader, J.: The relative power of SNPs and haplotype as genetic markers for association tests. Pharmacogenomics 2(1), 11–24 (2001)
3. Frazer, K., Ballinger, D., Cox, D., Hinds, D., Stuve, L., Gibbs, R., Belmont, J., Boudreau, A., Hardenbol, P., Leal, S., et al.: International HapMap Consortium, A second generation human haplotype map of over 3.1 million SNPs. Nature 449, 851–861 (2007)
4. Gudbjartsson, D.F., Thorvaldsson, T., Kong, A., Gunnarsson, G., Ingolfsdottir, A.: Allegro version 2. Nat. Genet. 37(10), 1015–1016 (2005)
5. Kruglyak, L., Daly, M.J., Reeve-Daly, M.P., Lander, E.S.: Parametric and nonparametric linkage analysis: a unified multipoint approach. Am. J. Hum. Genet. 58(6), 1347–1363 (1996)
6. Lander, E.S., Green, P.: Construction of multilocus genetic linkage maps in humans. Proc. Natl. Acad. Sci. USA 84(8), 2363–2367 (1987)
7. Li, J., Jiang, T.: Efficient inference of haplotypes from genotypes on a pedigree. J. Bioinform. Comput. Biol. 1(1), 41–69 (2003)
8. Li, X., Li, J.: An almost linear time algorithm for a general haplotype solution on tree pedigrees with no recombination and its extensions. J. Bioinform. Comput. Biol. 7(3), 521–545 (2009)

9. Li, X., Yin, X., Li, J.: Efficient identification of identical-by-descent status in pedigrees with many untyped individuals. Bioinformatics 26(12), i191 (2010)
10. Morris, R., Kaplan, N.: On the advantage of haplotype analysis in the presence of multiple disease susceptibility alleles. Genetic Epidemiology 23(3), 221–233 (2002)
11. Sobel, E., Lange, K.: Descent graphs in pedigree analysis: applications to haplotyping, location scores, and marker-sharing statistics. Am. J. Hum. Genet. 58(6), 1323–1337 (1996)
12. Xiao, J., Liu, L., Xia, L., Jiang, T.: Fast elimination of redundant linear equations and reconstruction of recombination-free mendelian inheritance on a pedigree. In: Proceedings of the Eighteenth Annual ACM-SIAM Symposium on Discrete Algorithms, p. 664. Society for Industrial and Applied Mathematics (2007)

Learning Cellular Sorting Pathways Using Protein Interactions and Sequence Motifs

Tien-ho Lin, Ziv Bar-Joseph, and Robert F. Murphy*

Lane Center for Computational Biology, School of Computer Science,
Carnegie Mellon University, Pittsburgh, PA, USA

Abstract. Proper subcellular localization is critical for proteins to perform their roles in cellular functions. Proteins are transported by different cellular sorting pathways, some of which take a protein through several intermediate locations until reaching its final destination. The pathway a protein is transported through is determined by carrier proteins that bind to specific sequence motifs. In this paper we present a new method that integrates sequence, motif and protein interaction data to model how proteins are sorted through these targeting pathways. We use a hidden Markov model (HMM) to represent protein targeting pathways. The model is able to determine intermediate sorting states and to assign carrier proteins and motifs to the sorting pathways. In simulation studies, we show that the method can accurately recover an underlying sorting model. Using data for yeast, we show that our model leads to accurate prediction of subcellular localization. We also show that the pathways learned by our model recover many known sorting pathways and correctly assign proteins to the path they utilize. The learned model identified new pathways and their putative carriers and motifs and these may represent novel protein sorting mechanisms.

Supplementary results and software implementation are available from http://murphylab.web.cmu.edu/software/2010_RECOMB_pathways/

1 Introduction

To perform their function(s), protein usually need to be localized to the specific compartment(s) in which they operate. Subcellular localization of proteins is typically achieved by targeting pathways involving carrier proteins. Disruption of these pathways leading to inaccurate localization plays an important role in several diseases, including cancer [8, 16, 13], Alzheimer's disease [9], hyperoxaluria [25] and cystic fibrosis [32]. Thus, an important problem in systems biology is to determine how proteins are localized to their target compartments, the carriers and motifs that govern this localization and the pathways that are being used.

Recent advances in fluorescent microscopy coupled with automated image-based analysis methods provide rich information about the compartments to

---

* To whom correspondence should be addressed.

V. Bafna and S.C. Sahinalp (Eds.): RECOMB 2011, LNBI 6577, pp. 204–221, 2011.

which proteins are localized in yeast [15, 6] and human [23, 3, 22]. Several computational methods have been developed to predict subcellular localization by integrating sequence data with other types of high throughput data [14, 11, 21, 28, 26, 2]. These methods either treat the problem as a one vs. all classification problem [11, 14] or utilize a tree that corresponds to the current knowledge regarding intermediate compartments, for example LOCtree [21], BaCelLo [24] and discriminative HMMs [18]. The tree based methods were shown to be superior to the one vs. all methods; however, these methods do not attempt to learn the sorting tree, relying instead on current (partial) knowledge.

A number of methods have learned decision trees for predicting subcellular localization. These include PSLT2 [28] which refines the location into subcompartments using a decision tree learned from data and YimLOC [30] which learns a decision tree for the mitochondrion compartment only using features that include predictions from SherLoc [29], an abstract-based localization classifier. While the decision trees generated by these methods are often quite accurate, they are not intended to reflect targeting pathways, and they utilize features that, while useful for classification, are not related to the biochemical process of protein sorting.

In contrast to the global localization prediction methods, several experimental researchers have focused on trying to assign a specific sorting pathway to a small number of proteins. For example, proteins containing a signal peptide are exported through the secretory pathway [19], while some proteins without a classical N-terminal signal peptide are found to be exported via the non-classical secretory pathway [27]. A number of computational methods were developed to use this information to predict, for a given pathway, whether a protein goes through that pathway or not based on its sequence (for example, SignalP [5] and SecretomeP [4]). However, these methods rely on the pathway as an input and cannot be used to infer new pathways.

While the above experimental methods provide some information on sorting pathways, no method exists to try and infer global sorting pathways from current localization information. In this paper, we show that by integrating sequence, motif and protein interaction data we can develop global models for the process in which proteins are localized to subcellular compartments. We use a hidden Markov model (HMM) to represent sorting pathways. Carrier proteins and motifs are used to define internal states in this model and the compartments serve as the final (goal) state. Using this model we identified several sorting pathways, the carrier proteins that govern them and the proteins that are being sorted according to these pathways. Simulation data indicates that the models we learn are accurate. Using data from yeast we show that our model leads to accurate classification of protein compartments while at the same time enabling us to recover many known pathways and the proteins that govern these pathways. Several new predictions are provided by the model representing new putative sorting pathways.

2   Methods

2.1   Modeling Targeting Pathway by Hidden Markov Models

We used a HMM to model the process of targeting proteins to their compartments. HMM is a generative model and thus provides the set of events that lead
to the observed localization of the proteins (see Figure 1). An allowed pathway
through the HMM state space structure represents a possible protein targeting
pathway. All proteins start at the same start state (representing their translation
in the cytoplasm). The assigned (final) compartment of a protein is represented
by a state in the model that does not have any outgoing transitions. Intermediate states correspond to intermediate compartments or to sorting events (for
example, interaction with a protein carrier). These internal states emit observed
features that are related to the sorting events, namely motifs (implying that the
targeted protein uses that motif to direct it to that state) and carrier proteins
that target proteins to the state. The emitted features of a protein are observed
and determine its path in the state space. Emission is probabilistic and so certain proteins can pass through states even if they do not contain any of the
motifs and do not interact with any of the carriers for that state. Note that
while the compartment information is available during training, we do not know
how many intermediate states should be included in the model (some sorting
pathways may be short and others long, and several compartments can share
parts of the pathways). Thus, unlike traditional HMM learning tasks that focus
on learning the transition and emission probabilities, for our model we also need
to learn the set of states that are used in the targeting HMM.

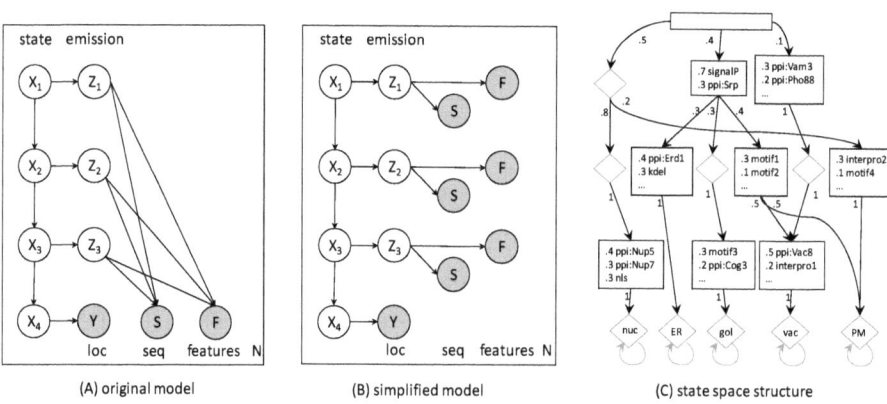

Fig. 1. (**A**) The graphical model representation of the original HMM for targeting
pathways and (**B**) the simplified HMM. (**C**) A sample state space: The top block
is the root and its outgoing arrows correspond to initial probabilities. Bottom nodes
are compartment states. The blocks are states and the arrows are transitions, with
transition probabilities labeled. The items listed inside a blocks are top features emitted
by the states, and emission probabilities are given on the left. Diamond-shaped blocks
are silent states that emit the background feature only.

2.2   A HMM for the Targeting Pathways Problem

We will discuss the likelihood of our HMM in detail here (see Figure 1). As discussed above, in our HMM model all proteins move from a single start state to their final compartment. For reasons that will become clear when talking about learning the parameters of the model, we associate each state in our model with a specific level. The root state is level 0, all compartment states are associated with the final level (T) and each intermediate state is associated with a specific level t ($0 < t < T$). We require that a state at level t can be reached from the root after exactly t transitions; connections that are more than one level apart move through several "silent" states so that transitions are only between adjacent levels. Silent states only emit a "background" feature (probabilities of the background feature are discussed later). Let X_t denote a hidden state at level t, $t = 1, 2, \cdots, T$ in a T-level model. The value of X_t can be one of J possible states, $X_t \in \{1, 2, \cdots, J\}$.

In addition to transition probabilities states are associated with emission probabilities. State X_t emits a feature index Z_t. Z_t can either be one of M motifs (represented as a likelihood score for each protein), or one of K binary features which include interactions with selected carriers, binary genome features based on UniProt (specifically, occurrences of deterministic motifs), or the background feature emitted by silent states. Hence $Z_t \in \{1, 2, \cdots M + K + 1\}$, where the motifs are indexed from 1 to M and the features are indexed from $M + 1$ to $M + K$.

Let S denote the sequence observed for each protein, F be the binary features from interaction databases and UniProt, and Y be the compartment assignments for a protein. The data likelihood of our HMM model (Figure 1), is defined as:

$$\Pr(S, F, Y | \Theta) = \sum_{X_1} \cdots \sum_{X_T} \sum_{Z_1} \cdots \sum_{Z_{T-1}} \Pr(S, F, Y, X_1, \cdots X_T, Z_1, \cdots Z_{T-1} | \Theta)$$

These joint probabilities can be decomposed based on the HMM independence assumptions as follows:

$$\Pr(S, F, Y, X_1, \cdots X_T, Z_1, \cdots Z_{T-1} | \Theta)$$
$$= \Pr(X_1) \prod_{t=1}^{T-1} \Pr(X_{t+1} | X_t) \Pr(Z_t | X_t) \Pr(S | Z_1, \cdots Z_{T-1}) \Pr(F | Z_1, \cdots Z_{T-1}) \Pr(Y | Z_T).$$
$$(1)$$

The parameters of our HMM are the initial, transition and emission probabilities, $\Theta = (\pi, A, B)$, defined as

$$\pi_i = \Pr(X_1 = i), \ A_{ij} = \Pr(X_{t+1} = j | X_t = i), \ B_{ik} = \Pr(Z_t = k | X_t = i).$$

where π_i is the initial probability of transition from the root to state i, A_{ij} is the transition probability between state i and state j, and B_{ik} is the emission probabilities from state i to emission k. Since each state only transits to a small number of states and emits a small number of features, these matrices are sparse.

2.3   Defining the Emission and Transition Probabilities for Our Model

Our input data is composed of the sequences of all proteins, their interactions and their compartments. Note that these observations are static and so may depend on all levels in the HMM. The emission probability for the sequence S is thus $\Pr(S|Z_1, \cdots Z_{T-1})$. Since probability depends on several motif models (one per level), which may be dependent (for example for overlapping motifs) and is thus computationally intractable given many combinations of motifs. As is commonly done [31] we approximate this term by the product of the conditional probabilities of the sequence given an individual emission at each level: $\prod_{t=1}^{T-1} \Pr(S|Z_t)$. Similarly we calculate the conditional probability of the binary features $\Pr(F|Z_1, \cdots Z_{T-1})$ using the product of the conditional probabilities of individual emissions (unlike for the sequence data this computation is exact since they are provided as independent events): $\prod_{t=1}^{T-1} \Pr(F|Z_t)$. This leads to the more typical HMM model shown in Figure 1B.

To translate the sequence information to a probability we use the likelihood of the sequence given the motif, $\Pr(S|\lambda_k)$, where λ_k is the motif model (we use a profile HMM model in this paper but any other probabilistic model including a PWM would work). This likelihood is termed the motif score, and indicates how well the sequence agrees with the motif model. For states emitting one of the binary features or the background feature, the likelihood of the sequence is $\Pr(S|\lambda_0)$, where λ_0 is the background model for which we use a 0th-order Markov model, which assumes that each position in the sequence are generated independently according to amino acid frequencies. Combined, the sequence likelihood is given by

$$\Pr(S|Z_t = k) = \begin{cases} \Pr(S|\lambda_k) \text{ if } 1 \leq k \leq M \\ \Pr(S|\lambda_0) \text{ if } M+1 \leq k \leq M+K+1 \end{cases} \tag{2}$$

The binary features observations, $F = (F_1, F_2, \cdots, F_K), F_k \in \{0,1\}$ correspond to observed protein interactions and deterministic motifs as discussed above. As mentioned above we assume independence between these features leading to:

$$\Pr(F|Z_t = k) = \prod_{j=1}^{K} \Pr(F_j|Z_t = k)$$

The conditional probability of observing a feature F_j given an emission Z_t is

$$\Pr(F_j = 1|Z_t = k) = \begin{cases} \nu_j \text{ if } k \neq M+j \\ \nu_0 \text{ if } k = M+j \end{cases}, \ 1 \leq j \leq K \tag{3}$$

where ν_j is probability of observing this interaction across all proteins in our dataset (background distribution) and $1 - \nu_0$ is the probability of false negatives, .i.e. proteins that should go through this state but do not have this interaction / motif. Note that we need to use ν_j since an interaction or a motif may be observed even if the corresponding feature is not emitted by one of the states

```
1. Estimate the associations between features and compartments using a hypergeometric test.
2. Select features significantly associated with at least one compartment.
2. Start with an initial structure estimated from associations between features and compartments.
3. While BIC score improves do
   a. For each level do
      i. Create a candidate structure as follows
      ii. Add a node (state) at this level
      iii. Link from all upper nodes and link to all lower nodes
      iv. Run EM to optimize parameters
      v. Remove edges (transitions) rarely visited based on the parameters
      vi. Remove emissions rarely used based on the parameters
      vi. Run EM again to adjust parameters
   b. Choose the candidate structure with highest BIC score
   c. If improving, update to that structure; otherwise stop
```

Fig. 2. Algorithm for structure search

since many interactions are not related to protein sorting but rather to another pathway in which this protein is a member.

The conditional probability of the compartment given the final state is denoted by: $Pr(Y|X_T)$. If a single compartment is given for a protein, the bottom state X_T is known for that protein and so this probability is 1 for that compartment and 0 for others. If the training data contains multiple compartments for a protein, it is reflected by the given compartment likelihood $Pr(Y = y|X_T = c)$, which is assumed to be uniform for all compartments listed for that protein.

2.4   Approximation and Feature Levels

Unlike a typical HMM learning problem, the emission data we observe (sequence and interaction data) is static and so cannot be directly associated with any sequence of events. In addition, since our features are static, they can be emitted multiple times along the *same* path. However, if this happens the independence assumptions of HMMs are violated. Specifically, if a feature is emitted by a state in level t and then again by a state in level $t + 1$ then it is not true anymore that the probability of emitting the feature given the state is independent of any emission events in previous states (since, if it was emitted before the protein can still emit it again). We thus constrain all features in our model so that each is only associated with a specific level and can only be emitted by states on that level. The level is determined in the initial structure estimation step discussed in the next section. Since no transitions are allowed between states on the same level no feature can thus be emitted more than once along the path and so the independence assumption holds. This requirement guarantees that the likelihood function obtained from the model presented in Figure 1B is a constant factor approximation of the likelihood function of our original model (Figure 1A). See Appendix for details.

2.5   Structure Learning

In addition to learning the parameters (emission and transition probabilities) we also need to learn the set of states that should be included in our model. The learning algorithm is formally presented in Figure 2. We start by associating potential features (protein interactions and known motifs) with compartments. For a potential feature, we use the hypergeometric distribution to determine the significance of this association (by looking at the overlap between proteins assigned to each compartment and proteins that are associated with each of the features). We next identify a set of significantly associated compartments (p-value < 0.01 with Bonferroni correction) for each potential feature. Features that are significantly associated with at least one compartment are selected and the remaining features are removed.

After feature selection, we estimate an initial structure by using the association between features and compartments. All features that correspond to the same set of associated compartments are grouped and assigned to a single state, such that this state emits these features with uniform probability. These features are fixed to the level corresponding to the number of compartments they are significantly associated with and can only be emitted by states on that level (we tried optimizing these feature levels as part of the iterative learning process but this did not improve performance while drastically increasing run time). Initial transition between states is determined from the inclusion relationship of the set of compartments (states for which features are associated with more compartments are assigned to higher levels). We initially only allow transitions between two states where the second state contains features that are associated with a subset of the compartments of the first state. The transition probability out of a state is also set to the uniform distribution.

Starting with this initial model, we use a greedy search algorithm which attempts to optimize the Bayes Information Criterion (BIC), which is the data log likelihood plus a penalty term for model selection.

$$BIC = -2 \log \Pr(\mathbf{S}, \mathbf{F}, \mathbf{Y} | \Theta) + |\Theta| \log N$$

where $\mathbf{S}, \mathbf{F}, \mathbf{Y}$ are the collection of sequences, feature observations, and compartments of the proteins in the training data. $\Theta = \pi, A, B)$ denote the parameters of the HMM. $|\Theta|$ is the number of parameters according to the structure, which is a function of the number of states and the number of transitions and emissions of each state. Complicated structures will have large $|\Theta|$ while simple structures will have small ones. N is the number of proteins in our training data.

To improve the initial structure described above we perform local moves in the following way. For each level we consider adding a state which is fully connected to all states in levels above and below it and emits all features on that level. We run standard EM algorithm [10] to optimize the parameters of the model for all states (transition and emission probabilities). Transitions and emissions with probabilities lower than a specific threshold are removed. Features not emitted by any states are also removed, so the feature set becomes smaller and smaller. Then we run EM algorithm again because the parameters are changed. A candidate model and structure is created by this process for each level, and the one with the highest BIC score is chosen. This procedure is repeated until the BIC score no longer improves.

3   Results

3.1   Simulated Data

We first tested our method using simulated data in order to determine how well it can recover a known underlying structure given only information on destinations, carriers and motifs. We manually created structures with 7, 14, 23, 25, and 31 states with multiple emitted features per state (see Supporting Website for the structure of these models). For each structure we simulate the probabilistic generative procedure and record the emitted features. 1,200 proteins are generated from the model, with varying levels of noise (leading to false positive and false negative features for proteins). We also tested various sizes of input sets with a fixed noise level.

Predicting Protein Locations. While it is not its primary goal, our method can provide predictions regarding the final localization of each protein. For each training dataset, we therefore generated a test dataset with 4,000 proteins from the same model and evaluated the accuracy of predicting protein localization for the test data using the structure and model learned by our method. Our method is compared to predictions made by the true model (note that due to noise, the true model can make mistakes as well) and by a linear support vector machine (SVM) learned from the training data using the features associated with each protein. Prediction accuracy on the 25-states dataset is shown in Figure 3 and the accuracy of other simulated datasets are available on the Supporting Website. As can be seen, when noise levels are low our model performs well and its accuracy is similar to that obtained by the true model for both simple and more complicated models. Both the learned model and the true model outperform SVM which does not try to model the generative process in which proteins are sorted in cells relying instead on a one vs. all classification strategy.

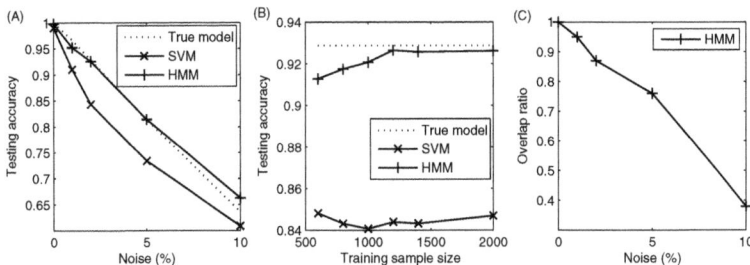

Fig. 3. (**A**) Testing error of simulated dataset generated from a structure with 25 states with varying levels of noise (false positive and false negative in features). The training sample size was fixed at 1400. (**B**) Testing error versus different training sample sizes. The noise level was fixed at 2%. (**C**) The ratio of overlapping nodes and edges between the learned model and the true model with varying levels of noise. The training sample size was fixed at 1400.

Recovering the True Structure. To quantitatively evaluate how well a learned structure resembles the true structure, we use the graph edit distance to measure their topological similarity [12]. First we need to match the nodes in a learned structure to a node in the true structure. We run the Viterbi algorithm on proteins in the testing data, and count the state co-occurrence matrix W whose elements W_{ij} is the co-occurrence of state i in the learned model and state j in the true model, i.e. the number of proteins in which the two states i and j occur in the Viterbi path inferred by the two models. The optimal one-to-one matching M, denoted as a set containing pairs of matched state indexes, can be found by running the Hungarian algorithm on the co-occurrence matrix W optimizing the objective function $\sum_{(i,j)\in M} W_{ij}$.

With the optimal matching we use the maximum common subgraph (MCS) and minimum common supergraph in the graph edit distance methodology to quantify similarity between two structures. Given two graphs G_1 and G_2, let $\hat{G}$ and $\check{G}$ be the MCS and minimum common supergraph of G_1 and G_2. Denote $|G|$ as the size, or the number of edges and nodes of a graph, we define the overlap rate as $|\hat{G}|/|\check{G}|$, i.e. the percentage of overlapping edges and nodes. The overlap rate comparing to the true model on the 25-states dataset is shown in Figure 3C. Structural comparison on other datasets is available on the supporting website. As can be seen, our algorithm successfully recovers the correct structure in all cases with 0% noise. As the noise increases the accuracy decreases. However, even for very high levels of noise the two models share a substantial overlap (around 40% of states and trnasitions could be matched).

3.2   Yeast Data

We next evaluated our method using subcellular locations of yeast proteins derived from fluorescence microscopy (the UCSF yeast GFP dataset [15]). This dataset contains 3,914 proteins that were manually annotated, based on imaging data, to 22 compartments. We used the following features to learn the model. Protein-protein interaction (PPI) data was downloaded from BioGRID (BiG) [33]. Protein sequences were downloaded from UniProt [1], and known motifs were downloaded from InterPro [20]. The above features are filtered by a hypergeometric test to identify features with a significant association with a final destination (p-value < 0.01 with Bonferroni correction) before learning the model. In addition to these known features, we applied the discriminative HMM motif finder we have previously described [18] to extract motifs present in one compartment but absent in other compartments. We extract 20 motifs for each compartment, and compared setting all to length 4 versus setting the length to range from 3 to 7. The performance in all following evaluations are similar and we show results based on motif length as 4. Furthermore, we use the occurrences of three signal sequences listed in UniProt.

1. Signal peptides: UniProt defines this sequence feature based on the literature or consensus vote of four programs, SignalP, TargetP, Phobius and Predotar.

2. Transmembrane region: UniProt annotates a sequence with this feature either based on literature or consensus vote of four programs, TMHMM, Memsat, Phobius and Eisenberg.
3. GPI anchor: UniProt annotation for this feature either relies on literature or prediction by the program big-PI.

Predicting Protein Locations. As with the simulated data, we first evaluated the accuracy of predicting the final subcellular location for each protein. This provides a useful benchmark for comparison to all other computational methods for which this is the end result. The performance is evaluated by 10-fold cross-validation. In each fold both feature selection and motif finding are restricted to the training data without accessing the testing data. The result is shown in Table 1. We compared our method with the k-Nearest Neighbors (kNN) from Lee *et al* [17] which was shown by the authors to outperform other methods. As can be seen in Table 1 PPI information (BiG) provides the major contribution for accurate predictions while InterPro motifs do not contribute as much. This agrees with previous studies [28, 17]. When adding more features the performance improves and the best result is achieved using all features. Note that the accuracy of our method is very close to that of the kNN method. However, it is important to note that our method performs the much harder task of simultaneously learning the sorting pathways as well as predicting locations.

Table 1. The accuracy of predicting the final subcellular location. For kNN we use the reported accuracy based on PPI information from BiG, deterministic InterPro motif annotation from UniProt, and amino acid composition of different length, gaps, and chemical properties [17]. For HMM we listed the mean and standard deviation of accuracy in 10-fold cross validation. The features for HMM also include InterPro and BiG, and three signal sequences from UniProt and novel motifs learned using discriminative HMM of length 4.

Methods and features	Accuracy
kNN BiG + InterPro + AA comp	65%
HMM InterPro	48% ±4
HMM BiG	61% ±2
HMM BiG + InterPro	61% ±2
HMM BiG + InterPro + Signals + DiscHMM 4	63% ±2

Evaluation of the Learned Structure. To evaluate the accuracy of the learned structure, we collected information about known targeting pathways from the literature. We were able to find information regarding 13 classical and non-classical targeting pathways (pathways followed by a minor fraction of proteins or that differ from the first discovered pathway are often referred to as non-classical pathways). For each of these pathways we identified a set of carriers or motifs that govern the pathway and, when available, the set of proteins that are predicted to use this pathway. Figure 4 presents the pathways we collected from the literature. For example the classical HDEL pathway into ER has two steps. In the first, proteins with signal peptide (SP) are introduced into this pathway by the SRP complex. In the second, proteins with the HDEL motif are retained in ER by interaction with proteins Erd1 and Erd2. The full list of carriers and motifs for these pathways is provided on the supporting website.

We first wanted to check if the databases we used for obtaining features contain the carrier information for the literature pathway. We filtered pathways for which carrier information in the BIG database did not contain the genes associated with the pathway (and thus no method can identify this pathway based in this input data) leaving 10 pathways that could, in principal, be recovered by computational models. Sorting steps that were filtered out in this way are represented as shaded links in Figure 4.

To determine whether we accurately recovered a pathway in our model we looked at the carriers and motifs that are associated with that pathway in the literature. A step in a literature pathway can be matched to a state if the state emits any carrier or motif in that step. A known pathway is considered recovered in a learned structure if its steps can be matched to the states along a path from the root to the compartment to which it leads. A pathway is partially recovered if only some of its steps can be matched. For example, the MVB pathway (Figure 4) is only partially recovered (66.7%) because the third step does not have a well-represented carrier in the data sources. The numbers of recovered pathways for different sets

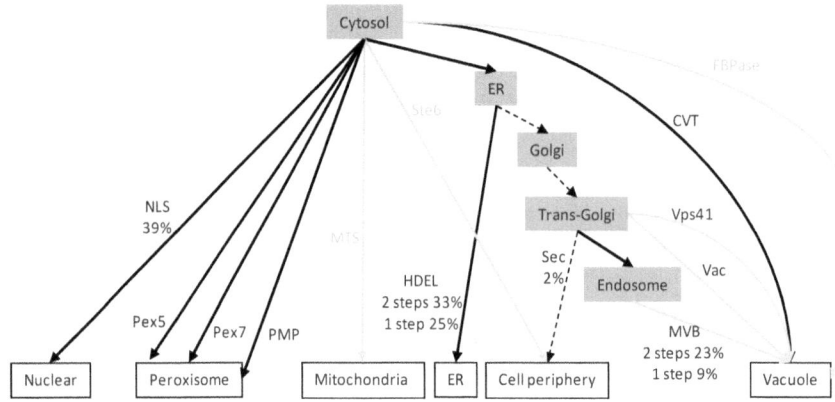

Fig. 4. Protein targeting pathways collected from the literature. Each pathway is a path from cytosol to a compartment at the bottom, consisting of one or more steps (the links) that transport proteins between intermediate locations. Each step has a list of carriers and motifs responsible for the transportation by which we can verify whether the pathway is recovered. Shaded links denote steps whose carriers are underrepresented on BiG (covering less than 5% of proteins transported to the corresponding compartment in the GFP dataset). Dashed lines denote steps taken by default without specific carriers. The percentage under pathway name is the protein sorting precision when the pathway is recovered, as described in Table 3.

of features are listed in Table 2. The ranges correspond to the different folds in our cross validation analysis. Fractions represent partial matches as discussed above. When using the full set of input features our algorithm is able to recover roughly 80% of known pathways. Most of these pathways are recovered in all 10 folds (Table 2). Note that because some carriers do not appear in our database not all steps in all pathways can be matched and the best possible recovery is 8.7. Thus, the 7.7 recovery obtained is very close to optimal.

For example, because of lack of evidence (the motif and carrier detection steps did not find the Vam3, Vam7, or the Vps41 features), the classical vacuole import pathway (Vac in Figure 4) and the alternative Vps41 pathway can only be 50% recovered (each missing a step). For both, the step of signal peptide (SP) is accurately found, but alternative motifs/carriers are selected to route proteins to the vacuole or cell periphery.

We further collected lists of proteins indicated as following specific pathways in the literature for 4 of the pathways, NLS, HDEL, Sec and MVB, and tested whether the recovered pathways indeed sort proteins on the correct path to the correct destination (allowing close compartments as above). For each protein, we use the Viterbi algorithm to infer the highest probability path of states the protein is expected to follow according to our learned model, and compare the Viterbi path to the known pathways. Again counting partial match of a multi-step pathway as above, on average using all features results in correctly assigning 21% of 63 proteins. Focusing on a representative feature set, detailed protein path results for each pathway are also given in Table 3. The recovered NLS pathway sorted 39% of proteins correctly, and the recovered HDEL pathway sorted 33% correctly but sorted the other 25% via SP. Similarly the recovered MVB pathway sorted 23% to go through two of the three steps (SP and MVB) and other 9% to one of the three steps. The recovered Sec pathway only sorted 2% of the proteins to go through SP and end at cell periphery. However, this was due to the fact that while 17 of the 28 proteins collected from literature as being secreted were included in the GFP dataset, the majority are labeled as ER and vacule and none are labeled as cell periphery. Overall the GFP dataset include 40 out of the 63 proteins whose pathway is known, of which only 28% are labeled in agreement with our lierature survey.

It is important to note that our analysis of the learned structure may underestimate its accuracy, since it may have recovered correct pathways that could not be verified due to insufficient detection of relevant motifs or carriers in the input data.

Table 2. Pathway recovery results of structure learned from different feature sets. The precision of inferred protein path is also listed here. Median, minimum and maximum among the 10 folds are shown.

Features	Pathway recovery	Inferred protein path
HMM BiG	5.9 (4.7 - 8.0)	9% (4% - 10%)
HMM BiG + InterPro + Signals	7.2 (5.7 - 8.7)	9% (6% - 11%)
HMM BiG + InterPro + Signals + DiscHMM 4	7.7 (6.7 - 8.7)	21% (16% - 23%)

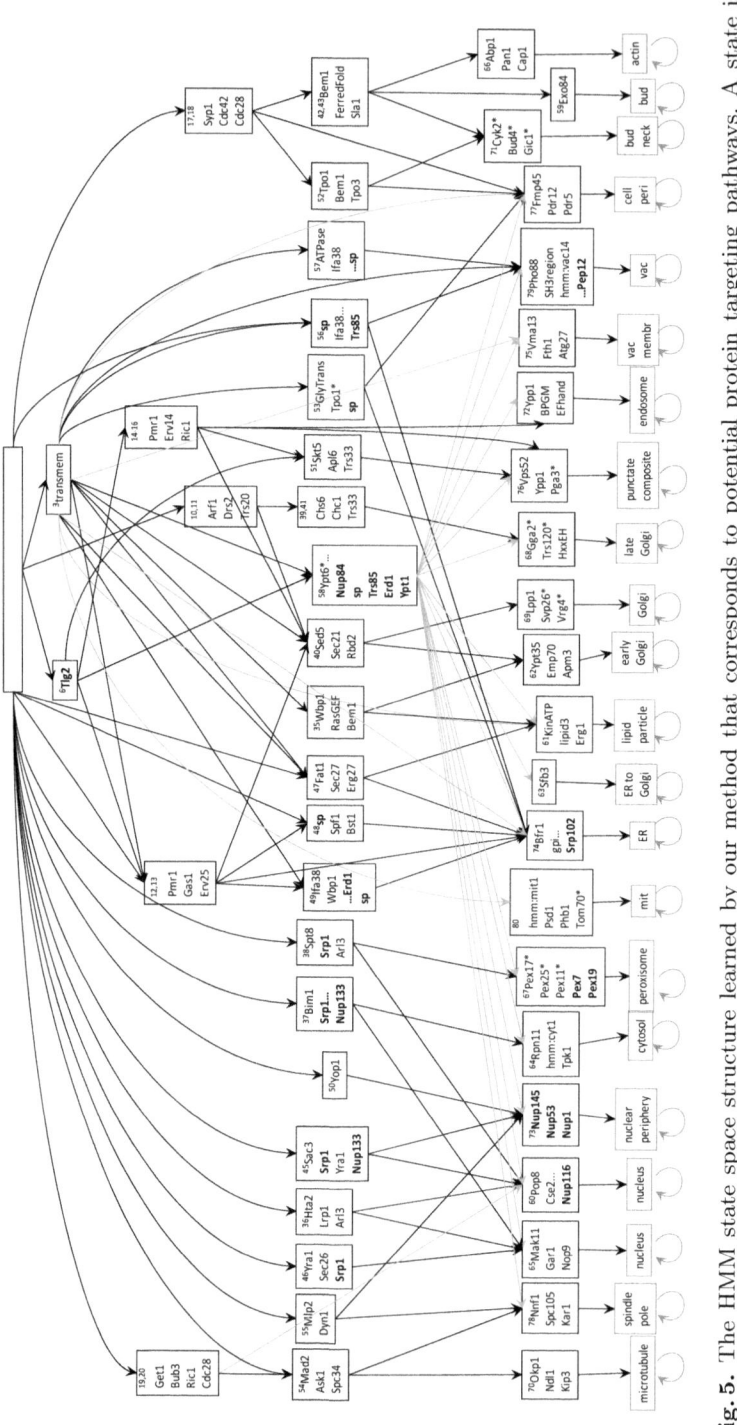

Fig. 5. The HMM state space structure learned by our method that corresponds to potential protein targeting pathways. A state is represented by a block; its transitions are shown as arrows and its top 3 emitting features are listed inside the block. The sparse transition and emission probabilities are omitted here. The initial state probabilities are denoted as arrows from the root block at the top. The bottom states are the final destination compartments. Some transitions are shaded only because of visual clarity, including transitions across levels or from and to the highly connected state (state 58). Carriers and motifs that matches our literature pathway collection are shown in boldface; other features potentially related to protein trafficking according to SGD are marked with an asterisk.

Table 3. Recovery and protein sorting results of each pathway using the features BiG + InterPro + Signals + DiscHMM 4

Compartment	Pathway (#proteins)	Recovery (folds)	Steps	Sorting
Nucleus	NLS(15)	10/10	all	39%
	Pex5	1/10	all	
Peroxisome	Pex7	10/10	all	
	PMP	9/10	all	
ER	HDEL(11)	10/10	SP+HDEL	33%
			SP	25%
Cell periphery	Sec(28)	10/10	SP	2%
	Vac	10/10	SP	
	MVB(9)	10/10	SP+MVB	23%
Vacuole			SP	9%
	Vps41	10/10	SP	
	CVT	10/10	all	

Figure 5 shows one of the learned structures obtained using all features. Besides carriers and motifs included in our literature pathway collection (marked as boldface), many other features were found that are also known to participate in protein trafficking as curated in SGD [7] (marked with an asterisk). For those compartments not covered by our collection of known pathways, the general topology of this structure agrees with our basic understanding of cell biology. For example microtubule share a step with spindle pole, which in turn share a step with nuclear periphery, and cell periphery share steps with bud neck, which in turn share steps with bud and actin.

4   Discussion

The goal of this research is to propose hypotheses about protein targeting mechanisms, not just to make predictions. We propose, for what we believe is the first time, a method to learn targeting pathways from protein localization annotation, based on co-occurrence of interacting partner and sequence motif. Our method is able to recover a significant part of known pathways collected from the literature, and to infer the correct path of proteins known to follow these pathways. Given that the sorting routes taken by many proteins are currently unknown, however, the most important part of our work is the potential to identify novel pathways. In this regard, we note that, just like hand-constructed pathways, any novel putative pathways contained in our learned model can be readily tested experimentally by perturbing motifs and/or carriers. An additional advantage of building comprehensive sorting models is that potential inconsistencies in canonical models can be identified and experiments performed to resolve them.

Acknowledgments. This work was supported in part by NIH grant R01 GM075205. The authors would like to thank Jennifer Bakal for programming support.

References

1. Bairoch, A., Apweiler, R., Wu, C.H., Barker, W.C., Boeckmann, B., Ferro, S., Gasteiger, E., Huang, H., Lopez, R., Magrane, M., Martin, M.J., Natale, D.A., O'Donovan, C., Redaschi, N., Su, L.: The Universal Protein Resource (UniProt). Nucleic Acids Res. 33(Database issue), D154–D159 (2005), http://dx.doi.org/10.1093/nar/gki070
2. Bannai, H., Tamada, Y., Maruyama, O., Nakai, K., Miyano, S.: Extensive feature detection of n-terminal protein sorting signals. Bioinformatics 18(2), 298–305 (2002)
3. Barbe, L., Lundberg, E., Oksvold, P., Stenius, A., Lewin, E., Björling, E., Asplund, A., Pontén, F., Brismar, H., Uhlén, M., Svahn, H.A.: Toward a confocal subcellular atlas of the human proteome. Mol. Cell Proteomics 7(3), 499–508 (2008), http://dx.doi.org/10.1074/mcp.M700325-MCP200
4. Bendtsen, J.D., Jensen, L.J., Blom, N., Von Heijne, G., Brunak, S.: Feature-based prediction of non-classical and leaderless protein secretion. Protein Eng. Des. Sel. 17(4), 349–356 (2004), http://view.ncbi.nlm.nih.gov/pubmed/15115854
5. Bendtsen, J.D., Nielsen, H., von Heijne, G., Brunak, S.: Improved prediction of signal peptides: SignalP 3.0. J. Mol. Biol. 340(4), 783–795 (2004), http://dx.doi.org/10.1016/j.jmb.2004.05.028
6. Chen, S.C., Zhao, T., Gordon, G.J., Murphy, R.F.: Automated image analysis of protein localization in budding yeast. Bioinformatics 23(13), i66–i71 (2007), http://dx.doi.org/10.1093/bioinformatics/btm206
7. Cherry, J.M., Adler, C., Ball, C., Chervitz, S.A., Dwight, S.S., Hester, E.T., Jia, Y., Juvik, G., Roe, T., Schroeder, M., Weng, S., Botstein, D.: SGD: Saccharomyces genome database. Nucleic Acids Research 26(1), 73–79 (1998), http://dx.doi.org/10.1093/nar/26.1.73
8. Cohen, A.A., Geva-Zatorsky, N., Eden, E., Frenkel-Morgenstern, M., Issaeva, I., Sigal, A., Milo, R., Cohen-Saidon, C., Liron, Y., Kam, Z., Cohen, L., Danon, T., Perzov, N., Alon, U.: Dynamic proteomics of individual cancer cells in response to a drug. Science 322(5907), 1511–1516 (2008), http://dx.doi.org/10.1126/science.1160165
9. De Strooper, B., Beullens, M., Contreras, B., Levesque, L., Craessaerts, K., Cordell, B., Moechars, D., Bollen, M., Fraser, P., St. George-Hyslop, P., Van Leuven, F.: Phosphorylation, subcellular localization, and membrane orientation of the Alzheimer's disease-associated presenilins. Journal of Biological Chemistry 272(6), 3590–3598 (1997), http://dx.doi.org/10.1074/jbc.272.6.3590
10. Dempster, A.P., Laird, N.M., Rubin, D.B.: Maximum likelihood from incomplete data via the em algorithm. Journal of the Royal Statistical Society. Series B (Methodological) 39(1), 1–38 (1977), http://dx.doi.org/10.2307/2984875, doi:10.2307/2984875
11. Emanuelsson, O., Nielsen, H., Brunak, S., von Heijne, G.: Predicting subcellular localization of proteins based on their N-terminal amino acid sequence. J. Mol. Biol. 300(4), 1005–1016 (2000), http://dx.doi.org/10.1006/jmbi.2000.3903
12. Gao, X., Xiao, B., Tao, D., Li, X.: A survey of graph edit distance. Pattern Analysis & Applications 13(1), 113–129 (2010), http://dx.doi.org/10.1007/s10044-008-0141-y
13. Gladden, A.B., Diehl, A.A.: Location, location, location: the role of cyclin D1 nuclear localization in cancer. Journal of cellular biochemistry 96(5), 906–913 (2005), http://dx.doi.org/10.1002/jcb.20613

14. Horton, P., Park, K.J., Obayashi, T., Fujita, N., Harada, H., Collier, C.J.A., Nakai, K.: WoLF PSORT: protein localization predictor. Nucleic Acids Res. 35(Web Server issue), W585–W587 (2007), http://dx.doi.org/10.1093/nar/gkm259

15. Huh, W.K., Falvo, J.V., Gerke, L.C., Carroll, A.S., Howson, R.W., Weissman, J.S., O'Shea, E.K.: Global analysis of protein localization in budding yeast. Nature 425(6959), 686–691 (2003), http://dx.doi.org/10.1038/nature02026

16. Kau, T.R., Way, J.C., Silver, P.A.: Nuclear transport and cancer: from mechanism to intervention. Nat. Rev. Cancer 4(2), 106–117 (2004), http://dx.doi.org/10.1038/nrc1274

17. Lee, K., Chuang, H.Y., Beyer, A., Sung, M.K., Huh, W.K., Lee, B., Ideker, T.: Protein networks markedly improve prediction of subcellular localization in multiple eukaryotic species. Nucleic Acids Research 36(20), e136+ (2008), http://dx.doi.org/10.1093/nar/gkn619

18. Lin, T.H., Murphy, R.F., Bar-Joseph, Z.: Discriminative motif finding for predicting protein subcellular localization. IEEE/ACM Trans. Comput. Biol. Bioinform. (2009) (to appear)

19. Lodish, H.F.: Molecular cell biology, 5threv. edn. W.H. Freeman and Company, New York (August 2003), http://www.worldcat.org/isbn/0716743663

20. Mulder, N.J., Apweiler, R., Attwood, T.K., Bairoch, A., Barrell, D., Bateman, A., Binns, D., Biswas, M., Bradley, P., Bork, P., Bucher, P., Copley, R.R., Courcelle, E., Das, U., Durbin, R., Falquet, L., Fleischmann, W., Jones, S.G., Haft, D., Harte, N., Hulo, N., Kahn, D., Kanapin, A., Krestyaninova, M., Lopez, R., Letunic, I., Lonsdale, D., Silventoinen, V., Orchard, S.E., Pagni, M., Peyruc, D., Ponting, C.P., Selengut, J.D., Servant, F., Sigrist, C.J.A., Vaughan, R., Zdobnov, E.M.: The InterPro database, 2003 brings increased coverage and new features. Nucleic Acids Res. 31(1), 315–318 (2003)

21. Nair, R., Rost, B.: Mimicking cellular sorting improves prediction of subcellular localization. J. Mol. Biol. 348(1), 85–100 (2005), http://dx.doi.org/10.1016/j.jmb.2005.02.025

22. Newberg, J.Y., Li, J., Rao, A., Pontén, F., Uhlén, M., Lundberg, E., Murphy, R.F.: Automated analysis of human protein atlas immunofluorescence images. In: Proceedings of the 2009 IEEE International Symposium on Biomedical Imaging, pp. 1023–1026 (2009)

23. Osuna, E.G., Hua, J., Bateman, N.W., Zhao, T., Berget, P.B., Murphy, R.F.: Large-scale automated analysis of location patterns in randomly tagged 3T3 cells. Ann. Biomed. Eng. 35(6), 1081–1087 (2007), http://dx.doi.org/10.1007/s10439-007-9254-5

24. Pierleoni, A., Martelli, P.L., Fariselli, P., Casadio, R.: Bacello: a balanced subcellular localization predictor. Bioinformatics 22 (2006), http://view.ncbi.nlm.nih.gov/pubmed/16873501

25. Purdue, P.E., Takada, Y., Danpure, C.J.: Identification of mutations associated with peroxisome-to-mitochondrion mistargeting of alanine/glyoxylate aminotransferase in primary hyperoxaluria type 1. J. Cell Biol. 111(6), 2341–2351 (1990), http://dx.doi.org/10.1083/jcb.111.6.2341

26. Rashid, M., Saha, S., Raghava, G.P.: Support Vector Machine-based method for predicting subcellular localization of mycobacterial proteins using evolutionary information and motifs. BMC Bioinformatics 8, 337 (2007), http://dx.doi.org/10.1186/1471-2105-8-337

27. Rubartelli, A., Sitia, R.: Secretion of mammalian proteins that lack a signal sequence. In: Unusual Secretory Pathways: From Bacteria to Man, pp. 87–104. RG Landes, Austin (1997)
28. Scott, M.S., Calafell, S.J., Thomas, D.Y., Hallett, M.T.: Refining protein subcellular localization. PLoS Comput. Biol. 1(6) (November 2005), http://dx.doi.org/10.1371/journal.pcbi.0010066
29. Shatkay, H., Höglund, A., Brady, S., Blum, T., Dönnes, P., Kohlbacher, O.: SherLoc: high-accuracy prediction of protein subcellular localization by integrating text and protein sequence data. Bioinformatics 23(11), 1410–1417 (2007), http://dx.doi.org/10.1093/bioinformatics/btm115
30. Shen, Y.Q., Burger, G.: 'unite and conquer': enhanced prediction of protein subcellular localization by integrating multiple specialized tools. BMC Bioinformatics 8, 420+ (2007), http://dx.doi.org/10.1186/1471-2105-8-420
31. Sinha, S.: On counting position weight matrix matches in a sequence, with application to discriminative motif finding. Bioinformatics 22(14), e454–e463 (2006), http://dx.doi.org/10.1093/bioinformatics/btl227
32. Skach, W.R.: Defects in processing and trafficking of the cystic fibrosis transmembrane conductance regulator. Kidney International 57(3), 825–831 (2000), http://dx.doi.org/10.1046/j.1523-1755.2000.00921.x
33. Stark, C., Breitkreutz, B.J., Reguly, T., Boucher, L., Breitkreutz, A., Tyers, M.: BioGRID: a general repository for interaction datasets. Nucleic Acids Research 34(suppl 1), D535–D539 (2006), http://dx.doi.org/10.1093/nar/gkj109

Appendix: Approximation of the Original Model

Here we will describe how to approximate the full model in Figure 1A by the simplified model in Figure 1B, given that each feature has a fixed level. Recall that the joint probabilities of the original model in Figure 1A is given in Equation (1). First we focus on the emission probabilities of the feature observations, and show that the likelihood ratio of the emission versus the background equals the product of this likelihood ratio on all levels.

$$\frac{\Pr(F_j = 1|Z_1, \cdots Z_{T-1})}{\nu_j} = \prod_{t=1}^{T-1} \frac{\Pr(F_j = 1|Z_t)}{\nu_j} \tag{4}$$

where ν_j is the likelihood given the background feature. From Equation (4) we can naturally obtain

$$\Pr(F_j = 1|Z_1, \cdots Z_{T-1}) = \nu_j^{2-T} \prod_{t=1}^{T-1} \Pr(F_j = 1|Z_t)$$

for each feature, and it is combined as

$$\Pr(F|Z_1, Z_2, \cdots Z_{T-1}) = \Big(\prod_j \nu_j^{2-T}\Big) \prod_{t=1}^{T-1} \Pr(F|Z_t) \tag{5}$$

The full emission probability for each feature, $\Pr(F_j|Z_1, Z_2, \cdots Z_{T-1})$, is defined as a noisy observation (with false positive and false negative) of the OR function over Z_t,

$$\Pr(F_j = 1 | Z_1 = k_1, Z_2 = k_2, \cdots Z_{T-1} = k_{T-1}) = \begin{cases} \nu_j \text{ if } \forall t \ k_t \neq M + j \\ \nu_0 \text{ if } \exists t \ k_t = M + j \end{cases}$$

However the OR function is unnecessary because we require feature F_j to have a fixed level, so only one level can emit the corresponding emission such that $Z_t = k_t = M + j$. Now to prove Equation (4), when one of the levels indeed emit the corresponding emission, we start from the right hand side of Equation (4) and apply Equation (3),

$$\prod_{t=1}^{T-1} \frac{\Pr(F_j = 1 | Z_t)}{\nu_j} = \frac{\nu_0 \nu_j^{T-2}}{\nu_j^{T-1}} = \frac{\nu_0}{\nu_j} = \frac{\Pr(F_j = 1 | Z_1, \cdots Z_{T-1})}{\nu_j}$$

and reach the left hand side of Equation (4). Similarly when none of the levels emit the corresponding emission,

$$\prod_{t=1}^{T-1} \frac{\Pr(F_j = 1 | Z_t)}{\nu_j} = \frac{\nu_j^{T-1}}{\nu_j^{T-1}} = \frac{\nu_j}{\nu_j} = \frac{\Pr(F_j = 1 | Z_1, Z_2, \cdots Z_{T-1})}{\nu_j}$$

Hence we have derived Equation (4) given the requirement that each feature must have a fixed level.

The above derivation for feature likelihood term is exact, but approximation is necessary for the sequence likelihood term. Similar to feature observations, we approximate the likelihood ratio of emission probabilities for sequence by a set of motifs over the background likelihood as the product of this likelihood at each level,

$$\frac{\Pr(S | Z_1, Z_2, \cdots Z_{T-1})}{\Pr(S | \lambda_0)} \approx \prod_{t=1}^{T-1} \frac{\Pr(S | Z_t)}{\Pr(S | \lambda_0)} \tag{6}$$

where λ_0 is the null model as in Equation (2). We assume that motifs are independent to each other since motif length is set to be short (either set to 4 peptides or 3 to 7 peptides) comparing to the sequence length, as is the case in most known targeting motifs. This is a common assumption (e.g. [31]) and necessary for avoiding overfitting. However as we discussed in section 2.4 this assumption requires that no motif is emitted twice in different levels, which is achieved by fixing the level of each feature. Similar to Equation (5) we also write the sequence likelihood term as

$$\Pr(S | Z_1, Z_2, \cdots Z_{T-1}) = \Pr(S | \lambda_0)^{2-T} \prod_{t=1}^{T-1} \Pr(S | Z_t). \tag{7}$$

By combining Equation (5) and (7), we show that the likelihood of the full model in Figure 1A and the likelihood of the simplified model in Figure 1B is approximately up to a constant factor, so that optimizing the simplified model also optimizes the original model.

A Geometric Arrangement Algorithm for Structure Determination of Symmetric Protein Homo-oligomers from NOEs and RDCs*

Jeffrey W. Martin[1], Anthony K. Yan[1,2], Chris Bailey-Kellogg[3],
Pei Zhou[2], and Bruce R. Donald[1,2,**]

[1] Department of Computer Science, Duke University, Durham, NC 27708, USA
[2] Department of Biochemistry, Duke University Medical Center, Durham,
NC 27710, USA
Tel.: 919-660-6583; Fax: 919-660-6519
brd+recomb11@cs.duke.edu
[3] Department of Computer Science, Dartmouth College, Hanover, NH 03755, USA

Abstract. Nuclear magnetic resonance (NMR) spectroscopy is a primary tool to perform structural studies of proteins in the physiologically-relevant solution-state. Restraints on distances between pairs of nuclei in the protein, derived from the nuclear Overhauser effect (NOE) for example, provide information about the structure of the protein in its folded state. NMR studies of symmetric protein homo-oligomers present a unique challenge. Current techniques can determine whether an NOE restrains a pair of protons across different subunits or within a single subunit, but are unable to determine in which subunits the restrained protons lie. Consequently, it is difficult to assign NOEs to particular pairs of subunits with certainty, thus hindering the structural analysis of the oligomeric state. Hence, computational approaches are needed to address this subunit ambiguity. We reduce the structure determination of protein homo-oligomers with cyclic symmetry to computing geometric arrangements of unions of annuli in a plane. Our algorithm, DISCO, runs in expected $O(n^2)$ time, where n is the number of distance restraints, and is guaranteed to report the exact set of oligomer structures consistent with ambiguously-assigned inter-subunit distance restraints and orientational restraints from residual dipolar couplings (RDCs). Since the symmetry axis of an oligomeric complex must be parallel to an eigenvector of the alignment tensor of RDCs, we can represent each distance restraint as a union of annuli in a plane encoding the configuration space of the symmetry axis. Oligomeric protein structures with the best restraint satisfaction correspond to faces of the arrangement contained in the greatest number of unions of annuli. We demonstrate our method using two symmetric protein complexes: the trimeric *E. coli* Diacylglycerol Kinase (DAGK), whose distance restraints possess at least two possible subunit

* This work was supported by the following grants from the National Institutes of Health: R01 GM-65982 to B.R.D. and R01 GM-079376 to P.Z.
** Corresponding author.

assignments each; and a dimeric mutant of the immunoglobulin-binding domain B1 of streptococcal protein G (GB1) using ambiguous NOEs. In both cases, DISCO computes oligomer structures with high accuracy.

1   Introduction

Structural characterization of proteins yields insight into their biological functions, which has become increasingly important for understanding the biochemical basis of human disease. Once the mechanism by which pathogens affect a host is better understood, one can begin to ask how it might be possible to alleviate the affects of infection, or prevent infection altogether. Determining the high-resolution 3D structures of proteins can enable design of molecules (drugs) to inhibit the native function of a pathogenic protein, or modify helpful proteins to perform a novel function to help stave off infection. One such protein redesign study modified a phenylalanine adenylation domain of the nonribosomal peptide synthetase enzyme gramicidin S synthetase A, an enzyme that originally manufactured the decapeptide gramicidin S, a strong antibiotic, to incorporate different substrates into the assembly line [1], thus showing it may be possible to use computational algorithms to engineer enzymes to produce new molecules of potential pharmacological interest. In addition, the same protein design methodology can help predict antibiotic resistance mutations in harmful pathogens such as methicillin-resistant *Staphylococcus aureus* (MRSA) [2], giving drug research the opportunity to keep one step ahead of its bacterial adversaries. Computational protein redesign is an increasingly popular tool for efficiently exploring possible modifications to protein sequence, but usually requires a structural model of the enzyme or protein of interest.

The majority of proteins assemble as symmetric homo-oligomers [3,4], including many membrane proteins, yet the symmetry complicates assignment of inter-subunit distance restraints, and hence oligomeric structure determination by NMR. The pace of structure determination of membrane proteins has lagged significantly behind soluble globular proteins [5], in part due to these challenges arising from symmetry. Structure determination of symmetric trimers and higher-order homo-oligomers is hindered by *subunit ambiguity* [6]: even if an NOE between two protons can be assigned as intra-subunit or inter-subunit through X-filtered NOESY [7], current experimental techniques are still unable to determine precisely in which subunits the restrained protons lie (see Figure 1).

Even with precise unambiguous distance restraint assignments, structure determination of monomeric proteins by NMR remains a difficult task. Structure determination protocols that rely only on local distance restraints have been proven strongly NP-Hard [8] and therefore vitiate guarantees of efficiency, accuracy, and completeness. Remarkably, the addition of global orientational constraints on internuclear vectors from residual dipolar couplings (RDCs) and a reduction to *sparse* distance restraints enabled a polynomial-time algorithm for monomeric structure determination [9]. For symmetric homo-oligomers of at least three subunits, subunit ambiguity complicates assignment of inter-subunit

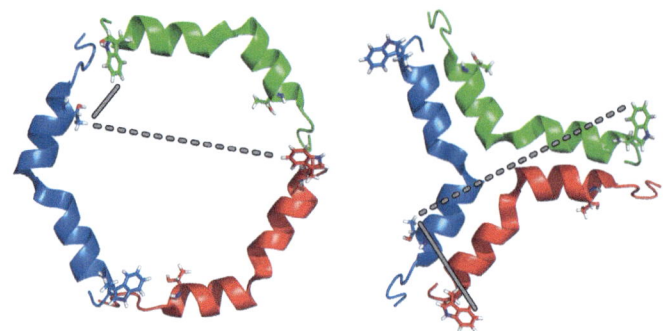

Fig. 1. Subunit ambiguity: An ambiguous inter-subunit NOE between two protons for a hypothetical symmetric trimer has two possible assignments. Without loss of generality, we can choose one proton to lie within the blue subunit. The other proton must lie in either the green or red subunits. The choice of assignment can potentially lead to vastly different overall folds for the trimer. Left: A ring-shaped scaffold satisfies the blue-green assignment (solid line), but not the blue-red assignment (dashed line). Right: A star-shaped scaffold satisfies the blue-red assignment (solid line), but not the blue-green assignment (dashed line).

distance restraints, and hence, calculation of oligomer structures, since naïvely enumerating possible assignment combinations requires exponential time. Potluri et al. [10,6] employed a branch and bound search algorithm which computed symmetric oligomer structures using ambiguously-assigned inter-subunit distance restraints and was guaranteed to return a superset of all oligomer structures satisfying the restraints. The algorithm avoided computing explicit distance restraint assignments, but provided no bound on running time. In this paper, we show how the addition of RDCs allows polynomial-time algorithms for structure determination of symmetric homo-oligomers, but with guarantees on solution quality as well as running time.

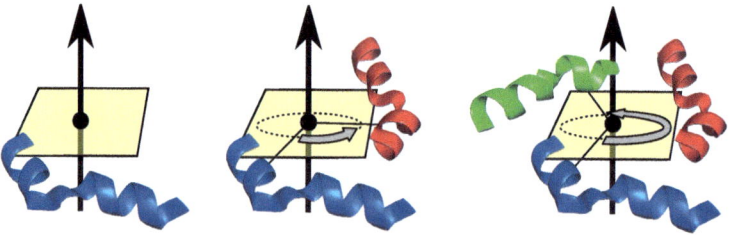

Fig. 2. Building a trimer structure using symmetry: Left: Compute the position and orientation of the symmetry axis (vertical arrow) relative to the subunit structure (blue α-helix). Middle: Copy the subunit structure and rotate by 120° about the symmetry axis to place the second subunit (red α-helix). Right: Copy the subunit structure again and rotate by 240° to place the final subunit (green α-helix).

In practice, approaches such as simulated annealing coupled with simplified molecular dynamics, which lack both combinatorial precision and guarantees on running time and solution quality, are used routinely for structure determination. These approaches require careful selection of annealing parameters, may not always converge, or can potentially miss structures consistent with the experimental restraints due to under-sampling. Since for every vector orientation that satisfies an RDC value, there exists an equally-satisfying inverse vector, restraints on internuclear vector orientation from RDCs are typically not included in the first annealing run. After an initial fold has been calculated from complementary restraints, RDCs are used to refine the structure with further annealing runs. We instead propose to incorporate RDCs into the beginning of the structure determination method, thereby creating a framework in which we analyze inter-subunit distance restraints without requiring a complete oligomer structure. Instead, the oligomer structure can be represented in terms of its axis of symmetry and the structure of its subunit (see Figure 2). Therefore, we perform structure determination in the configuration space of symmetry axes: two translational degrees of freedom (a plane, $\mathbb{R}^2$) and two rotational degrees of freedom (a unit sphere, $\mathbb{S}^2$).

Previous work [11,10,6] also formulated structure determination of homo-oligomers in a symmetry configuration space. Potluri et al. [10,6] computed the orientation and position of the symmetry axis using only inter-subunit distance restraints and a hierarchical subdivision of the configuration space ($\mathbb{R}^2 \times \mathbb{S}^2$). Regions of the space were pruned if geometric bounds proved they contained no satisfying symmetry axis configurations. Wang et al. [12] computed symmetry parameters for oligomer models using ambiguously-assigned distance restraints by partitioning Cartesian space instead of axis configuration space. After choosing three of the distance restraints as a geometric base, AMBIPACK computed symmetry axis parameters by computing the rigid transformation across the interface between two identical subunits. The three chosen distance restraints were used to define a coarse relative orientation between the subunits at the interface, which was iteratively refined against the remaining distance restraints. Due to the reliance on random sampling and local numerical optimization, the method may potentially miss structures that satisfy the distance restraints. Wang et al. [11] computed the orientation of the symmetry axis using just RDCs. The axis position was computed by generating putative dimer models on a grid over $\mathbb{R}^2$ and scoring the inter-subunit interface using a residue-pairing molecular mechanics function. Since dimer models were ranked only according to molecular mechanics scores, van der Waals energies, and RDC satisfaction, it was not necessary to assign or use inter-subunit NOEs. However, in doing so, the method misses the opportunity to incorporate the structural information provided by these distance restraints.

Work by Nilges [13] calculated oligomer models without explicit knowledge of the symmetry axis. Instead, structure calculation relied on symmetry potentials during runs of simulated annealing, and has been successfully employed in structure determination of homo-oligomers including a trimer [14] and a hexamer [15].

A non-crystallographic symmetry potential ensured subunits shared the same local conformation modulo relative placement and global orientation, while an additional potential arranged the subunits symmetrically by minimizing differences in distances for a chosen subset of the distance restraints. Building on the ambiguous distance restraint approach, Bardiaux et al. [16] implemented network anchoring into ARIA to simultaneously perform distance restraint assignment and oligomeric structure calculation.

To avoid the pitfalls of structure determination methods based on stochastic search, we instead propose algorithms that provide guarantees on the quality of the computed structures. In this paper, we describe a novel algorithm, DISCO, that computes oligomeric structures of protein complexes with cyclic symmetry (C_n) using RDCs and distance restraints such as NOEs, disulfide bonds, and distance restraints derived from paramagnetic relaxation enhancement (PREs). Along with returning the computed structural ensemble, DISCO guarantees the complete set of oligomer structures satisfying the RDCs and inter-subunit distance restraints can be computed exactly and in polynomial time. The following contributions are made in this paper:

- A novel geometric arrangement algorithm, DISCO, is presented to compute structures of homo-oligomeric protein complexes with C_n symmetry from RDCs, distance restraints such as NOEs, PREs, and disulfide bonds, and a structure of the subunit;
- DISCO guarantees all symmetric homo-oligomers satisfying the RDCs and the distance restraints are discovered, computed exactly, and computed in expected $O(n^2)$ time, where n is the number of distance restraints;
- DISCO can characterize the uncertainty in the position of the symmetry axis by computing the variance in atomic coordinates of oligomers sampled uniformly from the exact set of oligomer structures satisfying the RDCs and distance restraints;
- We introduce a technique to analyze ambiguous distance restraints that can discriminate between mutually consistent and inconsistent restraints; and
- We present results on the performance of DISCO on two symmetric proteins: E. coli Diacylglycerol Kinase (DAGK) [17] and a dimeric mutant of the immunoglobulin-binding domain B1 of streptococcal protein G (GB1) [18].

2   Methods

2.1   Solving for Protein Orientation

A single scalar RDC value r, measured experimentally, probes the orientation of an internuclear vector $\mathbf{v}$ through the following tensor equation: $r = D_{\max}\mathbf{v}^T S\mathbf{v}$ where $D_{\max}$ is the dipolar interaction constant, and S is the Saupe order matrix which represents an alignment tensor describing the average weak alignment of a protein in solution [19]. Our algorithm, DISCO, uses the observation that, for a C_n oligomer, one of the eigenvectors of the alignment tensor must be parallel to the symmetry axis of the complex [20,21], and hence computes the orientation of the symmetry axis from the RDCs. When it is possible to determine the high-resolution structure of the subunit based on entirely intra-molecular restraints [22,23,24], the alignment tensor can be least squares fit to the RDCs and the subunit structure using singular value decomposition [25].

For trimers and higher-order oligomers, we expect an alignment tensor with zero rhombicity. In this case, the symmetry axis is parallel to the eigenvector of the alignment tensor whose eigenvalue has the largest magnitude. For dimers, *which* eigenvector corresponds to the symmetry axis cannot be uniquely determined from a single set of RDCs alone, so all possibilities must be examined. If the alignment tensor has three distinct eigenvalues, then each corresponding eigenvector is evaluated by executing DISCO three times. We call any oligomer structure whose symmetry axis orientation has been computed from RDCs an *oriented oligomer structure*. Hence, the space of oriented oligomer structures corresponds to the space of symmetry axis positions; to build complete oligomer structures, all that remains is to choose a set of appropriate symmetry axis positions.

2.2   Solving for the Symmetry Axis Position

Inter-subunit distance restraints such as NOEs, PREs, and disulfide bonds can be used to restrict the position of the symmetry axis – even when precise subunit assignments are not known. Using the symmetry axis orientation computed from RDCs, each possible assignment for a distance restraint restricts the positions of the symmetry axis to an annulus in the plane ($\mathbb{R}^2$). However, distance restraints whose possible assignments include intra-subunit assignments will correspond to incorrect (or *decoy*) annuli if the true assignment of the restraint is intra-subunit. Therefore, distance restraints with possible intra-subunit assignments must not be used, unless DISCO is extended as described in Section 2.5.

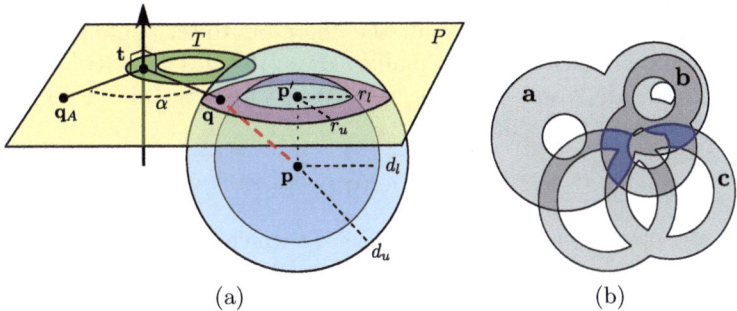

(a)   (b)

Fig. 3. (a) Symmetric distance restraint geometry. See Section 2.2 for complete explanation. (b) Unions of annuli (grey) for three hypothetical distance restraints (a, b, and c): This example shows two MSRs (blue).

First, chose a coordinate system so the z-axis ($\hat{z}$) is parallel to the symmetry axis. Then consider a single assignment for a strictly inter-subunit distance restraint with minimum and maximum distances d_l, d_u between atoms $\mathbf{p}, \mathbf{q} \in \mathbb{R}^3$ (see Figure 3(a)). Since the restraint must be inter-subunit, let $\mathbf{p}$ lie in subunit A and $\mathbf{q}$ lie in subunit B. If we assume the position and orientation of only subunit A are known, then $\mathbf{p}$ is known, but $\mathbf{q}$ is unknown. Let $\mathbf{q}_A$ be the position of the symmetric partner of $\mathbf{q}$ in subunit A. Due to the symmetry, $\mathbf{q}$ is related to $\mathbf{q}_A$ by a rotation about the symmetry axis (whose position $\mathbf{t}$ is also unknown):

$$\mathbf{q} = R(\mathbf{q}_A - \mathbf{t}) + \mathbf{t} \tag{1}$$

where R denotes a rotation about $\hat{z}$ by an angle $\alpha = \frac{2\pi}{m}$ and m is the oligomeric number of the protein. Therefore, to compute positions of the symmetry axis whose oligomer structures satisfy the distance restraint assignment, DISCO computes values of $\mathbf{t}$ such that distance restraint is satisfied: $d_l \leq |R(\mathbf{q}_A - \mathbf{t}) + \mathbf{t} - \mathbf{p}| \leq d_u$. Since we chose a coordinate system in which the symmetry axis is parallel to $\hat{z}$, we can simplify this problem to two dimensions instead of three. Construct a plane P perpendicular to $\hat{z}$ such that it contains $\mathbf{q}$ and $\mathbf{q}_A$. Let $A_3(\mathbf{p}, d_l, d_u)$ be a three-dimensional annulus centered at $\mathbf{p}$ whose radii d_l, d_u are equal to the lower and upper distance bounds of the distance restraint. The intersection of P with $A_3(\mathbf{p}, d_l, d_u)$ yields a two-dimensional annulus $A_2(\mathbf{p}', r_l, r_u)$ where $\mathbf{p}'$ is the projection of $\mathbf{p}$ along $\hat{z}$ onto P and the radii are: $r_l = \sqrt{d_l^2 - |\mathbf{p} - \mathbf{p}'|^2}$ and $r_u = \sqrt{d_u^2 - |\mathbf{p} - \mathbf{p}'|^2}$. Therefore, the distance restraint is satisfied when

$$\mathbf{q} \in A_2(\mathbf{p}', r_l, r_u). \tag{2}$$

By substituting (1) into (2), we relate the symmetry axis position $\mathbf{t}$ to satisfying positions of $\mathbf{q}$:

$$R(\mathbf{q}_A - \mathbf{t}) + \mathbf{t} \in A_2(\mathbf{p}', r_l, r_u). \tag{3}$$

To solve for $\mathbf{t}$, we return to (1) which can be rewritten: $(R - I)\mathbf{t} = R\mathbf{q}_A - \mathbf{q}$. Next, we substitute (2) for $\mathbf{q}$ and lift the operators to set operators to consider set membership in place of strict equality: $(R - I)\mathbf{t} \in R\mathbf{q}_A \ominus A_2(\mathbf{p}', r_l, r_u)$ where $\ominus$ represents the Minkowski difference [26]. We evaluate the Minkowski difference by simply translating the annulus:

$$(R - I)\mathbf{t} \in A_2(R\mathbf{q}_A - \mathbf{p}', r_l, r_u). \tag{4}$$

Consider all solutions to (4) for $\mathbf{t}$ as a set T, which represents the set of symmetry axis positions whose oligomer structures satisfy the distance restraint assignment:

$$T = \{\mathbf{t} \in \mathbb{R}^2 \mid (R - I)\mathbf{t} \in A_2(R\mathbf{q}_A - \mathbf{p}', r_l, r_u)\}. \tag{5}$$

To describe T, we make use of the following proposition which describes the matrix $(R - I) \in \mathbb{R}^{2 \times 2}$. The proof of this proposition is provided in the Supplementary Information (SI) [27], Section A.

Proposition 1. *The matrix* $(R - I)$ *is the composition of a 2D rotation* $W = \frac{1}{h}[\ \mathbf{u} - \hat{\mathbf{x}}\ \ \mathbf{v} - \hat{\mathbf{y}}\ ]$ *and a scaling* $h = |\mathbf{u} - \hat{\mathbf{x}}|$ *where* $R = [\ \mathbf{u}\ \ \mathbf{v}\ ]$ *and* $\hat{\mathbf{x}}, \hat{\mathbf{y}}$ *are the unit axes.*

Using Proposition 1 and (5), we can rewrite (4):

$$hWT = A_2(R\mathbf{q}_A - \mathbf{p}', r_l, r_u). \tag{6}$$

Since T represents a set and hW is invertible, we have replaced the set inclusion of (4) with strict equality. Solving for T, we see it must also be an annulus in two dimensions:

$$T = A_2\left(\frac{1}{h}W^{-1}(R\mathbf{q}_A - \mathbf{p}'), \frac{r_l}{h}, \frac{r_u}{h}\right). \tag{7}$$

Therefore, DISCO computes the annulus T exactly and in closed form using (7).

2.3   Analysis of Multiple Distance Restraints

If $\{(\mathbf{p}_i, \mathbf{q}_i)\}$ is a set of unambiguous distance restraints, DISCO evaluates (7) for each i to compute a set of annuli $\mathcal{T} = \{T_i\}$ that lies on a set of planes $\{P_i\}$. In the cases where $T_i = \emptyset$ (i.e., when $A_3(\mathbf{p}_i, d_l, d_u)$ and P_i do not intersect), the restraint cannot be satisfied by any oriented oligomer structure. Effectively, $T_i = \emptyset$ indicates the corresponding restraint is inconsistent with respect to the RDCs and the symmetry. Since all the P_i planes are mutually parallel and each P_i is perpendicular to $\hat{\mathbf{z}}$, each T_i is projected onto the xy-plane. In the ideal case with no noise and no incorrect assignments, the intersection of all annuli in $\mathcal{T}$ will result in a non-empty region of the xy-plane. Each symmetry axis position $\mathbf{t} \in \bigcap_i T_i$ corresponds to an oriented oligomer structure that satisfies every distance restraint. However, noise and incorrect assignments can cause $\bigcap_i T_i$ to be empty. Instead, DISCO computes the geometric arrangement of $\mathcal{T}$ using a randomized incremental algorithm [28] (implemented using the CGAL software library [29]), which returns all intersection points of the circles bounding the annuli, all edges between intersection points, and all faces bounded by the edges. Next, DISCO computes the faces of the arrangement contained in the greatest number of unions of annuli, the *maximally satisfying regions* (*MSRs*), which correspond to oriented oligomer structures that satisfy the maximal number of distance restraints (See Figure 3(b)).

If the distance restraints possess ambiguity, DISCO computes one annulus for each possible assignment. In one case, precise atom assignments are not known (*atom ambiguity*, often due to overlapping chemical shifts). Instead, each distance restraint has a set of possible assignments $\{(\mathbf{p}_j, \mathbf{q}_j)\}$ where $\mathbf{p}_j$ and $\mathbf{q}_j$ are the two atoms for assignment j. Instead of corresponding to a single annulus, the distance restraint corresponds to a *set* of annuli – one for each j. The restraint could be interpreted with any one of these possible assignments, and all of them are mutually exclusive. To avoid a combinatorial enumeration of assignment possibilities, we conservatively encode the choices using a logical OR operator by computing the set union of (7):

$$T_i = \bigcup_j A_2 \left(\frac{1}{h} W^{-1}(R\mathbf{q}_A^{(j)} - \mathbf{p}'_j), \frac{r_l}{h}, \frac{r_u}{h} \right) \tag{8}$$

where $\mathbf{q}_A^{(j)}$ represents the symmetric partner of $\mathbf{q}_j$ in subunit A. In the case where precise subunit assignments are not known (i.e., subunit ambiguity), DISCO computes an annulus for each of the $m - 1$ possible subunit assignments by varying the angle of rotation described by the matrix R in (7) to choose different subunits:

$$T_i = \bigcup_{j=1}^{m-1} A_2 \left(\frac{1}{h} W^{-1}(R_j\mathbf{q}_A - \mathbf{p}'), \frac{r_l}{h}, \frac{r_u}{h} \right) \tag{9}$$

where R_j is a rotation about the $\hat{\mathbf{z}}$ axis by an angle of $j\alpha$ and m is the oligomeric number of the protein. In both cases, DISCO computes the arrangement of the unions of annuli and selects as MSRs faces from the arrangement contained in the greatest number of unions of annuli. The following lemma describes the time complexity required to compute the MSRs. A proof of the complexity is provided in the SI [27], Section B.

Lemma 1. *For an oligomeric protein complex with cyclic symmetry and n distance restraints assigned ambiguously, the MSRs can be computed in expected $O(n^2)$ time.*

2.4   Evaluation

Once MSRs have been computed, DISCO evaluates the distance restraints using the continuous set of oligomer models described by the MSRs. We characterize a distance restraint as *inconsistent* if its corresponding union of annuli does not contain any of the MSRs. No oriented oligomer structure whose symmetry axis position was chosen from a MSR could satisfy an inconsistent restraint. Since the MSRs computed by DISCO represent continuous sets of symmetry axis positions, the corresponding oligomer structures are also continuous sets. To perform detailed structural analysis and for visualization, the MSRs are sampled on a uniform grid at a fine resolution to generate a discrete set of symmetry axis positions. One of the advantages of DISCO is that by computing the exact MSRs, it is unnecessary to sample the entire symmetry axis position configuration space. Instead, we can sample only within the MSR at a much finer resolution than would be possible using a grid search over the full configuration space. DISCO combines the sampled axis positions with the symmetry axis orientation computed from the RDCs to define a set of rigid transformations that, when applied to the subunit structure, generate symmetric oligomer structures. Figure 2 illustrates an example using a trimer. Each resulting structure is energy-minimized in XPLOR-NIH [30] using a fixed backbone, but flexible side chains to relieve minor steric clashes.

2.5   Extensions to DISCO

One restriction of DISCO as described above is that distance restraints must not have possible intra-subunit assignments. Since PREs have potential intra-subunit assignments as well as inter-subunit assignments, if the true assignment for a PRE is intra-subunit, then the annulus analysis presented in Section 2.2 will yield a decoy union of annuli. This union of annuli does not truly constrain the symmetry axis since an intra-subunit distance restraint cannot possibly describe the symmetry of the oligomer structure. One might hope to resolve the intra/inter-subunit assignment ambiguity directly, but no experimental or computational methods are currently known to perform such an assignment for PREs. However, DISCO's restriction can be relaxed if a set of distance restraints with no possible intra-subunit assignments are also available.

We therefore divide the available distance restraints into two classes: distance restraints with no possible intra-subunit assignments are considered *trusted*, and the remaining distance restraints are considered *untrusted*, since they may yield decoy unions of annuli. DISCO processes the trusted and untrusted distance restraints in two different phases of the algorithm. Phase one uses only the trusted restraints to compute MSRs (Section 2.2), which we refer to as *trusted MSRs*. In phase two, DISCO computes unions of annuli for the untrusted distance restraints, but does not immediately compute their arrangement. Instead, DISCO compares each of the untrusted unions of annuli to the trusted MSRs. If an untrusted union of annuli does not intersect the trusted MSRs, that union of annuli is discarded. The remaining unions of annuli that intersect the trusted MSRs are used along with the original trusted unions of annuli to compute a new arrangement, from which the *final MSRs* are selected. The final MSRs represent oligomer structures that are guaranteed to satisfy the trusted distance restraints, and also a subset of the untrusted distance restraints. This two-phase approach ensures all distance restraints contribute to the structure determination (despite some restraints having possible intra-subunit assignments), while avoiding the need to choose explicit intra/inter-subunit assignments.

3   Results

We evaluated the performance of DISCO on two proteins: DAGK [17] and a dimeric mutant of GB1 [18] (henceforth referred to simply as GB1). We compared structures computed by DISCO to known structures (i.e., *reference structures*) from the PDB [31] for DAGK: 2kdc, model 1 and GB1: 1q10, model 1. The subunit structure used by DISCO was the first subunit in the reference structure, which was determined using traditional protocols. This mirrors the experimental situation where the subunit structure can be determined with confidence [22,23,24], but the main bottleneck is subunit assignment and the assembly of subunit structures to form the oligomer structure.

We measured the structural similarity between structures computed by DISCO and the reference structures using the RMS deviation in backbone atom position. All computed structures were within 0.14 Å to the reference for DAGK, and 0.25

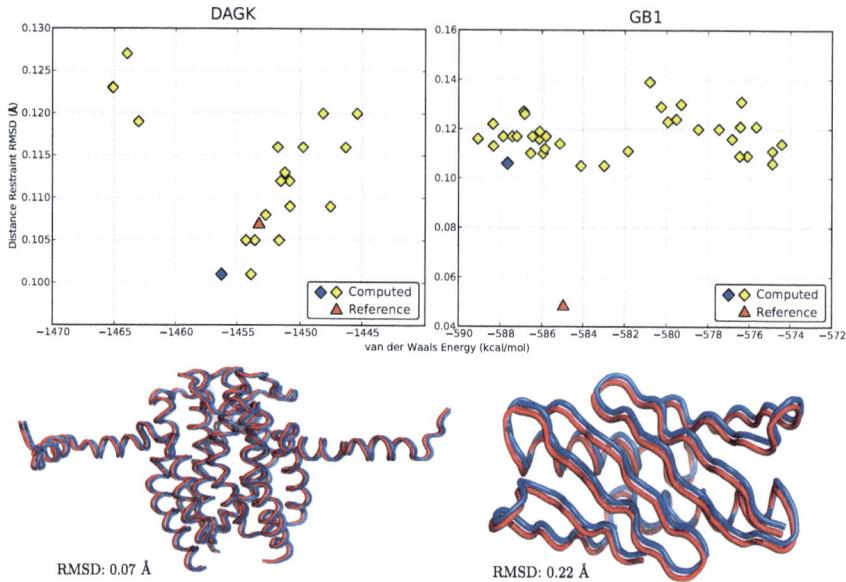

Fig. 4. Top: distance restraint satisfaction scores (lower is better) and van der Waals energies. Bottom: backbone alignment between the blue diamond structure from the top row (blue) and the reference structure (red). For this illustration, the structures are offset from one another by 1.0 Å so they appear distinct.

Å for GB1. After energy-minimization, we evaluated the RMS distance restraint violation and van der Waals energy (using the pairwise Lennard-Jones potential) of each oligomer structure [6]. Figure 4 shows the scores of the computed structures, which are comparable to those of the references. Since DISCO can compute the MSRs exactly and discrete structures are sampled uniformly from the MSRs, the variance in backbone atom position of the computed structural ensemble accurately represents uncertainty about the position of the symmetry axis inherent in the distance restraints. Statistics of the computed ensembles, including backbone RMSDs and the variance, are summarized in Table 1. Sections 3.1 and 3.2 describe in more detail the results for DAGK and GB1 respectively.

3.1   DAGK with Subunit Ambiguity

DAGK is a C_3 homo-trimeric membrane protein of 121 residues per subunit for which 67 NH RDCs, 200 PREs, and 24 disulfide bonds are available [17]. The PREs and disulfide bonds are inter-subunit distance restraints whose assignments are complicated by subunit ambiguity and therefore have two possible assignments each (See Figure 1). Additionally, it was not known whether the two PRE-related atoms were in the same subunit, or different subunits. Therefore, we used the two-phase extension to DISCO, labeling the disulfide bonds as trusted, and the PREs as untrusted.

Table 1. Statistics of oligomer structures computed by DISCO

	DAGK	GB1
Orientation difference[1]	0.16°	0.66°
MSR sample resolution[2]	0.025 Å	0.005 Å
Computed ensemble size	20	36
Average all-atom variance	9.2×10^{-3} Å^2	2.7×10^{-2} Å^2
Average backbone variance	5.8×10^{-3} Å^2	7.9×10^{-4} Å^2
Backbone atom RMSD[3]	0.05–0.14 Å	0.20–0.25 Å

[1]Difference in orientation between computed and reference symmetry axes. [2]Grid resolution at which symmetry axis positions were sampled from MSRs. [3]Range of RMSDs in the computed ensemble vs. reference.

During phase one, DISCO computed the arrangement of unions of annuli from the trusted disulfide bonds. DISCO allows sidechains to move during energy-minimization, but uses the rigid subunit structure to compute the annuli from the distance restraints. To account for motions of the sidechains during minimization that could potentially relieve violated distance restraints, we slightly increased the distances allowed by the restraint. We chose a *padding percentage* β such that the lower distance bound of each restraint is multiplied by $(1 - \beta)$ and the upper by $(1 + \beta)$. For these tests, we chose $\beta = 3\%$. Figure 5 (A) shows the arrangement and MSRs from phase one. The symmetry axis position of the reference structure is contained within the trusted MSR indicating the annulus analysis is able to correctly describe the satisfying symmetry axis positions of the oligomer structure for DAGK. Since DISCO is able to compute the MSRs exactly (and thus, the set of satisfying oriented oligomer structures exactly), the absence of any additional MSRs farther away rules out the possibility of a satisfying oligomer structure that is dissimilar to those already discovered by the algorithm. The MSRs for DAGK are sensitive to the padding percentage β chosen. With $\beta = 5\%$, oligomer structures sampled from the MSRs differed from the reference structure by as much as 2.7 Å backbone RMSD, but had a distance restraint RMSD of no worse than 0.38 Å.

The single trusted MSR computed satisfied 21 of the 24 disulfide bond restraints. The remaining 3 disulfide bond restraints were labeled inconsistent by DISCO, each resulting in small violations in the oligomer structures. For comparison, the same three disulfide bond restraints were also unsatisfied in the reference structure. Additional details of the inconsistent disulfide bonds are presented in the SI [27], Section C. Phase two of DISCO discarded 46 of the 200 PREs (also padded by $\beta = 3\%$) since their annuli did not intersect the trusted MSR. The remaining 154 untrusted PREs were combined with the original 24 trusted disulfide bonds to compute a new arrangement and the final MSRs which are shown in Figure 5 (B). As with the trusted MSR, DISCO also computed a single final MSR which again contains the symmetry axis position of the reference structure. If a large enough number of decoy unions of annuli still intersect the trusted MSR, it is possible for DISCO to be led astray and compute final MSRs that do not correctly describe the oligomer structure of the protein.

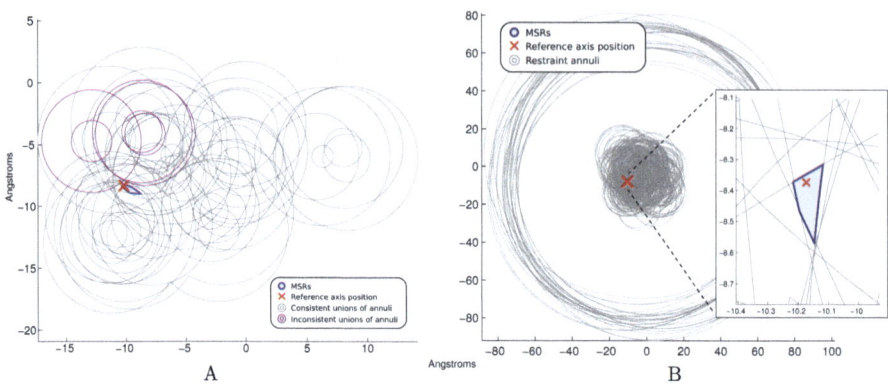

Fig. 5. (A): Unions of annuli from 24 disulfide bonds for DAGK. Even though an annulus marked consistent does not appear to intersect the MSR, it is the member of a union which does. (B): Unions of annuli from 24 disulfide bonds and 154 PREs. The outer ring represents outer boundaries for 27 PREs with large upper distances (~125 Å), for which only the lower bounds were meaningful. Inset: close-up of MSR and reference axis position.

However, the presence of the reference axis position within the final MSRs shows that the two-phase analysis is able to remove enough decoy unions of annuli to prevent them from conspiring to increase support for an incorrect answer.

3.2   GB1 with Simulated Atom Ambiguity

GB1 is a C_2 homo-dimer of 56 residues per subunit for which 56 NH RDCs and 296 experimental inter-subunit NOEs (assigned without subunit ambiguity, since GB1 is a dimer) are available [18]. We simulated atom ambiguity by expanding the published NOE assignments to include nuclei with similar chemical shifts, resulting in an average of 6.7 possible atom assignments per restraint. Window sizes of 0.05 ppm and 0.5 ppm were used for ^{1}H and ^{13}C/^{15}N shifts respectively. A search over the three alignment tensor eigenvectors revealed that MSRs computed from the D_{xx} eigenvector resulted in the

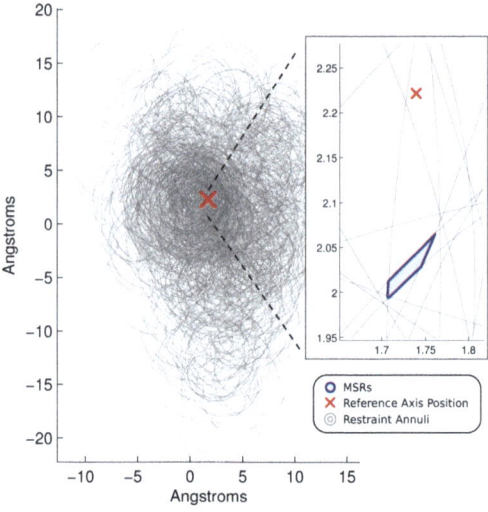

Fig. 6. Distance restraint unions of annuli for GB1 using 296 NOEs with simulated ambiguity. The six inconsistent unions are not labeled. Inset: close-up of the MSR and reference axis position.

greatest distance restraint satisfaction. Figure 6 shows the single MSR computed for GB1 (with $\beta = 0$) in comparison to the position of the symmetry axis of the reference structure. The ambiguity simulation resulted in 1993 total annuli. Remarkably, 32% of these annuli enclosed no points (i.e., are the empty set) and therefore, no satisfying symmetry axis positions exist, indicating these possible assignments are inconsistent with the computed symmetry axis orientation and ultimately the RDCs. DISCO also found six inconsistent inter-subunit NOEs whose unions of annuli did not intersect the MSR, each resulting in small violations in the oligomer structures. The reference structure violates 12 of its inter-subunit NOEs, although each to a lesser degree. Additional details of the inconsistent NOEs are presented in the SI [27], Section D.

4   Conclusion

DISCO can accurately determine the oligomer structures of proteins with C_n symmetry using RDCs and distance restraints. The MSRs are computed exactly and in expected $O(n^2)$ time, thus ensuring no satisfying oriented oligomer structures are missed by the algorithm. DISCO analyzes inter-subunit distance restraints, even when assigned ambiguously, but avoids enumerating explicit assignment combinations. A small number of distance restraints with low uncertainty work best, but DISCO performs well even when using a large number of noisy restraints. Using the two-phase protocol, DISCO can incorporate structural constraint provided by distance restraints with possible intra-subunit assignments in phase two. As a prerequisite, phase one of the protocol requires restraints with strictly inter-subunit assignments.

In practice, the D_{zz} eigenvector of the alignment tensor may differ slightly from the orientation of the true symmetry axis of the oligomer. To account for uncertainty in the symmetry axis position (possibly due to dynamics or experimental uncertainty), one can estimate the distribution of symmetry axis orientations described by the RDCs by considering the uncertainty of each RDC value. In future work, symmetry axis orientations sampled from this distribution can be analyzed by DISCO to select for the orientations whose resulting oligomer structures best satisfy the distance restraints. Since DISCO computes the exact set of oriented oligomer structures that satisfy the distance restraints, the variance in atom position (Table 1) yields a meaningful measure of the range of oligomer structures allowed by the distance restraints, whereas in methods that rely on stochastic search, the variance is merely an artifact of the sampling. The entire DISCO protocol has been completely automated in a software package that is freely available and open-source.

References

1. Chen, C.Y., Georgiev, I., Anderson, A.C., Donald, B.R.: Computational structure-based redesign of enzyme activity. Proceedings of the National Academy of Sciences 106, 3764–3769 (2009)

2. Frey, K.M., Georgiev, I., Donald, B.R., Anderson, A.C.: Predicting resistance mutations using protein design algorithms. Proceedings of the National Academy of Sciences 107, 13707–13712 (2010)

3. Goodsell, D.S., Olson, A.J.: Structural symmetry and protein function. Annual Review of Biophysics and Biomolecular Structure 29(1), 105–105 (2000), doi:10.1146/annurev.biophys.29.1.105

4. Levy, E.D., Erba, E.B., Robinson, C.V., Teichmann, S.A.: Assembly reflects evolution of protein complexes. Nature 453, 1262–1265 (2008), 10.1038/nature06942

5. White, S.H.: The progress of membrane protein structure determination. Protein Science 13, 1948–1949 (2004)

6. Potluri, S., Yan, A.K., Donald, B.R., Bailey-Kellogg, C.: A complete algorithm to resolve ambiguity for intersubunit NOE assignment in structure determination of symmetric homo-oligomers. Protein Science 16, 69–81 (2007)

7. Ikura, M., Bax, A.: Isotope-filtered 2D NMR of a protein-peptide complex: study of a skeletal muscle myosin light chain kinase fragment bound to calmodulin. Journal of the American Chemical Society 114, 2433–2440 (1992)

8. Saxe, J.: Embeddability of weighted graphs in k-space is strongly NP-hard. In: Proceedings of the 17th Allerton Conference in Communications, Control, and Computing, pp. 480–489 (1979)

9. Wang, L., Mettu, R.R., Donald, B.R.: A polynomial-time algorithm for de novo protein backbone structure determination from nuclear magnetic resonance data. Journal of Computational Biology 13, 1267–1288 (2006)

10. Potluri, S., Yan, A.K., Chou, J.J., Donald, B.R., Bailey-Kellogg, C.: Structure determination of symmetric homo-oligomers by a complete search of symmetry configuration space, using NMR restraints and van der Waals packing. Proteins: Structure, Function, and Bioinformatics 65, 203–219 (2006)

11. Wang, X., Bansal, S., Jiang, M., Prestegard, J.H.: RDC-assisted modeling of symmetric protein homo-oligomers. Protein Science 17, 899–907 (2008)

12. Wang, C.S.E., Lozano-Pérez, T., Tidor, B.: AmbiPack: A systematic algorithm for packing of macromolecular structures with ambiguous distance constraints. Proteins: Structure, Function, and Genetics 32, 26–42 (1998)

13. Nilges, M.: A calculation strategy for the structure determination of symmetric demers by ^{1}H NMR. Proteins: Structure, Function, and Genetics 17, 297–309 (1993)

14. Kovacs, H., O'Donoghue, S.I., Hoppe, H.J., Comfort, D., Reid, K.B.M., Campbell, I.D., Nilges, M.: Solution structure of the coiled-coil trimerization domain from lung surfactant protein D. Journal of Biomolecular NMR 24, 89–102 (2002), 10.1023/A:1020980006628

15. O'Donoghue, S.I., Chang, X., Abseher, R., Nilges, M., Led, J.J.: Unraveling the symmetry ambiguity in a hexamer: Calculation of the R6 human insulin structure. Journal of Biomolecular NMR 16(2), 93–108 (2000), 10.1023/A:1008323819099

16. Bardiaux, B., Bernard, A., Rieping, W., Habeck, M., Malliavin, T.E., Nilges, M.: Influence of different assignment conditions on the determination of symmetric homodimeric structures with ARIA. Proteins: Structure, Function, and Bioinformatics 75, 569–585 (2009)

17. Van Horn, W.D., Kim, H.J., Ellis, C.D., Hadziselimovic, A., Sulistijo, E.S., Karra, M.D., Tian, C., Sönnichsen, F.D., Sanders, C.R.: Solution nuclear magnetic resonance structure of membrane-integral diacylglycerol kinase. Science 324, 1726–1729 (2009)

18. Byeon, I.J.L., Louis, J.M., Gronenborn, A.M.: A protein contortionist: Core mutations of GB1 that induce dimerization and domain swapping. Journal of Molecular Biology 333, 141–152 (2003)

19. Donald, B.R., Martin, J.: Automated NMR assignment and protein structure determination using sparse dipolar coupling constraints. Progress in Nuclear Magnetic Resonance Spectroscopy 55, 101–127 (2009)
20. Al-Hashimi, H.M., Bolon, P.J., Prestegard, J.H.: Molecular symmetry as an aid to geometry determination in ligand protein complexes. Journal of Magnetic Resonance 142, 153–158 (2000)
21. Bewley, C.A., Clore, G.M.: Determination of the relative orientation of the two halves of the domain-swapped dimer of cyanovirin-N in solution using dipolar couplings and rigid body minimization. Journal of the American Chemical Society 122, 6009–6016 (2000)
22. Oxenoid, K., Chou, J.J.: The structure of phospholamban pentamer reveals a channel-like architecture in membranes. Proceedings of the National Academy of Sciences of the United States of America 102, 10870–10875 (2005)
23. Schnell, J.R., Chou, J.J.: Structure and mechanism of the M2 proton channel of influenza A virus. Nature 451(7178), 591–595 (2008), 10.1038/nature06531
24. Wang, J., Pielak, R.M., McClintock, M.A., Chou, J.J.: Solution structure and functional analysis of the influenza B proton channel. Nat. Struct. Mol. Biol. 16, 1267–1271 (2009), 10.1038/nsmb.1707
25. Losonczi, J.A., Andrec, M., Fischer, M.W.F., Prestegard, J.H.: Order matrix analysis of residual dipolar couplings using singular value decomposition. Journal of Magnetic Resonance 138, 334–342 (1999)
26. Lozano-Perez, T.: Automatic planning of manipulator transfer movements. IEEE Transactions on Systems, Man and Cybernetics 11, 681–698 (1981)
27. Martin, J.W., Yan, A.K., Bailey-Kellogg, C., Zhou, P., Donald, B.R.: Supplementary information: A geometric arrangement algorithm for structure determination of symmetric protein homo-oligomers from NOEs and RDCs (2011), http://www.cs.duke.edu/donaldlab/Supplementary/recombll/DISCO/
28. Halperin, D.: Arrangements. In: Goodman, J.E., O'Rourke, J. (eds.) Handbook of Discrete and Computational Geometry, 2nd edn., pp. 529–562. CRC Press, Inc., Boca Raton (1997)
29. Hanniel, I., Halperin, D.: Two-dimensional arrangements in CGAL and adaptive point location for parametric curves. In: Näher, S., Wagner, D. (eds.) WAE 2000. LNCS, vol. 1982, pp. 171–182. Springer, Heidelberg (2001)
30. Schwieters, C.D., Kuszewski, J.J., Tjandra, N., Clore, G.M.: The Xplor-NIH NMR molecular structure determination package. Journal of Magnetic Resonance 160, 65–73 (2003)
31. Berman, H.M., Westbrook, J., Feng, Z., Gilliland, G., Bhat, T.N., Weissig, H., Shindyalov, I.N., Bourne, P.E.: The protein data bank. Nucl. Acids Res. 28, 235–242 (2000)

Paired de Bruijn Graphs: A Novel Approach for Incorporating Mate Pair Information into Genome Assemblers

Paul Medvedev[1,*], Son Pham[1,*], Mark Chaisson[2],
Glenn Tesler[3], and Pavel Pevzner[1]

[1] Department of Computer Science and Engineering, Univ. of California, San Diego
[2] Pacific Biosciences of California, Menlo Park, CA
[3] Department of Mathematics, Univ. of California, San Diego

Abstract. The recent proliferation of next generation sequencing with short reads has enabled many new experimental opportunities but, at the same time, has raised formidable computational challenges in genome assembly. One of the key advances that has led to an improvement in contig lengths has been mate pairs, which facilitate the assembly of repeating regions. Mate pairs have been algorithmically incorporated into most next generation assemblers as various heuristic post-processing steps to correct the assembly graph or to link contigs into scaffolds. Such methods have allowed the identification of longer contigs than would be possible with single reads; however, they can still fail to resolve complex repeats. Thus, improved methods for incorporating mate pairs will have a strong effect on contig length in the future.

Here, we introduce the *paired de Bruijn graph*, a generalization of the de Bruijn graph that incorporates mate pair information into the graph structure itself instead of analyzing mate pairs at a post-processing step. This graph has the potential to be used in place of the de Bruijn graph in any de Bruijn graph based assembler, maintaining all other assembly steps such as error-correction and repeat resolution. Through assembly results on simulated error-free data, we argue that this can effectively improve the contig sizes in assembly.

1 Introduction

The recent proliferation of next generation sequencing with short reads has enabled new experimental opportunities, such as the 10K genomes project, which aims to sequence and assemble the genomes of approximately one species in every vertebrate genus [7]. At the same time, the short read length and sheer demand for powerful assemblers has raised formidable computational challenges. Thus, genome assembly continues to represent one of the most difficult and important algorithmic problems in bioinformatics.

The first generation of assemblers followed the overlap-layout-consensus paradigm, where overlaps were heuristically used to join reads together into

---

* These authors contributed equally to the work and are joint first authors.

V. Bafna and S.C. Sahinalp (Eds.): RECOMB 2011, LNBI 6577, pp. 238–251, 2011.

contigs [13,1]. Later, the introduction of de Bruijn graphs led to significant improvements in assembly [15,9,18]. In contrast to the overlap-layout-consensus approach, these assemblers first constructed a graph where the original genome is spelled by a series of walks through the graph, and non-branching walks correspond to substrings (contigs) of the genome. Compared to the earlier heuristic approaches, de Bruijn graphs produced longer contigs and gave rise to more powerful techniques for correcting errors and resolving repeats — identical, or nearly identical, stretches of DNA [19]. Their success led to the development of other types of graphs for sequence assembly: A-Bruijn graphs [17] and closely related string graphs [14], which together have become an essential part of most modern assembly tools, including EULER-SR [5], Velvet [22], ALLPATHS [3], ABySS [20], and others.

Despite these advances, the challenge of resolving repeats remains. When the length of a repeat is longer than twice the read length, it becomes difficult to correctly match its upstream and downstream regions. In order to alleviate this problem, sequencing technologies were extended to produce *mate pairs* [21] — pairs of reads between which the genomic distance (called the *insert size*) is well estimated. Because insert sizes could be much longer than the read length, mate pairs were able to span long repeats and could potentially match up the regions surrounding a repeat.

The challenge of algorithmically incorporating mate pair information into de Bruijn graph assemblers was first addressed by [16], who proposed a heuristic to look for a path between the two reads of a mate pair with a length of the insert size. If exactly one such path was found, then a *mate pair transformation* could be applied to "unwind" this path in the graph. Essentially, this amounted to transforming two mated reads into one long read where the gap between the mates was filled in with the nucleotide sequence representing the found path, thus potentially connecting the surrounding regions of a repeat. Several other heuristic approaches for utilizing mate pair information in the de Bruijn graph were developed [22,3,11].

Such methods had a great impact on genome assembly, allowing the construction of much longer contigs; however, they could still fail in complex repeat-rich regions, where there are multiple paths between the read pairs. Many current technologies (including Complete Genomics [6] and Helicos [8]) still generate very short reads (around 25 nt) for which the resulting de Bruijn graph is very tangled (even for bacterial genomes). In such cases, mate pair transformations often fail because of multiple paths. Additionally, the percentage of mate pairs that can be successfully transformed deteriorates when the insert size is high [4], and the search for paths between mates becomes prohibitively time-consuming. Unfortunately, these difficulties result in shorter contigs in complex repeat-rich regions. The limitations of the existing heuristics for analyzing mate pairs is thus a major hurdle towards assembling large contigs with short reads.

We believe that the shortcomings of current mate pair algorithms stem from the fact that they are heuristic approaches that are applied *after* the construction of the de Bruijn graph. The de Bruijn graph does an excellent job of incorporating the sequence information from the single reads; however, it ignores any mate pair information that is available. This information has to be recovered after the graph construction, and only then applied in a heuristic manner. In this paper, we propose the *paired de Bruijn graph*, a generalization of the de Bruijn graph that incorporates the mate pair information into the structure of the graph itself, as opposed to a post-processing step. Just as moving from the heuristic overlap-layout-consensus paradigm to the de Bruijn graph paradigm resulted in better assemblies, we believe that moving from heuristic mate pair algorithms to paired de Bruijn graphs could result in a more effective use of mate pair information. The paired de Bruijn graph is a potential replacement of the de Bruijn graph in existing de Bruijn graph based assemblers; existing assembly stages, including error correction and scaffolding, would not need to be substantially modified.

Through assembly results on simulated error-free data, we argue that when mate pair information is used in this manner, the read length (once above a small threshold) becomes much less relevant [4]. We find that the contig sizes in an assembly are largely dictated by the average insert size — when it exceeds 6000 nt, we can assemble all of *E. coli* into one contig and most of the human chromosome 22 into 15 contigs. Though this paper falls short of analyzing real data, we believe that, similar to how early error-free studies of de Bruijn graphs laid the foundation for their use in assembly [15], the paired de Bruijn graph can become the basis of practical assemblers.

2   From de Bruijn Graphs to Paired de Bruijn Graphs

2.1   Preliminaries

To simplify the presentation, we assume that the genome is a circular string, i.e., one circular, single-stranded chromosome, and that all reads have the same length ℓ; extending our approach for multiple linear chromosomes or varying read length is straightforward. Moreover, we assume that reads are error-free (see Section 5 for a discussion). In this setting, a mate pair is an ordered pair of strings of length ℓ drawn from the genome at positions i and j, respectively. Normally, the relative distance between reads is expressed in terms of the *insert size*, the number of nucleotides from the first nucleotide of a to the last nucleotide of b: $j-i+\ell$. However, for the purposes of our construction, it is more convenient to express it in terms of $d = j - i$, the difference in their leftmost coordinates. Note that d is the insert size minus one read length (see Fig. 1(a)).

As with any de Bruijn graph based approach, our algorithms have a parameter k that dictates the size of the substrings into which the reads are chopped up. Thus, though our input is a set of mate pairs of reads of any length, we immediately chop them up into smaller pieces. Formally, each mate pair of reads is replaced by its constituent $\ell - k$ (sub-)mate pairs, where the reads of each (sub-)mate pair now have length $k + 1$. Therefore, for the remainder of the paper, we will assume without loss of generality that the reads are immediately given with length $k + 1$. We now give some definitions.

A-Bruijn graphs: Let G be a directed graph on m vertices. The *gluing* of vertices v and w is defined by substituting v and w by a single vertex (called the *successor* of v and w) and retaining all edges incident to either v or w as edges incident to their successor. Let A be a boolean $m \times m$ matrix representing "glues" [17]. The *A-Bruijn graph* $A(G)$ is obtained by gluing all vertices v and w of G for which $A_{v,w} = 1$. One can execute these glues in an arbitrary order under the assumption that each gluing instruction $A_{v,w} = 1$ is applied to the successors of vertices v and w in the graph resulting from the previous gluing instructions.

Below we describe three A-Bruijn graphs: de Bruijn graphs (for unpaired reads); paired de Bruijn graphs (for mate pairs with an exact distance), and approximate paired de Bruijn graphs (for mate pairs with an approximate distance).

k-mers and labels: Define a k-*mer* as a string of length k. Below we assume that the parameter k is fixed. Given a circular string $S = s_1 \ldots s_n$, let $S_k(i)$ be the k-mer $s_i \ldots s_{i+k-1}$ (where the index is taken modulo n). The set of all k-mers $S_k(i)$ (for $1 \leq i \leq n$) is called the k-*spectrum* of S. For a k-mer $a = a_1 \ldots a_k$, we define two $(k-1)$-mers, $\text{prefix}(a) = a_1 \ldots a_{k-1}$ (remove last character) and $\text{suffix}(a) = a_2 \ldots a_k$ (remove first character). We say that k-mer a *aligns at position* i if $a = S_k(i)$.

(k,d)-mers and bilabels: A *bilabel* $(a|b)$ is a pair of strings, a and b, of equal length. Define $\text{left}(a|b) = a$ and $\text{right}(a|b) = b$. A k-*mer bilabel* indicates both a and b have length k. Define $\text{prefix}(a|b) = (a_1 \ldots a_{k-1}|b_1 \ldots b_{k-1})$ and $\text{suffix}(a|b) = (a_2 \ldots a_k|b_2 \ldots b_k)$. Given an integer d (usually $d \geq k$), a (k,d)-mer of S is a pair of k-mers $S_k(i)$ and $S_k(i+d)$ that start exactly d nucleotides apart. We use the bilabel notation $(S_k(i)|S_k(i+d))$ for (k,d)-mers. For a string S and parameters d and Δ, we say k-mer bilabel $(a|b)$ *aligns at position* i if $a = S_k(i)$ and $b = S_k(i+d+x)$ for some $-\Delta \leq x \leq \Delta$. A (k,d,Δ)-*mer* of S is a bilabel $(a|b)$ that aligns somewhere to S.

2.2 De Bruijn Graphs (Modelling Unpaired Reads)

Let C be a set of $(k+1)$-mers from a circular string S. We construct an A-Bruijn graph based on C as follows.

- First we define an initial graph G_0 consisting of $m = 2|C|$ vertices and $|C|$ isolated edges. For each $(k+1)$-mer $a \in C$, introduce two new vertices u, v and form an edge $u \to v$. Label the edge by the $(k+1)$-mer a; label u by the k-mer $\text{prefix}(a)$; and label v by the k-mer $\text{suffix}(a)$.
- Second, we glue certain vertices of G_0 together, by forming an $m \times m$ binary matrix A and setting $A_{i,j} = 1$ to indicate that vertices i and j should be glued together. For this construction, we set $A_{i,j} = 1$ when vertices i and j have the same label.

The labeled directed graph $G = \text{DB}(C, k)$ obtained from these gluings is the de Bruijn graph of C [17] (for an illustration, see Fig. 1(b,c,d)). It may be considered

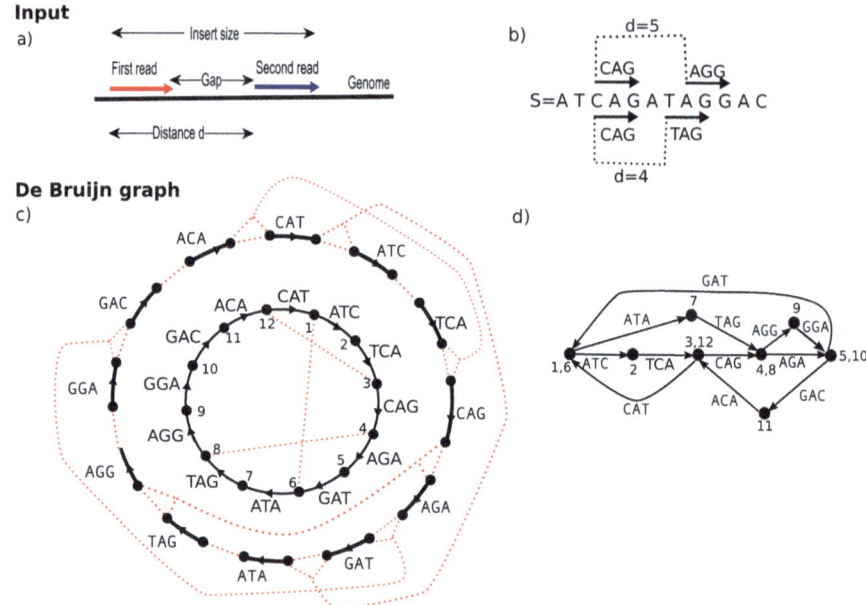

Fig. 1. An illustration of matepairs and of the de Bruijn graph. (a) A mate pair is a pair of reads with a distance of d between their start positions. (b) A circular genome S and two mate pairs, with $d = 4$ and $d = 5$. (c–d) The de Bruijn graph construction for $k = 2$. In (c), the outside circle shows a separate black edge for each 3-mer (equivalently, each element of the 3-spectrum). The dotted red lines indicate vertices that will be glued. The inner circle shows the result of applying some of the glues. Note that this is an intermediate step of the construction in which we only show the gluings of vertices arising from the same position of S. (d) The final de Bruijn graph, resulting from all the glues.

as either a simple graph (without parallel edges but with loops), or as a multigraph where the multiplicity of each edge is determined by the number of times the $(k+1)$-mer it represents is present in C. Consider a walk through edge/label sequence $e_1, \ldots, e_r$. The labels satisfy $\text{suffix}(e_i) = \text{prefix}(e_{i+1})$, and we may define the string of length $r + k$ spelled by this walk as $\text{walkword}(e_1, \ldots, e_r)$ by successively overlapping the labels with a shift of one character at a time.

Traditionally, the de Bruijn graph is also defined on a string S by setting the vertex set equal to the k-spectrum of S. For every $(k+1)$-mer a of S, define an edge $\text{prefix}(a) \rightarrow \text{suffix}(a)$ labeled by a. Explicitly, for each $S_{k+1}(i)$ of S, define an edge $S_k(i) \rightarrow S_k(i+1)$ labeled by $S_{k+1}(i)$ (for $i = 1, \ldots, n$).

In the case that C is the $(k+1)$-spectrum of S, the de Bruijn graph built on C using the gluing approach is identical to the one built directly on the genome S. Moreover, there is a covering cycle that spells S, where a *covering cycle* is a cyclical walk that visits every edge at least once. In this graph, the cycle is the sequence of edges $S_{k+1}(1), \ldots, S_{k+1}(n)$. The covering cycle property is crucial

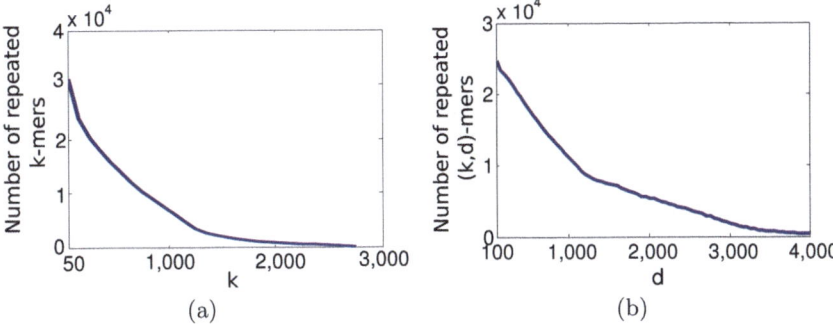

Fig. 2. (a) The number of repeated k-mers in the *E. coli* genome, for various values of k. (b) The number of repeated (k, d)-mers, for various values of d with $k = 50$.

for assembly because it implies that all walks whose interior vertices have just one out-neighbor must spell substrings in S (contigs).

2.3 Graph Complexity

The usefulness of a graph representation of a genome can vary widely. In general, the number of vertices can serve as a rough indicator of how useful the graph is — as the number of vertices grows (and the number of edges stays the same), the graph is likely to become less entangled, and the contigs are likely to become longer. Fig. 2(a) shows that in the de Bruijn graph, the number of repeated k-mers in *E. coli* drops as k increases, indicating that the de Bruijn graph has more vertices and likely becomes less entangled. Alternatively, consider pairs of k-mers, i.e., (k, d)-mers. Fig. 2(b) shows that, after fixing $k = 50$, the number of repeated (k, d)-mers drops as d increases. This is not surprising due to the repeat structure of genomes — the bigger the d, the less common it is to have pairs of repeats spaced a distance of d apart. Figs. 2(a) and (b) illustrate alternatives for improving contig lengths: increasing the read length (pursued by companies such as Pacific Biosciences) versus increasing the insert size (advocated by Chaisson et al., 2009 [4]). While the increase in the read length remains a difficult technological challenge, increasing the insert size (up to tens of thousands of nucleotides) is already within the power of current technologies. Thus, if we could build a graph whose vertices represent (k, d)-mers instead of k-mers, then the length of the contigs is likely to increase as the insert size grows. This is the basic motivation for the paired de Bruijn graph, and, as we will show in Section 3, the contig lengths in the paired de Bruijn graph do in fact increase with d.

2.4 Paired de Bruijn Graphs (Modelling Paired Reads with Exact Distance)

We now define a graph modelling mate pairs in the special case that all pairs are exactly the same distance d apart. This is an idealized case unachievable

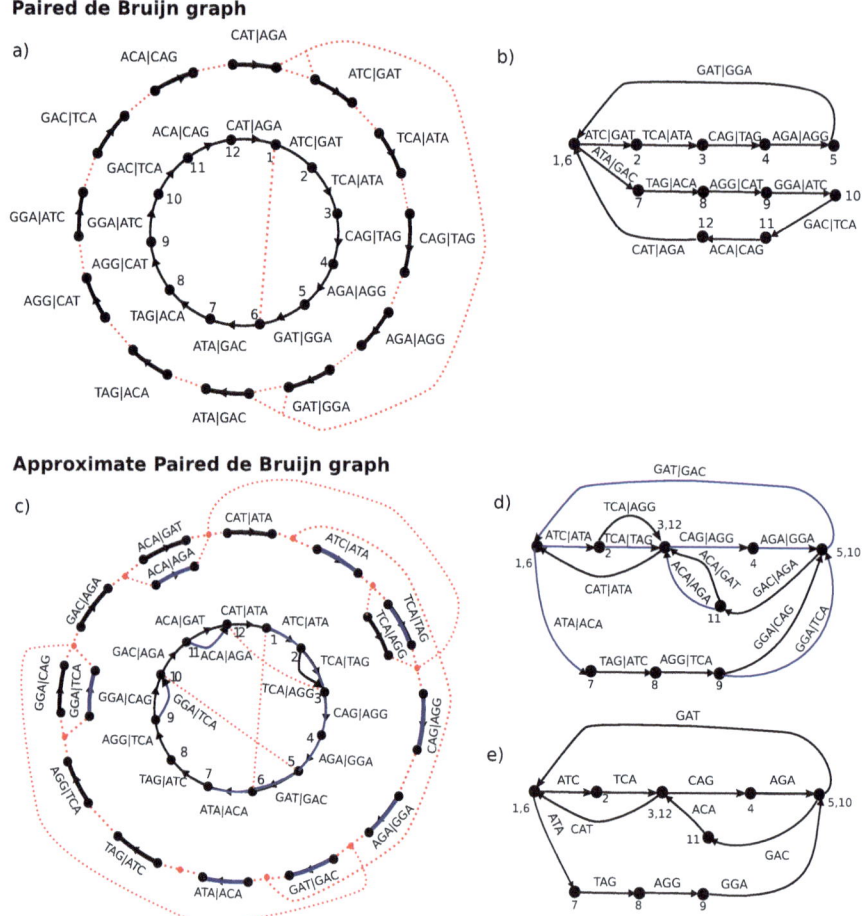

Fig. 3. (a–b) The paired de Bruijn construction for $k = 2$, $d = 4$ from the same string S as in Fig. 1. In (a), the outer circle has an edge from every element of the $(3, 4)$-spectrum. (b) The paired de Bruijn graph after all the gluings; notice that it has only one branching vertex, versus four in the de Bruijn graph (Fig. 1(d)). (c–e) The construction of the approximate paired de Bruijn graph for $k = 2$, $d = 5$, $\Delta = 1$. In (c), one possible covering spectrum is shown in the outside circle, with black edges for elements with mate pair distance 6 and blue edges for distance 5. Since $\Delta = 1$, we glue vertices if they have equal left labels and their right labels are a distance at most 2 apart from each other in the de Bruijn graph (Fig. 1(d)). The final multigraph after all vertex gluings is shown in (d), and the resulting simple graph, used to spell the contigs, is shown in (e). Notice that this graph now has three branching vertices.

with current sequencing technologies, but the next section will generalize the construction to varying distances. Given a set of $(k + 1, d)$-mers C (modelling mate pairs), construct an A-Bruijn graph as follows:

- Define an initial graph G_0 on $m = 2|C|$ vertices. For each bilabel $(a|b) \in C$ (representing a $(k + 1, d)$-mer), introduce two new vertices u, v and form an edge $u \rightarrow v$. Label the edge by $(a|b)$; label u by prefix$(a|b)$; and label v by suffix$(a|b)$.
- Glue vertices of G_0 together when they have the same label. The graph G so obtained is called the *paired de Bruijn graph* of C.

This procedure is illustrated in Fig. 3(a,b). An alternate construction of the paired de Bruijn graph is to define the vertex set as the (k, d)-mers present in C, and the edges as connecting prefix$(a|b)$ to suffix$(a|b)$ for every element of C.

As with the regular de Bruijn graph, in this construction, every vertex of G inherits the label common to all the vertices of G_0 that were glued together to form it, and this label is unique to that vertex. Any walk through the graph on edge sequence $e_1, \ldots, e_r$ spells out an $(r + k)$-mer bilabel $(L|R)$ where L is formed from the left labels, $L = $ walkword(left$(e_1), \ldots, $ left(e_r)), and R is formed from the right labels, $R = $ walkword(right$(e_1), \ldots, $ right(e_r)).

The (k, d)-spectrum of a string S is $\{(S_k(i)|S_k(i + d)) : i = 1, \ldots, n\}$. When C is the $(k + 1, d)$-mer spectrum of S, there is a covering cycle whose left labels spell S in G. The cycle consists of consecutive edges

$$(S_k(i)|S_k(i + d)) \rightarrow (S_k(i + 1)|S_k(i + d + 1)) \quad \text{for } i = 1, \ldots, n.$$

Just as with the de Bruijn graph, this is a key property that makes the paired graph useful for spelling contigs.

2.5 Approximate Paired de Bruijn Graphs (Modelling Inexact Distance)

We now define a graph modelling mate pairs where the distance between the two reads in each pair is only known to lie within some range $d \pm \Delta$. The parameter Δ can be estimated based on the mate pair generation protocol.

Let C be an arbitrary set of $(k + 1, d, \Delta)$-mers, representing the input data. The key insight is that if two (k, d, Δ)-mers $(a|b)$ and $(a|b')$ both arise from the same instance of a in S, then in the de Bruijn graph of S, there is a directed path from b to b', or vice-versa, with distance at most 2Δ. This insight was used for repeat resolution in [11], albeit as a post-construction modification step. We construct an A-Bruijn graph from C as follows:

- The initial graph G_0 consists of $|C|$ isolated edges on $2|C|$ vertices. For each $(a|b) \in C$, introduce an edge $u \rightarrow v$ on two new vertices. Label the edge by the $(k + 1)$-mer bilabel $(a|b)$. Label u by prefix$(a|b)$ and v by suffix$(a|b)$.
- For each k-mer α, glue together all vertices with labels $(\alpha|\beta), (\alpha|\beta')$ if there exists a directed path from β to β' (or vice-versa) in the de Bruijn graph

$D = \mathrm{DB}(C, k)$ of length at most 2Δ. Here, we assume that the construction of D implicitly breaks the $(k+1)$-mer bilabels of C into independent $(k+1)$-mers.

The graph $G = \mathrm{APDB}(C, k, d, \Delta)$ so obtained is the *approximate paired de Bruijn Graph* of C (Fig. 3(c,d,e)). The effect of this gluing is to merge all vertices (k-mer bilabels) that might align to the same position in the genome; vertices that align to the same position are thus guaranteed to be merged. However, the converse does not hold; vertices aligning to different positions in the genome are sometimes merged, either due to repeats that are not resolved by the given parameters, or due to chance short paths in D.

In the case that $k > 2\Delta$, we observed that if there is a directed path between β and β' in the de Bruijn graph D of length at most 2Δ, then β and β' should share an overlap of at least $k - 2\Delta$ characters. This observation leads to an alternate rule to glue vertices of G_0: for each k-mer α, glue together all vertices with labels $(\alpha|\beta), (\alpha|\beta')$ if β and β' share an overlap of at least $k-2\Delta$ characters. Note that this rule can only be used if $k > 2\Delta$ and may lead to a different graph; however, it is easier to implement.

Unlike our earlier constructions of the de Bruijn and paired de Bruijn graphs, the vertices of G do not inherit a single label from G_0; the vertices glued together have the same left label, but may have different right labels. In an edge walk $e_1, \ldots, e_r$ on G, the left labels spell the word walkword(left(e_1), ..., left(e_r)). However, the right labels typically do not successively overlap by $k-1$ characters as they did for the paired de Bruijn graph. Though we currently ignore these after gluing, we recognize that there is a potentially untapped benefit to using the right labels to later improve the assembly (see Section 5).

A set C of $(k+1)$-mer bilabels is a *covering spectrum of S* if for every position $i = 1, \ldots, n$, we have $(S_{k+1}(i)|S_{k+1}(i + d + x)) \in C$ for at least one x in the range $-\Delta \leq x \leq \Delta$. For each position i, there are $2\Delta+1$ choices of x. Note that there are many different covering spectra, and different choices of C may lead to different graphs. However, the graph will satisfy the key property of having a covering cycle that spells out S.

Theorem 1. *Let S be a circular string, and C a set of (k, d, Δ)-mers that is a covering spectrum of S. Then there is a covering cycle through the graph $G = \mathrm{APDB}(C, k, d, \Delta)$ that spells out S.*

Proof. For $i = 1, \ldots, n$, let $e_i \in C$ be any $(k + 1)$-mer bilabel in C aligning to position i in S. To prove $e_1, \ldots, e_n$ is a cycle in G, we need to show that consecutive edges $e_i = u_i \rightarrow v_i$ with label $(a|b)$, and $e_{i+1} = u_{i+1} \rightarrow v_{i+1}$ with label $(a'|b')$, share the connecting vertex, $v_i = u_{i+1}$. (Indices are taken modulo n.) Since C is a covering spectrum of S, the graph D is the ordinary de Bruijn graph of S. In G_0, v_i has label $(S_k(i + 1), \mathrm{suffix}(b))$ and u_i has label $(S_k(i + 1), \mathrm{prefix}(b'))$. Since these both align to position $i + 1$ in S, the distance between the start of b and b' in S is at most 2Δ. Thus in D, the directed distance from b to b' (or vice-versa) is at most 2Δ, so these vertices were glued together when forming G. □

3   Results

We implemented a prototype assembly program for mate paired data using the (approximate) paired de Bruijn graph approach and experimented with *E. coli* (4.6 Mbp) and *Human* chromosome 22 (49 Mbp). We removed all ambiguous bases (such as Ns) from chr22, resulting in 35 Mbp of sequence. The reads were generated with perfect coverage, meaning for every position in the genome we generated a single (k, d, Δ)-mer aligning to it. The insert size was picked uniformly at random from the specified range. We report as contigs the (left) words spelled by all maximal walks of the graph whose interior vertices have just one out-neighbor. We validated that any generated contigs mapped perfectly back to the original genome — this was the case for all the contigs.

Constructing the de Bruijn graph and finding all its non-branching paths takes time $O(n \log n)$, where n is the number of k-mers. The construction of the approximate paired de Bruijn graph has an additional cost of searching all neighbors within a distance 2Δ of each node. Therefore, the running time of the algorithms is $O(n \log n + n \min\{2^{\Delta}, n\})$, where n is the number of (k, d, Δ)-mers. However, since de Bruijn graphs are sparse, the searches in the graph are usually very fast, and in practice, even the run on chr22 with $\Delta = 200$ took less than 2 hours on a 8 core processor with 16G RAM. Moreover, the algorithm could be easily distributed over a large cluster to deal with larger Δ.

Our motivation for the paired de Bruijn graph approach was that the number of repeated (k, d)-mers quickly drops as d increases (Fig. 2(b)), and hence the contigs of the paired de Bruijn graph based on these (k, d)-mers could be longer. To test this hypothesis, we generated a set of mate pairs with varying insert sizes and plotted the length of the obtained contigs (Fig. 4(a)). To isolate the effect of the insert size, the coverage of the data was perfect (the (k, d)-spectrum), the insert sizes were perfect ($\Delta = 0$), and the read length was fixed to 50. We observed that contig lengths improved dramatically as the insert size increased. With an insert size of 6000 nt, all of *E. coli* was covered with just one contig, while for chr22, an insert size of 5000 nt enabled us to cover 98% of the chromosome with the 15 largest contigs. We thus believe that properly using mate pairs has a strong potential to increase contig lengths.

To explore the role that read length plays relative to the insert size, we generated sets of mate pairs with varying read lengths but with a fixed insert size (1000 nt). To isolate the effect of the read length, we had perfect coverage and no variation in the insert size. For *E. coli*, we found that, for an insert size of 1000 nt, once the read length grew over a small threshold of 10–20 nt, the contig lengths nearly reached the theoretical optimum that could be achieved by simply generating reads of length equal to the insert size (Fig. 4(b)). For *Human*, we needed to increase the read length to 300 nt in order to reach the optimum with 1000 nt insert size (Fig. 4(b)). However, for a longer insert size (5000 nt), a read length of 50 came close (Fig. 4(a)) to achieving the optimum (which, with 5000 nt reads, was a single contig). Therefore, by properly using mate pairs with large enough insert size, one can significantly reduce the limitations caused by short read length.

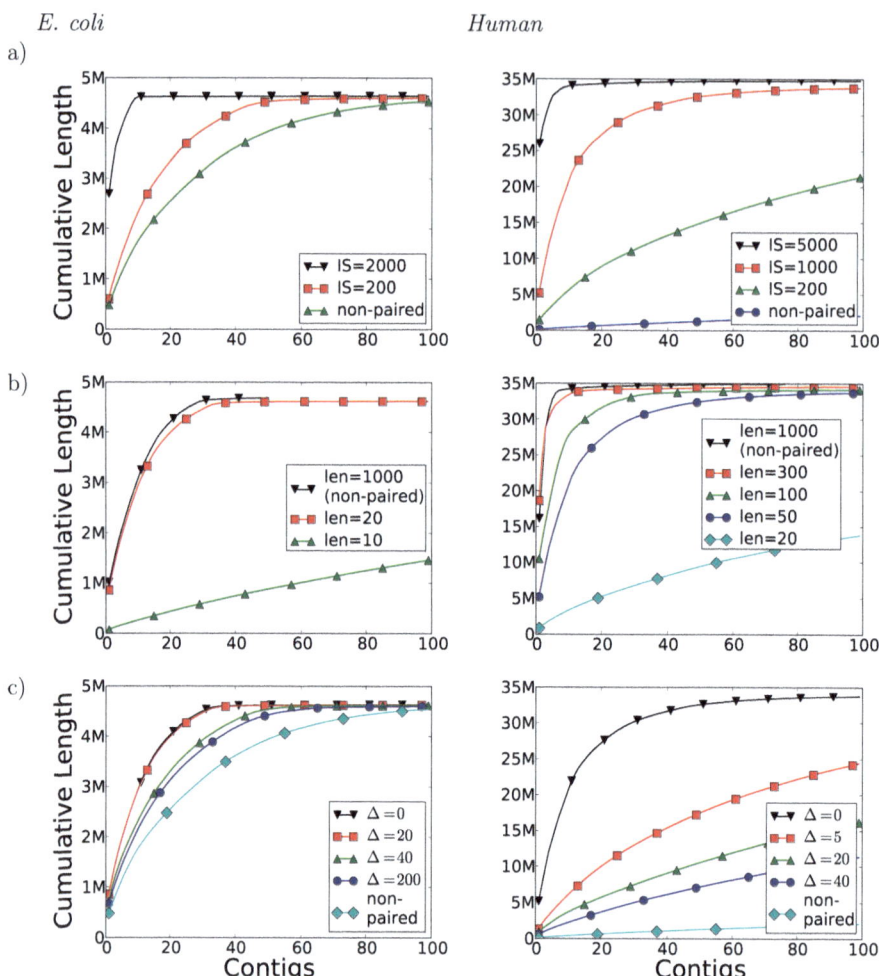

Fig. 4. Cumulative contig lengths (for standard and paired de Bruijn graphs) on simulated data with perfect coverage. Contigs are sorted in order from largest to smallest. Point (x, y) means the largest x contigs have cumulative length y. (a) To analyze the effect of the insert size (IS) on the assembly, we kept the read length fixed at 50, but varied the insert size. We also generated non-paired reads of length 50. For *E. coli*, the curve for insert size 6000 is not shown because there was only one contig, representing the whole genome. (b) To analyze the effect of read length on contig lengths, we fixed the insert size to 1000 but varied the read length. We also generated non-paired reads of length 1000, giving an upper bound on how good the assembly can be in this case. (c) To analyze the effect of variations in the insert size (Δ), we fixed the mean insert size (1000) and read length (50). We also show the baseline contig lengths in a non-paired dataset, with read length 50 and perfect coverage.

We measure the effect of increasing variability in the insert size (Δ) on the assembly. We fix the insert size to be 1000 nt and generate 50-long reads with perfect coverage, while varying Δ (Fig. 4(c)). We found that the assembly deteriorates with increasing Δ, especially for the *Human* genome. When Δ is large, the chance of two vertices of the de Bruijn graph being connected increases, and, hence, the number of vertices (bilabels) that do not align but nevertheless get glued together increases. In this situation, the read length is still important in determining the complexity of the (non-paired) de Bruijn graph. Some recent datasets achieve a small Δ, such as the Bentley et al. [2] human dataset with a mean insert size of 208 nt and a standard deviation of 13 nt. Nevertheless, we see robustness with respect to Δ as an important direction for improving the practical usefulness of our method.

4   Towards a Practical Paired de Bruijn Graph Assembler

We believe that, similarly to early studies of idealized fragment assembly with error-free k-mers [15], the (approximate) paired de Bruijn graphs can be of use in practical assemblers that utilize paired reads. Though this paper falls short of analyzing real data, we present here potential ways to remove our simplifications, and to move from the current de Bruijn graph assemblers to (approximate) paired de Bruijn graphs.

Base calling errors in reads. As with regular assembly, reads with base-calling errors may perturb the graph. Error correction algorithms for single reads may be used to improve the accuracy of the reads, while future error correction algorithms may also incorporate the mate pair information. Graph correction algorithms employed by current de Bruijn based assemblers [5,22] may also be applied to (approximate) paired de Bruijn graphs.

Insert size outliers. If a small percentage of read pairs are spaced outside the range $d \pm \Delta$, they will likely form isolated edges or terminal branches, which can be detected and discarded.

Double strandedness. The approximate paired de Bruijn graph is asymmetric in its treatment of the two reads $(a|b)$, and in the reverse complement, these are switched to $(b'|a')$ (where a', b' are the reverse complements of a and b). This makes existing methods [10,12,22] for accounting for double-strandedness difficult to apply. However, we may explicitly introduce the reverse complement of every read; perform assembly; match up reverse complement contigs after assembly; and reconcile any differences through a consensus stage.

5   Conclusion

In this paper, we introduced the paired de Bruijn graph and motivated its use in genome assembly. Instead of incorporating mate pairs into a post-graph-construction step, we have used them to construct the graph itself. Any procedures that could be performed on the regular de Bruijn graph (e.g., error correction) can be performed in the same manner on the paired de Bruijn graph. For instance, even when there are repeats that the paired de Bruijn graph does not resolve, mate pair transformations can still be applied to the graph to help resolve the remaining repeats.

By eliminating the need for mate pair transformations, the paired de Bruijn graph approach provides a potential method for assembly with short read mate pairs (like the ones generated by Complete Genomics [6] and Helicos [8]). By not requiring unique paths between paired reads in the de Bruijn graph, the paired approach could more efficiently resolve repeats despite the short read length. Moreover, the algorithms we describe can be extended to the *strobes* generated by Pacific Biosciences, which extend the notion of the mate pair to a set of multiple (more than two) reads separated by some distances.

A future direction lies in the use of the right labels on edges of the approximate paired de Bruijn graph. Currently, we spell out each contig using only the left label. The positions of the right labels are only known approximately, but this is often sufficient to form a righthand word displaced approximately d from the lefthand word. Moreover, after encountering an edge $(a|b)$ in a walk, we must encounter some edge $(b|c)$ approximately d edges away (unless it is past the end of the walk). This compatibility requirement may help to narrow the choice of valid paths when encountering branching vertices, thereby resolving longer repeats and improving contig lengths.

Acknowledgements. Glenn Tesler and Paul Medvedev were supported in part by NIH grant 3P41RR024851-02S1.

References

1. Batzoglou, S., Jaffe, D.B., Stanley, K., Butler, J., Gnerre, S., Mauceli, E., Berger, B., Mesirov, J.P., Lander, E.S.: ARACHNE: A Whole-Genome Shotgun Assembler. Genome Research 12(1), 177–189 (2002)
2. Bentley, D.R., Balasubramanian, S., Swerdlow, H.P., Smith, G.P., Milton, J., Brown, C.G., Hall, K.P., Evers, D.J., Barnes, C.L., Bignell, H.R., et al.: Accurate whole human genome sequencing using reversible terminator chemistry. Nature 456(7218), 53–59 (2008)
3. Butler, J., MacCallum, I., Kleber, M., Shlyakhter, I.A., Belmonte, M.K., Lander, E.S., Nusbaum, C., Jaffe, D.B.: ALLPATHS: de novo assembly of whole-genome shotgun microreads. Genome Research 18, 810–820 (2008)
4. Chaisson, M.J., Brinza, D., Pevzner, P.A.: De novo fragment assembly with short mate-paired reads: Does the read length matter? Genome Research 19, 336–346 (2009)
5. Chaisson, M.J., Pevzner, P.A.: Short read fragment assembly of bacterial genomes. Genome Research 18(2), 324–330 (2008)
6. Drmanac, R., Sparks, A.B., Callow, M.J., Halpern, A.L., Burns, N.L., Kermani, B.G., Carnevali, P., Nazarenko, I., Nilsen, G.B., Yeung, G., et al.: Human genome sequencing using unchained base reads on self-assembling DNA nanoarrays. Science 327(5961), 78 (2010)

7. Genome 10K Community of Scientists: Genome 10K: A proposal to obtain whole-genome sequence for 10000 vertebrate species. Journal of Heredity 100(6), 659–674 (2009)
8. Harris, T.D., Buzby, P.R., Babcock, H., Beer, E., Bowers, J., Braslavsky, I., Causey, M., Colonell, J., DiMeo, J., William Efcavitch, J., et al.: Single-molecule DNA sequencing of a viral genome. Science 320(5872), 106 (2008)
9. Idury, R.M., Waterman, M.S.: A new algorithm for DNA sequence assembly. Journal of Computational Biology 2, 291–306 (1995)
10. Kececioglu, J.D.: Exact and approximation algorithms for DNA sequence reconstruction. PhD thesis, University of Arizona, Tucson, AZ, USA (1992)
11. Medvedev, P., Brudno, M.: Ab initio whole genome shotgun assembly with mated short reads. In: Vingron, M., Wong, L. (eds.) RECOMB 2008. LNCS (LNBI), vol. 4955, pp. 50–64. Springer, Heidelberg (2008)
12. Medvedev, P., Georgiou, K., Myers, G., Brudno, M.: Computability of models for sequence assembly. In: Giancarlo, R., Hannenhalli, S. (eds.) WABI 2007. LNCS (LNBI), vol. 4645, pp. 289–301. Springer, Heidelberg (2007)
13. Myers, E.W.: Toward simplifying and accurately formulating fragment assembly. Journal of Computational Biology 2, 275–290 (1995)
14. Myers, E.W.: The fragment assembly string graph. Bioinformatics 21(suppl 2), ii79–ii85 (2005)
15. Pevzner, P.A.: L-Tuple DNA sequencing: computer analysis. J. Biomol. Struct. Dyn. 7(1), 63–73 (1989)
16. Pevzner, P.A., Tang, H.: Fragment assembly with double-barreled data. Bioinformatics 17(suppl 1), S223–S225 (2001)
17. Pevzner, P.A., Tang, H., Tesler, G.: De novo repeat classification and fragment assembly. Genome Research 14(9), 1786–1796 (2004)
18. Pevzner, P.A., Tang, H., Waterman, M.S.: An Eulerian path approach to DNA fragment assembly. Proceedings of the National Academy of Sciences of the United States of America 98(17), 9748–9753 (2001)
19. Schatz, M.C., Delcher, A.L., Salzberg, S.L.: Assembly of large genomes using second-generation sequencing. Genome Research 20(9), 1165–1173 (2010)
20. Simpson, J.T., Wong, K., Jackman, S.D., Schein, J.E., Jones, S.J.M., Birol, I.: ABySS: A parallel assembler for short read sequence data. Genome Research 6, 1117 (2009)
21. Weber, J.L., Myers, E.W.: Human whole-genome shotgun sequencing. Genome Research 7, 401–409 (1997)
22. Zerbino, D.R., Birney, E.: Velvet: algorithms for de novo short read assembly using de Bruijn graphs. Genome Research 18, 821–829 (2008)

An Optimization-Based Sampling Scheme for Phylogenetic Trees

Navodit Misra[1], Guy Blelloch[2], R. Ravi[3], and Russell Schwartz[4]

[1] Max Planck Institute for Molecular Genetics, Berlin, Germany
misra@molgen.mpg.de
[2] Computer Science Department, Carnegie Mellon University, Pittsburgh, USA
guyb@cs.cmu.edu
[3] Tepper School of Business, Carnegie Mellon University, Pittsburgh, USA
ravi@cmu.edu
[4] Department of Biological Sciences, Carnegie Mellon University, Pittsburgh, USA
russells@andrew.cmu.edu

Abstract. Much modern work in phylogenetics depends on statistical sampling approaches to phylogeny construction to estimate probability distributions of possible trees for any given input data set. Our theoretical understanding of sampling approaches to phylogenetics remains far less developed than that for optimization approaches, however, particularly with regard to the number of sampling steps needed to produce accurate samples of tree partition functions. Despite the many advantages in principle of being able to sample trees from sophisticated probabilistic models, we have little theoretical basis for concluding that the prevailing sampling approaches do in fact yield accurate samples from those models within realistic numbers of steps. We propose a novel approach to phylogenetic sampling intended to be both efficient in practice and more amenable to theoretical analysis than the prevailing methods. The method depends on replacing the standard tree rearrangement moves with an alternative Markov model in which one solves a theoretically hard but practically tractable optimization problem on each step of sampling. The resulting method can be applied to a broad range of standard probability models, yielding practical algorithms for efficient sampling and rigorous proofs of accurate sampling for some important special cases. We demonstrate the efficiency and versatility of the method in an analysis of uncertainty in tree inference over varying input sizes. In addition to providing a new practical method for phylogenetic sampling, the technique is likely to prove applicable to many similar problems involving sampling over combinatorial objects weighted by a likelihood model.

1 Introduction

Much of the theory and classic methods of phylogeny reconstruction were developed for a variety of optimization formulations of the problem (e.g., parsimony, likelihood or distance based). Optimization approaches have fallen into disfavor, however, due to the frequent presence of multiple optima or near-optima

V. Bafna and S.C. Sahinalp (Eds.): RECOMB 2011, LNBI 6577, pp. 252–266, 2011.

and a general desire to quantify uncertainty in the resulting trees. As a result, algorithms based on the idea of sampling over the space of phylogenetic trees for a given data set are now generally preferred to optimization approaches in order to provide better statistical support while answering questions such as whether a given bipartition is more likely than another conflicting bipartition. Popular sampling methods such as `MrBayes` [1] use Markov chains over a class of tree rearrangement moves such as nearest neighbor interchange (NNI), subtree pruning and re-grafting (SPR) and tree bisection and reconnection (TBR) [2] to estimate partition functions of trees under some implied probability distribution on tree topologies, branch lengths, ancestral sequences, and population genetic parameters. Despite their advantages, though, sampling-based approaches suffer from a comparatively poorly developed theoretical literature. In particular, there are few theoretical results regarding the mixing properties of their underlying Markov chains, particularly the number of steps for which one must run a model to sample accurately from its partition function. As a result, we rarely have any sound theoretical basis for concluding that a phylogeny sampling algorithm has been run sufficiently long to generate an accurate sample.

Among the few positive results are the methods of Diaconis and Holmes [3], for uniform sampling over all phylogenetic trees, and the recent result of Stefankovic and Vigoda [4], showing rapid mixing of SPR Markov chains when data is generated by phylogenies with sufficiently short branches. Mossel and Vigoda [5], and Stefankovic and Vigoda [6] have shown that Markov chains based on standard NNI or SPR moves do not always mix well. Their results are valid for a likelihood-based method on problem instances where input data can be represented by a mixture of two tree topologies. The question of a polynomial bound for mixing time on data generated from a single tree, with arbitrary branch lengths, is still open. There is, therefore, a need for either new theoretical insights into the mixing properties of the prevailing methods or the development of new sampling methods for which we can more readily analyze these properties.

In this paper, we pursue the latter approach, developing an alternative Markov chain-based phylogeny sampling algorithm that is more amenable to theoretical mixing time analysis and allows one to prove non-trivial mixing time bounds in important special cases. The key algorithmic insight of our method is that one can convert the hard sampling problem inherent in standard tree sampling into an easier sampling problem that uses, as a subroutine, the solution of a theoretically hard but practically tractable optimization problem (an instance of the minimum spanning tree problem with degree constraints). By repeatedly solving the embedded optimization problem provably to optimality, one can in turn solve the sampling problem with a small number of Monte Carlo steps. Our method can be used to sample from the likelihood distribution of labelled tree topologies, also known as the *ancestral likelihood*. We use this optimization-based method to theoretically bound the mixing time for the well known Cavender-Farris-Neyman (CFN) model [7,8,9], proving that our optimization based Markov chain mixes in time polynomial in the number of leaves (taxa) and the number of characters in the input for the CFN model. We then demonstrate the practical effectiveness of the method through a small empirical analysis of how uncertainty in tree topology increases with increasing numbers of taxa under a standard likelihood model. Our method can be readily generalized to sampling from the set of Steiner trees on arbitrary weighted graphs and might have applications to many similar problems involving sampling over combinatorial objects weighted by a likelihood function.

We begin by presenting some basic notation and background on likelihood models, and explain our new approach in their context. We then describe the Integer Linear Programming (ILP) formulation for the optimization subroutine in our sampler, and show how using this powerful procedure in each move, the mixing time of the CFN model of ancestral likelihood can be bounded by a polynomial number of steps in the input size. Finally, we close with some experimental results intended to demonstrate the practical use of our method.

2   Notation

We begin by defining some basic terminology and notation used throughout this manuscript. We refer the reader to the text by Felsenstein for a more thorough introduction to the topic of phylogenetics and the concepts and terminology presented below [2]. Let H be an input matrix that specifies a set χ of N taxa, over a set $C = \{c_1, \ldots c_M\}$ of M characters, such that H_{ij} represents the jth character of the ith taxon. Further, let n_k be the set of admissible states of the kth character c_k. The set of all possible states is the space $S \equiv n_1 \times n_2 \ldots \times n_M$. We will represent the ith character of any element $b \in S$, by $(b)_i$. The state space S can be represented as a graph $\mathcal{G} = (V_\mathcal{G}, E_\mathcal{G})$ with the vertex set $V_\mathcal{G} = S$ and edge set $E_\mathcal{G} = \{(u,v) | u, v \in S, d_h[u,v] = 1\}$, where $d_h[u,v]$ is the Hamming distance between u and v.

The set of all possible trees can be conveniently classified using the concept of phylogenetic X-Trees. A phylogenetic X-Tree $T(\chi)$ displaying a set of taxa χ is defined as follows: there is a bijection or labeling between the set of taxa χ and the leaves of $T(\chi)$. Furthermore, all internal nodes are of degree three or more. The latter requirement is equivalent to contracting the edges between any pair of degree-two internal nodes. Clearly, removing an edge in any tree disconnects the tree into two subtrees, each of which has a non-empty intersection with the set of taxa. Thus, each edge corresponds to a bipartition or a *split* of the taxa. The *topology* of a phylogenetic X-Tree is defined as the set of all splits obtained by removing an edge of $T(\chi)$. A popular approach to solving the phylogeny inference problem is to search through the space of all topologically distinct phylogenetic X-Trees for a given set of taxa. This search space is usually defined over the space of binary tree topologies (i.e., where all internal nodes are of degree three). Any instance of a phylogenetic X-Tree with an internal node with degree greater than three (also known as a *polytomy*) can be treated as a special case of a binary tree where two internal nodes represent the same vertex in the graph $\mathcal{G}$. From now on we will refer to such binary phylogenetic X-Trees simply as phylogenies. Each phylogeny on N leaves has $N - 2$ degree 3 internal nodes.

It is well known that there are $\frac{(2N-2)!}{2^{N-1}(N-1)!}$ distinct rooted tree topologies for phylogenies with N leaves. Diaconis and Holmes have shown that the set of tree topologies can be conveniently visualized using a connection between perfect matchings on $2N - 2$ points and phylogenies with N leaves [3]. Given a phylogeny T, we will use their method to assign a number to each internal node. We arbitrarily assign distinct labels to the leaves (from 1 to N) each of which corresponds to an element in χ. Since we are interested in unrooted phylogenies, we arbitrarily root the tree along any of the $2N - 3$ edges. Initially, all internal nodes are unlabeled. Each internal node is assigned a label between $N + 1$ and $2N - 2$ in the following sequence: 1. At each step, find an unlabeled internal node that has both its descendants labeled. In case there is more than one such internal node, choose the one that has a descendant with the lowest label; 2. Assign the lowest available label to this internal node; and 3. Recurse until all nodes are labeled. Diaconis and Holmes showed that this mapping is a bijection by showing how to transform any matching into a binary tree as follows. Assume we have a perfect matching P on $2N - 2$ points, such that the first N points represent the leaves of some binary tree T. Since, more than half of the points are leaves, at least one pair of leaves (say u and v) must be matched in P. We can represent this matched pair by a subtree where u and v are joined to an internal node with the smallest label (namely $N + 1$). If there is more than one pair of matched leaves, we take the pair that contains the leaf with the smallest label and connect them with the internal node $N + 1$. Now if we remove this pair of matched leaves and treat the internal node $N + 1$ as a new leaf, the remaining perfect matching on $2N - 4 = 2(N - 1) - 2$ points has $N - 2 + 1$ leaves along with a subtree associated with the new leaf node $N + 1$. If we iterate this process k times we reduce the matching to $2(N - k) - 2$ points and obtain a forest with $2N - 2 - 2k$ components on the vertices $\{1, \ldots, 2N - 1\}$, such that the leaf of each subtree has a label between 1 and N. After $N - 2$ steps, we join the final pair of nodes to get a binary tree with nodes 1 to N as leaves.

3   Likelihood Model

We will represent each phylogeny by a 4-tuple $T(\chi, \phi, \boldsymbol{\alpha}, \boldsymbol{\tau})$. We overload the symbol T to represent both the topology of the phylogeny as well as a bijection between leaves $l(T)$ and input taxa χ and a mapping from internal nodes of T onto a set $\phi \subseteq \mathcal{S}$ such that ϕ_i represents the label for the ith internal node. Next, we assign a likelihood to T assuming that the taxa have evolved via point mutations. Let $\boldsymbol{\alpha} = \{\alpha_k | \alpha_k[j, i] > 0 \ \forall i \neq j, c_k \in C\}$ be a set of rate matrices, such that $\alpha_k[j, i]$ represents the rate for a transition from state i to j for character c_k. We will assume $\boldsymbol{\alpha}$ is reversible with respect to $\pi_k[i]$ (representing the equilibrium frequency of state i at site k), such that $\alpha_k[i, j]\pi_k[j] = \alpha_k[j, i]\pi_k[i]$ and satisfies the conservation equation

$$\alpha_k[i, i] = -\sum_{j \neq i} \alpha_k[j, i] \tag{1}$$

The likelihood of each edge $e = (u, v) \in T$ is given by

$$L(e) = \prod_{c_k \in C} \exp(\tau_e \alpha_k)[(u)_k, (v)_k]$$

$$= \prod_{c_k \in C} \left(I[(u)_k, (v)_k] + \tau_e \alpha_k[(u)_k, (v)_k] + \ldots + \frac{\tau_e^n}{n!} \alpha_k^n[(u)_k, (v)_k] + \ldots \right) \quad (2)$$

where, $\tau_e \in [0, \infty)$ is the branch length representing relative time and I is the identity matrix. We root the tree arbitrarily at internal node $2N - 2$, represented by sequence r, and compute likelihood of edges directed away from the root. Let $\pi[r] = \prod_{c_k \in C} \pi_k[(r)_k]$ be the equilibrium density of the sequence representing the root. The ancestral likelihood of the phylogeny is then given by

$$L(T(\chi, \phi, \boldsymbol{\alpha}, \boldsymbol{\tau})) = \boldsymbol{\pi}[r] \prod_{e \in T} L(e) \quad (3)$$

The problem we want to solve is that of generating random samples from this likelihood distribution. We can simplify the problem somewhat by integrating out the set of branch lengths $\boldsymbol{\tau}$. Since we know the end points for each edge, this integral is easy to compute using a spectral decomposition of α_k in terms of its eigenvalues $\Lambda_k = \{-\lambda\}$ and corresponding eigenvectors $\{|\lambda\rangle\}$. However, since the smallest eigenvalue is zero (corresponding to the equilibrium distribution), we need to provide a suitable prior over branch lengths, in order to ensure that the integral is convergent. Choosing a suitable prior for an unbounded parameter requires care because a flat prior is not always an uninformative prior [2]. In practice, an exponentially decaying prior $Pr(\tau_e) = \eta e^{-\eta \tau_e}$ is usually recommended by popular methods such as MrBayes and we will follow the same convention in this paper. Combining this prior with our likelihood model gives us an expression for the posterior distribution:

$$L(T(\chi, \phi, \boldsymbol{\alpha})) = \boldsymbol{\pi}[r] \prod_{e \in T} \int_0^\infty L(e) Pr(\tau_e) d\tau_e$$

$$= \boldsymbol{\pi}[r] \prod_{(u,v) \in T} \int_0^\infty \prod_{c_k \in C} \left(\sum_{-\lambda \in \Lambda_k} \langle (u)_k | \lambda \rangle \langle \lambda | (v)_k \rangle e^{-\lambda \tau_e} \right) \eta e^{-\eta \tau_e} d\tau_e \quad (4)$$

Note that α_k is typically of dimension 20 or less, so the spectral decomposition is not a computational bottleneck and the likelihood $L(T(\chi, \phi, \boldsymbol{\alpha}))$ can be computed in $O(NM)$ time. We will not focus on sampling the branch lengths τ_e in this paper, however for completeness we note that τ_e can be sampled exactly and efficiently from the distribution represented by the integrand in the previous equation using rejection sampling [10]. In section 5, we will use a particularly simple closed form expression for the likelihood maxima for the standard CFN model to derive some theoretical results for our method. Note, that in contrast to our approach, methods based on fixing the tree topology followed by sampling branch lengths are known to get trapped in local optima.

Our New Model

Assume we are given the labels ϕ for the internal nodes and let $T_*(\chi, \phi, \boldsymbol{\alpha})$ be a most likely phylogeny. Let $\mathcal{T}(\chi, \boldsymbol{\alpha}) = \{T_*(\chi, \phi, \boldsymbol{\alpha}) : \phi \subset \mathcal{S}\}$. We will restrict ourselves to sampling from the likelihood distribution over the set $\mathcal{T}(\chi, \boldsymbol{\alpha})$. The intuition behind our approach is that this pruning might allow us to sample from the remaining phylogenies efficiently and reliably, if we can solve the optimization problem efficiently.

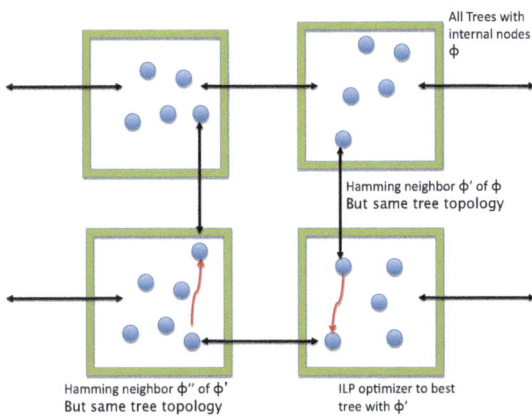

Fig. 1. A typical transition in our modified sampler: A move to a neighboring ancestral sequence is followed by an optimization step to reset the topology to be a most likely one

Now, consider the following family of distributions $L(v)^{\beta}$ for $\beta \in [0, \infty)$. Such distributions are usually called "heated" distributions and β, the inverse temperature, in analogy with the usual definition of temperature in physical processes. These distributions are consistent with our intuitive understanding that high β or low temperatures accentuate the "roughness" of the probability landscape. Such distributions are commonly used for approximately sampling from smoothed versions of distributions from which it is hard to sample (case with $\beta < 1$) or in simulated annealing for approximate optimization ($\beta > 1$). In this paper we will focus on the former scenario.

The rest of this paper is organized as follows — in section 4 we present an Integer Linear Program (ILP) for finding the optimal topology T_* given the labels ϕ for the set of internal nodes. In section 5 we present our Monte Carlo Markov Chain (MCMC) algorithm for sampling phylogenies in $\mathcal{T}(\chi, \boldsymbol{\alpha})$ and show that for the CFN model, the proposed Markov chain mixes in a number of optimization steps polynomial in the number of taxa and sequence length at sufficiently high temperatures. In section 6 we present results from experiments performed on simulated data sets. In section 7, we summarize the main contributions of the work and discuss future directions.

4 ILP for Solving the Degree Constrained Spanning Tree Problem

Each phylogeny is specified by the set ϕ of $N - 2$ internal nodes and the set χ of N taxa labeled according to the Diaconis-Holmes convention. Their convention provides valuable information regarding which nodes are potential descendants for a given internal node. We root the tree at internal node $2N - 2$ and initially connect each pair of vertices (both taxa and internal nodes) by a directed edge. Edges between taxa are removed and each edge between an internal node and a taxon is directed towards the taxon. Edges between two internal nodes are directed towards the node with smaller label. The internal node with the largest label $(2N - 2)$ has all edges directed away from it, from which we have to choose three, each taxon has all edges directed towards it, from which we have to choose 1. For all the remaining internal nodes we have to choose one edge directed towards it (from its parent) and two edges directed away from it (towards its children). An edge directed from vertex u to v corresponds to a Boolean variable $s_{u,v}$ with edge cost $w_{uv} = -\ln[\int_0^\infty L(e = (u, v))d\tau_e]$. The set of all such edges and the vertices form the graph G. Since the taxa are assigned labels in arbitrary order, we will try to find the minimum cost phylogeny over all possible orderings of the taxa. The following ILP finds the minimum cost tree compatible with these in and out degree constraints,

$$
\begin{aligned}
\text{Minimize} \quad & \sum_{(u,v) \in G} w_{uv} s_{u,v} \\
\text{subject to} \quad & \sum_v s_{v,u} = 1 \quad \forall u \in G \setminus \{2N - 2\} \\
& \sum_v s_{u,v} = 2 \quad \forall u \in \phi \setminus \{2N - 2\} \\
& \sum_v s_{2N-2,v} = 3 \\
& s_{u,v} \in \{0, 1\} \quad \forall (u, v) \in G
\end{aligned}
\tag{5}
$$

Lemma 1. *The ILP in equation 5 finds the minimum cost directed spanning tree given the edge costs w_{uv}.*

Proof. To prove the correctness of our method we show all feasible solutions to this ILP are acyclic. The degree constraints will then ensure that any feasible solution corresponds to a connected subgraph with no cycles, implying a tree. Suppose for a contradiction a feasible solution contains a cycle over a subset $A \subseteq G$. Since G is finite and elements of G are ordered (the directionality of the edge representing which vertex is a potential descendant of another vertex in G), A must contain a vertex v such that all vertices in $A \setminus \{v\}$ are ancestors of v. The only way to obtain a cycle over A is for v to be connected to two (or more) ancestors. But this would violate the in-degree constraint in the ILP.

5   Mixing Time for the Cavender-Farris-Neyman Model

In this section we use the Cavender-Farris-Neyman (CFN) model for binary sequences to establish some theoretical results regarding the convergence of the heated Markov chain. While we prove rapid mixing only for the CFN model, a special case of the class of likelihood model described above, we note that the proof will apply trivially to some generalizations of the CFN (e.g., non-binary data) and that the sampling technique itself applies to the full class of likelihood functions. The CFN model assigns an edge probability p_e to each edge, such that the likelihood for k mutations along e is $p_e^k (1-p_e)^{M-k}$. We will use the Hamming distance between two sets of internal nodes as a distance measure over the space $\mathcal{T}(\chi, \alpha)$. We will identify each set of internal nodes ϕ with the minimum cost tree $T(\phi)$ obtained by the method in section 4. Given a phylogeny $T(\phi)$, we will call the set of all phylogenies at a Hamming distance 1 the neighborhood of $T(\phi)$ (represented by $Nbd(\phi)$). We can think of the space $\mathcal{T}(\chi, \alpha)$ as a graph with each phylogeny as a vertex and edges connecting each pair of neighboring phylogenies. The Markov chain is defined by nearest neighbor moves over $\mathcal{T}(\chi, \alpha)$. We have the following bound on the change in cost function at each step of the Markov chain.

Lemma 2. *For any two neighboring phylogenies $u \in Nbd(v)$, $(eM)^{-3\beta} \leq \frac{L(u)}{L(v)} \leq (eM)^{3\beta}$*

Proof. Given any edge $e = (u, v)$ with branch length $l = -\ln(1 - 2p)$, and $d_h(u, v) = k$, the likelihood is given by $(p^k (1 - p)^{M-k})^\beta$. The optimal branch length maximizing this likelihood can be solved as $l = -\ln(1 - 2k/M)$. If we perturb one character for one internal node (say u), the maximum fractional change in edge likelihood is $\left((1 - 1/M)^{M-1}/M\right)^\beta > (1/eM)^\beta$ when $k = 0$. Since each internal node has three edges, the maximum change in likelihood for any tree topology is $(eM)^{3\beta}$. Also, the likelihood for each tree topology is a lower bound on the optimal likelihood. If we consider the topology that is optimal at u with likelihood $L(u)$ then we get an upper bound $(eM)^{-3\beta} \leq \frac{L(u)}{L(v)} \leq (eM)^{3\beta}$.

This previous result is sufficient to ensure rapid mixing at sufficiently high temperatures. We use path coupling method of Bubley and Dyer [11] to prove this result. We will use the Hamming distance $d_h(X, Y)$ between two phylogenies X and Y as a distance metric. Path coupling arguments are based on establishing a coupling of Markov chain moves between each pair of nearest neighbors such that the distance between them decreases on average at each iteration of the Markov chain. For completeness we state the main lemma from [11].

Lemma 3. *Let M be Markov chain over a graph $G(V, E)$ and let ρ define a set of edge distances over G such that $\rho(e) \geq 1 \ \forall e \in E$. Furthermore, let d be a distance metric over V such that given any pair of vertices $u, v \in V$,*

$$d(u, v) = min_{P(u,v)} \sum_{(x,y) \in P(u,v)} \rho(x, y) \tag{6}$$

where the minima is taken over the set of all paths $P(u, v)$ between u and v. Suppose that for all edges $\bar{E} = \{(u, v) | \rho(u, v) = d(u, v)\} \subseteq E$ there exists a coupling of Markov processes $\{u_t = M^t.u\}$ and $\{v_t = M^t.v\}$, such that the following bound holds

$$d(u, v) - E[d(M.u, M.v)] \geq \alpha d(u, v) \; \forall (u, v) \in \bar{E} \tag{7}$$

for some $\alpha > 0$, where $M.v$ represents the state after one step of the Markov chain starting at v. Then the total variation distance $\Delta(t) = max_{u \in V} |M^t.u - \pi| \leq e^{-\alpha t} \ln[D]$, where D is the diameter of G and π is the invariant probability measure on V.

Theorem 1. *For the CFN model, the heated Markov chain mixes in time $O(NM \ln[NM]/(1 - \tanh(3 \ln(M)\beta))$ for $\beta < -\ln[1 - 1/(NM + 1)]/3 \ln(M)$*

Proof. Suppose we have two random processes X_t and Y_t, evolving according to the Markov chain over phylogenies. We will concatenate the bit strings representing the internal nodes into one string of NM bits for each random variable. Suppose at time $t = 0$, X_0 and Y_0 differ in the kth bit. We define the coupling as follows — Select any bit b (representing a character for one of the internal nodes) uniformly at random. If the $b = k$, with probability $1/2$ flip the kth bit of variable X_0 and hold the state for Y_0 or vice versa; if $b \neq k$ with probability $1/2$ select identical proposal states for both random variables X_1 and Y_1 and with probability $1/2$ do nothing. If $b = k$, with probability $> 1/2$ both variables converge in one step. For each of the other choices, the Hamming distance either stays the same or increases by one. In each instance the probability that Hamming distance decreases is

$$d_h(X_0, Y_0) - E[d_h(X_1, Y_1)] \geq \frac{1/2 - (NM - 1)\frac{1}{2} |Pr[X_1 accepts] - Pr[Y_1 accepts]|}{NM} \tag{8}$$

Now, using lemma 2 we get the bounds

$$Pr[X_1 accepts] = \frac{L(X_1)}{L(X_1) + L(X_0)} \leq \frac{1}{1 + e^{-3 \ln(M)\beta}} \tag{9}$$

and

$$Pr[Y_1 accepts] = \frac{L(Y_1)}{L(Y_1) + L(Y_0)} \geq \frac{e^{-3 \ln(M)\beta}}{1 + e^{-3 \ln(M)\beta}} \tag{10}$$

combining these three equations we get

$$d_h(X_0, Y_0) - E[d_h(X_1, Y_1)] \geq 1/2 \left(1/NM - \left(\frac{1}{1 + e^{-3 \ln(M)\beta}} - \frac{e^{-3 \ln(M)\beta}}{1 + e^{-3 \ln(M)\beta}} \right) \right)$$
$$= 1/2 \left(1/NM - \tanh(3 \ln(M)\beta) \right) \tag{11}$$

This implies for $\beta < -\ln[1 - 1/(NM + 1)]/3 \ln(M)$, the distance between neighboring phylogenies decreases in expectation at a rate greater than $\alpha =$

$1/2NM(1 - NM \tanh(3 \ln(M)\beta)$. Finally, using the path coupling lemma 3 and the fact that our graph has diameter NM, we get the following condition for the total variation distance $\Delta(t)$ between the distribution of X_t and Y_t for $\alpha > 0$.

$$\Delta(t) = Pr[X_t \neq Y_t] \leq e^{-\alpha t} NM \qquad (12)$$

and the mixing time to reach a total variation distance $1/2$ is $\tau = \ln[2NM]/\alpha$.

While we have provided a proof just for the CFN model, the basic technique can be extended trivially to multi-state models, such as the Jukes-Cantor model.

6 Experiments

We implemented our method in C++ and used the **gnu** linear programming kit **GLPK** for solving the ILP. The Markov chain was simulated using the *replica sampling* heuristic as described next. In each of the experiments reported here, three Markov chains were simulated independently, at different values of β, for some user defined time steps N_1. One chain was always maintained at $\beta_0 = 1$ while the ith chain was heated to $\beta_i = (1 + i\delta)^{-1}$, for some user defined value of δ. At each optimization step the branch lengths were set to the maximum likelihood value, although in principle our method can sample from the full posterior distribution of branch lengths at each step. After N_1 steps of each chain, a pair of chains was picked uniformly at random and an exchange was proposed, followed by the usual Metropolis accept/reject criterion evaluated at the temperature of the colder of the two chains. After every N_{ex} attempts at exchanging states between the Markov chains, a measurement was made. We used data simulated using the CFN model on a user defined tree topology for our experiments, as described in the following section.

Our goals in validation were to verify that the model runs efficiently for moderate sized trees and to demonstrate its ability to ask questions about the ancestral likelihood function. For this purpose, we conducted a small study measuring the accuracy with which the true source tree of each data set can be inferred from the data for varying input sizes. We can assess this uncertainty by examining how often each bipartition in the source tree of a given data set occurs over the sample of trees. We quantify this measure of bipartition mismatch by the mean number of bipartitions that differ between observed tree and source tree across samples.

6.1 Data Sets

We report two sets of experiments on simulated data from 10, 25 and 35 taxa trees. Each set was prepared as follows: A tree topology with N leaves was generated by randomly choosing a matching by enumerating $2N - 2$ points in random order and matching successive points. Each edge was assigned a branch length by generating an exponentially distributed random number with user defined mean (mean was fixed by specifying the edge probability) and 100 characters were simulated using the CFN model starting from the root $2N - 2$. We initialized the

Markov chain simulator by the simple heuristic of starting with the set of leaves $S = \chi$ and true ancestral nodes. One tenth of the characters in each ancestral node were then randomly perturbed. This process was repeated independently for each chain participating in the replica exchange. All the experiments we report here had edge probabilities no more than 0.1, so these perturbations result in a fairly random initial state.

We first report results on experiments where we varied the number of taxa, keeping all other simulation and sampling parameters constant. In each case data was simulated on trees with each edge probability fixed at 0.1, so, on average, one in ten characters mutated along each edge of the tree. For the Monte Carlo sampling step, we used 3 coupled chains maintained at temperatures (or β^{-1}) = 1, 1.01 and 1.02. The temperatures were chosen heuristically [1]. After every 10 steps, two chains were picked at random and an attempt at swapping their states was made. These experiments were performed to assess the feasibility (both in run time performance and Markov chain convergence) of the proposed method.

For the second set of experiments, we fixed the number of taxa to 25 and simulated data for edge probabilities values of 0.01, 0.05 and 0.1. These experiments were performed to estimate the variation in the uncertainty in inferring the true topology as well as the rate of convergence of the Markov chain.

6.2   Results

Figure 2 shows inferred likelihoods per Monte Carlo step for each tree. The plot reveals that the sampler relaxes to a high likelihood tree in each case. Further, the number of steps until the likelihood plateaus increases monotonically with the number of taxa, as expected. However, due to the low temperature and our use of the replica exchange heuristic in these empirical tests, we cannot assert with certainty that these chains are well-mixed. The dashed horizontal lines, representing the ancestral likelihood for the source tree, seem to indicate that each chain is quite probably close to equilibrium.

Analysis of run times further shows the method to be very fast in practice despite the need for solving a hard optimization problem at each step. Table 1 shows mean run-times expressed in numbers of Monte Carlo steps solved per second. Run time does increase with numbers of taxa, but is still more than 1500 steps per minute for 35 taxa trees. The method is thus practical for tens of thousands of steps of Markov chain sampling on moderate problem sizes. For instance, each run presented in Figure 2 took less than 23 minutes.

Table 1 also shows the results of the uncertainty analysis. While the most likely ancestral tree is known to be statistically inconsistent in general, we see that the sampler is extremely efficient in identifying true bipartitions for these data sets.

The second set of experiments probes the ancestral likelihood landscape as we vary the mutation probabilities while keeping the number of taxa fixed at 25. Figure 3 shows the negative log-likelihood plots for three experiments with varying edge lengths. Once again we find that the sampled trajectories relax fairly rapidly to a high likelihood tree that is close to the likelihood of the source

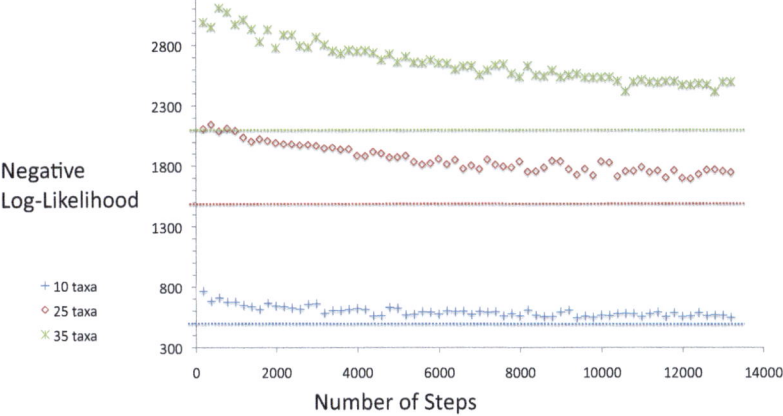

Fig. 2. Inferred negative log-likelihood as a function of Monte Carlo step for trees of 10, 25, and 35 taxa

Table 1. Run time and mismatch between true and sampled tree topologies for three input sizes

No. of taxa	ILP steps per second	Average mismatch
10	728.21	0.39
25	70.45	2.38
35	29.26	2.58

tree. Data sets with shorter branches (lower mutation probability) seem to relax faster, although we once again cannot assert that with certainty.

Table 2 gives us additional insight into the likely dynamics of the Markov chain. As the mutation probabilities along tree edges increase, so does the accuracy of inferring the source tree. On the other hand, looking at Figure 3 seems to suggest that the sampler relaxes more rapidly for high edge probability data. This set of experiments thus tends to strongly suggest that near equilibrium, the ability of the sampler to estimate the source tree deteriorates as edge probabilities become small. This agrees with our intuition that given two speciation events, it should be relatively easier to infer the order in which they occur if the sequences at the two branch points are "well separated," i.e., the branch length between the internal edges is large. At the same time, in the neighborhood of trees with long branches, the ancestral likelihood is comparably "flatter" (for the same reason that the peak of the likelihood curve in Lemma 2 is at the shortest branch). As a result, the Markov chain moves about the state space comparably faster (leading to a lower rejection ratio in Table 2), but at each new node the optimal tree does not differ much from the true tree. On the other hand, for extremely short branches, the most likely ancestral sequences are closely clustered together around a smaller region of the state space and

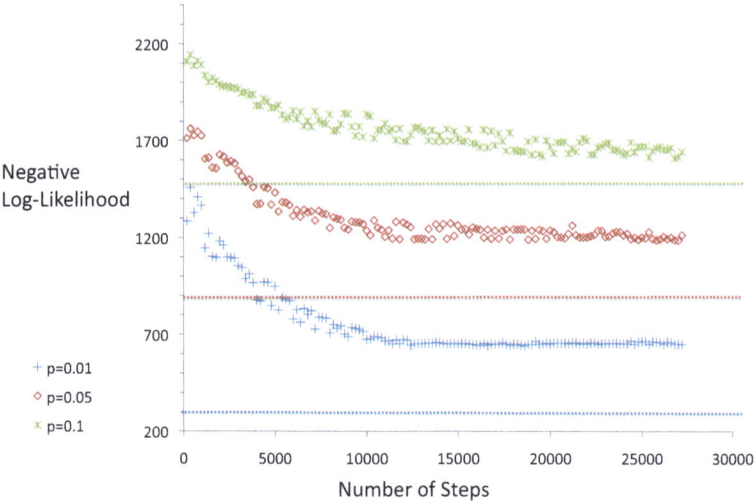

Fig. 3. Inferred negative log-likelihood as a function of Monte Carlo step for trees with 25 taxa and varying edge probabilities

Table 2. Rejection ratio and mismatch between true and sampled tree topologies for three different edge probabilities for 25 taxa data sets

Average edge probabilities	Average mismatch	Average rejection ratio
0.01	14.05	0.97
0.05	4.82	0.94
0.10	2.93	0.87

Markov chain rarely ventures out of this central region (leading to a high rejection ratio). However, since most ancestral nodes have largely similar sequences for the low edge probability case, it is relatively more common to swap the order of bifurcation events connecting two ancestral nodes, leading to a higher mismatch between the average estimated topology and the source tree. For the 0.01 edge probability case, the source tree had numerous higher degree internal nodes (or edges with zero branch lengths) and that may be the likely cause for the mismatch. In phylogenetics, it is well known that near polytomies or very closely spaced bifurcation events are generally harder to reconstruct from fixed length sequences [12] and our experiments seem to suggest a similar effect on our ancestral likelihood sampler.

6.3   Discussion

Our experiments show that our method provides an efficient sampler over the ancestral likelihood model for relatively large numbers of taxa. Although the method involved solving a formally intractable problem at each step of Markov chain sampling, in practice it proves extremely fast for even moderately large data sets. Furthermore, our novel formulation allows us to examine aspects of tree likelihood distribution inaccessible to standard samplers. Maximum ancestral likelihood has been known to be a statistically inconsistent estimator of the tree topology in general. Our experiments seem to indicate that the discrepancy between the topology of the average ancestral reconstruction and the source tree is small and robust to the number of leaves in the tree, except when the source tree has extremely short branch lengths.

7   Conclusions

We have developed a novel approach to phylogenetics designed to leverage methods for fast combinatorial optimization to efficiently sample over the space of phylogenetic trees under standard likelihood models. The method depends on an alternative formulation of phylogenetic likelihood to enable sampling over internal node states instead of tree topologies. The work establishes a new approach to performing efficient, accurate sampling over phylogenies and to establishing mixing time bounds for such sampling in practice. To demonstrate its theoretical value, we have and established mixing time bounds for the important practical case of the Cavender-Farris-Neyman model. These bounds can be extended to some generalizations of that model and provide a new strategy for establishing provable bounds on more general likelihood models. We further demonstrate the practical efficiency and utility of the method through a study of uncertainty in topology inference across samples under a standard likelihood function. The ideas developed here for efficient optimization-based sampling may be applicable to many similar problems involving sampling over likelihoods of combinatorial objects.

Acknowledgments

NM, GB, RR, and RS were supported in this work by U.S. National Science Foundation award #0612099. RS was additionally supported by U.S. National Institutes of Health award #1R01CA140214.

References

1. Huelsenbeck, J.P., Ronquist, F.: MRBAYES: Bayesian inference of phylogeny. Bioinformatics 17, 754–755 (2001)
2. Felsenstein, J.: Inferring Phylogenies. Sinauer Publications (2004)
3. Diaconis, P., Holmes, S.P.: Random walks on trees and matchings. Electronic Journal of Probability 7(6) (2002)
4. Stefankovic, D., Vigoda, E.: Fast Convergence of MCMC Algorithms for Phy- logenetic Reconstruction with Homogeneous Data on Closely Related Species arXiv:1003.5964 (2010)

5. Mossel, E., Vigoda, E.: Phylogenetic MCMC algorithms are misleading on mixtures of trees. Science 309(5744), 2207–2209 (2005)
6. Stefankovic, D., Vigoda, E.: Pitfalls of heterogeneous processes for phylogenetic reconstruction. Systematic Biology 56(1), 113–124 (2007)
7. Cavender, J.A.: Taxonomy with confidence. Mathematical Biosciences 40, 271–280 (1978)
8. Farris, J.S.: A probability model for inferring evolutionary trees. Systematic Zoology 22, 250–256 (1973)
9. Neyman, J.: Molecular studies of evolution: a source of novel statistical problems. In: Gupta, S.S., Yackel, J. (eds.) Statistical Decision Theory and Related Topics, pp. 1–27 (1971)
10. Misra, N., Schwartz, R.: Efficient stochastic sampling of first-passage time with applications to self-assembly simulation. Journal of Chemical Physics 129, 204109 (2008)
11. Bubley, R., Dyer, M.: Path coupling: A technique for proving rapid mixing in Markov chains. In: Proceedings of the 38th Annual Symposium on Foundations of Computer Science, pp. 223–231 (1997)
12. Rokas, A., Carroll, S.B.: Bushes in the Tree of Life. PLoS Biol. 4(11), e352 (2006)

Multiplex De Novo Sequencing of Peptide Antibiotics

Hosein Mohimani[1], Wei-Ting Liu[2], Yu-Liang Yang[3], Susana P. Gaudêncio[4], William Fenical[4], Pieter C. Dorrestein[2,3], and Pavel A. Pevzner[5,⋆]

[1] Department of Electrical and Computer Engineering, UC San Diego
[2] Department of Chemistry and Biochemistry, UC San Diego
[3] Skaggs School of Pharmacy and Pharmaceutical Sciences, UC San Diego
[4] Center for Marine Biotechnology and Biomedicine,
Scripps Institution of Oceanography, UC San Diego
[5] Department of Computer Science and Engineering, UC San Diego
ppevzner@cs.ucsd.edu

Abstract. Proliferation of drug-resistant diseases raises the challenge of searching for new, more efficient antibiotics. Currently, some of the most effective antibiotics (i.e., Vancomycin and Daptomycin) are cyclic peptides produced by non-ribosomal biosynthetic pathways. The isolation and sequencing of cyclic peptide antibiotics, unlike the same activity with linear peptides, is time-consuming and error-prone. The dominant technique for sequencing cyclic peptides is NMR-based and requires large amounts (milligrams) of purified materials that, for most compounds, are not possible to obtain. Given these facts, there is a need for new tools to sequence cyclic NRPs using picograms of material. Since nearly all cyclic NRPs are produced along with related analogs, we develop a mass spectrometry approach for sequencing all related peptides at once (in contrast to the existing approach that analyzes individual peptides). Our results suggest that instead of attempting to isolate and NMR-sequence the most abundant compound, one should acquire spectra of many related compounds and sequence all of them simultaneously using tandem mass spectrometry. We illustrate applications of this approach by sequencing new variants of cyclic peptide antibiotics from *Bacillus brevis*, as well as sequencing a previously unknown familiy of cyclic NRPs produced by marine bacteria.

1 Introduction

In 1939 Renê Dubos discovered that the peptide fraction *Tyrothricin*, isolated from the soil microbe *Bacillus brevis*, had an ability to inhibit the growth of *Streptococcus pneumoniae*, rendering it harmless. Although discovered 10 years after Penicillin, it was the first mass produced antibiotic deployed in Soviet hospitals in 1943. Unfortunately, the identification of amino acid sequences of cyclic peptides, once a heroic effort, remains difficult today. The dominant technique

---

⋆ Corresponding author.

V. Bafna and S.C. Sahinalp (Eds.): RECOMB 2011, LNBI 6577, pp. 267–281, 2011.
© Springer-Verlag Berlin Heidelberg 2011

for sequencing cyclic peptide antibiotics is 2D NMR spectroscopy, which requires large amounts of highly purified materials that, are often nearly impossible to obtain.

Tyrothricin is a classic example of a mixture of related cyclic decapeptides whose sequencing proved to be difficult and took over two decades to complete. By the 1970s, scientists had sequenced 5 compounds, Tyrocidine A-E, from the original mixture. However, these five are not the only peptides produced by *B. brevis* and even today it remains unclear whether *all* of the antibiotics produced by this bacterium have been documented (see reference [1] for a list of 28 known peptides from *B. brevis*).

Figure 1 (a) shows structure of Tyrocidine A. Table S1 illustrates that most cyclic decapeptides in the Tyrocidine/Tryptocidine family can be represented as shown (the rounded amino acid masses in daltons are also shown):

$$Val \begin{Bmatrix} Orn \\ Lys \end{Bmatrix} Leu\ Phe\ Pro \begin{Bmatrix} Phe \\ Trp \end{Bmatrix} \begin{Bmatrix} Phe \\ Trp \end{Bmatrix} Asn\ Gln \begin{Bmatrix} Tyr \\ Trp \\ Phe \end{Bmatrix}$$

$$99 \begin{Bmatrix} 114 \\ 128 \end{Bmatrix} 113\ 147\ 97 \begin{Bmatrix} 147 \\ 186 \end{Bmatrix} \begin{Bmatrix} 147 \\ 186 \end{Bmatrix} 114\ 128 \begin{Bmatrix} 163 \\ 186 \\ 147 \end{Bmatrix}$$

It may come as a surprise that there are no genes in *B. brevis* whose codons encode any of the Tyrocidine peptides! Tyrocidines, similar to many antibiotics such as Vancomycin or Daptomycin, represent cyclic *non-ribosomal* peptides (NRPs) that do not follow the central dogma "DNA produces RNA produces Protein". They are assembled by nonribosomal peptide synthetases that represent both the mRNA-free template and building machinery for the peptide biosynthesis [2]. Thus, NRPs are not directly inscribed in genomes and cannot be inferred with traditional DNA sequencing. Cyclic NRPs are of great pharmacological importance as they have been optimized by evolution for chemical defense and communication. Cyclic NRPs include antibiotics, antitumor agents, immunosuppressors, toxins, and many peptides with still unknown functions.

Most NRPs are cyclic peptides that contain nonstandard amino acids, increasing the number of possible building blocks from 20 to several hundreds. The now

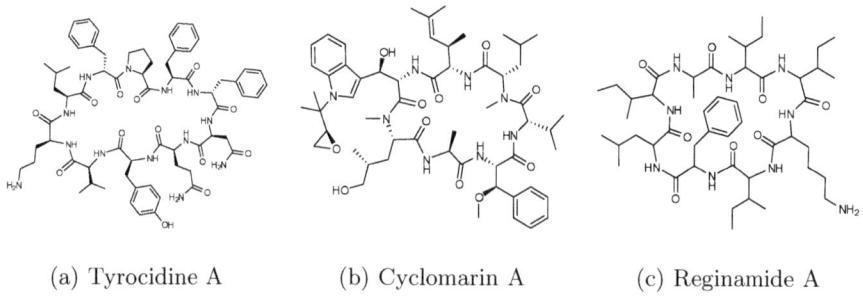

(a) Tyrocidine A   (b) Cyclomarin A   (c) Reginamide A

Fig. 1. Structures of Tyrocidine A (a), Cyclomarin A (b), and Reginamide A (c)

dominant 2D NMR-based methods for NRP characterization are time-consuming, error prone, and requires large amounts of highly purified material. Because NRPs are often produced by difficult to cultivable microorganisms, it may not be possible to get sufficient quantities for 2D structure elucidation, therefore it is important to develop a nmol scale structure elucidation approach [3,4]. Such methods promise to greatly accelerate cyclic NRP screening and may illuminate a vast resource for the discovery of pharmaceutical agents [5].

The first automated Mass Spectrometry (MS) based approach to sequencing cyclic peptides correctly sequenced 2 out of 6 Tyrocidines analyzed by Ng et al. [6]. While the correct sequences for 4 other Tyrocidines were highly ranked, Ng et al., 2009 [6] came short of identifying them as the *highest-scoring* candidates. Leao et al., [7], 2010, and Liu et al., [8], 2010, recently applied the algorithm from [6] for analyzing new cyclic peptides. In [7], the authors study peptides produced by the cyanobacterium *Oscillatoria sp.* that inhibit the growth of green algae and demonstrated that they function in a synergistic fashion, i.e., mixtures of these analogous peptides are needed to inhibit green algal growth. This observation emphasizes the importance of studying various peptide variants and calls for the development of a technology able to simultaneously sequence *all* peptides produced by a single organism.

Our first attempt to sequence cyclic NRPs from *Oscillatoria sp.* via MS using the algorithm described by Ng et al., [6] was inconclusive. We (Leao et al. 2010 [7]) resorted to purification of the most abundant peptide with the goal to sequence it via 2D NMR (purification of individual NRPs is often difficult since various NRP variants have similar physicochemical properties). This amounted to a large effort that involved applications of various NMR technologies (including HSQC, HMBC, COSY, and NOESY) but still failed to identify some inter-residue dependencies. Applications of both NMR and MS to finally sequence four compounds using NRP-Dereplication algorithm from [6] represented a large and time-consuming effort of a multidisciplinary team. A better approach would be to generate MS/MS spectra of *all* variant NRPs (without the need to purify large amounts of individual peptides) and to *multiplex* sequence them. By multiplex sequencing we mean simultaneous (and synergetic) sequencing of related peptides from their spectra.

Using this approach, we sequenced many known members of the Tyrocidine family as well as some still unknown Tyrocidine variants. Finding new Tyrocidine variants is surprising since this family has been studied for sixty years now. We further sequenced a previously unknown family of NRPs isolated from a bacterial strain that produces natural products with anti-asthma activities (named *Reginamides*). To validate these new sequences (obtained from a single mass spectrometry experiment) we analyzed one of them (named Reginamide A) using (rather time consuming) NMR experiments. The mass spectrometry approach revealed the sequence of masses with molecular composition (C_3H_5NO, $C_6H_{11}NO$, $C_6H_{11}NO$, $C_7H_{12}N_2O_2$, $C_6H_{11}NO$, C_9H_9NO, $C_6H_{11}NO$, $C_6H_{11}NO$) that was matched by NMR as the cyclic peptide AIIKIFLI with structure shown in Figure 1 (c). We emphasize that NMR confirmation of a compound with a known sequence (derived by MS) is much easier than NMR sequencing of a completely unknown compound. The crux of our approach is the analysis of the entire spectral network [9] of multiple Tyrocidines/Reginamides (Figure 4 (b-c) and Table 2 and 3) rather than analyzing each Tyrocidine/Reginamide isomer separately. The derived sequences of the Reginamides represent the first automated sequencing of a cyclic peptide family *before* NMR and highlights the future role that mass spectrometry may play in sequencing cyclic peptides. MS-CyclicPeptide software is available from the NCRR Center for Computational Mass Spectrometry at http://proteomics.ucsd.edu.

2 Results

Spectral datasets. We analyzed Tyrocidine, Cyclomarin, and Reginamide families of cyclic peptides (see Methods section for the detailed description of experimental protocols).

The Cyclomarins represent a family of cyclic heptapeptides with anti-inflammatory activity, isolated from a marine *Streptomyces* strain [10, 11, 12]. The structure of Cyclomarin A is shown in Figure 1 (b). We sequenced four variants of the Cyclomarins that differ in a single amino acid residue.

The Reginamides represent a newly isolated family of cyclic octapeptides isolated from a marine *Streptomyces* strain that also produces secondary metabolites with anti-asthma activities (*Splenocins*). Multiple variants of Reginamide isomers were sequenced using MS. Due to limited quantities of these cyclic peptides and severe separation challenges, it was only possible to purify one of the variants (named Reginamide A) for validating the derived sequences by NMR. Multi-dimensional NMR analysis confirmed the sequence of Reginamide A, derived by our multiplex sequencing algorithm.

Sequencing of individual peptides. Below we describe an algorithm for sequencing *individual cyclic peptides*. The goal of this algorithm is not improving the method of [6], but rather proposing the ground for multiplex peptide sequencing, something that the algorithm from [6] is not suited for.

Consider the cyclic peptide VOLFPFFNQY (Tyrocidine A) with integer masses (99, 114, 113, 147, 97, 147, 147, 114, 128, 163). We will interchangeably use the standard notation (VOLF...) and the sequence of rounded masses (99, 114, 113, 147, ...) to refer to a peptide. One may partition this peptide into three parts as OLF-PFF-NQYV with integer masses 374, 391 and 504 respectively. In general, a k-*partition* is a decomposition of a peptide P into k subpeptides with integer masses $m_1 \ldots m_k$ (we refer to $mass(P) = \sum_{i=1}^{k} m_i$ as the *parent mass* of peptide P). A k-*tag* of a peptide P is an arbitrary partition of $mass(P)$ into k integers. A k-tag of a peptide P is *correct* if it corresponds to masses of a k-subpartition of P, and *incorrect* otherwise. For example, (374, 391, 504) is a correct 3-tag, while (100, 1000, 169) is an incorrect 3-tag of Tyrocidine A.

A (linear) *subtag* of a cyclic k-tag $Tag = (m_1, \cdots, m_k)$ is a (continuos) linear substring $m_i \cdots m_j$ of the k-tag (we assume $m_i \cdots m_j = m_i \cdots m_k m_1 \cdots m_j$ in the case $j < i$). There are $k(k-1)$ subtags of a k-tag. The mass of a subtag is the sum of all elements of the subtag and the length of a subtag is the number of elements in the subtag. We define $\Delta(Tag)$ as the multiset of $k(k-1)$ subtag masses. For a peptide P, the *theoretical spectrum* of P is defined as $\Delta(P)$. For example, the theoretical spectrum of a cyclic peptide $AGPT = (71Da, 57Da, 97Da, 101Da)$ consists of 12 masses (57, 71, 97, 101, 128, 154, 172, 198, 225, 229, 255, and 269).

The problem of sequencing a cyclic peptide from a (complete and noiseless) spectrum corresponds to the *Beltway Problem* [13] and can be stated as follows:

Cyclic Peptide Sequencing Problem

- *Goal:* Given a spectrum, reconstruct the cyclic peptide[1] that generated this spectrum.
- *Input:* A spectrum S (a set of integers).
- *Output:* A cyclic peptide P, such that $\Delta(P) = S$.

While the Beltway Problem is similar to the well-studied Turnpike Problem [14, 15], the former is more difficult than the latter one [13]. Moreover, de novo sequencing of cyclic peptides is much harder than the (already difficult) Beltway Problem. Indeed, the real spectra are incomplete (missing peaks) and noisy (additional peaks). Table S2 represents an experimental spectrum of Tyrocidine A and illustrates that while the experimental spectrum captures many masses from the theoretical spectrum (45 out of 90 masses), it also contains 30 other masses (corresponding to noisy peaks and neutral losses). The limited correlation between the theoretical and experimental spectra makes the spectral interpretation difficult.

Given a tag Tag and an experimental spectrum S (represented as a set of integer masses), we define $Score(Tag, S)$ as the number of elements (masses) shared between $\Delta(Tag)$ and S (ignoring multiplicities of elements in $\Delta(Tag)$). For example, for the 3-tag $Tag = (374, 391, 504)$ of Tyrocidine A, $Score(Tag, S) = 5$, since the spectrum S contains 5 out of 6 elements in $\Delta(Tag) = (374, 391, 504, 765, 878, 895)$.

The problem of sequencing a cyclic peptide from an incomplete and noisy spectrum can be stated as follows:

Cyclic Peptide Sequencing Problem from Incomplete/Noisy Spectrum

- *Goal:* Given an incomplete and noisy spectrum, reconstruct the cyclic peptide that generated this spectrum.
- *Input:* A spectrum S (a set of integers) and an integer k (peptide length)
- *Output:* A cyclic peptide P of length k, such that $Spectrum$ and $\Delta(P)$ are as similar as possible, *i.e.* $Score(P, S)$ is maximized among all cyclic peptides of length k.

A tag is *valid* if all its elements are larger than or equal to 57 (minimal mass of an amino acid). A valid $(k + 1)$-tag derived from a k-tag Tag by breaking one

---

[1] We emphasize that the peptide might have amino acids with arbitrary masses, rather than the 20 standard amino acids.

of its masses into 2 masses is called an *extension* of Tag. For example, a 4-tag (374, 100, 291, 504) is an extension of a 3-tag (374, 391, 504). All possible tag extensions can be found by exhaustive search since for each k-tag $(m_1 \ldots m_k)$ there exist at most $\sum_{i=1}^{k} m_i$ extensions.

Our algorithm for sequencing individual peptides starts from scoring all 2-tags and selecting t top-scoring 2-tags, where t is a parameter. It further iteratively generates a set of all extensions of all top-scoring k-tags, combines all the extensions into a single list, and extracts t top scoring extensions from this list. Table 1 (a) shows the reconstructed 7-tags for the Tyrocidine family and illustrates that the highest-scoring tags are incorrect for most Tyrocidines. However, by *simultaneously* sequencing pairs of spectra of related peptides, one can achieve better results. For the sake of simplicity, we illustrate how our approach works with integer amino acid masses. However, with available high precision mass spectrometry data we are able to derive the elemental composition of each amino acid (see **Text S5**).

Furthermore, we describe an algorithm for combining information from all high scoring tags to generate a *spectral profile* (Figure 2) that compactly represents all high-scoring tags (similar to sequence logos [16]). Each $Tag = (m_1 \ldots m_k)$ with $\sum_{i=1}^{k} m_i = M$ defines an M-dimensional boolean vector $\overrightarrow{Tag}$ with 1s at k positions $\sum_{i=1}^{j} m_i$ for $1 \leq j \leq k$. For example, a tag (3,2,4) defines a vector 001010001. Given a vector $\mathbf{x} = x_1 \ldots x_M$, we define its i-*shift* as the vector $x_{M-i+1} x_{M-i+2} \ldots x_M x_1 \ldots x_{M-i}$ and its *reversal* as the vector $x_M x_{M-1} \ldots x_2 x_1$. We define the *reversed i-shift* as the reversal of the i-shift. For example, 2-shift of 001010001 is 010010100, and reversed 2-shift is 001010010. Given vectors $\mathbf{x}$ and $\mathbf{y}$, we define $alignment(\mathbf{x}, \mathbf{y})$ as a shift or reversed shift of $\mathbf{x}$ with maximum dot-product with $\mathbf{y}$. For $\mathbf{x} = 001010001$ and $\mathbf{y} = 101000000$, $alignment(\mathbf{x}, \mathbf{y}) = 101000100$.

Our algorithm for constructing the spectral profile (generated from a spectrum with parent mass M) starts from ordering t high-scoring k-tags $Tag_1 \ldots Tag_t$ in the decreasing order of their scores and defines T_0 as an M-dimensional vector with all zeros. It proceeds in t steps, at each step aligning the tag Tag_i against the vector T_{i-1}. At step i, it finds $alignment(\overrightarrow{Tag_i}, T_{i-1})$ between $\overrightarrow{Tag_i}$ and T_{i-1} and adds it to T_{i-1} to form $T_i = alignment(\overrightarrow{Tag_i}, T_{i-1}) + T_{i-1}$. After t steps, the algorithm outputs the vector $\frac{T_t}{t}$ as the spectral profile.

For example, for Tyrocidine A, the two 7-tags with the highest scores are $Tag_1 = (114, 147, 244, 260, 111, 119, 274)$ and $Tag_2 = (114, 147, 244, 291, 80, 133, 260)$. After the first step, we form a vector $T_1 = \overrightarrow{Tag_1}$ with 1s at positions 114, 261, 505, 765, 876, 995 and 1269. At the second step, we align $\overrightarrow{Tag_2}$ and T_1 and form a vector T_2 with 1s at positions 765, 995, 796, 1009 and 2s at positions 114, 261, 505, 876, and 1269. Repeating these steps for 100 high-scoring tags for Tyrocidine A results in the spectral profile shown in Figure 2(a). Table S4 provides the annotations of the spectral profiles for Tyrocidine A, B and C.

Sequencing of peptide pairs. We define a *spectral pair* as spectra S and S' of peptides P and P' that differ by a single amino acid. Consider a

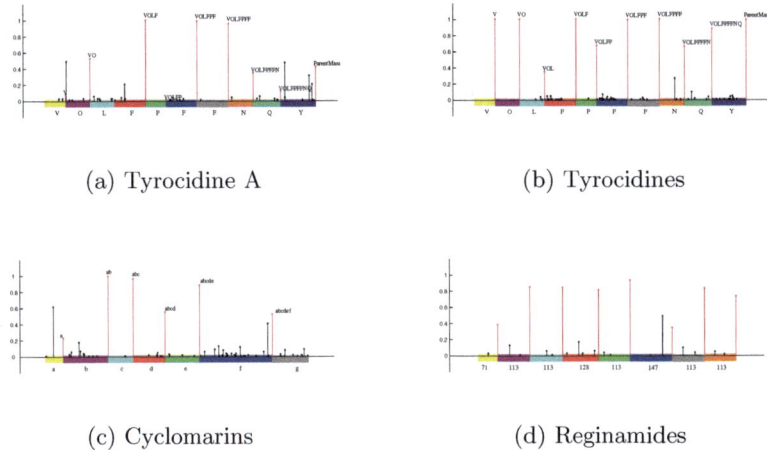

(a) Tyrocidine A

(b) Tyrocidines

(c) Cyclomarins

(d) Reginamides

Fig. 2. (a) Spectral profile of 100 highest scoring 7-tags for Tyrocidine A. Intensities of correct peaks account for 68% of total intensity. (b) Spectral profile of 100 highest scoring 10-tags for Tyrocidine A generated by multiplex sequencing of Tyrocidines. Intensities of correct peaks account for 86% of total intensity. (c) Spectral profile of 100 highest scoring 7-tags generated for Cycolmarin A by multiplex sequencing of four Cyclomarins (Cyclomarin A, Cyclomarin C, Dehydro Cyclomarin A and Dehydro Cyclomarin C). For Cyclomarin A, amino acids a, b, c, d, e, f and g stand for Alanine (71Da), β-methoxyphenylalanine (177Da), Valine (99Da), N-methylleucine (127Da), 2-amino-3,5-dimethylhex-4-enoic acid (139Da), N-(1,1-dimethyl-2,3-epoxyprophyl)-β-hydroxytryptophan (286Da) and N-methyl-δ-hydroxyleucine (143Da). In Cyclomarin C, f is replaced by N-prenyl-β-hydroxytryptophan (270Da). Dehydrations also occur on residue f. Intensities of correct peaks accounts for 59% of total intensites. (d) Spectral profile of 100 top scoring 8-tags of Reginamide A generated by multiplex sequencing of Reginamides. The top scoring 8-tag of Reginamide A, also verified by NMR, is $(71, 113, 113, 128, 113, 147, 113, 113)$. Intensities of correct peaks account for 81% of total intensity.

spectral pair (S, S') and set $\delta = Mass(S') - Mass(S)$. Given a k-tag $Tag = (m_1 \ldots m_k)$ of a spectrum S and an offset δ, we define a *corresponding* k-tag $Tag^i_{S \to S'} = (m_1 \ldots m_i + \delta \ldots m_k)$ of S' for each $1 \leq i \leq k$. For example for $Tag = (213, 260, 244, 147, 114, 128, 163)$ of Tyrocidine A, $Tag^1_{Tyc\ A \to Tyc\ A1} = (227, 260, 244, 147, 114, 128, 163)$ is the corresponding tag of Tyrocidine A1. Any k-tag of S corresponds to at most k k-tags of S', and any correct k-tag of S corresponds to (at least) one correct k-tag of S'. Given a k-tag Tag of a spectrum S, define its $PairwiseScore$ as

$$PairwiseScore(Tag, S, S') = \frac{Score(Tag, S) + \max_{1 \leq i \leq k} Score(Tag^i_{S \to S'}, S')}{2}$$

The algorithm for pairwise sequencing of the cyclic peptides is exactly the same as the algorithm for sequencing individual cyclic peptide but instead of using

Table 1. Individual (a), pairwise (b) and multiplex (c) de novo sequencing of Tyrocidines. The correct tag is selected from the set of 1000 top-scoring tags (the top scoring correct tag and its rank are shown). Table S3 shows the process of extensions of top scoring tags of Tyrocidine A from 2-tags to 7-tags. Rank 1···7 for the highest scoring tag of Tyrocidine A1 means that the seven highest scoring tags have equal score, and one of them is the correct tag. Composite masses such as [113+147] for Tyrocidine A mean that the sequencing algorithm returned 260Da instead of 113Da and 147Da corresponding to Leu and Phe. [99 + 114/128] for Tyrocidine A/A1 pair means that the mass 99 + 114 = 213 in the first position of Tyrocidine A is substituted by the mass 99 + 128 = 227 in Tyrocidine A1. Part (c) shows 10-tags resulting from multiplex sequencing of six Tyrocidines (projected to Tyrocidine A). Correct masses are shown in bold. MS stands for Multiplex Score, and WMS stands for weighted Multiplex Score (See **Text S2** for details).

Peptide	The highest-scoring correct 7-tag (among all generated tags)						Rank
Tyc A	[99+ 114]	[113+ 147]	97	147	147	114	[128+ 163] 384···1000
Tyc A1	[99+ 128]	[113+ 147]	[97+ 147]		147	114	128 163 1···7
Tyc B	[99+ 114]	113	147	97	[147+	186]	114 [128+ 163] 14···134
Tyc B1	99	128	[113+ 147]	[97+ 186]	147	[114+ 128]	163 2···13
Tyc C	99	114	[113+ 147]	[97+ 186]	[186+	114]	128 163 6···72
Tyc C1	99	128	[113+ 147]	[97+ 186]	186	114	[128+ 163] 4···38

(a) Individual

Pair	The highest-scoring correct 7-tag (among all generated tags)						Rank
Tyc A/A1	[99+ 114/128]	[113+ 147]	[97+ 147]	147	114	128	163 2···5
Tyc B/B1	99	114/128	[113+ 147]	[97+ 186]	147	[114+ 128]	163 1
Tyc C/C1	99	114/128	[113+ 147]	[97+ 186]	186	[114+ 128]	163 1
Tyc A/B	99	114	[113+ 147]	[97+ 147/186]	147	[114+ 128]	163 2···6
Tyc B/C	99	114	[113+ 147]	[97+ 186]	147/186	[114+ 128]	163 1
Tyc A1/B1	99	128	[113+ 147]	[97+ 147/186]	147	[114+ 128]	163 1···4
Tyc B1/C1	99	128	[113+ 147]	[97+ 186+	147/186]	114 128	163 43···82

(b) Pairwise

Family	Sequences (10-tags)	MS	WMS	Rank
Tyrocidines	**99 114 113 147 97 147 147 114 128 163** 232	29.14	1	
	99 114 113 147 97 147 147 69 173 **163** 228	28.78	2	
	99 114 141 119 **97 147 147 114 128 163** 222	28.14	3	
	99 114 113 147 97 147 147 114 111 180 222	27.85	4	

(c) Multiplex

Score(Tag, S) for scoring a single tag, it uses *PairwiseScore(Tag, S, S′)*. Table 1 (b) shows that while pairwise sequencing improves on sequencing of individual cyclic peptides, it does not lead to correct reconstructions of all Tyrocidines.

Identifying spectral pairs. While the described algorithm assumed that we know which spectra form spectral pair, i.e. which peptides differ by a single substitution, such an information is not available in *de novo* sequencing applications. The problem of whether spectra of two *linear* peptides form a spectral pair was investigated by Bandeira et al., [9]. In this section we address a more difficult problem of predicting whether the spectra of two *cyclic* peptides form a *spectral pair* based only on their spectra. Our approach extends the dereplication

goal: Given spectra of related cyclic peptides (of the same length), sequence of one of them, and their (estimated) spectral network, reconstruct all the cyclic peptides that generated this spectra.
input: Spectra $\mathbf{S} = (S_1, \cdots, S_m)$ of m related cyclic peptide, their (estimated) Spectral Network G, an integer k, a k-tag Tag of S_u for some $1 \leq u \leq m$, a scoring function $Score(Tag, S)$ for individual spectra.
output: an approximate solution $\mathbf{multitag}(Tag, u, \mathbf{S}, G)$ of constrained multiplex cyclic peptide sequencing problem.

> **for** $j = 1$ to m **do**
> $Tag_j \leftarrow$ **null**
> **end for**
> $Tag_u \leftarrow Tag$
> **repeat**
> $Change \leftarrow 0$
>   **for all** spectral pairs (S_j, S_r) in $E(G)$ **do**
> $\delta = ParentMass(S_r) - ParentMass(S_j)$
>   **if** $Tag_j \neq$ **null** and $r \neq u$ **then**
>   **for** $i = 1$ to k **do**
> $Tag'_r \leftarrow (i, \delta)$-modified Tag_j
>   **if** $Score(Tag'_r, S_r) > Score(Tag_r, S_r)$ **then**
> $Tag_r \leftarrow Tag'_r$
> $Change \leftarrow Change + 1$
>   **end if**
>   **end for**
>   **end if**
>   **end for**
> **until** $Change = 0$
> **return** $(Tag_1, \cdots, Tag_m)$

Fig. 3. Algorithm for generating multitags from a candidate Tag of a spectrum S_u in the spectral network formed by spectra $S_1, \ldots, S_m$ corresponding to the spectral network G. Given a k-tag Tag of the spectrum S_u, the algorithm initializes $Tag_u = Tag$ and $Tag_j = Null$ for all other $1 \leq j \leq m$. We assume that $Score(Null, S_i) = -\infty$ for all $1 \leq i \leq m$. $E(G)$ stands for the edge set of the spectral network G. Since the sum $\sum_{i=1}^{m} Score(Tag_i, S_i)$ is monotonically increasing, the algorithm converges (typically after few iterations).

algorithm from [6] by comparing spectra of mutated peptides (rather than comparing a spectrum against a sequence of a mutated peptide) and is based on the observation that related peptides usually have high-scoring corresponding tags. A simple measure of similarity between spectra is the number of (S, S')-shared peaks (see Table S6). In the following we introduce $\Delta(S, S')$ distance between spectra, that, in some cases, reveals the similarity between spectra even better than the number of (S, S')-shared peaks. Given a set of k-tags $TagList$ for a spectrum S, we define:

$$MaxScore(TagList, S) = \max_{Tag \in TagList} Score(Tag, S)$$

Given an additional spectrum S', we define:

$$MaxPairwiseScore(TagList, S, S') = \max_{Tag \in TagList} PairwiseScore(Tag, S, S')$$

Finally, given a set of k-tags $TagList$ for a spectrum S and a set of k-tags $TagList'$ for a spectrum S', define $\Delta(TagList, TagList', S, S')$ (or, simply,

Table 2. Reconstructed peptides from the spectra corresponding to vertices in the spectral network shown in Figure 4(b). The spectra were dereplicated using (known) Tyrocidines A, A1, B, B1, C and C1 by applying the **multitag** algorithm described in Figure 3. Four of the sequences are reported previously (see Table S12). For one spectrum with previously reported parent mass, 1292 Da, our reconstruction slightly differs from that of [1].

PM	Tag	Score	Comment
1269	99 114 113 147 97 147 147 114 128 163	21	Tyrocidine A
1283	99 128 113 147 97 147 147 114 128 163	26	Tyrocidine A1
1291	99 114 113 147 97 186 147 97 128 163	18	New
1292	99 114 113 147 97 186 131 114 128 163	22	PM matches Tryptocidine A[1]
1306	99 128 113 147 97 186 147 114 112 163	23	New
1308	99 114 113 147 97 186 147 114 128 163	25	Tyrocidine B
1322	99 128 113 147 97 186 147 114 128 163	32	Tyrocidine B1
1331	99 114 113 147 97 186 147 114 128 186	24	Tryptocidine B[1]
1345	99 128 113 147 97 186 147 114 128 186	27	previously reported[1]
1347	99 114 113 147 97 186 186 114 128 163	24	Tyrocidine C
1361	99 128 113 147 97 186 186 114 128 163	30	Tyrocidine C1
1370	99 114 113 147 97 186 186 114 128 186	26	Tyrocidine D[1]
1384	99 128 113 147 97 186 186 114 128 186	24	previously reported[1]

Table 3. Dereplication of Reginamide variants represented by the spectral network in the Figure 4(c)) from the Reginamide A, using **multitag** algorithm

PM	Peptide	Score
897	71 99 113 128 113 147 113 113	31
911	71 113 113 128 113 147 113 113	31
925	71 113 113 142 113 147 113 113	25
939	71 113 113 156 113 147 113 113	31
953	71 113 113 170 113 147 113 113	29
967	71 113 113 184 113 147 113 113	28
981	113 85 113 184 113 147 113 113	28
995	71 113 113 212 113 147 113 113	24
1009	113 113 113 184 113 147 113 113	26
1023	71 113 113 240 113 147 113 113	20

$\Delta(S, S')$) as the differences between the sum of scores of the best-scoring tags for S and S' and the sum of pairwise scores of the best-scoring tag of S/S' and S'/S pairs:

$$\Delta(S, S') = MaxScore(TagList, S) + MaxScore(TagList', S')$$
$$-MaxPairwiseScore(TagList, S, S') - MaxPairwiseScore(TagList', S', S)$$

It turned out that $\Delta(S, S')$ is a good indicator of whether or not peptides P and P' that produced S and S' are only one amino acid apart. Table S6 illustrates

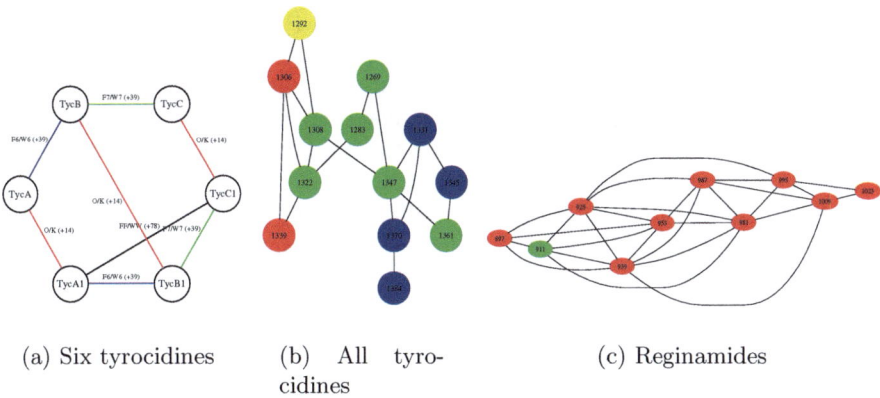

(a) Six tyrocidines

(b) All tyro-cidines

(c) Reginamides

Fig. 4. (a) The spectral network of six Tyrocidines analyzed in [6] reveals 7 (correct) spectral pairs differing by a single substitution and one (incorrect) spectral pair (Tyc A1 and Tyc C1) differing by two substitutions. (b) The spectral network of Tyrocidines after clustering similar spectra (see **Text S3** for details). The sequences were dereplicated from Tyrocidines A, A1, B, B1, C and C1 in Table 2 (green node) using the **multitag** algorithm. (c) The spectral network of Reginamides after clustering similar spectra (see **Text S4** for details). The sequences were dereplicated from Reginamide A in Table 3 (green node) using the **multitag** algorithm.

that all seven spectral pairs of Tyrocidines have Δ less than or equal to five, while for remaining pairs, Δ is greater than or equal to seven, with exception of Tyrocidine A1/C1 pair representing two substitutions at *consecutive* amino acids FF → WW. Such substitutions at consecutive (or closely located) positions are difficult to distinguish from single substitutions. For example, the theoretical spectrum for FF → WW substitutions (each with 39 Da difference in the mass of amino acids) is very similar to the theoretical spectrum of a peptide with a single substitution on either of Phe residues with 78 Da difference.

Spectral Network Construction. Given a set of peptides $P_1, \cdots, P_m$, we define their *spectral network* as a graph with m vertices $P_1, \cdots, P_m$ and edges connecting two peptides if they differ by a single amino acid substitution. In reality, we are not given peptides $P_1, \cdots, P_m$, but only their spectra $S_1, \cdots, S_m$. Nevertheless, one can approximate the spectral network by connecting vertices S_i and S_j if the corresponding peptides are predicted to differ by a single amino acid, i.e. if $\Delta(S, S')$ is less than a threshold. Figure 4(a) show the *spectral network* of six Tyrocidines analyzed in [6].

Multiplex sequencing of peptide families. We now move from pairwise sequencing to multiplex sequencing of *spectral networks* of (more than two) related cyclic peptides. While we use the notion of spectral networks from [9], the algorithm for sequencing linear peptides from spectral networks (as described in [9]) is not applicable for sequencing cyclic peptides.

In *multiplex sequencing of peptide families*, we are given a set of spectra of peptides of the same length n, without knowing their amino acid sequences, and without knowing which ones form spectral pairs. Sequencing of individual cyclic peptides is capable of generating a set of candidate k-tags, that typically contains a correct tag (at least for k smaller than n). However, sequencing of individual spectra typically fails to bring the correct peptide to the top of the list of high-scoring peptides or even, in some cases, fails to place it in this list. To alleviate this problem, we analyze all spectra in the spectral network and introduce a *multiplex scoring* that utilizes the information from all spectra.

Below we formulate the multiplex sequencing problem. Given a spectral network G of spectra $\mathbf{S} = (S_1, \cdots S_m)$, we call a set of peptides $(P_1, \cdots P_m)$ *G-consistent* if for every two spectra S_i and S_j connected by an edge in G, P_i and P_j differ by a single amino acid.

Multiplex Cyclic Peptide Sequencing Problem

- *Goal:* Given spectra of related cyclic peptides (of the same length) and their (estimated) spectral network, reconstruct all cyclic peptides that generated this spectra.
- *Input:* Spectra $\mathbf{S} = S_1, \cdots, S_m$, their (estimated) Spectral Network G, and an integer k.
- *Output:* A G-consistent[2] set of peptide $P_1, \cdots, P_m$ (each of length k) that maximizes $\sum_{i=1}^{m} Score(P_i, S_i)$ among all sets of G-consistent peptides of length k.

Let $\mathbf{S} = (S_1, \cdots S_m)$ be a set of spectra of m peptides forming a spectral network and let $\mathbf{Tag} = (Tag_1, \cdots, Tag_m)$ be a *multitag*, which is a set of tags such that Tag_i is a k-tag of spectrum S_i (for $1 \leq i \leq m$). In **Text S1** we describe multiplex scoring of multitags, taking into account dependencies between spectra in the spectral network. This is in contrast to scoring multitags as $\sum_{j=1}^{m} Score(Tag_j, S_j)$ that is equivalent to independent optimization of individual scores on all individual k-tags. This approach will not give any payoff in comparison to individual spectral sequencing.

MultiplexScore defined in **Text S1** scores a multitag against all spectra in the spectral network. However, generating a correct multitag from m lists of t top-scoring tags in spectra $S_1, \ldots, S_m$ is impractical since (i) the number of candidate multitags (t^m) is large, and (ii) some lists may not contain correct individual tags. We therefore generate candidate multitags from individual tags and score them against all spectra using *MultiplexScore*. Figure 3 describes the algorithm for generating a k-multitag from a single individual k-tag using the spectral network G. Given a candidate individual tag Tag

---

[2] Since we work with estimated (rather than exact) spectral networks, the multiplex cyclic peptide sequencing may not have a solution (i.e. a set of G-consistent peptides does not exist). Given a parameter u, a set of peptides is called (G, u)-consistent if for all but u edges (S_i, S_j), P_i and P_j differ by a single amino acid. The algorithm address finding (G, u)-consistent sets of peptides for a small parameter u.

of a spectrum S_u, $1 \leq u \leq m$, our algorithm generates a candidate multitag $\mathbf{multitag}(Tag, u, \mathbf{S}, G) = (Tag_1, \cdots, Tag_m)$, satisfying $Tag_u = Tag$. Note that given a tag $Tag = (m_1, \cdots, m_k)$, the (i, δ)-modification of Tag is defined as $(m_1, \cdots, m_i + \delta, \cdots, m_k)$.

We now define *multiplex score* on an individual tag Tag of a spectrum S_u as follows:

$$MultiplexScore(Tag, u, \mathbf{S}, G) = MultiplexScore(\mathbf{multitag}(Tag, u, \mathbf{S}, G), \mathbf{S}, G)$$

The multiplex sequencing algorithm (i) generates lists of individual tags for each spectrum in the spectral network, (ii) constructs the spectral network G, (iii) selects an individual Tag that maximizes $MultiplexScore(Tag, u, \mathbf{S}, G)$ among all individual tags, and (iv) outputs $\mathbf{multitag}(Tag, u, \mathbf{S}, G)$ as the solution of the multiplex sequencing problem.

Multiplex sequencing algorithm is exactly the same as the individual sequencing algorithm, with the only difference that we use $MultiplexScore$ here, instead of $Score$ (individual sequencing). Again we start with high scoring 2-tags (in $MultiplexScore$ sense), and extend them, keeping t highest scoring tags in each step. Table 1 (c) illustrates that the multiplex sequencing algorithm sequences all six Tyrocidines studied in [6] correctly.

Figure 2 (b-d) shows spectral profiles for $t = 100$ high scoring tags of multiplex sequencing of Q-TOF spectra of Tyrocidines, Cyclomarins, and Reginamides.

Figure 4(b) and Table 2 show spectral network and sequences of Tyrocidines, predicted by multiplex sequencing algorithm (using ESI-IT spectra, see **Text S3** for details). Figure 4(c) and Table 3 show similar results for Reginamides (see **Text S4** for details).

To analyze Reginamides, the Q-TOF and ESI-IT tandem mass spectrometry data was collected on both ABI QSTAR and ThermoFinnigan LTQ. In both cases, sequencing of Reginamide A resulted in a sequence of integer masses (71, 113, 113, 128, 113, 147, 113, 113). Using accurate FT spectra collected on ThermoFinnigan, we further derived amino acid masses as (71.03729, 113.08406, 113.08405, 128.09500, 113.08404, 147.06849, 113.08397, 113.08402) that pointed to amino acids Ala (71.03711), Ile/Leu (113.08406), Lys (128.09496) and Phe (147.06841) and revealed the elemental composition. These sequences were further confirmed by NMR (see **Text S6**).

3 Methods

Generating mass spectra. Q-TOF tandem mass spectrometry data for Tyrocidines, Cyclomarines, and Reginamides were collected on ABI-QSTAR. In addition, ESI-IT tandem mass spectrometry data were collected for Tyrocidines and Reginamides on a Finnigan LTQ-MS. All spectra were filtered as described in [6,17] by keeping five most intense peaks in each 50 dalton window. All masses were rounded after subtraction of charge mass and multiplication by 0.9995 as described in [18]. High resolution FT spectra of Reginamides were also collected on a Finnigan. Typical mass accuracy of IT instruments are between 0.1 to 1 Da, while typical accuracy of TOF and FT instruments are between 0.01 to 0.1Da, and 0.001 to 0.01Da respectively.

Isolation of Reginamide A. CNT357F5F5 sample was obtained from a cultured marine streptomyces in five 2.8 L Fernbach flasks each containing 1 L of a seawater-based medium and shaken at 230 rpm at 27°C. After seven days of cultivation, sterilized XAD-16 resin was added to adsorb the organic products, and the culture and resin were shaken at 215 rpm for 2 hours.

The resin was filtered through cheesecloth, washed with deionized water, and eluted with acetone. Pure Reginamide A eluted at 12.6 min to give 2.0 mg of pure material.

Generating NMR spectra. CD_3OD and C_5D_5N were purchased from Cambridge Isotope. ^1H NMR, ^{13}C NMR, $^1H - ^1H$ COSY, $^1H - ^1H$ TOCSY (mixing time 90 ms), HMBC (^2J or $^3J_{^1H-^{13}C} = 7$ Hz), HSQC ($^1J_{^1H-^{13}C} = 145$ Hz), and ROESY (mixing time = 400 ms) spectra were generated on the Bruker (AVANCE III 600) NMR spectrometer with 1.7 mm cryoprobe. All the NMR spectra are provided in the **Supplementary Information**.

Parameter Setting. Text S7 discusses setting of parameters of the algorithm.

Acknowledgement

SPG thanks Fundao para a Cincia e Tecnologia, Portugal, for a postdoctoral fellowship. This work was supported by US National Institutes of Health grants 1-P41-RR024851-01 and GM086283.

References

[1] Tang, X.J., Thibault, P., Boyd, R.K.: Characterization of the tyrocidine and gramicidin fractions of the tyrothricin complex from Bacillus brevis using liquid chromatography and mass spectrometry. Int. J. Mass Spectrom. Ion Processes 122, 153–179 (1992)

[2] Sieber, S.A., Marahiel, M.A.: Molecular Mechanisms Underlying Nonribosomal Peptide Synthesis: Approaches to New Antibiotics. Chem. Rev. 105, 715–738 (2005)

[3] Molinski, T.F., Dalisay, D.S., Lievens, S.L., Saludes, J.P.: Drug development from marine natural products. Nat. Rev. Drug Discovery 8 (1), 69–85 (2009)

[4] Molinski, T.F.: NMR of Natural Products at the Nanomole-Scale. Nat. Prod. Rep. 27 (3), 321–329 (2010)

[5] Li, J.W., Vederas, J.C.: Drug discovery and natural products: end of an era or an endless frontier? Science 325, 161–165 (2009)

[6] Ng, J., Bandeira, N., Liu, W., Ghassemian, M., Simmons, T.L., Gerwick, W.H., Linington, R., Dorrestein, P.C., Pevzner, P.A.: Dereplication and de novo sequencing of nonribosomal peptides. Nature Methods 6, 596–599 (2009)

[7] Leao, P.N., Pereirab, A.R., Liu, W.T., Ng, J., Pevzner, P.A., Dorrestein, P.C., Konig, G.M., Teresa, M., Vasconcelos, S.D., Vasconcelos, V.M., Gerwick, W.H.: Synergistic allelochemicals from a freshwater cyanobacterium (2010) (in press)

[8] Liu, W.T., Yang, Y.L., Xu, Y., Lamsa, A., Haste, N.M., Yang, J.Y., Ng, J., Gonzalez, D., Ellermeier, C.D., Straight, P.D., Pevzner, P.A., Pogliano, J., Nizet, V., Pogliano, K., Dorrestein, P.C.: Imaging mass spectrometry of intraspecies metabolic exchange revealed the cannibalistic factors of Bacillus subtilis. Proc. Nat. Acad. Sci. (2009) (in press)

[9] Bandeira, N., Tsur, D., Frank, A., Pevzner, P.A.: Protein identification by spectral networks analysis. Proc. Nat. Acad. Sci. 104(15), 6140–6145 (2007)

[10] Fenical, W.H., Jacobs, R.S., Jensen, P.R.: Cyclic heptapeptide anti-inflammatory agent, US Patent 5593960 (1995)

[11] Sugiyama, H., Shioiri, T., Yokokawa, F.: Synthesis of four unusual amino acids, constituents of cyclomarin A. Tetrahedron Letters 143, 3489–3492 (2002)

[12] Schultz, A.W., Oh, D.C., Carney, J.R., Williamson, R.T., Udwary, D.W., Jensen, P.R., Gould, S.J., Fenical, W., Moore, B.S.: Biosynthesis and structures of cyclomarins and cyclomarazines, prenylated cyclic peptides of marine actinobacterial origin. J. Am. Chem. Soc. 130(13), 4507–4516 (2008)

[13] Skiena, S.S., Smith, W.D., Lemke, P.: Reconstructing sets from interpoint distances. In: The Sixth Annual Symposium on Computational Geometry, Berkeley, California, pp. 332–339 (1990)

[14] Skiena, S.S., Sundaram, G.: A partial digest approach to restriction site mapping. Bulletin of Mathematical Biology 56(2), 275–294 (1994)

[15] Cieliebak, M., Eidenbenz, S., Penna, P.: Partial digest is hard to solve for erroneous input data. Theoretical Computer Science 349(3), 361–381 (2005)

[16] Schneider, T.D., Stephens, R.M.: Sequence Logos: A New Way to Display Consensus Sequences. Nucleic Acids Res. 18(20), 6097–6100 (1990)

[17] Liu, W.T., Ng, J., Meluzzi, D., Banderia, N., Gutirrez, M., Simmons, T.L., Schultz, A.W., Linington, R.G., Moore, B.S., Gerwick, W.H., Pevzner, P.A., Dorrestein, P.C.: Interpretation of Tandem Mass Spectra Obtained from Cyclic Nonribosomal Peptides. Anal. Chem. 81(11), 4200–4209 (2009)

[18] Kim, S., Gupta, N., Bandeira, N., Pevzner, P.: Spectral Dictionaries. Mol. Cell. Proteomics 8, 53–69 (2009)

Fractal and Transgenerational Genetic Effects on Phenotypic Variation and Disease Risk

Joe Nadeau

Institute for Systems Biology

To understand human biology and to manage heritable diseases, a complete picture of the genetic basis for phenotypic variation and disease risk is needed. Unexpectedly however, most of these genetic variants, even for highly heritable traits, continue to elude discovery for poorly understood reasons. The genetics community is actively exploring the usual explanations for missing heritability. But given the extraordinary work that has already been done and the exceptional magnitude of the problem, it seems likely that unconventional genetic properties are involved.

We made two surprising discoveries that may dramatically change our understanding of the genetic basis for phenotypic variation and disease risk, and that may also explain much of missing heritability. The first property involves fractal genetics, where a very large number of often closely linked genetic variants act in a remarkably strong, non-additive and context-dependent manner to control phenotypic variation, suggesting that networks of gene interactions are more important than the constant action of individual variants. The second property involves transgenerational genetic effects, where genetic variants acting in one generation affect phenotypes in subsequent generations. Because these transgenerational effects are common, strong and persistent across multiple generations, they rival the action of genetic variants that are inherited in the conventional manner. The search is ongoing to identify the molecular basis for this non-DNA inheritance. Together these discoveries in model organisms shed light on genetic phenomena that impact human biology but that are difficult to extremely detect directly in human populations. In particular, inheritance of traits and diseases without the corresponding genetic variants could revolutionize our understanding of inheritance.

V. Bafna and S.C. Sahinalp (Eds.): RECOMB 2011, LNBI 6577, p. 282, 2011.
© Springer-Verlag Berlin Heidelberg 2011

AREM: Aligning Short Reads from ChIP-Sequencing by Expectation Maximization

Daniel Newkirk[1,3,*], Jacob Biesinger[2,3,*], Alvin Chon[2,3,*],
Kyoko Yokomori[1], and Xiaohui Xie[2,3,**]

[1] Department of Biological Chemistry
[2] Department of Computer Science
[3] The Institute for Genomics and Bioinformatics,
University of California, Irvine, CA 92697
xhx@ics.uci.edu

Abstract. High-throughput sequencing coupled to chromatin immuno-precipitation (ChIP-Seq) is widely used in characterizing genome-wide binding patterns of transcription factors, cofactors, chromatin modifiers, and other DNA binding proteins. A key step in ChIP-Seq data analysis is to map short reads from high-throughput sequencing to a reference genome and identify peak regions enriched with short reads. Although several methods have been proposed for ChIP-Seq analysis, most existing methods only consider reads that can be uniquely placed in the reference genome, and therefore have low power for detecting peaks located within repeat sequences. Here we introduce a probabilistic approach for ChIP-Seq data analysis which utilizes all reads, providing a truly genome-wide view of binding patterns. Reads are modeled using a mixture model corresponding to K enriched regions and a null genomic background. We use maximum likelihood to estimate the locations of the enriched regions, and implement an expectation-maximization (E-M) algorithm, called AREM (aligning reads by expectation maximization), to update the alignment probabilities of each read to different genomic locations. We apply the algorithm to identify genome-wide binding events of two proteins: Rad21, a component of cohesin and a key factor involved in chromatid cohesion, and Srebp-1, a transcription factor important for lipid/cholesterol homeostasis. Using AREM, we were able to identify 19,935 Rad21 peaks and 1,748 Srebp-1 peaks in the mouse genome with high confidence, including 1,517 (7.6%) Rad21 peaks and 227 (13%) Srebp-1 peaks that were missed using only uniquely mapped reads. The open source implementation of our algorithm is available at http://sourceforge.net/projects/arem

Keywords: ChIP-Seq, Mixture Model, Expectation-Maximization, Cohesin, CTCF, Srebp-1, Repetitive Elements, High Throughput Sequencing, Peak-Caller.

---

[*] The first three authors contributed equally.
[**] Corresponding author.

V. Bafna and S.C. Sahinalp (Eds.): RECOMB 2011, LNBI 6577, pp. 283–297, 2011.

1   Introduction

In recent years, high-throughput sequencing coupled to chromatin immunopre-cipitation (ChIP-Seq) has become one of the premier methods of analyzing protein-DNA interactions [1]. The ability to capture a vast array of protein bind-ing locations genome-wide in a single experiment has led to important insights in a number of biological processes, including transcriptional regulation, epi-genetic modification and signal transduction [2,3,4,5]. Numerous methods have been developed to analyze ChIP-Seq data and typically work well for identi-fying protein-DNA interactions located within non-repeat sequences. However, identifying interactions in repeat regions remains a challenging problem since sequencing reads from these regions usually cannot be uniquely mapped to a reference genome. We present novel methodology for identifying protein-DNA interactions in repeat sequences.

ChIP-Seq computational analysis typically consists of two tasks: one is to iden-tify the genomic locations of the short reads by aligning them to a reference genome, and the second is to find genomic regions enriched with the aligned reads, which is often termed peak finding. Eland, MAQ, Bowtie, and SOAP are among the most popular for mapping short reads to a reference genome [6,7,8,9] and provide many or all of the potential mappings for a given sequence read. Once potential map-pings have been identified, significantly enriched genomic regions are identified us-ing one of several available tools [10,11,12,13,14,15,16,17,18]. Some peak finders are better suited for histone modification studies, others for transcription factor bind-ing site identification. These peak finders have been surveyed on several occasions [19,20,21].

Many short reads cannot be uniquely mapped to the reference genome. Most peak finding workflows throw away these non-uniquely mapped reads, and as a consequence have low power for detecting peaks located within repeat re-gions. While each experiment varies, only about 60% [in house data] of the sequence reads from a ChIP-Seq experiment can be uniquely mapped to a ref-erence genome. Therefore, a significant portion of the raw data is not utilized by the current methods. There have been proposals to address the non-uniquely mapped reads in the literature by either randomly choosing a location from a set of potential ones [22,23] or by taking all potential alignments [12], but most peak callers are not equipped to deal with ambiguous reads.

We propose a novel peak caller designed to handle ambiguous reads directly by performing read alignment and peak-calling jointly rather than in two separate steps. In the context of ChIP-Seq studies, regions enriched during immunopre-cipitation are more likely the true genomic source of sequence reads than other regions of the genome. We leverage this idea to iteratively identify the true ge-nomic source of ambiguous reads. Under our model, the true locations of reads and binding peaks are treated as hidden variables, and we implement an al-gorithm, AREM, to estimate both iteratively by alternating between mapping reads and finding peaks.

Two ChIP-Seq datasets were used in this study: 1) cohesin, a new dataset generated in house, and 2) Srebp-1, a previously published dataset [5].

We generated the cohesin dataset by performing ChIP-Seq using mouse embryonic fibroblasts and an antibody targeting Rad21 [24], a subunit of cohesin. Cohesin is an essential protein complex required for sister chromatid cohesion. In mammalian cells, cohesin binding sites are present in intergenic, promoter and 3' regions-especially in connection with CTCF binding sites [25,26]. It was found that cohesin is recruited by CTCF to many of its binding sites, and plays a role in CTCF-dependent gene regulation [27,28]. Cohesin has been shown to bind to repeat sequences in a disease-specific manner [24], making it a particularly interesting candidate for our study.

The second dataset is Srebp-1, a transcription factor important in allostatic regulation of sterol biosynthesis and membrane lipid composition [29]. This particular dataset [5] examines the genomic binding locations for Srebp-1 in mouse liver. Regulation of expression by Srebp-1 is important for regulation of cholesterol; repeat-binding for this transcription factor has not been shown previously [30,29]. We choose these datasets because both proteins have well characterized regulatory motifs, allowing us to directly test the validity of our peak finding method.

On a 2.8Ghz CPU, AREM takes about 20 minutes and 1.6GB RAM to call peaks from over 12 million alignments and about 30 minutes and 6GB RAM to call peaks from nearly 120 million alignments. Each dataset takes less than 40 iterations to converge. AREM is written in Python, is open-source, and is available at http://sourceforge.net/projects/arem.

2   Methods

2.1   Notations

Let $R = \{r_1, \cdots, r_N\}$ denote a set of reads from a ChIP-Seq experiment with read $r_i \in \Sigma^l$, where $\Sigma = \{\text{A}, \text{C}, \text{G}, \text{T}\}$, l is the length of each read, and N denotes the number of reads. Let $S \in \Sigma^L$ denote the reference sequence to which the reads will be mapped. In real applications, the reference sequence usually consists of multiple chromosomes. For notational simplicity, we assume the chromosomes have been concatenated to form one reference sequence.

We assume that for each read we are provided with a set of potential alignments to the reference sequence. Denote the set of potential alignments of read r_i to S by $A_i = \{(l_{ij}, q_{ij}) : j = 1, \cdots, n_i\}$, where l_{ij} and q_{ij} denote the starting location and the confidence score of the j-th alignment, and n_i is the total number of potential alignments. We assume $q_{ij} \in [0, 1]$ for all j, and use it to account for both sequencing quality scores and mismatches between the read and the reference sequence. There are several programs available to generate the initial potential alignments and confidence scores.

2.2   Mixture Model

We use a generative model to describe the likelihood of observing the given set of short reads from a ChIP-Seq experiment. Suppose the ChIP procedure results in the enrichment of K non-overlapping regions in the reference sequence S. Denote the K enriched regions (also called peak regions) by $\{(s_k, w_k) : k = 1, \cdots, K\}$, where s_k and w_k represent the start and the width, respectively, of the i-th enriched region in S. Let $E_k = \{s_k, \cdots, s_k + w_k - l\}$ denote the set of locations in the enriched region k that can potentially generate a read of length l. Let E_k^s, E_k^w denote the start and width of region k. We will use E_0 to denote all locations in S that are not covered by $\bigcup_{k=1}^K E_k$.

We use variable $z_i \in \{1, \cdots, n_i\}$ to denote the true location of read r_i, with $z_i = j$ representing that r_i originates from location l_{ij} of S. In addition, we use variable $u_i \in \{0, 1, \cdots, K\}$ to label the type of region that read r_i belongs to. $u_i = k$ represents that read r_i is from the non-enriched regions of S if $k = 0$, and is from k-th enriched region otherwise. Both z_i and u_i are not directly observable, and are often referred to as the hidden variables of the generative model.

Let $P(r_i|z_i = j, u_i = k)$ denote the conditional probability of observing read r_i given that r_i is from location l_{ij} and belongs to region k. Assuming different reads are generated independently, the log likelihood of observing R given the mixture model is then

$$\ell = \sum_{i=1}^N \log \left[\sum_{j=0}^{n_i} \sum_{k=0}^K P(r_i|z_i = j, u_i = k) P(z_i = j) P(u_i = k) \right],$$

where $P(z_i)$ and $P(u_i)$ represent the prior probabilities of the location and the region type, respectively, of read r_i. $P(z_i)$ is set according to the confidence scores of different alignments

$$P(z_i = j) = \frac{q_{ij}}{\sum_{k=1}^{n_i} q_{ik}}. \tag{1}$$

$P(u_i)$ depends on both the width and the enrichment ratio of each enriched region. Denote the enrichment ratio of the ChIP regions vs non-ChIP regions by α, which is often significantly impacted by the quality of antibodies used in ChIP experiments. We parametrize the prior distribution on region types as follows

$$P(u_i = k) = \frac{1}{(\alpha - 1) \sum_j w_j + L} \times \begin{cases} L - \sum_j w_j & \text{if } k = 0 \\ \alpha w_k & o.w. \end{cases} \tag{2}$$

2.3   Parameter Estimation

The conditional probability $P(r_i|z_i = j, u_i = k)$ can be modeled in a number of different ways. For example, bell-shaped distributions are commonly used to model the enriched regions. However, for computational simplicity, we will use a simple uniform distribution to model the enriched regions. If read r_i comes from one of the enriched regions, i.e., $k \neq 0$, we assume the read is equally likely to originate from any of the potential positions within the enriched region, that is,

$$P(r_i|z_i = j, u_i = k) = \frac{1}{w_k - l + 1} \mathbf{I}_{E_k}(l_{ij}), \tag{3}$$

where $\mathbf{I}_A(x)$ is the indicator function, returning 1 if $x \in A$ and 0 otherwise.

If the read is from non-enriched regions, i.e., $k = 0$, we use p_i^b to model the background probability of an arbitrary read originating from location i of the reference sequence. (We assume p_i^b has been properly normalized such that $\sum_{i=1}^{L} p_i^b = 1$.) Then the conditional probability $P(r_i|z_i = j, u_i = k)$ for the case of $k = 0$ is modeled by

$$P(r_i|z_i = j, u_i = 0) = \mathbf{I}_{E_0}(l_{ij}) \, p_{l_{ij}}^b. \tag{4}$$

Numerous ChIP-Seq studies have demonstrated that the locations of ChIP-Seq reads are typically non-uniform, significantly biased toward promoter or open chromatin regions [1]. The p_i^b's takes this ChIP and sequencing bias into account, and can be inferred from control experiments typically employed in ChIP-Seq studies.

Next we integrate out the u_i variable to obtain the conditional probability of observing r_i given only z_i

$$P(r_i|z_i = j) = P(u_i = 0)\mathbf{I}_{E_0}(l_{ij}) \, p_{l_{ij}}^b + \sum_{k=1}^{K} \frac{P(u_i = k)}{w_k - l + 1}\mathbf{I}_{E_k}(l_{ij}). \tag{5}$$

Note that because $E_0, E_1, \cdots, E_K$ are disjoint, only one term in the above summation can be non-zero. This property significantly reduces the computation for parameter estimation since we do not need to infer the values of u_i variables any more.

The log likelihood of observing R given the mixture model can now be written as

$$\ell(r_1, \cdots, r_n; \Theta) = \sum_{i=1}^{N} \log \left[\sum_{j=0}^{n_i} P(r_i|z_i = j)P(z_i = j) \right], \tag{6}$$

where $\Theta = (s_1, w_1, \cdots, s_K, w_K, \alpha)$ denotes the parameters of the mixture model. We estimate the values of these unknown parameters using maximum likelihood estimation

$$\hat{\Theta} = \arg\max_{\Theta} \ell(r_1, \cdots, r_n; \Theta). \tag{7}$$

2.4   Expectation-Maximization Algorithm

We solve the maximum likelihood estimation problem in Eq. (7) through an expectation-maximization (E-M) algorithm. The algorithm iteratively applies the following two steps until convergence:

Expectation step: Estimate the posterior probability of alignments under the current estimate of parameters $\Theta^{(t)}$:

$$Q^{(t)}(z_i = j|R) = \frac{1}{C}P(r_i|z_i = j, \Theta^{(t)})P(z_i = j), \tag{8}$$

where C is a normalization constant.

Maximization step: Find the parameters $\Theta^{(t+1)}$ that maximize the following quantity,

$$\Theta^{(t+1)} = \arg\max_{\Theta} \sum_{i=1}^{N} \sum_{j=0}^{n_i} Q^{(t)}(z_i = j|R) \log P(r_i|z_i = j, \Theta). \tag{9}$$

2.5   Implementation of E-M Updates

The mixture model described above contains $2K + 1$ parameters. Since K, the number of peak regions, is typically large, ranging from hundreds to hundreds of thousands, exactly solving Eq. (9) in the maximization step is nontrivial. Instead of seeking an exact solution, we identify the K regions from the data by considering all regions where the number of possible alignments is significantly enriched above the background.

For a given window of size w starting at s of the reference genome, we first calculate the number of reads located within the window, weighted by the current estimation of posterior alignment probabilities,

$$f(s, w) = \sum_{i=1}^{N} \sum_{j=1}^{n_i} Q^{(t)}(z_i = j|R) \, \mathbf{I}_{[s, s+w-l]}(l_{ij}). \tag{10}$$

We term this quantity the foreground read density. As a comparison, we also calculate a background read density $b(s, w)$, which is estimated using either reads from the control experiment or reads from a much larger extended region covering the window. Different ways of calculating background read density are discussed in [13].

Provided with both background and foreground read densities, we then define an enrichment score $\phi(s, w)$ to measure the significance of read enrichment within the window starting at position s with width w. For this purpose, we assume the number of reads are distributed according to a Poisson model with mean rate $b(s, w)$. If $f(s, w)$ is an integer, the enrichment score is defined to be $\phi(s, w) = -\log_{10}(1 - g(f, b))$, where

$$g(x, \lambda) = e^{-\lambda} \sum_{k=0}^{x} \frac{\lambda^k}{k!} \tag{11}$$

denotes the chance of observing at least x Poisson events given the mean rate of λ. However, if $f(s, w)$ is not an integer, the enrichment score cannot be defined this way. Instead, we use a linear extrapolation to define the enrichment score $\phi(s, w) = -\log_{10}(1 - \tilde{g}(f, b))$, where function $\tilde{g}$ is defined as

$$\tilde{g}(x, \lambda) = g(\lfloor x \rfloor, \lambda) + [g(\lceil x \rceil, \lambda) - g(\lfloor x \rfloor, \lambda)] (x - \lfloor x \rfloor). \tag{12}$$

If two potential alignments of a read have the same confidence score and are located in two peak regions with equal enrichment, the update of posterior alignment probabilities in Eq. (8) will assign equal weight to these two alignments.

This is so because we have assumed that peak regions have the same enrichment ratio as described in Eq. (2), which is not true as some peak regions are more enriched than others in real ChIP experiments. To address this issue, we have also implemented an update of the posterior probabilities that takes the calculated enrichment scores into account as

$$Q^t(z_i = j|R) \leftarrow \sum_{k=1}^{K} [\phi(E_k^s, E_k^w) P(z_i = j) \mathbf{I}_{E_k}(z_i)] \tag{13}$$

which is then normalized. In practice, we found this implementation usually behaves better than the one without using enrichment scores.

We use entropy to quantify the uncertainty of alignments associated with each read. For read i, the entropy at iteration t is defined to be

$$H_i^t = -\sum_{j=1}^{n_i} Q^t(z_i = j|R) \log Q^t(z_i = j|R). \tag{14}$$

We stop the E-M iteration when the relative square difference between two consecutive entropies is small, that is, when

$$\frac{\sum_{i=0}^{N}(H_i^t - H_i^{t-1})^2}{\sum_{i=0}^{N}(H_i^{t-1})^2} < \epsilon, \tag{15}$$

where $\epsilon = 10^{-5}$ for results reported in this paper.

AREM seeks to identify the true genomic source of multiply-aligning reads (also called multireads). Many of the multireads will map to repeat regions of the genome, and we expect repeats to be included in the K potentially enriched regions. To prevent repeat regions from garnering multiread mass without sufficient evidence of their enrichment, we impose a minimum enrichment score. Effectively, unique or less ambiguous multireads need to raise enrichment above noise levels for repeat regions to be called as peaks. The minimum enrichment score is a parameter of our model and its effect on called peaks is explored in Results.

3   Results

Building on the methodology of the popular peak-caller Model-based Analysis of ChIP-Seq (MACS) [13], we implement AREM, a novel peak caller designed to handle multiple possible alignments for each sequence read. AREM's peak caller combines an initial sliding window approach with a greedy refinement step and iteratively aligns ambiguous reads. We use two ChIP-Seq datasets in this study: Rad21 and Srebp-1. Rad21, a subunit of the structural protein cohesin, contained 7.2 million treatment reads and 7.4 million control reads (*manuscript in preparation*). Srebp-1, a regulator of cholesterol metabolism, had 7.7 million treatment reads and 6.4 million control reads [5].

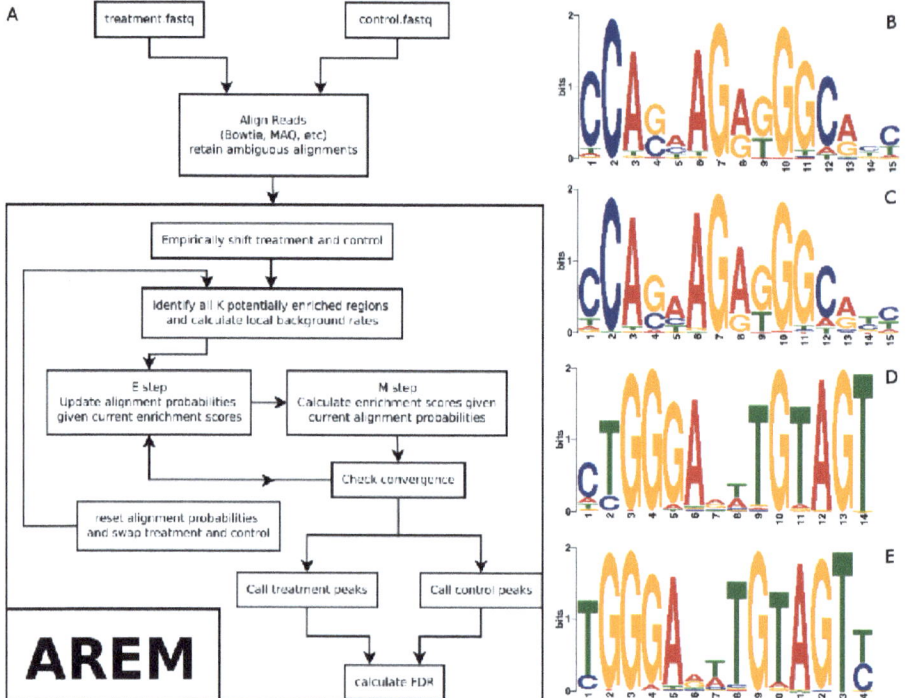

Fig. 1. A) AREM workflow diagram. **B-E** *de novo* discovery of motifs. From top to bottom: **B)** CTCF in MACS peaks from uniquely mapping reads, **C)** CTCF in AREM's peaks with multireads, **C)** Srebp-1 in MACS peaks from uniquely mapping reads and **D)** Srebp-1 in AREM's peaks with multireads.

Using AREM, we identify 19,935 Rad21 peaks covering more than 10 million base pairs at a low False Discovery Rate (FDR) of 3.7% and 1,474 Srebp-1 peaks covering nearly 1 million bases at a moderate FDR of 8%. For comparison, we also called peaks using MACS and SICER [15], another popular peak finding program. To compare our results, we use FDR and motif presence as indicators of *bona fide* binding sites.

3.1   AREM Identifies Additional Binding Sites

We seek to benchmark both AREM's peak-calling and its multiread methodology. To benchmark peak-calling, we limit all reads to their best alignment and run AREM, MACS and SICER. In the Rad21 dataset, AREM identifies 456 more peaks than MACS and 1920 more peaks than SICER but retains a similar motif presence (81.6% MACS, 82.5% SICER, 81.3% AREM) and has a lower FDR (2.8% MACS, 12.7% SICER, 1.9% AREM) (see Table 1). For Srebp-1, AREM identifies more than double the number of peaks compared to MACS and 816 more than SICER, though the FDR is slightly higher (4.85% MACS,

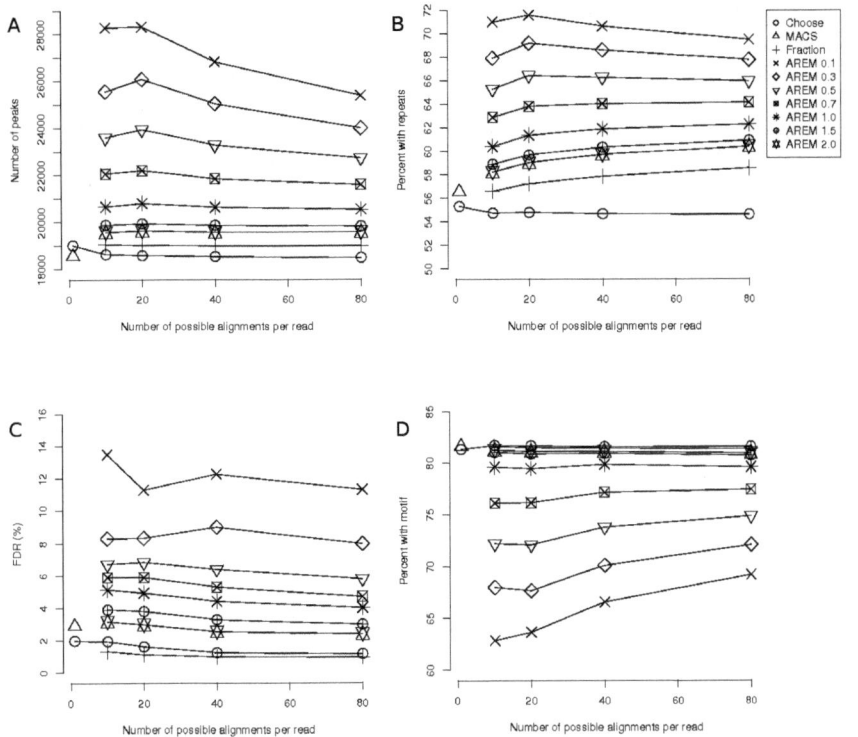

Fig. 2. Graphs displaying varying parameters and number of possible alignments per read. **A)** Total number of peaks discovered. **B)** Percentage of peaks with repetitive sequences. **C)** False Discovery Rate. **D)** Percentage of peaks with motif.

9% SICER, 8% AREM), and motif presence is slightly lower (46.6% MACS, 59% SICER, 39% AREM). In both datasets, AREM appears to be more sensitive to true binding sites, picking up more total sites with motif instances, although it trades off some specificity in Srebp-1.

To see if AREM can identify true sites that are not significant without multireads, we performed peak-calling with multireads, removing peaks that overlapped with those identified using AREM without multireads. Up to 1,546 (8.1%) and 272 (18.9%) previously unidentified peaks were called from Rad21 and Srebp-1, respectively. These new peaks have a similar motif presence compared to previous peaks but overlap with annotated repeat regions more often.

3.2 AREM's Sensitivity Is Increased with Ambiguous Reads

Several methods for dealing with ambiguous reads have been proposed, including retaining all possible mappings, retaining one of the mappings chosen at random, and distributing weight equally among the mappings. The first option will clearly lead to false positives, particularly in repeat regions as the number of

Table 1. Comparison of peak-calling methods for cohesin and Srebp-1. Three peak callers (MACS, SICER, and AREM) were run on both datasets. For AREM, the maximum number of retained alignments per read is varied (from 1 to 80). The total number of peaks and bases covered by peaks is reported as well as the FDR by swapping treatment and control. For both datasets, AREM's minimum enrichment score was fixed at 1.5 with 20 maximum alignments per read. For comparison, the motif background rate of occurence was 4.5% (CTCF) and 27% (Srebp-1) in 100,000 genomic samples, sized similarly to Rad21 MACS peaks and Srebp-1 MACS peaks, respectively.

Method	# Alignments	# Peaks	Peak Bases	FDR	New Peaks	Motif	Repeat
Cohesin							
MACS	2,368,229	18,556	9,546,641	2.8%	—	81.67%	56.55%
SICER	2,368,229	17,092	17,374,108	12.71%	—	82.55%	70.42%
AREM 1	2,368,229	19,012	9,353,567	1.9%	—	81.32%	55.30%
AREM 10	7,616,647	19,881	10,225,479	3.8%	1,404	81.04%	58.88%
AREM 20	12,312,878	19,935	10,531,465	3.7%	1,517	80.88%	59.66%
AREM 40	20,527,010	19,863	10,744,836	3.2%	1,546	80.93%	60.34%
AREM 80	34,537,311	19,820	10,972,796	2.9%	1,538	80.73%	60.91%
Srebp-1							
MACS	10,482,005	721	495,968	4.85%	—	46.60%	53.95%
SICER	10,482,005	622	963,778	9.0%	—	59.00%	77.33%
AREM 1	10,482,005	1,438	880,284	8.0%	—	39.08%	53.47%
AREM 10	28,347,869	1,815	996,346	10.5%	262	39.22%	56.04%
AREM 20	44,493,532	1,748	959,646	8.0%	227	39.95%	55.97%
AREM 40	72,453,642	1,685	983,459	8.2%	248	40.34%	56.46%
AREM 80	118,744,757	1,695	987,746	7.3%	272	40.66%	56.73%

retained mappings increases. We compare the latter two methods to our E-M implementation, varying the number of retained reads and summarizing the results in Table 1. Although both random selection and fractionating reads increases the number of peaks called, our E-M method outperforms them, yielding 1546 more peaks for Rad21, and 272 for Srebp-1 with comparable quality. As the number of retained alignments increases, the disparity gets smaller. AREM shows fairly consistent results across datasets with a large increase in total number of alignments (nearly 40-fold for Rad21, over 10-fold for Srebp-1).

For a given sample, the iterations show a continued shift of the max alignment probabilities to either 1 or 0. This shift is consistent across datasets with larger numbers of max alignments (data not shown), but does depend on other parameters. What is apparent is that AREM's E-M heuristic performs well, allowing for significant shift toward a "definitive" alignment; at the same time, it does not force a shift on reads with too little information, preventing misalignment and resulting spurious peak-calling.

3.3   AREM Is Sensitive to Repeat Regions

An important parameter in our model is the minimum enrichment score for all K regions. Since repeat regions have such similar sequence content, many reads will share the same repetitive elements. If one of the shared repeat elements has a slightly higher enrichment score by chance, the E-M method will iteratively shift probability into that repeat region, snowballing the region into what appears to be a full-fledged sequence peak. To distinguish repetitive peaks arising by small enrichment fluctuations from true binding sites within or adjacent to repetitive elements, we impose a minimum enrichment score on all regions. Using lower threshold scores, our method may include false positives from these random fluctuations. However, true binding peaks near repetitive elements may be missed if the score is too high.

To explore the effect of varying the minimum enrichment score, we varied the minimum score from 0.1 to 2, keeping the maximum number of alignments fixed at 20. For Rad21, we see a declining number of discovered peaks ranging from 28,305 to 19,634 peaks. In addition to a decline in discovered peaks as minimum enrichment score increases, we also see a decrease in the reported FDR and the percent of peaks in repeat regions from 11.28% to 2.95% FDR and 71.56% to 59.02%. Lastly, the percent of peaks with motif increases from 63.64% to 81.12%. These additional peaks appear to be of lower quality: motifs are largely absent from them and the FDR is much higher (see Figure 2).

For our method, detecting peaks near repeat regions is a tradeoff between sensitivity and specificity. As the minimum score increases, the method approaches the uniform or "fraction" distribution, in which only the initial mapping quality scores (and not the enrichment) affect alignment probabilities. The fraction method is explored explicitly, showing increased power compared to unique reads only, but decreased sensitivity to true binding sites compared to other AREM runs.

4   Discussion

Repetitive elements in the genome have traditionally been problematic in sequence analysis. Since sequenced reads are short and repetitive sequences are similar, many equally likely mappings may exist for a given read. Our method uses the low-coverage unique reads near repeat regions to evaluate which potential alignments for each read are the most likely. Our method's sensitivity to repeat regions is adjustable, but increasing sensitivity may introduce false positives. Further refinement of our methodology may lead to increased specificity.

Our results imply that functional CTCF binding sites exist within repeat regions, revealing an interesting relationship between repetitive sequence and chromatin structure. Another application of our method would be to explore the relationship between repetitive sequence and epigenetic modifications such as histone modifications. Regulation of and by transposable elements has been linked to methylation marks [31], and transposable elements have a major role in cancers [32]. Better identification of histone modifications in regions of repetitive DNA increases our understanding of key regulators of genome stability and diseases sparked by translocations and mutations.

Acknowledgement

The work was partly supported by NSF grant DBI-0846218 to XX, and NIH grant HD062951 to KY. DN and JB were supported by National Institutes of Health/NLM bioinformatics training grant (T15LM07443). We thank the Liu lab for releasing MACS as open-source, and R Chien, Y Chen, and N Infante for helpful discussions.

References

1. Park, P.: ChIP–seq: advantages and challenges of a maturing technology. Nature Reviews Genetics 10, 669–680 (2009)
2. Mikkelsen, T., Ku, M., Jaffe, D., Issac, B., Lieberman, E., Giannoukos, G., Alvarez, P., Brockman, W., Kim, T., Koche, R., et al.: Genome-wide maps of chromatin state in pluripotent and lineage-committed cells. Nature 448, 553–560 (2007)
3. Ouyang, Z., Zhou, Q., Wong, W.: ChIP-Seq of transcription factors predicts absolute and differential gene expression in embryonic stem cells. Proceedings of the National Academy of Sciences 106, 21521 (2009)
4. Blow, M., McCulley, D., Li, Z., Zhang, T., Akiyama, J., Holt, A., Plajzer-Frick, I., Shoukry, M., Wright, C., Chen, F., et al.: ChIP-Seq identification of weakly conserved heart enhancers. Nature Genetics 42, 806–810 (2010)
5. Seo, Y., Chong, H., Infante, A., Im, S., Xie, X., Osborne, T.: Genome-wide analysis of SREBP-1 binding in mouse liver chromatin reveals a preference for promoter proximal binding to a new motif. Proceedings of the National Academy of Sciences 106, 13765 (2009)
6. Cox, A.J.: Efficient Large-Scale Alignment of Nucleotide Databases. Whole genome alignments to a reference genome (2007),
   http://bioinfo.cgrb.oregonstate.edu/docs/solexa
7. Langmead, B., Trapnell, C., Pop, M., Salzberg, S.: Ultrafast and memory-efficient alignment of short DNA sequences to the human genome. Genome Biol. 10, R25 (2009)
8. Li, H., Ruan, J., Durbin, R.: Mapping short DNA sequencing reads and calling variants using mapping quality scores. Genome Research 18, 1851 (2008)
9. Li, R., Li, Y., Kristiansen, K., Wang, J.: SOAP: short oligonucleotide alignment program. Bioinformatics 24, 713 (2008)
10. Fejes, A., Robertson, G., Bilenky, M., Varhol, R., Bainbridge, M., Jones, S.: FindPeaks 3.1: a tool for identifying areas of enrichment from massively parallel short-read sequencing technology. Bioinformatics 24, 1729 (2008)
11. Ji, H., Jiang, H., Ma, W., Johnson, D., Myers, R., Wong, W.: An integrated software system for analyzing ChIP-chip and ChIP-seq data. Nature Biotechnology 26, 1293–1300 (2008)
12. Mortazavi, A., Williams, B., McCue, K., Schaeffer, L., Wold, B.: Mapping and quantifying mammalian transcriptomes by RNA-Seq. Nature Methods 5, 621–628 (2008)

13. Zhang, Y., Liu, T., Meyer, C., Eeckhoute, J., Johnson, D., Bernstein, B., Nussbaum, C., Myers, R., Brown, M., Li, W., et al.: Model-based analysis of ChIP-Seq (MACS). Genome Biology 9, R137 (2008)
14. Spyrou, C., Stark, R., Lynch, A., Tavaré, S.: BayesPeak: Bayesian analysis of ChIP-seq data. BMC Bioinformatics 10, 299 (2009)
15. Zang, C., Schones, D., Zeng, C., Cui, K., Zhao, K., Peng, W.: A clustering approach for identification of enriched domains from histone modification ChIP-Seq data. Bioinformatics 25, 1952 (2009)
16. Blahnik, K., Dou, L., O'Geen, H., McPhillips, T., Xu, X., Cao, A., Iyengar, S., Nicolet, C., Ludascher, B., Korf, I., et al.: Sole-Search: an integrated analysis program for peak detection and functional annotation using ChIP-seq data. Nucleic Acids Research 38, e13 (2010)
17. Qin, Z., Yu, J., Shen, J., Maher, C., Hu, M., Kalyana-Sundaram, S., Yu, J., Chinnaiyan, A.: HPeak: an HMM-based algorithm for defining read-enriched regions in ChIP-Seq data. BMC Bioinformatics 11, 369 (2010)
18. Salmon-Divon, M., Dvinge, H., Tammoja, K., Bertone, P.: PeakAnalyzer: Genome-wide annotation of chromatin binding and modification loci. BMC Bioinformatics 11, 415 (2010)
19. Kharchenko, P., Tolstorukov, M., Park, P.: Design and analysis of ChIP-seq experiments for DNA-binding proteins. Nature Biotechnology 26, 1351–1359 (2008)
20. Pepke, S., Wold, B., Mortazavi, A.: Computation for ChIP-seq and RNA-seq studies. Nature Methods 6, S22–S32 (2009)
21. Wilbanks, E., Facciotti, M.: Evaluation of Algorithm Performance in ChIP-Seq Peak Detection. PloS One 5, e11471 (2010)
22. Kagey, M., Newman, J., Bilodeau, S., Zhan, Y., Orlando, D., van Berkum, N., Ebmeier, C., Goossens, J., Rahl, P., Levine, S., et al.: Mediator and cohesin connect gene expression and chromatin architecture. Nature (2010)
23. Schmid, C., Bucher, P.: MER41 Repeat Sequences Contain Inducible STAT1 Binding Sites. PloS One 5, e11425 (2010)
24. Zeng, W., De Greef, J., Chen, Y., Chien, R., Kong, X., Gregson, H., Winokur, S., Pyle, A., Robertson, K., Schmiesing, J., et al.: Specific loss of histone H3 lysine 9 trimethylation and HP1γ/cohesin binding at D4Z4 repeats is associated with facioscapulohumeral dystrophy (FSHD) (2009)
25. Rubio, E., Reiss, D., Welcsh, P., Disteche, C., Filippova, G., Baliga, N., Aebersold, R., Ranish, J., Krumm, A.: CTCF physically links cohesin to chromatin. Proceedings of the National Academy of Sciences 105, 8309 (2008)
26. Liu, J., Zhang, Z., Bando, M., Itoh, T., Deardorff, M., Clark, D., Kaur, M., Tandy, S., Kondoh, T., Rappaport, E., et al.: Transcriptional dysregulation in NIPBL and cohesin mutant human cells. PLoS Biol. 7, e1000119 (2009)
27. Wendt, K., Yoshida, K., Itoh, T., Bando, M., Koch, B., Schirghuber, E., Tsutsumi, S., Nagae, G., Ishihara, K., Mishiro, T., et al.: Cohesin mediates transcriptional insulation by CCCTC-binding factor. Nature 451, 796–801 (2008)
28. Nativio, R., Wendt, K., Ito, Y., Huddleston, J., Uribe-Lewis, S., Woodfine, K., Krueger, C., Reik, W., Peters, J., Murrell, A.: Cohesin is required for higher-order chromatin conformation at the imprinted IGF2-H19 locus (2009)
29. Hagen, R., Rodriguez-Cuenca, S., Vidal-Puig, A.: An allostatic control of membrane lipid composition by SREBP1. FEBS Letters (2010)
30. Yokoyama, C., Wang, X., Briggs, M., Admon, A., Wu, J., Hua, X., Goldstein, J., Brown, M.: SREBP-1, a basic-helix-loop-helix-leucine zipper protein that controls transcription of the low density lipoprotein receptor gene. Cell 75, 187–197 (1993)

31. Huda, A., Jordan, I.: Epigenetic regulation of Mammalian genomes by transposable elements. Annals of the New York Academy of Sciences 1178, 276–284 (2009)
32. Chuzhanova, N., Abeysinghe, S., Krawczak, M., Cooper, D.: Translocation and gross deletion breakpoints in human inherited disease and cancer II: Potential involvement of repetitive sequence elements in secondary structure formation between DNA ends. Human Mutation 22, 245–251 (2003)
33. Rhead, B., Karolchik, D., Kuhn, R., Hinrichs, A., Zweig, A., Fujita, P., Diekhans, M., Smith, K., Rosenbloom, K., Raney, B., et al.: The UCSC genome browser database: update 2010. Nucleic Acids Research (2009)
34. Boeva, V., Surdez, D., Guillon, N., Tirode, F., Fejes, A., Delattre, O., Barillot, E.: De novo motif identification improves the accuracy of predicting transcription factor binding sites in ChIP-Seq data analysis. Nucleic Acids Research (2010)
35. Bailey, T., Elkan, C.: The value of prior knowledge in discovering motifs with MEME. In: Proc Int. Conf. Intell. Syst. Mol. Biol., vol. 3, pp. 21–29 (1995)

Appendix

Alignment

We aligned the data using Bowtie [7] with the Burrows-Wheeler index provided by the Bowtie website. The index is based on the unmasked MM9 reference genome from the UCSC Genome Browser [33]. We clipped the first base of all raw reads to remove sequencing artifacts and allowed a maximum of two mismatches in the first 28 bases of the remaining sequence. We generated several alignment collections for both Srebp-1 and Rad21 by varying k, the maximum number of reported alignments. We restricted our study to search the 1, 10, 20, 40, and 80 best alignments. Table 1 shows that the total number of alignments was only starting to plateau at k=80, indicating that many sequences have more than 80 possible alignments, for practicality we restricted our search as above. We calculated map confidence scores from Bowtie output as in [8]. We also provide an option for using the aligner's confidence scores directly rather than recalculating them from mismatches and sequence qualities. During preparation of the sequencing library, unequal amplification can result in biased counts for reads. To eliminate this bias, we limit the number of alignments to one for each start position on each strand. In particular, we choose the best alignment (based on quality score) for each position; in the event that all alignments have the same quality score, we choose a random read to represent that particular position.

Peak Finding

Our peak finding method is an adapted version of the MACS [13] peak finder. Like MACS, we empirically model the spatial separation between +/- strand tags and shift both treatment and control tags. We also continue MACS' conservative approach to background modeling, using the highest of three rates as the background (in this study, genome-wide or within 1,000 or 10,000 bases). As a divergence from MACS, we use a sliding window approach to identify large potentially enriched regions then use a smoothened greedy approach to refine called peaks. We call peaks within this large region by greedily adding reads to improve enrichment, but avoid local optima by always looking up to the full sliding window width away. The initial large regions correspond to the K regions used for the E-M steps of Section 2.5. During the E-M steps, local background rates are used as during final peak-calling. Peaks reported in this study are above a p-value of 10^{-5}. All enrichment scores and p-values are calculated using the poisson linear interpolation described in equation 12. Once E-M is complete on the treatment data and peaks are called, we reset the treatment alignment probabilities, swap treatment and control and rerun the algorithm, including E-M steps, to determine the False Discovery Rate (FDR). For all algorithms tested in this study, we define the FDR as the ratio of peaks called using control data to peaks called using treatment data. This method of FDR calculation is common in ChIP-Seq studies (e.g., [13,15]).

Motif Finding

Motif presence helps determine peak quality, as shown in [34]. To determine if our new peaks were of the same quality as the other peaks, we performed *de novo* motif discovery using MEME [35] version 4.4. Input sequence was limited to 150 bp (Rad21) and 200 bp (Srebp-1) around the summit of the peaks called by MACS from uniquely mapping reads. All sequences were used for Srebp-1, while 1,000 sequences were randomly sampled a total of 5 times for Rad21. The motif signal was strong in both datasets and we extracted the discovered motif position weight matrix (PWM) for further use. We also performed the motif search using Srebp-1 and CTCF motifs catalogued in Transfac 11.3, and found similar results. For the CTCF motif, we did genomic sampling (100,000 samples) to identify a threshold score corresponding to a z-score of 4.29. For Srebp-1, we used the threshold score reported by MEME (see Figure 1).

Blocked Pattern Matching Problem and Its Applications in Proteomics

Julio Ng[1], Amihood Amir[2,3], and Pavel A. Pevzner[4]

[1] Bioinformatics and Systems Biology Program, University of California San Diego
`jung@ucsd.edu`
[2] Department of Computer Science, Bar-Ilan University
[3] Department of Computer Science, Johns Hopkins University
`amir@cs.biu.ac.il`
[4] Department of Computer Science and Eng., University of California San Diego
`ppevzner@cs.ucsd.edu`

Abstract. Matching a *mass spectrum* against a text (a key computational task in proteomics) is slow since the existing text indexing algorithms (with search time independent of the text size) are not applicable in the domain of mass spectrometry. As a result, many important applications (e.g., searches for mutated peptides) are prohibitively time-consuming and even the standard search for non-mutated peptides is becoming too slow with recent advances in high-throughput genomics and proteomics technologies. We introduce a new paradigm – the Blocked Pattern Matching (BPM) Problem - that models peptide identification. BPM corresponds to matching a pattern against a text (over the alphabet of integers) under the assumption that each symbol a in the pattern can match a block of consecutive symbols in the text with total sum a.

BPM opens a new, still unexplored, direction in combinatorial pattern matching and leads to the Mutated BPM (modeling identification of mutated peptides) and Fused BPM (modeling identification of fused peptides in tumor genomes). We illustrate how BPM algorithms solve problems that are beyond the reach of existing proteomics tools.

1 Introduction

Matching a tandem mass *spectrum* (MS/MS) to a database is very slow as compared to matching a *pattern* to a database. The fundamental algorithmic advantage of the latter approach is that one can *index* the database (e.g., by constructing its suffix tree [24]) so that the complexity of the subsequent queries is not dependent on the database size. Since *efficient* indexing algorithms remain unknown in proteomics[1], many important applications, for example, database

---

[1] By efficient indexing we mean indexing that typically reduces spectral matching to a single look-up in the indexed database rather than a large number of look-ups (proportional to the database size). While there is no shortage of useful indexing approaches in proteomics (e.g., fast parent mass indexing like in [36] or peptide sequence tags like in [38]), such approaches may result in a large number of look-ups.

V. Bafna and S.C. Sahinalp (Eds.): RECOMB 2011, LNBI 6577, pp. 298–319, 2011.

searches for mutated peptides, remain extremely time-consuming. Moreover, even the standard applications, such as searching for non-modified peptides, are becoming prohibitively slow with recent advances in genomics and proteomics. On one hand, the total size of known sequenced bacterial proteomes (most of them sequenced using next generation DNA sequencing technologies) already amounts to billions of amino acids. On the other hand, Ion Mobility Separation (next generation proteomics technology) promises to increase the rate of spectra acquisition by two orders of magnitude.

Tandem mass spectrometry analyzes *peptides* (short 8-30 amino acid long fragments of proteins) by generating their *spectra*[2]. The still unsolved problem in computational proteomics is to reconstruct a peptide from its spectrum: even the advanced *de novo* peptide sequencing tools correctly reconstruct only 30 - 45% of the full-length peptides identified in MS/MS database searches [20]. After two decades of algorithmic developments, it seems that *de novo* peptide sequencing "hits a wall" and that accurate full-length peptide reconstruction is nearly impossible due to the limited information content of MS/MS spectra.

Recently, with the introduction of MS-GappedDictionary, Jeong et al., 2010 [26] advocated the use of *gapped peptides* to overcome the limitations of full-length *de novo* sequencing algorithms. Given a string of n integers $a_1, a_2, \ldots, a_n$ (a peptide) and k integers $1 \leq i_1 < \ldots < i_k < n$, a *gapped peptide* is a string of $(k + 1)$ integers $a_1 + \ldots + a_{i_1}, a_{i_1+1} + \ldots + a_{i_2}, \ldots, a_{i_k+1} + \ldots + a_n$. For example, if a peptide LNRVSQGK is represented as a sequence of its rounded amino acid masses 113, 114, 156, 99, 87, 128, 57, 128 then 113+114, 156+99+87,128+57, 128 represents a gapped peptide 227, 342, 185, 128. MS-GappedDictionary is a *database filtration* approach based on gapped peptides that are both long and accurate. Gapped peptides have higher accuracy and orders of magnitude higher filtering efficiency than traditional peptide sequence tags[3]. In contrast to a short peptide sequence tag, a gapped peptide typically has a single match in a proteome, reducing peptide identification to a single database look-up. MS-GappedDictionary generates 25-50 gapped peptides per spectrum (*Pocket Dictionary*) and guarantees that one of them is correct with high probability.

MS-GappedDictionary has a potential to be much faster than traditional proteomics tools because it matches *patterns* rather than *spectra* against a protein database, but algorithms to efficiently match such patterns to a database remain unknown. Therefore, this paper addresses the last missing piece in the series of recent developments aimed at the next generation of peptide identification algorithms [30,29,28,26], an efficient algorithm for matching gapped peptides against a proteome.

---

[2] Spectra are complex objects that, for the sake of brevity, are not formally described in this paper.

[3] A *peptide sequence tag* is a substring $a_i, \ldots, a_j$ of a peptide. While peptide sequence tags are used in many proteomics tools [38,37], the algorithms for generating *long* sequence tags remain inaccurate. As a result, in practice, applications of peptide sequence tags are limited to 3 amino acid long tags.

We describe two closely related MS/MS database search problems using gapped peptides. First, we describe algorithms to efficiently match a single gapped peptide to a database, and second, we present algorithms to solve the more general problem of matching multiple gapped peptides to a proteome. We present an indexing algorithm that is very relevant to approximate matching research because the indexing problem for gapped peptides falls under the category of a new recently proposed pattern matching paradigm – *pattern matching with address errors* [2,4,3,5,7,6,27]. In this model, the pattern *content* remains intact, but the relative positions (addresses) may change. Our work introduces a new direction in rearrangement matching by considering problems where "mass" (rather than the order of symbols) of the substring is important. Additionally, we present a novel pattern matching algorithm for the multiple gapped peptide matching problem that leads to a new peptide identification tool with search time that is nearly independent of the database size in practice. The resulting peptide identification software tool is so fast that its running time is dominated by spectral preprocessing rather than scanning the database[4]. In practice, it results in a peptide identification tool that is orders of magnitude faster than the state-of-the-art proteomics tools (four orders of magnitude speed-up over Sequest [19] and two orders of magnitude speed-up over InsPecT [38] in scanning the protein database).

In addition to exact matching, we describe *approximate matching* of gapped peptides that enables identification of mutated and fused peptides (characteristic for *fusion proteins* in cancer [18]). Such mutation tolerant framework can be adapted to detect annotation errors, programmed frame shifts, fusion proteins, and other features that remain beyond the reach of existing proteomics algorithms.

Due to the page limit, we describe only one of possible applications of the Blocked Pattern Matching paradigm to peptide identification. While traditional MS/MS searches assume that a proteome is known, *proteogenomics* searches use spectra to correct the proteome annotations [25,22,32,12]. The previous proteogenomics approaches searched spectra against the 6-frame translation of the genome in the *standard* genetic code. However, many species use non-standard genetic code [31,1] that is difficult to establish for a newly sequenced species. In particular, in addition to the standard ATG Start Codon, GTG and TTG also code for initial Methionine (rather than for Valine and Leucine as in the standard genetic code) in many bacterial genomes. The frequency of non-standard Start Codons varies widely: in *E. Coli*, GTG and TTG account for 14% and 3% of Start Codons (not to mention extremely rare ATG and CTG Start Codons), while in *Aeropyrum pernix* GTG and TTG are more common than ATG [35]. After a new bacterium is sequenced, the propensities of its Starts Codons are unknown making accurate gene predictions problematic [10]. Non-standard Start Codons GTG and TTG (or whatever other) can be discovered by finding mutated

---

[4] The running time of the existing proteomics algorithms [19,38] can be partitioned into *spectral preprocessing* time (approximated as $\alpha \cdot \#Spectra$) and database *scanning time* (approximated as $\beta \cdot \#Spectra \cdot ProteomeSize$).

peptides (GTG and TTG correspond to Valine and Leucine mutated to Methionine in the first peptide position). We illustrate applications of this approach to gene annotations in *Anthrobacter* sp.

2   The Blocked Pattern Matching Problem

Let $T = T[1], T[2], ..., T[n]$ be a *text* over a finite alphabet $\Sigma \subset \mathbb{N}$ and $P = P[1], P[2], ..., P[m]$ be a *pattern* over an alphabet $\mathbb{N}$ of all natural numbers. Let $S = T[i], T[i+1], ..., T[j]$ be a substring of T. Then $\overline{S}$ (the *mass* of substring S) $= \sum_{\ell=i}^{j} T[\ell]$.

Substrings $T[i], ..., T[j]$ and $T[j+1], ..., T[k]$ are called *consecutive*. A *block* in T is a sequence of consecutive substrings. The *mass of a block* B (denoted $\overline{B}$) is a string comprised of the masses of the consecutive substrings of B. Formally, if $B = S_1 \cdots S_k$ then $\overline{B} = \overline{S_1}, ..., \overline{S_k}$. We say that a pattern P *matches* a text T if there is a block B in T with $\overline{B} = P$.

Example. Let $T = 114, 77, 57, 112, 113, 186, 57, 99, 112, 112, 186, 113, 99$ be a text over an alphabet of 18 symbols that represents masses of 20 amino acids rounded to integers. The consecutive substrings $(57, 112, 113)$, $(186, 57)$, and $(99, 112, 112)$ define a block B in T with $\overline{B} = 282, 243, 323$. Thus, a pattern 282, 243, 323 matches the text T.

Definition 1. *The* Blocked Pattern Matching (BPM) Problem *is defined as follows:*

Input: A length-n text T over Σ and a length-m pattern P over $\mathbb{N}$.
Output: All blocks B in the text T such that $\overline{B} = P$.

Since BPM is a new algorithmic problem, we start from its theoretical analysis and describe an $O(nm)$ and $O(n \log n)$ BPM algorithms (Section 3). We further consider the *indexing* version of BPM, where the text preprocessed enable fast subsequent queries (Section 3.2). The indexing scheme handles a single-element pattern, but is a base for an efficient practical filter. We present an algorithm with a linear-time preprocessing and a query time, of $O(m \cdot (\sqrt{n \cdot tocc} + tocc))$ (*tocc* is the number of substrings matching the mass of the query in the text). The BPM problem is related to the pattern matching problem for spatial point sets in 1-dimension described in [11], but we are interested in an exact solution to the problem. The solution in [11] presents an algorithm that returns false positives. Additionally, a simpler version of the BPM problem is described in [21] for patterns of length 1. We generalize the problem for patterns of any length.

Modern mass spectrometers are capable of producing a million spectra per day and each spectrum corresponds to 25-50 gapped peptides in its Pocket Dictionary. While we describe efficient theoretical BPM algorithms, even these fast algorithms become too slow in practice when applied to billions of gapped peptides. In Section 4 we describe a modification of the classical *keyword tree* concept that leads to a practical (albeit memory demanding) BPM algorithm. In Section 4.2 we describe how constructing the keyword tree of patterns (rather than indexing the text) leads to a practical algorithm for solving the following problem:

Definition 2. *The* Multiple Blocked Pattern Matching (MBPM) Problem *is defined as follows:*

Input: A text T over Σ and a set $\mathcal{P}$ of patterns over $\mathbb{N}$.
Output: All blocks B in the text T such that $\overline{B} = P$, for some $P \in \mathcal{P}$.

3 BPM Algorithms

3.1 BPM Algorithms without Text Indexing

Finding all occurrences of a pattern of length 1 can be done in time $O(n)$ with a simple L1BPM algorithm (Fig. 1). Given a pattern $P[1], ..., P[m]$, separately find all occurrences of $P[1], P[1] + P[2], P[1] + P[2] + P[3], ..., \sum_{i=1}^{m} P[i]$. Every text location where all occurrences start is an occurrence of a pattern P. This results in a simple $O(nm)$ solution to the BPM problem that serves as a key to our indexing idea in Section 3.2. Below, we describe a faster BPM algorithm below.

ALGORITHM **Length-1 Blocked Pattern Matching (L1BPM)**
Input: A text T of length n, a pattern $P = P[1]$.
1   construct a new text $S = S[1], ..., S[n]$ where $S[i] = \sum_{\ell=1}^{i} T[\ell]$
2   initialize two pointers ℓ and r to location 1 on S
3   while $r \leq n$ do:
4   if $S[r] - S[\ell] = P[1]$ then:
5   there is a match at location ℓ
6   increment ℓ by 1
7   if $S[r] - S[\ell] < P[1]$ then increment r by 1
8   if $S[r] - S[\ell] > P[1]$ then increment ℓ by 1
9   endwhile

Fig. 1. The linear time BPM algorithm with a length-1 pattern

Given a natural number i, we define i^* as a boolean string of length i starting with 1 in the first position and ending with $i-1$ zeros (e.g., $5^* = 10000$). Given a string $T = T[1], T[2], ..., T[n]$, we define a boolean string B_T as the concatenation of strings $T[1]^*, T[2]^*, ...$ and $T[n]^*$.

We say that a binary array $B = B[1], ..., B[M]$ *point matches* a binary array $A = A[1], ..., A[N]$ at *location* i, if $A[i + \ell - 1] = 1$ for all ℓ such that $B[\ell] = 1$.

Example: Let $A = 001001000010101000100$ and $B = 10001$. B point matches in locations 11 and 15 of A because those are the only two locations where *every* 1 in B is aligned with a 1 in A.

Clearly, a pattern P matches a text T at location i iff there is a point matching of B_P in B_T at location $(\sum_{\ell=1}^{i-1} T[\ell]) + 1$. The following lemma shows that point matching can be computed by a *convolution* defined as $(T \otimes P)[j] = \sum_{i=1}^{m} T[j + i - 1] \cdot P[i]$, $j = 1, ..., n - m + 1$.

Lemma 1. *Let T and P be binary arrays of length N and M, respectively. The point matching of P in T can be computed in time $O(N \log M)$.*

Proof. Let $-T$ be a binary array $-T[1], ..., -T[N]$, where $-x$ stands for $1 - x$. The point matching locations between P and T are the locations j where $T[j + i - 1] \geq P[i]$, $i = 1, ..., m$. For every text location j, $(P \otimes -T)[j] = 0$ iff for every 1 in the pattern, there is also a 1 in the text matched to it, i.e. there is a point matching of P at location j of T. The convolution $T \otimes P$ can be computed by Fast Fourier Transform [15] in time $O(n \log m)$. $\square$

Below we consider patterns consisting of integers bounded by d (*d-bounded patterns*). The lemma above implies that we can solve the BPM problem for d-bounded patterns in time $O(n \log n)$ where n is the length of the text[5].

3.2   BPM Indexing for Length-1 Patterns

We would like to solve the BPM indexing problem with a fast preprocessing and with query time that does not depend on the text length. We start with the simpler problem of text indexing for length-1 d-bounded patterns. Extension to arbitrary mass values is described in the Appendix.

Split the text T into consecutive substrings $T_1^1, ..., T_{\lceil \frac{n}{2 \cdot d} \rceil}^1$ of size $2 \cdot d$ elements each except, possibly, the last one. Let $S_i^1 = S_i^1[1], ..., S_i^1[2 \cdot d]$ such that $S_i^1[j] = \sum_{\ell=1}^{j} T_i^1[\ell]$. Similarly, split T, starting from index d, into consecutive substrings $T_1^2, ..., T_{\lceil \frac{n}{2d} - \frac{1}{2} \rceil}^2$ of size $2 \cdot d$ elements each except, possibly, the last one, and construct $S_i^2 = S_i^2[1], ..., S_i^2[2 \cdot d]$ such that $S_i^2[j] = \sum_{\ell=1}^{j} T_i^2[\ell]$.

Example: Let $\Sigma = \{1, 2, 3\}$, $d = 3$, and $T = 1, 1, 2, 1, 3, 3, 2, 1, 3, 2, 3, 1, 1$. Then $T_1^1 = 1, 1, 2, 1, 3, 3$, $T_2^1 = 2, 1, 3, 2, 3, 1$, $T_3^1 = 1$ and $S_1^1 = 1, 2, 4, 5, 8, 11$, $S_2^1 = 2, 3, 6, 8, 11, 12$, $S_3^1 = 1$
$T_1^2 = 1, 3, 3, 2, 1, 3$, $T_2^2 = 2, 3, 1, 1$ and $S_1^2 = 1, 4, 7, 9, 10, 13$, $S_2^2 = 2, 5, 6, 7$

For $\lambda = 1, 2$, construct data structures I^λ that will enable us to quickly find, for a given mass ℓ, all T_i^λ whose prefixes have mass ℓ. For example, in the example above, since 8 appears in 5th position in S_1^1 and 4th position in S_2^1, I^1 will have a record $[8, < 1, 5 >, < 2, 4 >]$. Applying it to all masses, we arrive to the data structures:

$I^1 = [1, < 1, 1 >, < 3, 1 >], [2, < 1, 2 >, < 2, 1 >], [3, < 2, 2 >], [4, < 1, 3 >],$
$[5, < 1, 4 >], [6, < 2, 3 >], [8, < 1, 5 >, < 2, 4 >], [11, < 1, 6 >], [12, < 2, 6 >]$

---

[5] In the Appendix we describe an efficient algorithm for the infinite alphabet.

$$I^2 = [1, < 1, 1 >], [2, < 2, 1 >], [4, < 1, 2 >], [5, < 2, 2 >], [6, < 2, 3 >],$$
$$[7, < 1, 3 >, < 2, 4 >], \qquad [9, < 1, 4 >], [10, < 1, 5 >], [13, < 1, 6 >]$$

Our indexing algorithm below makes use of the *Set Intersection Problem*.

Definition 3. *Let U be a finite set, and let $S_1, ..., S_k$ be subsets of U. The* Set Intersection Problem *is that of constructing $S_i \cap S_j$, for a given pair, i, j.*

While computing the set intersection is difficult [17,16,9], Cohen and Porat [13] provide an algorithm that preprocesses the data in time $O(N)$, where $N = \sum_{i=1}^{k} |S_i|$, and, subsequently provides the intersection of sets S_i and S_j in time $O(\sqrt{N \cdot int} + int)$, where $int = |S_i \cap S_j|$.

Theorem 1. *Let T be a text of length n over a finite alphabet $\Sigma \subset \mathbb{N}$. Then it is possible to preprocess T in time $O(n)$ such that subsequent BPM queries for d-bounded patterns $P = P[1]$, can be answered in time $O(d \cdot f_{int}(n) + tocc)$, where tocc is the number of blocks in T with mass P and $f_{int}(\cdot)$ is the complexity of a* Set Intersection *algorithm.*

Proof. We construct data structures I^1 and I^2 as described above. It is easy to see that (i) the size of I^1 (I^2) is $O(n)$, (ii) the number of entries in I^1 (I^2) is $O(d)$, and (iii) $I^1(I^2)$ can be constructed in time $O(n)$. The advantage of the I^1, I^2 data structures is that finding a block with mass $ms \leq d$ in the text T (with n entries) is equivalent to finding it in the much smaller I^1, I^2 (with $O(d)$ entries).

The segmentation of T to the T^1 and T^2 substrings implies that for a block of mass ms starting at location $x = a \cdot d + i$ of T, there exists a block of mass ms starting at location i either in $T^1_{\frac{a}{2}+1}$ (for even a) or $T^2_{\frac{a+1}{2}}$ (for odd a). This observation enables us to run Algorithm **L1BPM** where, I^1 and I^2 play the role of array S (see Fig 1) and identify occurrences of a block of mass $ms \leq d$. However, now the length of this array is $O(d)$. The only problem is that I^1 (or I^2) are actually aggregates of all segments S_i^1 (or S_i^2) and thus we may declare a block when it really does not exist, since it "started" in one segment but "ended" in another.

This problem may be solved by ensuring that we indeed have a block that began and ended in the same segment. Line 4 in Algorithm **L1BPM**, needs to be modified so that not only $I[r] - I[\ell] = P[1]$ but also $I[r]$ and $I[\ell]$ have common segments in their lists. More precisely, the *intersection* of the set of segments in $I[r]$ and the set of segments in $I[\ell]$, with their appropriate indices, gives precisely the set of locations with blocks of the required mass.   □

Translated to our problem and combined with the Cohen and Porat [13] algorithm, we get[6]:

Theorem 2. *Let T be a text of length n over a finite alphabet $\Sigma \subset \mathbb{N}$. Then it is possible to preprocess T in time $O(n)$ such that subsequent BPM queries for*

---

[6] In Appendix, we present a scheme with the query time $O(P[1]^2)$ and the preprocessing time $O(n^{1.5})$.

patterns $P = P[1]$, can be answered in time $O(P[1] \cdot (\sqrt{n} \cdot tocc + tocc))$, where tocc is the number of blocks in T with mass P.

4   MBPM Algorithms

4.1   Transforming the Text into a Set of Patterns

Unfortunately, generalizing the previous scheme beyond length-1 patterns requires efficient set-intersection of more than two sets. However, from a practical point of view, it is possible to use our previous construction to index locations where all of $P[1], P[1] + P[2], ..., \sum_{\ell=1}^{m} P[\ell]$ match. While this results in a useful theoretical filter, below we describe text indexing for patterns of arbitrary length that further reduces query time at the expense of significantly increasing the memory requirements.

Since typical spectral searches identify peptides shorter than 30 amino acids, one can limit attention to k-mers in the proteome with $k \le 30$. From the perspective of the BPM, we are only interested in patterns that match no more than k symbols in the text. There exist 2^{k-1} ways to break a k-mer into its substrings $b_1, \ldots, b_n$ resulting in 2^{k-1} possible partitions $\overline{b_1}, \ldots, \overline{b_n}$ of each k-mer. For example, there exist 8 partitions arising from the 4-mer (114, 77, 99, 57): (114, 77, 99, 57), (114+77, 99, 57), (114, 77+99, 57), (114, 77, 99+57), (114+77,99+57) (114+77+99, 57), (114, 77+99+57), and (114+77+99+57).

Given a set of strings T, we define $KeywordTree(T)$ as the keyword tree of these strings [24]. One can generate all 2^{k-1} partitions for each k-mer in a text T, construct the keyword tree of these partitions, and match each pattern against the constructed keyword tree. While this approach is fast, it suffers from excessive memory requirements. Below we describe a more memory-efficient solution of the BPM Problem for d-bounded patterns. In practice, gaps in the gapped peptides typically do not exceed 500 Da, moreover MS-GappedDictionary can be run in such a way that it discriminates against large gaps [26]. While the number of d-bounded partitions that can be generated from a k-mer is large, below we describe a BPM algorithm that does not require generation of all d-bounded partitions.

Given a position i in a d-bounded pattern $P = p_1, \ldots, p_{i-1}, p_i, p_{i+1}, \ldots, p_n$ and a parameter $1 \le \delta \le n - i + 1$, (i, δ)-extension of P is a pattern $p_1, \ldots, p_{i-1}, p_i + p_{i+1} + \ldots + p_{i+\delta-1}, p_{i+\delta}, \ldots, p_n$. obtained from P by substituting δ symbols in P (starting from the i-th symbol) by their sum. Given a d-bounded pattern P, we define $P(i, d)$ as the set of all (i, δ)-extensions of P that result in d-bounded patterns. For example, for a pattern $P = (114, 77, 99, 55, 112), P(2, 300)$ consists of patterns $(114, 77, 99, 55, 112), (114, 176, 55, 112)$ and $(114, 231, 112)$.

A pattern P in the set of patterns T is called i-unique if no other pattern in T has the same prefix of length i (i-prefix) as P. We now describe a more memory efficient BPM (and MBPM) text indexing algorithm (Fig 2). Let T_0 be the set of all k-mers in text T, where each k-mer appears only once. We iteratively construct the set T_i from the set T_{i-1} by considering all non-i-unique patterns in T_{i-1} and substituting each such pattern P by the set of patterns $P(i, d)$. T_i is the resulting set of patterns with duplicates removed (i.e., each pattern appears only once). The MBPM algorithm iteratively generates the sets $\mathcal{T}_1, \ldots, \mathcal{T}_i = \mathcal{T}(T, d)$ and stops when all patterns in the set $\mathcal{T}_i$ become i-unique. We further construct $KeywordTree(\mathcal{T}_i)$ (denoted $KeywordTree(T, d)$) and classify each vertex in this tree as unique or non-unique according to the following rule. Let $q_1, \ldots, q_i$ be a pattern spelled by the path from the root to the vertex v in $KeywordTree(T, d)$. The vertex v is *unique* if the algorithm MBPM$(T, \mathcal{P}, d)$ classified $q_1, \ldots, q_i$ as a prefix of an i-unique pattern at some iteration, and *non-unique* otherwise.

ALGORITHM **Multiple Blocked Pattern Matching (MBPM)**
Input: A text T of length n, a set of d-bounded patterns $\mathcal{P}$, and a parameter k (k-mer size).
1   construct $\mathcal{T} \leftarrow$ set of all k-mers in T
2   initialize $i \leftarrow 1$
3   while there exist non-i-unique patterns in $\mathcal{T}$ do:
4   remove duplicates from $\mathcal{T}$
5   for each non-i-unique pattern $P \in \mathcal{T}$ do:
6   substitute P by $P(i, d)$ in $\mathcal{T}$
7   endfor
8   increment i by 1
9   endwhile
10 construct $KeywordTree(\mathcal{T})$
11 for each pattern $P \in \mathcal{P}$ do:
12   PartitionMatch$(P, KeywordTree(\mathcal{T}))$
13 endfor

Fig. 2. MBPM algorithm for matching a set of d-bounded patterns $\mathcal{P}$ against the text T. The PartitionMatch function works as follows. Let $p_1, \ldots, p_{i-1}$ be the longest prefix of $p_1, \ldots, p_n$ that matches the tree and let v be the last vertex of the path labeled by $p_1, \ldots, p_{i-1}$ in the tree (i.e., no outgoing edge from v is labeled by p_i). If v is a non-unique vertex, we declare that the pattern $p_1, \ldots, p_n$ does not match the the text. Otherwise, we attempt to match the suffix $p_i, \ldots, p_n$ against the path in the keyword tree that start at vertex v. Such matching simply amounts to checking whether the pattern $p_i, \ldots, p_n$ represents a partition of the string spelled by these path. If it is the case, the pattern $p_1, \ldots, p_n$ matches the tree, otherwise there is no match.

To solve the MBPM Problem, we match each pattern $p_1, \ldots, p_n$ against $KeywordTree(T, d)$. In standard searches with the keyword trees, a pattern $p_1, \ldots, p_n$ matches a tree if there exists a path in the keyword tree that spells $p_1, \ldots, p_n$, otherwise the pattern does not match the tree [24]. In contrast, for our application (with special processing of i-unique patterns), failure to find a path that spells $p_1, \ldots, p_n$ does not necessarily implies that the pattern $p_1, \ldots, p_n$ does not match the tree (see PartitionMatch function described in Fig.2).

For a "random" proteome T with 100,000 amino acids it takes only 4 iterations to stabilize $\mathcal{T}(T, 500)$ at ≈ 5 million patterns. However, the memory requirements significantly increase for real proteomes that typically contain repeats (Figure 3 in the Appendix). The Appendix describes practical MBPM solutions that further trade speed for memory.

4.2   Matching the Keyword Tree of Patterns against the Text

In the extreme case of trading memory for speed, one can construct the keyword tree of all patterns from the set $\mathcal{P}$ (rather than of all k-mers from the text T) while solving the MBPM Problem (this amortizes the time for scanning the text). One can further match all k-mers from the text against $KeywordTree(\mathcal{P})$. A pattern P (that does not belong to a set of patterns $\mathcal{P}$) is called $(i, \mathcal{P})$-unique with respect to the set $\mathcal{P}$ if its i-prefix matches at most one pattern in $\mathcal{P}$.

Matching $KeywordTree(\mathcal{P})$ against a text T amounts to constructing $KeywordTree(T, d, \mathcal{P})$ that we describe below. We use an analog of the MBPM algorithm to transform a text into a set of patterns and observe that there is no need to apply this transformation to a non-i-unique pattern that is $(i, \mathcal{P})$-unique. Thus, the only difference in constructing $KeywordTree(T, d, \mathcal{P})$ (as compared to $KeywordTree(T, d)$) is that it substitutes the notion of i-unique patterns by the notion of $(i, \mathcal{P})$-unique patterns in the **for** loop of the MBPM algorithm (Fig. 2).

The algorithm iteratively generates the sets $\mathcal{T}_1, \ldots, \mathcal{T}_i = \mathcal{T}(T, d, \mathcal{P})$ until all patterns in the set $\mathcal{T}_i$ become $(i, \mathcal{P})$-unique, constructs the keyword tree of $\mathcal{T}(T, d, \mathcal{P})$ (denoted $KeywordTree(T, d, \mathcal{P})$), and matches $KeywordTree(\mathcal{P})$ against $KeywordTree(T, d, \mathcal{P})$ [24] using the matching algorithm that is similar to the PartitionMatch algorithm described in the previous section. While $KeywordTree(T, d, \mathcal{P})$ can be large in practice, one can partition the text T into smaller segments so that their keyword trees fit into memory and solving the MBPM Problem for each of the resulting trees. In the extreme case, one can partition T into single k-mers resulting in a memory-efficient yet fast implementation (Fig. 4 in the Appendix).

4.3   Mutation-Tolerant Peptide Identification

Let $T_i(a)$ be a text obtained from a text $T = t_1, \ldots t_n$ by substituting a symbol a instead of the i-th symbol of T (for $1 \leq i \leq n$ and $a \in \mathcal{A}$). For example, if $T = 57, 112, 113, 113, 186$, $T_3(99) = 57, 112, 99, 113, 186$. To accommodate for insertions and deletions, we denote $T_i(\emptyset)$ as the deletion of the i-th symbol of T and $T_i(a^+)$ as the insertion of a symbol a before the i-th symbol of T. For example, $T_3(\emptyset) = 57, 112, 113, 186$ and $T_3(128^+) = 57, 112, 128, 113, 113, 186$. A *mutated block* in the text T is a block in $T_i(a)$ (for some i and a) that is not a block in T. For example, substrings $(112, 128)$ and $(113, 113, 186)$ form a mutated block in the text $T = 57, 112, 113, 113, 186$ because they form a block in $T_3(128^+)$. Let $\mathbb{N}_d$ be a set of natural numbers smaller than d and $\Sigma \subset \mathbb{N}_d$.

Definition 4. *The* Mutated Blocked Pattern Matching Problem *is defined as follows:*
Input: *A text T over the alphabet Σ and a pattern P over the alphabet $\mathbb{N}_d$.*
Output: *All mutated blocks B in the text T matching the pattern P.*

While the Mutated Blocked Pattern Matching Problem is limited to peptide identified with a *single* mutations, this is a reasonable limitation in proteomics. Indeed, Single Amino Acid Polymorphisms (SAAPs) are rarely clustered in the same region of proteins. Also, the false discovery rate in searches for peptides with two or more mutations becomes very high due to the Bonferroni correction to reflect the huge size of the (virtual) database of all mutated peptides [23].

Given a text $T = t_1, \ldots t_n$, we define a text $T_{i,j} = t_1, \ldots t_{i-1}, t_i, t_j, t_{j+1}, \ldots, t_n$. A *fused block* is a block in $T_{i,j}$ (for $1 \leq i, j \leq n$) that is not a block in T. E.g., substrings (57,112) and (113,186) form a fused block in the text $T = 57, 112, 113, 113, 186$ because they form a block in $T_{2,4}$.

Definition 5. *The* Fused Blocked Pattern Matching Problem *is defined as follows:*
Input: *A text T over the alphabet Σ and a pattern P over the alphabet $\mathbb{N}_d$.*
Output: *All fused blocks B in the text T matching the pattern P.*

Below we describe a simple approach to solving the Mutated and Fused Pattern Matching Problems that is based on the observation that for every pattern $p_1, \ldots, p_n$ matching a mutated/fused block in the text, either its prefix $p_1, \ldots, p_{n/2}$, or its suffix $p_{n/2+1}, \ldots, p_n$ of length $n/2$ matches a (non-mutated) block in the text. Therefore, in the case of the Mutated BPM Problem with a pattern $p_1, \ldots, p_n$, one can first solve the BPM problems for the prefix $p_1, \ldots, p_{n/2}$ and the suffix $p_{n/2+1}, \ldots, p_n$ and further check if some of the found matches can be locally extended into a match of the full pattern $p_1, \ldots, p_n$. The *local extension* can identify both mutated and post-translationally modified peptides as described in [38]. In the case of the Fused BPM Problem, the algorithm is a bit more involved since the found matches of prefixes and suffixes should be combined rather than locally extended as in [38]. In this case, we find all blocks $B_{prefix}(i)$ matching the prefix $p_1, \ldots, p_{i-1}$ and all blocks $B_{suffix}(i)$ matching the suffix $p_{i+1}, \ldots, p_n$, correspondingly (for all $1 < i < n$). Afterwards, we attempt to fuse pairs of blocks from $B_{prefix}(i)$ and $B_{suffix}(i)$ to match the full length pattern $p_1, \ldots, p_n$. In practice, the identification of fused peptides is typically limited to long peptides when both $p_1, \ldots, p_{i-1}$ and $p_{i+1}, \ldots, p_n$ are rather long [34] and thus the sets $B_{prefix}(i)$ and $B_{suffix}(i)$ are expected to be small making the algorithm practical.

4.4   Results

Benchmarking. We implemented various algorithms described in the main text and the Appendix and tested them on various spectral dataset. Due to page limit, below we only discuss the performance of the algorithm described in Section 4.2 (that we refer to as MS-BPM) on a single *Shewanella oneidensis* spectral dataset extensively studied in [22]. While other algorithms we described outperform MS-BPM on small proteomes and small spectral datasets, MS-BPM showed the best performance in searches against giant databases like the 6-frame translation of the human genome.

To evaluate the speed of various peptide identification tools, we measure the time to match a million spectra against a proteome consisting of a million amino acids (measured in *seconds per mil^2*) on a desktop machine (Intel Core i7-965, 3.20 Ghz with 24GB of RAM). For example, Sequest [19] (a popular peptide identification tool) takes $\approx 10^6$ seconds per mil^2, while InsPecT [38] (currently the fastest peptide identification algorithm) takes $\approx 1.7 \times 10^4$ seconds per $mil^{2,7}$. The running time of these algorithms can be partitioned into *spectral preprocessing* time (approximated as $\alpha \cdot \#Spectra$) and database *scanning time* (approximated as $\beta \cdot \#Spectra \cdot ProteomeSize$). Spectral preprocessing may include transforming raw spectra into its log-likelihood representation, parent mass correction, spectral calibration, etc. [29]. In MS-BPM, spectral preprocessing amounts to generating the Pocket Dictionary (using MS-GappedDictionary [26]) and takes ≈ 0.2 seconds per spectrum. Since the spectral preprocessing time is usually negligible as compared to scanning time (at least for large databases), it is usually ignored. However, MS-BPM turned out to be so fast (scanning time is 100 seconds per mil^2) that spectral preprocessing time is actually larger than the scanning time[8].

Identification of spectra of mutated peptides is a non-trivial algorithmic problem: such searches remain much slower than searches for non-mutated peptides[9]. Below we show that MS-BPM speeds up searches for mutated peptides by two orders of magnitude (as compared to MS-Alignment [39]) making them as fast as InsPecT searches for non-mutated peptides.

We performed a controlled experiment where we selected 5000 identified spectra from the *Shewanella oneidensis* spectral dataset corresponding to 5000 distinct non-overlapping peptides. For each identified peptide in the *Shewanella* proteome, we selected one position at random and randomly mutated the amino acid in this position into one of other 19 amino acids. MS-BPM (ran in the mode that solves the Mutated BPM Problem) took 246 seconds to match all gapped peptides generated for these spectra against the mutated proteome. This results in 3.8×10^4 seconds per mil^2 speed that is similar to InsPecT's speed in (much simpler) searches for non-mutated peptides. We therefore argue that MS-BPM has a potential to make searches for mutated and fused peptides as

---

[7] These estimates are derived from a proteogenomics search of the 6-frame translation of human genome (2.5 billion amino acids) as described in [29].

[8] We used MS-GappedDictionary [26] to generate gapped peptide reconstructions for 300000 mass spectra from *Shewanella oneidensis* (analyzed in [22]). The *Shewanella oneidensis* proteome is ≈ 1.3 million amino acids long. MS-GappedDictionary generated over 12×10^6 500-bounded gapped peptides (≈ 40 gapped peptides per spectrum). Constructing the keyword tree of all (12×10^6) gapped peptides took only 150 seconds. Searching 12 million gapped peptides against the *Shewanella* proteome took 40 seconds resulting in less than 100 seconds per mil^2 speed.

[9] MS-Alignment [39] tool for identification of mutated peptides, is two orders of magnitude slower than InsPecT (for non-mutated peptides). Since it takes on the order of 100 days (a single CPU time) to search 1 million spectra against the human proteome with MS-Alignment, such searches are rather time consuming even with large computing clusters.

routine as searches for non-mutated peptides are today thus opening possibilities for previously infeasible biological inquires.

Gene Annotations in *Arthrobacter*. The biological findings resulting from our analysis of various spectral datasets for mutated and fused peptides will be described elsewhere. Here, due to page limit, we only analyze a dataset of 221673 spectra from *Arthrobacter* sp. strain FB24 generated in Dick Smith's lab at PNNL using a LTQ-FT tandem mass spectrometer. 71760 non-modified spectra were identified in search against the 6-frame translation of the *Arthrobacter* genome of ($\approx$ 10.1 Mb amino acids) with a spectral probability threshold of 10^{-10} [30]. We analyzed the remaining 149913 spectra using the Mutated BPM algorithm. We transformed spectra into gapped peptides, searched them against the six-frame translation of *Arthrobacter* genome, and used an extremely selective spectral probability threshold 10^{-13} to report mutated peptides. This resulted in a small set of identified spectra that passed a stringent statistical significance threshold[10].

To identify alternative start codons, we collected all identified peptides with a mutation in the 1st position. Out of 123 identified mutated peptides, the most prevalent mutations can be alternatively explained as precursor mass errors (1 or 2 Da offsets), acetylation (42 Da offset) and oxidation (16 Da offset). For example, a Glu to Gln mutation ($\approx$ 1 Da offset) can be explained by a precursor mass error, while a Ser to Glu mutation ($\approx$ 42 Da offset) can be explained as acetylation. These alternative explanations represent useful peptide identifications (but not mutations) and account for $\approx$ 50% of the identified peptides.

The next class of the most prevalent mutations belong to amino acids that mutate into Methionine and reveal potential alternative Start Codons. The most common mutations in these peptides were Val into Met (4 peptides) and Leu into Met (5 peptides) in the first positions. All mutations from Val and Leu to Met represented GTG and TTG, potential Start Codons[11]. Seven of nine predictions were verified to be start codons in existing annotations. Additionally, we found the peptide LDTTVADTEVTMPEGQGPR which is not part of the annotated *Arthrobacter* sp proteome (mutation from initial Leu into Met). The closest match returned by BLAST is from an N-terminal peptide in *Arthrobacter aurescens* providing an additional evidence that this peptide represents a new coding region in *Arthrobacter* sp. that evaded the annotation. Lastly, MEQPIISGVAHDR is not at the beginning of the protein representing either an erroneously annotated Start

---

[10] The goal of this section is to illustrate the capabilities of the Mutated BPM algorithm rather than to provide a comprehensive re-annotation of *Arthrobacter*.

[11] The list of peptides identified with Val(GTG) to Met mutation at the first position are: MDSNDVQADLK, MLIAQRPTLSEEVVSENR, MIEETLLEAGDKMDK, and MSTVESLVGEWLPLPDVAEMMNVSITK. The list of peptides identified with Leu(TTG) to Met mutation at the first position are: MLTANAYAAPSADGDLVPTTIER, MEGPEIQFSEAVIDNGR, MLAEALEHLVR, MDTTVADTEVTMPEGQGPR and MEQPIISGVAHDR.

site, or a single nucleotide polymorphism, or DNA sequencing error, or false peptide identification.

5 Discussion

We introduced a new class of combinatorial pattern matching problems and proposed various algorithms for their solution. These algorithms represent both the initial attempts to study the theoretical complexity of BPM and practical applications of BPM in proteomics. Our results demonstrate that BPM has a potential to greatly speed up protein identifications, a key task in computational proteomics.

References

1. Abascal, F., Posada, D., Knight, R.D., Zardoya, R.: Parallel evolution of the genetic code in arthropod mitochondrial genomes. PLoS Biol. 4(5), e127 (2006)
2. Amir, A.: Asynchronous pattern matching. In: Lewenstein, M., Valiente, G. (eds.) CPM 2006. LNCS, vol. 4009, pp. 1–10. Springer, Heidelberg (2006)
3. Amir, A., Aumann, Y., Indyk, P., Levy, A., Porat, E.: Efficient computations of ℓ_1 and ℓ_∞ rearrangement distances. In: Ziviani, N., Baeza-Yates, R. (eds.) SPIRE 2007. LNCS, vol. 4726, pp. 39–49. Springer, Heidelberg (2007)
4. Amir, A., Aumann, Y., Benson, G., Levy, A., Lipsky, O., Porat, E., Skiena, S., Vishne, U.: Pattern matching with address errors: rearrangement distances. In: Proc. 17th ACM-SIAM Symposium on Discrete Algorithms (SODA), pp. 1221–1229 (2006)
5. Amir, A., Aumann, Y., Kapah, O., Levy, A., Porat, E.: Approximate string matching with address bit errors. In: Ferragina, P., Landau, G.M. (eds.) CPM 2008. LNCS, vol. 5029, pp. 118–129. Springer, Heidelberg (2008)
6. Amir, A., Eisenberg, E., Keller, O., Levy, A., Porat, E.: Approximate string matching with stuck address bits. In: Chavez, E., Lonardi, S. (eds.) SPIRE 2010. LNCS, vol. 6393, pp. 395–405. Springer, Heidelberg (2010)
7. Amir, A., Hartman, T., Kapah, O., Levy, A., Porat, E.: On the cost of interchange rearrangement in strings. In: Arge, L., Hoffmann, M., Welzl, E. (eds.) ESA 2007. LNCS, vol. 4698, pp. 99–110. Springer, Heidelberg (2007)
8. Amir, A., Kapah, O., Porat, E.: Deterministic length reduction: Fast convolution in sparse data and applications. In: Ma, B., Zhang, K. (eds.) CPM 2007. LNCS, vol. 4580, pp. 183–194. Springer, Heidelberg (2007)
9. Baeza-Yates, R.: A fast set intersection algorithm for sorted sequences. In: Sahinalp, S., Muthukrishnan, S., Dogrusoz, U. (eds.) CPM 2004. LNCS, vol. 3109, pp. 400–408. Springer, Heidelberg (2004)
10. Besemer, J., Lomsadze, A., Borodovsky, M.: Genemarks: a self-training method for prediction of gene starts in microbial genomes implications for finding sequence motifs in regulatory regions. Nucleic Acids Research 29(12), 2607–2618 (2001)
11. Cardoze, D.E., Schulman, L.J.: Pattern matching for spatial point sets. In: Proc. 39th Annu. IEEE Sympos. Found. Comput. Sci., pp. 156–165 (1998)
12. Castellana, N.E., Payne, S.H., Shen, Z., Stanke, M., Bafna, V., Briggs, S.P.: Discovery and revision of arabidopsis genes by proteogenomics. Proceedings of the National Academy of Sciences 105(52), 21034–21038 (2008)

13. Cohen, H., Porat, E.: Fast set intersection and two-patterns matching. In: López-Ortiz, A. (ed.) LATIN 2010. LNCS, vol. 6034, pp. 234–242. Springer, Heidelberg (2010)
14. Cole, R., Hariharan, R.: Approximate string matching: A simpler faster algorithm. SIAM J. Comput. 31(6), 1761–1782 (2002)
15. Cormen, T.H., Leiserson, C.E., Rivest, R.L.: Introduction to Algorithms. MIT Press and McGraw-Hill (1992)
16. Demaine, E.D., López-Ortiz, A., Munro, I.J.: Adaptive set intersections, unions and differences. In: Proc. 11th ACM-SIAM Symposium on Discrete Algorithms (SODA), pp. 743–752 (2000)
17. Dietz, P., Mehlhorn, K., Raman, R., Uhrig, C.: Lower bounds for set intersection queries. Algorithmica 14(2), 154–168 (1993)
18. Elenitoba-Johnson, K.S.J., Crockett, D.K., Schumacher, J.A., Jenson, S.D., Coffin, C.M., Rockwood, A.L., Lim, M.S.: Proteomic identification of oncogenic chromosomal translocation partners encoding chimeric anaplastic lymphoma kinase fusion proteins. Proceedings of the National Academy of Sciences 103(19), 7402–7407 (2006)
19. Eng, J., McCormack, A., Yates, J.: An approach to correlate tandem mass spectral data of peptides with amino acid sequences in a protein database. Journal of the American Society for Mass Spectrometry 5(11), 976–989 (1994)
20. Frank, A.M., Pevzner, P.A.: PepNovo: De Novo Peptide Sequencing via Probabilistic Network Modeling. Anal. Chem. 77, 964–973 (2005)
21. Guigó, R., Gusfield, D., Edwards, N., Lippert, R.: Generating peptide candidates from amino-acid sequence databases for protein identification via mass spectrometry. In: Guigó, R., Gusfield, D. (eds.) WABI 2002. LNCS, vol. 2452, pp. 68–81. Springer, Heidelberg (2002)
22. Gupta, N., Tanner, S., Jaitly, N., Adkins, J., Lipton, M., Edwards, R., Romine, M., Osterman, A., Bafna, V., Smith, R., Pevzner, P.: Whole proteome analysis of post-translational modifications: applications of mass-spectrometry for proteogenomic annotation. Genome Res. 17, 1362–1377 (2007)
23. Gupta, N., Pevzner, P.A.: False discovery rates of protein identifications: A strike against the two-peptide rule. Journal of Proteome Research 8(9), 4173–4181 (2009)
24. Gusfield, D.: Algorithms on strings, trees, and sequences: computer science and computational biology. Cambridge University Press, New York (1997)
25. Jaffe, J.D., Stange-Thomann, N., Smith, C., DeCaprio, D., Fisher, S., Butler, J., Calvo, S., Elkins, T., FitzGerald, M.G., Hafez, N., Kodira, C.D., Major, J., Wang, S., Wilkinson, J., Nicol, R., Nusbaum, C., Birren, B., Berg, H.C., Church, G.M.: The complete genome and proteome of mycoplasma mobile. Genome Research 14(8), 1447–1461 (2004)
26. Jeong, K., Bandeira, N., Kim, S., Pevzner, P.A.: Gapped spectral dictionaries and their applications for database searches of tandem mass spectra. Mol. Cell. Proteomics (2010) (in press)
27. Kapah, O., Landau, G.M., Levy, A., Oz, N.: Interchange rearrangement: The element-cost model. Theoretical Computer Science 410(43), 4315–4326 (2009)
28. Kim, S., Bandeira, N., Pevzner, P.A.: Spectral profiles: A novel representation of tandem mass spectra and its applications for de novo peptide sequencing and identification. Mol. Cell. Proteomics 8, 1391–1400 (2009)
29. Kim, S., Gupta, N., Bandeira, N., Pevzner, P.A.: Spectral dictionaries: Integrating de novo peptide sequencing with database search of tandem mass spectra. Mol. Cell. Proteomics 8(1), 53–69 (2009)

30. Kim, S., Gupta, N., Pevzner, P.A.: Spectral probabilities and generating functions of tandem mass spectra: A strike against decoy databases. Journal of Proteome Research 7(8), 3354–3363 (2008)
31. Knight, R.D., Freeland, S.J., Landweber, L.F.: Rewiring the keyboard: evolvability of the genetic code. Nat. Rev. Genet. 2(1), 49–58 (2001)
32. Merrihew, G.E., Davis, C., Ewing, B., Williams, G., Käll, L., Frewen, B.E., Noble, W.S., Green, P., Thomas, J.H., MacCoss, M.J.: Use of shotgun proteomics for the identification, confirmation, and correction of c. elegans gene annotations. Genome Research 18(10), 1660–1669 (2008)
33. Muthukrishnan, S.: New results and open problems related to non-standard stringology. In: Galil, Z., Ukkonen, E. (eds.) CPM 1995. LNCS, vol. 937, pp. 298–317. Springer, Heidelberg (1995)
34. Ng, J., Pevzner, P.A.: Algorithm for identification of fusion proteins via mass spectrometry. Journal of Proteome Research 7(1), 89–95 (2008)
35. Nielsen, P., Krogh, A.: Large-scale prokaryotic gene prediction and comparison to genome annotation. Bioinformatics 21(24), 4322–4329 (2005)
36. Park, C.Y., Klammer, A.A., Käll, L., MacCoss, M.J., Noble, W.S.: Rapid and accurate peptide identification from tandem mass spectra. Journal of Proteome Research 7(7), 3022–3027 (2008)
37. Shilov, I.V., Seymour, S.L., Patel, A.A., Loboda, A., Tang, W.H., Keating, S.P., Hunter, C.L., Nuwaysir, L.M., Schaeffer, D.A.: The paragon algorithm, a next generation search engine that uses sequence temperature values and feature probabilities to identify peptides from tandem mass spectra. Molecular & Cellular Proteomics 6(9), 1638–1655 (2007)
38. Tanner, S., Shu, H., Frank, A., Wang, L.C., Zandi, E., Mumby, M., Pevzner, P.A., Bafna, V.: Inspect: Identification of posttranslationally modified peptides from tandem mass spectra. Analytical Chemistry 77(14), 4626–4639 (2005)
39. Tsur, D., Tanner, S., Zandi, E., Bafna, V., Pevzner, P.: Identification of post-translational modifications by blind search of mass spectra. Nature Biotechnology 23(12), 1562–1567 (2005)

Appendix

A   The BPM Convolution Algorithm in the Infinite Alphabet Case

In the infinite alphabet case our reduction to bit-vectors B_T and B_P may be too large, even exponential in the input size. However, these vectors are quite sparse, with the vast majority of entries being 0. Therefore, we can overcome the exponential blowup in size by encoding the arrays as sets of the indices of the bit vector locations whose value is 1. The size of these sets is proportional to the original arrays.

The problem is that we are confronted with the problem of finding the convolution vector W of the two vectors B_T, B_P that are not given explicitly. While in the regular fast convolution the running time is $O(|B_T| \log |B_P|)$, the aim here is to compute W in time proportional to the number of non-zero entries in W. This problem was posed in by Muthukrishnan in [33] and solved by Cole and Hariharan in [14]:

Theorem 3. *The convolution of sparse arrays V_1 and V_2 (with n_1 and n_2 of non-zero entries, respectively) can be computed in time $O(w \log^2 n_2)$, by a Las Vegas randomized algorithm with failure probability that is inverse polynomial in n_2, where w is the number of non-zero entries in the convolution vector.*

The Cole-Hariharan algorithm, although randomized, is very fast in practice. However, our situation is even better. We are assuming a static database T where we can afford some preprocessing that subsequently will allow fast queries. In this case, there is a fast deterministic algorithm [8], that preprocesses the V_1 array in time $O(n_1^2)$ and subsequently achieves the following time for the sparse convolution:

Theorem 4. *Given two sparse arrays V_1 and V_2 (with n_1 and n_2 of non-zero entries, respectively), their convolution can be computed in time $O(w \log^3 n_2)$, where w is the number of non-zero entries in the convolution vector.*

B   BPM Text Indexing: Unbounded Patterns and the Expected Intersection Size

While Theorem 2 provides faster algorithm than the one based on the naive $O(n)$ time for Set Intersection, we are willing to spend more time in preprocessing but speedup the queries. It is clear that one can preprocess all pairs i, j, $i, j = 1, .., k$ and pre-compute all intersections in $O(|U| \cdot k^2)$ time. In our case, for a d-bounded pattern $P[1]$, we can consider all $O(d)$ entries of the I^1, I^2 data structures. Each one has $O(\frac{n}{d})$ elements. Thus, the total size of this preprocessed table is $O(n \cdot d)$. Since d is small in practice (smaller than 500 Da), a table of all intersections can be constructed in the preprocessing stage, making the query time $O(d + tocc)$.

In the case of unbounded pattern, $O(d \cdot n)$ may reach $O(n^2)$. Below we present a scheme where the query time never exceeds $O(P[1]^2)$ and the preprocessing time never exceeds $O(n^{1.5})$.

In the case of unbounded pattern, $O(d \cdot n)$ may reach $O(n^2)$. Below we present a scheme where the query time never exceeds $O(P[1]^2)$ and the preprocessing time never exceeds $O(n^{1.5})$.

Preprocessing: For each one of the mass values $1, 2, ..., \sqrt{n}$, we construct the I^1, I^2 data structures together with the intersection tables. It results in $O(n\sqrt{n})$ preprocessing time and space.

Query Processing: Given a query pattern $P[1]$. If $P[1] \leq \sqrt{n}$ then the BPM query is processed in time $O(P[1] + tocc)$ using the intersection tables. If $P[1] > \sqrt{n}$ then the intersections are computed in the naive way. This means that the query time is $O(P[1] \cdot smin + tocc)$, where $smin$ is the size of the smallest set whose intersection is computed. However, since $P[1] > \sqrt{n}$ then the size of each segment is less than $\frac{n}{P[1]}$, i.e., every segment size is less than $\sqrt{n}$. But since $P[1] > \sqrt{n}$, it means that every segment size does not exceed $P[1]$. We conclude:

Theorem 5. *Let T be a text over fixed finite alphabet $\Sigma \subset \mathbb{N}$. Then it is possible to preprocess T in time $O(n^{1.5})$ such that subsequent BPM queries for patterns*

$P = P[1]$, can be answered in time $O(P[1]^2 + tocc)$, where $tocc$ is the number of blocks in T with mass P, and $O(P[1])$ is the length of a block with mass P.

Above we considered the worst case complexity of calculating the intersection. In Appendix, we show that the particular situation in our application guarantees that even a naive implementation of the set intersection leads to an expected optimal time, since the intersections are generally "large".

In the main text, two options for the worst case of calculating the intersection were considered. We now show that the particular situation in our application guarantees that even a naive implementation of the set intersection leads to an expected optimal time, since the intersections are generally "large".

Lemma 2. Let $S_1, ..., S_k$ be subsets of $U = \{1, ..., n\}$, such that the subsets are constructed uniformly at random and such that there is a constant c for which we are guaranteed that the cardinality of every set S_k is at least $\frac{n}{c}$. Then for any $i, j \in \{1, ..., k\}$, $|S_i \cap S_j|$ is $\Omega(\frac{n}{c^2})$.

Proof. Fix sets S_i, S_j and $e \in U$. Because the sets are constructed uniformly at random and each set has at least $\frac{n}{c}$ elements, then the probability of e being in S_i is at least $\frac{1}{c}$, and the probability of e being in S_j is at least $\frac{1}{c}$. The probability of e being in *both* is at least $\frac{1}{c}^2$. $\square$

Theorem 6. Let T be a text over fixed finite alphabet $\Sigma \subset \mathbb{N}$. Then it is possible to preprocess T in time $O(n \log n)$ such that subsequent BPM queries for patterns $P = P[1]$, can be answered in expected time $O(P[1] + tocc)$, where $tocc$ is the number of blocks in T with mass P.

Proof. By Theorem 6, the query can be answered in worst case time $O(P[1] \cdot t_i + tocc)$, where t_i is the intersection time. Thus it is sufficient to show that the expected set-intersection time is the size of the intersection, thus $tocc$ will incorporate the set-intersection time.

If we assume that the alphabet is uniformly distributed in the text, then for every list of segments in I^λ, $\lambda = 1, 2$, the distribution of segments is uniformly random. In addition, since the text alphabet is $\Sigma = \{i_1, ..., i_c\}$, where $\Sigma \subset \mathbb{N}$, and $i_1 < i_2 < \cdots < i_c$, then for every substring of i_c entries in $I^1(I^2)$, every segment appears at least once. Thus the conditions of Lemma 2 hold and therefore the expected intersection size is $\Theta(\frac{n}{P[1] \cdot i_c^2})$, which is also the running time required for the naive intersection implementation.

However, our initial motivation was for T to be the proteome. Even without assumptions on the amino acid distribution, we can still prove that the expected intersection size is $\Theta(\frac{n}{P[1]})$. Let $P[1] = m$. Assume that in a set-intersection computation $S_i \cap S_{i+m}$ there are k entries (segments) in set S_i that are not elements of S_{i+m} (and $|S_i| - k$ entries are elements of S_{i+m}). Because of the construction of the S_j sets, it must be the case that all these k entries are elements of $\cup_{\ell=1}^{i_c-1} S_{i+m+\ell}$. This means that for the i_c queries of sizes $P[1] = m, ..., m + i_c$, the sum of the sizes of the intersections that S_i participates in throughout the running of the algorithm is $\Omega(|S_i|)$. Since this is true for every

S_i, and since i_c is a constant, then our theorem is proved for any distribution of the alphabet in T.   □

C   From Theory to Practice: Reducing Memory Requirements of MBPM Algorithms

C.1   Strongly Non-i-Unique Patterns

A non-i-unique pattern $P = p_1, \ldots, p_i, p_{i+1}, \ldots, p_n$ in the set of patterns T is called *weakly non-i-unique* if all patterns with i-prefix $p_1, \ldots, p_i$ have the same $(i+1)$-prefix $p_1, \ldots, p_i, p_{i+1}$. Otherwise, P is called *strongly non-i-unique*. A variation of the MBPM algorithm that substitutes "non-i-unique" by "strongly non-i-unique" in the **for** loop (Fig. 2) is more memory efficient in the case of real proteomes. The MATCH function becomes more involved in this case and is not discussed here. To further reduce memory, we use a directed acyclic graph structure (to be described elsewhere) rather than the keyword tree to index the text.

C.2 (i, w)-Unique Patterns

A pattern P in the set of patterns $\mathcal{P}$ is called (i, w)-*unique* if there are w or less patterns in $\mathcal{P}$ with the same i-prefix as P. The notion of (i, w)-unique patterns generalizes the notion of i-unique patterns (i-unique patterns are $(i, 1)$-unique). We now describe construction of *KeywordTree*(T, d, w) that requires significantly less memory than *KeywordTree*(T, d). The only difference in constructing *KeywordTree*(T, d, w) is that it substitutes the notion of i-unique patterns by the notion of (i, w)-unique patterns in the MBPM algorithm (Fig. 2).

The algorithm iteratively generates the sets $\mathcal{T}_1, \ldots, \mathcal{T}_i = \mathcal{T}(T, d, w)$ until all patterns in the set $\mathcal{T}_i$ become (i, w)-unique. We further construct the keyword tree of $\mathcal{T}(T, d, w)$ (denoted *KeywordTree*(T, d, w)) and classify each vertex in this tree as unique or non-unique. Let $q_1, \ldots, q_i$ be a pattern spelled by the path from the root to the vertex v in the keyword tree. The vertex v is *unique* if the MBPM algorithm classified $q_1, \ldots, q_i$ as a prefix of an (i, w)-unique pattern at some iteration, and *non-unique* otherwise. To solve the MBPM Problem, we match each pattern $p_1, \ldots, p_n$ against *KeywordTree*(T, d, t) using the matching algorithm that is similar to the matching algorithm described for *KeywordTree*(T, d) with the following difference. While in the matching for *KeywordTree*(T, d) we attempted to match the suffix $p_i, \ldots, p_n$ against the *path* in the keyword tree that start at vertex v, we now attempt to match the suffix $p_i, \ldots, p_n$ against the *paths* in the keyword tree that start at vertex v. Figure 3 shows the size of $\mathcal{T}(T, d, w)$ for various values of w and illustrates that (i, w)-unique patterns lead to a practical algorithm with a reasonable memory-speed trade-off that enable applications of the MBPM algorithm to entire bacterial proteomes (less than 10 million amino acid long). Large proteomes (e.g., eukaryotic proteomes with typical length under 100 million amino acids) can be processed by partitioning them into shorter (10 million amino acids long) segments.

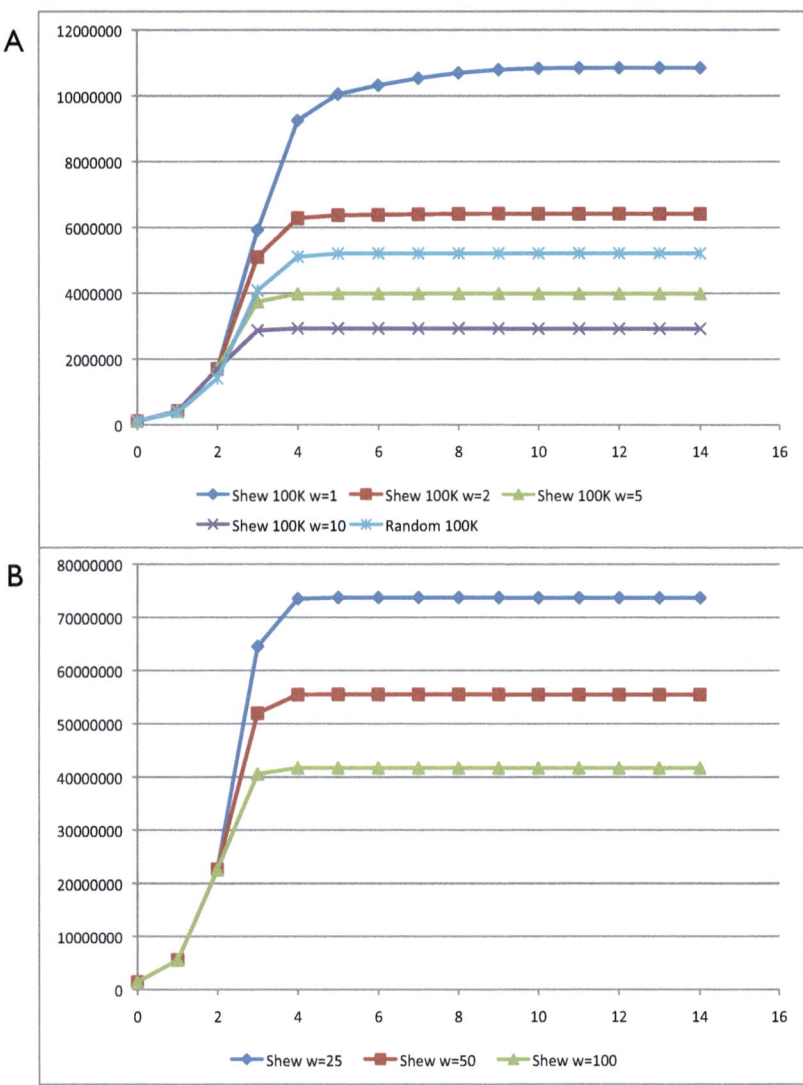

Fig. 3. Number of distinct patterns generated by the MBPM algorithm for various parameters and texts. The y-axis represents the number of patterns in the set $\mathcal{T}_i$ generated after the i-th iteration of the MBPM algorithm (x-axis represents the iteration number i). A) Size of $\mathcal{T}_i$ for $i = 0...14$ with $k = 15$ and varying parameter w for the first 100,000 amino acids of the *Shewanella* proteome compared to the size of set $\mathcal{T}_i$ for a random sequence of 100000 amino acids with $w = 1$. B) Size of $\mathcal{T}_i$ or $i = 0...14$ with $k = 15$ and varying w's for the complete *Shewanella* proteome.

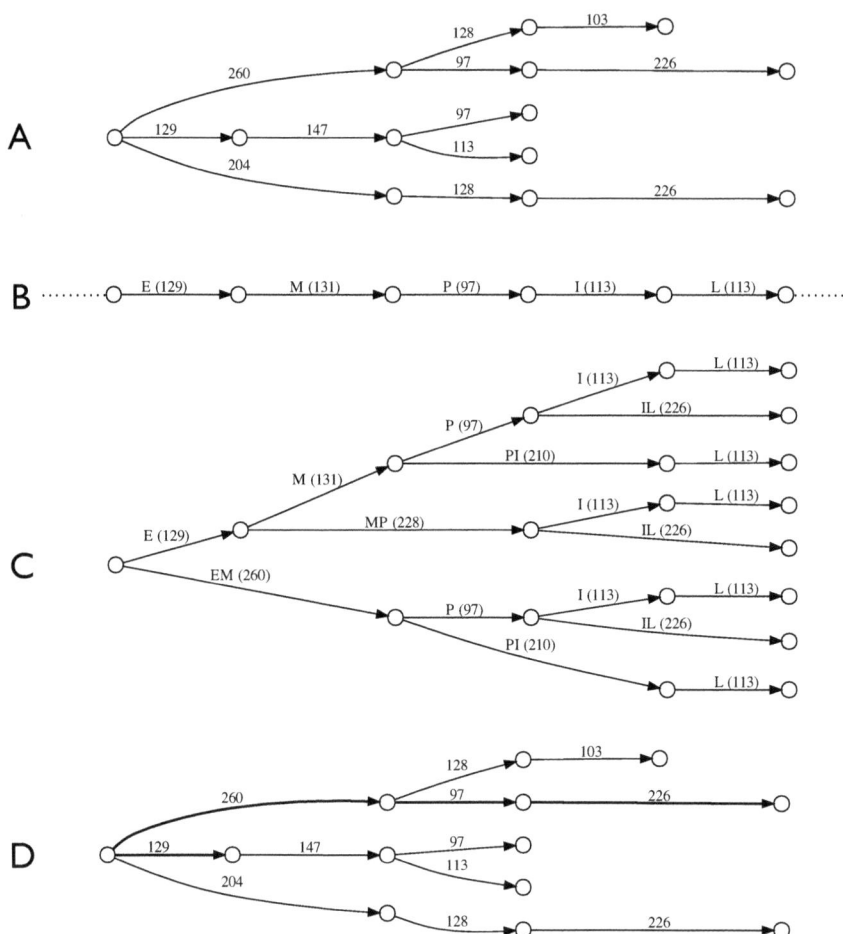

Fig. 4. Matching a keyword tree of patterns (*KeywordTree*($\mathcal{P}$)) against a text T. A) *KeywordTree*($\mathcal{P}$) built for a set $\mathcal{P}$ of 5 patterns: (129, 147, 97), (129, 147, 113), (204, 128, 226), (260, 97, 226) and (260, 128, 103). B) A text T containing a single k-mer EMPIL (129, 131, 97, 113, 113). C) The MBPM algorithm extends the original set of 5 patterns and constructs the keyword tree $Keyword(T, 300, \mathcal{P})$ of the resulting set of patterns. D) Matching *KeywordTree*($\mathcal{P}$) against $Keyword(T, 300, \mathcal{P})$ reveals the pattern matching the text. (shown as a bold path in the keyword tree).

C.3 i-Long Patterns

We previously described the memory overhead arising from processing non-i-unique k-mers
$p_1, \ldots, p_i, \ldots p_j, p_{j+1}, \ldots, p_n$ and $p_1, \ldots, p_i, \ldots p_j, p'_{j+1}, \ldots, p'_n$ sharing long common j-prefix (j is significantly larger than i). One way to reduce this overhead is not to extend the substring $p_i, \ldots, p_j$ of these patterns in the MBPM algorithm. This approach leads to some complications since in this case k-mers will represent a mosaic of processed and unprocessed substrings that have to be taken into account during the follow-up pattern matching step. However, it result in fast and memory efficient implementation capable of indexing the entire bacterial and even eukaryotic proteomes.

A substring $p_i, \ldots, p_j$ of a pattern $p_1, \ldots, p_i, \ldots, p_j, \ldots, p_n$ is called *long* if $\sum_{k=i}^{k=j} p_k \leq d$. Given a set of patterns $\mathcal{T}$, let $Patterns(\mathcal{T}, p_1, \ldots, p_i)$ be the subset of patterns in $\mathcal{T}$ with prefix $p_1, \ldots, p_i$. We call a position i in a pattern $P = p_1, \ldots, p_n$ a *break* if $Patterns(\mathcal{T}, p_1, \ldots, p_i) = Patterns(\mathcal{T}, p_1, \ldots, p_i, p_{i+1})$. A pattern P with k breaks can be divided into $k+1$ substrings between consecutive breaks that are classified into long and short. One can modify the MBPM algorithm to avoid processing long substrings, resulting in vastly reduced memory requirements (for real genomes). An actual implementation features a number of further refinements aimed at reducing memory. For example, while substituting a pattern $P = p_1, \ldots, p_i, \ldots, p_n$ by $P(i, d)$ in the MBPM algorithm, each pattern in $P(i, d)$ is represented by its i-prefix (rather than its full length) complemented by a pointer to the position of the keyword tree of the text where its $(i+1)$-prefix starts.

C.4 Matching the Keyword Tree of Patterns against the Text

Figure 4 illustrates matching a set of gapped peptides by building a keyword tree, to a text implicitly converted into a keyword tree. We can see that it is not necessary to explicitly construct the keyword tree for the patterns in the text.

A Three-Dimensional Model of the Yeast Genome

William Noble, Zhi-jun Duan, Mirela Andronescu, Kevin Schutz,
Sean McIlwain, Yoo Jung Kim, Choli Lee, Jay Shendure,
Stanley Fields, and C. Anthony Blau

University of Washington

Layered on top of information conveyed by DNA sequence and chromatin are higher order structures that encompass portions of chromosomes, entire chromosomes, and even whole genomes. Interphase chromosomes are not positioned randomly within the nucleus, but instead adopt preferred conformations. Disparate DNA elements co-localize into functionally defined aggregates or factories for transcription and DNA replication. In budding yeast, Drosophila and many other eukaryotes, chromosomes adopt a Rabl configuration, with arms extending from centromeres adjacent to the spindle pole body to telomeres that abut the nuclear envelope. Nonetheless, the topologies and spatial relationships of chromosomes remain poorly understood. Here we developed a method to globally capture intra- and inter-chromosomal interactions, and applied it to generate a map at kilobase resolution of the haploid genome of Saccharomyces cerevisiae. The map recapitulates known features of genome organization, thereby validating the method, and identifies new features. Extensive regional and higher order folding of individual chromosomes is observed. Chromosome XII exhibits a striking conformation that implicates the nucleolus as a formidable barrier to interaction between DNA sequences at either end. Inter-chromosomal contacts are anchored by centromeres and include interactions among transfer RNA genes, among origins of early DNA replication and among sites where chromosomal breakpoints occur. Finally, we constructed a three-dimensional model of the yeast genome. Our findings provide a glimpse of the interface between the form and function of a eukaryotic genome.

V. Bafna and S.C. Sahinalp (Eds.): RECOMB 2011, LNBI 6577, p. 320, 2011.
© Springer-Verlag Berlin Heidelberg 2011

Optimization of Combinatorial Mutagenesis

Andrew S. Parker[1], Karl E. Griswold[2], and Chris Bailey-Kellogg[1]

[1] Department of Computer Science, Dartmouth College
6211 Sudikoff Laboratory, Hanover, NH 03755, USA
{asp,cbk}@cs.dartmouth.edu
[2] Thayer School of Engineering, Dartmouth College
8000 Cummings Hall, Hanover, NH 03755, USA
karl.e.griswold@dartmouth.edu

Abstract. Protein engineering by combinatorial site-directed mutagenesis evaluates a portion of the sequence space near a target protein, seeking variants with improved properties (stability, activity, immunogenicity, etc.). In order to improve the hit-rate of beneficial variants in such mutagenesis libraries, we develop methods to select optimal positions and corresponding sets of the mutations that will be used, in all combinations, in constructing a library for experimental evaluation. Our approach, OCoM (Optimization of Combinatorial Mutagenesis), encompasses both degenerate oligonucleotides and specified point mutations, and can be directed accordingly by requirements of experimental cost and library size. It evaluates the quality of the resulting library by one- and two-body sequence potentials, averaged over the variants. To ensure that it is not simply recapitulating extant sequences, it balances the quality of a library with an explicit evaluation of the novelty of its members. We show that, despite dealing with a combinatorial set of variants, in our approach the resulting library optimization problem is actually isomorphic to single-variant optimization. By the same token, this means that the two-body sequence potential results in an NP-hard optimization problem. We present an efficient dynamic programming algorithm for the one-body case and a practically-efficient integer programming approach for the general two-body case. We demonstrate the effectiveness of our approach in designing libraries for three different case study proteins targeted by previous combinatorial libraries—a green fluorescent protein, a cytochrome P450, and a beta lactamase. We found that OCoM worked quite efficiently in practice, requiring only 1 hour even for the massive design problem of selecting 18 mutations to generate 10^7 variants of a 443-residue P450. We demonstrate the general ability of OCoM in enabling the protein engineer to explore and evaluate trade-offs between quality and novelty as well as library construction technique, and identify optimal libraries for experimental evaluation.

1 Introduction

Biotechnology is harnessing proteins for a wide range of significant applications, from medicine to biofuels [16,12]. In order to enable such applications, it is often

V. Bafna and S.C. Sahinalp (Eds.): RECOMB 2011, LNBI 6577, pp. 321–335, 2011.

necessary to modify extant proteins, developing variants with improved properties (stability, activity, immunogenicity, etc. [22,3,20]) for the task at hand. However, there is a massive space of potential variants to consider. Some protein engineering techniques (e.g., error-prone PCR [1] and DNA shuffling [28]) rely primarily on experiment to explore the sequence space, while others (e.g., structure-based protein redesign [11,2]) employ sophisticated models and algorithms in order to identify a small number of variants for experimental evaluation.

Computational design of combinatorial libraries [30,18,29,31,33] provides a middle ground between the primarily experimental and primarily computational approaches to development of improved variants. Library-design strategies seek to experimentally evaluate a diverse but focused region of sequence space in order to improve the likelihood of finding a beneficial variant. Such an approach is based on the premise that prior knowledge can inform generalized predictions of protein properties, but may not be sufficient to specify individual, optimal variants (resulting in both false positives and false negatives). Libraries are particularly appropriate when the prior knowledge does not admit detailed, robust modeling of the desired properties, but when experimental techniques are available to rapidly assay a pool of variants. Example scenarios would be instances where a three-dimensional structure is not available [13], or cases where definitive decisions regarding specific amino acid substitutions are non-obvious [22].

Nature employs both random mutation and recombination in generating diverse variants, and modern molecular biology has reconstituted these processes as highly controlled *in vitro* techniques. Here we develop library design methods for mutagenesis, wherein individual residue positions and corresponding mutations are first chosen, and then all possible combinations are constructed and subjected to screening or selection (Fig. 1). Most library optimization work has focused on recombination (i.e., selecting breakpoints), including approaches by Arnold and co-workers [30,15,17], Maranas and co-workers [26,25], and us [31,35,33,34]. Mayo and co-workers [29] have extended structure-based variant design to structure-based mutagenic library design, and applied it to the design of a library of green fluorescent proteins. Maranas and co-workers [18] have developed methods for optimizing both recombination and mutagenesis libraries, and applied them to the design of libraries of cytochrome P450s. LibDesign [14] is another useful tool for combinatorial mutagenesis, however it requires as input a predesigned library specification (positions and mutations). As we discuss further below, we develop here a more general method that encompasses different forms of computational library evaluation and optimization and experimental library construction, and explicitly optimizes both the quality and the novelty of the variants in the library.

Two techniques are commonly employed to introduce mutations in constructing combinatorial mutagenesis libraries (Fig. 1). When point mutagenesis is employed (Fig. 1, left), an individual oligonucleotide specific to a desired mutation is incorporated; there is a separate oligonucleotide for each such mutation. Combinatorial shuffling techniques [28] mix and match the mutated genes. When degenerate oligonucleotides are employed (Fig. 1, right), multiple amino acid-level mutations at a position are encoded by a single degenerate 3-mer (see [8]).

As with point mutagenesis, a library is generated by combinatorial shuffling. While the degenerate oligo approach is experimentally cheaper (a library costs about the same as a single variant), it can result in redundancy (multiple codons for the same amino acid) and junk (codons for undesired amino acids or stop codons), and is thus more appropriate when a larger library and lower hit rate are acceptable (e.g., when a high-throughput screen is available [5]).

Our method, *OCoM* (*O*ptimization of *Co*mbinatorial *M*utagenesis), encompasses both these approaches to experimental library construction. The key question is which mutations to introduce, given that the goal is isolation of functional variants with desired properties. A library-design strategy should therefore assess the predicted *quality* of prospective library members, e.g., by a sequence potential [18] or explicit structural evaluation [29]. We adopt a general sequence potential based on statistical analysis of a family of homologs to the target. The potential reveals both important residues (single-body conservation) and residue interactions (two-body coupling) for maintenance of protein stability and activity. Importantly, optimizing quality as a sole objective function might well result in libraries composed of sequences that are highly similar or even identical to extant proteins, an undesirable outcome. Thus it is necessary to balance quality assessment with *novelty* or diversity assessment. While this balance has been explicitly optimized for site-directed recombination [33], previous mutagenic library-design methods have only addressed this issue indirectly, e.g., by controlling factors such as the overall library size and the number of positions being mutated. Here we develop a new metric to explicitly account for the novelty of the variants compared to extant sequences, and we simultaneously optimize libraries for both novelty and quality.

While we have previously characterized the complexity of recombination library design for both quality [31] and diversity [35], to our knowledge, mutagenesis library design has never been similarly formalized or characterized. We show that, despite the combinatorial number of variants in the library, the OCoM design of an entire library is equivalent to the design of a single variant. Thus, like single-variant design, library optimization is NP-hard when accounting for a two-body potential. This stands in contrast to the polynomial-time algorithms for combinatorial recombination library design [31,35]. Consequently, we develop an integer programming approach that works effectively in practice on general OCoM problems, along with a polynomial-time dynamic programming approach that is appropriate for those without the two-body sequence potential.

To summarize the key contributions of OCoM, it supports a general scoring mechanism for variant quality, explicitly evaluates variant novelty, subsumes different approaches to library construction, accounts for bounds on library size and mutational sites, and evaluates the trade-offs between quality and diversity. While we focus on a statistical sequence potential for proposing and assessing mutations, our method is general and could employ a potential based on an initial round of experiments (e.g., from a randomization approach to remove phylogenetic bias [10]) or a list of high-quality results from structure-based design (e.g., [2]) from which it is desired to construct a library.

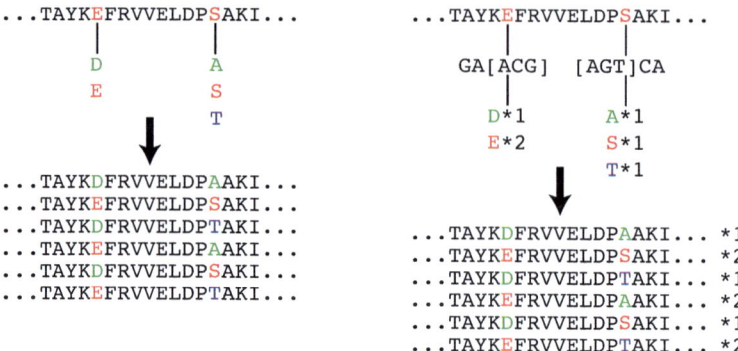

Fig. 1. Combinatorial mutagenesis libraries. (left) Specific point mutations at selected positions are introduced and shuffled to generate a library of all combinations of mutations. (right) Degenerate oligonucleotides (represented here in a regular expression-like notation, rather than IUPAC codes) are incorporated at selected positions, and shuffled to generate a library. Each degenerate oligo can code for a multiset of amino acids; consequently, some mutations may be represented more than others in the resulting library (e.g., in the first position, two codons for E vs. just one for D).

Our results illustrate the effectiveness of our approach. We show library plans for 3 proteins previously examined in combinatorial library experiments: a green florescent protein, a cytochrome P450, and a beta-lactamase. Our results span 6 orders of magnitude of library size, from 10^2 to 10^7 members. For each protein, libraries optimized under a range of constraints display distinct trade-offs between quality and novelty, as well as for the choice of library construction method (point mutations or degenerate oligos).

2   Methods

Given a target protein, our goal is to design an optimal combinatorial mutagenesis library, as measured by the overall quality and novelty of its variants.

Quality. To evaluate quality, we employ one- and two-body position-specific sequence potentials. Our current implementation uses potential scores derived from statistical analysis of an evolutionarily diverse multiple sequence alignment (MSA) of homologs of the target protein, but the method is generic to any potential of the same form. Details have been previously published [31,19]. The one-body term $\phi_i(a)$ for amino acid a position i captures conservation as the negative log frequency of a in the ith column of the MSA. Similarly, the two-body term $\phi_{ij}(a,b)$ for amino acid a at i and b at j captures correlated/compensating mutations as the negative log frequency of the pair (a,b) at the ith and jth columns, minus the independent terms $\phi_i(a)$ and $\phi_j(b)$ (to avoid double counting when summing the potentials). We filter the MSA to 90% sequence identity and restrict ϕ_{ij} to a relatively small, significant set of residue pairs by a χ^2 test of significant correlation (p-value 0.01).

$$\phi_i(a) = -\log \frac{|\{P \in \mathcal{S} : P[i] = a\}|}{|\mathcal{S}|} \qquad (1)$$

$$\phi_{i,j}(a,b) = -\log \frac{|\{P \in \mathcal{S} : P[i] = a \land P[j] = b\}|}{|\mathcal{S}|} - \phi_i(a) - \phi_j(b) \qquad (2)$$

The quality score of variant S is then $\sum_i \phi_i(S[i]) + \sum_{ij} \phi_{ij}(S[i], S[j])$ and the total quality score of a library is the sum of the quality scores of its variants. As these are based on negative logarithms, smaller is better.

Novelty. Given a whole sequence, we can assess its novelty in terms of how similar it is to the closest homolog (other than the target) in the MSA. That is, compute the minimum percent sequence identity to an extant sequence; the smaller the score, the more novel the variant. Without explicitly accounting for this, a library focused on quality could simply recapitulate natural sequences (which are of course high quality), wasting experimental effort.

To compute the percent sequence identity, we need an entire sequence. However, during the course of optimization, we want to be able to assess the impact on novelty of each mutation under consideration. Thus we introduce a position-specific novelty score $\nu_i(a)$ for amino acid a at position i, analogous to the quality score discussed above. The novelty contribution $\nu_i(a)$ assesses the sequence space distance between the mutant sequence containing a at i and homologs in the MSA.

$$\nu_i(a) = \min_{H \in \mathcal{S} \backslash S} \sum_{j=1}^{n} \frac{I\{S_{i \leftarrow a}[j] = H[j]\}}{n} \qquad (3)$$

where S is the target and $S_{i \leftarrow a}$ is the target with a mutation to amino acid a at position i, $\mathcal{S}$ is the MSA, n is the length of S and number of columns of $\mathcal{S}$, and $I\{\}$ the indicator function that returns 1 iff the predicate is true. Note that each $\nu_i(a)$ can be precomputed from the target and the MSA.

As with quality, the novelty score of variant S is then $\sum_i \nu_i(S[i])$, and the novelty score of a library sums the novelty scores of its variants. (Again, smaller is better.) The value for a variant is much like the percent sequence identity, except that each position does not account for mutations at other positions in computing the identity, and thus could underestimate the contribution. The value for a library is then much like the average percent sequence identity, and reduces the error in the total over the positions, since the library is comprised of the various combinations of mutations. While these thus are only approximations to the overall sequence identity, the error is independent of the actual mutations being made, and thus does not affect the optimization. We find in practice for the case studies presented in the results that the one-body potential is very highly correlated (over 0.99) with the full n-body one. Thus there is no need to go to a higher-order potential.

Library from tubes. Recall (Fig. 1) that there are two common molecular biology techniques for generating combinatorial mutagenesis libraries: point mutations and degenerate oligonucleotides. A convenient abstraction subsuming these two methods of library construction is to consider for each position a multiset of amino acids, which we call a *tube* (as in the experiment). For point mutation, a tube contains a selected set of amino acids to be incorporated at a position. For degenerate oligonucleotides, a tube contains a multiset of amino acids encoded by all codons represented by a degenerate oligonucleotide 3-mer. In this abstraction we always mean 3-mer. Note that the representation even supports multiple degenerate oligonucleotides (or a degenerate oligonucleotide and a specific one) at a position, which might be desirable to obtain the best balance of library quality, novelty, and size [8].

Given a set of tubes, one per position, the resulting library is defined by the cross-product of the tubes, with separate variants for each instance of an amino acid appearing multiple times in the multiset (Fig. 1). Note that in a multiset, every recurring appearance of an amino acid introduces redundancy, a scenario that is especially undesirable when screening is difficult. In optimizing a library, we select one tube for each position, from a preenumerated set of *allowed tubes*. These are in turn determined by the amino acids that should be considered as possible substitutions. Our current implementation only allows those appearing at expected uniform frequency 5% or greater in the MSA. This averages to 4 to 5 per position in our case studies, for at most $2^5 - 1 = 31$ tubes when considering all sets of point mutations. For degenerate oligos, we only allow tubes that have a ratio of at least 3 : 2 between codons for allowed substitutions and those for disallowed ones. We also eliminate tubes that code for the same proportions of amino acids in a larger multiset, for example, we would keep [GC]TC, coding for {L, V} instead of [GC]T[GC], coding equivalently but redundantly for {L, L, V, V}. Finally, we disallow tubes with STOP codons, though recognize that with a very high-throughput screen, those may still be acceptable. All combinations of the 4 nucleotides in each of 3 positions would yield 3375 possible degenerate oligos, but after our global filters there are fewer than 1000, which are further filtered for each position according to allowed substitutions, for an average of 10 in our case studies.

With the pieces in place, we can now formally define our problem.

Problem 1 (OCoM). Given a protein sequence S of length n and, for each position i a set $\mathcal{T}_i$ of allowed tubes, optimize a library $\mathcal{L} = T_1 \times T_2 \times \ldots \times T_n$ where for each i, $T_i \in \mathcal{T}_i$, so as to minimize

$$\sum_{S' \in \mathcal{L}} \alpha \left(\sum_{i=1}^{n} \phi_i(S'[i]) + \sum_{i=1}^{n-1} \sum_{j=i+1}^{n} \phi_{i,j}(S'[i], S'[j]) \right) + (1 - \alpha) \left(\sum_{i=1}^{n} \nu_i(S'[i]) \right)$$

The experimental cost can be constrained by the number of sites being substituted, the number of amino acids (including duplicates) in each tube, and the size of the library.

The parameter α controls the relative trade off between quality and novelty. For the results, we try a range of values, recognizing that in the future it is desirable to consider all trade-offs and select plans that are Pareto optimal [7].

Efficient library evaluation. Our optimization problem is expressed as a sum over all variants in the library. However, in practice, we do not want to enumerate all the variants in order to compute the value of the objective function. In previous work, we showed how to lift one- and two-body position-specific sequence potentials for single variants to corresponding potentials for recombination libraries [31,33]. We do the same here for combinatorial mutagenesis. For simplicity, consider just the one-body term ϕ_i; the two-body term ϕ_{ij} and the novelty ν_i work similarly.

$$\sum_{S' \in T_1 \times T_2 \times \ldots \times T_n} \sum_{i=1}^{n} \phi_i(S'[i]) = \sum_{i=1}^{n} \sum_{a \in T_i} \frac{|T_1| \cdot |T_2| \cdot \ldots \cdot |T_n|}{|T_i|} \phi_i(a)$$

$$= |\mathcal{L}| \sum_{i=1}^{n} \sum_{a \in T_i} \frac{\phi_i(a)}{|T_i|} \tag{4}$$

This follows by recognizing that amino acid type a at position i contributes $\phi_i(a)$ to each variant, i.e., each choice of amino acid types for the other positions.

Thus we develop a tube-based library potential by averaging over the set of amino acids in the tube:

$$\theta_i(T) = \alpha \frac{\sum_{a \in T} \phi_i(a)}{|T|} + (1 - \alpha) \frac{\sum_{a \in T} \nu_i(a)}{|T|} \tag{5}$$

$$\theta_{i,j}(T_i, T_j) = \alpha \frac{\sum_{a \in T_i} \sum_{b \in T_j} \phi_{i,j}(a, b)}{|T_i| \cdot |T_j|} \tag{6}$$

For simplicity of subsequent formulas, we assume that α is fixed before computing θ_i; recall that we do not have a two-body novelty term.

Note that our tube-based scores avoid a potential pitfall by automatically accounting for the relative frequencies of amino acids at a position, and their relative contribution to the library. That is, if one position has three amino acid types and another two (Fig. 1), then the contributions of the constituent amino acids are weighted by $1/3$ and $1/2$, respectively.

With the tube scores thus computed, our objective function is simplified:

$$f(T_1, \ldots, T_n) = \sum_{i=1}^{n} \theta_i(T_i) + \sum_{i=1}^{n-1} \sum_{j=i+1}^{n} \theta_{ij}(T_i, T_j) \tag{7}$$

Complexity. As Eq. 7 makes clear, once we have normalized tube scores, library optimization looks just like single-variant optimization, though over an "alphabet" of tubes rather than amino acids or rotamers. It immediately follows from the NP-hardness of protein design with a two-body potential [21] that OCoM-based combinatorial mutagenesis library design is NP-hard.

Dynamic programming. Without the two-body sequence potential, we can readily develop an efficient dynamic programming algorithm. Let $M(i, T)$ be the best score of a library optimized through position i, with tube T at position i. Because the one-body score allows for the choice of the optimal T at each position without consideration of any other position, the optimal library determined by the additional choice of T at i depends only on the library through $i - 1$. Thus

$$M(i, T) = \begin{cases} \theta_i(T) & i = 1 \\ \min_{T' \in T_{i-1}} M(i - 1, T') + \theta_i(T) & i > 1 \end{cases} \tag{8}$$

The time and space complexity is quadratic in the size of the input: $O(nm)$ for n the length of the sequence and m the maximum number of allowable tubes at any position. We can easily add a dimension to the DP matrix to count total mutational sites (up to M), for a total complexity of $O(nmM)$.

Integer programming. In order to solve the full library design problem, including the two-body potential, we develop an integer programming formulation that works well in practice using the IBM ILOG CPLEX solver.

Define singleton binary variable $s_{i,t}$ to indicate whether or not tube t is at position i. Similarly, define pairwise binary variable $p_{i,j,t,u}$ to indicate whether or not the tubes t, u are at i, j respectively.

We rewrite our objective function (Eq. 7) in terms of these binary variables:

$$\Phi = \sum_{i,t} s_{i,t} \cdot \theta_i(t) + \sum_{i,j,t,u} p_{i,j,t,u} \cdot \theta_{i,j}(t, u) \tag{9}$$

In order to guarantee that the variable assignments yield a valid combinatorial library, we impose the following constraints:

$$\forall i : \sum_t s_{i,t} = 1 \tag{10}$$

$$\forall i, t, j > i : \sum_u p_{i,j,t,u} = s_{i,t} \tag{11}$$

$$\forall j, u, i < j : \sum_t p_{i,j,t,u} = s_{j,u} \tag{12}$$

Eq. 10 ensures that exactly one tube is chosen at each position i. Eq. 11 and Eq. 12 maintain consistency between singleton and pairwise variables.

In order to specify desired properties of the mutated sites and library size, we impose the following additional constraints.

$$\log(\lambda) \leq \sum_i \sum_t s_{i,t} \log(|t|) \leq \log(\Lambda) \tag{13}$$

$$\mu \leq \sum_i \sum_{t \neq \{S[i]\}} s_{i,t} \leq M \tag{14}$$

The bounds on the library size (Eq. 13) and number of mutations per position (Eq. 14) may be set by the technology and resources available for library construction and screening. The expression $t \neq \{S[i]\}$ determines whether or not the tube has only the wild-type amino acid, and thereby whether or not that is a mutated position. We could likewise incorporate additional constraints on the number of mutated positions. We use these as constraints instead of terms in the objective function because there are likely to be a relatively small number of values to try, and the results can be compared and contrasted. Furthermore, our objective function incorporates an explicit novelty score; these terms somewhat implicitly affect diversity. A larger λ means more variants, which must be different from each other in some way, except in the case of redundant codons. A larger μ allows, but does not guarantee, greater site diversity.

3   Results

We applied OCoM to optimize libraries for three different proteins for which combinatorial libraries had previously been developed. We found that OCoM worked quite efficiently in practice, requiring only 1 hour even for the massive design problem of selecting 18 mutations to generate 10^7 variants for a 443-residue sequence. We demonstrate the general ability of OCoM in enabling the protein engineer to explore and evaluate trade-offs between quality and novelty as well as library construction technique, and identify optimal libraries for experimental evaluation.

Green Fluorescent Protein (GFP). GFP presents a valuable engineering target due to its widespread use in imaging experiments; the availability of distinct colors, some engineered, enables *in vivo* visualization of differential gene expression and protein localization and measurement of protein association by fluorescence resonance energy transfer [32]. Following the work of Mayo and colleagues, we targeted the wild type 238-residue GFP from *Aequorea victoria* (uniprot entry name GFP_AEQVI) with mutation S65T [29]. The sequence potential is derived from the 243 homologs in Pfam PF01353.

   Fig. 2(left) illustrates the trade-offs between library quality and novelty scores for fixed library size bounds and library construction techniques, over a range of α values (recall that higher α places more focus on quality). While we targeted 100- and 1000-member libraries, depending on the input and choice of parameters, not every exact library size is possible. Thus these numbers represent lower bounds on the library sizes; the upper bounds are slightly relaxed. The curves are fairly smooth but sometimes steep as a swift change in one property is made at relatively little cost to the other. Interestingly, the ≈100-member library curves intersect the ≈1000-member library curves. To the left of that point, the ≈100-member libraries yield better quality for a given novelty, while to the right, the ≈1000-member libraries yield a better novelty for a given quality, and thus would be preferred if that screening capacity is available. The curves intersect where the larger library approaches its maximum quality and the smaller library

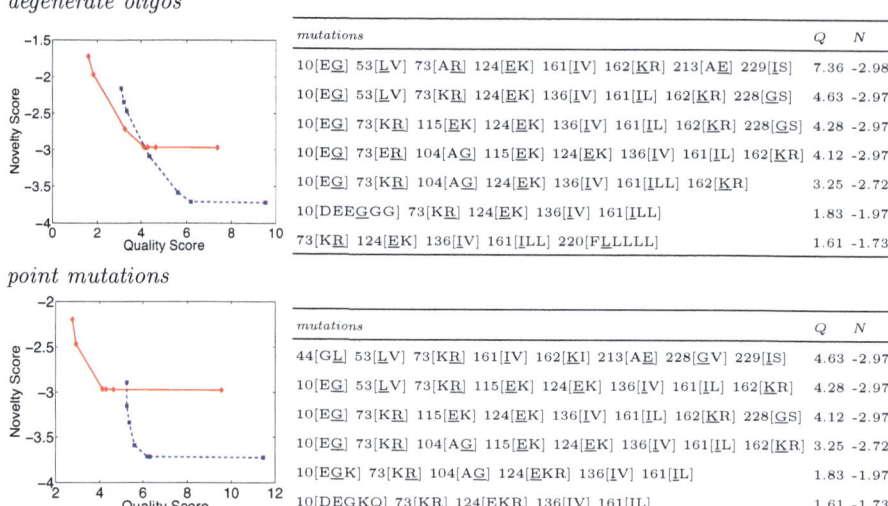

degenerate oligos

point mutations

Fig. 2. GFP plans under varying quality-novelty trade-offs, at fixed library size bounds, with two library construction techniques. Smaller scores are better. The left panels plot the scores of plans (one per point) for libraries of ≈100 members (red diamond solid) and ≈1000 members (blue square dash). The right panels detail the ≈100-member library plans, with selected positions and their wild-type amino acid types (underlined) and mutations.

reaches its maximum novelty; thus adjusting α only sends library plans along the vertical or horizontal.

The right panels of Fig. 2 summarize the mutations comprising each library. Within the degenerate oligo plans we notice single substitutions at each site, while within the point mutation plans we notice a set of different substitutions at the same site, including some that fall outside the natural degeneracy in the genetic code. We also notice that a number of mutations are attractive across a range of α values, and under both construction techniques. Several times both construction methods identify the same site and same mutation. And in both cases, we see concentration of mutations on less conserved sites (e.g., 124[EK] where Lysine is the consensus residue at 31%) for better quality, and spreading mutations over the sequence for better novelty.

Fig. 3(left) illustrates trends in planning GFP libraries of a wide range of sizes. The y-axis gives the total quality score summed over the unique variants in the library (lower is better). Compared to the number of variants in the library to be screened, this is a measure of the library efficiency. The point mutation libraries remain linear at an approximate slope of 1 on this log-log plot; essentially, each mutation is picking up a constant "penalty" against quality. While, as we also see in Fig. 2, degenerate oligo libraries tend to have better quality scores due to their multiset nature, the redundancy leads to fewer unique variants and

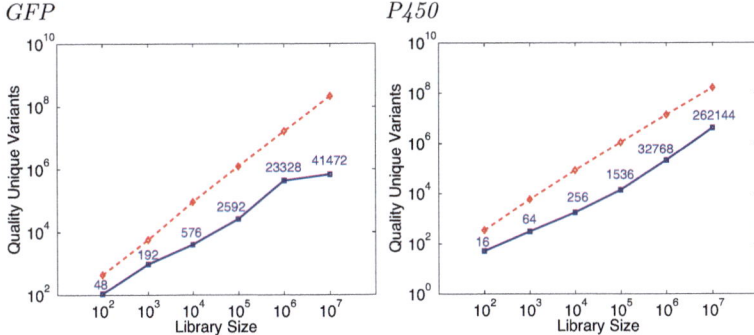

Fig. 3. Efficiency evaluation of plans for different GFP (left) and P450 (right) library sizes, under degenerate oligos (blue solid squares) and point mutations (red dashed diamonds). The y-axis plots the total quality score (ϕ; lower is better) of the *unique* variants in the library (i.e., removing duplicates from degenerate oligos). The degenerate oligo curve is labeled with the number of unique variants.

thus fewer expected "hits" for the same screening effort. Consequently, up to a factor of 10^3 more degenerate oligo variants than point mutation variants need to be screened to achieve the unique library size, consistent with trends in other studies [23]. On the other hand, degenerate oligo libraries are also cheaper to construct. These curves help elucidate the trade-offs. The degenerate oligo curve flattens out at 10^6 to 10^7 largely because the algorithm has reduced capacity to find more unique reasonable quality variants on this particular and relatively smaller protein.

To further study the use of degeneracy in library generation, we compared libraries using selected degenerate oligos with those using saturation mutagenesis, either with the NNK degenerate codon (coding all 20 amino acids) or the NDT degenerate codon (12 diverse amino acids). Reetz *et al.* [23] have studied the relative efficiency of the two saturation mutagenesis techniques, in the context of directed evolution. Using OCoM, we can further compare and contrast the selection of positions to mutate, at different levels of degeneracy. We separately optimized relatively conserved core residues (positions 57–72 [29]) and relatively less conserved surface ones. Fig. 4 shows the efficiency of libraries (using the total quality metric of the preceding paragraph) for different number of sites to mutate. As in our above library studies, there are sufficient degrees of freedom in any method, and both in the core and on the surface, to continue taking mutations at roughly the same penalty. Strikingly, the relative efficiency (ratio) of saturation, half saturation, and any choice is about the same in the core or on the surface, across the number of sites mutated. We also evaluated the use of "double-degenerate" oligos, combining two different degenerate oligos in a single tube. However, for these studies they yielded exactly the same plans as did the regular degenerate oligos. There was apparently insufficient motivation to select amino acids sufficiently different not to be naturally covered by the degeneracy in the genetic code.

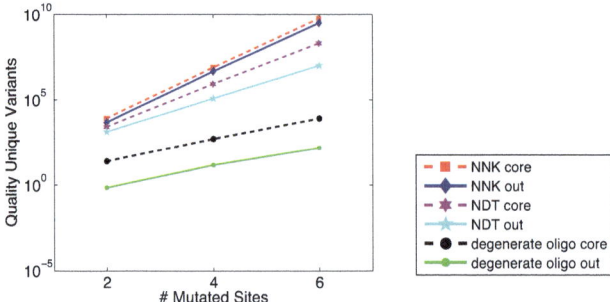

Fig. 4. Efficiency evaluation (as in Fig. 3) for GFP libraries optimized at different levels of degeneracy, for core or surface, at different numbers of mutated sites

Cytochrome P450. Cytochrome P450 is an essential enzyme at all levels of cellular life and thus extensively studied, especially given its significant engineering applications in biofuels [4]. We chose as a target a P450 from *Bacillus subtilis*, CYP102A2 (uniprot gene synonym cypD), used in previous library studies [18]. The P450 family is very diverse, so we identified a set of 194 homologs to our target by running PSI-BLAST for 3 iterations, and then multiply aligned them with ClustalW. As in the earlier studies, we focused on residues 6–449 because the remaining portions of the MSA were too sparse for meaningful statistics.

The trade-off curves (Fig. 5) are more distinct than those for GFP, and are quite sharp and sparse. This may be a result of looking here at a small library size with relatively few mutations, relative to the much larger size of this protein. The degenerate oligo plans focus on a few positions (an average of 4), while the point mutation plans are more spread over the sequence (an average of 7).

With increasing library size (Fig. 3(right)), we see similar trends as for GFP. As the library size increases, more and more screening effort (up to three orders of magnitude) is required to find fewer good unique variants in the degenerate oligo libraries. This illustrates a fundamental difference between the two library construction methods, and highlights a key advantage: using discrete oligos for each individual point mutation can always specifically target beneficial amino acids, even with increasing library size.

Beta Lactamase. The beta lactamase enzyme family hydrolyzes the beta lactam ring of penicillin-like drugs thereby conferring resistance to bacteria and presenting a potential drug target [6]. As it supports easy and inexpensive activity screening, beta lactamase is an ideal candidate for testing combinatorial library methods [9,15,31,33]. We took as target the TEM-1 beta lactamase from *E. coli*, and developed the sequence potential from an MSA of 149 homologs aligned to 263 residues used in our previous recombination work [31]. We found the trends too similar to our other case studies to merit repetition of detail here, but we note that in contrast to P450, but like GFP (of a more similar size), the trade-off curves are less sharp and more full. Like GFP and P450, the targeted mutation

degenerate oligos

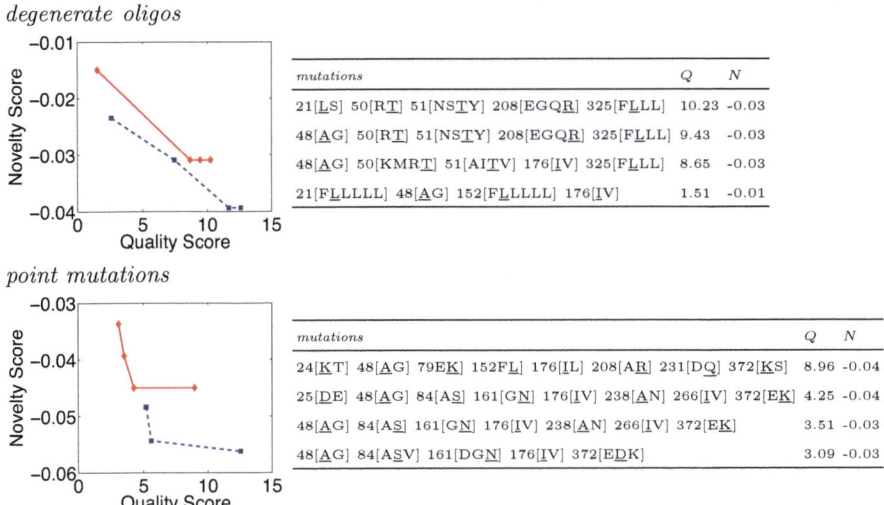

mutations	Q	N
21[LS] 50[RT] 51[NSTY] 208[EGQR] 325[FLLL]	10.23	-0.03
48[AG] 50[RT] 51[NSTY] 208[EGQR] 325[FLLL]	9.43	-0.03
48[AG] 50[KMRT] 51[AITV] 176[IV] 325[FLLL]	8.65	-0.03
21[FLLLLL] 48[AG] 152[FLLLLL] 176[IV]	1.51	-0.01

point mutations

mutations	Q	N
24[KT] 48[AG] 79EK] 152FL] 176[IL] 208[AR] 231[DQ] 372[KS]	8.96	-0.04
25[DE] 48[AG] 84[AS] 161[GN] 176[IV] 238[AN] 266[IV] 372[EK]	4.25	-0.04
48[AG] 84[AS] 161[GN] 176[IV] 238[AN] 266[IV] 372[EK]	3.51	-0.03
48[AG] 84[ASV] 161[DGN] 176[IV] 372[EDK]	3.09	-0.03

Fig. 5. P450 plans under varying quality-novelty trade-offs; see Fig. 2 for description

sites are similar, but the repertoire of substitutions can differ. For example, at Lys261 the degenerate oligo plans make D,E,N substitutions, while the point mutations make A,D,E,Q,V substitutions.

4   Discussion and Conclusion

OCoM provides a powerful and general mechanism to optimize combinatorial mutagenesis libraries so as to improve the "hit-rate" of novel variants with properties of interest. It enables protein engineers to study the trade-offs among predicted quality and novelty, library size, and expected success over two different approaches to library construction. While it readily allows effort to be focused on residues or regions of interest, that is not required; OCoM supports global design of a protein, accounting for interrelated effects of mutations. While the design problem is NP-hard in theory and clearly combinatorial in practice, our encoding of the constraints and homology-based filtering of poor choices, along with the power of the IBM ILOG solver, yielded an implementation that was able to compute the optimal 10^7 size library for each test case in under an hour.

As we have implemented here, 2-body quality scores are considered state-of-the art, and necessary for evaluation of stability and activity of new proteins [24,27]. However, there may be cases, such as large proteins (or complexes) with high degrees of sequence variability (and thus large tube sets), where only a 1-body potential will be practical because of the combinatorial explosion. In such cases, our dynamic programming formulation will still enable the optimization of libraries based on conservation statistics.

Since OCoM is modular, it is easily extensible to additional forms of variant and library evaluation and constraint, and those are key steps for our future work.

For example, rather than a general sequence potential and global design, it could be targeted to exploration of sequence space most affecting activity or stability, or it could be extended to incorporate evaluation of immunogenicity [20,19]. And as mentioned in the introduction, the potential could be derived from initial experiments or from structure-based analysis. Although beyond the scope of this paper, prospective application of OCoM in designing libraries for targets of engineering interest is of course the whole motivation of the work.

Acknowledgments. We thank Alan Friedman (Purdue) for helpful discussions on library design. This work was funded in part by NSF grant CCF-0915388 to CBK.

References

1. Cadwell, R.C., Joyce, G.F.: Randomization of genes by PCR mutagenesis. PCR Methods Appl. 2, 28–33 (1992)
2. Chen, C.Y., Georgiev, I., Anderson, A.C., Donald, B.R.: Computational structure-based redesign of enzyme activity. PNAS 106, 3764–3769 (2009)
3. Fox, R., et al.: Improving catalytic function by ProSAR-driven enzyme evolution. Nat. Biotechnol. 25, 338–344 (2007)
4. Fukuda, H., et al.: Reconstitution of the isobitene-forming reaction catalyzed by cytochrome p450 and p450 reductase from Rhodotorula minuta: Decarboxylation with the formation of isobutene. Biochem. Bioph. Res. Co. 201, 516–522 (1994)
5. Griswold, K.E., Aiyappan, N.S., Iverson, B.L., Georgiou, G.: The evolution of catalytic efficiency and substrate promiscuity in human theta class 1-1 glutathione transferase. J. Mol. Biol. 364, 400–410 (2006)
6. Harding, F.A., et al.: A beta-lactamase with reduced immunogenicity for the targeted delivery of chemotherapeutics using antibody-directed enzyme prodrug therapy. Mol. Cancer. Ther. 4, 1791–1800 (2005)
7. He, L., Friedman, A.M., Bailey-Kellogg, C.: Pareto optimal protein design. In: 3dsig: Structural Bioinformatics and Computational Biophysics, pp. 69–70 (2010)
8. Herman, A., Tawfik, D.S.: Incorporating synthetic oligonucleotides via gene reassembly (ISOR): a versatile tool for generating targeted libraries. Protein Eng. Des. Sel. 20, 219–226 (2007)
9. Hiraga, K., Arnold, F.: General method for sequence-independent site-directed chimeragenesis. J. Mol. Biol. 330, 287–296 (2003)
10. Jackel, C., Bloom, J.D., Kast, P., Arnold, F.H., Hilvert, D.: Consensus protein design without phylogenetic bias. J. Mol. Biol. 399, 541–546 (2010)
11. Jiang, L., et al.: De novo computational design of retro-aldol enzymes. Science 319(5868), 1387–1391 (2008)
12. la Grange, D.C., den Haan, R., van Zyl, W.H.: Engineering cellulolytic ability into bioprocessing organisms. Appl. Microbiol. Biotechnol. 87, 1195–1208 (2010)
13. Levin, A.M., Murase, K., Jackson, P.J., Flinspach, M.L., Poulos, T.L., Weiss, G.A.: Double barrel shotgun scanning of the Caveolin-1 scaffolding domain. ACS Chem. Biol. 2, 493–500 (2007)
14. Marco, A.M., Daugherty, P.S.: Automated design of degenerate codon libraries. Protein Eng. Des. Sel. 18, 559–561 (2005)
15. Meyer, M., Hochrein, L., Arnold, F.: Structure-guided SCHEMA recombination of distantly related beta-lactamases. Protein Eng. Des. Sel. 19, 563–570 (2006)

16. Nelson, A., Reichert, J.M.: Development trends for therapeutic antibody fragments. Nat. Biotech. 27, 331–337 (2009)
17. Otey, C., Landwehr, M., Endelman, J., Hiraga, K., Bloom, J., Arnold, F.: Structure-guided recombination creates an artificial family of cytochromes P450. PLoS Biol. 4, e112 (2006)
18. Pantazes, R., Saraf, M., Maranas, C.: Optimal protein library design using recombination or point mutations based on sequence-based scoring functions. Protein Eng. Des. Sel. 20, 361–373 (2007)
19. Parker, A.S., Griswold, K., Bailey-Kellogg, C.: Optimization of therapeutic proteins to delete T-cell epitopes while maintaining beneficial residue interactions. In: Proc. CSB, pp. 100–113 (2010)
20. Parker, A.S., Zheng, W., Griswold, K., Bailey-Kellogg, C.: Optimization algorithms for functional deimmunization of therapeutic proteins. BMC Bioinf. 11, 180 (2010)
21. Pierce, N., Winfree, E.: Protein design is *NP*-hard. Protein Eng. 15, 779–782 (2002)
22. Reetz, M.T., Carballira, J.: Iterative saturation mutagenesis (ISM) for rapid directed evolution of functional enzymes. Nat. Protocols 2, 891–903 (2007)
23. Reetz, M.T., Kahakeaw, D., Lohmer, R.: Addressing the numbers problem in directed evolution. ChemBioChem. 9, 1797–1804 (2008)
24. Russ, W.P., Lowery, D.M., Mishra, P., Yaffee, M.B., Ranganathan, R.: Natural-like function in artificial WW domains. Nature 437, 579–583 (2005)
25. Saraf, M.C., Gupta, A., Maranas, C.D.: Design of combinatorial protein libraries of optimal size. Proteins 60, 769–777 (2005)
26. Saraf, M.C., Horswill, A.R., Benkovic, S.J., Maranas, C.D.: FamClash: A method for ranking the activity of engineered enzymes. PNAS 12, 4142–4147 (2004)
27. Socolich, M., Lockless, S.W., Russ, W.P., Lee, H., Gardner, K.H., Ranganathan, R.: Evolutionary information for specifying a protein fold. Nature 437, 512–518 (2005)
28. Stemmer, W.P.C.: DNA shuffling by random fragmentation and reassembly: in vitro recombination for molecular evolution. PNAS 91, 10747–10751 (1994)
29. Treynor, T., Vizcarra, C., Nedelcu, D., Mayo, S.: Computationally designed libraries of fluorescent proteins evaluated by preservation and diversity of function. PNAS 104, 48–53 (2007)
30. Voigt, C.A., Martinez, C., Wang, Z.G., Mayo, S.L., Arnold, F.H.: Protein building blocks preserved by recombination. Nat. Struct. Biol. 9, 553–558 (2002)
31. Ye, X., Friedman, A.M., Bailey-Kellogg, C.: Hypergraph model of multi-residue interactions in proteins: Sequentially–constrained partitioning algorithms for optimization of site-directed protein recombination. J. Comput. Biol. 14, 777–790 (2007); In: Apostolico, A., Guerra, C., Istrail, S., Pevzner, P.A., Waterman, M. (eds.) RECOMB 2006. LNCS (LNBI), vol. 3909, pp. 15–29. Springer, Heidelberg (2006)
32. Zhang, J., Campbell, R., Ting, A., Tsien, R.: Creating new fluorescent probes for cell biology. Nat. Rev. Mol. Cell. Biol. 3, 906–918 (2002)
33. Zheng, W., Friedman, A., Bailey-Kellogg, C.: Algorithms for joint optimization of stability and diversity in planning combinatorial libraries of chimeric proteins. J. Comput. Biol. 16, 1151–1168 (2009); In: Vingron, M., Wong, L. (eds.) RECOMB 2008. LNCS (LNBI), vol. 4955, pp. 300–314. Springer, Heidelberg (2008)
34. Zheng, W., Griswold, K., Bailey-Kellogg, C.: Protein fragment swapping: A method for asymmetric, selective site-directed recombination. J. Comput. Biol. 17, 459–475 (2010); In: Batzoglou, S. (ed.) RECOMB 2009. LNCS, vol. 5541, pp. 321–338. Springer, Heidelberg (2009)
35. Zheng, W., Ye, X., Friedman, A., Bailey-Kellogg, C.: Algorithms for selecting breakpoint locations to optimize diversity in protein engineering by site-directed protein recombination. In: Proc. CSB, pp. 31–40 (2007)

Seeing More Is Knowing More: V3D Enables Real-Time 3D Visualization and Quantitative Analysis of Large-Scale Biological Image Data Sets

Hanchuan Peng and Fuhui Long

Howard Hughes Medical Institute

Everyone understands seeing more is knowing more. However, for large-scale 3D microscopic image analysis, it has not been an easy task to efficiently visualize, manipulate and understand high-dimensional data in 3D, 4D or 5D spaces. We developed a new 3D+ image visualization and analysis platform, V3D, to meet this need. The V3D system provides 3D visualization of gigabyte-sized microscopy image stacks in real time on current laptops and desktops. V3D streamlines the online analysis, measurement and proofreading of complicated image patterns by combining ergonomic functions for selecting a location in an image directly in 3D space and for displaying biological measurements, such as from fluorescent probes, using the overlaid surface objects. V3D runs on all major computer platforms and can be enhanced by software plug-ins to address specific biological problems. To demonstrate this extensibility, we built a V3D-based application, V3D-Neuron, to reconstruct complex 3D neuronal structures from high-resolution brain images. V3D-Neuron can precisely digitize the morphology of a single neuron in a fruitfly brain in minutes, with about a 17-fold improvement in reliability and tenfold savings in time compared with other neuron reconstruction tools. Using V3D-Neuron, we demonstrate the feasibility of building a high-resolution 3D digital atlas of neurite tracts in the fruitfly brain. V3D can be easily extended using a simple-to-use and comprehensive plugin interface.

V. Bafna and S.C. Sahinalp (Eds.): RECOMB 2011, LNBI 6577, p. 336, 2011.
© Springer-Verlag Berlin Heidelberg 2011

T-IDBA: A de novo Iterative de Bruijn Graph Assembler for Transcriptome[*]

(Extended Abstract)

Yu Peng, Henry C.M. Leung, S.M. Yiu, and Francis Y.L. Chin

Department of Computer Science, The University of Hong Kong,
Pokfulam Road, Hong Kong
{ypeng,cmleung2,smyiu,chin}@cs.hku.hk

Abstract. RNA-seq data produced by next-generation sequencing technology is a useful tool for analyzing transcriptomes. However, existing de novo transcriptome assemblers do not fully utilize the properties of transcriptomes and may result in short contigs because of the splicing nature (shared exons) of the genes. We propose the T-IDBA algorithm to reconstruct expressed isoforms without reference genome. By using pair-end information to solve the problem of long repeats in different genes and branching in the same gene due to alternative splicing, the graph can be decomposed into small components, each corresponds to a gene. The most possible isoforms with sufficient support from the pair-end reads will be found heuristically. In practice, our de novo transcriptome assembler, T-IDBA, outperforms Abyss substantially in terms of sensitivity and precision for both simulated and real data. T-IDBA is available at http://www.cs.hku.hk/~alse/tidba/

Keywords: de novo transcriptome assembly, de bruijn graph, alternative splicing, isoforms, next-generation sequencing.

1 Introduction

RNA sequencing (RNA-Seq) is a recently developed technique to sequence RNAs using the next-generation sequencing technologies. It is important in the analysis of transcriptomes and has been used successfully in multiple aspects [1-4]. Similar to the genome assembly problem, the de novo transcriptome assembly problem (the problem of reconstructing isoforms without a reference genome and annotated information) is very important. However, there has been little progress on the de novo transcriptome assembly problem. Most, if not all, existing approaches apply de novo genome assembly techniques (i.e. de Bruijn graph, string graph) directly to solve the de novo transcriptome assembly problem (e.g. [5, 6]) without fully utilizing the properties of transcriptomes. The performance of these approaches, in particular for the reconstruction of isoforms for the same gene, is not satisfactory.

There are three main difficulties for de novo transcriptome assembly. (1) Splicing nature of the genes: the same exon may appear in multiple isoforms which introducea

---

[*] This research is partly supported by RGC Grants.

V. Bafna and S.C. Sahinalp (Eds.): RECOMB 2011, LNBI 6577, pp. 337–338, 2011.
© Springer-Verlag Berlin Heidelberg 2011

lot of branches in the de Bruijn graph or the string graph. (2) Repeat patterns in different genes: subgraphs corresponding to different genes may merge together. (3) Different expression levels: it is difficult to identify low-expressed isoforms.

In this paper, we tackle the de novo transcriptome assembly problem. We analyze the properties of mammalian transcriptomes and observe that not too many genes (less than 1.4%) contain repeat patterns of length greater than 90 bp. This implies that if we can construct a de Bruijn graph using substrings of length 90 in the reads, subgraphs that correspond to different genes are more likely to be isolated. However, the current next-generation sequence technology may not produce such long reads and there are gap problems even if such long reads are available. To resolve this problem, we first build an accumulated de Bruijn graph [7] based on single-end reads up to say 50 bp (for reads of length 75 bp) and then extend to 90 bp based on pair-end reads. The graph will decompose into many connected components, most of which contain only a single or a few mRNAs. Finally, the branching problem introduced by the shared exons is resolved by a heuristic path finding algorithm based on pair-end reads to generate all possible isoforms.

T-IDBA can reconstruct 83.7% and 46.7% isoforms for simulated data and real mouse embryonic stem cells data [8] respectively. The precision of all predicted contigs are 93.4% and 79.7% respectively. It outperforms Abyss [9] which can only reconstruct 39.4% and 8.6% isoforms for the datasets and the precision of the contigs are 98.0% and 48.0% respectively.

References

1. Graveley, B.R.: Molecular biology: power sequencing. Nature 453, 1197–1198 (2008)
2. Nagalakshmi, U., Wang, Z., Waern, K., Shou, C., Raha, D., Gerstein, M., Snyder, M.: The transcriptional landscape of the yeast genome defined by RNA sequencing. Science 320, 1344–1349 (2008)
3. Trapnell, C., Pachter, L., Salzberg, S.L.: TopHat: discovering splice junctions with RNA-Seq. Bioinformatics 25, 1105–1111 (2009)
4. Jiang, H., Wong, W.H.: Statistical inferences for isoform expression in RNA-Seq. Bioinformatics 25, 1026–1032 (2009)
5. Birol, I., Jackman, S.D., Nielsen, C.B., Qian, J.Q., Varhol, R., Stazyk, G., Morin, R.D., Zhao, Y., Hirst, M., Schein, J.E., Horsman, D.E., Connors, J.M., Gascoyne, R.D., Marra, M.A., Jones, S.J.: De novo transcriptome assembly with ABySS. Bioinformatics 25, 2872–2877 (2009)
6. Jackson, B.G., Schnable, P.S., Aluru, S.: Parallel short sequence assembly of transcriptomes. BMC Bioinformatics 10 Suppl 1, S14 (2009)
7. Peng, Y., Leung, H.C.M., Yiu, S.M., Chin, F.Y.L.: IDBA- A Practical Iterative de Bruijn Graph De Novo Assembler. In: Berger, B. (ed.) RECOMB 2010. LNCS, vol. 6044, pp. 426–440. Springer, Heidelberg (2010)
8. Guttman, M., Garber, M., Levin, J.Z., Donaghey, J., Robinson, J., Adiconis, X., Fan, L., Koziol, M.J., Gnirke, A., Nusbaum, C., Rinn, J.L., Lander, E.S., Regev, A.: Ab initio reconstruction of cell type-specific transcriptomes in mouse reveals the conserved multi-exonic structure of lincRNAs. Nat. Biotechnol. 28, 503–510 (2010)
9. Simpson, J.T., Wong, K., Jackman, S.D., Schein, J.E., Jones, S.J., Birol, I.: ABySS: a parallel assembler for short read sequence data. Genome Res. 19, 1117–1123 (2009)

Experiment Specific Expression Patterns

Tobias Petri, Robert Küffner, and Ralf Zimmer

LMU Munich, Department of Informatics,
Amalienstr. 17, 80333 Munich, Germany
{petri,kueffner,zimmer}@bio.ifi.lmu.de
http://www.bio.ifi.lmu.de

Abstract. The differential analysis of genes between microarrays from
several experimental conditions or treatments routinely estimates which
genes change significantly between groups. As genes are never regulated
individually observed behavior may be a consequence of changes in other
genes. Existing approaches like co-expression analysis aim to resolve such
patterns from a wide range of experiments. The knowledge of such a background set of experiments can be used to compute expected gene behavior based on known links. It is particularly interesting to detect previously
unseen specific effects in other experiments. Here, a new method to spot
genes deviating from expected behavior (PAttern DEviation SCOring –
Padesco) is devised. It uses linear regression models learned from a background set to arrive at gene specific prediction accuracy distributions.
For a given experiment it is then decided whether each gene is predicted
better or worse than expected. This provides a novel way to estimate the
experiment specificityof each gene. We propose a validation procedure to
estimate the detection of such specific candidates and show that these
can be identified with an average accuracy of about 85 percent.

1 Introduction

Microarrays are often measurements of two or multiple conditions. A natural way
to analyze such data is through differential expression. It results in candidate
gene lists containing several hundreds of genes. Detailed biological downstream
studies are usually not feasible for all of these genes. Further filtering towards
more promising candidates is necessary. Moreover, most candidates are likely indirect targets of initially affected genes or, more generally, they follow a pattern
which can be observed similarly in other experiments. Such genes may not be
of immediate interest. In return, striking differences to known behavior indicate
specificity for a certain experiment and such genes are suited for further analysis.
We will now introduce how *Padesco* models common patterns and in which way
differences to known patterns are obtained.

Patterns. Hirsch et al. [15] noted that disease specific effects eventually trigger
core biological pathways and thus frequently lead to "a transcriptional signature that is common to a diverse set of human diseases". Such signatures can
be learned and used for experiment specific predictions. *Padesco* detects how

V. Bafna and S.C. Sahinalp (Eds.): RECOMB 2011, LNBI 6577, pp. 339–354, 2011.
© Springer-Verlag Berlin Heidelberg 2011

well the behavior of a gene can be derived from other genes. It allows to detect genes which show both differential and unexpected – and thereby interesting – behavior. The target gene patterns we learn are derived through Support Vector Regression (SVR) and basically constitute linear models describing its dependencies to other genes. Since we aim at unexpected *changes* rather than *states* we use fold-changes, not expression values to describe gene behavior.

Deviations. Not all differential genes detected by differential expression analysis are specific. A background set of experiments must be heterogeneous to assess this. *Padesco*'s key idea is that we can decide whether the behavior of a gene can be predicted worse than expected and is thus a specific gene. It is important to note that there may well be genes which are difficult or easy to predict in general. Our scoring scheme is designed to account for this individual prediction complexity by estimating an empirical distribution of deviations in a cross-validation (CV) setting.

Evaluation. Evaluation of differential expression results is difficult. Simulations may show methodological strengths and weaknesses, but biological evaluation is only possible through comparison to published knowledge or downstream experiments. We discuss *Padesco*'s performance both by means of an exhaustive simulation experiment and a detailed discussion of literature supporting genes found to be interesting by our approach. The simulation shows that genes deviating from their common behavior would be neglected due to differential expression analysis alone, since they often show only moderate differential expression.

Scoring. *Padesco* is trained on a background set of experiments consisting of 1437 microarrays from 25 experiments sharing 4117 genes. A leave-one-out cross-validation (LOOCV) across all experiments is done yielding predictions for a genes fold-change for all pairs of arrays within the omitted experiment. We estimate how well a gene can be predicted by deriving the empirical distribution of its deviations from the measured fold-changes. We then devise a score based on the median absolute deviation to score a gene in an unseen experiment. We assume that a gene in a differential setting is interesting if it exhibits a change in its gene expression. We thus use differentially expressed genes. Furthermore, a gene may be predicted better than expected, which points at stronger presence of a trained gene-gene dependency within this experiment. Although the problem is related we do not focus it in this work. If a gene is predicted worse than expected this suggests changes in a known dependency structure.

1.1   Related Work

Padesco is a natural extension to co-expression approaches like [8,27,22] or [35] as well as residual scoring schemes like [21] and [26]. Co-expression aims to construct gene sets by clustering, or construction of co-regulated gene sets across many samples. Our trained patterns are similar, yet not identical to these previously derived co-expression patterns. Mentionable differences are the use of fold-changes rather than raw measurements and our predictivity criterion. The idea that a gene can be predicted from other genes and the prediction error may model condition-specific changes is a concept previously applied as residual scoring ([21,26]). To achieve meaningful residual scores these approaches must be applied on homogeneous training data like a certain disease subtype. These scoring schemes fail on very heterogeneous experiment data. *Padesco* aims to bridge the gap between mere co-regulation detection and residual scoring. For heterogeneous training sets a background sensitive view on differential experiment results is provided. The most common framework for both residual scoring and differential analysis is provided by BioConductor ([12]). In general, identified sets of differentially expressed genes maximize sample discrimination. Such markers, or gene signatures, are promising targets for further analysis. [37] for instance, used gene expression profiling to identify transcription signatures for breast cancer prognosis classification. Gene expression profiles have been used by [27] to reveal pathway dysregulation in prostate cancer. Comprehensive resources like GEO (Gene Expression Omnibus,[7]) enable the analysis of co-expression across many experiments. The comparison can for instance be quantified by measures of reproducibility in-between experiments ([11]). *Padesco*'s linear models are loosely related to those used for imputation of missing values (IMV). Here, either data from single ([13]) or multiple experiments ([17]) is used. To our knowledge fold-change predictions have not been applied so far. SVR has been previously used by [20] for IMV.

2   Material and Methods

2.1   Data Sets

Padesco is trained on a set of experiments compiled by [22]. It consists of 3924 microarrays from 60 human data sets. These sets comprise 62.2 million expression measurements. They consist of at least 10 and up to 255 samples. Genes are filtered for a minimum amount of variance across samples. We restrict the data set to Affymetrix array platforms (HG-U95A, HG-U95Av2, HU6800, HuGeneFl, HG-U133A and HG-U133comb). There is no direct necessity for a comparability of absolute expression levels since our method works on fold-changes. We restrict to a subset G of 4117 genes which occur in more than 50% of all experiments and contain at least 75% valid (i.e. non-missing) measurements. 25 experiments with a total of $p=1437$ microarrays fulfill this constraint. We obtain a matrix of fold-changes F by sampling p pairs of arrays (Figure 1(b) and Section 2.5) from the space of all possible pairs within experiments.

For 13 selected experiments we combine arrays into sample groups, e.g. tumor vs. normal samples. This is required for differential expression analysis (Section 2.3) that compare the expression of genes between different sample groups. Some experiments, for instance time course measurements, can not be subdivided into sample groups but are used as background experiments. Overall, 62 sample groups have been analyzed. Comparisons are performed as one sample group versus the others from the same experiment, i.e. one comparison is performed for experiments with $n = 2$ sample groups and n comparisons otherwise. We thus conducted 59 comparisons in total.

2.2   Basic Protocol

Padesco uses a two step approach for the selection of candidate genes from expression measurements. Prior to the application of *Padesco*, expression patterns
must be trained on a background set *bs* of experiments (Section 2.1). We apply
Support Vector Regression (SVR, Sections 2.4 and 2.5) to train one model for
each gene using *bs*. Predicted labels are within-experiment fold-changes of this
gene. Training features are the fold-changes of all other genes. For a new experiment (not contained in *bs*), *Padesco* selects genes by two consecutive filter steps.
First, the measured genes are analyzed for differential expression (Section 2.3)
based on Wilcoxon's rank sum test ([41]). In general, any differential expression
approach can be applied as *Padesco* does not rely on a particular method. The
novel second filter step is based on an analysis of the trained regression models.
Here we discard genes that conform to the patterns learned from the background
experiments. We analyze the gene prediction errors using **residual scoring** (section 2.7) by comparing its predicted against the observed fold-changes. We then
assess its pattern conformance in terms of the distribution of its residuals across
the leave-one-out cross-validation (LOOCV) as described below. We show our
basic work flow in figure 1(a).

   To arrive at a background distribution of errors for each gene, we perform a
leave-one-out cross-validation omitting each experiment once. Each fold induces
$|G|$ models. The prediction performance can thus be evaluated independently for
each gene in each experiment. The background-training set for an (experiment
e, gene g)-pair contains all but this experiment using the gene as dependent
variable. Once trained, the application of *Padesco* involves only one prediction
per gene using the model trained on all known experiments.

2.3   Differential Analysis

We apply a common differential microarray analysis setup as primary filter step.
Different experimental sample groups are compared by the Wilcoxon rank sum
statistic. For a given gene, a p-value is computed for each sample group comparison as measure of significance of the differential expression. P-values are
corrected for multiple testing using the procedure of Benjamini-Hochberg ([2]).
In this primary filter step we assume genes as differentially expressed if they
exhibit significance-level α of 0.01.

2.4   Support Vector Regression

Padesco is based on the training of predictive regression models using ν-Support
Vector Regression (SVR, [36,34], Section 2.5). We focus Support Vector Machines (SVMs) which have acquired general acceptance for microarray applications and have been used for a wide range of tasks ([36,29]). In particular array
and tissue classification ([10,5]). They have been shown to yield very competitive
results for applications in molecular biology ([30]). In their most common formulation, SVMs work on a set D of real-valued vectors with associated labels L,

$D = \{\langle x, l(x)\rangle | x \in \mathbb{R}^n, l(x) \in L\}$. For Support Vector Regression (SVR) the labels are real numbers ($L = \mathbb{R}$). The ϵ-SVR estimates a function $f : \mathbb{R}^n \to L$ such that at most ϵ deviation ($|f(x) - l(x)| < \epsilon$) from the correct values of all training data is allowed. ν-SVR is an extension of ϵ-SVR which estimates an optimal value for ϵ. The ν parameter controls the number of support vectors and points outside of the ϵ-insensitive tube. A parameter C controls how strong deviations from the optimal model are penalized. In our experiments we apply the libSVM implementation of ν-SVR ([3]). A linear kernel is used in our experiments. We examined the linear kernel as well as the radial basis kernel. Final parameters have been chosen as $(C, \nu) = (100, 0.2)$ using an exhaustive grid search combined with leave-one-experiment out (LOO) for hyper-parameters $C \in \{10^i | i = -2, -1, \ldots, 4\}$ and $\nu \in \{0.2, \ldots, 0.8\}$. This initial screening has shown that on average the linear kernel provides similar performance to the radial basis kernel (screening $\gamma \in \{10^i | i = -4, -3, \ldots, 1\}$) and a stable prediction performance across a wide range of parameters in our setting rendering an additional CV unnecessary in our setting. We will now show how to use SVR to train models which are capable of predicting fold-changes from other genes.

2.5   Model Building

We derive SVR models to predict gene expression fold-changes. A model to predict the fold-change of a gene is trained on fold-changes of the other $(n-1)$ genes in a CV setting skipping an experiment at a time. Here, the linear kernel enables the immediate training on data sets containing missing values. We additionally examined an orthogonal coding scheme ([39]), 0-imputation and average-imputation but neither did increase the average performance of the method. Since imputation methods may introduce additional errors we omit instances carrying missing values (mis) as a label from both prediction and training. Raw data is represented as a matrix D of expression values $d_{ij} \in \mathbb{R} \cup \{mis\}$ with p rows (arrays) and n columns (genes). We denote the set of genes $G = \{g_1, \ldots, g_n\}$ and the set of arrays $A = \{a_1, \ldots, a_p\}$. Each array $a \in A$ belongs to an experiment $e(a) \in E, E = \{e_1, \ldots, e_k\}$ where each experiment e divides into sample groups $S_e = \{s_1^e, \ldots, s_{|S_e|}^e\}$. Thus, each array maps to one condition $s_e(a) \in S_e$. A single measurement is real-valued and written as m_a^g for gene $g \in G$ in array $a \in A$. *Padesco* is trained on a set of fold-change vectors derived from arrays $a, b \in A, a \neq b$ from the same experiment $e(a) = e(b)$ but different conditions i.e. $s_e(a) \neq s_e(b)$. The fold-change for gene g between array a and array b is computed as follows:

$$f_{a,b}^g = \begin{cases} mis, \text{ if } (m_a^g = mis) \text{ or } (m_b^g = mis) \\ log_2(\frac{m_a^g + c}{m_b^g + c}) \text{ otherwise.} \end{cases} \quad (1)$$

By adding a constant c, small expression values up to an estimated noise level will not lead to extreme fold-changes. The value of c was screened similar to SVR hyper-parameters during the initial parameter screening and thus set to 50. In a LOOCV, models are trained for the prediction of fold-changes of a gene g in an

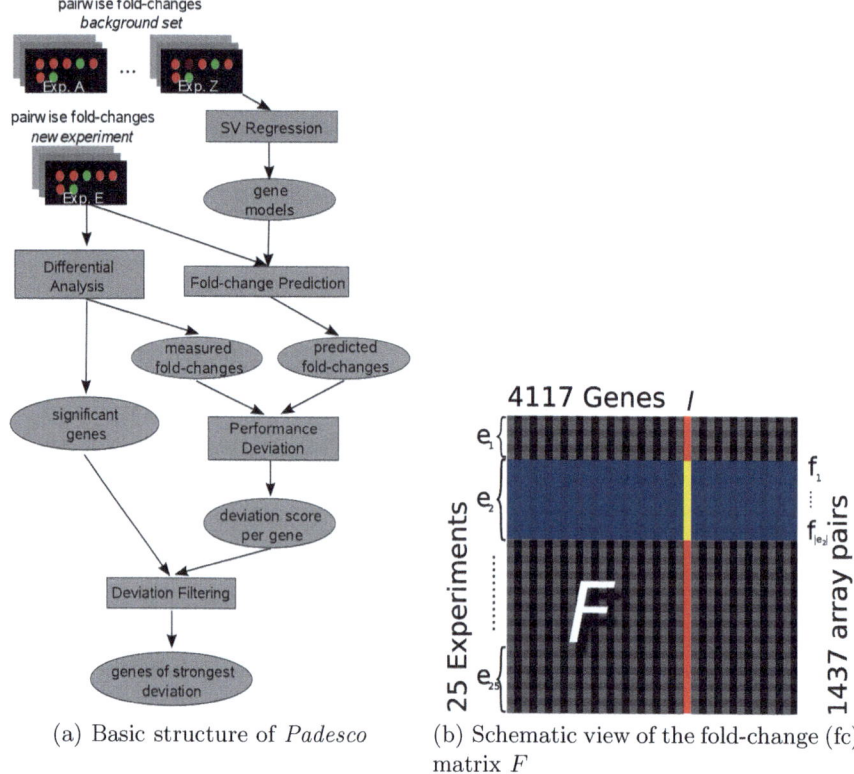

(a) Basic structure of *Padesco*   (b) Schematic view of the fold-change (fc) matrix F

Fig. 1. 1(a). A matrix F (see Figure 1(b)) of gene expression fold-changes (*fcs*) is computed. A leave-one-out cross validation (LOOCV) is performed for each gene and omitting each experiment once resulting in $|G| * |E|$ models. For a new experiment e, g's *fcs* are predicted with the model for (g,e) and compared to the measured *fcs* (*residual scoring*). Based on the LOOCV a deviation score (*padscore*) is derived. It enables deviation filtering of significant genes. **1(b). rows**: *fcs* of array pairs; **columns**: genes; *fcs* $f_1 \ldots f_p$ are computed among pairs from the same experiments. **Training**: linear models are derived from *fcs* in the training set (gray shades) with training labels of gene l (red). **Prediction** is done using the *fcs* in the test-set (blue areas) for the test labels of gene l (yellow). They result in predicted *fcs* which can be compared to measured *fcs* (*residual scoring*). **Deviation**: based on a leave-one-out validation each gene is assigned a background distribution of deviations. A median absolute deviation based score (*padscore*) is derived to estimate whether a prediction is better or worse than expected.

experiment e. For target gene g with associated label $l(g)$ the training is done using pairwise fold-change vectors of arrays within the same experiment. The induced fold-changes f_e^g constitute the target values. The experiment e to be predicted is omitted completely during training i.e. all its sample groups. Note that this setting avoids over-fitting more rigorously. It enables the estimation of prediction performance. The LOOCV results in $|E| = 25$ models for each gene and $|G| * |E|$ models overall. Note that the factor $|E|$ is only relevant for initial training purposes. The prediction requires the calculation of the prediction value for $|G|$ genes using existing models only. For our validation we obtain a model for g in each experiment e.

2.6   Fold-Change Prediction

We now simulate the real-world setting: the experiment e has never been seen during training and we like to do a prediction for a gene of interest g. First, we calculate the fold-changes for all genes among all t pairs in this particular experiment. This corresponds to a vector f_i^h for each $h \in G$ and each experiment pair $i = 1 \dots t$. We are especially interested in the values $\mathbf{f}^g = \{f_i^g | i = 1 \dots t\}$ holding the measured fold-changes for g and predict them using the model for g. This yields $\mathbf{f}'^g = \{f_i'^g | i = 1 \dots t\}$ constituting the vector of all predicted fold-changes for g in experiment e.

2.7   Scoring Performance Deviation

We now have a measured fold-change vector $\mathbf{f}^g$ and a predicted fold-change vector $\mathbf{f}'^g$ for an experiment e for a gene g. We compare them using the uncentered Pearson correlation $\rho_{g,e}$. Given the two n-dimensional vectors $\mathbf{f}^g$ and $\mathbf{f}'^g$ it is calculated as:

$$\rho_{g,e} = \frac{(\sum_{i=1}^{n} f_i^g \cdot f_i'^g)}{\sqrt{\sum_{i=1}^{n} f_i^{g2} \sum_{i=1}^{n} f_i'^{g2}}} \tag{2}$$

We also compute a discretized version of $\mathbf{f}^g$. We set all fold-changes above a threshold t_f to 1 and all below to 0. Thereby, we can compute an Area Under Curve (AUC) value by varying t_f for $\mathbf{f}'^g$ and leave the threshold for $\mathbf{f}^g$ fixed at 2.

We now know how well a gene performs in a single experiment. Yet, we cannot say whether this is more or less than we expected. To arrive at a deviation of known patterns we compute the empirical distribution D^g of all $\{\rho_{g,e} | e \in E\}$. The deviation for an experiment e and gene g is expressed in units of median absolute deviations (*padscore*) with respect to this distribution. Given the median med^g of D^g and the corresponding median absolute deviation mad^g the *padscore* is given by:

$$(med^g - \rho_{g,e})/mad^g \tag{3}$$

In order to be retained by *Padesco* genes must simultaneously satisfy the significance level of differential expression in a specific sample group given by a certain α-level as well as a minimum *padscore*. These genes are referred to as *specific*. Remaining *unspecific* genes are significantly regulated, but could be predicted by the SVR models.

2.8   Permutation Test for the Evaluation of *Padesco*

Gold standards on the experiment-specific expression of genes are not available. We use a permutation test to simulate genes deviating from their common behavior. By copying (*spiking*) the expression values from a significant gene g^+ to an insignificant gene g^- within an experiment e we force genes to violate their common behavior and trivially become significant. We have two choices in parameters here. First, we can choose a *z-score* level t_z for significance. Second, we can choose a *padscore* level t_p to select interesting genes. The following permutation test selects these thresholds and estimates the associated performance for spike-in controls. We sample from the significant genes, and spike into the insignificant genes. We then recompute the *padscore* for the previously insignificant gene. This process is repeated s times where s is the number of significant genes. $SPIKE$ denotes the set of spiked genes. After all repeats have been computed we obtain sensitivity=$tp/(tp + fn)$ and precision=$tp/(tp + fp)$ and repeat the evaluation for all possible thresholds t_z and t_p in the experiment. A gene g's recomputed *padscore* p determines whether it is tp, fn or fp based on these thresholds (see Table 1).

Table 1. Classification assignment for the evaluation. A threshold on *z-score* (t_z) and *padscore* (t_p) is chosen. After spiking (see Section 2.8) a gene g's recomputed *padscore* p determines its type.

Type	Abbreviation	Condition
true positive	tp	$g \in SPIKE \wedge p \geq t_p$
false positive	fp	$g \notin SPIKE \wedge p \geq t_p$
false negative	fn	$g \in SPIKE \wedge p < t_p$

3   Results

Evaluation of Expression Fold-change Predictions. We use the uncentered correlation to measure how well the regression models can predict expression fold-changes of gene g in experiment e. The prediction performance is significantly better than random for the majority of genes: the uncentered correlation achieved an average value of $\rho_u = 0.7$ in our experiment. 91% of the predictions exhibit a $\rho_u > 0$. After discretization we achieve an AUC of 81% on 15.7% cases with a fold-change of more than two. We argue that the remaining specific candidates that cannot be predicted well are particularly interesting because they exhibit an experiment-specific expression that could not be learned from the training data.

Prediction of Fold-changes for Individual Genes. Gene signatures, i.e. sets of genes that are differentially regulated between different cellular states are frequently published along with microarray experiments. Such signatures are expected to e.g. yield diagnostic markers that could help to differentiate between healthy and sick individuals. Here, we examine a gene signature that has

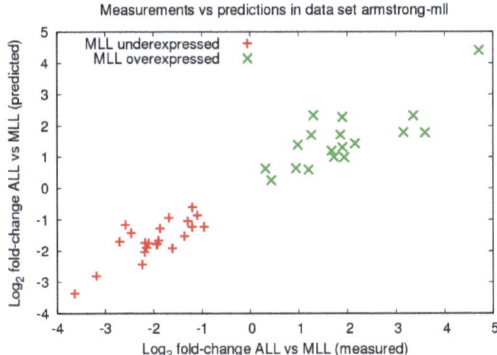

Fig. 2. Scatter plot of predicted vs. measured fold-changes. Microarray studies frequently derive gene signatures, i.e. sets of differentially expressed genes discriminating between experimental conditions. Gene signatures can be predicted well by our SVR models, as shown here for a gene signature distinguishing ALL and MLL leukemic genotypes published along with the data set of armstrong-mll ([1]). For every gene, our SVR models predict expression changes between ALL and MLL correctly, although the precise values of the measured fold-changes are not reproduced exactly. A gene is depicted as a single point, i.e. as average of all fold-changes of the given gene (section 2.5) over the array-pairs comparing the conditions ALL and MLL.

been compiled by [1] to distinguish a particular chromosomal translocation involving the MLL (mixed-lineage leukemia) gene from the regular ALL (acute lymphoblastic leukemia) genotype. The MLL translocation is significant as it frequently leads to an early relapse after chemotherapy. In Figure 2 we compare our predictions against the experimentally determined expression fold-changes for this gene signature. For all genes, the direction of differential expression can be correctly derived from our predictions, although the values of the measured fold-changes are not reproduced exactly. Similar results have been obtained for other published signatures.

Permutation Test Based Evaluation. *Padesco* filters genes based on a standard differential expression *z-score* (Wilcoxon test, Section 2.3) and a novel *padscore* (Section 2.7) indicating experiment-specific expression. This second score indicates whether genes can be predicted less well than expected from the training (background) set of experiments. They are selected due to their *padscore* since they do not conform to their trained patterns. Cutoffs on the two scores are required for the selection of specific candidate genes that exhibit differential as well as experiment specific expression. The permutation test introduced in Section 2.8 generates artificial pattern deviations through spiked genes. As shown in figure 3 at a *padscore* cutoff of 2.0 specific candidates are accurately detected (85% precision). The precision increases for differentially expressed genes ($z > 3$). Based on the evaluation we picked a *padscore* threshold

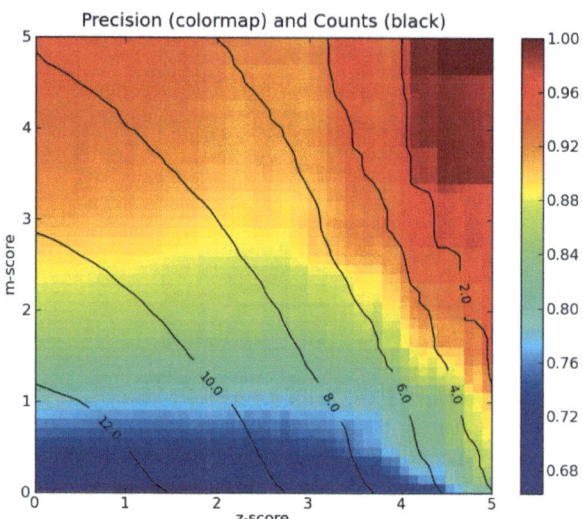

Fig. 3. Precision (percent, colormap) vs. number of detected genes (log base 2, contour). As gold standards are not available, we estimate the performance of specific candidate gene detection by a permutation test. This test evaluates how well known spike-in controls can be recovered by *Padesco* for arbitrary z-score (differential expression) and *padscore*/m-score (experiment specific expression) thresholds. Specific candidate genes can be reliably identified (85% precision) using a *padscore* above 1.5 even if they exhibit only moderate levels of differential expression (*z-score* < 4). By combined *z-score* and *padscore* thresholds candidate gene lists can efficiently be reduced for follow-up studies. Analyzed here are 59 condition comparisons from 13 experiments. At the chosen p-score (2.0) and z-score (3.0) thresholds, some 250 specific candidates (contour) are detected by *Padesco*.

of 2.0 and a *z-score* threshold of 3.0 to receive a moderate number of candidates exhibiting a high precision. Thereby we selected some 250 specific candidates.

Specific Candidates. In this section we discuss sample results in two experiments that examine prostate cancer ([32,40]) and one study on leukemia ([1]). As a further example we provide results from a *Toxoplasma gondii* infection study by [4]. *Padesco* does not aim to whole relevant pathways but allows to focus on a small subset of interesting genes for further analysis. [38] discuss key pathways which are likely to be disturbed to promote cancer in almost any cell type. Development of cancer is strongly coupled to the perturbation of one or more such pathways. The interleukin 2 pathway has been shown to be deregulated in many cancer types and was also described to be involved in prostate cancer. IL2RB dimerization with the α-subunit leads to a higher affinity towards interleukin-2. IL-2 treatment was previously shown to lead to reduced prostate tumor growth in rats ([14,24]). Another cancer therapy using IL-2 has been developed by [25].

Eicosanoids are known to interact with immune messengers like interleukins. In the first experiment by [40] tumor samples were compared against normal and HUVEC (Human Umbilical Vein Endothelial Cells) samples, where 27 genes have been detected as differentially expressed. For an initial screening of functional significance we apply a gene ontology over-representation analysis (DAVID, [6,18]) on the differentially expressed genes, i.e. without *padscore* filter. As for [40] a screening for significance (Benjamini-Hochberg corrected scores, $\alpha = 0.01$) shows no significant enrichment, yet 3 genes (IPR, EPR3R and CYT450J) are found to belong to the eicosanoid metabolism (p=0.08). With *padscore* filter Interleukin 2 receptor β (IL2RB) is the only gene found to be interesting in this experiment. *Padesco* reported an unusual expression of IL2RB in the tumor samples, which could explain the decoupling of eicosanoid pathway members from IL2RB regulation.

The second examined experiment on prostate cancer is described in [32]. Here, 60 differentially expressed genes were identified. DAVID analysis shows no significant over-representation. After *Padesco*-filtering, 4 genes remain that we discuss in the following. HCK (*padscore*=7.2), an src related tyrosin kinase is most interesting in terms of the *padscore*. [33] describes its association with gpl130 and the formation of a complex with IL-6R which promotes high affinity binding of IL-6. In prostate cancer, IL-6 is a key protein. It has been suggested to contribute to prostate cancer progression towards an androgen-independent state. We observe MGC17330 (PIK3IP1, *padscore*=2.6) to be the second *padscore* relevant gene. It is a negative regulator of PI3K. Src kinases are upstream mediators for the PI3K signaling pathway with important roles in proliferation, migration and survival. It is described to be a tumor suppressor in heptacellular carcinomas ([9]). It shows only a weak positive fold-change in this experiment which may explain why it fails as a suppressor here. Mutations within the PI3K pathway have been described by [38] to be involved in a number of tumor types. The third gene found is ATP2B1 (PMCA1, *padscore*=2.6). It is a Ca^{2+} ATPase subunit. An unusual reduction in gene expression can be observed in our data. This reduction is discussed by [28]. Ca^{2+} pumps are likely to provide good therapeutic targets for anticancer drug development as suggested by [23] and [28]. They emphasize the role of Ca^{2+} (intracellular calcium) in the life-and-death decisions of the cell such that disturbed control of Ca^{2+} may lead to an inappropriate cell fate. The fourth candidate – C2orf3 (*padscore*=2.03) – has also been identified by an approach by [16]. They analysed three prostate cancer sets but since more than a hundred genes are identified and C2orf3 is no top-ranking gene this candidate has not been subject to further discussion. The transcription repressor binds GC-rich sequences of the epidermal growth factor receptor, beta-actin and calcium-dependent protease promoters.

[4] analyze diverse parasite infections on human macrophages and dendritic cells. We find CCR1 to be the most prominent gene (*padscore* = 9.29) in the Toxoplasma infection subgroup. Mice lacking this chemokine receptor CCR1 have shown dramatically increased mortality after Toxomplasma gondii infection [19].

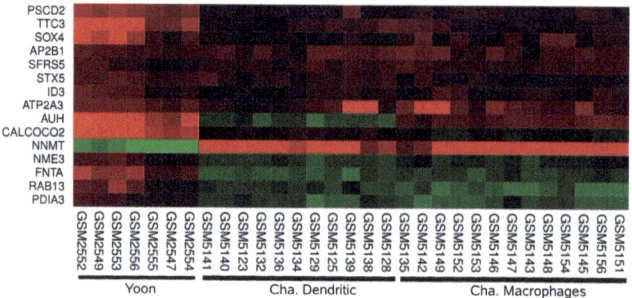

Fig. 4. Common Regulation and Experiment Deviation. The heat map shows a cluster of genes which are usually correlated (pairwise Pearson's Correlation above 0.8, 'yoon-p53 (Yoon)' [42] is given as reference). Here, AUH with a *padscore* of 5.09 is detected as interesting in 'chaussabel-parasite (Cha)' [4].

4   Discussion

Genes are not regulated individually ([22,35]). Frequently, patterns of co- or anti-regulation can be observed such that the up-regulation of a gene A is a good hint that another gene B will also be up-regulated while a third gene C will rather be down-regulated. The disruption of such patterns pinpoints genes with experiment-specific expressions. We call such genes *specific* candidates in contrast to the remaining *unspecific* candidates that exhibit only generic expression patterns. *Padesco* detects *specific* candidates by analyzing fold-change based co-expression patterns with Support Vector Regression (SVR) models trained on a background set of microarray experiments. After training, we select *specific* candidate genes via a two stage filter. The first filter step is a routine analysis of differential expression (significant genes). A novel second filter selects genes that show deviations from generic expression patterns predictable by linear models (interesting genes).

In order to avoid the predictions of false *specific* candidates *Padesco* depends on a good prediction performance of the underlying SVR models. The prediction performance can be evaluated rigorously as the prediction target experiment is excluded from training in a leave one out cross-validation setting where all conditions of particular experiments are left out. This not only excludes condition specific but also experiment specific biases. We examined 4117 genes across 25 experiments consisting of 1437 individual microarrays. Predictions by *Padesco* are better than expected by chance in 91% of the cases. It has been criticized that gene signatures rarely help to identify the involved biological processes or the causal regulatory mechanisms ([31]). [15] further argued that a gene signatures frequently do not represent specific attributes of the measured biological conditions. We analyzed gene signatures published together with the corresponding microarray experiments. These signatures were selected by the authors of the corresponding studies to discriminate between experimental conditions (sample groups). We found that expression changes for genes in signatures are predicted well by our SVR models trained on other, unrelated experiments. An example is the signature distinguishing ALL and MLL ([1]). Although the ALL/AML signature certainly provides discriminating marker genes, it does not capture experiment specific expression patterns according to *Padesco*.

The extend of differential expression alone does not indicate experiment *specific* involvement of genes. Based on the prediction performance we identified specific candidates genes that exhibit experiment *specific* expression, i.e. expression changes that cannot be explained (predicted) by our models. This analysis is related to co-expression studies and complements differential expression analysis. It enables to focus on concise candidate lists for follow-up studies that consist of experiment-specific candidates only. We screened for filter thresholds and estimated *Padesco*'s performance from permutation tests as comprehensive gold standards for the experiment specific expression of genes are not available. This newly devised simulation approach suggests that *specific* candidates are identified reliably by *Padesco* ($> 85\%$ precision at *padscore* > 1.5) even if they show only marginal levels of differential expression. On the other hand, more than 90% of the genes selected by differential expression alone exhibit only generic expression patterns and could thus be excluded from further studies. *Specific* candidates are likely to represent characteristic features of the corresponding experimental conditions.

We evaluated *Padesco* selected genes for two data sets on prostate cancer. Besides interesting new candidates, we found several genes with a known involvement in the disease. Some of them, such as IL-2RB, have already been reported as promising drug targets. We demonstrated that such examples are more difficult to detect by differential expression analysis alone. Instead, differential expression tends to pick up genes that act similarly in other, biologically unrelated experiments. Thus, in combination with differential expression analysis, *Padesco* is a promising protocol for the detection and analysis of particularly distinctive features of microarray experiments.

References

1. Armstrong, S.A., Staunton, J.E., Silverman, L.B., Pieters, R., den Boer, M.L., Minden, M.D., Sallan, S.E., Lander, E.S., Golub, T.R., Korsmeyer, S.J.: MLL translocations specify a distinct gene expression profile that distinguishes a unique leukemia. Nature Genetics 30, 41–47 (2002)
2. Benjamini, Y., Hochberg, Y.: Controlling the false discovery rate: a practical and powerful approach to multiple testing. Journal of the Royal Statistical Society. Series B (Methodological) 57(1), 289–300 (1995)
3. Chang, C.-C., Lin, C.-J.: LIBSVM: a library for support vector machines (2001), software http://www.csie.ntu.edu.tw/~cjlin/libsvm
4. Chaussabel, D., Semnani, R.T., McDowell, M.A., Sacks, D., Sher, A., Nutman, T.B.: Unique gene expression profiles of human macrophages and dendritic cells to phylogenetically distinct parasites. Blood 102(2), 672–681 (2003)
5. Chen, Z., Li, J., Wei, L.: A multiple kernel support vector machine scheme for feature selection and rule extraction from gene expression data of cancer tissue. Artif. Intell. Med. 41(2), 161–175 (2007)

6. Dennis, G., Sherman, B.T., Hosack, D.A., Yang, J., Gao, W., Lane, H.C., Lempicki, R.A.: DAVID: Database for Annotation, Visualization, and Integrated Discovery. Genome Biology 4(5), P3 (2003)

7. Edgar, R., Domrachev, M., Lash, A.E.: Gene Expression Omnibus: NCBI gene expression and hybridization array data repository. Nucl. Acids Res. 30(1), 207–210 (2002)

8. Eisen, M., Spellman, P., Brown, P., Botstein, D.: Cluster analysis and display of genome-wide expression patterns. Proc. Natl. Acad. Sci. U.S.A 95, 14863–14868 (1998)

9. Fabregat, I.: Dysregulation of apoptosis in hepatocellular carcinoma cells. World Journal of Gastroenterology: WJG 15(5), 513 (2009)

10. Furey, T.S., Cristianini, N., Duffy, N., Bednarski, D.W., Schummer, M., Haussler, D.: Support vector machine classification and validation of cancer tissue samples using microarray expression data. Bioinformatics 16(10), 906–914 (2000)

11. Garrett-Mayer, E., Parmigiani, G., Zhong, X., Cope, L., Gabrielson, E.: Cross-study validation and combined analysis of gene expression microarray data. Biostatistics 9, 333–354 (2008)

12. Gentleman, R., Carey, V., Bates, D., Bolstad, B., Dettling, M., Dudoit, S., Ellis, B., Gautier, L., Ge, Y., Gentry, J., Hornik, K., Hothorn, T., Huber, W., Iacus, S., Irizarry, R., Leisch, F., Li, C., Maechler, M., Rossini, A., Sawitzki, G., Smith, C., Smyth, G., Tierney, L., Yang, J., Zhang, J.: Bioconductor: open software development for computational biology and bioinformatics. Genome Biology 5, R80 (2004)

13. Hastie, T., Tibshirani, R., Sherlock, G., Eisen, M., Brown, P., Botstein, D.: Imputing Missing Data for Gene Expression Arrays. Technical report, Stanford Statistics Department (1999)

14. Hautmann, S.H., Huland, E., Huland, H.: Local intratumor immunotherapy of prostate cancer with interleukin-2 reduces tumor growth. Anticancer Res. 19(4A), 2661–2663 (1999)

15. Hirsch, H.A., Iliopoulos, D., Joshi, A., Zhang, Y., Jaeger, S.A., Bulyk, M., Tsichlis, P.N., Liu, X.S., Struhl, K.: A transcriptional signature and common gene networks link cancer with lipid metabolism and diverse human diseases. Cancer Cell 17(4), 348–361 (2010)

16. Hong, D., Lee, J., Hong, S., Yoon, J., Park, S.: Extraction of Informative Genes from Integrated Microarray Data. Springer, Heidelberg (2008)

17. Hu, J., Li, H., Waterman, M., Zhou, X.: Integrative missing value estimation for microarray data. BMC Bioinformatics 7, 449 (2006)

18. Huang, D.W., Sherman, B.T., Lempicki, R.A.: Systematic and integrative analysis of large gene lists using DAVID bioinformatics resources. Nat. Protoc. 4(1), 44–57 (2009)

19. Khan, I., Murphy, P., Casciotti, L., Schwartzman, J., Collins, J., Gao, J., Yeaman, G.: Mice lacking the chemokine receptor CCR1 show increased susceptibility to Toxoplasma gondii infection. The Journal of Immunology 166(3), 1930 (2001)

20. Kim, H., Golub, G.H., Park, H.: Missing value estimation for DNA microarray gene expression data: local least squares imputation. Bioinformatics 21(2), 187–198 (2005)

21. Kostka, D., Spang, R.: Finding disease specific alterations in the co-expression of genes. Bioinformatics 20(Suppl 1), i194–i199 (2004)

22. Lee, H.K., Hsu, A.K., Sajdak, J., Qin, J., Pavlidis, P.: Coexpression analysis of human genes across many microarray data sets. Genome Research 14(6), 1085–1094 (2004)

23. Monteith, G., McAndrew, D., Faddy, H., Roberts-Thomson, S.: Calcium and cancer: targeting Ca2+ transport. Nature Reviews Cancer 7(7), 519–530 (2007)
24. Moody, D.B., Robinson, J.C., Ewing, C.M., Lazenby, A.J., Isaacs, W.B.: Interleukin-2 transfected prostate cancer cells generate a local antitumor effect in vivo. Prostate 24(5), 244–251 (1994)
25. Otter, W.D., Jacobs, J.J.L., Battermann, J.J., Hordijk, G.J., Krastev, Z., Moiseeva, E.V., Stewart, R.J.E., Ziekman, P.G.P.M., Koten, J.W.: Local therapy of cancer with free IL-2. Cancer Immunol. Immunother. 57(7), 931–950 (2008)
26. Prieto, C., Rivas, M.J., Sánchez, J.M., López-Fidalgo, J., Rivas, J.D.L.: Algorithm to find gene expression profiles of deregulation and identify families of disease-altered genes. Bioinformatics 22(9), 1103–1110 (2006)
27. Rhodes, D.R., Barrette, T.R., Rubin, M.A., Ghosh, D., Chinnaiyan, A.M.: Meta-Analysis of Microarrays: Interstudy Validation of Gene Expression Profiles Reveals Pathway Dysregulation in Prostate Cancer. Cancer Res. 62(15), 4427–4433 (2002)
28. Roderick, H.L., Cook, S.J.: Ca2+ signalling checkpoints in cancer: remodelling Ca2+ for cancer cell proliferation and survival. Nat. Rev. Cancer 8(5), 361–375 (2008)
29. Schölkopf, B., Smola, A.J.: Learning with Kernels: Support Vector Machines, Regularization, Optimization, and Beyond (Adaptive Computation and Machine Learning). The MIT Press, Cambridge (December 2001)
30. Schölkopf, B., Tsuda, K., Vert, J.: Kernel Methods in Computational Biology (Computational Molecular Biology). The MIT Press, Cambridge (2004)
31. Segal, E., Friedman, N., Koller, D., Regev, A.: A module map showing conditional activity of expression modules in cancer. Nat. Genet. 36(10), 1090–1098 (2004)
32. Singh, D., Febbo, P.G., Ross, K., Jackson, D.G., Manola, J., Ladd, C., Tamayo, P., Renshaw, A.A., D'Amico, A.V., Richie, J.P., Lander, E.S., Loda, M., Kantoff, P.W., Golub, T.R., Sellers, W.R.: Gene expression correlates of clinical prostate cancer behavior. Cancer Cell 1(2), 203–209 (2002)
33. Smith, P., Hobisch, A., Lin, D., Culig, Z., Keller, E.: Interleukin-6 and prostate cancer progression. Cytokine & Growth Factor Reviews 12(1), 33–40 (2001)
34. Smola, A., Schölkopf, B.: A tutorial on support vector regression. Technical Report NC2-TR-1998-030, NeuroCOLT2 (1998)
35. Ucar, D., Neuhaus, I., Ross-MacDonald, P., Tilford, C., Parthasarathy, S., Siemers, N., Ji, R.-R.: Construction of a reference gene association network from multiple profiling data: application to data analysis. Bioinformatics 23(20), 2716–2724 (2007)
36. Vapnik, V.N.: Statistical Learning Theory. Wiley-Interscience, Hoboken (September 1998)
37. Veer, V.L., Dai, H., van de Vijver, M.J., He, Y.D., Hart, A.A., Mao, M., Peterse, H.L., van der Kooy, K., Marton, M.J., Witteveen, A.T., Schreiber, G.J., Kerkhoven, R.M., Roberts, C., Linsley, P.S., Bernards, R., Friend, S.H.: Gene expression profiling predicts clinical outcome of breast cancer. Nature 415(6871), 530–536 (2002)
38. Vogelstein, B., Kinzler, K.W.: Cancer genes and the pathways they control. Nature Medicine 10(8), 789–799 (2004)
39. Wang, X., Li, A., Jiang, Z., Feng, H.: Missing value estimation for DNA microarray gene expression data by Support Vector Regression imputation and orthogonal coding scheme. BMC Bioinformatics 7(1), 32 (2006)

40. Welsh, J.B., Sapinoso, L.M., Su, A.I., Kern, S.G., Wang-Rodriguez, J., Moskaluk, C.A., Frierson, J., Henry, F., Hampton, G.M.: Analysis of Gene Expression Identifies Candidate Markers and Pharmacological Targets in Prostate Cancer. Cancer Res. 61(16), 5974–5978 (2001)
41. Wilcoxon, F.: Individual Comparisons by Ranking Methods. Biometrics Bulletin 1(6), 80–83 (1945)
42. Yoon, H., Liyanarachchi, S., Wright, F.A., Davuluri, R., Lockman, J.C., de la Chapelle, A., Pellegata, N.S.: Gene expression profiling of isogenic cells with different TP53 gene dosage reveals numerous genes that are affected by TP53 dosage and identifies CSPG2 as a direct target of p53. PNAS USA 99(24), 15632–15637 (2002)

Geometric Interpretation of Gene Expression by Sparse Reconstruction of Transcript Profiles

Yosef Prat[1], Menachem Fromer[1], Michal Linial[2,*], and Nathan Linial[1,3]

[1] School of Computer Science and Engineering, The Hebrew University of Jerusalem, Israel
[2] Dept of Biological Chemistry, Institute of Life Sciences,
The Hebrew University of Jerusalem, Israel
[3] Sudarsky Center for Computational Biology, The Hebrew University of Jerusalem, Israel

Large-scale data collection technologies have come to play a central role in biological and biomedical research in the last decade. Consequently, it has become a major goal of functional genomics to develop, based on such data, a comprehensive description of the functions and interactions of all genes and proteins in a genome. Most large-scale biological data, including gene expression profiles, are usually represented by a matrix, where n genes are examined in d experiments. Here, we view such data as a set of n points (vectors) in d-dimensional space, each of which represents the profile of a given gene over d different experimental conditions. Many known methods that have yielded meaningful biological insights seek geometric or algebraic features of these vectors.

To properly appreciate our approach, it is useful to discuss the activity in the machine-learning area as a modern-day approach to the classical questions of statistics. The data at hand is considered as being sampled from some distribution and the question is to get a comprehensive description of that distribution. When data items are (or can be naturally viewed as) points in space, it is possible to utilize any "unexpected" geometric properties that this set of points (corresponding to data items) has. For instance, unanticipated pairs of points (even nearly) in the same direction from the origin are likely a reflection of an interesting property in the domain from which the data set came. This is our interpretation of correlation analysis. Likewise, a generic point set in Euclidean space is not expected to be stretched in any special directions. Therefore if your data set, viewed geometrically, is stretched in certain directions, it can often be used to discover interesting phenomena, this is our interpretation of SVD analysis, and its principal component analysis (PCA) implementation.

Correlations and stretch are only two of the numerous properties that one may consider in a point set in Euclidean space. Our work considers another very basic property that we know not to exist in generic sets: (Nearly) linearly-dependent sets of points of cardinality that is substantially smaller than the dimension of the host space. When such an unexpected property of the data set is discovered, two questions suggest themselves: (i) Is this phenomenon only coincidental? and (ii) How can this geometric property of the data help us learn something about the system which it represents? In this study, we aim to address these questions. We confirm the robustness of this property under multiple

---

[*] Corresponding author.

V. Bafna and S.C. Sahinalp (Eds.): RECOMB 2011, LNBI 6577, pp. 355–357, 2011.
© Springer-Verlag Berlin Heidelberg 2011

perturbations and the generality for multiple model organisms, and further demonstrate the significance of the biological insights it unveils.

The geometric principle that underlies this is close in spirit, and inspired by, recent advances in compressive sensing and sparse signal recovery (Candes and Tao, 2005; Donoho, 2006). A conceptually new method that we call SPARCLE (SPArse ReCovery of Linear combinations of Expression) is introduced. In practice, we wish to discover linear dependencies within groups of expression profiles, using full transcriptome measured under multiple environmental conditions. To this end, for each

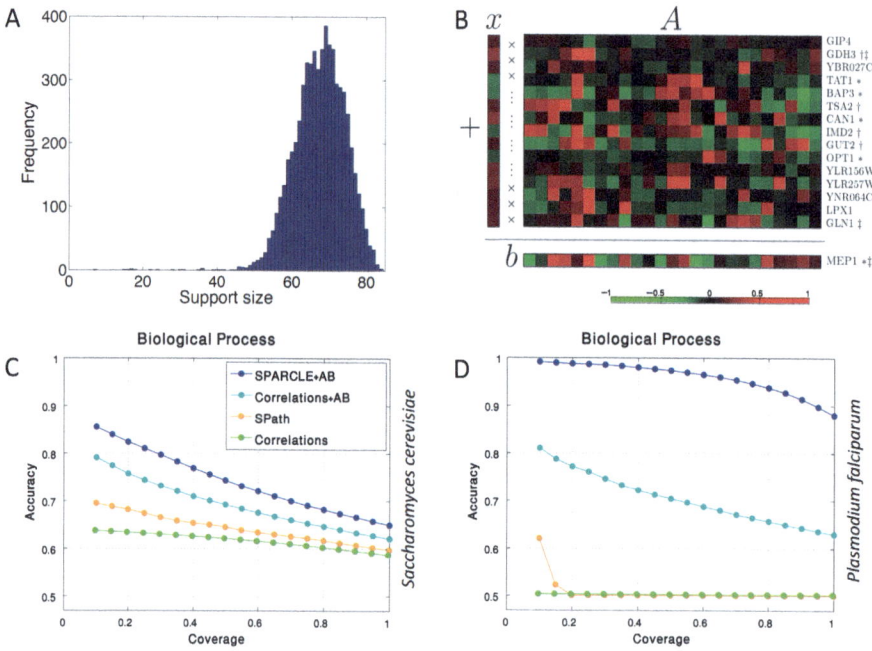

Fig. 1. Sparse reconstruction of *S. cerevisiae* genes expression profiles, and SPARCLE-based machine learning prediction. (A) Number of expression profiles needed to linearly reconstruct each profile in the yeast transcriptome. 6254 yeast genes in 85 experiments analyzed. The common representation uses considerably less genes than the rank of the matrix (85). (B) The expression profile reconstruction for MEP1 (ammonium transporter) as recovered by SPARCLE. The expression profile of the gene (bottom) is displayed as a linear combination of the profiles of its supporting genes, with their corresponding coefficients (left). For comprehensibility, only the 15 genes with the largest absolute value coefficients are shown, as well as a third of the 85 conditions. (C) Prediction of *S. cerevisiae* genes' associations according to GO Biological Process, where accuracy is traded off with coverage. A comparison of SPARCLE-based AdaBoost learning (SPARCLE+AB), correlation-based AdaBoost learning (Correlations+AB), correlations-based shortest path (SPath) method (Zhou, et al., 2002), and pairwise correlations for the raw data (Correlations). (D) Similar predictions of *P. falciparum* genes' associations.

gene, we seek the smallest number of expression profiles, whose linear span contains the expression profile of that gene. We exemplified SPARCLE on *Saccharomyces cerevisiae* transcriptome that consists of 170 conditions and ~6200 genes (Knijnenburg, et al., 2009), recovering the expression profiles using short (Fig. 1A) representations. Specifically, each profile reconstructed as a linear combination of a small subset of other profiles (Fig. 1B). We further confirmed the stability and robustness of SPARCLE results under perturbations to the data. The value of the information retrieved by the SPARCLE approach was demonstrated by using its results as a basis for machine learning classification of gene associations. A comprehensive evaluation was performed on the immensely explored budding yeast and the poorly annotated malaria-parasite *Plasmodium falciparum* transcriptomes (Knijnenburg, et al., 2009). The evaluation covers the PPI networks and all resolution levels of the GO annotation database, and included accurate prediction of pairwise genes' associations (Fig. 1 C,D), using the AdaBoost platform. We demonstrate, applying the SPARCLE based machine learning method, the large potential of using such a poorly studied geometric approach to extract meaningful insights from raw high-throughput data. It is a first step of utilizing the above-described geometric perspective toward a deeper understanding of the complexity of gene associations.

A detailed study is available in Prat *et al.* (2011) Bioinformatics, in press. This study is partially supported by the ISF grant and the Prospects EU Framework VII.

References

Candes, E.J., Tao, T.: Decoding by linear programming. IEEE Transactions on Information Theory 51, 4203–4215 (2005)

Donoho, D.L.: For most large underdetermined systems of linear equations the minimal l(1)-norm solution is also the sparsest solution. Communications on Pure and Applied Mathematics 59, 797–829 (2006)

Zhou, X., et al.: Transitive functional annotation by shortest-path analysis of gene expression data. Proc. Natl. Acad. Sci. USA 99, 12783–12788 (2002)

Hu, G., et al.: Transcriptional profiling of growth perturbations of the human malaria parasite Plasmodium falciparum. Nat. Biotechnol. 28, 91–98 (2009)

Knijnenburg, T.A., et al.: Combinatorial effects of environmental parameters on transcriptional regulation in Saccharomyces cerevisiae: A quantitative analysis of a compendium of chemostat-based transcriptome data. Bmc Genomics 10, 53 (2009)

A Ribosome Flow Model for Analyzing Translation Elongation

(Extended Abstract)

Shlomi Reuveni[1,2,*], Isaac Meilijson[1], Martin Kupiec[3],
Eytan Ruppin[4,5], and Tamir Tuller[6,7,*,**]

[1] Statistics and Operations Research
[2] School of Chemistry
[3] Molecular Microbiology and Biotechnology Department
[4] School of Computer Sciences
[5] School of Medicine Tel Aviv University, Ramat Aviv, Israel
[6] Faculty of Mathematics and Computer Science, Weizmann Institute of Science
[7] Department of Molecular Genetics, Weizmann Institute of Science. Rehovot, Israel
tamir.tuller@weizmann.ac.il

We describe the first genome wide analysis of translation based on a model aimed at capturing the physical and dynamical aspects of this process. The Ribosomal Flow Model (RFM) is a computationally efficient approximation of the Totally Asymmetric Exclusion Process (TASEP) model (*e.g.* see [1]). The RFM is sensitive to the order of codons in the coding sequence, the tRNA pool of the organism, interactions between ribosomes and their size (see Figure 1). The RFM predicts fundamental outcomes of the translation process, including translation rates, protein abundance and ribosomal densities [2] and the relation between all these variables, better than alternative ('non-physical') approaches (*e.g.* see [3,4]). In addition, we show that the RFM model can be used for accurate inference of initiation rates, the effect of codon order on protein abundance and the cost of translation. All these variables could not be inferred by previous predictors.

In the RFM, mRNA molecules are coarse-grained into sites of C codons; (in the Figure $C = 3$); in practice we use $C = 25$, a value that is close to various geometrical properties of the ribosome. Ribosomes arrive at the first site with initiation rate λ but are only able to bind if this site is not occupied by another ribosome. A ribosome that occupies the ith site proceeds, with rate λ_i, to the consecutive site provided the latter is not occupied by another ribosome. Transition rates are determined by the codon composition of each site and the tRNA pool of the organism (Figure 1). Denoting the probability that the ith site is occupied at time t by $p_i(t)$, the rate of ribosome flow into/out of the system is given by: $\lambda(1 - p_1(t))$ and $\lambda_n p_n(t)$ respectively. The rate of ribosome flow from site i to site $i + 1$ is given by: $\lambda_i p_i(t)(1 - p_{i+1}(t))$.

---

* SR and TT contributed equally to this work.
** Corresponding author.

V. Bafna and S.C. Sahinalp (Eds.): RECOMB 2011, LNBI 6577, pp. 358–360, 2011.
© Springer-Verlag Berlin Heidelberg 2011

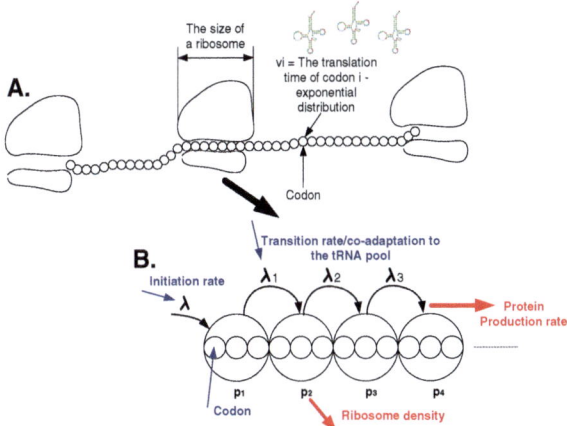

Fig. 1. An illustration of the RFM. *A*. The RFM is based on the TASEP and is sensitive to the order of codons in the coding sequence, the tRNA pool of the organism, interactions between ribosomes and their size. *B*. To accelerate simulation times we approximate the stochastic dynamics of the TASEP by a set of differential equations describing the flow of ribosomes.

From applying the RFM in a systems biological study of translation it follows that increasing the number of available ribosomes (or equivalently the initiation rate) increases the genomic translation rate and the mean ribosome density only up to a certain point, beyond which both saturate. Strikingly, assuming that the translation system is tuned to work at the pre-saturation point leads to the best predictions of experimental data. This implies that, in all the analyzed organisms (from bacteria to Human), ribosome allocation (initiation rate) is optimized to the pre-saturation point. The fact that similar results were not observed for heterologous genes indicates that this feature is under selection.

Remarkably, the gap between the performances of the RFM in comparison with alternative predictors is strikingly large in the case of heterologous genes [5,6], testifying to the model's promising biotechnological value in predicting the protein abundance of heterologous proteins before expressing them in the desired host.

Acknowledgment. We thank Prof. Elchanan Mossel, Prof. Eitan Bachmat and Prof. Yitzhak Pilpel for helpful discussions. T.T. is a Koshland Scholar at Weizmann Institute of Science and is travel supported by EU grant PIRG04-GA-2008-239317. S.R. acknowledges support from the Converging Technologies program of the Israeli Council for higher education. MK is supported by grants from the Israel Science Foundation and the James S. McDonnell Foundation.

References

1. Shaw, L.B., et al.: Totally asymmetric exclusion process with extended objects: a model for protein synthesis. Phys. Rev. E Stat. Nonlin. Soft Matter Phys. 68, 021910 (2003)

2. Ingolia, N.T., et al.: Genome-wide analysis in vivo of translation with nucleotide resolution using ribosome profiling. Science 324, 218–223 (2009)
3. dos Reis, M., et al.: Solving the riddle of codon usage preferences: a test for translational selection. Nucleic Acids Res. 32, 5036–5044 (2004)
4. Tuller, T., et al.: Determinants of protein abundance and translation efficiency in s. cerevisiae. PLoS Comput. Biol. 3, e248 (2007)
5. Kudla, G., et al.: Coding-sequence determinants of gene expression in escherichia coli. Science 324, 255–258 (2009)
6. Welch, M., et al.: Design parameters to control synthetic gene expression in escherichia coli. PLoS One 4, e7002 (2009)

Design of Protein-Protein Interactions with a Novel Ensemble-Based Scoring Algorithm[*]

Kyle E. Roberts[1], Patrick R. Cushing[3], Prisca Boisguerin[4],
Dean R. Madden[3], and Bruce R. Donald[1,2,**]

[1] Department of Computer Science, Duke University, Durham, NC 27708, USA
[2] Department of Biochemistry, Duke University Medical Center,
Durham, NC 27708, USA
Tel.: 919-660-6583; Fax: 919-660-6519
brd+recomb11@cs.duke.edu
[3] Department of Biochemistry, Dartmouth Medical School, Hanover, NH 03755, USA
[4] Institute for Medical Immunology, Charite Universitätsmedizin,
10115 Berlin, Germany

Abstract. Protein-protein interactions (PPIs) are vital for cell signaling, protein trafficking and localization, gene expression, and many other biological functions. Rational modification of PPI targets provides a mechanism to understand their function and importance. However, PPI systems often have many more degrees of freedom and flexibility than the small-molecule binding sites typically targeted by protein design algorithms. To handle these challenging design systems, we have built upon the computational protein design algorithm K^* [8,19] to develop a new design algorithm to study protein-protein and protein-peptide interactions. We validated our algorithm through the design and experimental testing of novel peptide inhibitors.

Previously, K^* required that a complete partition function be computed for one member of the designed protein complex. While this requirement is generally obtainable for active-site designs, PPI systems are often much larger, precluding the exact determination of the partition function. We have developed proofs that show that the new K^* algorithm combinatorially prunes the protein sequence and conformation space and guarantees that a provably-accurate ε-approximation to the K^* score can be computed. These new proofs yield new algorithms to better model large protein systems, which have been integrated into the K^* code base.

K^* computationally searches for sequence mutations that will optimize the affinity of a given protein complex. The algorithm scores a single protein complex sequence by computing Boltzmann-weighted partition functions over structural molecular ensembles and taking a ratio of the partition functions to find provably-accurate ε-approximations to the K^* score, which predicts the binding constant. The K^* algorithm uses several provable methods to guarantee that it finds the gap-free optimal

---

[*] This work is supported by the following grants from the National Institutes of Health: R01 GM-78031 to B.R.D. and R01 DK075309 to D.R.M.

[**] Corresponding author.

V. Bafna and S.C. Sahinalp (Eds.): RECOMB 2011, LNBI 6577, pp. 361–376, 2011.
© Springer-Verlag Berlin Heidelberg 2011

sequences for the designed protein complex. The algorithm allows for flexible minimization during the conformational search while still maintaining provable guarantees by using the minimization-aware dead-end elimination criterion, minDEE. Further pruning conditions are applied to fully explore the sequence and conformation space.

To demonstrate the ability of K^* to design protein-peptide interactions, we applied the ensemble-based design algorithm to the CFTR-associated ligand, CAL, which binds to the C-terminus of CFTR, the chloride channel mutated in human patients with cystic fibrosis. K^* was retrospectively used to search over a set of peptide ligands that can inhibit the CAL-CFTR interaction, and K^* successfully enriched for peptide inhibitors of CAL. We then used K^* to prospectively design novel inhibitor peptides. The top-ranked K^*-designed peptide inhibitors were experimentally validated in the wet lab and, remarkably, all bound with μM affinity. The top inhibitor bound with seven-fold higher affinity than the best hexamer peptide inhibitor previously available and with 331-fold higher affinity than the CFTR C-terminus.

Abbreviations used: PPI, protein-protein interaction; CFTR, Cystic fibrosis transmembrane conductance regulator; CAL, CFTR-associated ligand; DEE, Dead-end elimination; MC, Monte Carlo; CF, cystic fibrosis; NHERF1, Na^+/H^+ Exchanger Regulatory Factor 1; GMEC, global minimum energy conformation; BLU, biochemical light unit; ROC, receiver operating curve; AUC, area under the curve.

1   Introduction

The ability to accurately design protein-protein interactions (PPIs) would provide a powerful method to create new interactions or disrupt existing ones. PPI binding is very difficult to fully characterize due to a PPI's large, flexible and energetically shallow binding surface. This complexity is evidenced by the difficulty that has been encountered when trying to design small molecules to disrupt a target PPI [22,52]. Protein design methods provide a promising approach to rationally design PPIs, but most protein design studies have focused on smaller protein-small molecule systems. The intricacies of PPIs require improved accuracy in protein design algorithms. In light of these challenges, we have developed the computational protein design algorithm, K^*, to study protein-protein and protein-peptide interactions.

Typically, protein design algorithms focus on finding the single global minimum energy conformation (GMEC) by searching over a reduced conformation space using a fixed backbone and discrete side-chain rotamers [10]. The new design algorithm, K^*, improves the scoring of mutation sequences by including continuous rotamer flexibility during the conformational search, and using an ensemble-based (as opposed to single structure-based) score to better predict the binding affinity of each mutated protein sequence [8,19]. We build upon the previous K^* algorithm by developing new proofs that show that K^* can be applied to protein-protein interactions and implementing the necessary changes in the existing software. We highlight the benefits of using this approach by successfully predicting and designing peptide inhibitors of the protein complex of the CFTR-associated ligand (CAL) bound to the cystic fibrosis transmembrane conductance regulator (CFTR).

1.1 Protein Design and K^*

Structure-based computational protein design seeks to find amino-acid sequences that are compatible with a specific protein fold. Often, additional functional constraints are applied to the problem in order to design a protein with a given catalytic or binding activity. Because the conformational space of a protein is large, design algorithms often assume a fixed backbone conformation and reduce side-chain configuration space by using discrete conformations called *rotamers* [14,27,41,46]. Thus, the traditional design problem can be defined as: for a given fixed backbone conformation, find the side chain rotamers such that the energy function for the entire protein structure is minimized, also known as finding the GMEC [10,11,12,34]. As stated, the protein design problem has been shown to be NP-hard and NP-hard to approximate [7,44].

The complexity of the protein design problem has motivated the development of several different algorithms. Heuristic algorithms try to explore the most relevant parts of protein conformational space by employing self-consistent mean field theory [33], genetic algorithms [12,30], Monte Carlo (MC) and simulated annealing protocols [28,35,40], or belief propagation [16,55]. These algorithms have no guarantee on how close their results are to the optimal solution so it is possible to miss good candidate structures. Provable algorithms do not suffer from this deficiency. These algorithms include branch-and-bound techniques [23,26,38], integer linear programming [1,32], tree decomposition [39], or dead-end elimination (DEE), the most commonly used provable technique [13,26]. DEE uses provable pruning criteria to remove rotamers from the search space that cannot be part of the GMEC [13,18,21,36,43].

These methods strive to find the GMEC for a given design problem, but proteins exist as a thermodynamic ensemble and not just a single low-energy structure [20]. There is evidence that accounting for this ensemble can help find true native protein structures [3,31,56]. The new design algorithm, K^*, uses this ensemble by computing Boltzmann-weighted partition functions over structural molecular ensembles to find provably-accurate approximations to the binding constant for a protein complex [8,19]. K^* first prunes the conformational space with minimized DEE (minDEE) [19], and then uses the A* algorithm to enumerate conformations in order of their lower energy bounds to compute the partition functions. The K^* algorithm has been shown previously to design a switch in enzyme specificity for an enzyme in the non-ribosomal peptide synthetase pathway [8] and to predict resistance mutations for antibiotic targets [15]. We develop new proofs for the K^* algorithm and enhance the ability of the software to handle large systems so novel protein-protein interactions can be designed.

1.2   CFTR and CAL

CFTR is an epithelial chloride channel that is mutated in cystic fibrosis (CF) patients. The most common disease-associated mutation is a single amino acid deletion that causes CFTR misfolding and endoplasmic reticulum retention. If rescued from folding errors, mutant CFTR still maintains residual channel activity, but is rapidly degraded. The PDZ domain-containing proteins CAL and Na^+/H^+ Exchanger Regulatory Factor 1 (NHERF1) competitively bind the C-terminus of CFTR. They are involved in the endocytic trafficking of CFTR to either degradation or recycling pathways, respectively (Fig. 1C) [9,25].

The PDZ domain is a common structural motif in proteins and generally binds to the C-termini of other proteins. When bound to the PDZ domain, the C-terminus of a protein essentially forms another β-strand with the existing PDZ domain β-sheet (Fig. 1A). The specific β-sheet interactions and major binding pocket can be seen in Fig. 1B. PDZ domains can be separated into classes that are known to bind different amino acid sequence motifs [42]. The CAL PDZ domain is a class I PDZ domain which means it recognizes the C-terminal sequence X-S/T-X-I/V/L (where X is any amino acid), consistent with the CFTR C-terminus sequence DTRL.

RNA interference experiments have demonstrated that knocking down CAL increases expression of mutant CFTR in the membrane [53]. This suggests that inhibiting the CAL-CFTR interaction could also enhance CFTR membrane expression and could provide a potential avenue to ameliorating cystic fibrosis symptoms. In this study we seek to use computational design to find a peptide inhibitor to disrupt CAL-CFTR binding.

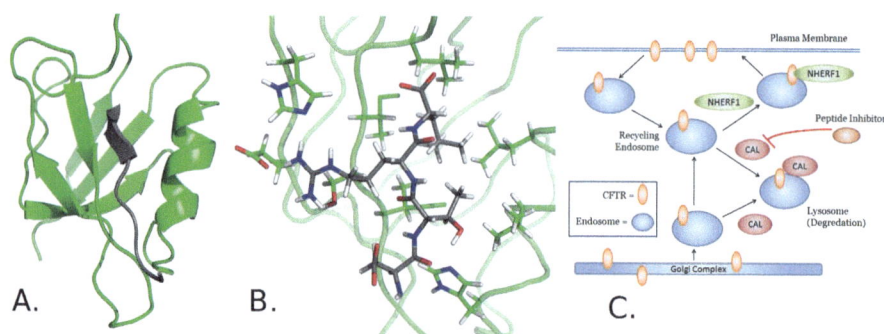

Fig. 1. (A) Cartoon representation of the CAL PDZ domain (green) bound to a CFTR C-terminus mimic (gray) (Structure from: [45]) (B) Detailed view of the CAL PDZ domain binding site. (C) CFTR trafficking pathway adapted from [25].

1.3   Previous Work

Design methodologies have been previously developed to study protein-protein interactions and, more specifically, PDZ domain interactions. These methods can be divided into sequence- and structure-based methods. Sequence-based methods rely on large sequence alignments in order to determine the likelihood that a given peptide binds to a certain protein from a given protein family. Two previous methods have specifically been applied to PDZ domains [4,50]. The methods analyze the sequence alignments as well as known protein-peptide interactions for the PDZ domain family to predict which peptides are likely to bind a given PDZ domain.

Previous structure-based computational techniques have been applied to protein-protein interactions. These techniques have used either MC methods [29], a combination of MC and DEE [48], charge optimization with discrete rigid-rotamer DEE/A* [2] or belief propagation [31] to predict and/or improve the binding of protein-protein complexes. One study successfully used a self-consistent mean field approach to improve the binding of two PDZ domains to their respective peptide ligand [47]. That study focused on finding mutations to the PDZ domain that could increase binding affinity, while our work focuses on methods to find a peptide inhibitor to bind a specific PDZ domain.

The methodology presented here complements these methods while obtaining some significant advantages over what has been done previously. Compared to the sequence-based methods, no structural information is lost in our approach. Also, the sequence-based methods require a large number of sequences for proteins in a given family as well as large amounts of binding data on the family. Our methods do not require this multitude of data in order to improve the binding of a protein complex. Our methods are general, requiring only a starting template structure to design other PPI systems. All but one of the structure-based methods mentioned above focus on finding the single GMEC conformation. In addition, only the work of [2] utilizes provable techniques, but none use both provable techniques and protein ensembles. Our statistical mechanics-based method has provable guarantees for finding the optimal sequence, and also scores each sequence with a partition function-based ranking, which better reflects natural binding affinity. Finally, the K^* method uses the minDEE pruning criteria [19] in order to allow for continuous minimization of rotamers during the mutation search.

In this paper we present a novel ensemble-based algorithm for the design of protein-protein interactions. The K^* algorithm uses a statistical mechanics-derived ranking to score protein complexes that will improve binding. The following contributions are made in this work:

- Introduction of an ensemble-based algorithm, K^*, for protein-protein interface design;
- Development of proofs showing that the sequence ranking score can be computed with provable accuracy;
- The use of K^* to retrospectively predict CAL-CFTR peptide-array data;
- The use of K^* to prospectively search 2166 peptide inhibitor sequences (approximately 10^{15} possible conformations) and predict novel peptides that successfully inhibit the CAL-CFTR protein complex;
- Experimental (wet-lab) testing of the top 11 novel designed inhibitors. Inhibitors were found with up to a 331-fold improvement in binding over the wildtype CFTR sequence.

2   Methods

2.1 K^* Algorithm

K^* computationally searches over protein sequence mutations for a given protein-protein or protein-peptide complex and assigns each sequence a score, called a K^* *score*. To compute the score for a given protein-protein complex sequence, K^* evaluates the low-energy conformations for the sequence and uses them to compute a Boltzmann-weighted partition function. Unfortunately, it is not possible to compute exact partition functions for a given protein complex, because that would require integrating an exact energy function over a protein's entire conformation space. Instead, partition functions are computed for each protein binding partner using rotamer-based ensembles defined as follows:

$$q_A = \sum_{a \in A} \exp(-E_a/RT), \; q_B = \sum_{b \in B} \exp(-E_b/RT), \; q_{AB} = \sum_{ab \in AB} \exp(-E_{ab}/RT)$$

where q_{AB} is the partition function for protein A bound to protein B, and q_A and q_B are the partition functions for the unbound proteins, A and B. The K^* score is defined as the ratio of partition functions: $K^* = \frac{q_{AB}}{q_A q_B}$, which is an approximation of the protein complex binding constant, K_A [19]. Sequences are ranked based on their K^* score where sequences with a higher K^* score are considered to have a better binding constant for the bound complex.

Fig. 2 shows the general framework for the K^* algorithm. In order to efficiently search protein conformation space, K^* searches over discrete side-chain conformations called rotamers. Side-chain rotamers are statistically overrepresented discrete side-chain conformations that are used to efficiently search the continuous side-chain conformation space. During a partition function calculation, K^* uses dead-end elimination (DEE) to prune side-chain conformations that provably cannot be part of low-energy structures. Originally, DEE was only able to prune over rigid side-chain rotamers [13]. K^* utilizes the newer DEE pruning criteria, minDEE [19], in order to prune side-chains where the discrete rotamers are allowed to minimize during the search in order to relieve potential clashes that can arise from only allowing rigid side-chain placements. After minDEE pruning, the branch-and-bound algorithm A* [38] is used to enumerate low-energy conformations [19]. These conformations are then Boltzmann weighted and incorporated into the partition function. The partition function is computed with respect to the input model described above, so the accuracy of the partition function is bounded by the accuracy of the input model.

Most of the steps in K^* are provable. That is, guarantees can be made on the optimality of the results from a given step. The minDEE pruning criteria provably maintains that all rotamers that cannot be a part of the lowest energy conformations are pruned from the protein search space. The A* algorithm provably enumerates the protein conformations gap-free in order of increasing lower bounds on energy. Finally, K^* uses these low energy conformations to compute an ε-approximation to the partition function, where ε is a user-selected parameter. The ability to compute this partition function approximation hinges on the first two provable steps. If we were to use heuristic steps to find the low energy conformations, it could not be guaranteed that all the low energy conformations are found and we would lose the ability to calculate a provably-good ε-approximation to the partition function. Because of the provable aspects of K^*, if K^* makes an errant prediction, we can be certain that it is due to an inaccuracy in the input model and not a problem (such as inadequate optimization) with our search algorithm. This makes it substantially easier to improve the model based on experimental feedback, as we show in Sec. 3.2.

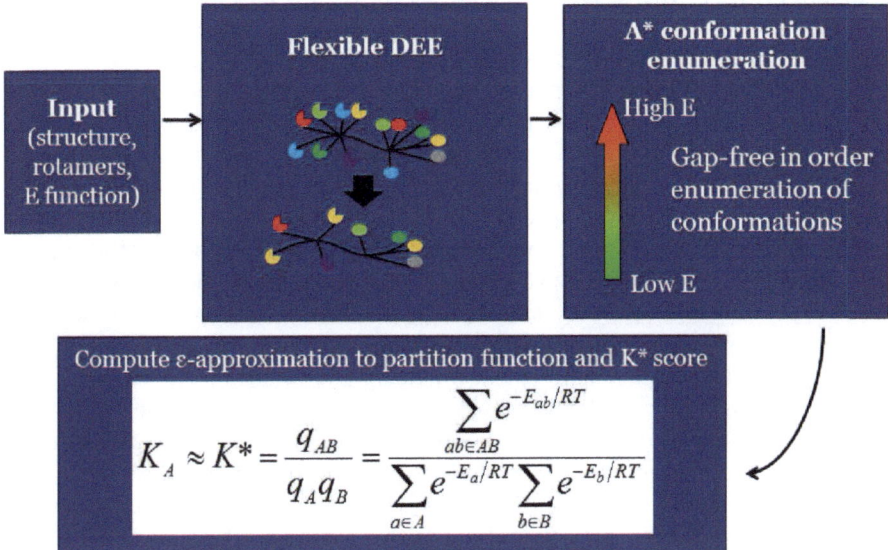

Fig. 2. Overview of K^* Algorithm. K^* takes as input an initial protein structure, a rotamer library to search over side-chain conformations, and an energy function to evaluate conformations. Minimization-aware DEE is used to prune rotamers that are not part of the lowest energy conformations for a given sequence. The remaining conformations from DEE are enumerated in order of increasing energy lower bounds using A*. Finally, the conformations are Boltzmann-weighted and used to compute partition functions and ultimately a K^* score for each sequence.

2.2 Extension of K^* to Mutations/Flexibility on Two Protein Strands

K^* has previously been successfully applied to alter enzyme active site specificity [8] and predict resistance mutations of a highly adaptive protein [15]. These studies show the great potential that K^* has to design biological systems. However, protein-protein systems can be much larger than these previous systems K^* has studied. When applied to PPI systems, the original K^* design methodology does not produce the provable guarantees it obtained for enzyme active sites. As described above, these guarantees are a key part of the K^* algorithm. Most importantly, the provable properties allow the calculation of the ε-approximation of the K^* score, which predicts the K_A for the protein complex.

According to previous work, the guarantees for calculating the K^* score require that one of the partition functions in the protein complex be computed exactly. While this requirement is generally true for active-site designs, PPI systems are often much larger and the full partition function calculation is not computationally tractable. Specifically, the previous K^* proofs [19] for intermutation pruning and calculation of an ε-approximation to the K^* score relied on this requirement. We now show that it is possible to improve the K^* algorithm to maintain these critical provable guarantees. As a result, systems where both binding partners in the protein complex are flexible or mutable during the search can be accurately studied by K^*.

The idea behind intermutation pruning is that it is possible to provably show that, in some cases, a K^* score that is currently being computed for one sequence will never be better than a K^* score that has already been found for another sequence. This pruning step significantly reduces the number of K^* scores that must be fully computed and speeds up the algorithm. Since in a positive design we only desire the top few sequences, we use a parameter (γ) to limit the number of sequences for which we must find an ε-approximation. Below, we develop the new pruning condition, by maintaining that the partition function for each protein complex partner is an ε-approximation.

We seek to show that there exists a halting condition for the conformation enumeration such that we know we have an ε-approximation to the bound partition function for a given protein complex. Given that $K_i^* \geq \gamma K_0^*$, where K_i^* is the K^* score of the current sequence, K_0^* is the best score observed so far, and γ is a user-specified parameter defining the number of top scoring sequences for which we want an ε-approximation, there exists an intermutation pruning criteria for PPI designs. In the following lemma, note that n is the number of conformations in the search that remain to be computed, k is the number of conformations that have been pruned from the search with DEE, E_0 is the lower energy bound on all pruned conformations, R is the universal gas constant, and T is the temperature. The full partition function for the protein-protein complex, and unbound proteins are q_{AB}, q_A, and q_B respectively, while q_{AB}^*, q_A^*, and q_B^* denote the current calculated value of the partition functions during the computational search.

Lemma 1. *If the lower bound E_t on the minimized energy of the $(m+1)^{th}$ conformation returned by A^* satisfies $E_t \geq -RT(\ln(\gamma \varepsilon K_0^* q_A^* q_B^* - k \exp(-E_0/RT)) - \ln n)$, then the partition function computation can be halted, with q_{AB}^* guaranteed to be an ε-approximation to the true partition function, q_{AB}, for a mutation sequence whose score K_i^* satisfies $K_i^* \geq \gamma K_0^*$.*

This lemma shows that even when designing for protein-protein interactions, there exists a provable sequence pruning criterion during the K^* search. The proof of Lemma 1 is provided in the supplementary information (SI) [49].

Now we show that we can obtain a provable guarantee on the accuracy of the K^* score for each protein sequence. Since both partition functions are

ε-approximations, we no longer obtain an ε-approximation to the K^* score but rather the following:

Lemma 2. *When mutations (or flexible residues) are allowed on both strands in a computational design, the computed K^* score is a σ-approximation to the actual K^* score, where $\sigma = \varepsilon(2 - \varepsilon)$.*

Since neither of the protein complex partition functions are calculated fully, the K^* score approximation is an 2ε-approximation as opposed to the ε-approximation of the previous method. This implies that we must compute better partition function approximations than before to maintain the same level of K^* score approximation. Nevertheless, the fact that the K^* score can still be provably approximated, confers all the advantages of a provable algorithm as stated in Sec. 2.1. The proof of Lemma 2 is provided in the SI [49].

3   Results

Using the results of Sec. 2.2, we applied the updated K^* algorithm to the CAL-CFTR system. The CAL PDZ domain has been implicated in the trafficking of CFTR. We seek to find peptide inhibitors of the CAL PDZ domain to disrupt CAL-CFTR binding, and in this section we describe how we used K^* to discover successful peptide inhibitors. First, in order to obtain accurate designs we trained the energy term weights using known inhibitory constants for natural CAL PDZ domain peptide inhibitors. Next, the K^* algorithm was applied retrospectively to predict peptide-array binding data to validate the design methodology. The retrospective test showed we were able to enrich for peptide inhibitors, so we then used K^* to prospectively find new peptide inhibitors of CAL. Top predicted sequences were then synthesized and experimentally validated and we determined that they all bind CAL with μM affinity.

3.1   Computational Designs with K^*

The previously determined NMR structure of the CAL PDZ domain bound to the C-terminus of CFTR was used to study the binding of CAL to CFTR [45]. The CFTR peptide in the NMR structure was truncated to the six most C-terminal amino acids and mutated to the amino acid sequence WQTSII to mimic the best peptide hexamer for CAL discovered thus far (unpublished data). An acetyl group was modeled onto the N-terminus of the peptide using restrained molecular dynamics and minimization where the N-terminus of the peptide was allowed to move, while the remainder of the protein complex was restrained using a harmonic potential [6]. An 8 Å shell around the peptide hexamer was used as the input structure to K^*. The four most C-terminal residues, TSII, were allowed to mutate to the following residues during the design search: Thr (all amino acids except Pro), Ser (T/S), Ile (all amino acids except Pro), and Ile (I/L/V). In addition, the Probe program [54] was used to determine the side-chains on CAL that interact with the CFTR peptide mimic. These nine residues that interact with the peptide, as well as the two most N-terminal residues on the peptide, were allowed to be flexible during the design search. The peptide was also allowed to rotate and translate as a rigid body during the search, as previously described for small molecules [8,15,19]. To explore the feasibility of our new algorithms, unless otherwise noted, full partition functions were not computed and a maximum of 10^3 conformations were allowed to contribute to each partition function. The system size was addressed in part by efficiently parallelizing the partition function calculation in the design software.

Rotamer values were taken from the Penultimate Rotamer Library modal values [41]. The energy function used to evaluate protein conformations has been previously described [8,15]. The energy function, Energy = vdW + Coul + EEF1, consists of a van der Waals term, a Coulombic electrostatics term, and an EEF1 implicit solvation term [37]. For all but one of the design runs, the `Amber` [51] forcefield terms were used, with the `Charmm` [5] forcefield parameters used for the other run.

3.2   Training of Energy Function Weights

In order to obtain accurate predictions for the CAL-CFTR system, scaling parameters for the three energy function components (See 3.1) were determined. In order to determine how best to weight each of these terms, we trained the scaling terms using experimental data from the CAL-CFTR system. Sixteen K_i values for natural ligands of CAL measured in [9] were used for the training. K^* scores were computed for each of the 16 natural ligands values. Note, for this training, the CAL-CFTR structure only included the four most C-terminal residues of the peptide inhibitor. The energy weights were searched using a gradient descent method to optimize the correlation between the K^* scores and the experimental K_i^{-1} values.

The best correlation found through the parameter search that maintains reasonable K^* scores is shown in Fig. 3A. The correlations throughout the search range from 0.0 to 0.75 which highlights the importance of choosing the correct weighting factors. The resulting parameters used for the retrospective and prospective studies (design runs) described in Secs. 3.3-3.4 is as follows: van der Waals scaling = 0.9, dielectric constant for electrostatic scaling = 20, and a solvation scaling = 0.76. These parameters are reasonable and similar to parameters used in previous designs. The dielectric constant might appear high (typically the interior of a protein is thought to have a dielectric of 2-4), but since the peptide design occurs at the surface of the protein, this might necessitate a higher dielectric constant.

3.3   Retrospective Test against Peptide Array Data

Using the energy weights discovered through the training process, K^* peptide inhibitor designs were conducted for sequences from a CAL peptide-array. The peptide-array data from [9] was used to validate the K^* peptide inhibitor predictions. Briefly, the peptide array consisted of 6223 C-termini (11-mers) of

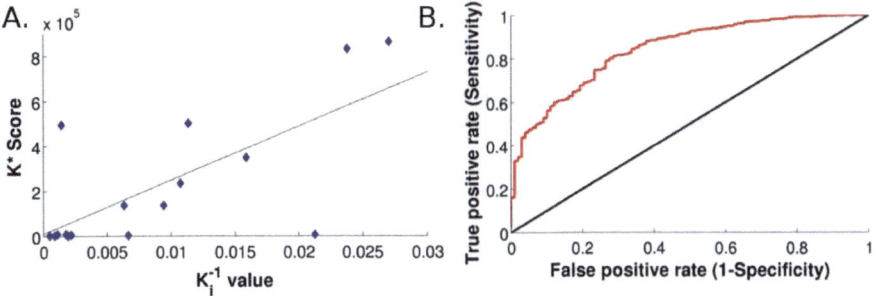

Fig. 3. (A) Correlation between K^* score and experimental K_i^{-1} values for CAL PDZ peptide inhibitors. Pearson Correlation of 0.75. (B) ROC curve showing the ability of K^* designs to enrich for peptide sequences that can bind the CAL PDZ domain. The AUC is 0.84.

human proteins. The array was incubated with the CAL PDZ domain in order to determine binding of CAL to the 11-mers. The K^* algorithm was used to evaluate 4-mer structural models of the peptide-array sequences to verify the accuracy of the predictions.

Since the peptide array data is somewhat noisy, to compare the array data with the K^* predictions, the quantitative array data, measured in biochemical light units (BLUs), was converted into a binary CAL binding event. In other words, by setting a binding cutoff on the peptide array, we classify each sequence as either a CAL binder or non-binder. The cutoff value was determined as three standard deviations away from the average BLU value of the array. Since K^* is being compared against a binary binding event, the design algorithm can be viewed as a filter enriching for sequences that bind CAL.

Fig. 3B shows the resulting receiver operating curve (ROC) when comparing the K^* scores to the CAL binding event of the peptide array. The ROC curve has an area under the curve (AUC) of 0.84 which shows that K^* greatly enriches for peptides that bind CAL. Consider if we were conducting this as a prospective test and we were to test the top 30 K^* ranked sequences. According to the peptide array, 11 of the top 30 sequences would be found to bind CAL. Notably, this is a 20-fold increase over the number of binders that would be expected to be found if the binding sequences were distributed randomly in the rankings.

Class I PDZ domains like CAL are known to bind the canonical sequence motif: X-S/T-X-L/V/I. Therefore a much more stringent test of the K^* design algorithm is to determine the degree to which K^* enriches for binders if we restrict the peptide array to sequences that have the class I motif. With this new restriction, K^* is still able to significantly enrich for CAL peptide binders producing a ROC curve with an AUC of 0.71. When considering the top 30 K^* ranked sequences, 17 of the 30 sequences are binders, which results in a 2-fold increase over the expected random distribution.

3.4   Prospective validation of the K^* Algorithm

Since K^* was able to successfully enrich for CAL binders based on peptide array data, we took the analysis a step further by using K^* to prospectively find novel CAL peptide inhibitors.

Computational Predictions. K^* was used to search over all peptide sequences within the class I PDZ domain sequence motif to find new CAL peptide inhibitors. For computational efficiency the number of conformations enumerated by A* for each partition function was limited to 10^3 conformations. In order to choose the most promising peptide inhibitors, a second K^* design was done where K^* scores for the top 30 sequences were re-calculated with the number of enumerated conformations per partition function increased to 10^5. Several top-ranked sequences were chosen to be experimentally tested. First, the top 7 ranked sequences from the second run were chosen. In addition, two sequences that significantly increased in ranking from the first to second run (rank 29 to 9, and rank 28 to 11) were chosen as well. Finally, a K^* run was conducted using Charmm forcefield parameters instead of Amber parameters. Two sequences that scored high on both the Amber and Charmm runs were chosen to be experimentally tested as well. In total 11 K^*-predicted peptide inhibitor sequences were experimentally tested (Fig. 4A).

Comparison with minDEE and DEE. To determine the importance of the ensemble-based K^* rankings we compared the predictions to two single-structure GMEC-based methods, minDEE [19], and rigid-rotamer DEE (rigid-DEE) [24]. Both minDEE, and rigidDEE were run with the same parameters as the K^* designs, except that they used reference energies. We compared the top 30 sequences from minDEE and rigidDEE and found they had no sequences in common. This supports our findings that in over 69 protein design systems minDEE finds low energy sequences that rigidDEE discards because rigidDEE does not allow minimization [17,19]. In addition, when we compare the top 30 rigidDEE and minDEE results to the top K^* designs we find that they have only three and four sequences in common, respectively. If we were to have used only GMEC-based approaches instead of K^*, we would not have predicted most of the experimentally-successful sequences that K^* found. In addition, the overall sequence rankings show a very poor correlation between the minDEE and K^* predictions; the same is true of the rigidDEE and K^* predictions ($R^2 = 0.1$ and 0.09 respectively).

Experimental Procedure. To test the ability of K^* to predict CAL-CFTR peptide inhibitors, the inhibitory constants of top-ranking peptide sequences from the K^* CAL-CFTR design were experimentally determined. As a control, the best known peptide hexamer was also retested. The corresponding N-terminally acetylated peptides were purchased from NEO BioScience and the K_i values for the peptides were detected using fluorescence polarization, using the method previously described in [9]. Briefly, the CAL PDZ domain was

Sequence	K^* Ranking (out of 2166)	Experimental K_i (μM)
Ac-WQVTRV	9	2.1
Ac-WQFTRL	$1^\dagger$	6.7
Ac-WQKTRL	2	8.6
Ac-WQRTRL	5	10.0
Ac-WQKTRI	4	11.6
Ac-WQKTRV	1	12.9
Ac-WQRTRI	7	14.4
Ac-WQFTKL	$2^\dagger$	15.0
Ac-WQLTKL	11	15.6
Ac-WQKTKL	6	18.2
Ac-WQRTRV	3	18.4

A.

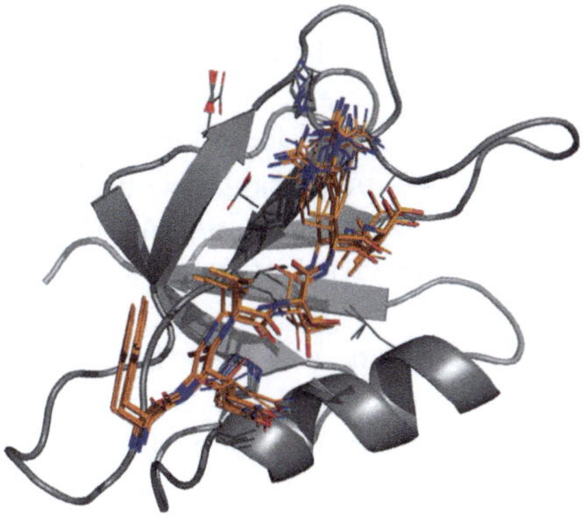

B.

Fig. 4. (A) Top-ranked K^* predictions that we experimentally tested by fluorescence polarization. K_i values marked in green denote that the binding affinity was higher than the best previously known hexamer (14.8 μM). †Sequence rank obtained by ordering the quantity: $\frac{R_A+R_C}{2}$, where R_A is the sequence rank from a design run using the **Amber** forcefield and R_C is the sequence rank from a run using the **Charmm** forcefield. (B) Ensemble of top 100 conformations for the peptide with tightest binding to CAL (WQVTRV).

incubated with a labeled peptide of known binding affinity. Each top ranking peptide was serially diluted and the protein-peptide mixture was added to each dilution. Finally, the amount of competitive inhibition was tracked using residual fluorescence polarization.

Experimental Validation. All of our designed inhibitors are novel and none had been predicted or experimentally tested before. Remarkably, all of the predicted peptides bind CAL with high affinity (Fig. 4A). The previously best known peptide hexamer (WQTSII) to bind CAL has a K_i of 14.8 μM. Seven of the eleven tested sequences show an improvement in binding compared to the best-known peptide hexamer, and the best peptide displayed 2.1 μM affinity, representing a 7-fold improvement over the previously best-known hexamer. For comparison, note that the K_i for the wild-type CFTR sequence (TEEEVQDTRL) is 690 μM and the highest known affinity natural ligand (ANGLMQTSKL) for CAL is 37 μM. The goal of this experiment was to find a peptide inhibitor for CAL binding to CFTR. Using the K^* design algorithm we successfully found a peptide inhibitor with 331-fold higher affinity than the interaction we were trying to inhibit. Thus, we have successfully designed peptide inhibitors of the CAL PDZ domain.

4   Conclusions

We presented a novel, provable, ensemble-based protein design algorithm for protein-peptide and protein-protein interactions. As a demonstration of K^*'s design capabilities we showed that K^* can accurately predict sequences that will tightly bind the CFTR trafficking regulator, CAL. In order to make accurate predictions we first trained 3 energy weighting parameters for the energy function used to evaluate each conformation. The gradient descent search for the optimal parameters is important because a poor choice of parameters can lead to a bad correlation with experimental results. With training we were able to obtain a Pearson correlation of 0.75 which is very good for the current state of protein design.

We have demonstrated that K^* can accurately enrich for and predict peptide inhibitors for the CAL-CFTR system. The retrospective tests comparing K^* scores to the CAL peptide-array demonstrate our strong ability to enrich for sequences that bind CAL over the human protein sequence space. Even when the search is made more difficult by adding stringent restrictions to the sequence space by searching only within the class I PDZ sequence motif, K^* is able to enrich for binding peptides. Finally, the experimental validation in the wet lab of top-ranked K^* sequences confirms that K^* is able to predict novel CAL peptide inhibitors. Compared to the inhibitory constant of the natural CFTR C-terminus, the designed sequences are much stronger binders. Indeed, our approach was able to find peptide sequences that bound tighter than the previously best known hexamer sequence.

It has been demonstrated that knockdown of CAL using RNA interference can lead to increased CFTR membrane expression and that CAL-mediated degradation of CFTR requires an intact CAL PDZ domain [53]. Peptides that are able to inhibit CAL-CFTR binding may provide a method of treatment to increase mutant CFTR membrane expression. Ultimately, this approach could lead to the development of a drug to treat cystic fibrosis. Since peptides themselves often make poor drugs, we are continuing to extend the K^* software to incorporate non-natural amino acids into the design search space. We can design compounds that inhibit CAL, but cannot be broken down inside the human body as easily as peptides. In addition, we plan to use negative design (as we showed was possible in [15]) in order to ensure that the peptides predicted by K^* do not bind other PDZ domain containing proteins. Specifically, the PDZ domain containing protein NHERF1 is involved in the endocytic recycling of CFTR back to the membrane [25]. So the goal of future studies is to ensure that designed peptides inhibit CAL but do not inhibit NHERF1.

The K^* algorithm is a general algorithm that can now be applied to many protein-protein interface systems. In addition, PDZ domains are prevalent in human proteins and are a very important structural motif to understand. The application of K^* to CAL-CFTR demonstrates the potential that this method has to analyze many other PDZ domains and protein-protein interfaces.

Acknowledgments

We thank all members of the Donald Lab, in particular Mr. Pablo Gainza for helpful discussions and comments.

References

1. Althaus, E., et al.: Journal of Computational Biology 9(4), 597–612 (2002)
2. Altman, M.D., et al.: Proteins 70(3), 678–694 (2008)
3. Berezovsky, I.N., et al.: PLoS Comput. Biol. 1(4), e47 (2005)
4. Brannetti, B., Helmer-Citterich, M.: Nucleic Acids Res. 31(13), 3709–3711 (2003)
5. Brooks, B.R., et al.: Journal of Computational Chemistry 4(2), 187–217 (1983)
6. Case, D.A., et al.: Journal of Computational Chemistry 26(16), 1668–1688 (2005)
7. Chazelle, B., et al.: INFORMS Journal on Computing 16(4), 380–392 (2004)
8. Chen, C., et al.: Proc. Natl. Acad. Sci. USA 106(10), 3764–3769 (2009)
9. Cushing, P.R., et al.: Biochemistry 47(38), 10084–10098 (2008)
10. Dahiyat, B.I., Mayo, S.L.: Protein Science 5(5), 895–903 (1996)
11. Dahiyat, B.I., Mayo, S.L.: Science 278(5335), 82–87 (1997)
12. Desjarlais, J.R., Handel, T.M.: Protein Science 4(10), 2006–2018 (1995)
13. Desmet, J., et al.: Nature 356(6369), 539–542 (1992)
14. Dunbrack, R.L., Karplus, M.: J. Mol. Biol. 230(2), 543–574 (1993)
15. Frey, K.M., et al.: Proc. Natl. Acad. Sci. USA 107(31), 13707–13712 (2010)
16. Fromer, M., Yanover, C.: Bioinformatics 24(13), i214–i222 (2008)
17. Gainza, P., Roberts, K.E., Donald, B.R. (2011) (submitted)
18. Georgiev, I., et al.: Bioinformatics 22(14), e174–e183 (2006)
19. Georgiev, I., et al.: Journal of Computational Chemistry 29(10), 1527–1542 (2008)
20. Gilson, M., et al.: Biophysical Journal 72(3), 1047–1069 (1997)
21. Goldstein, R.: Biophysical Journal 66(5), 1335–1340 (1994)
22. Gorczynski, M.J., et al.: Chemistry & Biology 14(10), 1186–1197 (2007)
23. Gordon, D.B., Mayo, S.L.: Structure 7(9), 1089–1098 (1999)
24. Gordon, D.B., et al.: Journal of Computational Chemistry 24(2), 232–243 (2003)
25. Guggino, W.B., Stanton, B.A.: Nature Reviews. Molecular Cell Biology 7(6), 426–436 (2006)
26. Hong, E., et al.: Journal of Computational Chemistry 30(12), 1923–1945 (2009)
27. Janin, J., Wodak, S.: Journal of Molecular Biology 125(3), 357–386 (1978)
28. Jiang, X., et al.: Protein Science 9(2), 403–416 (2000)
29. Joachimiak, L.A., et al.: Journal of Molecular Biology 361(1), 195–208 (2006)
30. Jones, D.T.: Protein Science 3(4), 567–574 (1994)
31. Kamisetty, H., et al.: Proteins 79(2), 444–462 (2011)
32. Kingsford, C.L., et al.: Bioinformatics 21(7), 1028–1039 (2005)
33. Koehl, P., Delarue, M.: Journal of Molecular Biology 239(2), 249–275 (1994)
34. Koehl, P., Levitt, M.: Journal of Molecular Biology 293(5), 1161–1181 (1999)
35. Kuhlman, B., Baker, D.: Proc. Natl. Acad. Sci. USA 97(19), 10383–10388 (2000)
36. Lasters, I., et al.: Protein Eng. 8(8), 815–822 (1995)
37. Lazaridis, T., Karplus, M.: Proteins 35(2), 133–152 (1999)
38. Leach, A.R., Lemon, A.P.: Proteins 33(2), 227–239 (1998)
39. Leaver-Fay, A., et al.: Pacific Symposium on Biocomputing 10, 16–27 (2005)
40. Lee, C., Subbiah, S.: Journal of Molecular Biology 217(2), 373–388 (1991)
41. Lovell, S.C., et al.: Proteins 40(3), 389–408 (2000)
42. Nourry, C., et al.: Sci. STKE 2003(179), re7 (2003)
43. Pierce, N.A., et al.: Journal of Computational Chemistry 21(11), 999–1009 (2000)
44. Pierce, N.A., Winfree, E.: Protein Eng. 15(10), 779–782 (2002)
45. Piserchio, A., et al.: Biochemistry 44(49), 16158–16166 (2005)
46. Ponder, J.W., Richards, F.M.: Journal of Molecular Biology 193(4), 775–791 (1987)
47. Reina, J., et al.: Nat. Struct. Mol. Biol. 9(8), 621–627 (2002)

48. Reynolds, K.A., et al.: Journal of Molecular Biology 382(5), 1265–1275 (2008)
49. Roberts, K.E., Cushing, P.R., Boisguerin, P., Madden, D.R., Donald, B.R.: Supplementary material: Design of protein-protein interactions with a novel ensemble-based scoring algorithm (2011), available online
    `http://www.cs.duke.edu/donaldlab/Supplementary/recomb11/kstar-ppi`
50. Thomas, J., et al.: Proteins 76(4), 911–929 (2009)
51. Weiner, S.J., et al.: Journal of Computational Chemistry 7(2), 230–252 (1986)
52. Wells, J.A., McClendon, C.L.: Nature 450(7172), 1001–1009 (2007)
53. Wolde, M., et al.: Journal of Biological Chemistry 282(11), 8099–8109 (2007)
54. Word, J.M., et al.: Journal of Molecular Biology 285(4), 1711–1733 (1999)
55. Yanover, C., Weiss, Y.: Advances in Neural Information Processing Systems, pp. 84–86 (2002)
56. Zhang, J., Liu, J.S.: PLoS Comput. Biol. 2(12), e168 (2006)

Computing Fragmentation Trees from Metabolite Multiple Mass Spectrometry Data

Kerstin Scheubert[1], Franziska Hufsky[1,2],
Florian Rasche[1], and Sebastian Böcker[1]

[1] Lehrstuhl für Bioinformatik, Friedrich-Schiller-Universität Jena,
Ernst-Abbe-Platz 2, Jena, Germany
{kerstin.scheubert,franziska.hufsky,florian.rasche,
sebastian.boecker}@uni-jena.de
[2] Max Planck Institute for Chemical Ecology, Beutenberg Campus, Jena, Germany

Abstract. Since metabolites cannot be predicted from the genome sequence, high-throughput *de-novo* identification of small molecules is highly sought. Mass spectrometry (MS) in combination with a fragmentation technique is commonly used for this task. Unfortunately, automated analysis of such data is in its infancy. Recently, fragmentation trees have been proposed as an analysis tool for such data. Additional fragmentation steps (MS^n) reveal more information about the molecule.

We propose to use MS^n data for the computation of fragmentation trees, and present the Colorful Subtree Closure problem to formalize this task: There, we search for a colorful subtree inside a vertex-colored graph, such that the weight of the transitive closure of the subtree is maximal. We give several negative results regarding the tractability and approximability of this and related problems. We then present an exact dynamic programming algorithm, which is parameterized by the number of colors in the graph and is swift in practice. Evaluation of our method on a dataset of 45 reference compounds showed that the quality of constructed fragmentation trees is improved by using MS^n instead of MS^2 measurements.

Keywords: metabolomics, computational mass spectrometry, multiple-stage mass spectrometry, hardness results, parameterized algorithms.

1 Introduction

The phenotype of an organism is strongly determined by the small chemical compounds contained in its cells. These compounds are called metabolites; their mass is typically below 1000 Da. Unlike biopolymers such as proteins and glycans, the chemical structure of metabolites is not restricted. This results in a great variety and complexity in spite of their small size. Except for primary metabolites directly involved in growth, development, and reproduction, most metabolites remain unknown. Plants, filamentous fungi, and marine bacteria synthesize huge numbers of secondary metabolites, and the number of metabolites in any higher

V. Bafna and S.C. Sahinalp (Eds.): RECOMB 2011, LNBI 6577, pp. 377–391, 2011.

eukaryote is currently estimated between 4 000 and 20 000 [9]. Unlike for proteins, the structure of metabolites usually cannot be deduced by using genomic information, except for very few metabolite classes like polyketides.

Mass spectrometry (MS) is one of the key technologies for the identification of small molecules. Identification is usually achieved by fragmenting the molecule, and measuring masses of the resulting fragments. The fragmentation mechanisms of electron ionization (EI) during gas chromatography MS (GC-MS) are well described [12]. Unfortunately, only thermally stable and volatile compounds can be analyzed by this technique. Liquid chromatography MS (LC-MS) can be adapted to a wider array of (even thermally unstable) molecules, including a range of secondary metabolites [9]. LC-MS uses the more gentle electrospray ionization (ESI) and a selected compound is fragmented in a second step using collision-induced dissociation (CID), resulting in MS^2 spectra. Different from peptides where CID fragmentation is generally well understood, this understanding is in its infancy for metabolites. The manual interpretation of CID mass spectra is cumbersome and requires expert-knowledge. Even searching spectral libraries is problematic, since CID mass spectra are limited in their reproducibility on different instruments [14]. Additionally, compound libraries to search against are vastly incomplete. For these reasons, automated *de novo* interpretation of CID mass spectra is required as an important step towards the identification of unknowns.

Multiple-stage mass spectrometry (MS^n) allows to further fragment the products of the initial fragmentation step. To this end, fragments of the MS^2 fragmentation are selected as precursor ions, and subjected to another fragmentation reaction. Several precursor ions can be selected successively. Selection can either be performed automatically for a fixed number of precursor ions with maximal intensity, or manually by selecting precursor ions. Fragments from MS^3 fragmentations can, in turn, again be selected as precursor ions, resulting in MS^4 spectra. Typically, the quality of mass spectra is reduced with each additional fragmentation reaction. Furthermore, measuring time is increased, reducing the throughput of the instrument. Hence, for untargeted analysis by LC-MS, analysis is usually limited to few additional fragmentation reactions beyond MS^2.

In the past years some progress has been made in searching of spectral and compound libraries using CID spectra [14,15,11], and there exist some pioneering studies towards the automated analysis of such spectra [16,10,18]. Recently, a method for *de novo* interpretation of metabolite MS^2 data has been developed [6, 17]. It helps to identify metabolite sum formulas and further to interpret the fragmentation processes, resulting in hypothetical fragmentation trees. These fragmentation trees can be compared to each other to identify compound classes of unknowns [17]. In fact, applying this method of computing fragmentation trees to MS^n data is possible, but dependencies between different fragmentation steps are not taken into account. For peptide sequencing, MS^3 spectra have been used to increase the accuracy of *de novo* peptide sequencing algorithms [2].

Here, we present a method for automated interpretation of MS^n data. We adjust the fragmentation model for MS^2 data from [6] to MS^n data to reflect the succession of fragmentation reactions. This results in the COLORFUL SUBTREE CLOSURE problem that has to be solved in conjunction with the original MAXIMUM COLORFUL SUBTREE problem [6]. We show that the COLORFUL SUBTREE CLOSURE problem is NP-hard, and present intractability results regarding the approximability of this and the MAXIMUM COLORFUL SUBTREE problem. Despite these negative results, we present an exact algorithm for the combined problem: This fixed-parameter algorithm, based on dynamic programming, has a worst-case running time with exponential dependence only on the number of peaks k in the spectrum. In application, we choose some fixed k' such as $k' = 15$, limit exact calculations to the k' most intense peaks in the mass spectra and attach the remaining peaks heuristically. We apply our algorithm to a set of 185 mass spectra from 45 compounds, and show that adding MS^n information to the analysis improves quality of results but does not affect the running time in comparison to the algorithm for MS^2 data from [6].

2   Constructing Fragmentation Trees from MS^2 and MS^n Data

Fragmentation of glycans and proteins is generally well understood, but this is not the case for metabolites and small molecules. That makes it difficult both to predict the fragmentation process, and to interpret metabolite MS data. Böcker *et al.* [6] propose fragmentation trees to interpret MS^2 data: In a fragmentation tree nodes are annotated with molecular formulas of fragments, and edges represent fragmentation reactions or *neutral losses*.

The algorithm to compute a fragmentation tree proceeds as follows [6]: Each fragment peak is assigned one or more molecular formulas with mass sufficiently close to the peak mass [5]. The resulting molecular formulas including the parent molecular formula, are considered vertices of a directed acyclic graph (DAG). We assume that the parent molecular formula is either given or can be calculated from isotope pattern analysis. Vertices in the graph are colored, such that vertices that explain the same peak receive the same color. Edges represent neutral losses, that is, fragments of the molecule that are not observed, as they were not ionized. Two vertices u, v are linked by a directed edge if the molecular formula of v is a sub-molecule of the molecular formula of u. Edges are weighted, reflecting that some edges are more likely to represent true neutral losses than others. Also, peak intensities and mass deviations are taken into account in these weights. Now, each subtree of the resulting graph corresponds to a possible fragmentation tree. To avoid the case that one peak is explained by more than one molecular formula, only *colorful* subtrees that use every color at most once are considered. In practice, it is very rare that a peak is indeed created by two different fragments, whereas our optimization principle without restriction would always choose all explanations of a peak. Therefore, searching for a colorful subtree of maximum weight means searching for the best explanation of the observed fragments:

Maximum Colorful Subtree problem. Given a vertex-colored DAG $G = (V, E)$ with colors $\mathcal{C}$ and weights $w : E \to \mathbb{R}$. Find the induced colorful subtree $T = (V_T, E_T)$ of G of maximum weight $w(T) := \sum_{e \in E_T} w(e)$.

We now modify this problem to take into account MS^n data when constructing fragmentation trees. From the experimental data, we construct a DAG $G = (V, E)$ together with a vertex coloring $c : V \to \mathcal{C}$, called *fragmentation graph*. Recall that the vertices of V correspond to potential molecular formulas of the fragments, colors $\mathcal{C}$ correspond to peaks in the mass spectra, and molecular formulas corresponding to the same peak mass have the same color. In contrast to the fragmentation graph where each edge indicates a direct succession, the MS^n data does not only hint to direct but also to indirect successions. So, in the graph constructed from the MS^n data we also have to score the transitive closure of the induced subtrees. The *transitive closure* $G^+ = (V, E^+)$ of a DAG $G = (V, E)$ contains the edge $(u, v) \in E^+$ if and only if there is a directed path in G from u to v. In case G is a tree, the transitive closure can be computed in time $O(|V|^2)$ using Nuutila's algorithm [13]. The MS^n data gives additional information about the provenience of certain peaks/colors, but does not differentiate between different explanations of these peaks via molecular formulas, so we will score not edges but pairs of colors.

To score the closure, let $w^+ : \mathcal{C}^2 \to \mathbb{R}$ be a weighting function for pairs of colors. We define the *transitive weight* of an induced tree $T = (V_T, E_T)$ with transitive closure $T^+ = (V_T, E_T^+)$ as:

$$w^+(T) := \sum_{(u,v) \in E_T^+} w^+\big(c(u), c(v)\big) \tag{1}$$

Again, we limit our search to colorful trees, where each color is used at most once in the tree. Scoring the transitive closure of an induced colorful subtree, we reach the following problem definition:

Colorful Subtree Closure problem. Given a vertex-colored DAG $G = (V, E)$ with colors $\mathcal{C}$ and transitive weights $w^+ : \mathcal{C}^2 \to \mathbb{R}$. Find the induced colorful subtree T of G of maximum weight $w^+(T)$.

We will see in the next section that this is again a computationally hard problem. But the problem we are interested in is even harder as it combines the two above problems:

Combined Colorful Subtree problem. Given a vertex-colored DAG $G = (V, E)$ with colors $\mathcal{C}$, edge weights $w : E \to \mathbb{R}$, and transitive weights $w^+ : \mathcal{C}^2 \to \mathbb{R}$. Find the induced colorful subtree T of G of maximum weight $w^*(T) = w(T) + w^+(T)$.

3   Hardness Results

Fellows *et al.* [8] and Böcker and Rasche [6] independently showed that the MAXIMUM COLORFUL SUBTREE problem is NP-hard. It turns out that the COLORFUL SUBTREE CLOSURE problem is NP-hard even for unit weights:

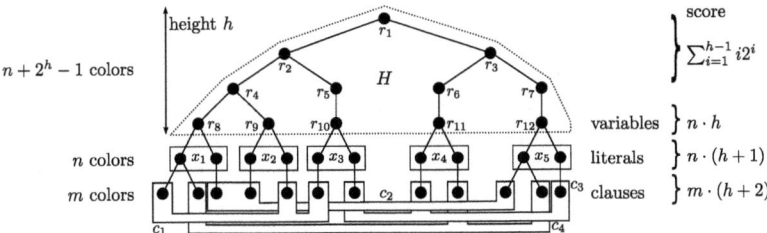

Fig. 1. Proof of Theorem 1: Example for the construction of G for $\Phi = (x_1 \vee \overline{x_2} \vee x_3) \wedge (\overline{x_1} \vee x_2 \vee x_5) \wedge (\overline{x_3} \vee x_4 \vee \overline{x_5}) \wedge (\overline{x_4} \vee x_5 \vee x_1)$

Theorem 1. *The* COLORFUL SUBTREE CLOSURE *problem is NP-hard even if the input graph is a binary tree with unit weights* $w^+ \equiv 1$.

Proof. To prove NP-hardness we use a reduction from the NP-hard 3-SAT* problem [3]:

3-SAT*. Given a Boolean expression in conjunctive normal form (CNF) consisting of a set of length three clauses, where each variable occurs at most three times in the clause set. Decide whether the expression is satisfiable.

Given an instance of 3-SAT* as a CNF formula $\Phi = c_1 \wedge \cdots \wedge c_m$ over variables $x_1, \ldots, x_n$ we construct an instance of COLORFUL SUBTREE CLOSURE. Since variables occurring only with one literal are trivial, we assume that the formula contains both literals of each variable. We first construct a colorful binary tree H with root vertex r and n leaves, that has height $h := \lceil \log_2 n \rceil$ and is a perfect binary tree up to height $h - 1$. This tree uses $p := n + 2^h - 1$ colors, namely $r_1, \ldots, r_p$. To each leaf i, $1 \leq i \leq n$, we connect two vertices using the same color x_i and representing the different truth assignments for x_i. One vertex in the color x_i represents $x_i = true$, the other one $x_i = false$. If a truth assignment to x_i satisfies clause c_j we connect a vertex colored c_j to the vertex in the color x_i, that corresponds to this truth assignment (Fig. 1). The resulting tree G possesses $n + 2^h - 1 + n + m$ colors, namely $r_1, \ldots, r_p, x_1, \ldots, x_n, c_1, \ldots, c_m$. The tree is binary, since each variable occurs in at most three clauses and we assumed that both literals are contained in the formula. Finally, we define unit weights $w^+ \equiv 1$.

The resulting tree G has as many leaves as there are literals in Φ, hence the construction is polynomial. We claim that Φ is satisfiable if and only if the colorful subtree T of G with maximum transitive closure has score $\sum_{i=1}^{h-1} i2^i + nh + n(h+1) + m(h+2)$. To prove the forward direction, assume a truth assignment ϕ that satisfies Φ. Define $A \subseteq V(G)$ to be the subset of vertices in the colors x_i that correspond to the assignment ϕ. Then, for every $1 \leq j \leq m$ there exists at least one vertex colored c_j in the neighborhood of A. Add an arbitrary representative of these vertices colored c_j to the set $B \subseteq V(G)$. The union of the sets $A \cup B \cup \{r_1, \ldots, r_p\}$ forms a colorful subtree T of G with transitive closure that has score $\sum_{i=1}^{h-1} i2^i + nh + n(h+1) + m(h+2)$, as $\sum_{i=1}^{h-1} i2^i$ corresponds to the score of the perfect binary tree up to height $h-1$, nh is the additional score from the leaves of H, $n(h+1)$ the additional score induced by the colors x_i and $m(h+2)$ the additional score induced by the colors c_j. No colorful subtree with higher score of the transitive closure can exist.

To prove the backward direction assume there is a colorful subtree T of G with maximum transitive closure that has score $\sum_{i=1}^{h-1} i2^i + nh + n(h+1) + m(h+2)$. Any optimal solution uses all colors from H, all colors x_i and all colors c_j. The truth assignment corresponding to the vertices of T colored x_i satisfies Φ, as for all $1 \le j \le m$ exactly one vertex colored c_j is connected to these vertices, otherwise T would not contain all colors.   □

We now turn to the inapproximability of the above problems. Dondi *et al.* [7] show that the MAXIMUM MOTIF problem, that is closely related to the MAXIMUM COLORFUL SUBTREE problem, is APX-hard even if the input graph is a binary tree. In fact, Proposition 8 in [7] implies that the MAXIMUM COLORFUL SUBTREE problem is APX-hard for such trees. We infer that there exists no Polynomial Time Approximation Scheme (PTAS) for the problem unless P = NP [1]. In Proposition 10 Dondi *et al.* [7] prove the even stronger result that there is no constant-factor approximation for MAXIMUM LEVEL MOTIF problem, unless P = NP.

Lemma 1. *The* MAXIMUM COLORFUL SUBTREE *problem is APX-hard even if the input graph is a binary tree with unit weights* $w \equiv 1$.

Lemma 2. *The* MAXIMUM COLORFUL SUBTREE *problem has no constant-factor approximation unless* P = NP, *even if the input graph is a tree with unit weights* $w \equiv 1$.

We now concentrate on the COLORFUL SUBTREE CLOSURE problem. We show that the problem is MAX SNP-hard even for unit weights, but we have to drop the requirement that the tree is binary in this case. We infer the non-existence of a PTAS unless P = NP [1].

Theorem 2. *The* COLORFUL SUBTREE CLOSURE *problem is MAX SNP-hard even if the input graph is a tree with unit weights* $w^+ \equiv 1$.

The construction used in the proof of Theorem 2 is very similar to that of Theorem 1. We defer the details to the full version of the paper.

We infer that the COMBINED COLORFUL SUBTREE problem is computationally hard and also hard to approximate, as it generalizes the above two problems. Note that the input graphs in our application are transitive graphs, whereas we assume trees in our hardness proofs. One might argue that the problem is actually simpler for transitive graphs; but for a given tree $T = (V, E)$ with unit weights, its transitive closure $G := T^+$ can be complemented with a binary weighting $w : E^+ \to \{0, 1\}$ such that $w(e) = 1$ if and only if $e \in E$. So, the COLORFUL SUBTREE PROBLEM remains hard for transitive input graphs. Also note that the input graphs constructed from mass spectra, possess a topological sorting that respects colors. Again, one might argue that the problem is actually simpler for such graphs. It turns out that this is not the case, either: Dondi *et al.* [7] show that the MAXIMUM MOTIF problem is APX-hard even for leveled trees. All the trees constructed in our reductions are leveled, and all our results are also valid on leveled trees. Similar to above, leveled trees can be encoded in "color-sorted" graphs using binary weightings. Thus, the problems remain hard for graphs with this property.

The MAXIMUM COLORFUL SUBTREE problem becomes tractable if the input graph is a colorful graph with non-negative edge weights. But the COLORFUL SUBTREE CLOSURE problem remains hard, even in this case:

Theorem 3. *The* COLORFUL SUBTREE CLOSURE *problem is MAX SNP-hard even if the input graph is a colorful DAG with a single source and binary weights.*

As we consider a colorful DAG, we can discard all colors and search for a subtree with maximum transitive closure. The transitive closure need not to be defined on colors, but can also be defined on vertices. So, each transitive edge has individual 0/1 weight. We defer the proof of Theorem 3 to the full version of the paper. This proof can be easily adapted to a DAG with maximal vertex degree three.

Surprisingly, we can still find a swift and exact algorithm for the COLORFUL SUBTREE CLOSURE problem, presented in the next section.

4   An Exact Algorithm for the Combined Colorful Subtree Problem

Several heuristics for the simpler MAXIMUM COLORFUL SUBTREE problem have been evaluated both regarding quality of scores [6] and quality of fragmentation tree [17]. Results of using only the heuristics were of appalling quality, so we refrain from using only heuristics to solve the COMBINED COLORFUL SUBTREE problem. Furthermore, no constant-factor approximation can exist, unless P = NP. But despite the hardness of the problem, we will now present an exact algorithm with reasonable running time in applications. The algorithm is fixed-parameter tractable with respect to the number of colors $k = |C|$, and uses dynamic programming to find the optimum. Note that in application, we can choose k arbitrarily, see below. Let $n := |V|$ and $m := |E|$ be the number of vertices and edges in the input graph $G = (V, E)$, respectively.

Let $W^*(v, S)$ be the maximum score $w^*(T)$ of a colorful subtree with root v using colors $S \subseteq C$. Then $W^*(v, S)$ can be calculated as

$$W^*(v, S) = \max \begin{cases} \displaystyle\max_{u:c(u)\in S\setminus\{c(v)\}} \left\{ \begin{array}{l} W^*(u, S\setminus\{c(v)\}) + w(v, u) \\ + \displaystyle\sum_{c'\in S\setminus\{c(v)\}} w^+\big(c(v), c'\big) \end{array} \right\} \\ \displaystyle\max_{\substack{(S_1,S_2):S_1\cap S_2=\{c(v)\} \\ S_1\cup S_2=S}} W^*(v, S_1) + W^*(v, S_2) \end{cases} \tag{2}$$

where, obviously, we have to exclude the cases $S_1 = \{c(v)\}$ and $S_2 = \{c(v)\}$ from the computation of the second maximum. We initialize $W^*(v, \{c(v)\}) = 0$, and set the weight of nonexistent edges to $-\infty$. To prove the correctness of recurrence (2), we note that we only have to differentiate three cases: A subtree root v can have no children, one child, or two or more children. The case of no children, that is v is a leaf, is covered by the initialization. If v has one child u, we add the score of the tree rooted in u, the score of the new edge (v, u), and scores of all new transitive edges. This is done in the first line of the recurrence. If v has two or more children, we can "glue together" two trees rooted in v, where we arbitrarily distribute the children of v and the colors of S to the two trees.

We now analyze the running time of recurrence (2). Extending a tree by a single vertex takes $O(2^k m)$ time, as we can calculate the sum in constant time by going over the 2^k partitions in a reasonable order. Gluing together two trees, the k colors are partitioned into three groups: those not contained in S, elements of S_1, and elements of S_2. There are 3^k possibilities to perform this partition, so running time is $O(3^k n)$. This results in a total running time of $O(3^k n + 2^k m)$. Running time can be improved to $O^*(2^k)$ using subset convolutions and the Möbius transform [4], but this is of theoretical interest only. In comparison to the algorithm presented in [6], the worst-case running time is not affected by scoring the transitive closure. The necessary space is $O(2^k n)$. In our implementation, we only iterate over defined values in W^*: An entry is not defined if there exists no subtree rooted in v using exactly the colors in S. This algorithm engineering technique does not improve worst-case running times and memory consumption, but greatly improves them in practice. To decrease memory consumption, we use hash maps instead of arrays.

Unfortunately, the above method is limited by its memory and time consumption. In application, exact calculations are limited to $k' \leq k$ colors for some moderate k', such as $k' = 15$. These colors correspond to the k' most intense peaks in the mass spectra, as these contribute most to our scoring. The remaining peaks are added in descending intensity order by a greedy heuristic: For each vertex v with an unassigned color, we try to attach v to every vertex u of the tree constructed so far, where some or all of the children of u in the tree may become children of v. We omit the technical details, and just note that our heuristic is inspired by Kruskal's algorithm for computing a maximum spanning tree.

5   Scoring

Particularly in fragmentation spectra, the charge of metabolites is mostly ± 1, so we may assume that m/z and mass are equal. Note that our calculations are not limited to a charge of one, though.

The transitive closure score w^+ is defined using the MS^n data. Recall that this score is defined for peaks or, equivalently, colors. In detail, we score three cases, see Fig. 2:

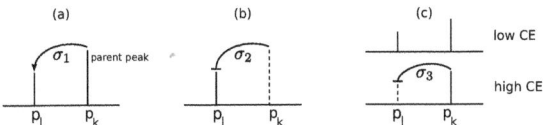

Fig. 2. Scoring of the transitive closure referring to the three cases. A dashed peak is not occurring in the spectrum drawn but in another one (typically the MS^2 spectrum). A connection $p_k \rightarrow p_l$ indicates that p_l is a fragment of p_k while a connection $p_k \dashv p_l$ indicates that peak p_l is unlikely to be a fragment of p_k.

- A spectrum with parent peak p_k and peak p_l indicates that the fragment corresponding to p_l has evolved from the fragment corresponding to p_k. To reward this, we increase the transitive score of the tree by $\sigma_1 \geq 0$ if the fragment corresponding to p_k is a direct or indirect ancestor of the fragment corresponding to p_l, see Fig. 2(a).
- Given a spectrum that contains peak p_l but not peak p_k, and mass $p_l <$ mass p_k. This indicates that the fragment corresponding to p_l has not evolved from the fragment corresponding to p_k. To penalize this, we add $\sigma_2 \leq 0$ to the score if the fragment corresponding to p_k is a direct or indirect ancestor of the fragment corresponding to p_l, see Fig. 2(b).
- Given two spectra with different collision energies and two peaks p_k and p_l with mass $p_l <$ mass p_k. If the spectrum with higher collision energy contains only p_k but the spectrum with lower collision energy contains both peaks, the fragment corresponding to p_l has probably not evolved from the fragment corresponding to p_k. To penalize this case, we add $\sigma_3 \leq 0$ to the score if the fragment corresponding to p_k is a direct or indirect ancestor of the fragment corresponding to p_l, see Fig. 2(c).

In all cases, σ_1, σ_2, and σ_3 are not used to score edges of the fragmentation tree but instead, edges of the transitive closure of the tree. Two peaks are identified to correspond to the same fragment if their masses differ in less than 0.1 Da. For each fragment only the peak with maximum intensity is taken into account for further calculations.

The scoring scheme of the fragmentation graph is the same as introduced in [6], taking the following properties into account: peak intensities, mass deviation between explanation and peak, chemical properties, collision energies and neutral losses. First, every peak is given a base score of b, $b \geq 0$. To score the mass deviation we evaluate the logarithmized Gaussian probability density function with SD σ at the measuring error value. Further we use the density function of the normal distribution with mean 0.59 and SD 0.56 to score the hetero atom to carbon ratio of the decompositions. Due to the collision energies of the different spectra, some peaks cannot represent fragments of other peaks. A fragment appearing at lower collision energy than its predecessor is penalized with $\log(\alpha)$, $\alpha \ll 1$. If there is no spectrum containing both, neither containing none of the peaks we add only a penalty of $\log(\beta)$, $\alpha < \beta < 1$. Common neutral losses are rewarded with $\log(\gamma)$, $\gamma > 1$, while radical neutral losses are penalized by $\log(\delta)$, $\delta < 1$, and large neutral losses by $\log(1 - \frac{\text{mass(neutral loss)}}{\text{parent mass}})$. In addition to the scoring from [6], we use an extension that takes into account *rare* neutral losses: If a rare neutral loss occurs in a fragmentation step we penalize it by adding $\log(\eta), \eta \ll 1$. We also penalize neutral losses that consists carbon or only nitrogen atoms by adding $\log(\epsilon), \epsilon \ll 1$. In contrast, radical losses are not penalized, since they sometimes occur in fragmentation reactions. Due to space constraints, we defer a list of all rare neutral losses and radical losses to the full version of the paper.

6   Results

To evaluate our work we implemented the algorithm in Java 1.6. As test data we used 185 mass spectra of 45 compounds, mostly representing plant secondary metabolites. The 185 mass spectra are composed of 45 MS^2 spectra, 128 MS^3 spectra and twelve MS^4 spectra (unpublished). All spectra were measured on a Thermo Scientific Orbitrap XL instrument, we omit the experimental details. Peak picking was performed using the Xcalibur software supplied with the instrument. The data set was analyzed with the following options: For decomposing peak masses we use a relative error of 20 ppm and the standard alphabet containing carbon (C), hydrogen (H), nitrogen (N), oxygen (O), phosphorus (P), and sulfur (S). For the construction of the fragmentation graph, we use the collision energy scoring parameters $\alpha = 0.1$, $\beta = 0.8$, the neutral loss scoring parameters $\gamma = 10$, $\delta = 10^{-3}$, $\epsilon = 10^{-4}$, $\eta = 10^{-3}$, the intensity scoring parameter $\lambda = 0.1$, the base score $b = 0$ and the standard deviation of the mass error $\sigma = 20/3$ as described in [6, 17]. We can identify the molecular formulas of the compounds from isotope pattern analysis and by calculating the fragmentation trees for all candidate molecular formulas [17]. In this paper, we assume that this task has been solved beforehand, and that all molecular formulas are known.

Comparing Trees. We evaluate the impact of using MS^n instead of MS^2 data, as well as the influence of scoring parameters $\sigma_1, \sigma_2, \sigma_3$ from Sec. 5, using pairwise tree comparison. In each fragmentation tree, vertices are implicitly labeled by molecular formulas of the corresponding fragments. We limit our comparison to those fragments that appear in both trees, and discard orphan fragments. We distinguish four cases:

- A fragment is *identically placed*, if its parent fragments are identical in both trees.
- A fragment is *pulled up*, if its parent fragment in the second tree is one of its predecessors in the first tree (and the fragment is not identically placed).
- A fragment is *pulled down*, if its parent fragment in the first tree is one of its predecessors in the second tree (and the fragment is not identically placed).
- A fragment is *regrafted*, if it is not identically placed, pulled up or pulled down.

The obvious way to evaluate our method would be to compare our results against some gold standard. Unfortunately, such gold standard is not available for our study. Rasche *et al.* [17] have evaluated the method from [6] by expert annotation of MS^2 fragmentation trees for a subset of the compounds used in this paper. Unfortunately, the input data (that is, fragments observed in the MS^2 and MS^n mode of the instrument) differ strongly. Hence, a comparison against these expert-annotated fragmentation trees is impossible.

As mentioned in Sec. 4 the exact algorithm is memory and time consuming. So, we use the exact algorithm for only the k' most intense peaks, and attach the remaining peaks using a greedy heuristic. We find that decreasing k' from 20 to 15, has a comparatively small effect on the computed fragmentation trees: 97.1% of the fragments were identically placed, 0.4% were pulled up, 0.6% were pulled down, and only 1.9% were regrafted. On the other hand, average running time per compound was decreased from 30.8 min to 3.97 s. In the remainder of this section, we set $k' = 15$ and use only the 15 most intensive peaks for exact computations. Choosing a moderate $k' = 15$ has a much stronger effect here, than it was observed for the Maximum Colorful Subtree problem [6], where constructed fragmentation trees were practically identical for $k' = 15$ and $k' = 20$. We attribute this to the transitive scoring, which appears to be harder to grasp by the heuristic.

To show the effect of evaluating MS^n data, we individually varied the three score parameters, and compared the resulting trees to the trees constructed without scoring the transitive closure, see Fig. 3. As σ_1 increases, the fraction of changes in the trees (pull ups, pull downs and regrafts) converges to about 14%. A similar behavior is observed as σ_2, σ_3 are decreased. The main difference between the bonus score σ_1 and the penalty scores σ_2 and σ_3 is that increasing σ_1 results in more pull downs than pull ups, while decreasing penalty scores σ_2, σ_3 produces more pull ups than pull downs. This can be explained as follows: Reward scores can rather be realized if fragments are inserted deep, that is, far from the root. In contrast, negative penalty scores are avoided if the fragments are inserted "shallow", that is, close to the root. So, $\sigma_1 \gg 0$ tends to deepen the trees, whereas $\sigma_2, \sigma_3 \ll 0$ tends to broaden the trees.

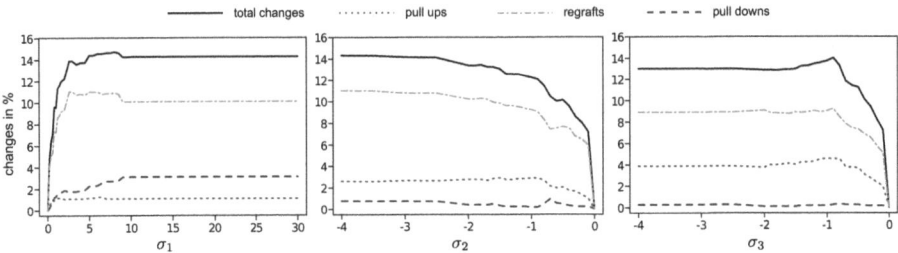

Fig. 3. Percentage of pull ups, pull downs, regrafted fragments, and total changed fragments when varying score parameters σ_1, σ_2, and σ_3. Left: Varying σ_1 with $\sigma_2 = 0$ and $\sigma_2 = 0$ fixed. Middle: Varying σ_2 with $\sigma_1 = 0$ and $\sigma_3 = 0$ fixed. Right: Varying σ_3 with $\sigma_1 = 0$ and $\sigma_2 = 0$ fixed.

Based on the above analysis, we decided to use the following parameter values: $k' = 15$, $\sigma_1 = 3$, $\sigma_2 = -0.5$, and $\sigma_3 = -0.5$. We choose a large σ_1 as the underlying MS^n observation is a clear signal that some fragment should be placed as a successor of another fragment. In comparison, the MS^n reasoning behind σ_2 and σ_3 is somewhat weaker, so we choose smaller absolute values for these parameters that less influence the trees. The crucial comparison is now between the fragmentation trees computed without scoring the transitive closure and the fragmentation trees computed with the above scores. As we have only one MS^2 spectrum per compound and one spectrum contains too few peaks to calculate a reasonable tree, we transform the MS^n data to "pseudo-MS^2" data by merging all fragmentation spectra of a compound into one. This simulates MS^2 spectra with different collision energies. By merging all spectra into one, we loose all information about dependencies between peaks/colors. This is implicitly achieved by setting $\sigma_1, \sigma_2, \sigma_3 = 0$. Between these trees 76.21% of the fragments are identically connected, 4.90% are pull ups, 1.79% pull downs and 17.11% regrafted fragments. Hence, almost one quarter of all fragments are changed due to the information from MS^n data.

We cannot evaluate whether these changed neutral losses are true or false and, hence, whether MS^n fragmentation trees are truly better than the MS^2 trees. But we will now show an example where the MS^n tree agrees well with the observed MS^n data: To this end, we consider the fragmentation trees of Phenylalanine, with and without scoring the transitive closure, see Fig. 4. The two fragmentation trees are almost identical, with the single exception of fragment C_7H_9 at 93.1 Da: This fragment is connected to $C_9H_9O_2$ at 149.0 Da in the MS^2 tree, and to $C_8H_{10}N$ at 120.1 Da in the MS^n tree. In the MS^2 interpretation, the neutral loss C_2O_2 is explained as two common neutral losses CO and, hence, it is preferred over the neutral loss CHN (hydrogen cyanide). Using MS^n data, we can resolve this: the peak at 93.1 Da *does* occur in the MS^3 spectrum with parent peak at 120.1 Da, therefore C_7H_9 at 93.1 Da probably resulted (directly or indirectly) as a fragment of $C_8H_{10}N$ at 120.1 Da. This is rewarded by our algorithm, adding $\sigma_1 = +3$ to the score of the modified tree. The fact that the peak at 107.0 Da is missing in the MS^3 spectrum with parent peak at 120.1 Da, does not change the score: In the MS^2 analysis, fragment C_7H_7O cannot be a successor of $C_8H_{10}N$ at 120.1 Da, nor are 91.1 Da, 93.1 Da, or 103.1 Da assumed to be its successor. Another example where the MS^n tree agrees well with the observed MS^n data is tryptophan. Due to space constraints, we defer the details of this analysis to the full version of the paper.

As shown in Sec. 3, the theoretical worst case running time of our algorithm is identical with that of the MAXIMUM COLORFUL SUBTREE algorithm in [6]. We investigated whether this also holds in application. Running times were measured on an Intel Core 2 Duo, 2.4 GHz with 4 GB memory, with parameter $k' = 15$. We find that total running times of the algorithm, with and without using MS^n data, are practically identical: Average running time is about 3.8 s, and the maximal running time for one compound was 17.6 s. We omit further details.

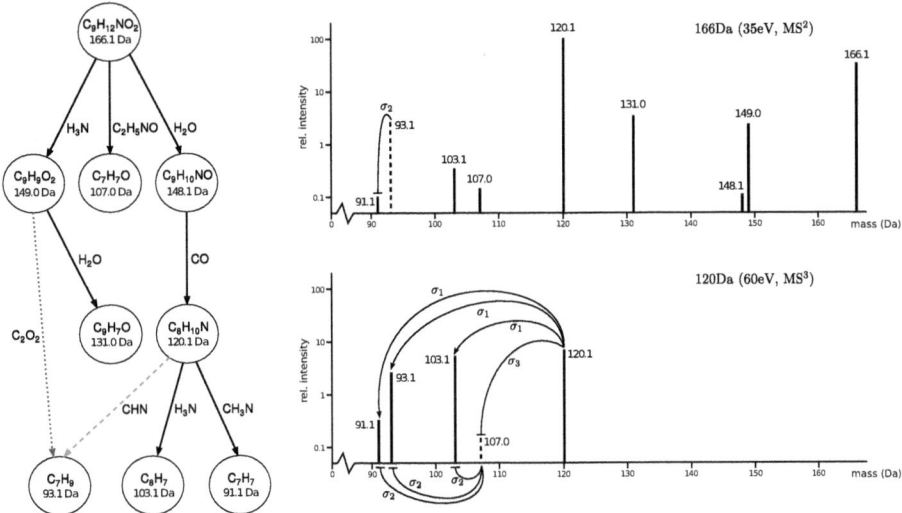

Fig. 4. Left: Fragmentation trees of phenylalanine. Solid edges are neutral losses present in both trees, the red dotted (green dashed) edge is present in the MS^2 (or MS^n) tree only, respectively. Right: MS^2 spectrum of the parent peak (top) and MS^3 spectrum of the 120.1 Da fragment (bottom). Dashed peaks are not contained in the particular spectrum.

7   Conclusion

In this paper, we have presented a framework for computing metabolite fragmentation trees using MS^n data. Our fragmentation model results in the COMBINED COLORFUL SUBTREE problem, a conjunction of the MAXIMUM COLORFUL SUBTREE problem and the COLORFUL SUBTREE CLOSURE problem. Both problems are NP-hard, and no PTAS can exist for either problem. The latter problem remains MAX SNP-hard even if the input graph is colorful.

We have presented an exact dynamic programming algorithm for the COMBINED COLORFUL SUBTREE problem, showing that the problem is fixed-parameter tractable with respect to the parameter "number of colors". We have introduced a scoring scheme based on the dependencies between the different fragmentation steps. To reduce memory and time requirements, we limit exact computations to the $k' \leq k$ most intense peaks in the spectrum. Although the COMBINED COLORFUL SUBTREE problem is computationally hard, the resulting algorithm is fast in practice.

For our application, the score of the transitive closure $w^+ : C^2 \to \mathbb{R}$ is defined on *pairs of colors*. From the theoretical standpoint, one can modify the problem such that the score $w^+ : E^+ \to \mathbb{R}$ is defined on *edges of the transitive closure* of the fragmentation graph $G = (V, E)$. In this case, our algorithm from Sec. 4 cannot be used, and it remains an open problem whether this modified version of the COLORFUL SUBTREE CLOSURE problem is fixed-parameter tractable with respect to the number of colors. Clearly, the problem is in FPT for unit weights.

We have seen that using additional information from MS^n data does change the computed fragmentation trees. In our experiments, one quarter of fragments were differently inserted when including MS^n information. As our scoring scheme is "chemically reasonable", we argue that the trees are actually improved using MS^n data. Unfortunately, MS^n is less suited for high-throughput measurements, as individual measurements are more time-consuming. On the other hand, for about three quarters of the fragments, trees remain identical between MS^2 and MS^n. Thus, calculating fragmentation trees from MS^2 data extracts valuable information concealed in these spectra and results in largely reasonable trees.

In the future, we want to increase the speed and decrease the memory consumption of our exact algorithm. Also, we want to use MS^n fragmentation trees to fine-tune the scoring parameters for computing MS^2 fragmentation trees. The next step of the analysis pipeline is a method for automated comparison of fragmentation trees.

Acknowledgments. We thank Aleš Svatoš and Ravi Kumar Maddula from the Max Planck Institute for Chemical Ecology in Jena, Germany for supplying us with the test data. K. Scheubert was funded by Deutsche Forschungsgemeinschaft, project "IDUN". F. Hufsky was supported by the International Max Planck Research School Jena.

References

1. Arora, S., Lund, C., Motwani, R., Sudan, M., Szegedy, M.: Proof verification and the hardness of approximation problems. J. ACM 45(3), 501–555 (1998)
2. Bandeira, N., Olsen, J.V., Mann, J.V., Mann, M., Pevzner, P.A.: Multispectra peptide sequencing and its applications to multistage mass spectrometry. Bioinformatics 24(13), i416–i423 (2008)
3. Berman, P., Karpinski, M., Scott, A.D.: Computational complexity of some restricted instances of 3-SAT. Discrete Appl. Math. 155, 649–653 (2007)
4. Björklund, A., Husfeldt, T., Kaski, P., Koivisto, M.: Fourier meets Möbius: fast subset convolution. In: Proc. of ACM Symposium on Theory of Computing (STOC 2007), pp. 67–74. ACM Press, New York (2007)
5. Böcker, S., Lipták, Z.: A fast and simple algorithm for the Money Changing Problem. Algorithmica 48(4), 413–432 (2007)
6. Böcker, S., Rasche, F.: Towards de novo identification of metabolites by analyzing tandem mass spectra. Bioinformatics 24, 149–155 (2008); Proc. of European Conference on Computational Biology (ECCB 2008)
7. Dondi, R., Fertin, G., Vialette, S.: Complexity issues in vertexcolored graph pattern matching. J. Discrete Algorithms (2010) (in press), doi:10.1016/j.jda, 09.002
8. Fellows, M., Fertin, G., Hermelin, D., Vialette, S.: Sharp tractability borderlines for finding connected motifs in vertex-colored graphs. In: Arge, L., Cachin, C., Jurdziński, T., Tarlecki, A. (eds.) ICALP 2007. LNCS, vol. 4596, pp. 340–351. Springer, Heidelberg (2007)
9. Fernie, A.R., Trethewey, R.N., Krotzky, A.J., Willmitzer, L.: Metabolite profiling: from diagnostics to systems biology. Nat. Rev. Mol. Cell Biol. 5(9), 763–769 (2004)

10. Heinonen, M., Rantanen, A., Mielikäinen, T., Kokkonen, J., Kiuru, J., Ketola, R.A., Rousu, J.: FiD: a software for ab initio structural identification of product ions from tandem mass spectrometric data. Rapid Commun. Mass Spectrom. 22(19), 3043–3052 (2008)
11. Hill, D.W., Kertesz, T.M., Fontaine, D., Friedman, R., Grant, D.F.: Mass spectral metabonomics beyond elemental formula: Chemical database querying by matching experimental with computational fragmentation spectra. Anal. Chem. 80(14), 5574–5582 (2008)
12. McLafferty, F.W., Tureček, F.: Interpretation of Mass Spectra, 4th edn. University Science Books, Mill valley (1993)
13. Nuutila, E.: An efficient transitive closure algorithm for cyclic digraphs. Inform. Process. Lett. 52(4), 207–213 (1994)
14. Oberacher, H., Pavlic, M., Libiseller, K., Schubert, B., Sulyok, M., Schuhmacher, R., Csaszar, E., Köfeler, H.C.: On the inter-instrument and inter-laboratory transferability of a tandem mass spectral reference library: 1. results of an Austrian multicenter study. J. Mass Spectrom. 44(4), 485–493 (2009)
15. Oberacher, H., Pavlic, M., Libiseller, K., Schubert, B., Sulyok, M., Schuhmacher, R., Csaszar, E., Köfeler, H.C.: On the inter-instrument and the inter-laboratory transferability of a tandem mass spectral reference library: 2. optimization and characterization of the search algorithm. J. Mass Spectrom. 44(4), 494–502 (2009)
16. Pelander, A., Tyrkkö, E., Ojanperä, I.: In silico methods for predicting metabolism and mass fragmentation applied to quetiapine in liquid chromatography/time-of-ight mass spectrometry urine drug screening. Rapid Commun. Mass Spectrom. 23(4), 506–514 (2009)
17. Rasche, F., Svatoš, A., Maddula, R.K., Böttcher, C., Böcker, S.: Computing fragmentation trees from tandem mass spectrometry data. Anal. Chem. (December 2010) (in press), doi:10.1021/ac101825k
18. Sheldon, M.T., Mistrik, R., Croley, T.R.: Determination of ion structures in structurally related compounds using precursor ion fingerprinting. J. Am. Soc. Mass Spectrom. 20(3), 370–376 (2009)

Metric Labeling and Semi-metric Embedding for Protein Annotation Prediction

Emre Sefer[1,2] and Carl Kingsford[1,2]

[1] Department of Computer Science, University of Maryland, College Park
[2] Center for Bioinformatics and Computational Biology,
Institute for Advanced Computer Studies
University of Maryland, College Park
{esefer,carlk}@cs.umd.edu

Abstract. Computational techniques have been successful at predicting protein function from relational data (functional or physical interactions). These techniques have been used to generate hypotheses and to direct experimental validation. With few exceptions, the task is modeled as multi-label classification problems where the labels (functions) are treated independently or semi-independently. However, databases such as the Gene Ontology provide information about the similarities between functions. We explore the use of the METRIC LABELING combinatorial optimization problem to make use of heuristically computed distances between functions to make more accurate predictions of protein function in networks derived from both physical interactions and a combination of other data types. To do this, we give a new technique (based on convex optimization) for converting heuristic semimetric distances into a metric with minimum least-squared distortion (LSD). The METRIC LABELING approach is shown to outperform five existing techniques for inferring function from networks. These results suggest METRIC LABELING is useful for protein function prediction, and that LSD minimization can help solve the problem of converting heuristic distances to a metric.

1 Introduction

Networks encoding pairwise relationships between proteins have been widely used for protein function prediction and for data aggregation and visualization. Sometimes these networks are derived from a single data source such as protein-protein interactions [17,37,30]. In other instances, they are constructed from integration of large collection of experiments involving different data types, such as gene expression [11], protein localization [16], etc. The precise meaning of an edge can differ, but a common feature of these networks is that two proteins connected by an edge often have similar functions. By extension, these networks generally have the property that two proteins that are "close" in the network are more likely to have closely related functions. This correlation has given rise to a number of computational approaches to extract hypotheses for protein function from relational data [34,33,14,29,18,7].

V. Bafna and S.C. Sahinalp (Eds.): RECOMB 2011, LNBI 6577, pp. 392–407, 2011.
© Springer-Verlag Berlin Heidelberg 2011

Nearly all of these computational methods treat the function prediction problem as a labeling problem, where the labels are drawn from a vocabulary of biological functions or processes. They typically ignore any relationships between the functions, treating them as independent labels. However, there are usually known relationships among functions that ought to be useful to make more accurate predictions of protein function. For example, the Gene Ontology (GO) [36] is a manually curated database of biological functions and processes that represents the hierarchical relationships among different functions as a DAG. However, most prediction methods have ignored such a structure.

In the few cases it has been done, integrating Gene Ontology knowledge into protein function prediction methods [1,7] and clustering [5] has resulted in improved predictions. For example, Barutcuoglu et al. [1] developed a Bayesian framework for combining multiple SVM classifiers based on the GO constraints to obtain the most probable, consistent set of predictions. Their approach used a hierarchy of support vector machine (SVM) classifiers trained on multiple data types. This method also exploits the relationship between functions in GO but does not exploit distances between functions directly. Taking another approach, Deng et al. [7] uses the correlations between which proteins are labeled with each functions but they estimate these correlations from training data and do not consider GO structure.

1.1   Metric Labeling for Function Prediction

Here, we propose to integrate Gene Ontology relationships with relational data by modeling the protein function prediction problem as an instance of METRIC LABELING [20] which is a special case of MRF [21] in which the distance function among labels is a metric. The METRIC LABELING problem seeks to assign labels (here, protein functions) to nodes in a graph (here, proteins or genes) to minimize the distance (in the metric) between labels assigned to adjacent nodes in addition to the cost of assigning labels to nodes. The advantage of this formulation is that rather than treating function labels as independent, unrelated entities, their similarities can be directly incorporated into the objective function. A more detailed description of the METRIC LABELING problem is given in Section 2.1.

The METRIC LABELING formulation can be seen as an generalization of minimum multiway cut [39], which implicitly assigns distance 0 between two identical functions and distance 1 between any pair of distinct functions. METRIC LABELING softens this to account for varied levels of similarities between the functions. METRIC LABELING can also be seen as special case of Markov Random Field (MRF). MRFs encode the same combinatorial problem, but the distance function is not restricted to metrics or semimetrics [21]. However, optimization with such arbitrary distance functions is NP-Hard [6,21], and there is no approximation algorithm that can approximate the global optimum within a non-trivial bound. In contrast, there are practical approximation algorithms for METRIC LABELING with logarithmic approximation guarantees [4,20]. In this paper, we will use the integer programming formulation by [4] which yields an $O(\log k)$ approximation algorithm for METRIC LABELING where k is number of labels.

1.2   Constructing a Metric Distance between GO Functions

METRIC LABELING (and MRF models) have typically been used in applications related to computer vision [21,26,2] where often the distance between the labels naturally can be expressed by metrics. In the case of function prediction from relational data, while heuristic relationships between functions can be readily computed from the structure of the Gene Ontology graph, it is more difficult to make these distances obey the requirements of a metric. Recall that a **metric** $d(\cdot, \cdot)$ over items X satisfies the following 4 properties for all x, y, z in X:

$$d(x, y) \geq 0 \qquad \qquad \text{(Nonnegativity)} \qquad (1)$$
$$d(x, y) = 0 \text{ if and only if } x = y \qquad \qquad (2)$$
$$d(x, y) = d(y, x) \qquad \qquad \text{(Symmetry)} \qquad (3)$$
$$d(x, z) \leq d(x, y) + d(y, z) \qquad \text{(Triangle Inequality)} \qquad (4)$$

Typically, properties (1)–(3) can be easily satisfied, but often natural distance measures do not satisfy the triangle inequality (4). When d satisfies (1)-(3) but not the triangle inequality (4), it becomes a **semimetric**.

To apply METRIC LABELING when the distance function on the labels is merely a semimetric, we will first convert the semimetric into a metric that is as similar to the semimetric as possible. Approximating a semimetric by a close metric and MRF optimization when the distances are semimetric are topics of recent interest, and Kumar and Koller [24] have recently suggested an algorithm based on minimizing the distortion. If $\mathcal{S}$ is a semimetric, and $\mathcal{M}$ is a metric approximating $\mathcal{S}$, the *contraction* of this mapping is the maximum factor by which distances are shrunk in $\mathcal{M}$ and the *expansion* or stretch of this mapping is the maximum factor by which distances are stretched in $\mathcal{M}$. The *distortion* of this approximation is the product of the contraction and the expansion. Although distortion minimization has traditionally been used in metric embeddings, distortion considers the error introduced in the largest outlier and does not take into account the distribution of the error over all the distances. For imperfect data that is far from a metric, intuition indicates that minimizing the error introduced in the other distances would yield a better metric.

To design metric approximations to semimetrics that better preserve all distances, we propose a least squared minimization algorithm that tries to minimize the total squared error among all distances. To contrast it with traditional distortion, we call this approach *least squared distortion* (LSD). This problem can easily be solved in polynomial time because it is a convex case of quadratic programming. Thus, to apply METRIC LABELING in cases when the distances among the labels are not metric, we first map the semimetric to a close metric using the LSD algorithm and then run METRIC LABELING on the new metric. Experiments on protein function prediction suggest this is a good metric approximation method. The issue of converting a set of heuristically estimated distances to a metric arises in many practical contexts and the LSD approach may also be useful for other applications.

1.3   Improvement in Function Prediction

We test the LSD algorithm and the METRIC LABELING approach for function prediction on relational data for 7 species: *S. cerevisae*, *A. thaliana*, *D. melanogaster*, *M. musculus*, *C. elegans*, *S. pombe* and Human. For *S. cerevisae*, we apply the algorithms to an integrated data set that derives pairwise relationships between proteins from several lines of evidence such as gene expression, protein localization data, and known protein complexes. For all 7 species, we also test the approaches on networks derived from high-throughput protein-protein interaction experiments.

The algorithms are tested in a variety of settings. The set of functional labels are drawn from the Gene Ontology's Biological Process sub-ontology. The number of considered GO terms is varied between 90 and 300 in order to evaluate the effect of the size and specificity of the label set on performance. Various metrics and semimetrics relating the GO terms are also tested. A simple shortest-path metric is compared with two other semimetrics derived from lowest common ancestor in the Gene Ontology DAG, semimetrics computed from a training set of labels, and semimetrics computed from both training set and GO. See Section 2.4.

1.4   Our Contributions

We introduce the use of METRIC LABELING for protein function prediction from relational data and show that under many reasonable metrics it outperforms the approaches based on Markov Random Fields [25], Functional Flow [29], minimum multiway cut [39,19], neighborhood enrichment [14], and simple majority rule [33]. We test on 7 species in both protein-protein and integrated networks using several different collections of GO terms. The results indicate that the clean METRIC LABELING formulation is useful for automated function prediction.

In addition, we introduce the LSD objective function for finding a metric that approximates a semimetric with the goal of preserving many distances rather than just limiting the maximum distortion. The convex optimization approach for this problem may be useful in other contexts where reasonable heuristic distances do not satisfy the triangle inequality. We compare the performance of running first our LSD metric approximation algorithm and then running METRIC LABELING on the LSD's output metrics with a recent algorithm by Kumar and Koller [24] and METRIC LABELING with LSD metric approximation results in better predictions.

2   Methods

2.1   The Metric Labeling Problem

The METRIC LABELING problem has been extensively investigated from a theoretical point of view [20,4]. Formally, we have a graph $G = (P, E)$ over a set P of n nodes (here, proteins), E of edges and a set L of k possible labels (here, functions) that we want to assign to objects. We have a metric $d(\cdot, \cdot)$ satisfying properties (1)–(4) defined between any labels in L. We are also given a function $c(p, \ell)$ that provides the cost of assigning label $\ell \in L$ to $p \in P$. METRIC LA-BELING seeks an assignment $f : P \to L$ of labels to proteins that minimizes the objective function:

$$Q(f) = \sum_{p \in P} c(p, f(p)) + \sum_{(p,q) \in E} w(p, q)d(f(p), f(q)). \qquad (5)$$

where $w(p, q) = w(q, p)$ is the weight of the edge between proteins p and q in the graph. The first summation is called the *assignment costs* and depends only on individual choice of label we make for each protein and second summation is called the *separation costs* and is based on the pair of choices we make for two interacting proteins.

The intuition is that pairs of proteins that are highly related ($w(p.q)$ is high) ought to be assigned labels that are highly similar ($d(f(p), f(q))$ is low). The assignment costs prevent the problem from becoming trivial by forbidding the assignment of the same label to every protein. For a protein p with a known function b, typically $c(p, b)$ will be 0 and $c(p, \ell) = \infty$ for all $\ell \in L$ except b.

2.2   Integer Programming Formulation of Metric Labeling

The METRIC LABELING problem defined above can be written as an ILP [4]. In this formulation, $x(u, i)$ is binary variable indicating that vertex u is labeled with i and $x(u, i, v, j)$ is binary variable indicating that vertex u is labeled with i and vertex v is labeled with j for edge $(u, v) \in E$. The objective is then to

$$\text{minimize} \sum_{v \in V} \sum_{i \in L} c(u, i)x(u, i) + \sum_{(u,v) \in E} \sum_{i \in L} \sum_{j \in L} w(u, v)d(i, j)x(u, i, v, j). \qquad (6)$$

The variables are subject to the following constraints:

$$\sum_{i \in L} x(u, i) = 1 \qquad\qquad \forall u \in V \qquad (7)$$
$$\sum_{j \in L} x(u, i, v, j) = x(u, i) \qquad \forall u \in V, v \in N(u), i \in L \qquad (8)$$
$$x(u, i, v, j) = x(v, j, u, i) \qquad \forall u, v \in V, i, j \in L \qquad (9)$$
$$x(u, i) \in \{0, 1\} \qquad\qquad \forall u \in V, i \in L \qquad (10)$$
$$x(u, i, v, j) \in \{0, 1\} \qquad \forall (u, v) \in E, i, j \in L \qquad (11)$$

Constraints (7) mean each vertex must receive some label. Constraints (8) force consistency in the edge variables: if $x(u, i) = 1$ and $x(v, j) = 1$, they force $x(u, i, v, j)$ to be 1. Constraints (9) express the fact that (u, i, v, j) and (v, j, u, i) refer to the same edge.

Solving this integer programming instance optimally is NP-complete. Since we are dealing with large networks, we use the $O(\log k)$ approximation algorithm given by Chekuri et al. [4] that is based on solving the linear programming relaxation to identify a deterministic HST metric [9] of the given metric such that the cost of our fractional solution on this HST metric is at most $O(\log k)$ times the LP cost on the original metric. We implemented and ran the LP formulation in GLPK [13].

2.3 Metric Approximation via Least Square Distortion Minimization

The algorithms suggested above have guaranteed performance bounds when the distance d is a metric. However, finding a metric distance in practical contexts can be difficult. Ideally, the distance encodes a large amount of knowledge about the relationship between protein functions. It is likely that such as distance will not satisfy the triangle inequality. We define a novel metric approximation algorithm, called LSD, based on minimizing the total least squared error between a given semimetric set of distances and the computed metric distances. Least squared error approximation is intuitive because the error of every distance contributes to the total error of the metric approximation instead of only the maximum expansion and contraction as in distortion case.

The LSD algorithm is defined as a quadratic program below, where $S = \{s_1, \ldots, s_{\binom{n}{2}}\}$ is the given set of semimetric distances between each pair of n items, and $M = \{m_1, \ldots, m_{\binom{n}{2}}\}$ is corresponding set of metric distances, where for all i, s_i and d_i represent distances between the same pair of proteins. Let $I = \{1, \ldots, \binom{n}{2}\}$ be the set of indices of distances.

To find a good approximation to the distances in S we seek values for the $\{m_i\}$ variables to

$$\text{minimize} \sum_{i \in I} (s_i - m_i)^2. \tag{12}$$

We require that the m_i values satisfy the following constraints for all $i, j, k \in I$ that should be related by the triangle inequality:

$$m_i + m_j - m_k \geq 0 \tag{13}$$
$$m_i + m_k - m_j \geq 0 \tag{14}$$
$$m_k + m_j - m_i \geq 0 \tag{15}$$

The objective function can be written as $(1/2)x^T Q x + c^T x$ where $n \times n$ matrix Q is symmetric, and c is any $n \times 1$ vector. In our case, the matrix Q is positive definite and if the problem has a feasible solution then the global minimizer is unique. In this case, the problem can be solved by interior point methods in polynomial time. We implemented and ran the problem in CPLEX [38].

2.4 Metrics and Semimetrics

We test 4 different distance measures between protein functions:

1. $d_{SP}(x, y)$ = the shortest path distance in the GO DAG between x and y divided by diameter of GO. This is a metric and intuitively simple.

2. $d_{\mathrm{LCA}}(x, y) = (b+c)/(2a+b+c)$, where a is shortest path distance from the root of the ontology to the lowest common ancestor u of x and y and b is the shortest distance from x to u and c is the shortest distance from y to u. The LCA distance measure does not satisfy triangle inequality and is only a semimetric.

3. $d_{\mathrm{Lin}}(x, y) = (\log \mathrm{Pr}(x) + \log \mathrm{Pr}(y))/(2 \log \mathrm{Pr}(\mathrm{lca}(x, y)))$, where $\mathrm{Pr}(x)$ is the empirical probability (computed from the training annotations) that a protein is annotated with x, and $\mathrm{lca}(x, y)$ is the LCA of x and y. This is defined in [28] as a similarity measure, and we take its reciprocal as a distance. It is similar to the LCA distance above but uses the probabilities of each annotation instead of GO distances. It has mostly been used in NLP applications [3,27]. However, it has recently been used in other applications of Gene Ontology distances [32,8]. It is a semimetric.

4. $d_{\mathrm{KB}}(x, y) = \sum_{p_1 \in P_x} \sum_{p_2 \in P_y} \mathrm{sp}(p_1, p_2)/(\mathrm{diameter} \cdot |P_x| \cdot |P_y|)$, where P_x and P_y are sets of proteins in the training set annotated with x and y respectively, $\mathrm{sp}(x, y)$ is the shortest path distance between x and y, $\mathit{diameter}$ is the diameter of network.

We also consider the combination of a structure-based $d \in \{d_{\mathrm{SP}}, d_{\mathrm{LCA}}, d_{\mathrm{Lin}}\}$ with the knowledge-based d_{KB} using the formula:

$$d_{\mathrm{comb}}(x, y) = (1 - \alpha)d(x, y) + \alpha d_{\mathrm{KB}}(x, y), \tag{16}$$

where α is a weight of contribution of training set estimations. For $\alpha < 1$, none of the combined distances are metric (but are semimetric).

When the distance is not a metric, we first run the LSD metric approximation algorithm (Section 2.3) to obtain a metric and then run METRIC LABELING on those metric distances. When it is a metric, we just run METRIC LABELING.

In addition, we test two schemes for the assignment costs $c(u, i)$ of assigning label i to node u. Either they are chosen to be uniformly 1 or non-uniformly according to the density of a label in a particular region of the graph as follows: We estimated for each protein p and label i cost $c(p, i) = n_p/(n_{pi}n_p) = 1/n_{pi}$ where n_p and n_{pi} are number of neighbors of p and number of neighbors of p in the training set that have function i respectively. In the case where p has no neighbors with function i, $c(p, i) = 2$. When a function of protein is known, the cost of assigning that function is 0 whereas assigning any other function is ∞.

2.5   Network Data

We tested our algorithm on the protein-protein interaction (PPI) networks of 7 species obtained from BIOGRID [35]: *S. cerevisiae*, *C. elegans*, *D. melanogaster*, *A. thaliana*, *M. musculus*, *H. sapiens*, and *S. pombe*. We used all physical interactions in BIOGRID. Duplicate edges were counted as single edges. We consider only the largest connected component. We used GO annotations downloaded from the Gene Ontology as our true annotations. Only non-EIA annotations are considered. When considering only PPI networks, weight of every edge is 1.

For *S. cerevisiae*, we also considered an integrated network derived from several data sources, including gene expression [11], protein localization [16], protein complexes [12,15], and protein interaction [35]. We used protein complex dataset by assigning binary interactions between any two proteins participating in the same complex, yielding 49313 interactions. For gene expression data, we assigned binary interactions between genes whose correlation in [11] is greater than 0.8 or smaller than -0.8. We assigned binary interactions between any proteins that are annotated to the same location in [16].

We combined these data sources into one network by using noisy-or with their reliability scores, where the interaction score between nodes u and v is taken to be $w(u, v) = \text{Score}(u, v) = 1 - \prod_{i \in E_{uv}} (1 - r_i)$ where E_{uv} are the experiments in which u and v were observed to interact. The reliability r_i of each source i was estimated by the percent of edges from i that connect proteins of shared function in the training data.

2.6 Comparison to Other Methods

We ran the algorithms on a Mac that had 2 GHz Intel Core 2 Duo processor and 2 Gb memory. The METRIC LABELING algorithm took approximately 15 minutes to run. We compared METRIC LABELING predictions with several well-known direct function prediction methods:

Majority: Each protein is annotated with the function that occurs most often among its neighbors as described in [33]. The main disadvantage of this method is that the full topology of network is not considered.

Neighborhood: For each protein, we consider all other proteins within a radius $r = 2$ as described in [14] and a χ^2-test is used to determine if each function is overrepresented.

GenMultiCut: This approach is described in [39] and [19]. It tries to cluster the network by minimizing the number of edges between clusters. This algorithm is a simpler version of our algorithm in which distance between two functions are either 1 (if they are not the same) or 0 (if they are equal). Hence, it cannot take the relations among functions into account. We followed the same approach as [29] and ran an ILP formulation for this problem 50 times, each time perturbing the weights by a very small offset drawing from uniform distribution on $(-w_{\max}10^{-5}, w_{\max}10^{-5})$ where $w_{\max}$ is the maximum edge weight in the graph. Then probability of assigning a function to a protein will be the fraction of number of annotations of this protein with that function. We implemented this by using MathProg and GLPK. It runs in < 1 minute on yeast.

FunctionalFlow: Each function is independently flowed through the whole network according to an update rule and each node is assigned to functions depending on the amount of flow it receives [29].

MRF: This method is from [25]. It is based on kernel logistic regression which is the improvement over previous MRF models [7,23]. This method also tries to exploit the relation between different functions by identifying a set of functions that are correlated with the function of interest. However, it does not use the structure of GO when estimating the correlation. This approach takes < 5 minutes to run on yeast.

We also compared LSD with a recent approach for MAP estimation under a semimetric:

Semimetric MAP Estimation Algorithm: This algorithm from [24] tries to approximate a given semimetric distance function using a mixture of r-hierarchically well-separated tree (r-HST) metrics [9]. Then, it solves each resulting r-HST metric labeling problem. We followed the same approach as in GenMultiCut, run the formulation 50 times by perturbing the edges and assign the fraction of number of annotations of this protein with that function as probability of annotating this protein with that function. We modified code provided by the authors to work on our data sets. It ran in less than a 1 minute on yeast.

Solving LSD optimally takes an hour to three hours depending on number of elements in the ontology. However, we only run that once to come up with metrics. This time can easily be reduced to several minutes by considering an iterative approach that starts with point set which elements satisfy triangle inequality and adding other points iteratively by minimizing the total distance modifications made so that current set of points after each iteration will keep satisfying triangle inequality. However, solution of this iterative approach is not guaranteed to be optimal anymore.

2.7   Evaluating Performance

We use fivefold cross-validation to compare the predictive performance of the algorithms. The d_{KB} distance and the non-uniform assignment costs are computed using only the remaining 80% of annotated proteins each time. All performance measurements are the average of the 5 runs. Each method described in Section 2.6 yields a score, and we assess performance at different false positive rates by varying the score thresholds from 0 to 1 by 0.05 increments. We varied the number of considered functions from 90 to 300. The GO terms are selected for each species to match sets of terms used in previous publications [22] and the annotations known for each species. Depending on the number of annotations required, those annotations that are seen more than others and also that are not ancestors of each other are selected. The annotations for each case and for each species can be found in the online supplementary material. We counted each annotation seperately as a separate example.

3   Results

3.1   Function Prediction in Yeast Using a PPI Network

Predictive performance on the yeast PPI network is shown in Fig. 1. The curves show that METRIC LABELING combined with our LSD metric approximation

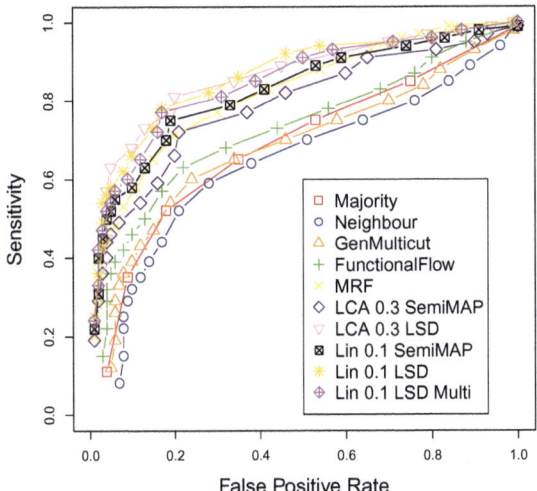

Fig. 1. ROC curves comparing various algorithms with METRIC LABELING approaches using 90 ontology terms. **SemiMap** indicates the semimetric-to-metric conversation algorithm by Kumar and Koller [24] is run; **LSD** means we first run our LSD minimization algorithm and then METRIC LABELING. The trade-off α between the GO-based distance and the training distance (Eqn. 16) is either 0.1 or 0.3 as indicated.

algorithm performs better than the other tested algorithms. METRIC LABELING is more accurate than GenMultiCut in every case since GenMultiCut ignores the effect of distances between functions. FunctionalFlow also does not perform as well as METRIC LABELING, which again may be due to its independence assumption between functions. METRIC LABELING still performs well when number of elements in the ontology is increased to 150 and 300 (Fig. 2).

METRIC LABELING also outperforms the MRF-based algorithm [25]. This may be because the correlation estimations between functions used in that approach depend solely on training data whereas our distances are estimated from both the training set and the structure of the GO DAG. This indicates that, while the Gene Ontology is an imperfect, incomplete, manually edited resource, the distances between annotations in the ontology do contain useful information that can be exploited to make more accurate predictions.

Among various distance heuristics we used, the LCA and Lin distances are better in general since they take the lowest common ancestor into account. The d_{Lin} and d_{LCA} distances perform about the same but they both perform better than the d_{SP} metric (Fig. 3a). This further indicates that lowest common ancestor is a good distance estimator when there are hierarchical relations among points as shown previously in WordNet [10]. This also echos results in several other

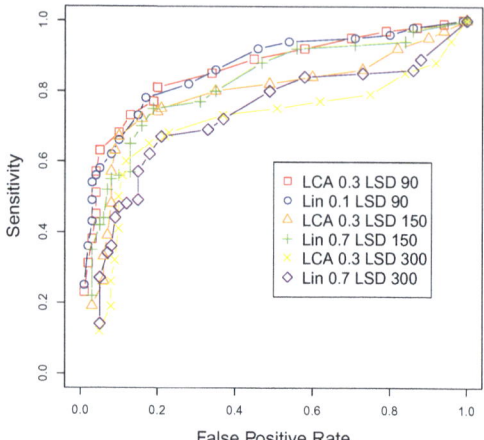

Fig. 2. Performance degrades as the number of terms increases

papers [27,3,31] in terms of showing effectiveness of lowest common ancestor as a measure between ontology terms. In addition, in almost all cases the nonuniform assignment costs performs slightly better than uniform assignment costs, although the effect is not large, and if nonuniform assignment costs are not available, uniform assignment costs can be nearly as effective.

Running the LSD minimization for semimetrics and then running METRIC LABELING performs better than Semimetric MAP Estimation algorithm [24] in most of the cases. This shows optimizing least squared error, rather than the classical distortion, for metric approximation is effective in the protein function prediction application.

3.2   Trade-off between GO-Distances and Network Distances

We also investigate how performance varies as the tradeoff between a distance computed from the GO structure $(d_{SP}, d_{LCA}, d_{Lin})$ and a distance computed from proximity in the network (d_{KB}) is varied. Figure 3b shows the performance of METRIC LABELING with LSD metric approximation and the LCA distance for different trade-offs α between the GO-based structural distance (d_{LCA}) and the trained distances d_{KB} as described in Eqn. 16. In almost all cases, using distances based solely on GO performs better than using only d_{KB} but using estimations both from training set and Gene Ontology structure performs better than using either one alone.

Combining the Gene Ontology knowledge with training set estimations using low values of α ($\alpha = 0.1$ or $\alpha = 0.3$) achieves the best performance by a slight

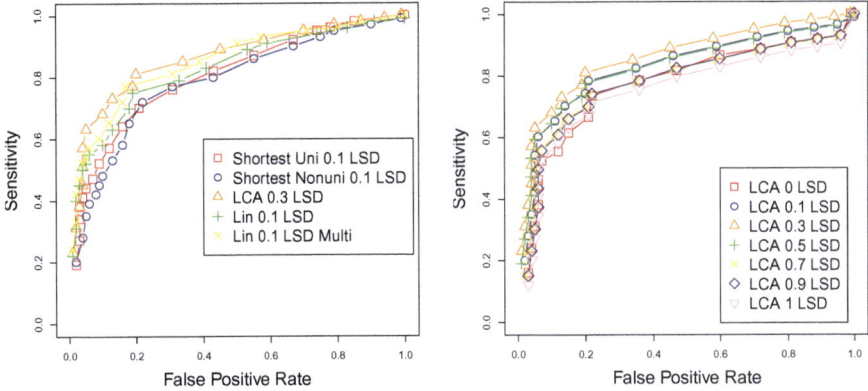

Fig. 3. (a) Performance of METRIC LABELING combined with the LSD minimization using various distance measures. **Uni** indicates the assignment costs of METRIC LABELING are uniformly 1 except for known annotations and **Nonuni** means assignment costs are nonuniform as described in Section 2.4. **Shortest, LCA, Lin** indicate the different distances functions in Section 2.4. (b) Performance of the d_{LCA} distance combined with the d_{KB} distance with various α using the LSD algorithm.

margin for most of the cases. When the number of elements in ontology increases, best performance is achieved by running METRIC LABELING with combination of d_{Lin} distances and training set estimates when $\alpha = 0.7$. After the initial benefit of using some of the d_{KB} distances, the performance starts to decrease as the weight α is increased. This may mean that d_{KB} is most effective when it operates as a tie-breaker between terms that have the same GO distances. (The dependence on α of the performance of the other GO-based distances d_{SP}, d_{Lin} is similar).

3.3   Robustness on the Yeast PPI Network

METRIC LABELING combined with LSD metric approximation is more robust to noise in both misannotations and edge removal. We tested for robustness of the predicted results in two ways. First, we removed various percentages of edges randomly from the PPI network and re-run our algorithm. Performance clearly decreased but even when 50% and 40% of the PPI edges are removed on 90 and 150 element ontologies respectively a METRIC LABELING approach performs as well as other algorithms run on the true PPI network. The fewer elements the ontology has, the more robust it is in terms of edge removal. The Lin and LCA distance measures again outperform shortest path distance and running LSD minimization for semimetrics and then METRIC LABELING does better than using the Semi-Metric MAP Estimation algorithm [24]. The LSD minimization may handle the noise in the data better during its error minimization.

Secondly, we also tested robustness by misannotating various percentages of protein annotations and then running our algorithm. Performance even when 30% of the proteins are misannotated on both 90 and 150 element ontologies is still comparable with its performance with the true labels, and it is not worse than other algorithms on the true labels. However, in the case of misannotations, combining GO knowledge with training set estimations ($\alpha = 0.1$ or $\alpha = 0.3$) no longer performs the best. Rather, the GO structure-based distances in isolation perform the best as expected.

3.4   Performance on Other Networks

When we created integrated network from multiple sources as described in Section 2.5, the performance increases slightly (last curve in Fig. 1). This shows that the METRIC LABELING approach is also useful on relational data other than PPI networks. We also tested our algorithm on several species. Among those species, performance strongly depends on how complete PPI network is, with sparser networks generally exhibiting worse performance. Again, the METRIC LABELING approach performs competitively with existing methods. Due to space limitations, the complete results for the 7 considered species are available at http://www.cbcb.umd.edu/kingsford-group/metriclabeling.

4   Conclusions

We show that GO structural information can be exploited to achieve better protein function prediction. We also show that the clean, combinatorial problem of METRIC LABELING can effectively use these distances and produce accurate predictions in a reasonable amount of computational time.

Our novel LSD metric approximation algorithm combined with METRIC LABELING performs better than the semimetric MAP estimation algorithm in most cases. This is interesting since distortion defined as in Section 1 has nearly always been used as the performance measure for metric embeddings. However, as mentioned, distortion does not consider the distribution of the error on all points. Its minimization considers just the minimization of the boundary cases (of maximum contraction and expansion). LSD minimization instead tries to minimize the total least squared error which makes sense both intuitively and experimentally as we have seen on protein function prediction. Its effectiveness on different application domains is an open question, but the LSD approach is likely to be useful for the common problem of converting a set of heuristic distances into a metric for subsequent processing with an algorithm (such as that for METRIC LABELING) that assume a metric. The LSD metric approximation is completely independent of METRIC LABELING. Either of these algorithms can be changed without affecting the other. However, this is not the case for Semimetric MAP Estimation algorithm, for which the two phases of metric estimation and prediction are not independent.

References

1. Barutcuoglu, Z., Schapire, R.E., Troyanskaya, O.G.: Hierarchical multi-label prediction of gene function. Bioinformatics, 830–836 (2006)
2. Boykov, Y., Veksler, O., Zabih, R.: Fast approximate energy minimization via graph cuts. IEEE T. on Pat. Anal. Mach. Intell. 23(11), 1222–1239 (2001)
3. Budanitsky, A., Hirst, G.: Semantic distance in WordNet: An experimental, application-oriented evaluation of five measures. In: Workshop on WordNet and Other Lexical Resources, Second Meeting of The North American Chapter of The Association For Computational Linguistics (2001)
4. Chekuri, C., Khanna, S., Naor, J., Zosin, L.: A linear programming formulation and approximation algorithms for the metric labeling problem. SIAM J. Discret. Math. 18(3), 608–625 (2005)
5. Cheng, J., Cline, M., Martin, J., Finkelstein, D., Awad, T., Kulp, D., Siani-Rose, M.A.: A knowledge-based clustering algorithm driven by Gene Ontology. J. Biopharm. Stat. 14(3), 687–700 (2004)
6. Chuzhoy, J., Naor, J.S.: The hardness of metric labeling. In: 45th Annual IEEE Symp. Foundations of Computer Science, pp. 108–114. IEEE Computer Society, Washington, DC (2004)
7. Deng, M., Tu, Z., Sun, F., Chen, T.: Mapping gene ontology to proteins based on protein–protein interaction data. Bioinformatics 20(6), 895–902 (2004)
8. Dotan-Cohen, D., Kasif, S., Melkman, A.A.: Seeing the forest for the trees: using the Gene Ontology to restructure hierarchical clustering. Bioinformatics 25(14), 1789–1795 (2009)
9. Fakcharoenphol, J., Rao, S., Talwar, K.: A tight bound on approximating arbitrary metrics by tree metrics. In: Proc. 35th Annual ACM Symp. on Theory of Computing, pp. 448–455 (2003)
10. Fellbaum, C. (ed.): WordNet: An Electronic Lexical Database (Language, Speech, and Communication). The MIT Press, Cambridge (1998)
11. Gasch, A.P., Spellman, P.T., Kao, C.M., Carmel-Harel, O., Eisen, M.B., Storz, G., Botstein, D., Brown, P.O., Silver, P.A.: Genomic expression programs in the response of yeast cells to environmental changes. Mol. Biol. Cell 11, 4241–4257 (2000)
12. Gavin, A.C., Bosche, M., Krause, R., et al.: Functional organization of the yeast proteome by systematic analysis of protein complexes. Nature 415(6868), 141–147 (2002)
13. GNU Linear Programming Kit (2010), http://www.gnu.org/software/glpk/
14. Hishigaki, H., Nakai, K., Ono, T., Tanigami, A., Takagi, T.: Assessment of prediction accuracy of protein function from protein–protein interaction data. Yeast 18(6), 523–531 (2001)
15. Ho, Y., Gruhler, A., Heilbut, A., et al.: Systematic identification of protein complexes in Saccharomyces cerevisiae by mass spectrometry. Nature 415(6868), 180–183 (2002)
16. Huh, W.K., Falvo, J.V., Gerke, L.C., Carroll, A.S., Howson, R.W., Weissman, J.S., O'Shea, E.K.: Global analysis of protein localization in budding yeast. Nature 425(6959), 686–691 (2003)
17. Ito, T., Chiba, T., Ozawa, R., Yoshida, M., Hattori, M., Sakaki, Y.: A comprehensive two-hybrid analysis to explore the yeast protein interactome. Proc. Natl. Acad. Sci. USA 98(8), 4569–4574 (2001)

18. Jensen, L.J., Gupta, R., Strfeldt, H.H., Brunak, S.: Prediction of human protein function according to Gene Ontology categories. Bioinformatics 19(5), 635–642 (2003)
19. Karaoz, U., Murali, T.M., Letovsky, S., Zheng, Y., Ding, C., Cantor, C.R., Kasif, S.: Whole-genome annotation by using evidence integration in functional-linkage networks. Proc. Natl. Acad. Sci. USA 101(9), 2888–2893 (2004)
20. Kleinberg, J., Tardos, E.: Approximation algorithms for classification problems with pairwise relationships: Metric labeling and markov random fields. In: Proc. 40th Annual IEEE Symp. on Foundations of Computer Science, pp. 14–23 (1999)
21. Komodakis, N., Tziritas, G.: Approximate labeling via graph-cuts based on linear programming. IEEE T. Pat. Anal. Mach. Intell. 29(8), 1436–1453 (2007)
22. Kourmpetis, Y.A., van Dijk, A.D., Bink, M.C., van Ham, R.C., Ter Braak, C.J.: Bayesian markov random field analysis for protein function prediction based on network data. PloS One 5(2), e9293+ (2010)
23. Kui, M.D., Zhang, K., Mehta, S., Chen, T., Sun, F.: Prediction of protein function using protein-protein interaction data. J. Computat. Biol. 10, 947–960 (2002)
24. Kumar, M.P., Koller, D.: MAP estimation of semi-metric MRFs via hierarchical graph cuts. In: UAI 2009: Proc. Twenty-Fifth Conf. on Uncertainty in Artificial Intelligence, pp. 313–320. AUAI Press, Arlington (2009)
25. Lee, H., Tu, Z., Deng, M., Sun, F., Chen, T.: Diffusion kernel-based logistic regression models for protein function prediction. OMICS 10(1), 40–55 (2006)
26. Li, S.Z.: Markov random field modeling in computer vision. Springer, London (1995)
27. Lin, D.: Automatic retrieval and clustering of similar words. In: Proc. 17th Internat. Conf. on Computational Linguistics, pp. 768–774. Association for Computational Linguistics, Morristown (1998)
28. Lin, D.: An information-theoretic definition of similarity. In: Proc. 15th Internat. Conf. Machine Learning, pp. 296–304. Morgan Kaufmann, San Francisco (1998)
29. Nabieva, E., Jim, K., Agarwal, A., Chazelle, B., Singh, M.: Whole-proteome prediction of protein function via graph-theoretic analysis of interaction maps. Bioinformatics 21(Suppl 1), i302–i310 (2005)
30. Rain, J.C., Selig, L., De Reuse, H., Battaglia, V., Reverdy, C., Simon, S., Lenzen, G., Petel, F., Wojcik, J., Schachter, V., Chemama, Y., Labigne, A., Legrain, P.: The protein-protein interaction map of *Helicobacter pylori*. Nature 409(6817), 211–215 (2001)
31. Resnik, P.: Semantic Similarity in a Taxonomy: An Information-Based Measure and its Application to Problems of Ambiguity in Natural Language. J. Artificial Intelligence Research 11, 95–130 (1999)
32. Schlicker, A., Domingues, F., Rahnenfuhrer, J., Lengauer, T.: A new measure for functional similarity of gene products based on gene ontology. BMC Bioinformatics 7(1), 302 (2006)
33. Schwikowski, B., Uetz, P., Fields, S.: A network of protein-protein interactions in yeast. Nat. Biotechnol. 18(12), 1257–1261 (2000)
34. Sharan, R., Ulitsky, I., Shamir, R.: Network-based prediction of protein function. Mol. Syst. Biol. 3, 88 (2007)
35. Stark, C., Breitkreutz, B.J., Reguly, T., Boucher, L., Breitkreutz, A., Tyers, M.: BioGRID: a general repository for interaction datasets. Nucl. Acids Res. 34(suppl 1), D535–D539 (2005)

36. The Gene Ontology Consortium: Gene ontology: tool for the unification of biology. Nat. Genetics 25(1), 25–29 (2000)
37. Uetz, P., Giot, L., Cagney, G., et al.: A comprehensive analysis of protein-protein interactions in *Saccharomyces cerevisiae*. Nature 403(6770), 623–627 (2000)
38. ILOG CPLEX (2010), `http://www.ibm.com/software/integration/optimization/cplex-optimizer`
39. Vazquez, A., Flammini, A., Maritan, A., Vespignani, A.: Global protein function prediction from protein-protein interaction networks. Nat. Biotechnol. 21(6), 697–700 (2003)

Efficient Traversal of Beta-Sheet Protein Folding Pathways Using Ensemble Models

Solomon Shenker[1], Charles W. O'Donnell[2], Srinivas Devadas[2], Bonnie Berger[2,3,⋆], and Jérôme Waldispühl[1,2,⋆]

[1] School of Computer Science & McGill Centre for Bioinformatics, McGill University, Montreal, Canada
[2] Computer Science and AI Lab, MIT, Cambridge, USA
[3] Department of Mathematics, MIT, Cambridge, USA
jeromew@cs.mcgill.ca, bab@mit.edu

Abstract. Molecular Dynamics (MD) simulations can now predict ms-timescale folding processes of small proteins — however, this presently requires hundreds of thousands of CPU hours and is primarily applicable to short peptides with few long-range interactions. Larger and slower-folding proteins, such as many with extended β-sheet structure, would require orders of magnitude more time and computing resources. Furthermore, when the objective is to determine only which folding events are necessary and limiting, atomistic detail MD simulations can prove unnecessary. Here, we introduce the program tFolder as an efficient method for modelling the folding process of large β-sheet proteins using sequence data alone. To do so, we extend existing ensemble β-sheet prediction techniques, which permitted only a fixed anti-parallel β-barrel shape, with a method that predicts arbitrary β-strand/β-strand orientations and strand-order permutations. By accounting for all partial and final structural states, we can then model the transition from random coil to native state as a Markov process, using a master equation to simulate population dynamics of folding over time. Thus, all putative folding pathways can be energetically scored, including which transitions present the greatest barriers. Since correct folding pathway prediction is likely determined by the accuracy of contact prediction, we demonstrate the accuracy of tFolder to be comparable with state-of-the-art methods designed specifically for the contact prediction problem alone. We validate our method for dynamics prediction by applying it to the folding pathway of the well-studied Protein G. With relatively very little computation time, tFolder is able to reveal critical features of the folding pathways which were only previously observed through time-consuming MD simulations and experimental studies. Such a result greatly expands the number of proteins whose folding pathways can be studied, while the algorithmic integration of ensemble prediction with Markovian dynamics can be applied to many other problems.

⋆ Corresponding authors.

V. Bafna and S.C. Sahinalp (Eds.): RECOMB 2011, LNBI 6577, pp. 408–423, 2011.

1   Introduction

Protein folding and unfolding is a key mechanism used to control biological activity and molecule localization [1]. The simulation of folding pathways is thus helpful to decipher the cell behavior. Classical molecular dynamics (MD) methods [2] can produce reliable predictions but unfortunately the heavy computational load required by these techniques limits their application to inputs tens of amino acids long and prevents their application to large sequences (i.e. hundreds of amino acids). Recently, P. Faccioli *et al.* proposed an effective solution of the Fokker-Planck equation to compute dominant protein folding pathways [3], but the same size limitations remain.

The development of distributed computing technologies has dramatically extended the range of application of MD techniques. For instance, Pande and co-workers achieved a 1.5 millisecond folding simulation of a 39 residue protein NTL9 [4]. In spite of this achievement, this strategy still seems limited to small polypeptides (about 50 residues) and, more importantly, requires several months of parallel computing and typically thousands of GPU's.

In this paper, we introduce a complete methodology to address these computational complexity limitations. Our approach aims to complement the range of techniques already offered. Unlike MD simulations which use an all-atom description of structures together with a fine-tuned energy force field, here we use a residue-level representation of the structure with a statistical residue contact potentials. This simplification enable us to sample intermediate structures and build a coarse-grained model of the energy landscape and subsequently simulate folding processes.

Since the seminal work of Levitt and Warshel [5], it is widely acknowledged that simplified representations of protein structures and motions are required to circumvent computational limitations. A conceptual breakthrough came when Amato and co-workers applied motion planning techniques to the protein folding problem [6,7]. The method is much faster than classical MD techniques and enables the study of the folding of large proteins. However, this approach does not predict structures, rather it requires the three-dimensional structure of the native state to compute potential intermediate structures and unfolding pathways, on which the folding simulations are performed. It follows that the methodology cannot be applied to proteins with unknown structures and cannot be relied upon to study misfolding processes.

In fact, all the methods previously described face a difficulty common with MD: efficient sampling of the conformational landscape. MD algorithms explore the landscape through force-directed local search and progressive modification of the structure. However, the scalability and numerical efficiency when modeling large molecular structures remains problematic, limiting their application to small molecular systems. On the other hand, motion planning algorithms use a 3D structure of the native fold to predict distant structural intermediates. Accordingly, the accuracy of the method can suffer when intermediates sampled are far away from the native state. Recently, Hosur *et al.* [8] have combined efficient motion planning techniques with machine-learning to model proteins as an ensemble, but this approach is effective only in the local neighborhood of the input structure.

Such obstacles have been addressed for RNA molecules by the development of structural ensemble prediction algorithms [9,10], and the derivation of a finely-tuned energy model based on experimental data [11]. Combined together, these techniques enable us to compute the RNA secondary structure energy landscapes and sample structures from sequence information alone. Wolfinger *et al.* [12] further demonstrated how an RNA energy landscape can be constructed by connecting these samples together and estimating the transition rates between pairs of interconverting states. The resulting ordinary differential equation (ODE) system can be solved to predict and characterize RNA folding pathways. The method has since been improved to analyze the motion of large RNAs [13].

In this paper, we propose to expand the methodology developed for RNAs to the more complicated case of proteins. First, we design an algorithm to sample the complete conformational landscape of large protein sequences given sequence data alone. Then, we use this sampling algorithm to build a coarse-grain representation of the energy landscape of a protein, from which we construct an ODE system modeling transition rates between folding intermediates that we solve to simulate protein folding.

We choose to address specifically β-sheet structures. The folding of these structures is particularly difficult to simulate. Indeed, β-sheets are stabilized by inter-strand residue interactions, and thus the folding and assembly of these structures is largely influenced by long-range interactions and global conformational rearrangements. For instance, Voeltz *et al.* recently showed that the rate-limiting step in the NTL9 fold was beta-sheet hairpin formation [4].

Since the original work of Mamitsuka and Abe [14], several groups have proposed models to predict general β-sheets [15,16,17]. However, none of these methods are capable of computing *ensembles* of β-sheet structures (i.e. perform an exact enumeration of all β-structures without duplicates) and therefore cannot be used to sample the β-sheet energy landscape.

We recently introduced a structural ensemble predictor for transmembrane β-barrel (TMB) proteins [18], continuing earlier work on molecular structure modeling [19,20]. However, TMBs are a special case of β-sheets where each strand pairs with its two sequence neighbors via an anti-parallel interaction (except the "closing" pair which involves the first and last strands). Here, we expand these techniques to allow any β-strand organization in the β-sheet, with parallel and anti-parallel orientations, and enable the sampling of general β-sheets. This algorithm is implemented in the program tFolder.

We use tFolder to sample the β-sheet conformational landscape and build a coarse-grain model of the energy landscape. More specifically, we cluster protein configurations according to contact distance metrics, and associate each cluster with an intermediate folding state. We use the difference between the ensemble free energies of the clusters to compute the transition rates and build an ODE system that models the energy landscape. Finally, we solve this system to estimate the distribution of conformations over folding time.

This methodology reconciles the MD and motion planning approaches for studying folding pathways. Using `tFolder`, we are now able to simulate in a couple of minutes on a single desktop the folding of large proteins, and to predict the folding pathways (as well as possible misfolding pathways) of proteins with unknown structures. Thus we are able to provide a broader range of applications, while offering computational efficiency comparable with motion planning techniques. Although we focus on β-sheet proteins, our method in principle could be extended to describe the folding pathways of a wider class of protein structures.

This paper is organized as follows. In section 2 we describe the `tFolder` algorithm and explain how we construct the coarse grained energy landscape model. Then, in section 3, we benchmark our methods. First, we evaluate the accuracy of `tFolder` for simple inter-strand residue contact prediction and show that it performs comparably with more sophisticated techniques specifically designed for this task. Importantly, our contact predictions are not dependent on the separation between the residue indices, which means an improved "very" long-range contact prediction accuracy. Then, we illustrate the insights provided by our methods by analyzing the energy landscape of the extensively studied Protein G. We show that `tFolder` predicts the correct folding pathways, and interestingly, our simulation reveals a possible off-pathway structure. All these simulations can be performed on query sequences using our program `tFolder`, available at `http://csb.cs.mcgill.ca/tFolder`

2   Methods

To predict realistic protein folding pathways, we exploit well-established ensemble prediction algorithms [18] for their ability to accurately predict the energy scores of millions of feasible structural conformations from sequence alone. Our approach proceeds in two steps: (1) Given an arbitrary peptide sequence, we produce ensemble predictions of the energetic weight for all possible β-sheet structures and sub-structures, utilizing an enhancement to standard ensemble predictors which allows permutation. (2) Using each conformation's energetic score and metrics of conformational similarity, we derive the likelihood of dynamic state-to-state transitions and assemble a set of complete folding paths. In this way, we can identify and rank the most likely pathways from an unfolded conformation to a fully folded conformation based on predicted energy landscapes.

Modelling β-Sheet Ensembles

We model the set of all possible β-sheet conformations a peptide can attain using a statistical-mechanical framework. Conceptually, each structure is described by the set of residue/residue contacts that form hydrogen bonds between β-strand backbones, and is assigned a Boltzmann-distributed pseudo-energy, determined by the specific residues involved in contacts. To characterize the energetic landscape of this ensemble, a partition function Z can be calculated over all structural states $S = \{1...n\}$ such that

$$Z = \sum_{i=1}^{n} e^{-\frac{E_{S_i}}{RT}},$$

with energies E_{S_i}, temperature T, and the Boltzmann constant R. For example, from this the relative abundance of a structure S_i can be easily derived:

$$p(S_i) = \frac{e^{-E_{S_i}}}{Z}.$$

Our energy model is based on statistical potentials and follows directly from prior prediction tools that have been shown to be accurate [18][21]. An energy $E_{i,j}$ is given to each residue/residue pair within the β-sheet fold following $E_{i,j} = -RT[\log(p(i,j)) - Z_c]$, where Z_c is a statistical recentering constant and $p(i,j)$ is the probability of these two residues appearing in a β-sheet environment, as observed across all non-sequence-homologous solved structures in the PDB [18]. Further, we assign separate probabilities based on the hydrophobicity of the environment on either face of a β-sheet.

A naive approach to computing the partition function would thus be to enumerate all possible structures and compute each structure's contribution to the sum individually. However, as was previously shown for the special case of anti-parallel β-strands in transmembrane β-barrel proteins, a much more efficient method exists using dynamic programming [18]. We have generalized this approach to enable the computation of arbitrary single β-sheet fold topologies.

Permutable β-Templates

We introduce the concept of permutable β-templates to enable the calculation of the partition function of a β-sheet with arbitrary β-strand topologies. This extends existing ensemble prediction techniques by allowing any combination of parallel and anti-parallel β-strands to be including within a single β-sheet fold, and by removing any sequence dependency between β-strand/β-strand pairing partners. Prior methods supported only all-anti-parallel β-strands and required β-strand/β-strand interactions to be separated only by coil (and not other strands) [18].

To efficiently encode these generic shapes, each strand is labeled $\{1...n\}$ to allow a stepwise permutation through β-strand ordering, and a signed permutation is defined such that each β-strand is assigned to be parallel or anti-parallel relative to the first strand in the sheet (Figure 1). Algorithmically, tFolder is capable of constructing a dynamic program over all such permutations to calculate the partition function. In practice, since such an encoding can result in unrealistic combinations of β-strand/β-strand pairings (such as if β-strands 1 and 4 had too short a coil between them in Figure 1), we impose that valid foldings must satisfy steric and biologically derived constraints. These include a minimum and maximum β-strand length, maximum shear between neighboring β-strands (the amount of inclination that causes the β-sheet to deviate from a perfect rectangle), and minimum inter-strand loop size. These constraints serve

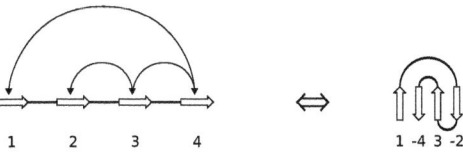

Fig. 1. An illustration of how a permutable β-template can be encoded as a signed permutation. The permutation lists the strands in the order that they occur in the sheet, with the sign indicating whether the strand is parallel (+) or anti-parallel (-) to the first strand.

to limit the exploration of unrealistic conformations, minimizing excess computation and allowing directed investigation into specific motifs.

The energy of a structure with n strands, can be recursively defined as $E(S_n) = E(S_{n-1}) + Pairing(s_{n-1}, s_n)$, where $E(S_{n-1})$ is the interaction energy between the first $n-1$ strands, and $Pairing(s_{n-1}, s_n)$ is the energy of the pairing of strand $n-1$ with strand n (See Figure 2(a)). tFolder exploits the shared structure between instances in the ensemble by computing this recursion using a dynamic programming algorithm. The result of each recursive call is stored in a table indexed by the parameters of the call. Subsequent recursive calls made with the same parameters perform a table lookup instead of re-computing the value of the recursion.

For a sheet of n strands, the table has n rows, where the kth row has entries corresponding to valid configurations of the first k strands. For the kth strand, these configurations are partitioned by the location of four indices k_1, k_2, k_3, k_4, which denote the boundaries of the region occupied by the k strands (Figure 2(b)). To begin, the algorithm enumerates all possible positions of the first two strands, and for each stores the strand pair interaction energy in entry $E_{2_1 2_2 2_3 2_4}$ of the table. For each subsequent strand k, the value of $E_{k_1 k_2 k_3 k_4}$ is computed as:

$$E_{k_1 k_2 k_3 k_4} = \sum_{i_1 i_2 i_3 i_4} E_{i_1 i_2 i_3 i_4} + Pairing(i, k),$$

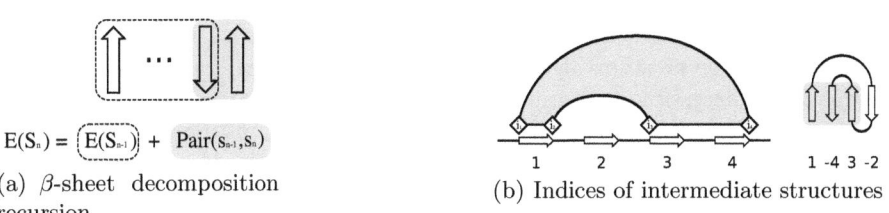

$E(S_n) = \boxed{E(S_{n-1})} + \boxed{Pair(s_{n-1}, s_n)}$

(a) β-sheet decomposition recursion

(b) Indices of intermediate structures

Fig. 2. (a) Illustrates how the energy function of a β-sheet can be recursively defined as the sum of the contribution of the last two strands with the contribution of the remaining structure. (b) Indicates the indices used to store the energies of intermediate structures for the recursion.

where i_1, i_2, i_3, i_4 are enumerated for all valid settings for the boundaries of the preceding strands, given the boundaries of the kth strand. Once the recursion has filled the table, the partition function Z is calculated by summing over all possible settings of n_1, n_2, n_3, n_4:

$$Z = \exp\left(-\sum_{n_1 n_2 n_3 n_4} E_{n_1 n_2 n_3 n_4}\right).$$

The table constructed to calculate the partition function can be used to sample the distribution of configurations of a given topology, utilizing the approach established by Ding and Lawrence for RNA secondary structure [22], and successfully applied previously by Waldispühl *et al.* to sample conformations of β-barrel proteins[20]. To do this, we perform a traceback through the table and, at *ith* step, sample the indices within which the first i strands are contained, according to the Boltzmann representation of these i-stranded structures (Figure 3).

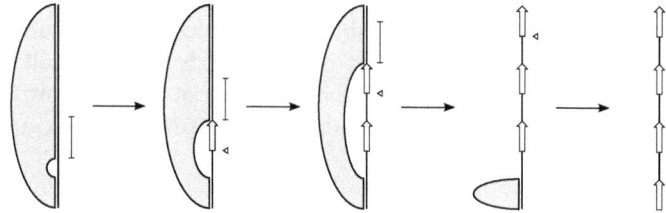

Fig. 3. Illustration of of how the sampling procedure performs a traceback through the table, over the indices of intermediate structures. During each step of the sampling procedure, the location of a single strand is sampled from the region indicated by the vertical bars. The triangles denote the location of the strand sampled during the previous step.

2.1   Predicting Folding Dynamics

Conceptually, we model the folding process as a path through a graph of varyingly folded conformations of a protein. In this graph, different protein conformations are represented as states, and two states that inter-convert in a folding pathway are connected by an edge, analogous to work with RNA described previously [12]. The tFolder algorithm provides a means to efficiently sample the energetically accessible conformations that make up the states of this graph. We further propose a means to determine the connectivity between states and demonstrate how this can be applied to calculate the dynamics of the folding process.

Since we do not know the final structure, we begin by sampling configurations from all possible permutations of β-sheet topology, as described above. For every pair of states, we add an edge between two states if (1) the states have compatible topologies, and further, (2) the states show structural similarity.

Two templates are compatible if they are identical to each other, modulo the addition or removal of a single strand pairing. This operation can result in the growth of a core structure, or the nucleation of an independent strand pair (see Figure 4). Note that the requirements for satisfying the second criterion of structural similarity depends on the metric used to estimate structural similarity between two conformations. In practice, we use a contact based metric and deem two structures to be structurally similar if the metric is below the transition threshold.

Given the graph constructed according to these two criteria, the change in the probability of the system being in state i at time t is calculated from the total flux into and out of state i,

$$\frac{dp_i}{dt} = \sum_{j \in X} r_{ij} p_j(t),$$

where p_i is the probability of state i, X is the state space, and r_{ij} is the rate of transition from state i to state j. Given that two states are connected in the graph, the rate at which two states inter-convert is proportional to the difference between free energies of the states(ΔG); the system tends toward energetically favorable states. We calculate the transition rate r_{ij} between states i and j using the Kawasaki rule (with parameter r_0 to scale the time dimension):

$$r_{ij} = r_0 \exp\left(-\Delta G_{ij}/2RT\right).$$

The dynamics of the system are calculated by treating the folding process as a continuous time discrete state Markov process. Given the matrix of folding rates R, where $R_{ij} = r_{ij}$ and initial state density $p(0)$, the distribution over states $p(t)$ of the system at time t is given by the explicit solution to the system of linear differential equations,

$$p(t) = \exp\left(Rt\right) p(0).$$

Since we sample hundreds of states from each β-strand topology, we partition the state space into macro states using clustering, in order to work with a tractably sized system. Under this approximation, we consider two clusters the graph to be connected if the minimum distance between any two states from each cluster are connected. We define the ensemble free energy difference ΔG_{ij} between two macrostates i and j by summing over the states from which they are composed.

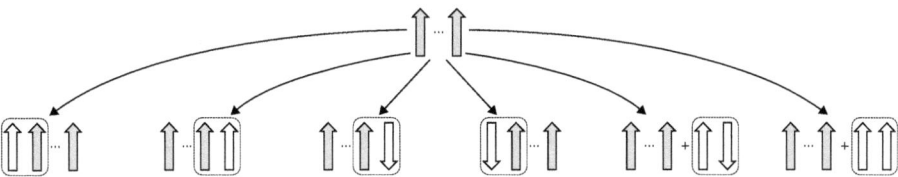

Fig. 4. The topologies that are compatible with a given state (shaded gray) result from the addition of a single pairing between strands (dashed box). The '+' indicates that there is no pairing between the gray structure and the white strand pair.

$$\Delta G_{ij} = E(\chi_i) - E(\chi_j) = \sum_{x \in \chi_i} E(x) - \sum_{x \in \chi_j} E(x).$$

Although this approximation lessens the computational burden, it represents a trade off. The granularity achievable by our simulation is at the level of the macrostates. Note, energy barriers are not explicitly incorporated into the model, since entire β-strands are either added or removed between states without partially-formed intermediates.

3   Results

Evaluation of Contact Prediction

To evaluate the contact prediction performance of tFolder, we tested it using a 31 protein benchmark[1]. Proteins were selected from the Protein Data Bank that had dominantly beta structure and low sequence homology. From each of these the β-topology was extracted and used as input for tFolder, along with the amino-acid sequence and fixed strand length of 4–6 residues. Since predicting folding dynamics involves a permutation over all β-topologies, this demonstrates the expected accuracy of each folding state along the pathway. We sample 500 configurations of each protein, and use these ensembles to compute a stochastic contact map and distribution of strand locations (See Figure 5(a) and 5(b) for example). The contact map represents the probability of observing a given contact, and predicted contacts are the set of all contacts with probability above a threshold value t. The selection of t influences the measured performance (Figure 5(c)), so to objectively set the threshold, we chose a t that maximizes the F-measure. We evaluated the quality of our contact maps based on the Accuracy ($\frac{no.\ of\ correctly\ predicted\ contacts}{no.\ of\ predicted\ contacts}$), Coverage ($\frac{no.\ of\ correctly\ predicted\ contacts}{no.\ of\ observed\ contacts\ contacts}$), and F-measure ($\frac{2 \cdot Accuracy \cdot Coverage}{Accuracy + Coverage}$) of our predictions. We calculated these measures in terms of β-contacts, which we defined as residues located withing β-strands less than 8Å apart (between C_α atoms) in the PDB structure. A summary of tFolder performance on Protein G, as well as average performance on the 31 protein dataset, is presented in Table 1. Here we distinguish between results for long range contacts, greater than 0, 12, or 24 residues apart. Thus, tFolder maintains reasonable predictive accuracy even with large contact separations

In order to evaluate the performance of tFolder with respect to other approaches for contact prediction, results on this protein dataset are presented in Table 2 along with a comparison with two leading contact prediction alogrithms, SVMcon and BETApro. The method SVMcon used ten of the proteins in this dataset for the training of their SVM, so they were excluded from the evaluation for the comparison of methods. It can be seen that tFolder is able to perform comparably, in particular for the F-measure of contacts with sequence separation greate than 24 residues. Although these methods sometimes perform better

---

[1] Complete benchmark results available at the tFolder website.

Table 1. The performance of tFolder for contact prediction is evaluated based on the Accuracy, Coverage, and F-measure of experimentally observed contacts. These performance metrics are reported for contacts that are more than 0, 12, and 24 residues apart, showing that tFolder maintains reasonable predictive accuracy even with large contact separations. Additionally, these metrics are evaluated when predicted contacts are within ±2 residues of an observed contact.

| | Protein G | | | | | | 31 protein benchmark | | | | | |
| | Exact | | | ± 2 | | | Exact | | | ± 2 | | |
	≥ 0	≥ 12	≥ 24	≥ 0	≥ 12	≥ 24	≥ 0	≥ 12	≥ 24	≥ 0	≥ 12	≥ 24
Accuracy	13.3	10.6	14.0	52.1	54.1	58.3	8.6	7.0	8.4	24.2	32.1	39.8
Coverage	56.3	53.8	37.5	97.9	61.5	87.5	9.1	11.6	11.9	43.5	44.5	51.1
F-measure	21.5	17.7	20.4	68.0	57.6	70.0	8.3	9.3	19.0	27.3	29.7	45.2

for contact prediction, it is important to note that the predictive performance of tFolder is less sensitive to the distance of contact separation. Since critical protein folding steps can involve both short-range and long-range β-sheet contacts, it is especially important for long-range contacts to be predicted correctly to allow an accurate folding pathways to be reconstructed. Furthermore, since we cannot apply cross validation techniques to BETApro and SVMcon, we also indicate their performance for CASP 7.

Table 2. Comparison of the performance of tFolder contact prediction with contact prediction algorithms SVMcon and BETApro. The methods are evaluated based on their ability to perform contact prediction for contacts greater than 12 and 24 residue separation respectively. The metric values for contacts within ±2 residues of an observed contact are reported in parentheses.

| | ≥ 12 | | | ≥ 24 | | |
Method	F-measure	Accuracy	Coverage	F-measure	Accuracy	Coverage
tFolder	9.0 (24.0)	6.7 (29.4)	8.8 (37.2)	20.1 (41.8)	11.7 (41.0)	15.1 (44.8)
BETApro	10.8 (28.1)	41.5 (78.1)	4.8 (16.0)	6.2 (22.8)	28.0 (57.7)	1.1 (7.2)
(CASP 7)		35.4	5.1		19.7	3.2
SVMcon	27.8 (55.7)	26.7 (69.7)	32.9 (48.4)	19.9 (40.0)	15.6 (54.0)	29.1
(CASP 7)		27.7	4.7		13.1	2.8

Predicting the Folding Pathways of the B1 Domain of Protein G

To demonstrate the efficacy of our techniques for predicting protein folding pathways, we reconstruct the folding landscape of the B1 domain of Protein G — a well-studied protein for which the pathway has been elucidated through many experimental studies and MD simulations. To do this, all possible permutations of a 4-strand β-sheet topology were sampled and clustered. For each of these sets of structures, the cluster with the highest probability of being observed was selected to be representative of each topology.

The graph of the folding pathway was constructed by considering all pairs of clusters. If the minimum distance beween two clusters was less than the transition treshold, we considered that there was exchange between the two states. We tried several metrics, including segment overlap, mountain metric, and a contact based metric [24] [25], selecting the contact based metric, because it performed best empirically. The resulting graph of protein conformations is illustrated in Figure 6(a). Inspection of this graph, along with the folding dynamics computed from this graph in Figure 6(b), reveals folding intermediates consistent with those

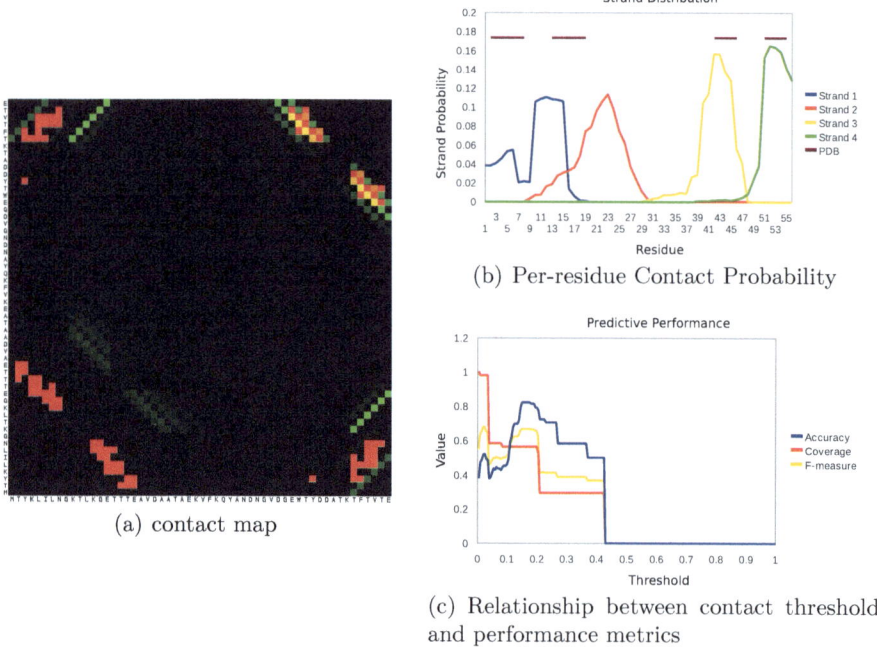

(a) contact map

(b) Per-residue Contact Probability

(c) Relationship between contact threshold and performance metrics

Fig. 5. Summary of the distribution of structures predicted by **tFolder** for Protein G (a) Shows the contact probability predicted by **tFolder** between all pairs of amino acids. Green squares indicate contacts predicted by **tFolder**, whereas red squares represent pairs of amino acids that are less than 8Å from each other in the observed structure. A higher intensity of green indicates a higher predicted probability of the contact, and yellow squares are an indication of agreement between prediction and observed contacts. (b) Shows the probability of the location of each strand, computed from the ensemble of sampled structures.The bars at the top of the plot indicate the location of the strands from the experimentally determined structure. (c) Shows the relationship between the threshold used to determine a contact from the contact map, and the values of the three metrics Accuracy, Coverage, and F-measure. The threshold can be set to maximize the value of the F-measure, representing a reasonable trade-off between Coverage and Accuracy.

previously reported by Song *et al.* [26]. It should also be noted that although we compute other configurations of the sequence that are energetically favorable (faded states), they are not predicted to form because they are unreachable from the unfolded state. Interestingly, a four-stranded off-pathway structure is predicted to form, which has not been observed previously. Furthermore, our results agree with the work of Hubner *et al.*, who show that the anti-parallel beta-hairpin, predicted to form an interaction between residues 39–44 and 50–55, center around known nucleation points W43, Y50, F54 [27].

Algorithm Running Time

The computational bottleneck of our approach is the computation of the partition function of a template. The primary factors influencing this calculation are the length of sequence and the number of strands in the β-topology (the depth of the recursion). The partition function for sequences between 40–130 residues and 4–6 strands was calculated using a single 2.66GHz processor with 512 MB of RAM. The effect of these two parameters on the computation time is depicted in Figure 7 below. Further, computing the parition function across multiple β-templates is trivially parallelizable. The ability to formulate quick, coarse-grained predictions in a matter of minutes, rather than days of atomistic-detail simulation, is a fundamental benefit of our technique.

4   Discussion

We present tFolder, a novel approach for quickly predicting protein folding pathways through the accurate prediction of the conformational landscape of arbitrary β-sheet proteins. What distinguishes tFolder from other computational approaches that attempt to probe protein folding processes is that tFolder does not require vast computational resources; in fact, it can be ran on a single personal computer. To achieve this performance we use a simplified model for protein folding, allowing us to very rapidly compute a coarse-grained picture of the folding of a protein from sequence information alone. This contrasts with methods that attempt to determine folding mechanisms by trying to unfold proteins from their native state. Such methods require the *a priori* knowledge of the native structure, and as such are not applicable to study protein sequences with unknown structures. When computing protein folding pathways, our method explores all possible β-sheet configurations, and thus does not face such limitations. Interestingly, this independence from known structures could provide insights into off-pathway kinetics, such as the aggregation of proteins into amyloid structures.

Although tFolder only predicts coarse folding pathway transitions in β-sheet proteins, its strength lies in its ability to quickly separate conformational transitions that are critical to folding from those transitions that could simply result from minor structural fluctuations. This complements the use of MD simulations as the MD can be used to explore the nuanced structural interactions

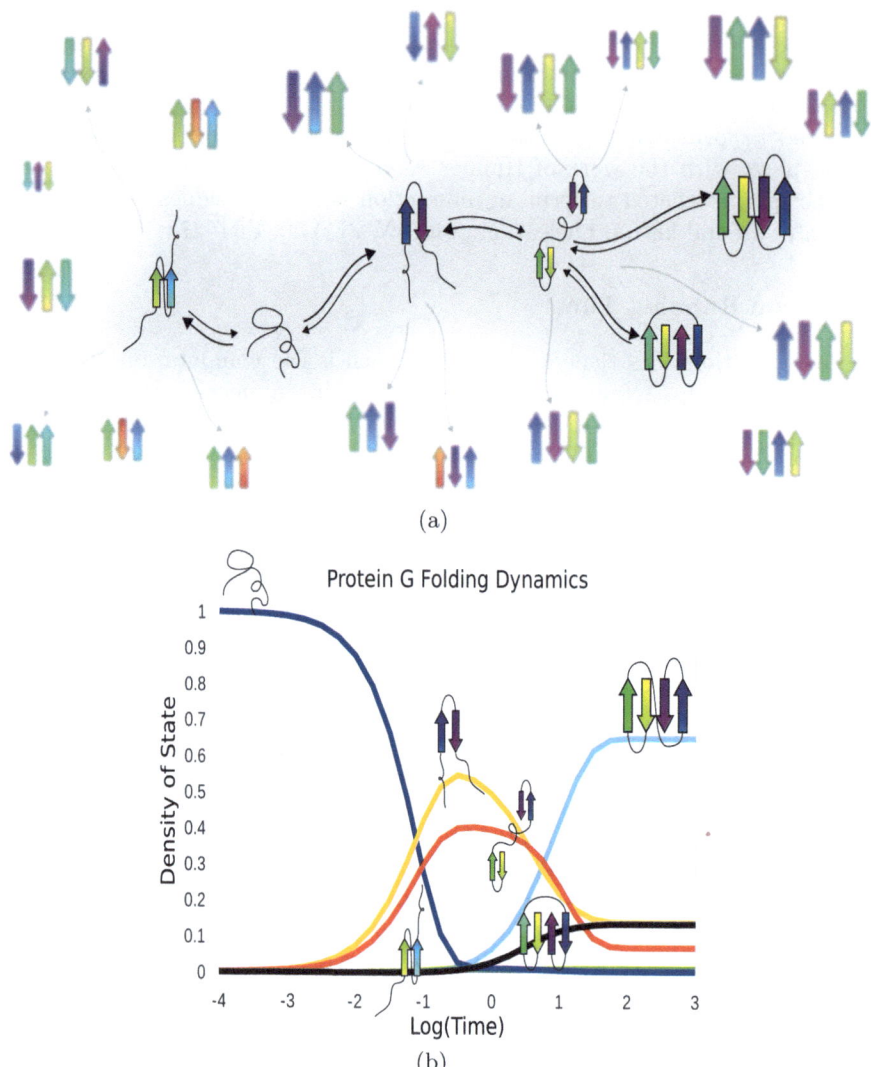

Fig. 6. (a)The graph of the folding landscape of Protein G predicted by `tFolder` is illustrated above. The gray shaded region indicates the states predicted to be reachable from the unfolded state. The dark arrows indicate transitions between states, and the size of the arrow indicates the favored direction of transition along each edge. Faded arrows are drawn between states that have compatible topologies but do not reach the transition threshold. The size of each state indicates its relative representation at equilibrium. The faded structures indicate states that are unreachable from the unfolded state. (b)The folding dynamics of Protein G shows how the probability of observing any of the reachable states changes over the time the protein folds. Each line is annotated with an image of the state it represents.

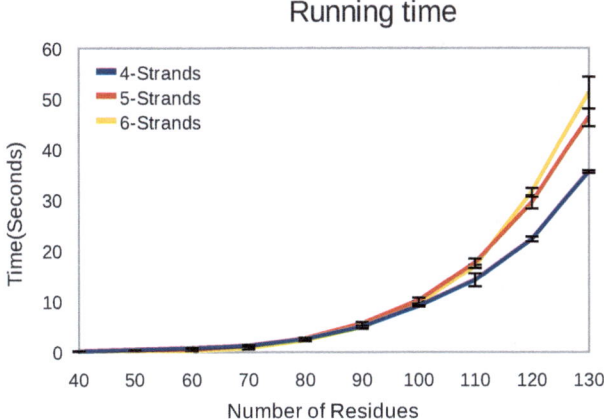

Fig. 7. The time required to compute the partition function increases with increasing size of amino acid sequence, and number of strands. The time was computed by averaging over n=3 trials, for sequences ranging from 40–130 residues in length, with 4–6 strands.

that certainly occur near a transition highlighted by **tFolder**. Further, although we are able to produce good results using a fairly simplistic energy model, a more complicated formulation, such as one including entropic forces, would clearly improve **tFolder**'s analysis. More advanced heuristics also exist [13] that more efficiently extract folding pathway information, which could be applied to **tFolder**.

Understanding the folding dynamics of β-sheet proteins, especially which β-strand contacts drive folding and conformational stability, could help create better models of hierarchical folding, protein aggregation, and evolutionary pressure. Significant overlap likely exists between many proteins' folding pathways to even permit a classification of common transition elements (e.g. [28]); however, creating such a database would only be possible with sufficiently fast and accurate algorithms. **tFolder** takes a step toward this end by demonstrating techniques for efficiently predicting ensembles of arbitrary β-sheet proteins, and for combining these predictions to construct accurate protien folding transition landscapes.

References

1. Dobson, C.M.: Protein folding and misfolding. Nature 426(6968), 884–890 (2003)
2. Karplus, M., McCammon, J.A.: Molecular dynamics simulations of biomolecules. Nat. Struct. Biol. 9(9), 646–652 (2002)
3. Faccioli, P., Sega, M., Pederiva, F., Orland, H.: Dominant pathways in protein folding. Phys. Rev. Lett. 97(10), 108101 (2006)
4. Voelz, V.A., Bowman, G.R., Beauchamp, K., Pande, V.S.: Molecular simulation of ab initio protein folding for a millisecond folder NTL9(1-39). J. Am. Chem. Soc. 132(5), 1526–1528 (2010)

5. Levitt, M., Warshel, A.: Computer simulation of protein folding. Nature 253(5494), 694–698 (1975)
6. Tapia, L., Thomas, S., Amato, N.M.: A motion planning approach to studying molecular motions. Communications in Information and Systems 10(1), 53–68 (2010)
7. Amato, N.M., Song, G.: Using motion planning to study protein folding pathways. J. Comput. Biol. 9(2), 149–168 (2002)
8. Hosur, R., Singh, R., Berger, B.: Sparse estimation for structural variability. Algorithms Mol. Biol. (2011)
9. McCaskill, J.: The equilibrium partition function and base pair binding probabilities for RNA secondary structure. Biopolymers 29, 1105–1119 (1990)
10. Ding, Y., Lawrence, C.E.: A bayesian statistical algorithm for RNA secondary structure prediction. Comput. Chem. 23(3-4), 387–400 (1999)
11. Turner, D.H., Mathews, D.H.: NNDB: the nearest neighbor parameter database for predicting stability of nucleic acid secondary structure. Nucleic Acids Res. 38(Database issue), 280–282 (2010)
12. Wolfinger, M.T., Andreas Svrcek-Seiler, W.A., Flamm, C., Hofacker, I.L., Stadler, P.F.: Efficient computation of RNA folding dynamics. Journal of Physics A: Mathematical and General 37(17) (2004)
13. Tang, X., Thomas, S., Tapia, L., Giedroc, D.P., Amato, N.M.: Simulating RNA folding kinetics on approximated energy landscapes. J. Mol. Biol. 381(4), 1055–1067 (2008)
14. Mamitsuka, H., Abe, N.: Predicting location and structure of beta-sheet regions using stochastic tree grammars. In: ISMB, pp. 276–284 (1994)
15. Chiang, D., Joshi, A.K., Searls, D.B.: Grammatical representations of macromolecular structure. J. Comput. Biol. 13(5), 1077–1100 (2006)
16. Kato, Y., Akutsu, T., Seki, H.: Dynamic programming algorithms and grammatical modeling for protein beta-sheet prediction. J. Comput. Biol. 16(7), 945–957 (2009)
17. Tran, V.D., Chassignet, P., Sheikh, S., Steyaert, J.M.: Energy-based classification and structure prediction of transmembrane beta-barrel proteins. In: Proceedings of the First IEEE International Conference on Computational Advances in Bio and medical Sciences (ICCABS) (2011)
18. Waldispühl, J., O'Donnell, C.W., Devadas, S., Clote, P., Berger, B.: Modeling ensembles of transmembrane beta-barrel proteins. Proteins 71(3), 1097–1112 (2008)
19. Waldispühl, J., Steyaert, J.M.: Modeling and predicting all-alpha transmembrane proteins including helix-helix pairing. Theor. Comput. Sci. 335(1), 67–92 (2005)
20. Waldispühl, J., Berger, B., Clote, P., Steyaert, J.M.: Predicting transmembrane beta-barrels and interstrand residue interactions from sequence. Proteins 65(1), 61–74 (2006)
21. Cowen, L., Bradley, P., Menke, M., King, J., Berger, B.: Predicting the beta-helix fold from protein sequence data. J. Comput. Bio.l, 261–276 (2001)
22. Ding, Y., Lawrence, C.E.: A statistical sampling algorithm for RNA secondary structure prediction. Nucleic Acids Res. 31, 7280–7301 (2003)
23. Cheng, J., Baldi, P.: Improved residue contact prediction using support vector machines and a large feature set. BMC Bioinformatics 8, 113 (2007)
24. Zemla, A., Venclovas, C., Fidelis, K., Rost, B.: A modified definition of sov, a segment-based measure for protein secondary structure prediction assessment. Proteins 34(2), 220–223 (1999)
25. Moulton, V., Zuker, M., Steel, M., Pointon, R., Penny, D.: Metrics on RNA secondary structures. J. Comput. Biol. 7, 277–292 (2000)

26. Song, G., Thomas, S., Dill, K.A., Scholtz, J.M., Amato, N.M.: A path planning-based study of protein folding with a case study of hairpin formation in protein G and L. Pac. Symp. Biocomput., 240–251 (2003)
27. Hubner, I.A., Shimada, J., Shakhnovich, E.I.: Commitment and nucleation in the protein G transition state. J. Mol. Biol. 336, 745–761 (2004)
28. Fulton, K.F., Devlin, G.L., Jodun, R.A., Silvestri, L., Bottomley, S.P., Fersht, A.R., Buckle, A.M.: PFD: a database for the investigation of protein folding kinetics and stability. Nucleic Acids Res. 33(Database issue), D279–D283 (2005)

Optimally Orienting Physical Networks

Dana Silverbush[1,*], Michael Elberfeld[2,*], and Roded Sharan[1]

[1] Blavatnik School of Computer Science, Tel Aviv University, Tel Aviv 69978, Israel
{danasilv,roded}@post.tau.ac.il
[2] Institute of Theoretical Computer Science,
University of Lübeck, 23538 Lübeck, Germany
elberfeld@tcs.uni-luebeck.de

Abstract. In a network orientation problem one is given a mixed graph, consisting of directed and undirected edges, and a set of source-target vertex pairs. The goal is to orient the undirected edges so that a maximum number of pairs admit a directed path from the source to the target. This problem is NP-complete and no approximation algorithms are known for it. It arises in the context of analyzing physical networks of protein-protein and protein-DNA interactions. While the latter are naturally directed from a transcription factor to a gene, the direction of signal flow in protein-protein interactions is often unknown or cannot be measured en masse. One then tries to infer this information by using causality data on pairs of genes such that the perturbation of one gene changes the expression level of the other gene. Here we provide a first polynomial-size ILP formulation for this problem, which can be efficiently solved on current networks. We apply our algorithm to orient protein-protein interactions in yeast and measure our performance using edges with known orientations. We find that our algorithm achieves high accuracy and coverage in the orientation, outperforming simplified algorithmic variants that do not use information on edge directions. The obtained orientations can lead to better understanding of the structure and function of the network.

Keywords: network orientation, protein-protein interaction, protein-DNA interaction, integer linear program, mixed graph.

1 Introduction

High-throughoutput technologies are routinely used nowadays to detect physical interactions in the cell, including chromatin immuno-precipitation experiments for measuring protein-DNA interactions (PDIs) [10], and yeast two-hybrid assays [6] and co-immunoprecipitation screens [8] for measuring protein-protein interactions (PPIs). These networks serve as the scaffold for signal processing in the cell and are, thus, key to understanding cellular response to different genetic or environmental cues.

---

* These authors contributed equally to this work.

V. Bafna and S.C. Sahinalp (Eds.): RECOMB 2011, LNBI 6577, pp. 424–436, 2011.

While PDIs are naturally directed (from a transcription factor to its regulated genes), PPIs are not. Nevertheless, many PPIs transmit signals in a directional fashion, with kinase-substrate interactions (KPIs) being one of the prime examples. These directions are vital to understanding signal flow in the cell, yet they are not measured by most current techniques. Instead, one tries to infer these directions from perturbation experiments. In these experiments, a gene (cause) is perturbed and as a result other genes change their expression levels (effects). Assuming that each cause-effect pair should be connected by a directed pathway in the physical network, one can predict an orientation (direction assignments) to the undirected part of the network that will best agree with the cause-effect information.

The resulting combinatorial problem can be formalized by representing the network as a mixed graph, where undirected edges model interactions with unknown causal direction, and directed edges represent interactions with known directionality. The cause-effect pairs are modeled by a collection of *source-target vertex pairs*. The goal is to orient (assign single directions to) the undirected edges so that a maximum number of source-target pairs admit a directed path from the source to the target.

Previous work on this and related problems can be classified into theoretical and applied work. On the theoretical side, Arkin and Hassin [1] studied the decision problem of orienting a mixed graph to admit directed paths for a given set of source-target vertex pairs and showed that this problem is NP-complete. The problem of finding strongly connected orientations of graphs can be solved in polynomial time [3,5]. For a comprehensive discussion of the various kinds of graph orientations (not necessarily reachability preserving), we refer to the textbook of Bang-Jensen and Gutin [2].

For the special case of an undirected network (with no pre-directed edges), the orientation problem was shown to be NP-complete and hard to approximate to within a constant factor of $11/12$ [12]. On the positive side, Medvedovsky et al. [12] provided an ILP-based algorithm, and showed that the problem is approximable to within a ratio of $O(1/\log n)$, where n is the number of vertices in the network. The approximation ratio was recently improved to $O(\log \log n / \log n)$ [7]. The authors considered also the more general problem on mixed graphs, but the polylogarithmic approximation ratio attained was not satisfying as its power depends on some properties of the actual paths.

On the practical side, several authors studied the orientation problem and related annotation problems using statistical approaches [16,13]. However, these approaches rely on enumerating all paths up to a certain length between a pair of nodes, making them infeasible on large networks.

Our main contribution in this paper is a first efficient ILP formulation of the orientation problem on mixed graphs, leading to an optimal solution of the problem on current networks. We implemented our approach and applied it to a large data set of physical interactions and knockout pairs in yeast. We collected interaction and cause-effect pair information from different publications and integrated them into a physical network with 3,658 proteins, 4,000 PPIs, 4,095

PDIs, along with 53,809 knockout pairs among the molecular components of the network. We carried out a number of experiments to measure the accuracy of the orientations produced by our method for different input scenarios. In particular, we studied how the portion of undirected interactions and the number of cause-effect pairs affect the orientations. We further compared our performance to that of two layman approaches that are based on orienting undirected networks, ignoring the edge directionality information. We demonstrate that our method retains more information to guide the search, achieving higher numbers of correctly oriented edges.

The paper is organized as follows: In the next section we provide preliminaries and define the orientation problem. In Section 3 we present an ILP-based algorithm to solve the orientation problem on mixed graphs. In Section 4 we discuss our implementation of this algorithm and in Section 5 we report on its application to orient physical networks in yeast. For lack of space, some proofs are shortened or omitted.

2   Preliminaries

We focus on simple graphs with no loops or parallel edges. A *mixed graph* is a triple $G = (V, E_U, E_D)$ that consists of a set of vertices V, a set of *undirected edges* $E_U \subseteq \{e \subseteq V \mid |e| = 2\}$, and a set of *directed edges* $E_D \subseteq V \times V$. We assume that every pair of vertices is either connected by a single edge of a specific type (directed or undirected) or not connected. For convenience, we also use the notations $V(G)$, $E_U(G)$, and $E_D(G)$ to refer to the sets V, E_U, and E_D, respectively.

Let G_1 and G_2 be two mixed graphs. The graph G_1 is a *subgraph* of G_2 iff the relations $V(G_1) \subseteq V(G_2)$, $E_U(G_1) \subseteq E_U(G_2)$, and $E_D(G_1) \subseteq E_D(G_2)$ hold; in this case we also write $G_1 \subseteq G_2$. Similarly, an *induced subgraph* $G[V']$ is a subset $V' \subseteq V$ of the graph's vertices and all their pairwise relations (directed and undirected edges).

A *path* in a mixed graph G of length m is a sequence $p = v_1, v_2, \ldots, v_m$, v_{m+1} of distinct vertices $v_i \in V(G)$ such that for every $i \in \{1, \ldots, m\}$, we have $\{v_i, v_{i+1}\} \in E_U(G)$ or $(v_i, v_{i+1}) \in E_D(G)$. It is a *cycle* iff $v_1 = v_{m+1}$. Given $s \in V(G)$ and $t \in V(G)$, we say that t *is reachable from* s iff there exists a path in G that goes from s to t. In this case we also say that G *satisfies* the pair (s, t). The *transitive closure* $C(G)$ of a mixed graph G is the set of all its satisfied vertex pairs. A mixed graph with no cycles is called a *mixed acyclic graph* (MAG).

Let G be a mixed graph. An *orientation* of G is a directed graph $G' = (V(G), \emptyset, E_D(G'))$ over the same vertex set whose edge set contains all the directed edges of G and a single directed instance of every undirected edge, but nothing more. We are now ready to state the main optimization problem that we tackle:

*Problem 2.1 (*MAXIMUM-MIXED-GRAPH-ORIENTATION*).*

Input: A mixed graph G, and a set of vertex pairs $P \subseteq V(G) \times V(G)$.
Output: An orientation G' of G that satisfies a maximum number of pairs from P.

3   An ILP Algorithm for Orienting Mixed Graphs

In this section we present an integer linear program (ILP) for optimally orienting a mixed graph. The inherent difficulty in developing such a program is that a direct approach, which represents every possible path in the graph with a single variable (indicating whether, in a given orientation, this path exists or not), leads to an exponential program. Below we will work toward a polynomial size program.

Many algorithms for problems on directed graphs first solve the problem for the graph's strongly connected components independently and, then, work along the directed acyclic graph (DAG) of strongly connected components to produce a solution for the whole instance. Our ILP-based approach for orienting mixed graphs has the same high level structure: In Section 3.1 we define a generalization of strongly connected components to mixed graphs, called strongly orientable components, and show how the computation of a solution for the orientation problem can be reduced to the mixed acyclic graph of strongly orientable components. For MAGs, in turn, we present (in Section 3.2) a polynomial-size ILP that optimally solves the orientation problem.

3.1   A Reduction to a Mixed Acyclic Graph

Let G be a mixed graph. The graph G is *strongly orientable* iff it has a strongly connected orientation. The *strongly orientable components* of G are its maximal strongly orientable subgraphs. It is straightforward to prove that a graph can be partitioned into its strongly orientable components (by noting that if the vertex sets of two strongly orientable graphs intersect, then their union is also strongly orientable). The *strongly orientable component graph*, or *component graph*, G_{SOC} of G is a mixed graph that is defined as follows: Its vertices are the strongly orientable components $C_1, \ldots, C_n$ of G. Its edges are constructed as follows: There is a directed edge (C_i, C_j) in G_{SOC} iff $(v, w) \in E_D(G)$ for some $v \in V(C_i)$ and $w \in V(C_j)$. There is an undirected edge $\{C_i, C_j\}$ in G_{SOC} iff $\{v, w\} \in E_U(G)$ for some $v \in V(C_i)$ and $w \in V(C_j)$. Note that G_{SOC} must be acyclic. The strongly orientable components of a mixed graph G and, hence, the graph G_{SOC}, can be computed in polynomial time as follows: Repeatedly identify cycles in the graph and orient their undirected edges in a consistent direction. After orienting all cycles the strongly connected components that are made up by the directed edges are exactly the strongly orientable components of the initial graph.

To complete the reduction we need to specify the new set of source-target pairs. This also involves a slightly more general definition of the orientation problem where the collection of input pairs is allowed to be a multi-set. Let

P be the input multi-set for the original graph G. The multi-set P_{SOC} for the reduced graph is constructed as follows: for every pair $(s,t) \in P$ we insert a pair (C, C') into P_{SOC}, where C and C' are the strongly orientable components that contain s and t, respectively. The following lemma establishes the correctness for the reduction from instances (G, P) to $(G_{\text{SOC}}, P_{\text{SOC}})$.

Lemma 3.1. *Let G be a mixed graph and P a set of vertex pairs from G. For every $k \in \mathbb{N}$ there exists an orientation G' of G that satisfies k pairs from P iff there exists an orientation G'_{SOC} of G_{SOC} that satisfies k pairs from P_{SOC}.*

A mixed acyclic graph $G_{\text{SOC}} = (V, E_{\text{U}}, E_{\text{D}})$ is, in general, neither a forest nor a directed acyclic graph. Its structure inherits from both of these concepts: The undirected graph $(V, E_{\text{U}}, \emptyset)$ is a forest whose trees are connected by the directed edges E_{D} without producing cycles. This observation gives rise to the following definition of topological sortings for mixed graphs: A mixed graph G *admits a topological sorting* if (1) the connected components of $(V, E_{\text{U}}, \emptyset)$ are trees and (2) they can be arranged in a linear order $T_1, \ldots, T_n$, such that directed edges from E_{D} can only go from a vertex in T_i to a vertex in T_j if $i < j$. The linear order $T_1, \ldots, T_n$ of the trees is called a *topological sorting* of G. Note that the definition of topological sortings for MAGs also works for DAGs – with every tree being a single vertex. Moreover, similar to DAGs, every MAG admits a topological sorting.

3.2 An ILP Formulation for Mixed Acyclic Graphs

Given an instance of a MAG G and a multiset of vertex pairs P, our ILP consists of a set of binary *orientation* variables, describing the edge orientations, and binary *closure* variables, describing reachability relations in the oriented graph. The objective of satisfying a maximum number of vertex pairs can then be phrased as summing over closure variables for all pairs from P.

   The ILP relies on a topological sorting $T_1, \ldots, T_n$ of the input MAG, which allows formulating constraints that force a consistent assignment of values to the orientation and closure variables. The formulation is built iteratively on growing parts of the graph following the topological sorting. Specifically, for every $i \in \{1, \ldots, n\}$, we define $G_i = G[V(T_1) \cup \cdots \cup V(T_i)]$ and $P_i = P \cap (V(G_i) \times V(G_i))$, and for every $i \in \{2, \ldots, n\}$, we define $E_i = E_{\text{D}}(G) \cap (V(G_{i-1}) \times V(T_i))$. We will first define the variables of the ILP and discuss their intuitive meaning. Then we will define the constraints and the objective function of the ILP, followed by a discussion about the correctness. The ILP I for G and P is made up by the variable set variables(I) that is the union of the binary variables:

$$\{o_{(v,w)} \mid \{v,w\} \in E_{\text{U}}(G)\} \tag{1}$$

$$\{c_{(v,w)} \mid (v,w) \in V(G) \times V(G)\} \tag{2}$$

$$\{p_{(v,v',w',w)} \mid \exists\, 2 \le i \le n : (v,w) \in V(G_{i-1}) \times V(T_i) \wedge$$
$$(v',w') \in E_i\} \tag{3}$$

The orientation variables (1) are used to encode orientations of the edges: an assignment of 1 to $o_{(v,w)}$ means that the undirected edge $\{v, w\}$ is oriented from v to w. The closure variables (2) are used to represent which vertex pairs of the graph are satisfied: an assignment of 1 to $c_{(v,w)}$ will imply that there exists a directed path from v to w in the constructed orientation. During the construction we will set closure variables $c_{(v,w)}$ with $(v, w) \in E_D(G)$ to 1, and closure variables $c_{(v,w)}$ where w is not reachable from v in G to 0. *Path variables* are used to describe the satisfaction of a vertex pair (v, w) by using an intermediate directed edge (v', w'): an assignment of 1 to $p_{(v,v',w',w)}$ will imply that there exists a directed path from v to w that goes through the directed edge (v', w').

The ILP contains the constraints

$$o_{(v,w)} + o_{(w,v)} = 1 \quad \text{for all } \{v, w\} \in E_U(G) \tag{4}$$

$$c_{(v,w)} \leq o_{(x,y)} \quad \begin{array}{l} \text{for all } v, w \in V(T_i), \text{ and all } x, y \in V(T_i) \\ \text{where } y \text{ comes directly after } x \text{ on the} \\ \text{unique path from } v \text{ to } w \text{ in } T_i, 1 \leq i \leq n \end{array} \tag{5}$$

$$c_{(v,w)} \leq \sum_{(v',w') \in E_i} p_{(v,v',w',w)}$$
$$\text{for all } (v, w) \in V(G_{i-1}) \times V(T_i), 2 \leq i \leq n \tag{6}$$

$$p_{(v,v',w',w)} \leq c_{(v,v')}, c_{(w',w)}$$
$$\text{for all } (v, w) \in V(G_{i-1}) \times V(T_i), (v', w') \in E_i, 2 \leq i \leq n \tag{7}$$

and the objective

$$\text{maximize} \sum_{(s,t) \in P} c_{(s,t)} \tag{8}$$

Constraints (4) force that each undirected edge is oriented in exactly one direction. The remaining constraints (5) to (7) are used to connect closure variables to the underlying orientation variables. They force that every closure variable $c_{(v,w)}$ can only be set to 1 if the orientation variables describe a graph that has a directed path from v to w. Whenever v and w are in the same undirected component (which is a tree since the whole graph is a MAG), they can only be connected via an orientation of the unique undirected path between them. For vertex pairs of these kind constraint (5) ensures the above property. Next we consider the case where v and w are in different components T_i and T_j with $i < j$. We need to associate $c_{(v,w)}$ with all possible paths from v to w; this is done by using the path variables: If there is a path from v to w then it must visit a directed edge (v', w') that starts in some component that precedes T_j and ends at T_j (Constraint (6)). Path variables are, in turn, constrained by (7). The objective function maximizes the number of closure variables with assignment 1 that correspond to pairs from P. The above discussion contains the basic ideas to prove the following lemma, which formally implies the correctness of the ILP.

Lemma 3.2. *The following properties hold:*

Completeness: *For every orientation G' of G there exists an assignment $a :$*
  variables$(I) \rightarrow \{0,1\}$ with $\{(v,w) \in V(G) \times V(G) \mid a(c_{(v,w)}) = 1\} = C(G')$
  that satisfies the constraints (4) to (7).
Soundness: *For every assignment $a :$ variables$(I) \rightarrow \{0,1\}$ that satisfies the*
  constraints (4) to (7) there exists an orientation G' of G with $\{(v,w) \in$
  $V(G) \times V(G) \mid a(c_{(v,w)}) = 1\} \subseteq C(G')$.

The ILP has polynomial size and can be constructed in polynomial time: The
construction starts by sorting the MAG topologically. Constant length constraints
(4) are constructed for all undirected edges. For every ordered pair (v,w) of
vertices v and w that are inside the same undirected component T_i, we construct
at most $|E_U|$ constraints of type (5) using reachability queries to T_i. The sum
constraints (6) are constructed for all ordered vertex pairs (v,w) where the
undirected component of v comes before the undirected component of w in the
topological sorting of the MAG. Each sum iterates over the directed edges that
lead into the component of w. Thus, each sum's length is bounded by $O(|E_D|)$
and it can be written down in polynomial time. The constraints (7) of constant
length are constructed by iterating over the same vertex pairs and directed edges.
In total, the size of the ILP is asymptotically bounded by $O(|V(G)|^2(|E_D| + |E_U|))$.

One may ask if it is possible to apply the ILP construction to general mixed
graphs instead of MAGs. The MAG-based construction explores the graph iteratively by using a topological sorting. It relies on the fact that connecting paths in
MAGs are either unique (inside the undirected components) or can only go from
a component T_i to a component T_j if $i < j$. In a mixed graph $G = (V, E_U, E_D)$
that contains cycles, connecting paths in (V, E_U) are, in general, not unique
and there may be directed edges going back and forth between components of
(V, E_U). This prevents the iterative construction and implies a construction that
needs to revise already constructed parts of the formulation instead of just appending new constraints at each step. We are not aware of any method that
directly produces polynomial size ILP formulations for general graphs.

4 Implementation Details

Our implementation is written in C++ using BOOST C++ libraries (version number 1.43.0) and the commercial IBM ILOG CPLEX optimizer (version number 12)
to solve ILPs. The input of our program consists a mixed graph $G = (V, E_U, E_D)$
and a collection P of vertex pairs from G. The program predicts an orientation
G' for G that satisfies a maximum number of pairs from P.

The program starts by computing strongly connected orientations for all
strongly orientable components of the input graph. This can be done in polynomial time, as described in Section 3.2. Our program implements a linear time
approach for this step that is based on combined ideas from [15] and [5]. Next,
the program computes the acyclic component graph G_{SOC} of G and transforms

the collection of pairs P into the collection of pairs P_{SOC}. Finally, the program computes an optimal orientation for the resulting instance $(G_{\text{SOC}}, P_{\text{SOC}})$ via the ILP approach from Section 3.2. This results in an orientation for all undirected edges that are not inside strongly orientable components and the number of satisfied pairs, which is optimal. Altogether, the program outputs an optimal orientation for the input instance and, if desired, the satisfied pairs and their number.

Due to the combinatorial nature of our approach, there is possibly more than one orientation that results in an optimal number of satisfied pairs. To determine if an undirected edge $e = \{v, w\}$ has the same orientation in all maximum solutions, one can utilize our computational pipeline as follows: First compute the number of satisfied pairs in an optimal solution s_{opt}. Let (v, w) be the orientation of e in this solution. Then run the experiment again, but this time with $\{v, w\}$ replaced by (w, v) in the input network. After that set a *confidence value* $c_e = s_{\text{opt}} - s_e$, where s_e is the maximum number of satisfied pairs for the modified instance. The edge e is said to be oriented with *confidence* iff $c_e \geq 1$; in this case its direction is the same in all optimal orientations of the input.

5   Experimental Results

5.1   Data Acquisition and Integration

We gathered physical interactions (PPIs, PDIs, and KPIs) and cause-effect pair information for *Saccharomyces cerevisiae* from different sources. We used the PPI data set "Y2H-union" from Yu et al. [17], which contains 2,930 highly-reliable undirected interactions between 2,018 proteins. The PDI data were taken from MacIsaac et al. [11], an update of which can be found at http://fraenkel.mit.edu/improved_map/. We used the collection of PDIs with $p < 0.001$ conserved over at least two other yeast species, which consists of 4,113 unique PDIs spanning 2,079 proteins. The KPIs were collected from Breitkreutz et al. [4] by taking the directed kinase-substrate interactions out of their data set. This results in 1361 KPIs among 802 proteins. A set of 110,487 knockout pairs among 6,228 proteins where taken from Reimand et al. [14].

We integrated the data to obtain a physical network of undirected and directed interactions. We removed self loops and parallel interactions; for the latter, whenever both a directed and an undirected edge were present between the same pair of vertices, we maintained the former only. Pairs of edges that are directed in opposite directions were maintained, and will be contracted into single vertices in later phases of the preprocessing. The resulting physical network, which we call the *integrated network* spans 3,658 proteins, 2,639 PPIs, 4,095 PDIs and 1,361 KPIs. For some of the following experiments we want to control the amount of directed edges better, to investigate their contribution in a purified manner. To this end we will also use the subnetwork of 2,579 proteins of the integrated network that is obtained by taking only the directed PDIs and PKIs, leaving the PPIs out; we call it the *refined network*. To orient the physical networks, we use the set of 110,487 knockout pairs and consider the subset of pairs with endpoints being in the physical network. The integrated network contains 53,809 of the pairs; the refined network contains 34,372 of the pairs.

5.2   Application and Performance Evaluation

To study the behavior and properties of our algorithm, we apply it to the physical networks and monitor properties of the instance from the intermediate steps and the resulting orientations. For the former, we examine the contraction step, monitoring the size of the component graph obtained (number of vertices, directed edges and undirected edges), and the number of cause-effect pairs after the contraction. For the latter, we run the algorithm in a cross-validation setting, hiding the directions of some of the edges and testing our success in orienting them.

The component graph for the integrated network contains 763 undirected edges and 2,910 directed edges among 2,569 vertices. We filter from the corresponding set of pairs P_{SOC} those pairs that have the same source and target vertices; these pairs lie inside strongly orientable components and are already satisfied. About 85% (44,825) of the initial pairs from the large knockout pair data remain in the contracted graph and can be used to guide the orientation produced by our ILP algorithm. Considering the whole integrated network with the large set of knockout pairs, the component orientation and component contraction steps take 3 seconds and the solution of the ILP takes 70 seconds. Considering only the refined network, the preprocessing as elaborated takes 5 seconds, as there are eventually more components in the contracted graph, and 57 seconds for the solution of the ILP, as there are less choices to be made. Computing confidence scores for undirected edges requires rerunning the steps of the computational pipeline for each of these edges, resulting in about 3.4 hours for the integrated network and 5 hours for the refined network (in which more test edges remain after the cycle contraction).

Next, we wished to evaluate the orientations suggested by our algorithm. To this end, we defined a subset of the directed edges in the input graph (KPIs or PDIs) as undirected *test edges*. Guided by the set of knockout pairs, our program computes orientations for all undirected edges, including the test edges. In the evaluation of the orientation we focus on test edges that survive the contraction and *remain* in the component graph, as the orientation of the other test edges depends only on the cycles they lie in and not on the input cause-effect pairs. We further focus on confident orientations, as other orientations are arbitrary. We define the *coverage* of an orientation as the percent of remaining test edges that are oriented with confidence. The *accuracy* of an orientation is the percent of confidently oriented test edges whose orientation is correct.

When using the integrated network and all 1,361 KPIs as test edges, 166 (12%) of them remain after cycle contraction. The algorithm covers 158 (95%) of the remaining test edges, orienting correctly 137 (86%) of the covered ones. In the refined network 290 (21%) of the KPIs remain, 264 (91%) of them are covered, and 228 (86%) of those are oriented correctly. When using the integrated network and the 4,095 PDIs as test edges 712 (17%) of the test edges remain, 634 (89%) of them are covered, and 614 (96%) of those are oriented correctly. In the refined

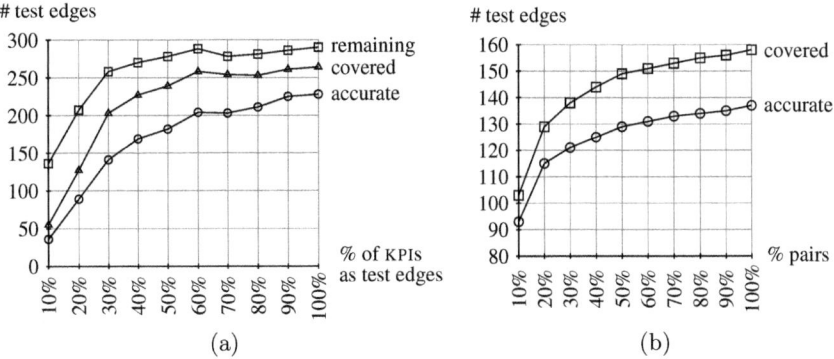

Fig. 1. (a) Remaining, covered and accurate test edges as a function of the percentage of input test edges. x-axis: percentage of KPIs that are used as test edges. y-axis: numbers of test edges that remain (squares), are covered (triangles) and accurately oriented (circles). (b) The number of covered and accurately oriented test edges as a function of the percentage of cause-effect pairs guiding the orientation.

network 996 (24%) of the PDIs remain, 895 (90%) of them are covered, and 868 (97%) of those are oriented correctly. Expectedly, more test edges remain in the refined network; coverage and accuracy are high in all these experiments.

Effects of the portion of undirected edges. The previous results hint that to obtain a higher percentage of remaining edges, it is helpful to consider networks with a smaller number of undirected edges and larger number of directed edges. To test the effect of the portion of undirected edges more systematically, we focused on the refined network and used different portions (chosen at random) of KPIs as test sets. All KPIs that are not test edges are deleted from the network. The results are depicted in Figure 1(a), demonstrating that the percentage of remaining test edges increases when we consider small fractions of them. This stems from the fact that a smaller number of test edges gives rise to fewer component contractions in the input graph. Interestingly, the coverage and accuracy go up when considering larger amounts of test edges. The reason is that while many parts of the graph are contracted, the initial large number of input test edges leads to a large number of test edges after the contractions. As a result, there is more information to guide the orientation compared to the smaller test sets.

Effects of the amount of cause-effect pairs. Next, we wished to study the effect of the amount of cause-effect pairs on the orientation. We used as input the integrated network, with the KPIs serving as test edges, and applied the algorithm with increasing portions (chosen at random) of pairs. Out of the 166 test edges that remain after contraction, different numbers of covered and accurate edges were attained depending on the input pairs. As evident from Figure 1(b), the more pairs the higher the number of covered edges, albeit with similar accuracy (86-90%), supporting our use of the cause-effect pairs to guide the orientation. Although this is not our objective, it is interesting to note that a high percentage (approximately 85%) of the knockout pairs are satisfied throughout the experiments. The much smaller fraction of unsatisfied pairs may be due to noise in the expression data, incomplete interaction data or molecular events that are not covered by the physical interactions considered here.

5.3   Comparison to Layman Approaches

To the best of our knowledge, there exists no previous method to orient mixed graphs, but one can try to adapt methods for undirected graphs to the mixed graph case. The only previous method for orienting large undirected graphs is the one from Medvedovsky et al. [12]. In our terminology, it first computes the graph's component graph, which is a tree for undirected input graphs. It then applies an ILP-formulation, using the fact that there is at most one path between any two vertices. We consider two ways of transforming mixed graphs into undirected graphs to which this method can be applied. Both approaches take their action after the construction of the component graph for the mixed input graph. While our approach, which we call MIXED, uses an ILP at this point, the DELETION approach removes all directed edges from the component graph, yielding a forest of its undirected components to which an ILP is applied. The UNDIRECTED approach considers all directed edges as being undirected and applies a second component contraction step to produce a forest to which the ILP is applied. The same forest can be obtained by starting from the input graph, making all directed edges undirected, and applying a single contraction step.

Table 1. (a) Properties of the intermediate steps of the three orientation approaches. (b) A comparison of the three orientation approaches with cross-validation experiments using different fractions of cause-effect pairs.

(a)

	DELETION	UNDIRECTED	MIXED
# of undirected edges in the input	2639	8089	2639
# of directed edges in the input	5450	0	5450
# of vertices in the component graph	2569	1483	2569
# of undirected edges in the component graph	763	1445	763
# of directed edges in the component graph	0	0	2910
# of pairs between different vertices in P_{soc}	44825	24423	44825
# of pairs between different vertices in P_{soc} that are satisfied in G_{soc}	4705	23587	29792

(b)

	100% cause-effect pairs			10% cause-effect pairs		
	DELETION	UNDIRECTED	MIXED	DELETION	UNDIRECTED	MIXED
# of test edges						
that remain	290	226	290	290	226	290
that are covered	240	215	265	102	133	144
that are accurate	212	187	229	87	112	121

The behaviors of the intermediate steps of the three approaches when applied to the integrated network are shown in Table 1(a). In comparison to UNDIRECTED, MIXED maintains a higher number of vertices in the component graph, as less cycles are contracted. In comparison to DELETION, MIXED maintains a much higher amount (6 fold) of pairs that are satisfied in the component graph and, therefore, potentially affect the orientation process. This is due to the fact that the edge deletion separates large parts of the graph. Overall, one can see that MIXED retains more information for the ILP step in the form of vertices in the component graph and causal information from the knockout pairs.

To compare the orientations produced by the three approaches, we applied them to the refined network using the KPIs as test edges and different portions of the cause-effect pairs. As the baseline for computing the coverage of the three approaches should be the same – the number of test edges after the initial contraction – we report in the following the absolute numbers of covered (confidently oriented) and correctly oriented interactions, rather than the relative coverage and accuracy measures. Table 1(b) present these results, comparing the numbers of remaining, covered and correctly oriented test edges among the three approaches. Evidently, MIXED yields higher numbers of test edges, covered edges, and correctly oriented edges.

6   Conclusions

We presented an ILP algorithm that efficiently computes optimal orientations for mixed graph inputs. We implemented the method and applied it to the orientation of physical interaction networks in yeast. Depending on the input the method yields very high coverage and accuracy in the orientation. Our experiments further show that the algorithm works very fast in practice and produces orientations that cover (accurately) larger portions of the network compared to the ones produced by previous approaches that ignore the directionality information and operate on undirected versions of the networks.

While in this paper we concentrated on the computational challenges in network orientation, the use of the obtained orientations to gain biological insights on the pertaining networks is of great importance. As demonstrated by [9], the directionality information facilitates pathway inference. It may also contribute to module detection; in particular, it is intriguing in this context to map the correspondence between contracted edges (under our method) and known protein complexes.

Acknowledgements

M.E. was supported by a research grant from the Dr. Alexander und Rita Besser-Stiftung. R.S. was supported by a research grant from the Israel Science Foundation (grant no. 385/06).

References

1. Arkin, E.M., Hassin, R.: A note on orientations of mixed graphs. Discrete Applied Mathematics 116(3), 271–278 (2002)
2. Bang-Jensen, J., Gutin, G.: Digraphs: Theory, Algorithms and Applications, 2nd edn. Springer, Heidelberg (2008)
3. Boesch, F., Tindell, R.: Robbins's theorem for mixed multigraphs. The American Mathematical Monthly 87(9), 716–719 (1980)
4. Breitkreutz, A., et al.: A global protein kinase and phosphatase interaction network in yeast. Science 328(5981), 1043–1046 (2010)
5. Chung, F.R.K., Garey, M.R., Tarjan, R.E.: Strongly connected orientations of mixed multigraphs. Networks 15(4), 477–484 (1985)
6. Fields, S., Song, O.-K.: A novel genetic system to detect protein-protein interactions. Nature 340(6230), 245–246 (1989)
7. Gamzu, I., Segev, D., Sharan, R.: Improved orientations of physical networks. In: Moulton, V., Singh, M. (eds.) WABI 2010. LNCS, vol. 6293, pp. 215–225. Springer, Heidelberg (2010)
8. Gavin, A., et al.: Functional organization of the yeast proteome by systematic analysis of protein complexes. Nature 415(6868), 141–147 (2002)
9. Gitter, A., Klein-Seetharaman, J., Gupta, A., Bar-Joseph, Z.: Discovering pathways by orienting edges in protein interaction networks. Nucleic Acids Research (2010), doi:10.1093/nar/gkq1207
10. Lee, T.I., et al.: Transcriptional regulatory networks in saccharomyces cerevisiae. Science 298(5594), 799–804 (2002)
11. MacIsaac, K., et al.: An improved map of conserved regulatory sites for saccharomyces cerevisiae. BMC Bioinformatics 7(1), 113 (2006)
12. Medvedovsky, A., Bafna, V., Zwick, U., Sharan, R.: An algorithm for orienting graphs based on cause-effect pairs and its applications to orienting protein networks. In: Crandall, K.A., Lagergren, J. (eds.) WABI 2008. LNCS (LNBI), vol. 5251, pp. 222–232. Springer, Heidelberg (2008)
13. Ourfali, O., Shlomi, T., Ideker, T., Ruppin, E., Sharan, R.: SPINE: a framework for signaling-regulatory pathway inference from cause-effect experiments. Bioinformatics 23(13), i359–i366 (2007)
14. Reimand, J., Vaquerizas, J.M., Todd, A.E., Vilo, J., Luscombe, N.M.: Comprehensive reanalysis of transcription factor knockout expression data in saccharomyces cerevisiae reveals many new targets. Nucleic Acids Research 38(14), 4768–4777 (2010)
15. Tarjan, R.: Depth-first search and linear graph algorithms. SIAM Journal on Computing 1(2), 146–160 (1972)
16. Yeang, C., Ideker, T., Jaakkola, T.: Physical network models. Journal of Computational Biology 11(2-3), 243–262 (2004)
17. Yu, H., et al.: High-Quality binary protein interaction map of the yeast interactome network. Science 322(5898), 104–110 (2008)

Opera: Reconstructing Optimal Genomic Scaffolds with High-Throughput Paired-End Sequences

Song Gao[1], Niranjan Nagarajan[2], and Wing-Kin Sung[2,3]

[1] NUS Graduate School for Integrative Sciences and Engineering,
Centre for Life Sciences, 28 Medical Drive, Singapore
[2] Computational and Systems Biology, Genome Institute of Singapore,
60 Biopolis Street, Singapore
[3] School of Computing, National University of Singapore,
21 Lower Kent Ridge Road, Singapore
sungk@gis.a-star.edu.sg

Abstract. Scaffolding, the problem of ordering and orienting contigs, typically using paired-end reads, is a crucial step in the assembly of high-quality draft genomes. Even as sequencing technologies and mate-pair protocols have improved significantly, scaffolding programs still rely on heuristics, with no gaurantees on the quality of the solution. In this work we explored the feasibility of an exact solution for scaffolding and present a first fixed-parameter tractable solution for assembly (Opera). We also describe a graph contraction procedure that allows the solution to scale to large scaffolding problems and demonstrate this by scaffolding several large real and synthetic datasets. In comparisons with existing scaffolders, Opera simultaneously produced longer and more accurate scaffolds demonstrating the utility of an exact approach. Opera also incorporates an exact quadratic programming formulation to precisely compute gap sizes.

Keywords: Scaffolding, Genome Assembly, Fixed-parameter Tractable, Graph Algorithms.

1 Introduction

With the advent of second-generation sequencing technologies, while the cost of sequencing has decreased dramatically, the challenge of reconstructing genomes from the large volumes of fragmentary *read* data has remained daunting. Newly developed protocols for second-generation sequencing can generate *paired-end reads* (reads from the ends of a fragment of known approximate length) for a range of library sizes [1] and third-generation *strobe sequencing* protocols [2] provide linking information that, in principle, can be valuable for assembling a genome. In recent work, the importance of paired reads has been further highlighted, with some authors even questioning the need for long reads in the presence of libraries with large insert lengths [4,5].

V. Bafna and S.C. Sahinalp (Eds.): RECOMB 2011, LNBI 6577, pp. 437–451, 2011.
© Springer-Verlag Berlin Heidelberg 2011

Scaffolding, the problem of using the connectivity information from paired reads to order and orient partially reconstructed *contig* sequences in the genome, has been well-studied in the assembly literature with several algorithms proposed in recent years [7,8,9,6,10,5,3]. In work in 2002, Huson et al [6] presented a natural formulation of this problem (in terms of finding an ordering of sequences that minimizes paired read violations) and showed that its decision version is NP-complete. Several related problems are also known to be NP-complete [8,10] and hence, to maintain efficiency and scalability, existing algorithms have relied on various heuristic solutions. For instance, in [6], the authors proposed a greedy solution that iteratively merges scaffolds connected by the most paired reads. Similarly, the algorithm proposed in the Phusion assembler [11] relies on a greedy heuristic based on the distance contraints imposed by the paired reads. Other approaches, used in assemblers such as ARACHNE and JAZZ [12,13], also employ error-correction steps to minimize the potential impact of misjoins from heuristic searches.

In addition to paired reads, similarity to a reference genome [14,15,16] and restriction-map based approaches [8,17] have been used to order contigs, partly because they can lead to a more computationally tractable problem. However, while reference-guided assembly uses potentially misleading synteny information, restriction-map based approaches can produce an ambiguous order and find it hard to place small contigs [17]. Paired-end reads therefore remain the most general source of information for generating high-quality scaffolds.

In this work, we focus on the problem of scaffolding of a set of contigs using paired-end reads, though similar ideas could be extended to multi-contig constraints from sources such as strobe sequencing and restriction maps [2]. Unlike existing solutions which use heuristics, we provide a combinatorial algorithm that is guaranteed to find the optimal scaffold under a natural criterion similar to [6]. By exploiting the fixed-parameter tractability of the problem and a contraction step that leverages the structure of the graph, our scaffolder (Opera) effectively constructs scaffolds for large genomic datasets. The fundamental advantages of this approach are twofold: Firstly, the algorithm provides a solution that explains/uses as much of the paired read data as possible (as we show, this also translates into a more complete and reliable scaffold in practice). Secondly, the algorithm provides a clear guarantee on the quality of the assembly and avoids overly aggressive assembly heuristics that can produce large scaffolds at the expense of assembly errors.

While libraries from new sequencing technologies generate a vast amount of paired-end reads that provide detailed connectivity information, assembly and mapping errors from shorter read lengths and an abundance of chimeric mate-pairs in some protocols [1] can complicate the scaffolding effort. We show how these sources of error can be handled in our optimization framework in a robust fashion. We also employ a quadratic programming formulation (and an efficient solver) to compute gap sizes that best agree with the mate-pair library derived contraints. Our experiments with several large real and synthetic datasets suggest that these theoretical advantages in Opera do translate to larger, more reliable and well-defined scaffolds when compared to existing programs.

2   Methods

2.1   Definitions

In a typical whole-genome shotgun sequencing project, randomly sheared fragments of DNA are sequenced, using one or more of the several sequencing technologies that are now available. The resulting reads are then assembled *in silico* to produce longer contig sequences [18]. In addition, the reads are often generated from the ends of long fragments (of known approximate sizes and from one or more libraries) and this information is used to link together contigs and order and orient them (see Fig. 1).

Consider a set of contigs $C = \{c_1, \ldots, c_n\}$. For every $c_i \in C$, we denote the two possible orientations as c_i and $-c_i$. A *scaffold* is then given by a signed permutation of the contigs as well as a list of gap sizes between adjacent contigs (see Fig. 1(d)). Given two contigs c_i and c_j linked by a paired-read (i.e. one end falls on c_i and the other end on c_j), the relative orientation of the contigs suggested by the paired-read can be encoded as a bidirected edge in a graph (see Fig. 1(a, c)). We then say that a paired-read is *concordant* in a scaffold if the

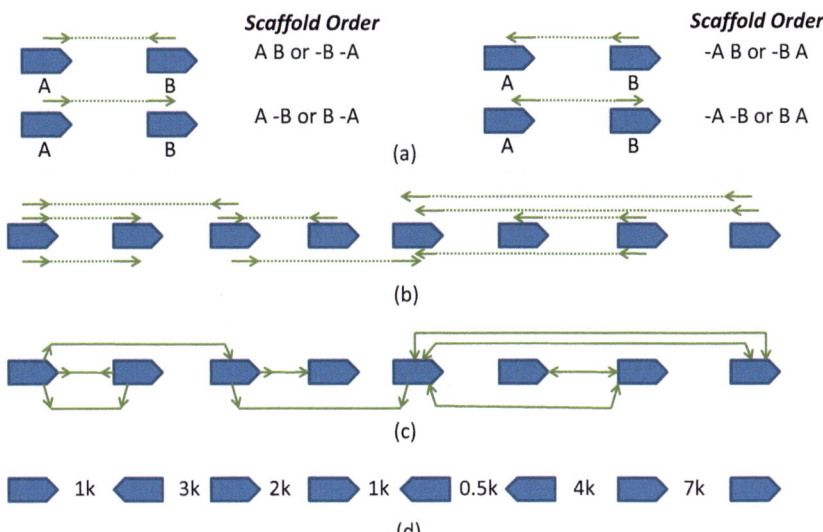

Fig. 1. Paired-read and Scaffold Graph. (a) Paired-read constraints on order and orientation of contigs (pointed boxes) (b) A set of paired-read and contigs (c) The resulting scaffold graph (d) A scaffold for the graph in (c).

suggested orientation is satisfied and the distance between the reads is less then a specified maximum[1] library size τ.

2.2 Scaffold Graph

Given a set of contigs and a mapping of paired-reads to contigs, we use an "edge bundling" step as described in [6] to construct a scaffold graph (actually bidirected multi-graph) where contigs are nodes and are connected by scaffold edges representing multiple paired-reads that suggest a similar distance and orientation for the contigs (see Fig. 1(b, c)). After the bundling step, existing scaffolders typically filter edges with reads less than an arbitrary (sometimes user-specified) threshold. This is done to reduce the number of incorrect edges in the graph from chimeric paired-reads. Instead of setting this threshold arbitrarily and independent of genome size or sequence coverage, we use the following simulation to determine an appropriate threshold: we simulate chimeric reads by selecting paired-reads at random and exchanging their partners. This is then repeated till a significant proportion of the reads (say 10%) are chimeric. We then bundle the chimeric reads as before and repeat the simulation a 100 times to determine the scaffold edge with most chimeric reads supporting it (say d) and set the threshold to be one more than that (i.e. $d + 1$). This then effectively removes the "stochastic noise" introduced by chimeric constructs and allows the main scaffolding algorithm to focus on systematic assembly and mapping errors that lead to incorrect scaffold edges.

Extrapolating the notion of *concordance* to scaffold edges we get the following natural formulation of the Scaffolding Problem:

Definition 1 (Scaffolding Problem). *Given a scaffold graph G, find a scaffold S of the contigs that maximizes the number of concordant edges in the graph.*

As this problem is analogous to that in [6] (where the optimality criterion is the number of concordant paired-reads) it is easy to modify their proof to show that the decision version of the scaffolding problem is NP-complete.

2.3 Fixed-Parameter Tractability

The scaffolding problem that we defined (as well as the one in [6]) does not specifically delineate a structure for the scaffold graph. In practice, however, the scaffold graph is constrained by the fact that paired-read libraries have an upperbound τ and contigs have a minimum length, l_{min}. This defines an upper-bound on the number of contigs that can be spanned by a paired-read i.e. the *width* of the library (or w, where $w \leq \frac{\tau}{l_{min}}$). Here we show that considering width as a fixed parameter, we can indeed construct an algorithm that is polynomial in the size of the graph. This is similar to the work in [19], where the focus is on a bounded version of the graph bandwith problem. The scaffolding problem can

---

[1] In principle, a lower-bound can also be determined and used in defining concordant edges.

be seen as a generalization of a bounded version of the graph bandwith problem where nodes and edges in the graph have orientations and lengths and not all edge constraints have to be satisfied.

For ease of exposition we first consider the special case where the optimal scaffold in a scaffold graph has no discordant edges (i.e. a *bounded-width graph*). We consider the case of discordant edges in Section 2.4. Also, without loss of generality, we assume that the graph is connected (otherwise, we can compute optimal scaffolds for each component independently). Finally, it is easy to see that we can limit our search to scaffolds where all gap sizes are 0 (we show how more appropriate gap sizes can be computed in Section 2.7). We begin with a few definitions: For a scaffold graph $G = (V, E)$, a *partial scaffold* S' is a scaffold on a subset of the contigs and the *dangling set*, $D(S')$, is the set of edges from S' to $V - S'$. The *active region* $A(S')$ is then the shortest suffix of S' such that all dangling edges are adjacent to a contig in $A(S')$. A partial scaffold S' is said to be *valid* if all edges in the induced subgraph are concordant.

We now describe a dynamic-programming based search over the space of scaffolds to find the optimal scaffold. Note that a naive search for an optimal scaffold would enumerate over all possible signed permutations of the contigs and count the number of concordant scaffold edges. Since there are $2^{|V|}|V|!$ possible signed permutations, this approach is clearly not feasible. Instead, we can limit our search over an equivalence class of partial scaffolds as shown in the following lemma.

Lemma 1. *Consider two valid partial scaffolds S'_1 and S'_2 of the scaffold graph G. If $(A(S'_1), D(S'_1)) = (A(S'_2), D(S'_2))$ then (1) S'_1 and S'_2 contain the same set of contigs; and (2) both or neither of them can be extended to a solution.*

Proof. For (1), suppose there exists a contig c which appears in S'_2 but not in S'_1. Since $G = (V, E)$ is a connected graph, there exists a path (in an undirected sense) $z_1 = y, z_2, \ldots, z_i = c$ in $G = (V, E)$ where y is the first contig in the active region of both S'_1 and S'_2 while $z_2, \ldots, z_i \in V - S'_1$. For S'_2, since y is the first contig which appears in the active region of S'_2, we have $z_2, \ldots, z_i \in S'_2$. Hence, (z_1, z_2) is a dangling edge of S'_1 but not a dangling edge of S'_2 which gives us a contradiction.

For (2), let S'' be any scaffold of $V - S'_1 = V - S'_2$. Since S'_1 and S'_2 have the same active region, $S'_1 S''$ has no discordant edges if and only if $S'_2 S''$ has no discordant edges. □

Based on the above lemma, the algorithm in Figure 2 starts from an empty scaffold $S = \emptyset$, extends it a contig at a time to search over the equivalence class of partial scaffolds, and finds a scaffold with no discordant edges (if it exists). The proof of correctness of the algorithm follows directly from Lemma 1.

We prove the runtime complexity of the algorithm in the following theorem.

Theorem 1. *Given a scaffold graph $G = (V, E)$ and an empty scaffold, the algorithm Scaffold-Bounded-Width runs in $O(|E||V|^w)$ time.*

Scaffold-Bounded-Width(S')
Require: A scaffold graph $G = (V, E)$ and a valid partial scaffold S'
Ensure: Return a scaffold S of G with no discordant edges and where S' is a prefix
  of S
1: **if** S' is a scaffold of G **then**
2:    return S'
3: **end if**
4: **for** every $c \in V - S'$ in each orientation **do**
5:    Let S'' be the scaffold formed by concatenating S' and c;
6:    Let A be the active region of S'';
7:    Let D be the dangling set of S'';
8:    **if** (A, D) is unmarked **then**
9:       Mark (A, D) as processed;
10:      **if** S'' is valid **then**
11:         $S''' \leftarrow$ Scaffold-Bounded-Width(S'');
12:         If $S''' \neq$ FAILURE, return S''';
13:      **end if**
14:   **end if**
15: **end for**
16: Return FAILURE;

Fig. 2. An algorithm for generating a scaffold for a bounded-width scaffold graph

Proof. The number of contigs in an active region is bounded by w and each contig has two possible orientations. Hence, the set of possible active regions is $O((2|V|)^w)$. Every contig in an active region has $\leq w$ dangling edges. Thus, a given active region has at most $O(2^{w^2})$ possible dangling sets. The number of equivalence classes is therefore bounded by $O(2^{w^2}(2|V|)^w) = O(|V|^w)$ (treating w as a constant). For each equivalence class, updating the active region and the dangling set in steps 6 and 7 takes $O(|E|)$ time.   □

The runtime analysis presented here is clearly a coarse-grained analysis and with some more work, tighter bounds can be proven (for example, since we extend the scaffold in only one direction, we do not need to keep track of the dangling set). However, the main point here is that for a fixed w, the worst-case runtime of the algorithm is polynomial in the size of the graph, i.e. we have a fixed-parameter tractable algorithm for the problem. In the next section, we discuss how this analysis can be extended to the case where not all edges in the optimal scaffold are concordant.

2.4   Minimizing Discordant Edges

Treating the width parameter as a fixed constant is a special case of the scaffolding problem and correspondingly the NP-completeness result discussed in Section 2.2 does not hold. In the following theorem (with proof in the appendix) we show that the decision version of the scaffolding problem (allowing for discordant edges) is NP-complete even when the width of the paired-end library is treated as a constant.

Theorem 2. *Given a scaffold graph G and treating the library width w as a fixed-parameter, the problem of deciding if there exists a scaffold S with less than p discordant edges is NP-complete.*

Theorem 2 suggests that we cannot hope to design an algorithm that is polynomial in p, the number of discordant edges. However, treating p as a fixed-parameter as well, we can extend the algorithm in Section 2.3 and still maintain a runtime polynomial in the size of the graph. The basic idea here is that we need to extend the notion of equivalence class by keeping track of discordant edges from the partial scaffold (denoted by $X(S')$ for a partial scaffold S'). Also, we redefine the dangling set to contain only concordant edges and note that as the scaffold is only extended in one direction, the dangling set is completely determined by the active region and the set of discordant edges. Then the following lemma is a straightforward extension of Lemma 1:

Lemma 2. *Consider two partial scaffolds S_1' and S_2' of G with less than p discordant edges. If $(A(S_1'), X(S_1')) = (A(S_2'), X(S_2'))$ then (1) S_1' and S_2' contain the same set of contigs; and (2) both or neither of them can be extended to a solution.*

Based on this lemma an extension of the algorithm in Figure 2 that handles discordant edges is presented in Figure 3. In addition, we extend the runtime analysis in the following lemma:

Lemma 3. *Consider a scaffold graph $G = (V, E)$ and let p be the maximum allowed number of discordant edges. The algorithm Scaffold runs in $O(|V|^w |E|^{p+1})$ time.*

Proof. As before, the set of possible active regions is $O(|V|^w)$. Also, there are at most $O(|E|^p)$ possible sets of discordant edges. Finally, for each equivalence class, updating the active region and the set of discordant edges in steps 6 and 7 takes $O(|E|)$ time. ☐

To convert this algorithm into one that optimizes over p, we can rely on a branch-and-bound approach where (1) a quick heuristic search is used to find a good solution and define an upper-bound on p and (2) the upper-bound is refined as better solutions are found and the search is not terminated till all extensions have been explored in step 4. We implemented such an approach but found that in some cases our heuristic search would return a poor upper-bound and thus affect the runtime of the algorithm. To get around this, our current implementation tries each value of p (starting from 0) and stops when a scaffold can be constructed (the total runtime is still $O(|V|^w |E|^{p+1})$).

Note that while the worst-case runtime bound suggests that if p increases by one, runtime would increase by a factor proportional to the size of the graph, in practice, we observe only a constant factor increase (i.e. runtime growth is C^p where $C \leq 5$). For real datasets, we can further exploit the structure of the graph and one idea that improves runtimes significantly is detailed below.

Scaffold(S', p)

Require: A scaffold graph $G = (V, E)$ and a partial scaffold S' with at most p
   discordant edges

Ensure: Return a scaffold S of G with at most p discordant edges and where S' is
   a prefix of S

1: **if** S' is a scaffold of G **then**
2:    return S'
3: **end if**
4: **for** every $c \in V - S'$ in each orientation **do**
5:    Let S'' be the scaffold formed by concatentating S' and c;
6:    Let A be the active region of S'';
7:    Let X be the set of discordant edges of S'';
8:    **if** (A, X) is unmarked **then**
9:       Mark (A, X) as processed;
10:       **if** $|X| \le p$ **then**
11:          $S''' \leftarrow$ Scaffold(S'', p);
12:          If $S''' \ne$ FAILURE, return S''';
13:       **end if**
14:    **end if**
15: **end for**
16: Return FAILURE;

Fig. 3. An algorithm for generating a scaffold with at most p discordant edges

2.5   Graph Contraction

Contigs assembled from a whole-genome shotgun sequencing data come in a
range of sizes and often a successful assembly produces several contigs longer
than paired-read library thresholds (τ). For a particular library size, we label
such contigs as *border* contigs and note the fact that a scaffold derived from
such a scaffold graph will not have concordant library edges spanning a border
contig. For a scaffold graph $G = (V, E)$, we then define $G' = (V', E')$ as a *fenced*
subgraph of G if edges in E from $V - V'$ to V' are always adjacent to a border
contig. For example, Figure 4(b) shows a fenced subgraph of the scaffold graph
in Figure 4(a).

   We now prove a lemma on the relationship between optimal scaffolds of G'
and G.

Lemma 4. *Given a scaffold graph $G = (V, E)$, let $G' = (V', E')$ be a fenced
subgraph of G. Suppose $\mathcal{S}' = \{S'_1, \ldots, S'_n\}$ forms the optimal scaffold set of G'
(disconnecting scaffolds connected by discordant edges). There exists an optimal
scaffold set $\mathcal{S}$ of G where every S'_i is a subpath of some scaffold of $\mathcal{S}$.*

Proof. Let $\mathcal{S}$ be an optimal scaffold set of G that does not contain $\mathcal{S}'$ as subpaths.
We construct a new scaffold set that does, by first removing all contigs in V'.
For each remaining partial scaffold whose end was adjacent to a border contig
b, we append that end to the corresponding scaffold S'_i (with b on its end and

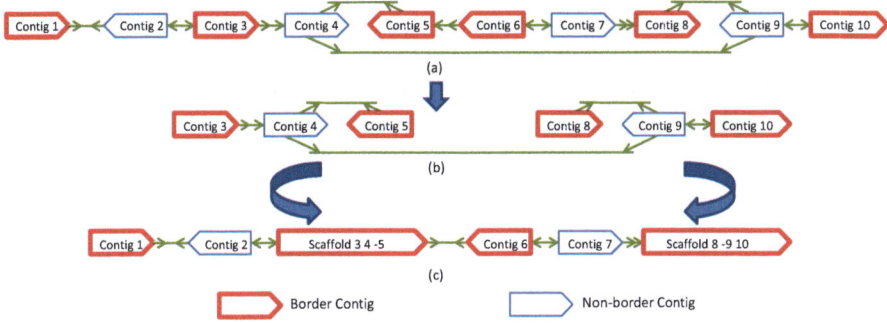

Fig. 4. Contracting the Scaffold Graph. (a) Original scaffold graph G. (b) A fenced subgraph of G (with optimal scaffolds "3 4 -5" and "8 -9 10"). (c) The new graph after contraction of optimal scaffolds for the subgraph in (b).

in the right orientation). This new scaffold is at least as optimal as $\mathcal{S}$. To see this, note that the number of concordant edges between nodes in $V - V'$ as well as those between nodes in V and $V - V'$ has remained the same. Also, since $\mathcal{S}'$ is optimal for G' the number of concordant edges in V' could only have gone up. □

Based on the lemma, we devise a recursive, graph contraction based algorithm to compute the optimal scaffold and this is outlined in Figure 5 and illustrated by an example in Figure 4.

2.6   Handling of Repeat Contigs

Repeat regions in the genome are often assembled (especially by short-read assemblers) as a single contig and in the scaffolding stage, information from paired-reads

---

ContractAndScaffold(G)
Require: A Scaffold Graph G
Ensure: An optimal scaffold $\mathcal{S}$ of G
 1: Identify a minimal fenced subgraph G' of G using a traversal from a border contig.
 2: Solve the scaffolding problem for G' to obtain a set of scaffolds $S'_1, \ldots, S'_n$ (see Section 2.4).
 3: From G, form a new scaffold graph G'' by contracting all contigs in S'_i to a node s_i, for each $i = 1, \ldots, n$.
 4: Call ContractAndScaffold(G') to obtain the scaffold $\mathcal{S}''$ of G''.
 5: From $\mathcal{S}''$ construct $\mathcal{S}$ by replacing every s_i by S'_i, for each $i = 1, \ldots, n$.

Fig. 5. A recursive graph contraction based algorithm to compute an optimal scaffold for a scaffold graph G

could help place such contigs in multiple scaffold locations. The optimization algorithm described here can be naturally extended to handle such cases but due to space constraints we do not explore this extension here and instead filter such contigs (based on read coverage greater than 1.5 times the genomic mean) before scaffolding with Opera.

2.7   Determination of Gap Sizes

After the order and the orientation of contigs in a scaffold have been computed, the constraints imposed by the paired-reads can also be used to determine the sizes of intervening gaps between contigs. This then serves as an important guide for *genome finishing* efforts as well as downstream analysis. Since scaffold edges can span multiple gaps and impose competing constraints on their sizes, we adopt a maximum likelihood approach to compute gap sizes:

$$\max_{G} p(E|G) = \max_{G} \Pi_{i \in E} \frac{1}{\sqrt{2\pi\sigma_i^2}} e^{-\frac{(s_i(G)-\mu_i)^2}{\sigma_i^2}} \tag{1}$$

where, E is the set of scaffold edges (whose sizes follow normal distribution with parameters μ_i, σ_i), G is the set of gap sizes, and $s_i(G)$ is the observed separation for scaffold edge i determined from the gap sizes. If c_i is the total length of contig sequences spanned by a scaffold edge and G_i is the set of gaps spanned, then we can reformulate this as the minimization of the following quadratic function:

$$\sum_{i \in E} \frac{((c_i + \sum_{j \in G_i} g_j) - \mu_i)^2}{\sigma_i^2} \tag{2}$$

where g_j are the gap sizes. The resulting quadratic program (with gap sizes bounded by τ) can be shown to have a positive definite Q matrix with a unique solution that can be found by the Goldfarb-Idnani active-set dual method in polynomial time [20]. This procedure thus efficiently computes gap sizes that optimize a clear likelihood function while taking all scaffold edges into account. As we show below, this also leads to improved estimates for gap sizes in practice.

3   Experimental Results

3.1   Datasets

To evaluate Opera, we compared it against existing programs (Velvet [5], Bambus [10] and SOPRA [3]) on a dataset for *B. pseudomallei* [23] as well as synthetic datasets for *E. coli*, *S. cerevisiae* and *D. melanogaster* (chromosome X). The synthetic datasets were generated using Metasim [21] (with default Illumina error models) and the reference genome in each case was downloaded from the NCBI website. Similar to the real dataset, for the synthetic sets, we simulated a high-coverage read library as well as a low-coverage paired-read library. Detailed information about the datasets can be found in Table 1.

Table 1. Test datasets and sequencing statistics. Note that for a library where insert size is $\mu\pm\sigma$, τ was set to $\mu+6\sigma$ and in all cases l_{min} was set to 500bp. For the synthetic datasets, 10% of the large-insert library reads were made chimeric by exchanging read-ends at random.

	E. coli	B. pseudomallei	S. cerevisiae	D. melanogaster
Size (Mbp)	4.6	7	12.1	22.4
Chromosomes	1	2	16	1
Reads (length, insert size, coverage)	80bp, 300±30bp 40X	100bp (454 reads) 20X	80bp, 300±30bp 40X	80bp, 300±30bp 40X
Paired-reads (length, insert size, coverage)	50bp, 10±1Kbp 2X	20bp, 10±1.5Kbp 2.8X	50bp, 10±1Kbp 2X	50bp, 10±1Kbp 2X

In all cases (except for *B. pseudomallei*), the reads were assembled and scaffolded using Velvet (with default parameters and $k = 31$). For Bambus and Opera, contigs assembled by Velvet were provided as input and scaffolded with the aid of the paired-read library. For SOPRA, we used the combined Velvet-SOPRA pipeline as described in [3]. In the case of *B. pseudomallei*, the 454 reads were assembled using Newbler (http://www.454.com) and scaffolded using Bambus and Opera (Velvet and SOPRA cannot directly take contigs as input).

3.2 Scaffold Contiguity

For each dataset and for each method, we assessed the contiguity of the reported set of scaffolds, by the N50 size (the length ℓ of the longest scaffold such that at least half of the genome is covered by scaffolds longer than ℓ) and the length of the longest scaffold. We also report the total number of scaffolds as well as the number of scaffolds with more than one contig (see Table 2). As can be seen from Table 2, Opera consistently produces the smallest number of scaffolds, the largest N50 sizes and the largest single scaffold.

Table 2. A comparison of scaffold contiguity for different methods

		E. coli	B. pseudomallei	S. cerevisiae	D. melanogaster
Scaffolds (non-singletons)	Velvet	241 (2)	-	1131 (26)	2148 (23)
	Bambus	200 (9)	183 (62)	1085 (39)	2062 (42)
	SOPRA	545 (90)	-	2171 (308)	4927 (149)
	Opera	3 (2)	3 (2)	31 (22)	36 (15)
N50 (Mbp)	Velvet	3.02	-	0.55	1.88
	Bambus	0.73	0.25	0.36	1.05
	SOPRA	0.05	-	0.04	0.03
	Opera	3.02	3.81	0.65	3.18
Max. Length (Mbp)	Velvet	3.02	-	0.96	4.31
	Bambus	1.35	0.47	0.72	2.33
	SOPRA	0.14	-	0.15	0.82
	Opera	3.02	3.81	1.04	7.69

Table 3. Comparison of scaffold correctness for different methods. Breakpoints were not assessed for *B. pseudomallei* due to the lack of a finished reference for the sequenced strain.

		E. coli	*B. pseudomallei*	*S. cerevisiae*	*D. melanogaster*
No. of breakpoints	Velvet	3	-	6	11
	Bambus	31	-	57	107
	SOPRA	1	-	67	10
	Opera	0	-	1	4
No. of discordant edges	Velvet	4	-	7	16
	Bambus	19	673	55	423
	Opera	1	19	3	4

3.3   Scaffold Correctness

To check the correctness of the reported scaffolds, we aligned the corresponding contigs to the reference genome using MUMmer [24]. Consecutive contigs in a scaffold that do not have the same order and orientation in the reference genome were then counted as *breakpoints* in the scaffold (see Table 3). In all datasets, Opera reports scaffolds with fewer breakpoints and therefore with greater agreement with the reference genome. Table 3 also reports the number of discordant edges seen in the scaffolds for the various methods (SOPRA is not compared as it uses a different set of contigs) and as expected Opera produces the best results under this criteria.

3.4   Running Time and Gaps

The current implementation of Opera is in JAVA (for ease of programming) and has not been optimized for runtime. However, despite this Opera had favorable runtimes on all datasets (see Table 4). This is likely due to the fact that it can effectively contract the scaffold graph while it searches for the optimal scaffold. We also compared the gap sizes estimated by the scaffolders and, in general, Velvet and Opera had the most consistent scaffolds and gap sizes. For gaps ($\geq$ 1Kbp) shared by their scaffolds, both Velvet and Opera produced accurate gap size estimates for *S. cerevisiae*, but Velvet had more gaps with relative error $> 10\%$ (13 compared to 7 for Opera). For *D. melanogaster*, both scaffolders had

Table 4. Runtime Comparison. Note that we do not report results for Velvet as it does not have a stand-alone scaffolding module

		E. coli	*B. pseudomallei*	*S. cerevisiae*	*D. melanogaster*
Time	Bambus	50s	16m	2m	3m
	SOPRA	49m	-	2h	5h
	Opera	4s	7m	11s	30s

many more gaps with relative error > 10%, but Opera was slightly better (31 versus 36 for Velvet).

4   Discussion

In this paper we explored a formal approach to the problem of scaffolding of a set of contigs using a paired-read library. As we describe in the methods, despite the computational complexity of the problem, we can devise a fixed-parameter tractable algorithm for scaffolding. Furthermore, by exploiting the structure of the scaffold graph (using a graph contraction procedure), this method can scaffold large graphs and long paired-read libraries (for example, the *B. pseudomallei* graph has more than 900 contigs).

Our experimental results, while limited, do suggest that Opera can more fully utilize the connectivity information provided by paired-reads. When compared with existing heuristic approaches, Opera simultaneously produces longer scaffolds and with fewer errors. This highlights the utility of minimizing the number of discordant edges in the scaffold graph and suggests that good approximation algorithms for this problem could achieve similar results with better scalability.

We plan to explore several promising extensions to Opera including the use of strobe-sequencing reads and information from overlaping contigs to improve the scaffolds. Another extension is to incorporate additional quality metrics (such as a lower-bound for the library size) to help differentiate between solutions that are equally good under the current optimality criterion. A C++ version of Opera (handling repeat contigs as well) will be publicly available soon.

Acknowledgments. This work was supported in part by the Biomedical Research Council / Science and Engineering Research Council of A*STAR (Agency for Science, Technology and Research) Singapore and by the MOEs AcRF Tier 2 funding R-252-000-444-112. S.G. is supported by a NUS graduate scholarship. We would also like to thank Mihai Pop for stimulating interest in this topic and Pramila Nuwantha Ariyaratne for providing advice on datasets.

References

1. Ng, P., Tan, J.J., Ooi, H.S., et al.: Multiplex sequencing of paired-end ditags (MS-PET): A strategy for the ultra-high-throughput analysis of transcriptomes and genomes. Nucleic Acids Research 34, e84 (2006)
2. Eid, J., Fehr, A., Gray, J., et al.: Real-time DNA sequencing from single polymerase molecules. Science 323(5910), 133–138 (2009)
3. Dayarian, A., Michael, T.P., Sengupta, A.M.: SOPRA: Scaffolding algorithm for paired reads via statistical optimization. BMC Bioinformatics 11(345) (2010)
4. Chaisson, M.J., Brinza, D., Pevzner, P.A.: De novo fragment assembly with short mate-paired reads: does the read length matter? Genome Research 19, 336–346 (2009)

 5. Zerbino, D.R., McEwen, G.K., Marguiles, E.H., Birney, E.: Pebble and rock band: heuristic resolution of repeats and scaffolding in the velvet short-read de novo assembler. PLoS ONE 4(12) (2009)
 6. Huson, D.H., Reinert, K., Myers, E.W.: The greedy path-merging algorithm for contig scaffolding. Journal of the ACM 49(5), 603–615 (2002)
 7. Myers, E.W., Sutton, G.G., Delcher, A.L., et al.: A whole-genome assembly of Drosophila. Science 287(5461), 2196–2204 (2000)
 8. Kent, W.J., Haussler, D.: Assembly of the working draft of the human genome with GigAssembler. Genome Research 11, 1541–1548 (2001)
 9. Pevzner, P.A., Tang, H.: Fragment assembly with double-barreled data. Bioinformatics 17(S1), 225–233 (2001)
10. Pop, M., Kosack, S.D., Salzberg, S.L.: Hierarchical scaffolding with bambus. Genome Research 14, 149–159 (2004)
11. Mullikin, J.C., Ning, Z.: The phusion assembler. Genome Research 13, 81–90 (2003)
12. Jaffe, D.B., Butler, J., Gnerre, S., et al.: Whole-genome sequence assembly for mammalian genomes: Arachne 2. Genome Research 13, 91–96 (2003)
13. Aparicio, S., Chapma, J., Stupka, E., et al.: Whole-genome shotgun assembly and analysis of the genome of Fugu rubripes. Science 297, 1301–1310 (2002)
14. Pop, M., Phillipy, A., Delcher, A.L., Salzberg, S.L.: Comparative genome assembly. Briefings in Bioinformatics 5(3), 237–248 (2004)
15. Richter, D.C., Schuster, S.C., Huson, D.H.: OSLay: optimal syntenic layout of unfinished assemblies. Bioinformatics 23(13), 1573–1579 (2007)
16. Husemann, P., Stoye, J.: Phylogenetic comparative assembly. Algorithms for Molecular Biology 5(3) (2010)
17. Nagarajan, N., Read, T.D., Pop, M.: Scaffolding and validation of bacterial genome assemblies using optical restriction maps. Bioinformatics 24(10), 1229–1235 (2008)
18. Pop, M.: Shotgun sequence assembly. Advances in Computers 60 (2004)
19. Saxe, J.: Dynamic programming algorithms for recognizing small-bandwidth graphs in polynomial time. SIAM J. on Algebraic and Discrete Methodd 1(4), 363–369 (1980)
20. Goldfarb, D., Idnani, A.: A numerically stable dual method for solving strictly convex quadratic programs. Mathematical Programming 27 (1983)
21. Richter, D.C., Ott, F., Schmid, R., Huson, D.H.: Metasim: a sequencing simulator for genomics and metagenomics. PloS One 3(10) (2008)
22. MacCallum, I., Przybylksi, D., Gnerre, S., et al.: ALLPATHS2: small genomes assembled accurately and with high continuity from short paired reads. Genome Biology 10, R103 (2009)
23. Nandi, T., Ong, C., Singh, A.P., et al.: A genomic survey of positive selection in Burkholderia pseudomallei provides insights into the evolution of accidental virulence. PLoS Pathogens 6(4) (2010)
24. Kurtz, S.A., Phillippy, A., Delcher, A.L., et al.: Versatile and open software for comparing large genomes. Genome Biology 5, R12 (2004)

Appendix: Proof of Theorem 2

Proof. Given a scaffold, it is easy to see that it can be checked in polynomial time and hence the problem is in NP.

We now show a reduction from the $(1, 2)$-traveling salesperson problem. Given a complete graph $H = (V, E)$ whose edges are of weight either 1 or 2, the $(1, 2)$-traveling salesperson problem asks if a weight k path exists that visits all vertices.

To construct a scaffold graph $G = (V', E')$, we set $V' = V$ and E' to a subset of E in which all edges with weight 2 are discarded (for every pair of such nodes $(u, v) \in E$, there are actually two bidirected edges in E' corresponding to the permutations uv and $-u - v$). Note that the graph G can be constructed from H in polynomial time and while the reduction is for the case $w = 0$ it extends in a straightforward fashion for other values (by inserting a path of w contig nodes between every pair of nodes from V).

We now show that H has a path of weigth $L + 2(|V| - 1 - L)$ if and only if G has a solution which omits $|E'| - L$ edges, where L is the number of weight-1 edges in a scaffold of G.

Suppose H has a path of length $L + 2(|V| - 1 - L)$, i.e., the path has L edges of weight 1 and $(|V| - 1 - L)$ edges of weight 2. Then, in G, we can construct a scaffold S which consists of these L edges of weight 1 (by choosing the appropriate bidirected edge). S is a valid scaffold which omits $|E'| - L$ edges in G.

Suppose G has a scaffold which omits $|E'| - L$ edges (for multiple independent scaffolds, consider them in any order). Since the weights of all edges in G are 1, all edges in the solution connect two adjacent nodes. As H is a clique, if there is no edge in a pair of adjacent nodes in the solution, there must be an edge of weight 2 in H. Then, a travelling-salesperson path of length $L + 2(|V| - 1 - L)$ can be constructed by selecting all such missing edges from H.   □

Increasing Power of Groupwise Association Test with Likelihood Ratio Test

Jae Hoon Sul, Buhm Han, and Eleazar Eskin*

Computer Science Department, University of California, Los Angeles, California 90095, USA
eeskin@cs.ucla.edu

Abstract. Sequencing studies have been discovering a numerous number of rare variants, allowing the identification of the effects of rare variants on disease susceptibility. As a method to increase the statistical power of studies on rare variants, several groupwise association tests that group rare variants in genes and detect associations between groups and diseases have been proposed. One major challenge in these methods is to determine which variants are causal in a group, and to overcome this challenge, previous methods used prior information that specifies how likely each variant is causal. Another source of information that can be used to determine causal variants is observation data because case individuals are likely to have more causal variants than control individuals. In this paper, we introduce a likelihood ratio test (LRT) that uses both data and prior information to infer which variants are causal and uses this finding to determine whether a group of variants is involved in a disease. We demonstrate through simulations that LRT achieves higher power than previous methods. We also evaluate our method on mutation screening data of the susceptibility gene for ataxia telangiectasia, and show that LRT can detect an association in real data. To increase the computational speed of our method, we show how we can decompose the computation of LRT, and propose an efficient permutation test. With this optimization, we can efficiently compute an LRT statistic and its significance at a genome-wide level. The software for our method is publicly available at http://genetics.cs.ucla.edu/rarevariants.

1   Introduction

Current genotyping technologies have enabled cost-effective genome-wide association studies (GWAS) on common variants. Although these studies have found numerous variants associated with complex diseases [1,2,3], common variants explain only a small fraction of disease heritability. This has led studies to explore effects of rare variants, and recent studies report that multiple rare variants affect several complex diseases [4,5,6,7,8,9,10,11,12,13,14]. However, the traditional statistical approach that tests each variant individually by comparing the frequency of the variant in individuals who have the disease (cases) with the frequency in individuals who do not have the disease (controls) yields low statistical power when applied to rare variants due to their low occurrences.

---

* Corresponding author.

V. Bafna and S.C. Sahinalp (Eds.): RECOMB 2011, LNBI 6577, pp. 452–467, 2011.

Identifying genes involved in diseases through multiple rare variants is an important challenge in genetics today. The main approach currently proposed is to group variants in genes and detect associations between a disease and these groups. The rationale behind this approach is that multiple rare variants may affect the function of a gene. By grouping variants, we may observe a larger difference in mutation counts between case and control individuals and hence, power of studies increases. Recently, several methods have been developed for the groupwise approach such as the Cohort Allelic Sums Test (CAST) [15], the Combined Multivariate and Collapsing (CMC) method [16], a weighted-sum statistic by Madsen and Browning (MB) [17], a variable-threshold approach (VT) [18], and Rare variant Weighted Aggregate Statistic (RWAS) [19].

In combining information from multiple rare variants, a groupwise association test faces two major challenges. The first is unknown effect sizes of variants on the disease phenotype. To address this challenge, MB and RWAS discuss a disease risk model in which rarer variants are assumed to have higher effect sizes than common variants [17,19]. This model provides a simulation framework that would be appropriate for testing the groupwise tests on rare variants because it describes associations usually not found in traditional GWAS. RWAS is shown to outperform other grouping methods under this disease risk model [19]. The second challenge is that only a subset of the rare variants in the gene will have an effect on the disease and which of these variants are causal is unknown. Including non-causal variants in a groupwise association test may reduce power because it may weaken effects of causal variants. RWAS and VT attempt to overcome this challenge by utilizing prior information of which variants are likely causal, and prior information can be obtained from bioinformatics tools such as Align-GVGD [20] , SIFT [21] and PolyPhen-2 [22]. By incorporating prior information into the methods, RWAS and VT reported that they achieved higher power [18,19].

These methods do not achieve the best performance even under the assumptions of their disease model (as we show below) and we improve on the previous methods by taking advantage of the following ideas. First, observational data can give us a clue to which variants are causal in data because casual variants occur more frequently in cases than in controls. Hence, a method that infers causal variants from data would outperform methods that do not, and previous methods fall into the latter category. In addition, previous methods such as RWAS, MB, and VT compute their statistics using a linear sum of mutation counts. In these methods, a large discrepancy in mutation counts between cases and controls has the same effect on a statistic as a sum of two small discrepancies with half the size of the large one. However, the large discrepancy should contribute more than the sum of small discrepancies because a variant that causes the large difference in mutation counts is more likely to be involved in a disease. To emphasize the large discrepancy, a nonlinear combination of mutation counts is necessary. Finally, the set of rare variants in the gene and their distribution among cases and controls can be used to estimate the effect sizes of the rare variants on the disease. This estimate can then be used to improve the statistical power of the method.

In this paper, we present a novel method for the groupwise association test based on a likelihood ratio test (LRT). LRT computes and compares likelihoods of two models; the null model that asserts no causal variants in a group and the alternative model that asserts at least one causal variant. To compute likelihoods of the models, LRT assumes that some variants are causal and some are not (called "causal statuses of variants") and computes the likelihood of the data under each possible causal status. This allows LRT to compute likelihoods of the null and alternative models, and a statistic of LRT is a ratio between likelihoods of the two models.

LRT takes advantage of both prior information and data to compute likelihoods of underlying models, and hence it uses more information than previous methods to identify a true model that generated data. Simulations show that LRT is more powerful than previous methods such as RWAS and VT using the same set of prior information. We also show by using real mutation screening data of the susceptibility gene for ataxia telangiectasia that LRT is able to detect an association previously reported by [23] and [19].

Unfortunately, to compute the LRT statistic directly, we must consider a number of possible models exponential in the number of rare variants in the gene. In addition, we must perform this computation once in each permutation and we must perform millions of permutations to guarantee that we control false positives when trying to obtain genome-wide significance. We address these computational challenges by decomposing the computation of LRT and developing an efficient permutation test. Unlike the standard approach to compute the LRT statistics which requires exponential time complexity, we make a few assumptions and derive a method for computing the LRT statistic whose time complexity is linear. For the permutation test, we further decompose the LRT statistic and take advantage of the distribution of allele frequency. These techniques allow us to compute a statistic of each permutation efficiently, and hence we can perform a large number of permutations to obtain genome-wide significance. We provide the software package for LRT at http://genetics.cs.ucla.edu/rarevariants.

2   Material and Methods

2.1   Computation of Likelihoods of Haplotypes

We consider likelihoods of two models under LRT; the likelihood of the null model (L_0) and the likelihood of the alternative model (L_1). The null model assumes that there is no variant causal to a disease while the alternative model assumes there is at least one causal variant. To compute the likelihood of each model, let D^+ and D^- denote a set of haplotypes in case and control individuals, respectively. We assume there are M variants in a group, and let V^i be the indicator variable for the "causal status" of variant i; $V^i = 1$ if variant i is causal, and $V^i = 0$ if not causal. Let $V = \{V^1, ..., V^M\}$ represent the causal statuses of M variants, and there exist 2^M possible values for V. Among them, let $v_j = \{v_j^1, ..., v_j^M\}$ be jth value, consisting of 0 and 1 that represent one specific scenario of causal statuses [19]. We use c_i to denote the probability of variant i being causal to a disease. Then, we can compute the prior probability of each scenario v_j as

$$P(v_j) = \prod_{i=1}^{M} c_i^{v_j^i} (1 - c_i)^{1-v_j^i} . \tag{1}$$

We define $L(D^+, D^-|v_j)$ as the likelihood of observing case and control haplotypes given jth scenario. Then, L_0 and L_1 can be defined as

$$L_0 = L(D^+, D^-|v_0)P(v_0) \tag{2}$$

$$L_1 = \sum_{j=1}^{2^M-1} L(D^+, D^-|v_j)P(v_j) \tag{3}$$

where v_0 is a scenario where $v_0^i = 0$ for all variants; no causal variants. In the Appendix we describe how we can compute $L(D^+, D^-|v_j)$.

The statistic of LRT is a ratio between L_1 and L_0, L_1/L_0, and we perform a permutation test to compute a p-value of the statistic.

2.2 Computation of Likelihoods of Variants

We decompose L_0 and L_1 in (Equations 2, 3) such that we compute likelihoods of variants instead of likelihoods of haplotypes to reduce the computational complexity. To compute L_1 in (Equation 3), we need to compute likelihoods of 2^M scenarios of causal statuses, which is computationally expensive if there are many rare variants in a group. To decompose likelihoods of haplotypes, we make two assumptions. The first assumption is low disease prevalence, and the second assumption is no linkage disequilibrium between rare variants [16,24,25].

Assume there are $N/2$ case and $N/2$ control individuals. Let $H_k = \{H_k^1, H_k^2, \ldots, H_k^M\}$ denote kth haplotype, where $H_i^k = \{0, 1\}$. $H_k^i = 1$ if ith variant in kth haplotype is mutated, and $H_k^i = 0$ if not. Let p_i denote population minor allele frequency (MAF) of variant i, and p_i^+ and p_i^- represent the true MAF of case and control individuals, respectively. We denote relative risk of variant i by γ_i. Then, L_0 and L_1 of (Equations 2, 3) can be decomposed into (see the Appendix for the derivation)

$$L_0 = \prod_{i=1}^{M} \left\{ (1 - c_i) \prod_{H_k \in D^+} p_i^{H_k^i}(1 - p_i)^{1-H_k^i} \prod_{H_k \in D^-} p_i^{H_k^i}(1 - p_i)^{1-H_k^i} \right\} \tag{4}$$

$$L_0 + L_1 = \prod_{i=1}^{M} \left\{ (1 - c_i) \prod_{H_k \in D^+} p_i^{H_k^i}(1 - p_i)^{1-H_k^i} \prod_{H_k \in D^-} p_i^{H_k^i}(1 - p_i)^{1-H_k^i} \right.$$

$$\left. + c_i \prod_{H_k \in D^+} {p_i^+}^{H_k^i}(1 - p_i^+)^{1-H_k^i} \prod_{H_k \in D^-} p_i^{H_k^i}(1 - p_i)^{1-H_k^i} \right\} \tag{5}$$

where p_i^+ and p_i^- are

$$p_i^+ = \frac{\gamma_i p_i}{(\gamma_i - 1)p_i + 1} \tag{6}$$

$$p_i^- = p_i \quad \text{(assuming the disease prevalence is very small)} \tag{7}$$

We estimate the population MAF of a variant (p_i) using an observed overall sample frequency.

$$p_i = \frac{\hat{p}_i^+ + \hat{p}_i^-}{2}$$

where $\hat{p}_i^+$ and $\hat{p}_i^-$ represent observed case and control MAF, respectively.

This decomposition reduces the time complexity of computing L_1 from exponential to linear, substantially increasing the computational efficiency.

2.3   Efficient Permutation Test for LRT

We propose a permutation test that is substantially more efficient than a naive permutation test that permutes case and control statuses in each permutation. The naive permutation test is computationally expensive because every haplotype of case and control individuals needs to be examined in each permutation, and hence it requires more computation as the number of individuals increases. Moreover, to compute a p-value at a genome-wide level, more than 10 million permutations are necessary assuming a significance threshold of 2.5×10^{-6} (computed from the overall false positive rate of 0.05 and the Bonferroni correction with 20,000 genes genome-wide). It is often computationally impractical to perform this large number of permutations with the naive permutation test. Hence, we develop a permutation test that does not permute case and control statuses, and this makes the time complexity independent of the number of individuals and allows the permutation test to be capable of performing more than 10 million permutations.

First, we reformulate L_0 and L_1 (Equations 4, 5) such that they are composed of terms that do not change and terms that change per each permutation (see the Appendix for the derivation).

$$L_0 = \prod_{i=1}^{M} X_i \tag{8}$$

$$L_0 + L_1 = \prod_{i=1}^{M} \left\{ X_i + K_i Y_i^{N\hat{p}_i^+} \right\} \tag{9}$$

where

$$X_i = (1 - c_i) p_i^{2Np_i} (1 - p_i)^{2N - 2Np_i}$$

$$K_i = c_i (1 - p_i^+)^N (1 - p_i)^N \left(\frac{p_i}{1 - p_i} \right)^{2Np_i}$$

$$Y_i = \left(\frac{p_i^+}{1 - p_i^+} \cdot \frac{1 - p_i}{p_i} \right)$$

In (Equations 8 and 9), it is only a $\hat{p}_i^+$ term that changes when the dataset is permuted because p_i and p_i^+ are invariant per permutation, meaning X_i, K_i, and Y_i are constant. $N\hat{p}_i^+$ follows the hypergeometric distribution with the mean equal to Np_i and the variance equal to $\frac{N}{2} p_i (1 - p_i)$ under permutations. Hence, we sample $N\hat{p}_i^+$ from the hypergeometric distribution, and since this sampling strategy does not permute and examine haplotypes of individuals, it is more efficient than the naive permutation test when studies have a large number of individuals.

To speed up sampling from the hypergeometric distribution, we pre-compute hypergeometric distributions of all rare variants (e.g. variants whose MAF are less than 10%) before performing the permutation test. Computing the hypergeometric distribution requires several factorial operations, which is computationally expensive. The pre-computation of distributions allows the permutation test to avoid having the expensive operations repeatedly per permutation, and the number of pre-computed distributions is limited due to the small range of MAF. For common variants, we sample $N\hat{p}_i^+$ from the normal distribution, which approximately follows the hypergeometric distribution when $\hat{p}_i^+$ is not close to 0 or 1. Sampling from the normal distribution is substantially more efficient than sampling from the hypergeometric distribution, and hence there is no need to pre-compute the normal distribution.

We find that our permutation test is efficient enough to calculate a p-value of the LRT statistic at a genome-wide level. For example, using a dataset that contains 1000 cases and 1000 controls with 100 variants, 10 million permutations take about 10 CPU minutes using one core of a Quad-Core AMD 2.3 GHz Opteron Processor. Note that the time complexity of our method is $O(N + kMP)$ where N is the total number of individuals, M is the number of variants, P is the number of permutations, and k is the number of iterations in the local search algorithm discussed below (see "Estimating PAR of a group of variants using LRT" section for more details). We find that k is very small in permutations and $MP \gg N$ for a large number of permutations (e.g. 100 millions). Thus, the time complexity of our method becomes approximately $O(MP)$, and this shows that the amount of computation our method needs mostly depends on the number of variants and the number of permutations.

We note that our permutation test can also be applied to previous grouping methods such as RWAS [19]. RWAS assumes that its statistic (a weighted sum of z-scores of variants) approximately follows the normal distribution, and the p-value is obtained accordingly. Since the permutation test does not make any assumptions on the distribution of a statistic, it may provide a more accurate estimate of a p-value and improve the power of previous methods.

2.4   Power Simulation Framework

The effect sizes and the causal statuses of variants are two major factors that influence the power of the groupwise association test. To simulate these two factors, we adopt the same simulation framework as one discussed in Sul et al. and Madsen and Browning [17,19]. In this framework, population attributable risk (PAR) defines the effect sizes of variants, and we assign the predefined group PAR to a group of variants. The group PAR divided by the number of causal variants is the marginal PAR, denoted as ω, and every variant has the same ω.

The effect size of a variant also depends on its population MAF in this simulation framework. We assign each variant population MAF (p_i) sampled from Wright's formula [26,27], and we use the same set of parameter values for the formula as discussed in Sul et al. and Madsen and Browning (see [17,19] for details). Using ω and population MAF, we can compute relative risk of variant i (γ_i) as following.

$$\gamma_i = \frac{\omega}{(1 - \omega)p_i} + 1 \tag{10}$$

(Equation 10) shows that a rarer variant has the higher effect size. Given relative risk and population MAF of a variant, we compute the true case and control MAF of the variant according to (Equations 6 and 7). We then use the true case and control MAF to sample mutations in case and control individuals, respectively.

To simulate the causal status of a variant, we assign each variant the probability of being causal to a disease. Let c_i denote this probability for variant i, and in each dataset, a variant is causal with the probability c_i, and not causal with the probability $1 - c_i$. Relative risk of a causal variant is defined in (Equation 10) while that of non-causal variant is 1.

Given all parameters of variants, we generate 1,000 datasets, and each dataset has 1,000 case and 1,000 control individuals with 100 variants. Since we are interested in comparing power of the groupwise tests, we only include datasets that have at least two causal variants. The number of significant datasets among the 1,000 datasets is used as an estimate of power with the significance threshold of 2.5×10^{-6}.

2.5   Estimating PAR of a Group of Variants Using LRT

We need a few model parameters to compute the LRT statistic, and we use data, prior information, and the LRT statistic itself to estimate the parameters. More specifically, we need to know relative risk of variant i, γ_i, to compute p_i^+ in (Equation 6). According to (Equation 10), γ_i depends on population MAF (p_i) and the marginal PAR (ω) which is the group PAR divided by the number of causal variants. We can estimate p_i from observational data, and we use prior information (c_i) of variants to compute the expected number of causal variants, which we use as an estimate of the number of causal variants.

To estimate the group PAR, we use the LRT statistic because we are likely to observe the greatest statistic when LRT is given the group PAR that generated observational data. We apply a local search algorithm to find the value of PAR that maximizes the LRT statistic; we compute the statistic assuming a very small PAR value (0.1%), and iteratively compute statistics using incremental values of PAR (0.2%, 0.3%, etc.) until we observe a decrease in the LRT statistic. After we find the maximum LRT statistic, we perform the permutation test with the same local search algorithm to find the significance of the statistic.

3   Results

3.1   Type I Error Rate of LRT

We examine the type I error rate of LRT by applying it to "null" datasets that contain no causal variants. We measure the type I error rates under three significance thresholds; $0.05, 0.01$, and 2.5×10^{-6} (the significance threshold for the power simulation). A large number of null datasets are necessary to accurately estimate the type I error rate under the lowest significance threshold (2.5×10^{-6}). Thus, we create 10 million datasets, and each dataset contains 1000 case and 1000 control individuals with 100 variants. We estimate the type I error rate as the proportion of significant datasets among the 10 million datasets.

To efficiently measure the type I error rates of LRT, we use the following approach. We first test LRT on all 10 million datasets with 100,000 permutations. This small number of permutations makes it possible to test LRT on all null datasets and allows us to estimate the type I error rates under the 0.05 and 0.01 significance thresholds. As for the lowest significance threshold, we need to test LRT with a very large number of permutations (e.g. 100 million) to obtain a genome-wide level p-value. To reduce the amount of computation, we exclude datasets whose p-values cannot be lower than 2.5×10^{-6} with 100 million permutations. More specifically, to obtain a p-value less than 2.5×10^{-6}, the number of significant permutations (permutations whose LRT statistics are greater than the observed LRT statistic) must be less than 250 with 100 million permutations. We exclude datasets having more than 250 significant permutations after the 100,000 permutations. We then apply the adaptive permutation test on the remaining datasets; we stop the permutation test when the number of significant permutations is greater than 250. The proportion of datasets whose permutation tests do not stop until 100 million permutations is the type error rate under the 2.5×10^{-6} threshold.

We find that the type I error rates of LRT are 0.0500946, 0.0100042, and 2.6×10^{-6} for the significance thresholds of 0.05, 0.01, and 2.5×10^{-6}, respectively. This shows that the type I error rates are well controlled for LRT under the three different thresholds.

3.2   Power Comparison between LRT and Previous Grouping Methods

We compare power between LRT and previous methods using two simulations. We design these simulations to observe how LRT's implicit inference of which variants are causal affects the power compared to methods which do not make this kind of inference. In the first simulation, we generate datasets in which variants have true $c_i = 0.1$. This means that only a subset of variants is causal, and causal statuses of variants vary per datasets. In the second simulation, all 100 variants in datasets are causal; true c_i of all variants is 1.

We test four different methods in this experiment; LRT, Optimal Weighted Aggregate Statistic (OWAS), MB, and VT. OWAS computes a difference in mutation counts between case and control individuals for each variant, or z-score of a variant, and assigns weights to z-scores according to the non-centrality parameters of z-scores [19]. Sul et al. reported that OWAS achieves slightly higher power than RWAS [19]. Thus, we test OWAS instead of their proposed method, RWAS, to compare power between a weighted sum of z-scores approach and the LRT approach. Since OWAS needs to know the effect sizes of variants, we give OWAS the true group PAR that generated data. OWAS divides the true group PAR by the expected number of causal variants to compute the marginal PAR (ω) and then compute relative risk of variants (Equation 10). We also apply our permutation test for LRT (see Material and Methods) to OWAS to estimate its p-value more accurately. To test VT, we use an R package available online [18]. LRT, OWAS, and VT are given prior information that is equivalent to true c_i of datasets, and we perform 10 million permutations to estimate p-values of their statistics.

Results of the two simulations show that LRT outperforms the previous groupwise tests in the first simulation, and it has almost the same power as OWAS in the second simulation. In the first simulation, LRT has higher power than other tests at all group PAR values (Figure 1A); at the group PAR of 5%, LRT achieves 94.5% power while OWAS and VT achieve 53.7% and 83.6% power, respectively. This shows that data may provide useful information about causal statuses of variants, and a method that takes advantage of data achieves higher power than those that do not. When prior information, however, can alone identify which variants are causal as in the second simulation, LRT and OWAS have almost the identical power (Figure 1B). This is because both methods know which variants are causal from prior information. Hence, this experiment demonstrates that LRT is generally a more powerful approach than the weighted sum of z-scores approach because it achieves higher power in studies where prior information cannot specify which variants are causal.

3.3   LRT on Real Mutation Screening Data of ATM

We apply LRT to real mutation screening data of the susceptibility gene for ataxia telangiectasia [23]. This gene, called *ATM*, is also an intermediate-risk susceptibility gene for breast cancer. Tavtigian *et al.* conducted mutation screening studies and collected data from 987 breast cancer cases and 1021 controls. Tavtigian *et al.* increased the number of cases and controls to 2531 and 2245, respectively, by collecting data from seven published *ATM* case-control mutation screening studies. This dataset is called "bona fide case-control studies," and 170 rare missense variants are present in this dataset. Sul *et al.* also analyzed the dataset with RWAS [19].

To obtain prior information of variants in the dataset, Tavtigian *el al.* used two missense analysis programs, Align-GVGD [20] and SIFT [21]. A difference between the two programs is that while SIFT classifies a variant as either deleterious (SIFT scores ≤ 0.05) or neutral (SIFT scores > 0.05), Align-GVGD classifies a variant into seven grades from C0 (most likely neutral) to C65 (most likely deleterious). To convert the seven grades of Align-GVGD into c_i values, we arbitrarily assign c_i values from 0.05 to 0.95 in increments of 0.15 to the seven grades. As for converting SIFT scores into c_i values, we assign c_i value of 1 to variants whose SIFT scores are ≤ 0.05 and c_i of 0 to other variants. This is the same conversion used in [19].

When LRT uses prior information from Align-GVGD, it yields a p-value of 0.0058, which indicates a significant association between a group of rare variants and the

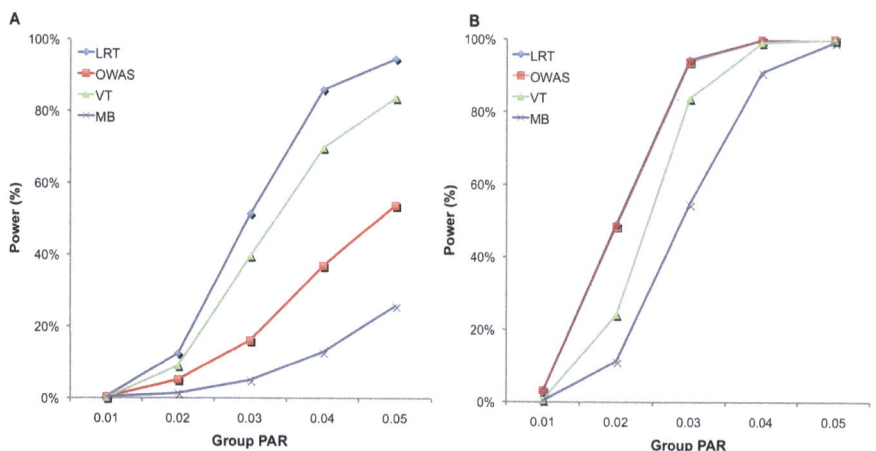

Fig. 1. Power comparison among four different groupwise association tests on datasets where $c_i = 0.1$ for all variants (A) and $c_i = 1$ (B) over different group PAR values

disease. This result is consistent with the findings of [23] and [19]; Tavtigian *et al.* and Sul *et al.* both obtained significant p-values when they used outputs of Align-GVGD as prior information. The result shows that we can apply LRT to real data to discover an association.

LRT yields a non-significant p-value of 0.39341 when it does not use prior information, and this is also consistent with results of [23] and [19]; Tavtigian *et al.* and Sul *et al.* reported non-significant p-values when they analyzed the data without prior information. When SIFT scores are used as prior information, LRT similarly reports a non-significant p-value of 0.08384, and Sul *et al.* also obtained a non-significant p-value [19]. However, the analysis of Tavtigian *et al.* with SIFT scores showed a significant association [23]. According to [19], the reason for this difference may be that LRT and RWAS need to know the relative degree of how deleterious a variant is to better detect an association. However, it may be difficult to know this relative deleteriousness of variants with SIFT scores because variants are either deleterious or neutral. Thus, this experiment shows that more informative prior information such as the seven grades of Align-GVGD may yield better results with LRT.

4 Discussion

We developed a likelihood ratio test (LRT) to increase power of association studies on a group of rare variants. The power of statistical methods that group rare variants depends on which rare variants to group or to exclude from the group because including non-causal variants in the group decreases power [19]. Although prior information provides useful information of how likely each variant is casual to a disease, determining whether a variant is causal or not in data only from prior information is often infeasible. LRT takes advantage of data to identify causal variants, and when it is not possible to identify causal variants from prior information, we showed that LRT outperforms previous methods.

To evaluate LRT on real data, we used mutation screening data of the *ATM* gene [23]. Tavtigian *et al.* and Sul *et al.* both found the significant association in the data [23,19], and we showed that LRT also detected the association using the output of Align-GVGD as prior information of variants. This shows that LRT can be applied to detect an association in real association studies.

One of the two assumptions that we made to efficiently compute the LRT statistic and its p-value is the independence between variants. Several studies suggest that there would be very low linkage disequilibrium between rare variants due to their low occurrences [16,24,25]. However, if non-negligible LD is expected between variants, especially when common variants are in linkage disequilibrium in the group, we can change our permutation test as follows to take into account LD and to correctly control the false positive rate. Instead of separately sampling $N\hat{p}_i^+$ of each common variant from the normal distribution, we sample $N\hat{p}_i^+$ of all common variants from the multivariate normal distribution (MVN). This approach is similar to the approach of Han *et al* who used the MVN framework to correct for multiple testing on correlated markers [28]. The covariance matrix of the MVN we create consists of correlations (r) between common variants, and hence $N\hat{p}_i^+$ sampled from this MVN takes into account LD between variants. For rare variants, we use our proposed method that samples $N\hat{p}_i^+$ of each rare variant from the hypergeometric distribution because LD between rare variants is expected to be very low.

The other assumption of our method is the low disease prevalence, and this assumption does not influence the false positive rate of our method while it may affect the power. The false positive rate of LRT is controlled even though the disease we consider is highly prevalent because we perform the permutation test. Therefore, LRT can still be applied to association studies involving diseases with high prevalence while its power may not be as high as the power it achieves on diseases with low prevalence.

The software package for computing the LRT statistic and performing the proposed permutation test is publicly available online at http://genetics.cs.ucla.edu/rarevariants.

References

1. Corder, E.H., Saunders, A.M., Strittmatter, W.J., Schmechel, D.E., Gaskell, P.C., Small, G.W., Roses, A.D., Haines, J.L., Pericak-Vance, M.A.: Gene dose of apolipoprotein e type 4 allele and the risk of alzheimer's disease in late onset families. Science 261(5123), 921–923 (1993)
2. Bertina, R.M., Koeleman, B.P.C., Koster, T., Rosendaal, F.R., Dirven, R.J., de Ronde, H., Van Der Velden, P.A., Reitsma, P.H., et al.: Mutation in blood coagulation factor v associated with resistance to activated protein c. Nature 369(6475), 64–67 (1994)
3. Altshuler, D., Hirschhorn, J.N., Klannemark, M., Lindgren, C.M., Vohl, M.C., Nemesh, J., Lane, C.R., Schaffner, S.F., Bolk, S., Brewer, C., et al.: The common pparγ pro12ala polymorphism is associated with decreased risk of type 2 diabetes. Nature Genetics 26(1), 76–80 (2000)
4. Gorlov, I.P., Gorlova, O.Y., Sunyaev, S.R., Spitz, M.R., Amos, C.I.: Shifting paradigm of association studies: value of rare single-nucleotide polymorphisms. Am. J. Hum. Genet. 82(1), 100–112 (2008)
5. Kryukov, G.V., Pennacchio, L.A., Sunyaev, S.R.: Most rare missense alleles are deleterious in humans: implications for complex disease and association studies. Am. J. Hum. Genet. 80(4), 727–739 (2007)
6. Cohen, J.C., Kiss, R.S., Pertsemlidis, A., Marcel, Y.L., McPherson, R., Hobbs, H.H.: Multiple rare alleles contribute to low plasma levels of hdl cholesterol. Science 305(5685), 869–872 (2004)
7. Fearnhead, N.S., Wilding, J.L., Winney, B., Tonks, S., Bartlett, S., Bicknell, D.C., Tomlinson, I.P.M., Mortensen, N.J.M., Bodmer, W.F.: Multiple rare variants in different genes account for multifactorial inherited susceptibility to colorectal adenomas. Proc. Natl. Acad. Sci. USA 101(45), 15992–15997 (2004)
8. Ji, W., Foo, J.N., O'Roak, B.J., Zhao, H., Larson, M.G., Simon, D.B., Newton-Cheh, C., State, M.W., Levy, D., Lifton, R.P.: Rare independent mutations in renal salt handling genes contribute to blood pressure variation. Nat. Genet. 40(5), 592–599 (2008)
9. Bodmer, W., Bonilla, C.: Common and rare variants in multifactorial susceptibility to common diseases. Nature Genetics 40(6), 695–701 (2008)
10. Romeo, S., Pennacchio, L.A., Fu, Y., Boerwinkle, E., Tybjaerg-Hansen, A., Hobbs, H.H., Cohen, J.C.: Population-based resequencing of angptl4 uncovers variations that reduce triglycerides and increase hdl. Nat. Genet. 39(4), 513–516 (2007)
11. Blauw, H.M., Veldink, J.H., van Es, M.A., van Vught, P.W., Saris, C.G.J., van der Zwaag, B., Franke, L., Burbach, J.P.H., Wokke, J.H., Ophoff, R.A., van den Berg, L.H.: Copy-number variation in sporadic amyotrophic lateral sclerosis: a genome-wide screen. Lancet Neurol. 7(4), 319–326 (2008)

12. Consortium, I.S.: Rare chromosomal deletions and duplications increase risk of schizophrenia. Nature 455(7210), 237–241 (2008)
13. Xu, B., Roos, J.L., Levy, S., Van Rensburg, E.J., Gogos, J.A., Karayiorgou, M.: Strong association of de novo copy number mutations with sporadic schizophrenia. Nature Genetics 40(7), 880–885 (2008)
14. Walsh, T., McClellan, J.M., McCarthy, S.E., Addington, A.M., Pierce, S.B., Cooper, G.M., Nord, A.S., Kusenda, M., Malhotra, D., Bhandari, A., Stray, S.M., Rippey, C.F., Roccanova, P., Makarov, V., Lakshmi, B., Findling, R.L., Sikich, L., Stromberg, T., Merriman, B., Gogtay, N., Butler, P., Eckstrand, K., Noory, L., Gochman, P., Long, R., Chen, Z., Davis, S., Baker, C., Eichler, E.E., Meltzer, P.S., Nelson, S.F., Singleton, A.B., Lee, M.K., Rapoport, J.L., King, M.C.C., Sebat, J.: Rare structural variants disrupt multiple genes in neurodevelopmental pathways in schizophrenia. Science 320(5875), 539–543 (2008)
15. Morgenthaler, S., Thilly, W.G.: A strategy to discover genes that carry multi-allelic or mono-allelic risk for common diseases: a cohort allelic sums test (cast). Mutat. Res. 615(1-2), 28–56 (2007)
16. Li, B., Leal, S.M.: Methods for detecting associations with rare variants for common diseases: application to analysis of sequence data. Am. J. Hum. Genet. 83(3), 311–321 (2008)
17. Madsen, B.E., Browning, S.R.: A groupwise association test for rare mutations using a weighted sum statistic. PLoS Genet. 5(2), e1000384 (2009)
18. Price, A.L., Kryukov, G.V., de Bakker, P.I.W., Purcell, S.M., Staples, J., Wei, L.J.J., Sunyaev, S.R.: Pooled association tests for rare variants in exon-resequencing studies. Am. J. Hum. Genet. 86(6), 832–838 (2010)
19. Sul, J.H., Han, B., He, D., Eskin, E.: An optimal weighted aggregated association test for identification of rare variants involved in common diseases. Genetics (in press)
20. Tavtigian, S.V., Deffenbaugh, A.M., Yin, L., Judkins, T., Scholl, T., Samollow, P.B., de Silva, D., Zharkikh, A., Thomas, A.: Comprehensive statistical study of 452 brca1 missense substitutions with classification of eight recurrent substitutions as neutral. J. Med. Genet. 43(4), 295–305 (2006)
21. Ng, P.C., Henikoff, S.: Sift: Predicting amino acid changes that affect protein function. Nucleic Acids Res. 31(13), 3812–3814 (2003)
22. Adzhubei, I.A., Schmidt, S., Peshkin, L., Ramensky, V.E., Gerasimova, A., Bork, P., Kondrashov, A.S., Sunyaev, S.R.: A method and server for predicting damaging missense mutations. Nature Methods 7(4), 248–249 (2010)
23. Tavtigian, S.V., Oefner, P.J., Babikyan, D., Hartmann, A., Healey, S., Le Calvez-Kelm, F., Lesueur, F., Byrnes, G.B., Chuang, S.C.C., Forey, N., Feuchtinger, C., Gioia, L., Hall, J., Hashibe, M., Herte, B., McKay-Chopin, S., Thomas, A., Vallée, M.P., Voegele, C., Webb, P.M., Whiteman, D.C., Study, A.C., (BCFR), B.C.F.R., for Research into Familial Aspects of Breast Cancer (kConFab), K.C.F.C., Sangrajrang, S., Hopper, J.L., Southey, M.C., Andrulis, I.L., John, E.M., Chenevix-Trench, G.: Rare, evolutionarily unlikely missense substitutions in atm confer increased risk of breast cancer. Am. J. Hum. Genet. 85(4), 427–446 (2009)
24. Pritchard, J.K., Cox, N.J.: The allelic architecture of human disease genes: common disease-common variant..or not? Hum. Mol. Genet. 11(20), 2417–2423 (2002)
25. Pritchard, J.K.: Are rare variants responsible for susceptibility to complex diseases? The American Journal of Human Genetics 69(1), 124–137 (2001)
26. Wright, S.: Evolution in mendelian populations. Genetics 16(2), 97–159 (1931)
27. Ewens, W.J.: Mathematical population genetics, 2nd edn. Springer, Heidelberg (2004)
28. Han, B., Kang, H.M., Eskin, E.: Rapid and accurate multiple testing correction and power estimation for millions of correlated markers. PLoS Genet. 5(4) (2009)

Appendix

Computation of $L(D^+, D^-|v_j)$ in LRT

We show how the likelihood of haplotypes under certain causal statuses of variants, $L(D^+, D^-|v_j)$, can be computed. Let H_k denote kth haplotype, and $H_k = \{H_k^1, H_k^2, \ldots, H_k^M\}$. $H_k^i = 1$ if ith variant in kth haplotype is mutated, and $H_k^i = 0$ otherwise. Let p_i denote population minor allele frequency (MAF) of ith variant, and we can compute the probability of a haplotype H_k as $P(H_k) = \prod_{i=1}^M p_i^{H_k^i}(1-p_i)^{1-H_k^i}$.

Then, we define the likelihood of haplotypes as

$$L(D^+, D^-|v_j) = \prod_{H_k \in D^+} P(H_k|+, v_j) \prod_{H_k \in D^-} P(H_k|-, v_j) \qquad (11)$$

where $+$ and $-$ denote case and control statuses. In order to compute $P(H_k|+/-, v_j)$, we first denote F as disease prevalence and $\gamma_{v_j}^{H_k}$ as the relative risk of kth haplotype under v_j. We define $\gamma_{v_j}^{H_k}$ as $\gamma_{v_j}^{H_k} = \prod_{i=1}^M \gamma_i^{v_j^i H_k^i}$. Let H_0 denote the haplotype with no variants, and using Bayes' theorem and independence between H_k and v_j, and between disease status ($+$ and $-$) and v_j, we can define the $P(H_k|+/-, v_j)$ as

$$P(H_k|+, v_j) = \frac{P(H_k, +|v_j)}{P(+)} = \frac{P(+|H_k, v_j)P(H_k)}{F} = \frac{\gamma_{v_j}^{H_k} P(+|H_0, v_j)P(H_k)}{F} \qquad (12)$$

$$P(H_k|-, v_j) = \frac{P(H_k, -|v_j)}{P(-)} = \frac{(1 - P(+|H_k, v_j))P(H_k)}{1 - F} \qquad (13)$$

$P(+|H_0, v_j)$, or the probability of having a disease given no variants in the haplotype under jth causal statuses, can be computed as

$$\sum_{k=0}^{2^M-1} \gamma_{v_j}^{H_k} P(+|H_0, v_j)P(H_k) = F$$

$$P(+|H_0, v_j) = \frac{F}{\sum_{k=0}^{2^M-1} \gamma_{v_j}^{H_k} P(H_k)}$$

Decomposition of Likelihoods of Haplotypes into Likelihoods of Variants in LRT

First, we consider two variants case. We have 4 possible causal statuses, denoted as $v_{00}, v_{01}, v_{10}, v_{11}$ and 4 possible haplotypes, denoted as $H_{00}, H_{01}, H_{10}, H_{11}$. Let p_1 and p_2 denote population MAF of two variants and p_1^+ and p_2^+ are MAF of case individuals at two variants. The original LRT statistic based on (2) and (3) compute the following likelihoods

$$L_0 = (1-c_1)(1-c_2) \prod_{H_k \in D^+} P(H_k|+, v_{00}) \prod_{H_k \in D^-} P(H_k|-, v_{00})$$

$$L_0 + L_1 = (1-c_1)(1-c_2) \prod_{H_k \in D^+} P(H_k|+, v_{00}) \prod_{H_k \in D^-} P(H_k|-, v_{00})$$

$$+ (1-c_1)c_2 \prod_{H_k \in D^+} P(H_k|+, v_{01}) \prod_{H_k \in D^-} P(H_k|-, v_{01})$$

$$+ c_1(1-c_2) \prod_{H_k \in D^+} P(H_k|+, v_{10}) \prod_{H_k \in D^-} P(H_k|-, v_{10})$$

$$+ c_1 c_2 \prod_{H_k \in D^+} P(H_k|+, v_{11}) \prod_{H_k \in D^-} P(H_k|-, v_{11}) \tag{14}$$

Our first assumption for decomposition is that F or disease prevalence is very small. Then, we can decompose $P(H_k|-, v_j)$ for all causal statuses j, as

$$P(H_k|-, v_j) = p_1^{H_k^1}(1-p_1)^{1-H_k^1} \times p_2^{H_k^2}(1-p_2)^{1-H_k^2} = P(H_k|+, v_{00}) \tag{15}$$

Then, we decompose $P(H_k|+, v_j)$ for different v_j, and first, let's consider v_{11} where two variants are both causal. We make another assumption here, which is the independence between rare variants; there is no linkage disequilibrium (LD) [16,24,25]. If variants are independent, $P(H_{00}|+, v_{11})$ can be formulated as

$$\begin{aligned} P(H_{00}|+, v_{11}) &= \frac{P(H_{00})}{P(H_{00}) + P(H_{10})\gamma_1 + P(H_{01})\gamma_2 + P(H_{11})\gamma_1\gamma_2} \\ &= \frac{(1-p_1)(1-p_2)}{(1-p_1)(1-p_2) + p_1(1-p_2)\gamma_1 + (1-p_1)p_2\gamma_2 + p_1 p_2 \gamma_1 \gamma_2} \\ &= \frac{(1-p_1) \times (1-p_2)}{((1-p_1) + p_1\gamma 1) \times ((1-p_2) + p_2\gamma 2)} \\ &= (1-p_1^+)(1-p_2^+) \end{aligned}$$

The last derivation comes from (6) where $p_i^+ = \frac{p_i \gamma_i}{(1-p_i)+p_i\gamma_i}$. Similarly, we can define the probabilities of other haplotypes (H_{01}, H_{10}, H_{11}) as

$$P(H_{01}|+, v_{11}) = (1-p_1^+)p_2^+$$
$$P(H_{10}|+, v_{11}) = p_1^+(1-p_2^+)$$
$$P(H_{11}|+, v_{11}) = p_1^+ p_2^+$$

Combining these probabilities, we have the following decomposition of $P(H_k|+, v_{11})$.

$$P(H_k|+, v_{11}) = p_1^{+H_k^1}(1-p_1^+)^{1-H_k^1} \times p_2^{+H_k^2}(1-p_2^+)^{1-H_k^2} \tag{16}$$

Using the similar derivation, decomposition of $P(H_k|+, v_{01})$ and $P(H_k|+, v_{10})$ is

$$P(H_k|+, v_{01}) = p_1^{H_k^1}(1-p_1)^{1-H_k^1} \times p_2^{+H_k^2}(1-p_2^+)^{1-H_k^2} \tag{17}$$

$$P(H_k|+, v_{10}) = p_1^{+\,H_k^1}(1 - p_1^+)^{1-H_k^1} \times p_2^{\,H_k^2}(1 - p_2)^{1-H_k^2} \tag{18}$$

By the 4 decompositions (15,16,17,18), we can finally decompose the likelihoods of haplotypes (14) as

$$
\begin{aligned}
L_0 = \quad & (1 - c_1)(1 - c_2) \prod_{H_k \in D^+} p_1^{\,H_k^1}(1 - p_1)^{1-H_k^1} p_2^{\,H_k^2}(1 - p_2)^{1-H_k^2} \\
& * \prod_{H_k \in D^-} p_1^{\,H_k^1}(1 - p_1)^{1-H_k^1} p_2^{\,H_k^2}(1 - p_2)^{1-H_k^2} \\[2mm]
L_0 + L_1 = \quad & (1 - c_1)(1 - c_2) \prod_{H_k \in D^+} p_1^{\,H_k^1}(1 - p_1)^{1-H_k^1} p_2^{\,H_k^2}(1 - p_2)^{1-H_k^2} \\
& * \prod_{H_k \in D^-} p_1^{\,H_k^1}(1 - p_1)^{1-H_k^1} p_2^{\,H_k^2}(1 - p_2)^{1-H_k^2} \\
+ \quad & (1 - c_1)c_2 \prod_{H_k \in D^+} p_1^{\,H_k^1}(1 - p_1)^{1-H_k^1} p_2^{+\,H_k^2}(1 - p_2^+)^{1-H_k^2} \\
& * \prod_{H_k \in D^-} p_1^{\,H_k^1}(1 - p_1)^{1-H_k^1} p_2^{\,H_k^2}(1 - p_2)^{1-H_k^2} \\
+ \quad & c_1(1 - c_2) \prod_{H_k \in D^+} p_1^{+\,H_k^1}(1 - p_1^+)^{1-H_k^1} p_2^{\,H_k^2}(1 - p_2)^{1-H_k^2} \\
& * \prod_{H_k \in D^-} p_1^{\,H_k^1}(1 - p_1)^{1-H_k^1} p_2^{\,H_k^2}(1 - p_2)^{1-H_k^2} \\
+ \quad & c_1 c_2 \prod_{H_k \in D^+} p_1^{+\,H_k^1}(1 - p_1^+)^{1-H_k^1} p_2^{+\,H_k^2}(1 - p_2^+)^{1-H_k^2} \\
& * \prod_{H_k \in D^-} p_1^{\,H_k^1}(1 - p_1)^{1-H_k^1} p_2^{\,H_k^2}(1 - p_2)^{1-H_k^2}
\end{aligned}
$$

$$
\begin{aligned}
L_0 + L_1 = \Bigg(& (1 - c_1) \prod_{H_k \in D^+} p_1^{\,H_k^1}(1 - p_1)^{1-H_k^1} \prod_{H_k \in D^-} p_1^{\,H_k^1}(1 - p_1)^{1-H_k^1} \\
& + c_1 \prod_{H_k \in D^+} p_1^{+\,H_k^1}(1 - p_1^+)^{1-H_k^1} \prod_{H_k \in D^-} p_1^{\,H_k^1}(1 - p_1)^{1-H_k^1} \Bigg) \times \\
\Bigg(& (1 - c_2) \prod_{H_k \in D^+} p_2^{\,H_k^2}(1 - p_2)^{1-H_k^2} \prod_{H_k \in D^-} p_2^{\,H_k^2}(1 - p_2)^{1-H_k^2} \\
& + c_2 \prod_{H_k \in D^+} p_2^{+\,H_k^2}(1 - p_2^+)^{1-H_k^2} \prod_{H_k \in D^-} p_2^{\,H_k^2}(1 - p_2)^{1-H_k^2} \Bigg)
\end{aligned} \tag{19}
$$

If we generalize (19) to M variants, we have the likelihood of M variants as in (4, 5)

Reformulation of L_0 and L_1 in LRT for an Efficient Permutation Test

First, computation of L_0 (4) can be reformulated as

$$
L_0 = \prod_{i=1}^{M} \left\{ (1 - c_i) \prod_{H_k \in D^+} p_i^{H_k^i} (1 - p_i)^{1 - H_k^i} \prod_{H_k \in D^-} p_i^{H_k^i} (1 - p_i)^{1 - H_k^i} \right\}
$$

$$
= \prod_{i=1}^{M} \left\{ (1 - c_i) \prod_{H_k \in D^{\pm}} p_i^{H_k^i} (1 - p_i)^{1 - H_k^i} \right\}
$$

$$
= \prod_{i=1}^{M} \left\{ (1 - c_i) p_i^{2Np_i} (1 - p_i)^{2N - 2Np_i} \right\} \triangleq \prod_{i=1}^{M} X_i
$$

Similarly, we can reformulate L_1 as

$$
L_1 = \prod_{i=1}^{M} \left\{ (1 - c_i) \prod_{H_k \in D^+} p_i^{H_k^i} (1 - p_i)^{1 - H_k^i} \prod_{H_k \in D^-} p_i^{H_k^i} (1 - p_i)^{1 - H_k^i} \right.
$$

$$
\left. + c_i \prod_{H_k \in D^+} p_i^{+ H_k^i} (1 - p_i^+)^{1 - H_k^i} \prod_{H_k \in D^-} p_i^{- H_k^i} (1 - p_i^-)^{1 - H_k^i} \right\}
$$

$$
= \prod_{i=1}^{M} \left\{ X_i + c_i p_i^{+ N\hat{p}_i^+} (1 - p_i^+)^{N - N\hat{p}_i^+} p_i^{- N\hat{p}_i^-} (1 - p_i^-)^{N - N\hat{p}_i^-} \right\}
$$

$$
= \prod_{i=1}^{M} \left\{ X_i + c_i (1 - p_i^+)^N \left(\frac{p_i^+}{1 - p_i^+} \right)^{N\hat{p}_i^+} (1 - p_i^-)^N \left(\frac{p_i^-}{1 - p_i^-} \right)^{N\hat{p}_i^-} \right\}
$$

Using the fact that $N\hat{p}_i^+ + N\hat{p}_i^- = 2Np_i$ under permutations,

$$
L_1 = \prod_{i=1}^{M} \left\{ X_i + c_i (1 - p_i^+)^N (1 - p_i^-)^N \left(\frac{p_i^+}{1 - p_i^+} \right)^{N\hat{p}_i^+} \left(\frac{p_i^-}{1 - p_i^-} \right)^{2Np_i - N\hat{p}_i^+} \right\}
$$

$$
= \prod_{i=1}^{M} \left\{ X_i + c_i (1 - p_i^+)^N (1 - p_i^-)^N \left(\frac{p_i^-}{1 - p_i^-} \right)^{2Np_i} \left(\frac{p_i^+}{1 - p_i^+} \cdot \frac{1 - p_i^-}{p_i^-} \right)^{N\hat{p}_i^+} \right\}
$$

$$
\triangleq \prod_{i=1}^{M} \left\{ X_i + K_i Y_i^{N\hat{p}_i^+} \right\}
$$

Conservative Extensions of Linkage Disequilibrium Measures from Pairwise to Multi-loci and Algorithms for Optimal Tagging SNP Selection[*]

Ryan Tarpine[1], Fumei Lam[2,**], and Sorin Istrail[1]

[1] Center for Computational Molecular Biology, Department of Computer Science, Brown University, Providence, RI 02912
[2] Department of Computer Science, University of California, Davis, CA 95616
{ryan,sorin}@cs.brown.edu, flam@cs.ucdavis.edu

Abstract. We present results on two classes of problems. The first result addresses the long standing open problem of finding unifying principles for Linkage Disequilibrium (LD) measures in population genetics (Lewontin 1964 [10], Hedrick 1987 [8], Devlin and Risch 1995 [5]). Two desirable properties have been proposed in the extensive literature on this topic and the mutual consistency between these properties has remained at the heart of statistical and algorithmic difficulties with haplotype and genome-wide association study analysis. The first axiom is *(1) The ability to extend LD measures to multiple loci as a conservative extension of pairwise LD.* All widely used LD measures are pairwise measures. Despite significant attempts, it is not clear how to naturally extend these measures to multiple loci, leading to a "curse of the pairwise". The second axiom is *(2) The Interpretability of Intermediate Values.* In this paper, we resolve this mutual consistency problem by introducing a new LD measure, *directed informativeness* $\overrightarrow{\mathcal{I}}$ (the directed graph theoretic counterpart of the *informativeness* measure introduced by Halldorsson et al. [6]) and show that it satisfies both of the above axioms. We also show the maximum informative subset of tagging SNPs based on $\overrightarrow{\mathcal{I}}$ can be computed exactly in polynomial time for realistic genome-wide data. Furthermore, we present polynomial time algorithms for optimal genome-wide tagging SNPs selection for a number of commonly used LD measures, under the bounded neighborhood assumption for linked pairs of SNPs. One problem in the area is the search for a quality measure for tagging SNPs selection that unifies the LD-based methods such as LD-select (implemented in Tagger, de Bakker et al. 2005 [4], Carlson et al. 2004 [3]) and the information-theoretic ones such as informativeness. We show that the objective function of the LD-select algorithm is the *Minimal Dominating Set (MDS)* on r^2-SNP graphs and show that we can compute MDS in polynomial time for this class of graphs. Although

---

[*] Supported by National Science Foundation.
[**] Work done while in the Department of Computer Science and Center for Computational Molecular Biology, Brown University.

V. Bafna and S.C. Sahinalp (Eds.): RECOMB 2011, LNBI 6577, pp. 468–482, 2011.
© Springer-Verlag Berlin Heidelberg 2011

in LD-select the "maximally informative" solution is obtained through a greedy algorithm, and therefore better referred to as "locally maximally informative," we show that in fact, Tagger (LD-select) performs very close to the global maximally informative optimum.

1   Desiderata for Linkage Disequilibrium Measures

Linkage Disequilibrium (LD) is of fundamental importance for population genetics and evolutionary studies. In the past 10 years, there has been extensive literature on the study of patterns of LD in the human genome; one of the major driving hopes is that these LD patterns and measurements hold the key to developing powerful computational tools for mapping complex disease loci through genome-wide association studies.

The authors of published criticism about measures of LD have often accompanied their criticism with commentary on criteria or axioms that an ideal measure should satisfy. Some of these criteria are presented as mathematical properties and some are presented only informally. The aim of our approach is to develop an axiomatic framework by formalizing two such desiderata about LD measures proposed in the literature and then to study the problem of finding LD measures satisfying both of them.

1.1   Axiom 1: The Extendability of LD Measures to Multi-loci as a Unique Conservative Extension of the Pairwise Values

Currently, all extensively used LD measures are defined pairwise. In order to measure LD for a genomic region with three of more loci, an aggregation function is then required to combine the pairwise values in the region. The choice of the aggregation function is necessarily ad hoc, with many options on general principles, and the effect of this ad hoc choice is hard to evaluate. For example, one could use a weighted sum of all the pairwise values, the min-max of pairwise values, or a graph theoretic objective (e.g., all connected nodes within a threshold) [3]. Our formulation of this axiom is to obtain an LD measure that generalizes to many sites, and is a conservative extension of the values for pairwise LD. To our knowledge, none of the measures in use have such a conservative extension.

We call this major unresolved difficulty, the *curse of pairwise*. The name is inspired by discussions with Andy Clark during our collaboration on the Minimum Informative Subset of SNPs Problem [7]. In this problem, we would like to genotype the minimum set of informative SNPs for a region with the property such that the remaining SNPs can be inferred from the subset of the SNPs typed. Solving this problem required a new information theoretic measure, called *Informativeness*, that quantifies how much information a set of SNPs S captures about a target SNP t. The optimization problem asks for the minimum subset of SNPs $S' \subset S$ that contains the same amount of information as the entire set of SNPs S. This problem reduces to a well-studied problem in computer science, the

Set Cover Problem. Based on that insight, exact algorithms were obtained in [7] that work in polynomial time for genome-wide data, and compute the globally minimum informative subset of SNPs. We present in this paper a contribution of the same type.

This paper introduces a sister measure to Informativeness, called *Directed Informativeness*. The approach is based on graph theoretic and algorithmic results. Finding the Minimum Informative Subset under Directed Informativeness is also NP-complete, but as in the case of informativeness, we have a polynomial time and practical algorithm that works for genome-wide data. However, the new measure has unexpected properties unifying a number of aspects of the extensively used LD measures, and as a consequence, it is consistent with the two axioms we propose. In particular, like Informativeness, Directed Informativeness can be extended to multiple loci as a conservative pairwise-to-all extension.

1.2   Axiom 2: The Interpretability of Intermediary Values

Criticism of D':
"Because the magnitude of D' depends strongly on sample size, samples are difficult to compare ... intermediate values of D' should not be used for comparison of the strength of LD between studies, or to measure the extent of LD." Ardlie, Kruglyak, Seielstad [1]

Intermediate values of r^2 are easily interpretable.
Pritchard and Przeworski show the relationship between r^2, Pearson correlation coefficient χ^2 and effective population size N [11]. We show that directed informativeness can be related to r^2, thus obtaining as a corollary the relationship of directed informativeness to χ^2 and effective population size.

2   Directed Informativeness and the Minimum Informative Subset Tagging SNPs Problem

A *Single Nucleotide Polymorphism* (SNP) is a position in the genome at which two or more different alleles occur in the population, each with frequency above a certain threshold. The goal of association studies is to correlate genetic variation with the occurrence of disease. The difficulty arises in the large number of candidate sites and the combinatorial explosion in the number of subsets of SNPs when multiple sites are considered. In chromosome-wide studies, whole genome scans are performed and it is desirable for cost efficiency reasons to select only a subset of SNPs which accurately represent the genetic variation of the entire population [2, 7, 9].

We introduce a measure for association which extends the *minimum informative subset*, a concept from data compression that has been studied in tagging and in the analysis of haplotype block robustness [2, 12, 13].

2.1 Informativeness

We first introduce the graph theoretic measure of informativeness from [7], which aims to capture how well a set of SNPs can predict a target SNP (our notation follows closely that of [7]). The central idea is to define a measure of informativeness to quantify the accuracy of predicting an untyped SNP by a set of proximal SNPs. The input is an $n \times m$ matrix M representing n binary $(0/1)$ haplotypes of length m, together with an $n \times 1$ disease vector t. The vector t takes values 0 and 1 to distinguish between 'case' and 'control' individuals. Consider the $n \times (m+1)$ matrix M', obtained from the matrix M by appending column t. The value in the ith row and jth column of M' is denoted by $M'_{i,j}$. For a column s of M' and haplotypes i, j, let $D^s_{i,j}$ denote the event that SNP s differs in positions i and j ($M_{i,s} \neq M_{j,s}$).

Definition 1. *The* informativeness *of SNP s with respect to the disease vector t is*

$$I(s,t) = Prob_{i \neq j}(D^s_{i,j} \mid D^t_{i,j}).$$

We can also interpret the informativeness of a SNP with respect to the disease vector in terms of graph theory. Associate with each column s of M' a complete bipartite graph G_s on n vertices, with bipartition defined by the alleles of s. Each vertex in G_s corresponds to a row of M', and there are edges between any two rows which differ at column s, i.e.,

$$E(G_s) = \{\{i,j\} \mid M_{i,s} \neq M_{j,s}\}.$$

Let $V(G_{s,0})$ denote the vertices in graph G_s corresponding to allele 0 and $V(G_{s,1})$ denote the vertices in G_s corresponding to allele 1. For $1 \leq i < j \leq n$, let

$$\delta_{ij}(s,t) = \begin{cases} 1 & \text{if } M_{i,s} \neq M_{j,s} \text{ and } M_{i,t} \neq M_{j,t} \\ 0 & \text{otherwise.} \end{cases}$$

Then the informativeness of s with respect to t can be expressed in terms of the bipartite graphs G_s and G_t as

$$I(s,t) = \frac{\sum_{1 \leq i < j \leq n} \delta_{ij}(s,t)}{|E(G_t)|}.$$

In [7], this measure was used to detect how well a target SNP can be predicted from a set of tagging SNPs and to solve the *k most informative SNPs problem* by observing that the minimum informative subset problem is equivalent to the minimum set cover problem. The advantage of this measure is that it can easily be generalized to define informativeness for multi-locus SNPs [7], therefore satisfying Axiom (1) of the desired properties for a linkage disequilibrium measure.
   For $S' \subseteq S$, let

$$I(S',t) = Prob_{i \neq j}\left(D^{S'}_{i,j} \mid D^t_{i,j}\right) = \frac{|E(S') \cap E(t)|}{|E(t)|}$$

2.2   Directed Informativeness

Using the graph-theoretic interpretation of informativeness as a starting point, we modify the graph under consideration to define a measure we call *directed informativeness* and relate this measure to existing measures of linkage disequilibrium. For site s, denote the major allele by 0 and the minor allele by 1. We create a *directed* bipartite graph $\overrightarrow{G_s}$ for site s, with vertex set $\{1, 2, \dots n\}$ and directed edge set

$$E(\overrightarrow{G_s}) = \{(i, j) \mid M_{i,s} = 0 \text{ and } M_{j,s} = 1\}.$$

In this directed bipartite graph, all the edges are between different alleles and are oriented from the major allele to the minor allele. This addition of edge orientations in the graph will allow us to make a connection between the graph theoretic interpretation of informativeness with existing linkage disequilibrium measures. Note that by considering the underlying undirected graph of $\overrightarrow{G_s}$ for each site s, we obtain the undirected graph G_s defined in [7]. However, the pattern of intersection between the directed edges will play an important role in our extended definition of directed informativeness.

For $1 \le i < j \le n$, let

$$\overrightarrow{\delta_{ij}}(s, t) = \begin{cases} 1 & \text{if } M_i^s = M_i^t = 0 \text{ and } M_j^s = M_j^t = 1 \\ -1 & \text{if } M_i^s = M_j^t = 0 \text{ and } M_j^s = M_i^t = 1 \\ 0 & \text{otherwise.} \end{cases}$$

Definition 2. *The* directed informativeness *of SNP s with respect to SNP t is defined as*

$$\overrightarrow{\mathcal{I}}(s, t) = \frac{\sum_{1 \le i < j \le n} \overrightarrow{\delta_{ij}}(s, t)}{|E(G_t)|}$$

2.3   Directed Informativeness and the Conservative Extension to Multi-loci Axiom

One problem with the LD measures r^2 and D' is that they are not adequate for SNP subset selection/tagging SNPs and do not extend to multiple SNPs in a canonical way. In contrast, we demonstrate how the directed informativeness measure can be extended to multiple sites while satisfying the desired properties.

Let $S = \{s_1, s_2, \dots s_k\}$ and $T = \{t_1, t_2, \dots t_l\}$ be disjoint subsets of loci. For each site $s_i \in S$ and $t_j \in T$, consider the associated directed graphs G_{s_i} and G_{t_j}. Each of the graphs G_{s_i} and G_{t_j} are defined on vertex set $\{1, 2, \dots n\}$. Let G_T be the graph with vertex set $\{1, 2, \dots n\}$ and with edge set $E(G_{t_1}) \cup E(G_{t_2}) \cup \dots \cup E(G_{t_l})$. Note that G_T is a graph (not a multigraph), and any edge (i, j) appearing in two or more graphs $G_{t_j} (1 < i < l)$ appears only once in G_T. Let $E(G_T)$ denote the edge set of graph G_T. Then the directed informativeness of SNP subset S with respect to SNP t is

$$\overrightarrow{\mathcal{I}}(S,t) = \frac{\sum_{e\in E(G_s)} \overrightarrow{\delta}_e(S,t)}{|E(G_t)|}$$

3   Directed Informativeness and the Interpretability of Intermediary Values Axiom

We now relate the directed informativeness measure $\mathcal{I}$ with the widely used linkage disequilibrium measure r^2. This direct relationship provides insight into the observation in [7] that "optimizing for either [informativeness or r^2] optimizes very well for the other. These results thus suggest that informativeness and haplotype r^2, although distinct measures, are closely related in practice."

Note that while r^2 is a symmetric measure with respect to the pair of sites considered, the directed informativeness measure is not (i.e., $\overrightarrow{\mathcal{I}}(s,t) \neq \overrightarrow{\mathcal{I}}(t,s)$). The following theorem shows that the two measures are related by a natural product symmetrization.

Theorem 1. *For any two SNPs s and t, linkage disequilibrium measure r^2 between s and t is equal to*

$$r^2(s,t) = \overrightarrow{\mathcal{I}}(s,t)\overrightarrow{\mathcal{I}}(t,s).$$

Proof. Recall that $r^2(s,t) = \frac{(p_{00}p_{11} - p_{01}p_{10})^2}{p_{0+}p_{+0}p_{1+}p_{+1}}$. For each $i, j \in \{0, 1\}$, let $C_{ij} = p_{ij}n$, where n is the size of the sample. We will show that

(1) $C_{00}C_{11} - C_{01}C_{10} = \sum_{1\leq i<j\leq n} \overrightarrow{\delta}_{ij}(s,t)$
(2) $C_{0+}C_{1+}C_{+0}C_{+1} = |E(G_s)||E(G_t)|$

To prove (1), observe that each haplotype occurrence of 00 and 11 in columns i and j contribute $+1$ to both the left and right hand sides of the equation, while each haplotype occurrence of 01 and 10 contribute -1 to both the left and right hand sides of the equation. Furthermore, all remaining haplotype pairs contribute 0 to both equations.

To prove (2), note that

$$C_{0+}C_{1+} = |E(G_s)|$$

and

$$C_{+0}C_{+1} = |E(G_t)|.$$

This proves the theorem.   □

Pritchard and Przeworski show the relationship between LD measure r^2, Pearson correlation coefficient χ^2 and effective population size N [11]. As a corollary, the directed informativeness between two SNPs can also be related to the χ^2 Pearson correlation coefficient and effective population size.

Corollary 1. *For any two SNPs s and t,*

$$\chi^2(s,t) = \frac{\overrightarrow{\mathcal{I}}(s,t)\,\overrightarrow{\mathcal{I}}(t,s)}{N}.$$

4 Formalizing LD-Select/Tagger

A SNP Threshold Graph (STG) is constructed as follows: Given a set S of n SNPs ($|S| = n$), construct a set V of vertices where each vertex corresponds to a single SNP in S ($|V| = n$). There exists an edge between two vertices u and v if and only if there is linkage disequilibrium (LD) above a certain threshold τ between the SNPs represented by u and v.

LD-Select [3] and Tagger (without multimarker tests) [4] are essentially greedy heuristics which attempt to find the minimum dominating set for this graph. A dominating set for a graph $G = (V, E)$ is a subset $V' \subseteq V$ such that every vertex not in V' is connected to at least one member of V' by an edge. The minimum dominating set problem is to find the smallest such set. In the context of SNP selection, finding the smallest possible set V' means finding the smallest set of tagging SNPs such that every SNP is either in the set or is in LD above τ with at least one SNP in the set.

In general graphs, the dominating set problem is NP-complete. However, SNP Threshold Graphs have certain properties which allow them to be solved optimally in polynomial time.

4.1 LD Measure Assumptions

In order to construct the graph, we make the following assumptions about the LD measure under consideration:

1. The measure is symmetric: $\forall s, \forall t, LD(s, t) = LD(t, s)$. Pairwise measures usually fulfill this criterion; e.g., r^2.
2. Long-range LD (LD between SNPs hundreds of kilobases apart) is not meaningful and can be ignored. This induces a "neighborhood" around each SNP, with SNPs beyond the edges of the neighborhood ignored because they are too far away. To simplify things, a fixed neighborhood size w is often chosen [6].

4.2 SNP Selection Graph Properties

These assumptions yield a graph with the following properties:

1. The graph is undirected
2. There is a linear ordering on the vertices. Each vertex uniquely represents one SNP, so the ordering on the SNPs gives an ordering on the vertices.

3. Because we are not interested in long-range LD, there are restrictions on which vertices may be connected by edges. This is more meaningful than just a limit to the degree of nodes; based on the vertex ordering, we know that if the neighborhood is centered on SNP i and w is odd, then vertex i can only have edges to vertices $i - (w - 1)/2$ through $i + (w - 1)/2$.

This final property is key in finding the minimum dominating set efficiently. LD-Select and Tagger perform well in practice because the greedy approximation algorithm they use is known to find an $O(\log d)$-approximation for graphs of maximum degree d, and the degree is bounded by both the window size and the actual extent of LD between SNPs.

The final property allows us to apply the dynamic programming (DP) algorithm given by Halldorsson et al. [6] with a novel measure of informativeness to compute the minimum dominating set. We call our method MIS-DS because it utilizes the minimum informative subset (MIS) algorithm to solve the dominating set problem.

4.3   Minimum Dominating Set Algorithm

The DP algorithm of Halldorsson et al. requires an upper bound k on the number of tagging SNPs in order to run (since it computes a dynamic programming matrix, it must know the dimensions of the matrix in advance). An obvious upper bound would be n, the total number of SNPs, but this is unnecessarily high. Instead, we first run a greedy heuristic such as LD-Select [3] to establish a tighter upper bound. We then use the size of the resulting set as the value of the parameter k. We run the DP algorithm with the following informativeness measure I:

$$I_\tau(S', t) = \begin{cases} 1 & \exists s \in S' \text{ such that } LD(s, t) \geq \tau \\ 0 & \text{otherwise} \end{cases}$$

where $LD(s, t)$ computes r^2 between SNPs s and t.

I_τ is a binary measure of whether SNP t is in LD above τ with at least one of the SNPs in the set S'. The DP algorithm finds in time $O(nk2^w)$ the set S' of size k which maximizes $\iota_k = \sum_t I_\tau(S', t)$, which in this case counts the number of SNPs which are either in the set S' or have LD above τ with a SNP in S'. As long as k is equal to or greater than the size of the minimum dominating set, this value will be n, the total number of SNPs. This allows us to find the size of the minimum dominating set regardless of how well the greedy heuristic performed.

The DP matrix contains implicitly the maximum informativeness ι' possible for all upper bounds $0 \leq k' \leq k$. Therefore, we can check whether $k - 1$ would have gotten the same result ι, and if so, whether $k - 2$ would have done the same, etc. Finding the maximum informativeness takes time $O(2^w)$ for each k' (because we must examine the DP matrix for each assignment A_s, of which there are 2^w), so in total time $O(k2^w)$ we can find the smallest $k^\star$ such that $\iota_{k^\star} = n$. Then backtracking in the DP matrix will yield the set $S^\star$ such that $|S^\star| = k^\star$ and $S^\star$ is a dominating set, i.e., the smallest possible set of tagging SNPs which captures all of the SNPs with $r^2 \geq \tau$.

The MIS-DS algorithm is a globally optimal solver for the dominating set problem that works on arbitrarily large genome-wide data and runs in polynomial time. It is not restricted to r^2; it can be used with any pairwise measure of LD that fulfills the criteria above, such as D', Q, and so on.

4.4   Comparison of LD-Select/Tagger vs. Optimal

To test the performance of MIS-DS we used as out dataset the first 100 SNPs with MAF ≥ 0.05 of chromosome 22 from the HapMap3 release 2 phasing data. The following tables show the number of tagging SNPs required to capture all of the SNPs above the r^2 threshold:

r^2 threshold	0.9	0.8	0.7	0.6	0.5	0.4	0.3	0.2	0.1
Tagger pairwise	64	55	49	45	39	34	27	18	13
MIS-DS $w = 17$	67	61	54	50	45	39	32	24	18

First 50 SNPs:

r^2 threshold	0.9	0.8	0.7	0.6	0.5	0.4	0.3	0.2	0.1
Tagger pairwise	40	34	31	29	26	22	17	13	9
MIS-DS $w = 17$	40	35	31	30	27	24	19	14	10

First 40 SNPs:

r^2 threshold	0.9	0.8	0.7	0.6	0.5	0.4	0.3	0.2	0.1
Tagger pairwise	33	28	26	25	23	20	16	14	9
MIS-DS $w = 17$	35	33	29	28	26	24	20	15	12

Considering that the MIS-DS algorithm is optimal within the window size, we immediately assumed that Tagger's smaller result set was due to a lack of this constraint. Examining some of Tagger's results confirmed this to be true: in more than one case, a tagging SNP was capturing another SNP with 11 SNPs in between. This would require a centered window size of 25 (12 to the left, 12 to the right, and the center position), which is not feasible with our current implementation. In terms of physical distance, the SNPs are about 20 kb apart, so the LD is certainly significant. Future research will involve expanding the algorithm to handle larger window sizes.

4.5   Tagger "Best N"

It is well understood that by choosing tagging SNPs using an r^2 threshold τ the same power is approximately achieved by increasing the sample size by a factor of $1/\tau$ [11]. In this sense, the experimenter understands the trade-off between using a low threshold versus a high threshold—a low threshold means fewer tagging SNPs to genotype but requires a larger sample size. However, this is only true when every SNP of interest is in LD above the threshold with a tagging SNP. When using Tagger's "best N" method (the "Max tags" feature of Haploview), the experimenter is told exactly how many SNPs were captured within the threshold, but nothing is known about the SNPs that were not captured. It is possible that some of them are in LD just below the threshold with the chosen tagging SNPs. It is also possible that they are not in significant LD with any of them; moreover, it is possible that if the experimenter had set the threshold slightly lower, then the "best N" tagging SNPs would have been very different. This is because lowering the threshold adds edges to the SNP Threshold Graph, which can considerably increase the degrees of certain nodes, and this changes the minimum dominating set—the set could be much smaller and contain different vertices.

A better algorithm would not require a binary measure of whether LD between two SNPs is above a certain level or not—it would take into account the fractional LD value as is. One possibility is using the MIS DP algorithm with the informativeness measure:

$$I_{\max}(S', t) = \max_{s \in S'} LD(s, t)$$

$I_{\max}$ computes the maximum LD between SNP t and any SNP in set S'. Maximizing $\sum_t I_{\max}(S', t)$ would then attempt to make sure every SNP is in some level LD with tagging SNPs, rather than focusing entirely on LD above an arbitrary threshold.

5   Comparison of Directed Informativeness with (Undirected) Informativeness

5.1   Prediction vs. Distinguishability

The measure of informativeness was defined in order to capture distinguishability between haplotypes. Given a sample of haplotypes, find the smallest set of SNPs such that these SNPs are capable of distinguishing between the original haplotypes—if two haplotypes were different when all the known SNPs were genotyped, then these haplotypes will still be different when only considering the tagging SNPs. For a single SNP s, its associated graph G_s can be used to infer which allele each individual has because it is a complete bipartite graph—it is clear which individuals possess one allele and which contain the other. However, once the measure is extended to compute the informativeness of a set of SNPs S', the union of the edges $E(S') = \bigcup_{s \in S'} E(s)$ is no longer sufficient to infer the alleles of the SNP t for which $I(S', t)$ is near 1, because the graph is no longer bipartite.

Directed informativeness, on the other hand, captures the ability to impute non-genotyped SNPs directly from the tagging SNPs. The directed edges of $\overrightarrow{G}_s$ make clear exactly which haplotypes contain 0 and which contain 1. For a set of SNPs S', the directedness of the edges in $E(S')$ makes it possible to determine for each vertex whether its allele is 0 or 1 simply by comparing its outdegree and indegree: if the outdegree is greater (or equal), then the allele is 0. This is because each edge represents a SNP $s \in S'$ for which the allele was 0 at this vertex and 1 at its endpoint. If the outdegree is greater, then there is more evidence that the allele is 0 than 1. If the indegree and outdegree are equal, then the predicted allele is still 0 since 0 is the major allele, by definition more likely if no other information is known.

5.2 Non-monotonicity vs. Monotonicity

Because informativeness is calculated by counting edges in an undirected graph, adding more tagging SNPs can never decrease informativeness. That is, $I(S', t) \leq I(S' \cup \{s\}, t)$ for all s. This is not the case with directed informativeness. If the edges contributed by an additional tagging SNP s have the opposite orientation with respect to edges in t's graph, then those edges will decrease the directed informativeness, not increase it. This means that we cannot assume every tagging SNP in the neighborhood of a non-genotyped SNP t should be used to predict it, because if any of those tagging SNPs have a different allele pattern than t then they could potentially decrease the directed informativeness and hinder imputing the SNP. Otherwise, even tagging every SNP would fail to produce total directed informativeness because each SNP would interfere with its dissimilar neighbors.

Ramifications to Minimum Informative Subset problem. Non-monotonicity means that we must change the definition of the minimum informative subset problem [6] in order to make it relevant to directed informativeness. We define it as:

$$\overrightarrow{\mathcal{I}}(S', T) = \sum_{t \in T} \max_{S'_t \subseteq S' \cap N(t)} \overrightarrow{\mathcal{I}}(S'_t, t)$$

Rather than using all the tagging SNPs in the neighborhood of t to predict t, we use the subset of the tagging SNPs in the neighborhood which maximizes directed informativeness. We modified the algorithm of Halldorsson et al. to use this metric to compute the minimum directed informative subset.

6   Empirical Results

6.1 Haplotype Discrimination

One measure-agnostic method for testing the level of information captured by a set of tagging SNPs is haplotype discrimination. Given a set of m unique haplotypes, are the tagging SNPs sufficient to distinguish between all m, or do they collapse into a smaller set? For this test we used as our dataset the 189 unique haplotypes generated by taking the first 40 SNPs with a minor allele frequency of at least 0.05 of chromosome 22 of the CEU population in HapMap3 release 3. Since the Tagger algorithm involves some degree of randomness, for each test we ran Tagger 10 times and used the results of the run which yielded the highest mean r^2 as reported by Tagger.

Figure 1(a) shows what fraction of those haplotypes are still unique after viewing only the tagging SNPs chosen by different algorithms.

We chose 33 to be the upper bound for the number of tagging SNPs because this was the number of SNPs selected by Tagger when run without a limit with the threshold $r^2 \geq 0.9$. Lower thresholds and aggressive tagging both require fewer tags.

The measure $\mathcal{I}^2$ is defined by $\mathcal{I}^2(S, t) = \overrightarrow{\mathcal{I}}(S, t)\overrightarrow{\mathcal{I}}(t, S)$. This is the measure created by taking the pairwise identity with r^2 and extending it to multiple loci. Tagging SNPs for $\mathcal{I}^2$ are chosen by using the minimum directed informative subset algorithm with $\mathcal{I}^2$ as the measure. We discuss $\mathcal{I}^2$ below.

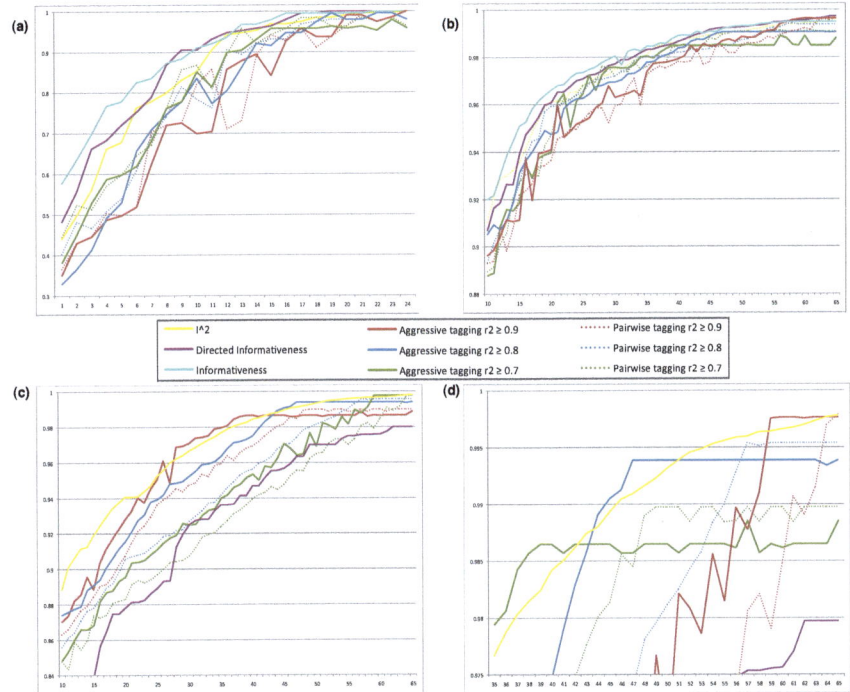

Fig. 1. For all graphs the x-axis shows the number of tagging SNPs chosen. **(a)** A comparison of haplotype discrimination according to the tagging SNPs chosen by different measures. **(b)** A comparison of the ability to impute non-genotyped alleles by looking into a dataset where the alleles are known using the tagging SNPs chosen by different measures. The Y axis shows the fraction of the alleles of the original dataset correctly imputed. **(c)** A comparison of the ability to impute non-genotyped alleles using only the tagging SNPs chosen by different measures. **(d)** Similar to (c) but restricted to large numbers of tagging SNPs.

6.2   Haplotype Imputation

Another measure of how representative tagging SNPs are is whether by looking at only the tagging SNPs, can the untyped SNPs be inferred. There are two methods for imputation: looking up the missing SNPs in a reference panel where these SNPs are known, and computing the missing alleles using only the genotyped SNPs themselves. We tested both types of imputation, using the 207 unique haplotypes generated by taking the first 100 SNPs with a minor allele frequency of at least 0.05 of chromosome 22 of the CEU population in HapMap3 release 3.

Using a Reference Panel. For the first type of imputation we tested each measure in a method similar to Halldorsson et al. [6]: an untyped SNP of an individual is inferred by looking at the typed SNPs in its neighborhood. If there are haplotypes in the training set that had the same allele as the individual on all the tagging SNPs, then a majority vote is taken to estimate which allele the individual has. If the votes are split 50/50, or if there are no training haplotypes which share every allele, then votes are taken from the haplotypes which differ by at most one allele. If even these votes are split or if no training haplotypes match, then the process continues by counting votes from haplotypes which differ on up to two alleles (and so on).

Given the tagging SNPs assigned by each measure, we tested the percentage of SNPs correctly imputed by this method. The results can be seen in figure 1(b).

(Undirected) informativeness is the best measure for imputation using a reference panel because of its ability to distinguish haplotypes. This property is exactly what is needed in order to look up the correct haplotype to observe the missing SNPs. The SNPs which are tagged maximizing informativeness are like a key into the reference panel: they are sufficient to find the exact haplotype, if present, because they distinguish between all of the different known haplotypes.

Tagging SNPs Only. To impute non-genotyped SNPs using only the tagging SNPs, for $\mathcal{I}^2$ we used the method detailed above in Section 5.1. For tagging SNPs chosen by Tagger, there are two cases. If a SNP t is captured by a single tagging SNP s, then we assume for each individual that his allele for t is the same as the allele genotyped for s. If t is captured by a haplotype, then we compare the individual's haplotype (i.e., the alleles of the test SNPs) to the test haplotype given by Tagger. If they are equal, then the predicted SNP is said to have the allele 1, otherwise 0. The results of this analysis can be seen in figure 1(c).

It is evident that no measure is best for all numbers of tagging SNPs. $\mathcal{I}^2$ does the best for few tagging SNPs, but aggressive tagging with Tagger with an r^2 threshold of 0.7 results in more accurate imputation for a large range (28-42) of tagging SNPs. However, 39 SNPs is sufficient to capture all of the SNPs according that threshold, so allowing for more SNPs beyond that point does not result in any improvement—any increase or decrease beyond 39 SNPs is only due to randomness in the Tagger algorithm. More tagging SNPs allow higher r^2 thresholds to perform better, but each performs optimally only within a certain range. A zoomed-in view of the performance with high numbers of tagging SNPs can be seen in figure 1(d). $\mathcal{I}^2$ performs consistently, and near-optimally, across all ranges.

Directed informativeness alone is not a suitable measure for tagging SNP selection, due to an aspect of its asymmetric nature. If $\overrightarrow{\mathcal{I}}(S,\,t)$ is near 1.0, then most of the edges of G_t are present in G_S. However, it is possible that there are edges present in G_S which are not in G_t—this has no effect on directed informativeness, because its definition involves counting only edges present in G_t. Our algorithm for SNP imputation is negatively affected by these additional edges, because it uses these edges to infer what the alleles of t would be if it were genotyped. A measure which makes sure that $E(G_S)$ is *similar* to $E(G_t)$, not only a (near) superset, is $\mathcal{I}^2$. The additional factor of $\overrightarrow{\mathcal{I}}(t,\,S)$ ensures that $E(G_S)$ does not contain too many additional edges. We find that $\mathcal{I}^2$ is an excellent measure for tagging SNP selection.

However, in certain circumstances $\mathcal{I}^2$ does not perform as well as the aggressive mode of Tagger. We believe that this is due to the fact that $\mathcal{I}^2$, like r^2, can only capture SNPs using tagging SNPs which are "similar" to them. Haplotype markers are more powerful and enable the capture of SNPs which are not similar to any one of the SNPs in the haplotype.

7   Computational Complexity of the Minimum Directed Informative Set of SNPs/Tagging SNPs Problem

In this section, we establish the complexity of the Minimum Directed Informative SNPs problem.

Lemma 1. *The Minimum Directed Informative SNPs problem is NP-complete.*

Proof. The proof follows the proof for the complexity of the Minimum (undirected) Informative SNPs problem, with a reduction from the set cover problem. Given a collection C of subsets of a finite set X, and positive integer $k \leq |C|$, the set cover problem asks if there exist $C' \subseteq C$ with $|C'| \leq k$ such that every element of X belongs to at least one member of C'. We construct a SNP matrix M with $|X| + 1$ haplotypes and $|C| + 1$ SNPs. For each subset $C_j \in C$, define a SNP $M[*, j]$ such that

$$M[i, j] = \begin{cases} 0 \text{ if } i \leq |X| \text{ and } X_i \in C_j \\ 1 \text{ otherwise} \end{cases}$$

The SNP $t = M[*, |C| + 1]$ is dened by the vector $[0, 0, \ldots, 0, 1]$ with exactly $|X|$ zeros and a single one. Then $C' \subseteq C$ covers X if and only if the corresponding subset of SNPs S' are informative with respect to t.

A polynomial time algorithm – that is practical for genome-wide data sets – for Directed Informativeness is obtained by generalizing the algorithm used for the Informativeness measure [6] (and available from the authors).

References

1. Ardlie, K., Kruglyak, L., Seielstad, M.: Patterns of linkage disequilibrium in the human genome. Nature Reviews, Genetics 3, 299–309 (2002)
2. Bafna, V., Halldrsson, B.V., Schwartz, R., Clark, A.G., Istrail, S.: Haplotypes and informative snp selection algorithms: don't block out information. In: RECOMB, pp. 19–27 (2003)
3. Carlson, C., Eberle, M., Reider, M., Yi, Q., Kruglyak, L., Nickerson, D.: Selecting a maximally informative set of single-nucleotide polymorphisms for association analyses using linkage disequilibrium. Am. J. Hum. Genet. 74, 106–120 (2004)
4. de Bakker, P., Yelensky, R., Peer, I., Gabriel, S., Day, M., Altshuler, D.: Efficiency and power in genetic association studies. Nature Genetics 37, 1217–1223 (2005)
5. Delvin, B., Risch, N.: A comparison of linkage diseqilibrium measures for fine-scale mapping. Genomics 29, 311–322 (1995)
6. Halldorsson, B., Bafna, V., Lippert, R., Schwartz, R., De La Vega, F., Clark, A., Istrail, S.: Optimal haplotype block-free selection of tagging snps for genome-wide association studies. Genome Research 14, 1633–1640 (2004)
7. Halldrsson, B.V., Bafna, V., Edwards, N., Lippert, R., Yooseph, S., Istrail, S.: Combinatorial problems arising in snps and haplotype analysis. In: Calude, C.S., Dinneen, M.J., Vajnovszki, V. (eds.) DMTCS 2003. LNCS, vol. 2731, pp. 26–47. Springer, Heidelberg (2003)
8. Hedrick, P.: Gametic disequilibrium measures: Proceed with caution. Genetics 117, 331–341 (1987)
9. Lancia, G., Bafna, V., Istrail, S., Lippert, R., Schwartz, R.: SNPs problems, complexity, and algorithms. In: Meyer auf der Heide, F. (ed.) ESA 2001. LNCS, vol. 2161, pp. 182–193. Springer, Heidelberg (2001)
10. Lewontin, R.: On measures of gametic disequilibrium. Genetics 120, 849–852 (1988)
11. Pritchard, J.K., Przeworski, M.: Linkage disequilibrium in humans: Models and data. The American Journal of Human Genetics 69, 1–14 (2001)
12. Schwartz, R., Clark, A.G., Istrail, S.: Methods for inferring block-wise ancestral history from haploid sequences. In: Guigó, R., Gusfield, D. (eds.) WABI 2002. LNCS, vol. 2452, pp. 44–59. Springer, Heidelberg (2002)
13. Schwartz, R., Halldrsson, B.V., Bafna, V., Clark, A.G., Istrail, S.: Robustness of inference of haplotype block structure. Journal of Computational Biology 10, 13–20 (2003)

Protein Loop Closure Using Orientational Restraints from NMR Data[*]

Chittaranjan Tripathy[1], Jianyang Zeng[1],
Pei Zhou[2], and Bruce Randall Donald[1,2,**]

[1] Department of Computer Science, Duke University, Durham, NC 27708
[2] Department of Biochemistry, Duke University Medical Center, Durham, NC 27710
Tel.: 919-660-6583; Fax: 919-660-6519
brd+recomb11@cs.duke.edu

Abstract. Protein loops often play important roles in biological functions such as binding, recognition, catalytic activities and allosteric regulation. Modeling loops that are biophysically sensible is crucial to determining the functional specificity of a protein. A variety of algorithms ranging from robotics-inspired inverse kinematics methods to fragment-based homology modeling techniques have been developed to predict protein loops. However, determining the 3D structures of loops using global orientational restraints on internuclear vectors, such as those obtained from residual dipolar coupling (RDC) data in solution Nuclear Magnetic Resonance (NMR) spectroscopy, has not been well studied. In this paper, we present a novel algorithm that determines the protein loop conformations using a minimal amount of RDC data. Our algorithm exploits the interplay between the sphero-conics derived from RDCs and the protein kinematics, and formulates the loop structure determination problem as a system of low-degree polynomial equations that can be solved exactly and in closed form. The roots of these polynomial equations, which encode the candidate conformations, are searched systematically, using efficient and provable pruning strategies that triage the vast majority of conformations, to enumerate or prune all possible loop conformations consistent with the data. Our algorithm guarantees completeness by ensuring that a possible loop conformation consistent with the data is never missed. This data-driven algorithm provides a way to assess the structural quality from experimental data with minimal modeling assumptions. We applied our algorithm to compute the loops of human ubiquitin, the FF Domain 2 of human transcription elongation factor CA150 (FF2), the DNA damage inducible protein I (DinI) and the third IgG-binding domain of Protein G (GB3) from experimental RDC data. A comparison of our results versus those obtained by using traditional structure determination protocols on the same data shows that our algorithm is able to achieve higher accuracy: a 3- to 6-fold improvement in backbone RMSD. In addition, computational experiments on synthetic RDC data for a set of protein loops of length 4, 8 and 12 used

---

[*] This work is supported by the following grants from National Institutes of Health: R01 GM-65982 to B.R.D. and R01 GM-079376 to P.Z.
[**] Corresponding author.

V. Bafna and S.C. Sahinalp (Eds.): RECOMB 2011, LNBI 6577, pp. 483–498, 2011.

in previous studies show that, whenever sparse RDCs can be measured, our algorithm can compute longer loops with high accuracy. These results demonstrate that our algorithm can be successfully applied to compute loops with high accuracy from a limited amount of NMR data. Our algorithm will be useful to determine high-quality complete protein backbone conformations, which will benefit the nuclear Overhauser effect (NOE) assignment process in high-resolution protein structure determination.

1   Introduction

Protein loops are the segments of polypeptide chain that connect two secondary structure elements (SSEs) such as α-helices or β-strands. In addition to serving as linkers between SSEs, loops often play crucial roles in protein folding and stability pathways, and in many other important biological functions such as binding, recognition, catalytic activities and allosteric regulation [42,7,55,27].

While the *global fold*, i.e., the conformations and orientations of the SSEs of a protein, can often be determined with high accuracy via traditional experimental techniques such as X-ray crystallography or nuclear magnetic resonance (NMR) spectroscopy, modeling loops that seamlessly close the gap between two consecutive SSEs by satisfying the geometric, biophysical, and data constraints remains a difficult and open problem. In X-ray crystallography, for instance, the disorder in a protein crystal can render interpretation of the resulting electron density for loops difficult. As a result, protein structures found in the Protein Data Bank (PDB) [3] often have missing loops or disordered loops. The problem of computing loops that are biophysically reasonable and geometrically valid is called the *loop closure problem*. Since its introduction four decades ago in the classic paper by Gō and Scheraga [26], the loop closure problem has been an active area of research. In fact, modeling of loops can be regarded as an *ab initio* protein folding problem at a smaller scale. It is also an important problem in *de novo* protein structure prediction. Therefore, solutions and algorithms for accurate modeling of loops are highly desirable for understanding of the physical-chemical principles that determine protein structure and function.

Exploring the conformation space of a protein loop to identify low energy loop conformations is a difficult computational problem. Methods to identify such loops include database search and homology modeling [64,60,20], *ab initio* methods based on the   minimization of empirical molecular mechanics energy functions [22,54,30], and robotics-inspired inverse kinematics and optimization-based methods [16,40,69,17,8,31]. These techniques work in two phases: first, the protein conformation space is explored to find a set of candidate loop constructs, which are then evaluated in the second phase using an appropriate empirical energy function to select the most promising set of loops.

Database methods [64,60,20] identify a set of candidate loops from a library of fragments derived from a protein structure database such as the PDB [3] that fit the anchor residues on either end of a loop. These loops are further ranked using criteria such as the sequence homology and conformational energy. Since these methods heavily rely on the statistical diversity of the structure database, the accuracy of loop predictions depends on how well the loop is represented in the database. However, in general, database methods suffer from limited sampling of the loop conformations by the fragments in the database.

Ab initio loop modeling methods sample the conformation space randomly or use robotics-based sampling algorithms to generate a large number of loop conformations. Loop closure and energy minimization are done by using methods such as random tweak [54,21], analytical loop closure techniques [17,69], molecular dynamics simulation [5], Markov Chain Monte Carlo (MCMC) simulated annealing [13,22], and other optimization techniques [30]. The accuracy of loop prediction here depends on the efficacy of the conformational space exploration techniques used, and on the quality and proper parameterization (e.g., implicit or explicit solvent effects) of the force field employed to evaluate the conformational energy. These algorithms are computationally expensive due to a large number of random moves accompanied by repeated energy computations.

The protein loop closure problem is an *inverse kinematics* (IK) problem in computational biology, i.e., given the poses of terminal anchor residues, it asks to find all possible values of the degrees of freedom (DOFs) (i.e., the dihedrals ϕ and ψ) for which the fragment connects both the anchor residues. This problem has been studied widely in robotics and biology [16,40,69,17,8,31]. Tri-peptide loop closure, for which the number of DOFs is six and exactly six geometric constraints are stipulated due to the closure criterion, can be solved analytically [69,17,39,11] using exact IK solvers to give at most 16 possible solutions. For longer loops, the loop closure problem is underconstrained, so a continuous family of solutions are possible without additional constraints. Optimization-based IK solvers such as random tweak [54,21], and the cyclic coordinate descent (CCD) algorithm [8] have been successful in dealing with a large number of DOFs, and have found many applications [52,29,65]. These methods iteratively solve for the DOFs until the loop closure constraints are satisfied. However, the problem of loop closure subject to orientational restraints (e.g., from NMR data) has not been studied rigorously in the robotics or computational biology literature, and no practical deterministic algorithm exists to our knowledge.

Protein structure determination using nuclear Overhauser effect (NOE) distance restraints is NP-hard [50]. Traditional protein structure determination from solution NMR data starts with an elongated polypeptide backbone chain, and uses NOEs and dihedral angle restraints in a simulated annealing/simplified molecular dynamics (SA/MD) protocol [12,28,41,32,51] to compute the protein structure. Residual dipolar coupling (RDC) restraints are only incorporated in the final stages of the structure computation to refine the structures [6,51]. NOE-based structure determination protocols are known to be prone to local minima or lead to wrong convergence. To overcome the shortcomings of NOE-based methods, approaches in [18,46,4,56,25,1] have been proposed that primarily use RDC data, which provides precise global orientational restraints on internuclear vector orientations, to determine protein backbone structure. However, most of these approaches employ stochastic search, and therefore lack any algorithmic guarantee on the quality of the solution or running time. In recent work from our lab [66,68,19,71], polynomial-time algorithms have been proposed for high-resolution backbone global fold determination from a minimal amount of RDC data. These algorithms represent the RDC equations and protein kinematics in algebraic form, and use exact methods in a divide-and-conquer framework to compute the global fold. In addition, these algorithms use a sparse set of RDC measurements (e.g., only two RDCs per residue), with the goal of minimizing the number of NMR experiments, hence the time and cost to perform them.

A high-resolution protein backbone is often a starting point for structure-based protein design [23,10,24,35]. An accurate backbone structure facilitates the assignment of NOESY spectra (i.e., *NOE assignment*), which is a prerequisite for high-resolution structure determination protocols, including side-chains. For example, the algorithms in [66,68,19,71] have been used in [67,73,72] to develop new algorithms for NOE assignment, based on which in [72] we recently developed a new framework for high-resolution protein structure determination, which was used prospectively to solve the solution structure of the FF Domain 2 of human transcription elongation factor CA150 (FF2) (PDB id: 2kiq). The global folds obtained by [66,68,19,71] have all the loops missing which requires a new algorithm that can compute the missing loops from RDCs. To alleviate this problem, a heuristic local minimization approach [51] for loops was used in [72].

In this paper, we give a solution to the loop closure problem. We present an efficient deterministic algorithm, POOL, that computes the missing loops from RDC data. Our algorithm exploits the interplay between protein backbone kinematics and the global orientational restraints derived from RDC data to naturally discretize the conformation space by polynomial-root solutions, and represents the candidate conformations using a tree. A systematic depth-first search of the conformation tree is used to enumerate all possible loop conformations that are consistent with the data. POOL uses efficient pruning strategies (Section 2.6) capable of pruning the majority of the conformations that are provably not part of a valid loop, thereby achieving a huge reduction in the search space. Unlike other algorithms, e.g. [4], that attempt to compute backbone structure using as many as 15 RDCs per residue recorded in two alignment media, our algorithm uses as few as 2 RDCs per residue in one alignment medium, which is often experimentally feasible. As we will show in Section 3.2, when given the same data, our algorithm performs better than traditional SA/MD-based approaches, e.g., [51]. Additional RDCs, and other data that provide constraints in torsion-angle space (e.g., TALOS [14,53] dihedral restraints) or in Euclidean space (e.g., sparse NOEs), whenever available, can directly be incorporated into our algorithm. In summary, we make the following contributions in this paper:

1. Derivation of quartic equations for backbone dihedrals ϕ and ψ from experimentally-recorded RDC sphero-conics and backbone kinematics, that can be solved *exactly* and in closed form;
2. Systematic search of the roots of the polynomial equations that encode the conformations, using efficient pruning methods to prune the vast majority of conformations;

3. Design and implementation of an efficient algorithm to determine the loop conformations from a limited amount of experimental RDC data;
4. Promising results from the application of our algorithm both on experimental NMR data sets for four proteins, and on simulated data sets for protein loops studied previously in [36,17,8].

2   Theory and Methods

2.1   Overview

POOL solves the following loop closure problem. Let the residues of the protein be numbered from 1 to n (from N- to C-terminus). Suppose the global fold of the protein has been determined from RDCs in a principal order frame (POF) of RDCs, as we showed was feasible in [66,68,19,72,71]. In principle, the global fold of proteins could also be computed using protein structure prediction [2], or homology modeling [33,34]; alternatively, X-ray structures (with missing loops) can be used. Given two consecutive SSEs with n_1 and n_2 being the last residue of the first SSE and first residue of the second SSE, respectively, the missing loop $[n_1, n_2]$ is defined as the fragment between residues n_1 and n_2 with both end residues included. The residues n_1 and n_2 that are part of the SSEs will be called the *stationary anchors*, and those of a candidate loop will be called the *mobile anchors*. We assume that the n_1 mobile anchor of the loop is attached to the n_1 stationary anchor of the first SSE. Then the loop closure problem is stated as follows: in the POF, given the poses of the stationary anchors n_1 and n_2 (points in $\mathbb{R}^3 \times SO(3)$), compute a complete set of conformations of fragments $[n_1, n_2]$ so that n_2 mobile anchor of each fragment in the set assumes the pose of the stationary anchor n_2, while satisfying the RDC data and standard protein geometry.

Our algorithm builds upon the initial work from our lab [19,68,72,71], where the authors developed polynomial time algorithms to compute high-resolution backbone global fold *de novo* from N-H^N and C^α-H^α RDCs in one alignment medium. These sparse-data algorithms have been extended to incorporate combinations of different types of RDCs (see Table 1) in one or two alignment media. The new generalized framework is called RDC-ANALYTIC [72,71]. POOL implements a novel algorithm to determine protein loop backbone structures from a minimal amount of RDC data, and is a crucial addition to the RDC-ANALYTIC suite, which did not compute loops before.

Table 1 describes the RDC types that POOL uses to compute the backbone dihedrals exactly and in closed form (Section 2.3). A ϕ-*defining* RDC is used to compute the backbone dihedral ϕ, and a ψ-*defining* RDC is used to compute the backbone dihedral ψ. The input data to POOL include: (1) the global fold of the

Table 1. A ϕ-defining RDC is used to compute the backbone dihedral ϕ, and a ψ-defining RDC is used to compute the backbone dihedral ψ exactly and in closed form

ϕ-defining RDC	C^α-H^α, C^α-C', C^α-C^β
ψ-defining RDC	N-H^N, C'-N, C'-H^N

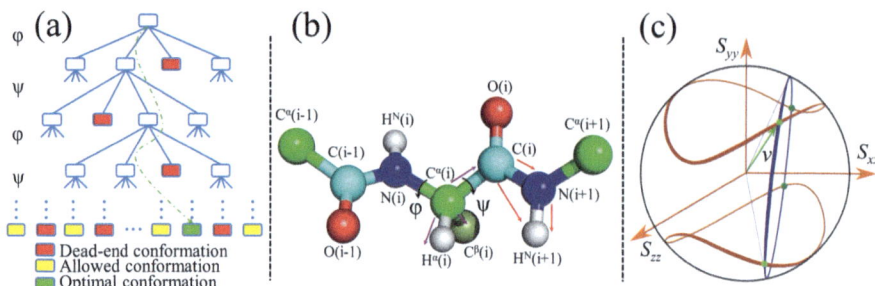

Fig. 1. (a) An example conformation tree. (b) The internuclear vectors (shown using arrows) for which RDCs are possible to measure. The magenta and red arrows represent ϕ-defining and ψ-defining RDCs, respectively. (c) The pringle-shaped RDC sphero-conic curves inscribed on a unit sphere constrain the internuclear vector **v** (green arrow) to lie on one of them. The kinematic circle (shown in blue almost edge-on) of **v** intersects the sphero-conic curves in at most four points (green dots) leading to a maximum of four possible orientations for the internuclear vector **v**.

protein computed by [68,19,72]; (2) the alignment tensor, which generally can be computed from the global fold using [37,66]; (3) at least one ϕ-defining and one ψ-defining RDCs per residue, and optionally other data, e.g., TALOS [14,53] dihedral restraints and sparse NOEs; and (4) the primary sequence of the protein.

Solving a system of equations from RDCs, protein kinematics and loop closure constraints simultaneously is a difficult computational problem since it leads to solving a high-degree polynomial system. However, since RDCs are very precise measurements, an algorithm which is able to compute protein fragments by inductively solving low-degree polynomial equations derived from RDCs and backbone kinematics, and drives the computation to satisfy the loop closure criterion, will achieve the desired objective. Our algorithm POOL is based on this key insight. Starting from a stationary anchor, it solves each DOF sequentially using the equations derived in Sections 2.3 and 2.4. The discrete values of the DOFs computed from the polynomial roots, are represented by a conformation tree grown recursively as we solve for the DOFs progressively. An internal (i.e., non-leaf) node in the tree represents the conformation of a part of a candidate loop, and a leaf node represents a candidate loop conformation computed from RDCs. Figure 1 (a) illustrates a conformation tree for a loop. As each node is visited in a depth-first traversal of the tree, if the conformation represented by that node fails the conformation filters (Section 2.6), it is called a *dead-end* node, and the sub-tree rooted at that node is pruned. Dead-end nodes identified at lower levels (i.e., closer to the root) of the conformation tree prune more conformations than those identified at higher levels. Finally, all remaining unpruned conformations (leaf nodes) already close to the stationary anchor (since they satisfy the reachability criterion; see Section 2.6), are evaluated for loop closure. At this stage minimization techniques can be applied to improve the closure. Conformations satisfying the closure criterion are added to the final ensemble of loops. POOL enumerates all loop conformations that satisfy the RDC data and pass the conformation filters; therefore, it guarantees completeness.

2.2 RDC Sphero-Conics

The residual dipolar coupling r between two spin-$\frac{1}{2}$ nuclei a and b is given by

$$r = D_{\max}\mathbf{v}^T\mathbf{S}\mathbf{v}, \tag{1}$$

where $\mathbf{v}$ is the unit internuclear vector between a and b, $D_{\max}$ is the dipolar interaction constant, $\mathbf{S}$ is the *Saupe order matrix* [49], or *alignment tensor*, that specifies the ensemble-averaged anisotropic orientation of the protein in the laboratory frame. $\mathbf{S}$ is a 3×3 symmetric, traceless, rank 2 tensor with five independent elements [57,58,43,19]. The constant $D_{\max}$ is given by

$$D_{\max} = \frac{\mu_0 \hbar \gamma_a \gamma_b}{4\pi^2} \left\langle r_{ab}^{-3} \right\rangle, \tag{2}$$

where μ_0 is the magnetic permeability of vacuum, $\hbar$ is Planck's constant, γ_a and γ_b are the gyromagnetic ratios of the nuclei a and b, respectively, and $\left\langle r_{ab}^{-3} \right\rangle$ represents the vibrational ensemble-averaged inverse cube of the distance between the two nuclei. Letting $D_{\max} = 1$ (i.e., scaling the RDCs appropriately), and considering a global coordinate frame that diagonalizes the alignment tensor $\mathbf{S}$, often called the *principal order frame* (POF), Eq. (1) can be written as

$$r = S_{xx}x^2 + S_{yy}y^2 + S_{zz}z^2, \tag{3}$$

where S_{xx}, S_{yy} and S_{zz} are the three diagonal elements of a diagonalized alignment tensor $\mathbf{S}$, and x, y and z are, respectively, the x, y and z components of the unit vector $\mathbf{v}$ in a POF that diagonalizes $\mathbf{S}$. Since $\mathbf{v}$ is a unit vector, i.e.,

$$x^2 + y^2 + z^2 = 1, \tag{4}$$

an RDC constrains the corresponding internuclear vector $\mathbf{v}$ to lie on the intersection of a concentric unit sphere (Eq. (4)) and a quadric (Eq. (3)) [44]. This gives a pair of closed curves inscribed on the unit sphere that are diametrically opposite to each other (see Figure 1 (b), (c)). These curves are known as *sphero-conics* or *sphero-quartics* [9,47].

Using Eq. (4) in Eq. (3), we can rewrite Eq. (3) in the following form:

$$ax^2 + by^2 = c, \tag{5}$$

where $a = S_{xx} - S_{zz}$, $b = S_{yy} - S_{zz}$, and $c = r - S_{zz}$. Henceforth, we refer to Eq. (5) as the *reduced RDC equation*. For background on RDCs and RDC-based structure determination, the reader is referred to [57,58,43,19].

2.3   Analytic Solutions for Peptide Plane Orientations from ϕ-Defining and ψ-Defining RDCs in One Alignment Medium

The derivation below assumes standard protein geometry, which is exploited in the kinematics [66]. We choose to work in an orthogonal coordinate system defined at the peptide plane P_i with z-axis along the bond vector $\mathrm{N}(i) \to \mathrm{H}^{\mathrm{N}}(i)$, where the notation $a \to b$ means a vector from the nucleus a to the nucleus b. The y-axis is on the peptide plane i and the angle between y-axis and the bond vector $\mathrm{N}(i) \to \mathrm{C}^{\alpha}(i)$ is fixed. The x-axis is defined based on the right-handedness. Let $\mathbf{R}_{i,\mathrm{POF}}$ denote the orientation (rotation matrix) of P_i with respect to the POF. Then $\mathbf{R}_{1,\mathrm{POF}}$ denotes the relative rotation matrix between the coordinate system defined at the first residue of the current SSE and the principal order frame. $\mathbf{R}_{i,\mathrm{POF}}$ is used to derive $\mathbf{R}_{i+1,\mathrm{POF}}$ inductively after we compute the dihedral angles ϕ_i and ψ_i. $\mathbf{R}_{i+1,\mathrm{POF}}$, in turn, is used to compute the $(i+1)^{st}$ peptide plane.

Proposition 1. *Given the diagonalized alignment tensor components S_{xx} and S_{yy}, the peptide plane P_i, and a ϕ-defining RDC r for the corresponding internuclear vector of residue i, there exist at most 4 possible values of the dihedral angle ϕ_i that satisfy the RDC r. The possible values of ϕ_i can be computed exactly and in closed form by solving a quartic equation.*

Proof. Let the unit vector $\mathbf{v}_0 = (0,0,1)^T$ represent the N-H^{N} bond vector of residue i in the local coordinate frame defined on the peptide plane P_i. Let $\mathbf{v}_1 = (x,y,z)^T$ denote the internuclear vector for the ϕ-defining RDC for residue i in the principal order frame. We can write the forward kinematics relation between $\mathbf{v}_0$ and $\mathbf{v}_1$ as follows:

$$\mathbf{v}_1 = \mathbf{R}_{i,\mathrm{POF}} \; \mathbf{R}_l \; \mathbf{R}_z(\phi_i) \; \mathbf{R}_r \; \mathbf{v}_0. \tag{6}$$

Here $\mathbf{R}_l$ and $\mathbf{R}_r$ are constant rotation matrices that describe the kinematic relationship between $\mathbf{v}_0$ and $\mathbf{v}_1$. $\mathbf{R}_z(\phi_i)$ is the rotation about the z-axis by ϕ_i.

   Let c and s denote $\cos\phi_i$ and $\sin\phi_i$, respectively. Using this while expanding Eq. (6) we have

$$x = A_0 + A_1 c + A_2 s, \quad y = B_0 + B_1 c + B_2 s, \quad z = C_0 + C_1 c + C_2 s, \tag{7}$$

in which A_i, B_i, C_i for $0 \le i \le 2$ are constants. Using Eq. (7) in the reduced RDC equation Eq. (5) and simplifying we obtain

$$K_0 + K_1 c + K_2 s + K_3 cs + K_4 c^2 + K_5 s^2 = 0, \tag{8}$$

in which K_i, $0 \le i \le 5$ are constants. Using half-angle substitutions

$$u = \tan(\frac{\phi_i}{2}), \quad c = \frac{1-u^2}{1+u^2}, \quad \text{and} \quad s = \frac{2u}{1+u^2} \tag{9}$$

in Eq. (8) we have

$$L_0 + L_1 u + L_2 u^2 + L_3 u^3 + L_4 u^4 = 0, \tag{10}$$

in which L_i, $0 \le i \le 4$ are constants.

Eq. (10) is a quartic equation which can be solved exactly and in closed form. Let $\{u_1, u_2, u_3, u_4\}$ denote the set of (at most) four real solutions of Eq. (10). For each u_i, the corresponding ϕ_i value can be computed using Eq. (9).   □

Proposition 2. *Given the diagonalized alignment tensor components S_{xx} and S_{yy}, the peptide plane P_i, the dihedral ϕ_i, and a ψ-defining RDC r for the corresponding internuclear vector on peptide plane P_{i+1}, there exist at most 4 possible values of the dihedral angle ψ_i that satisfy the RDC r. The possible values of ψ_i can be computed exactly and in closed form by solving a quartic equation.*

Proof. The proof is provided in the supporting information (SI) **Appendix A** available online [61].   □

Proposition 3. *Given the diagonalized alignment tensor components S_{xx} and S_{yy}, the peptide plane P_i, a ϕ-defining RDC and a ψ-defining RDC for ϕ_i and ψ_i, respectively, there exist at most 16 orientations of the peptide plane P_{i+1} with respect to P_i that satisfy the RDCs.*

Proof. This follows directly from Proposition 1 and Proposition 2.   □

2.4   Analytic Solutions for the ϕ Angle of Glycine from C^α-H^α RDC

The amino acid residue glycine (Gly) has two H^α atoms which we denote by $H^{\alpha 2}$ and $H^{\alpha 3}$, respectively. The C^α-H^α RDC measured for Gly is the sum of the RDCs for these two bond vectors. We show that given the C^α-H^α RDC for a Gly residue we can compute all possible solutions for the dihedral ϕ.

Proposition 4. *Given the diagonalized alignment tensor components S_{xx} and S_{yy}, the peptide plane P_i, and the C^α-H^α RDC r for residue i which is a glycine, there exist at most 4 possible values of the dihedral angle ϕ_i that satisfy the C^α-H^α RDC r. The possible values of ϕ_i can be computed exactly and in closed form by solving a quartic equation.*

Proof. The proof is provided in the SI **Appendix B** available online [61].   □

2.5   Sampling the DOFs When RDCs Are Missing

Theoretically, for a loop with n (> 6) DOFs, $n - 6$ DOFs are redundant. Therefore, $n - 6$ equality constraints are necessary to solve for the loop conformations so that the number of conformations is discrete. We systematically sample (at 5° resolution) the dihedrals from the Ramachandran map (and TALOS dihedral restraints if available) for the DOFs for which RDCs are missing, and use analytic equations to solve for the other dihedrals for which RDCs are available, to compute an ensemble of loops complete to the resolution of sampling. If RDCs can be recorded for the missing ones in a second alignment medium, POOL can use them (see the online SI **Appendix C** [61]). Table 2 shows that when as many as 5 RDCs are missing in a loop, POOL still could compute the loops accurately.

2.6   Pruning with Conformation Filters

Loop conformations are generated by traversing a conformation tree in a depth-first search order (Section 2.1). At each node, conformation filters are applied as *predicates*. If the node passes all the filters, then the subtree rooted at that node is visited; otherwise, the subtree is pruned. Failing a predicate at lower levels (closer to the root) of the conformation tree prunes more conformations than that detected at higher levels (farther from the root). In fact, pruning at depth i eliminates $O(b^{n-i})$ conformations, where b is the average number of branches in the conformation tree, and n is the height of the conformation tree. For loops with constrained work-space, substantial pruning can be achieved resulting in significant speedup. POOL uses the following conformation filters.

Real Solution Filter. While solving the equations derived in Sections 2.3 and 2.4 to compute the dihedrals, all non-real roots with the imaginary parts greater than a chosen threshold are discarded [72]. Also, multiplicities of the roots are eliminated, thereby pruning the subtrees rooted at the eliminated-roots.

Ramachandran and TALOS Filters. There exist regions in the Ramachandran map (*Rama-map*) that are forbidden for any biophysically relevant (ϕ, ψ) values for a given residue type. Therefore, any disallowed value for a dihedral suggested by the Rama-map, whenever it appears in the conformation tree, is pruned. We used the data from [38], and implemented a *residue-specific* Ramachandran filter. Our implementation considers four residue types: Gly, Pro, pre-Pro, and other general amino acid types (called *general*). It has been specifically optimized for $O(1)$-time queries for the *favored* or *allowed* intervals for ϕ, and ψ given ϕ. If $M_{\mathcal{T}}$ is the Rama-map for residue type $\mathcal{T}$, and $I_{\mathcal{T}}$ is the set of all allowed ϕ-intervals for $\mathcal{T}$, we evaluate if $\phi \in I_{\mathcal{T}}$ for a computed ϕ. Similarly, when a ψ is computed, we evaluate if $\psi \in I_{\mathcal{T}}|_\phi$. TALOS [14,53] dihedral information, whenever available, are used as follows. If for the dihedral ϕ_i of the residue i of type $\mathcal{T}$, $I_{\mathcal{L}}$ is the TALOS-predicted interval, then for a computed ϕ for the residue i, we evaluate if $\phi \in I_{\mathcal{T}} \cap I_{\mathcal{L}}$. Similarly, for a computed ψ, the predicate $\psi \in I_{\mathcal{T}}|_\phi \cap I_{\mathcal{L}}$ is evaluated. The subtree rooted at the node representing the dihedral is pruned if any of these predicates fail. Further, in the absence of RDC data for a dihedral, finite-resolution uniform sampling of the Rama-map is used for that dihedral.

Steric Filter. We use our in-house implementation of the steric checker similar to that in [70]. During the depth-first search of the conformation tree, at each node corresponding to a newly added residue, the steric check is performed for (i) self-collision, i.e., if the fragment clashes with itself, and (ii) collision with the rest of the protein. If the clash score [70] is greater than a user-defined threshold, then the branch is pruned and the search backtracks.

Reachability Criterion. As each node of the conformation tree is visited, we test if the rest of the fragment, if grown using the best possible kinematic chain, can ever reach the stationary anchor. The node is pruned if this test fails. For long loops, this test prunes a large fraction of conformations, especially at the tree nodes at higher level (farther from the root).

Closure Criterion. When the distance between the mobile anchor (i.e., the conformation at a leaf node), and the stationary anchor is less than a user-specified threshold (chosen to be 0.2 Å), called the *closure distance*, and defined as the root-mean-square distance between the N, C^α and C' atoms of the mobile anchor and stationary anchor, the conformation is accepted and added to the ensemble of computed loops. Otherwise, the conformation is subject to a gradient-descent minimization over the last few dihedrals to improve the closure distance to below 0.2 Å while maintaining the user-defined RDC RMSD thresholds. If after minimization the closure is achieved, the conformation is accepted; otherwise, rejected. The RDC RMSD between back-computed and experimental RDCs is computed using the equation $\mathrm{RMSD}_x = \sqrt{\frac{1}{n}\sum_{i=1}^{n}(r_{x,i}^b - r_{x,i}^e)^2}$, where x is either a ϕ-defining or a ψ-defining RDC type, n is the number of RDCs, $r_{x,i}^e$ is the experimental RDC, and $r_{x,i}^b$ is the corresponding back-computed RDC.

Pruning using Unambiguous NOEs. When unambiguous backbone NOEs are available, they can be used as predicates to prune unsatisfying conformations.

3   Results and Discussion

To study the effectiveness of our algorithm, we applied it on experimental NMR data sets for four proteins. Further, to study the robustness of our algorithm to

Table 2. (*a*) The anchor residues are always included. (*b*) number of residues. (*c*) experimental RDCs used. The C^α-H^α, C^α-C' and N-H^N RDC RMSDs of loops computed by POOL are less than 2.0, 0.2 and 1.0 Hz, respectively. (*d*) Missing means unavailable. (*e*) Backbone RMSD computed vs. the NMR reference loops. The results show that the loops computed by POOL are more accurate than those computed by XPLOR-NIH [51].

Protein Loop[a]	Length[b]	Types of RDCs[c]	RDCs missing[d]	RMSD[e] (Å) (POOL)	RMSD[e] (Å) (XPLOR-NIH)
Ubiquitin 7-12	6	C^α-H^α, N-H^N	2	0.64	1.40
Ubiquitin 17-23	7	C^α-H^α, N-H^N	2	0.60	2.25
Ubiquitin 33-41	9	C^α-H^α, N-H^N	2	0.89	2.07
Ubiquitin 45-48	4	C^α-H^α, N-H^N	0	0.27	1.58
Ubiquitin 50-65	16	C^α-H^α, N-H^N	2	0.66	3.94
Ubiquitin 7-12	6	C^α-C', N-H^N	3	0.37	0.67
Ubiquitin 17-23	7	C^α-C', N-H^N	3	0.60	3.54
Ubiquitin 33-41	9	C^α-C', N-H^N	5	0.58	3.11
Ubiquitin 45-48	4	C^α-C', N-H^N	0	0.11	1.02
Ubiquitin 50-65	16	C^α-C', N-H^N	4	1.06	4.48
FF2 18-27	10	C^α-H^α, N-H^N	3	1.41	3.20
FF2 33-38	6	C^α-H^α, N-H^N	3	0.34	1.09
FF2 42-48	7	C^α-H^α, N-H^N	4	1.31	2.14
DinI 8-17	10	C^α-H^α, N-H^N	5	1.57	4.17
DinI 32-39	8	C^α-H^α, N-H^N	3	0.61	3.45
DinI 45-49	5	C^α-H^α, N-H^N	2	0.28	2.27
DinI 53-58	6	C^α-H^α, N-H^N	2	0.42	2.62
GB3 8-13	6	C^α-H^α, N-H^N	0	0.43	1.07
GB3 19-23	5	C^α-H^α, N-H^N	0	0.34	0.23
GB3 36-42	7	C^α-H^α, N-H^N	1	0.27	1.34
GB3 46-51	6	C^α-H^α, N-H^N	0	0.65	3.61

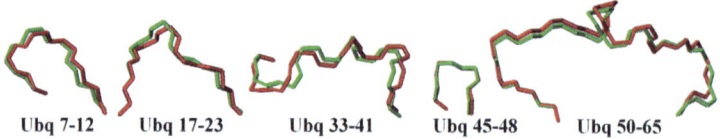

Fig. 2. Overlay of the loops (green) of ubiquitin computed by POOL using C^α-H^α and N-H^N RDCs vs. the corresponding loops (red) in the NMR reference structure (1d3z model 1) without any structural alignment

the variations in standard peptide geometry, we tested it on synthetic datasets for three sets of canonical loops of length 4, 8 and 12 residues that were investigated by three other protein loop closure algorithms [8,17,36].

3.1   Tests on Experimental NMR Data

We applied POOL to compute the loops of four proteins: FF2 (PDB id: 2kiq) [72], human ubiquitin (PDB id: 1d3z) [15], the DNA damage inducible protein I (DinI) (PDB id: 1ghh) [45], and the third IgG-binding domain of Protein G (GB3) (PDB id: 2oed) [62]. The RDC data for FF2 was recorded using Varian 600 and 800 MHz spectrometers at Duke University. Details of the NMR experimental procedures are provided in the SI **Appendix D** available online [61]. For ubiquitin, DinI and GB3, NMR data were obtained from BioMagResBank (BMRB) [63]. For each of these proteins, we used the NMR model 1 with loops removed as the respective test structures. RDCs were perturbed within the experimental-error window [66] to account for experimental errors.

Table 2 summarizes the results computed by POOL. For ubiquitin we used two different combinations of RDCs, viz. (C^α-H^α, N-H^N) and (C^α-C', N-H^N) to test the performance of our algorithm on different types of RDC data. In most cases, *sub-angstrom* RMSD loops were computed by POOL. Figure 2 shows the overlay of the loops computed for ubiquitin using C^α-H^α and N-H^N RDCs with the corresponding loops from the NMR reference structure. For FF2, DinI and GB3, the results show that POOL is able to compute accurate loops when as many as 5 RDCs are missing.

The run-time analysis of POOL is similar to that in [68]. In practice, for short loops, POOL runs in minutes, and for longer loops (e.g., ubiquitin 50-65) it runs in hours on a 2.5 GHz dual-core processor Linux workstation.

3.2   Comparison vs. Traditional Structure Determination Protocols

To investigate whether traditional SA/MD-based structure determination protocols can compute accurate loop conformations using sparse data, we ran XPLOR-NIH [51] on the same input used by POOL for ubiquitin, FF2, DinI and GB3. Table 2 summarizes the results. In Figure 3, a comparison is made between the results obtained by applying POOL versus those obtained by applying XPLOR-NIH. The loops computed by POOL have much smaller (3- to 6-fold less for longer loops) backbone RMSD vs. the reference structures than those computed using

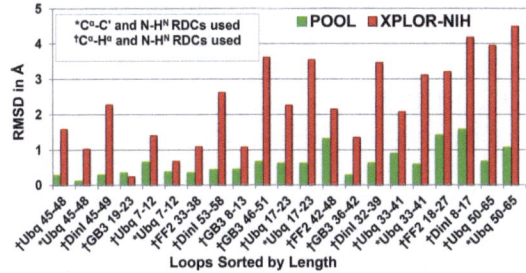

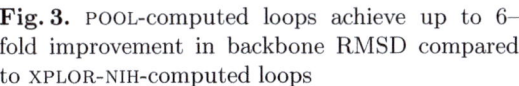

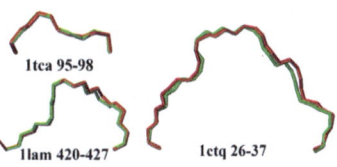

Fig. 3. POOL-computed loops achieve up to 6–fold improvement in backbone RMSD compared to XPLOR-NIH-computed loops

Fig. 4. Overlay of the lowest RMSD loop (green) computed by POOL for 4, 8 and 12-residue loops vs. the X-ray structures of the reference loops (red) without any structural alignment

Table 3. The minimum RMSD (Å) from X-ray structures for these four algorithms. The loops computed by POOL using only one ϕ-defining and one ψ-defining RDC per residue simulated using an alignment tensor estimated using PALES [76,75]. SOS, CSJD and CCD results were obtained from Table 1, Table 1 and Table 2 of [36], [17] and [8], respectively. These three methods do not use any experimental NMR data.

4-residue loops				8-residue loops				12-residue loops			
Loop	POOL	SOS	CSJD	CCD	Loop	POOL	SOS	CSJD	CCD	Loop	POOL	SOS	CSJD	CCD
1dvjA_20	0.74	0.23	0.38	0.61	1cruA_85	0.72	1.48	0.99	1.75	1cruA_358	1.54	2.39	2.00	2.54
1dysA_47	0.25	0.16	0.37	0.68	1ctqA_144	0.91	1.37	0.96	1.34	1ctqA_26	0.65	2.54	1.86	2.49
1eguA_404	0.42	0.16	0.36	0.68	1d8wA_334	0.28	1.18	0.37	1.51	1d4oA_88	1.83	2.44	1.60	2.33
1ej0A_74	0.18	0.16	0.21	0.34	1ds1A_20	0.70	0.93	1.30	1.58	1d8wA_46	0.93	2.17	2.94	4.83
1i0hA_123	0.27	0.22	0.26	0.62	1gk8A_122	0.87	0.96	1.29	1.68	1ds1A_282	1.50	2.33	3.10	3.04
1id0A_405	0.63	0.33	0.72	0.67	1i0hA_122	0.45	1.37	0.36	1.35	1dysA_291	0.76	2.08	3.04	2.48
1qnrA_195	0.47	0.32	0.39	0.49	1ixh_106	0.68	1.21	2.36	1.61	1eguA_508	1.25	2.36	2.82	2.14
1qopA_44	0.36	0.13	0.61	0.63	1lam_420	0.42	0.90	0.83	1.60	1f74A_11	0.76	2.23	1.53	2.72
1tca_95	0.12	0.15	0.38	0.39	1qopB_14	0.87	1.24	0.69	1.85	1qlwA_31	1.27	1.73	2.32	3.38
1thfD_121	0.25	0.11	0.36	0.50	3chbD_51	0.96	1.23	0.96	1.66	1qopA_178	0.87	2.21	2.18	4.57
Average	0.37	0.20	0.40	0.56	Average	0.69	1.19	1.01	1.59	Average	1.14	2.25	2.34	3.05

XPLOR-NIH. For example, for ubiquitin loop 50-65, the loop computed by POOL has backbone RMSD 0.66 Å, a 6-fold decrease vs. the loop computed by XPLOR-NIH (3.94 Å). This shows that when given sparse data, our algorithm is able to compute more accurate loop conformations than the SA/MD-based protocols.

3.3 Comparison with Loop Prediction Algorithms

We compared the performance of POOL with three other loop prediction algorithms including the CCD method by Canutescu and Dunbrack [8], the CSJD algorithm by Coutsias *et al.* [17], and the self-organizing superimposition (SOS) algorithm by Liu *et al.* [36]. Unlike these algorithms, which do not use any data, POOL is a sparse data-driven algorithm. While CCD, CSJD and SOS algorithms have applications in protein structure prediction, none of them is specifically designed to incorporate geometric restraints from experimental NMR data. Our algorithm POOL provides an approach to fill this gap by being able to compute loops using sparse NMR data, specifically, RDCs.

In our study, we used the same test set as in [36,17,8]. This set consists of 10 loops each with 4, 8 and 12 residues long chosen from a set of nonredundant X-ray crystallographic structures from the PDB. Since there is no experimental RDC data available for these proteins, we simulated the RDCs using PALES [76,75]. Details of the RDC simulation are described in the SI **Appendix E** available online [61]. The alignment tensor, the RDC data, and the two anchor peptide planes of the loop were used by POOL to compute the loop conformations.

Table 3 summarizes the results for POOL, CCD, CSJD and SOS algorithms. In Figure 4, examples of minimum RMSD loop conformations determined by POOL are shown. For 4-residue loops the average minimum RMSD of the computed loops by POOL is larger than that for SOS, but smaller than that for CSJD and CCD. This can be explained by the fact that SOS allows slight deviations from standard protein geometry. For 8 and 12-residue loops POOL computes more accurate loops than other algorithms. For example, for 12-residue loops, the average minimum RMSD of the loops are 1.14, 2.25, 2.34 and 3.05 Å for POOL, SOS, CSJD and CCD, respectively, which shows a 2-fold improvement in accuracy by POOL. For five of these loops, POOL computed loops with *sub-angstrom accuracy*. Further, the reference loops in Table 3 have deviations from standard protein geometry; therefore, the RDCs simulated on them inherits these deviations, in addition to a Gaussian noise of 1 Hz added to account for experimental errors. These results suggest that POOL is robust to both experimental uncertainties in RDCs, and minor deviations from standard protein geometry assumptions. Therefore, POOL can be useful to compute longer loops with high accuracy using a minimal amount of RDC data.

4   Conclusions

While the global fold of a protein can often be determined from experimental NMR data [25,66,68,72,71], determining loop conformations from sparse experimental RDCs is a difficult problem. We described a novel, efficient, and practical deterministic algorithm, POOL, that determines accurate loop conformations from sparse RDC data. Empirical comparison with traditional structure determination protocols [51] demonstrates that POOL is able to achieve up to 6-fold improvement over the latter methods under sparse-data settings.

Since an accurate and complete protein backbone is a prerequisite for NOE-assignment algorithms [28,72] and side-chain resonance assignment methods [74] in traditional NMR structure determination protocols, POOL will be useful in high-resolution protein structure determination. Whenever RDCs can be collected for proteins with known X-ray structures containing missing loops, POOL can be used to determine the loop conformations.

Since RDCs also provide sensitive probes to protein conformational dynamics [59,48] over nano- to millisecond timescales, it will be interesting to extend our algorithm to capture and characterize the motional fluctuations, and deconvolve the dynamics from measured RDCs. In such cases, the ensemble of loops computed by POOL will effectively define a normal distribution of conformations centered at the experimentally-measured RDCs, and as such encode a unimodal dynamic ensemble about a protein's native fold. Our algorithm can even be a stepping stone to computing ensembles reflecting more complex dynamics.

Availability. The source code of our algorithm is available open-source under the GNU Lesser General Public License (Gnu, 2002).

Acknowledgments. We thank professors Jane and Dave Richardson, Dr. A. Yershova, Dr. A. Yan, Mr. V. Chen, Mr. J. MacMaster, Mr. C.-Y. Chen, Mr. J. Martin, Mr. P. Gainza, Mr. M. Hallen and Ms. S. Jain for many valuable suggestions. We thank all members of the Donald, Zhou and Richardson Labs for helpful discussions.

References

1. Andrec, M., et al.: J. Biomol. NMR 21, 335–347 (2004)
2. Baker, D., Sali, A.: Science 294, 93–96 (2001)
3. Berman, H.M., et al.: Nucleic Acids Res. 28(1), 235–242 (2000)
4. Bouvignies, G., et al.: Angewandte Chemie 118, 8346–8349 (2006)
5. Bruccoleri, R.E., Karplus, M.: Macromolecules 29, 1847–1862 (1990)
6. Brünger, A.T.: Yale University Press, New Haven (1992)
7. Buchbinder, J.L., Fletterick, R.J.: J. Biol. Chem. 271(37), 22305–22309 (1996)
8. Canutescu, A.A., Dunbrack Jr., R.L.: Protein Sci. 12(5), 963–972 (2003)
9. Casey, J.: Proceedings of the Royal Society of London XIX, 495–497 (1871)
10. Chen, C.-Y., et al.: Proc. Natl. Acad. Sci. USA 106(10), 3764–3769 (2009)
11. Chirikjian, G.S.: In: Proceedings of IROS, vol. 2, pp. 1067–1073 (1993)
12. Clore, G.M., et al.: J. Magn. Reson. 131, 159–162 (1998)
13. Collura, V., et al.: Protein Sci. 2, 1502–1510 (1993)
14. Cornilescu, G., et al.: J. Biomol. NMR 13, 289–302 (1999)
15. Cornilescu, G., et al.: J. Am. Chem. Soc. 120, 6836–6837 (1998)
16. Cortés, J., et al.: J. Comput. Chem. 25(7), 956–967 (2004)
17. Coutsias, E.A., et al.: J. Comput. Chem. 25, 510–528 (2004)
18. Delaglio, F., et al.: J. Am. Chem. Soc. 122, 2142–2143 (2000)
19. Donald, B.R., Martin, J.: Prog. NMR Spectrosc. 55(2), 101–127 (2009)
20. Du, P., et al.: Protein Engineering 16(6), 407–414 (2003)
21. Fine, R.M., et al.: Proteins 1(4), 342–362 (1986)
22. Fiser, A., et al.: Protein Sci. 9(9), 1753–1773 (2000)
23. Frey, K.M., et al.: Proc. Natl. Acad. Sci. USA 107(31), 13707–13712 (2010)
24. Georgiev, I., et al.: J. Comput. Chem. 29, 1527–1542 (2008)
25. Giesen, A.W., et al.: J. Biomol. NMR 25, 63–71 (2003)
26. Gō, N., Scheraga, H.A.: Macromolecules 3, 178–187 (1970)
27. Gorczynski, M.J., et al.: Chemistry & Biology 14(10), 1186–1197 (2007)
28. Güntert, P.: Prog NMR Spectrosc. 43, 105–125 (2003)
29. Hu, X., et al.: Proc. Natl. Acad. Sci. USA 104(45), 17668–17673 (2007)
30. Koehl, P., Delarue, M.: Nat. Struct. Biol. 2, 163–170 (1995)
31. Kolodny, R., et al.: Int. J. Robot Res. 24, 151–163 (2005)
32. Kuszewski, J., et al.: J. Am. Chem. Soc. 126(20), 6258–6273 (2004)
33. Langmead, C.J., Donald, B.R. In: Proceedings of CSB, pp. 209–217 (2003)
34. Langmead, C.J., Donald, B.R. In: Proceedings of CSB, pp. 278–289 (2004)
35. Lilien, R.H., et al.: J. Comput. Biol. 12(6), 740–761 (2005)
36. Liu, P., et al.: PLoS Comput. Biol. 5(8), e1000478 (2009)
37. Losonczi, J.A., et al.: J. Magn. Reson. 138, 334–342 (1999)
38. Lovell, S.C., et al.: Proteins 50, 437–450 (2003)

39. Manocha, D., Canny, J.F.: IEEE T. Robotic Autom. 10, 648–657 (1994)
40. Milgram, R.J., et al.: J. Comput. Chem. 29(1), 50–68 (2008)
41. Mumenthaler, C., et al.: J. Biomol. NMR 10(4), 351–362 (1997)
42. Pesce, S., Benezara, R.: Mol. Cell. Biol. 13(12), 7874–7880 (1993)
43. Prestegard, J.H., et al.: Chemical Reviews 104, 3519–3540 (2004)
44. Ramirez, B.E., Bax, A.: J. Am. Chem. Soc. 120, 9106–9107 (1998)
45. Ramirez, B.E., et al.: Protein Sci. 9, 2161–2169 (2000)
46. Rohl, C.A., Baker, D.: J. Am. Chem. Soc. 124, 2723–2729 (2002)
47. Salmon, G.: Longmans, Green and Company, London (1912)
48. Salmon, L., et al.: Angew Chem. Int. Edit. 48(23), 4154–4157 (2009)
49. Saupe, A.: Angewandte Chemie 7(2), 97–112 (1968)
50. Saxe, J.B.: In: Proc. 17th Allerton Conf. Comm., Ctrl. Comput., pp. 480–489 (1979)
51. Schwieters, C.D., et al.: J. Magn. Reson. 160, 65–73 (2003)
52. Shehu, A., et al.: Proteins 65(1), 164–179 (2006)
53. Shen, Y., et al.: J. Biomol. NMR 44, 213–223 (2009)
54. Shenkin, P.S., et al.: Biopolymers 26(12), 2053–2085 (1987)
55. Shi, L., Javitch, J.A.: Proc. Natl. Acad. Sci. USA 101(2), 440–445 (2004)
56. Tian, F., et al.: J. Am. Chem. Soc. 123, 11791–11796 (2001)
57. Tjandra, N., Bax, A.: Science 278, 1111–1114 (1997)
58. Tolman, J.R., et al.: Proc. Natl. Acad. Sci. USA 92, 9279–9283 (1995)
59. Tolman, J.R., et al.: Nat. Struct. Biol. 4(4), 292–297 (1997)
60. Tosatto, S.C.E., et al.: Protein Engineering 15(4), 279–286 (2002)
61. Tripathy, C., Zeng, J., Zhou, P., Donald, B.R.: Supporting Information (2011),
    http://www.cs.duke.edu/donaldlab/Supplementary/recomb11/pool/
62. Ulmer, T.S., et al.: J. Am. Chem. Soc. 125, 9179–9191 (2003)
63. Ulrich, E.L., et al.: Nucleic Acids Res. 36(Database issue), D402–D408 (2008)
64. van Vlijmen, H.W.T., Karplus, M.: J. Mol. Biol. 267, 975–1001 (1997)
65. Wang, C., et al.: J. Mol. Biol. 373(2), 503–519 (2007)
66. Wang, L., Donald, B.R.: J. Biomol. NMR 29(3), 223–242 (2004)
67. Wang, L., Donald, B.R.: In: Proceedings of CSB, pp. 189–202 (2005)
68. Wang, L., et al.: J. Comput. Biol. 13(7), 1276–1288 (2006)
69. Wedemeyer, W.J., Scheraga, H.A.: J. Comput. Chem. 20(8), 819–844 (1999)
70. Word, J.M., et al.: J. Mol. Biol. 285, 1711–1733 (1999)
71. Yershova, A., et al.: In: Proceedings of WAFR, vol. 68, pp. 355–372 (2010)
72. Zeng, J., et al.: J. Biomol. NMR 45(3), 265–281 (2009)
73. Zeng, J., et al.: In: Proceedings of CSB, pp. 169–181 (2008) ISBN 1752–7791
74. Zeng, J., et al.: A markov random field framework for protein side-chain resonance
    assignment. In: Berger, B. (ed.) RECOMB 2010. LNCS, vol. 6044, pp. 550–570.
    Springer, Heidelberg (2010)
75. Zweckstetter, M.: Nat. Protoc. 3, 679–690 (2008)
76. Zweckstetter, M., Bax, A.: J. Am. Chem. Soc. 122(15), 3791–3792 (2000)

De Novo Discovery of Mutated Driver Pathways in Cancer

Fabio Vandin[1,2,*,**], Eli Upfal[1,**], and Benjamin J. Raphael[1,2,**]

[1] Department of Computer Science, Brown University, Providence, RI
[2] Center for Computational Molecular Biology, Brown University, Providence, RI
{vandinfa,eli,braphael}@cs.brown.edu

Motivation. Next-generation DNA sequencing technologies are enabling genome-wide measurements of somatic mutations in large numbers of cancer patients. A major challenge in interpretation of this data is to distinguish functional *driver* mutations that are important for cancer development from random, *passenger* mutations. A common approach to identify driver mutations is to find genes that are mutated at significant frequency in a large cohort of cancer genomes. This approach is confounded by the observation that driver mutations target multiple cellular signaling and regulatory pathways. Thus, each cancer patient may exhibit a different combination of mutations that are sufficient to perturb the necessary pathways. However, the current understanding of the somatic mutational process of cancer [3,5,6] places two additional constraints on the expected patterns of somatic mutations in a cancer pathway. First, an important cancer pathway should be perturbed in a large number of patients. Thus we expect that with genome-wide measurements of somatic mutations a driver pathway will exhibit high *coverage*, where most patients will have a mutation in some gene in the pathway. Second, since driver mutations are relatively rare and typically a single driver mutation is sufficient to perturb a pathway, a reasonable assumption is that most patients have a single driver mutation in a pathway. Thus, the genes in a driver pathway exhibit a pattern of *mutually exclusive* driver mutations, where driver mutations are observed in exactly one gene in the pathway in each patient. There are numerous examples of sets of mutually exclusive mutations [5,6].

Methods. Motivated by these biological observations and the availability of somatic mutation data on large sets of patients, we introduce the problem of finding sets of genes with the following properties: (i) *coverage*: most patients have at least one mutation in the set; (ii) *exclusivity*: nearly all patients have no more than one mutation in the set. We define a measure on sets of genes that quantifies how much sets exhibit both properties and show that finding sets of genes that

---

* Corresponding author.
** This work is supported by NSF grant IIS-1016648, the Department of Defense Breast Cancer Research Program, the Alfred P. Sloan Foundation, and the Susan G. Komen Foundation. BJR is also supported by a Career Award at the Scientific Interface from the Burroughs Wellcome Fund.

V. Bafna and S.C. Sahinalp (Eds.): RECOMB 2011, LNBI 6577, pp. 499–500, 2011.
© Springer-Verlag Berlin Heidelberg 2011

optimize this measure is NP-hard. In contrast, we prove that a straightforward greedy algorithm produces an optimal solution with high probability when given a sufficiently large sets of patients and subject to some statistical assumptions on the distribution of the mutations.

Since the number of patients in currently available datasets is only in the hundreds and the statistical assumptions for the greedy algorithm may be too restrictive (e.g. they are not satisfied by copy number aberrations), we also introduce a second approach. We use a Markov Chain Monte Carlo (MCMC) algorithm to sample from sets of genes with a distribution that gives significantly higher probability to sets of genes with high coverage and exclusivity. Our MCMC algorithm is based on the Metropolis-Hastings method. Although the Metropolis-Hastings method defines a chain that is guaranteed to converge to the desired stationary distribution, there is in general no guarantee how rapidly the chain will converge. While there has been significant progress in recent years in developing mathematical tools for analyzing the convergence time [4], our ability to analyze useful chains is still limited, and in practice most MCMC algorithms rely on simulations to provide evidence of convergence to stationarity [2]. Nevertheless, we prove that our MCMC algorithm converges rapidly to the equilibrium distribution.

Results. We apply the MCMC algorithm to analyze sequencing data from 623 genes in 188 lung adenocarcinoma patients and 601 genes in 84 glioblastoma patients. In both datasets, we find sets of 2-3 genes that are mutated in large subsets of patients and are largely exclusive. These sets include genes in the Rb, p53, and mTor signaling pathways, all pathways known to be important in cancer. In glioblastoma, the set of three genes that we identify was shown to be associated with shorter survival [1]. Finally, we show that the MCMC algorithm efficiently identifies sets of six genes with high coverage and exclusivity in simulated mutation data with thousands of genes and patients.

References

1. Backlund, L.M., Nilsson, B.R., Goike, H.M., Schmidt, E.E., Liu, L., Ichimura, K., Collins, V.P.: Short postoperative survival for glioblastoma patients with a dysfunctional Rb1 pathway in combination with no wild-type PTEN. Clin. Cancer Res. 9, 4151–4158 (2003)
2. Gilks, W.: Markov Chain Monte Carlo in Practice. Chapman and Hall, London (1998)
3. McCormick, F.: Signalling networks that cause cancer. Trends Cell Biol. 9, M53–M56 (1999)
4. Randall, D.: Rapidly mixing markov chains with applications in computer science and physics. Computing in Science and Engineering 8(2), 30–41 (2006)
5. Vogelstein, B., Kinzler, K.W.: Cancer genes and the pathways they control. Nat. Med. 10, 789–799 (2004)
6. Yeang, C.H., McCormick, F., Levine, A.: Combinatorial patterns of somatic gene mutations in cancer. The FASEB Journal (2008)

An Unbiased Adaptive Sampling Algorithm for the Exploration of RNA Mutational Landscapes under Evolutionary Pressure

Jérôme Waldispühl[1],* and Yann Ponty[2],*

[1] School of Computer Science & McGill Center for Bioinformatics,
McGill University, Montreal, Canada
[2] Laboratoire d'Informatique, École Polytechnique, Palaiseau, France
jeromew@cs.mcgill.ca, yann.ponty@lix.polytechnique.fr

Abstract. The analysis of the impact of mutations on folding properties of RNAs is essential to decipher principles driving molecular evolution and to design new molecules. We recently introduced an algorithm called RNAmutants which samples RNA sequence-structure maps in polynomial time and space. However, since the mutation probabilities depend of the free energy of the structures, RNAmutants is bias toward G+C-rich regions of the mutational landscape. In this paper we introduce an unbiased adaptive sampling algorithm that enables RNAmutants to sample regions of the mutational landscape poorly covered by previous techniques. We applied the method to sample mutations in complete RNA sequence-structures maps of sizes up to 40 nucleotides. Our results indicate that the G+C-content has a strong influence on the evolutionary accessible structural ensembles. In particular, we show that low G+C-contents favor the apparition of internal loops, while high G+C-contents reduce the size of the evolutionary accessible mutational landscapes.

1 Introduction

Our understanding of the mechanisms regulating cell activity has considerably improved over the last two decades. Ribonucleic acids (RNAs) have emerged as one of the most important biomolecules, playing key roles in various aspects of the gene transcription and regulation processes. For instance, ribozymes are involved in the cleavage of messenger RNAs (mRNAs), and riboswitches undergo structural changes to regulate gene expression.

To achieve their functions, RNAs use sophisticated structures which are mainly determined by their sequence. Any modification of the sequence may result in a change in its structure and a loss (or an improvement) in function. The development of tools to estimate the effect of mutations on structures, or conversely the influence of structure conservation on the mutational process, is essential for understanding the mechanisms of molecular evolution [1], the origin of genetic

---

* Corresponding authors.

V. Bafna and S.C. Sahinalp (Eds.): RECOMB 2011, LNBI 6577, pp. 501–515, 2011.

diseases [2] or to develop bioengineering applications such as the design of RNA molecules (a.k.a. inverse folding) [3].

To understand the role of specific nucleotides, mutagenesis experiments proceed to point-wise mutations in order to observe putative changes in the expression profile of the experiments (i.e. the experimental observation) revealing a modification of the functionality of the molecule. Such experiments are critical to identify mutations modifying the function and structure of RNAs. However, all experiments are time-consuming and have a substantial cost, and it follows that exhaustive experimental studies are impossible.

While it is not realistic to conduct large scale experimental studies on the complete RNA mutational landscape, this limitation could be circumvented in computational studies. Indeed, the structure of RNAs can be predicted from sequence data only [4,5]. More importantly, the secondary structure can be predicted with dynamic programming techniques in polynomial time and space [4,6] using a nearest neighbor energy model [7]. These algorithms are implemented in various programs [8,9,10] and enable to predict the secondary structures of thousands of sequences in a short time. Therefore it has becomes possible to compute the complete mutational landscape small RNA sequences [11] and to simulate the evolution of the structure of populations of RNAs [12,13].

Several groups intended to explore the mutational landscape of RNAs and to quantify the dependences between sequences and structures. The most representative work in this area has been achieved by P. Schuster and co-worker on the sequence-structure maps and neutral networks [14,15]. So far, all these studies were limited by brute force approaches requiring to compute individually the structure of a number of mutants growing exponentially with the length of the sequence (e.g. there is 4^n sequences of length n), thus making an exhaustive exploration of the mutational landscape intractable on sequences with more than 20 nucleotides.

To address this issue, we have developed the program RNAmutants which, from an input sequence, computes the structural ensembles of all sequences with k mutations in *polynomial time and space* [16]. To achieve this algorithmic advance, we expanded the seminal dynamic programming rules introduced 30 years ago by Zuker and Stiegler [4]. The dramatic improvement of the algorithmic complexity (from an exponential to a polynomial running time) enabled us to investigate problems that could not have been addressed with previous techniques. For instance, we provided evidences that the complete sequence of the 3'UTR of the GB RNA virus C has been optimized to preserve its secondary structure from the deleterious effect of mutations [16]. RNAmutants has been developed upon a formal grammar-based model [17,18] which, in particular, can be used to compute k-mutants (i.e. sequences with exactly k mutations) with the lowest free energy structure.

Formally, RNAmutants takes an input sequence and computes the minimum free energy (MFE) structure and the Boltzmann partition function of all k-mutants sequences in the k-Hamming neighborhood (i.e. ensemble of sequences with exacly k mutations) of the input sequence. In addition, it samples k-mutants together with a secondary structure. This naturally extends the seminal Zuker and Stiegler's [4], McCaskill's [6] and Ding and Lawrence's algorithms [19] which do not consider sequence variations.

In this model, the probabilities of the sequences in the k-mutants ensembles are determined by their ensemble free energies. It follows that these ensembles are dominated by sequences with high G+C-contents and that RNAmutants has a bias towards A/U→C/G mutations. This bias is a serious drawback for complete and rigorous analysis of RNA sequence-structure maps or the prediction of mutations altering the native structure. Indeed, sampling sequences with a large number of mutations will necessarily produce mutants with high G+C-content folding into long single stem structures, while in reality a broader range of structures are observed. The nucleotide distribution can also be used to indirectly control the folding and functional properties on RNAs. For instance, Chan *et al.* showed a correlation between the G+C-content and the RNAi efficiency [20]. Finally, RNAs may also experience a stabilizing selection, independent of their structural stability, acting directly on their nucleotide composition [21]. But, such scenarios are impossible to model using the original RNAmutants algorithm.

In this paper, we develop an unbiased adaptive sampling algorithm enabling to control of the nucleotide composition of the sequences sampled by RNAmutants. These techniques alleviate RNAmutants from its previous limitations and enable us to study mutational processes at a finer resolution level. Importantly, this algorithmic advance is achieved at a minimal computational cost and can be generalized to sample any regions of the mutational and structural landscapes which are difficult to reach with classical algorithms.

This article is organized as follows. In section 2, we formally define the problem addressed, explain why a brute force approach fails, and show how a multivariate Boltzmann model can be integrated into RNAmutants to control the nucleotide composition of sampled sequences. Then, in section 3 we illustrate the efficiency of our techniques by providing an analysis of complete sequences-structure maps of RNAs of sizes up to 40 nucleotides (while previous exhaustive studies were limited to sizes of 20). Our computational experiments reveal interesting properties of RNA sequence-structure maps that can be parameterized by the G+C-content. In particular, we find that low G+C-contents favor the apparition of bulges and internal loops, thus the possible insertion of non-canonical interactions and tertiary structure motifs (Section. 3). We also show that the diversity of mutants improving the stability of the fold is effectively optimal for medium G+C-contents (around 50%) and that high G+C-contents reduce the size of the evolutionary reachable mutational landscape (Section. 3). These finding suggest that the G+C-content is essential to balance the competition between the evolutionary accessibility (i.e. the sequence diversity) and the structural stability.

2   Methods

Notations, Definitions and Existing Works. Throughout this document, we will abstract an RNA molecule ω as a sequence of bases over an alphabet

$\mathbb{B} := \{A, C, G, U\}$. The length of an RNA sequence will be denoted by $n = |\omega|$. Following standard notations, we will denote by ω_i the base at position i, and by $\omega_{i,j}$ the portion of ω delimited by positions i and j inclusive. A secondary structure s for an RNA ω is defined as a set of base pairs of the form $(i, j) \in [1, n]^2$ with $i < j$, such that any two base pairs $\{(i, j), (k, l)\} \subset s$ do not share an extremity ($\{i, j\} \cap \{k, l\} = \varnothing$), and are either non-overlapping ($[i, j] \cap [k, l] = \varnothing$) or stricly inclusive ($[i, j] \subset [k, l]$ or $[k, l] \subset [i, j]$). Moreover in order to avoir steric clashes, a minimal number of bases θ is usually required between the two extremities of a base pair (i, j) ($i + \theta < j$). Let us denote by $\mathcal{S}_{\omega,\theta}$ the set of all secondary structures compatible with a given RNA ω under the θ constraint.

Free-Energy model. For the sake of clarity, we will illustrate our claims and algorithms on a generalization of the energy model proposed by Nussinov and Jacobson [22], assigning additive free-energy contributions to each base-pair. This model may appear overly simplistic in comparison with the Turner model [4], but it is sufficient to capture the key algorithmic elements while remaining easier to grasp. It should however be noted that the implementations used for our experiments make use of the full Turner model, as was described in the initial presentation of RNAMutants [16].

In this section, each base-pair $(a, b) \in s$ within a sequence ω is associated with a free-energy contribution $\Delta_{\omega_a, \omega_b}$ and unpaired bases are not taken into account by the model. Consequently the overall free-energy $E(\omega, s)$ of a structure s over a sequence ω is given by $E(\omega, s) = \sum_{(i,j) \in s} \Delta_{\omega_i, \omega_j}$. Note that this energy model captures the incompatibility of a base-pair $(x, y) \in \mathbb{B}^2$ upon setting $\Delta_{x,y} = +\infty$.

Partition Function. Following McCaskill [6], one can define a Boltzmann distribution and assign to each structure s a Boltzmann factor $\mathcal{B}_\omega(s) := e^{\frac{-E(\omega,s)}{RT}}$ where is T the temperature and R the universal gas constant. This induces a Boltzmann probability distribution on the set $\mathcal{S}_{\omega,\theta}$ of structures compatible with ω such that

$$P(s \mid \omega) = \frac{\mathcal{B}_\omega(s)}{\mathcal{Z}_\omega} \tag{1}$$

where $\mathcal{Z}_\omega$ is the partition function defined as $\mathcal{Z}_\omega = \sum_{s \in \mathcal{S}_{\omega,\theta}} \mathcal{B}_\omega(s)$.

Restricting our attention to an interval $[i, j]$ of ω, we can easily observe that within a secondary structure on $[i, j]$, the first position i is either unpaired and is followed by a secondary structure on $[i + 1, j]$, or paired to some position $l \in [i + \theta + 1, j]$, in which case the non-crossing condition forces the existence of two independent structures on intervals $[i + 1, l - 1]$ and $[l + 1, j]$. Furthermore this case decomposition is complete as shown by Waterman [23].

A restricted version of the partition function, only considering the subinterval $\omega_{i,j}$ of sequence ω, can then be defined as $\mathcal{Z}_{[i,j]} = \sum_{s \in \mathcal{S}_{\omega_{i,j},\theta}} \mathcal{B}_\omega(s)$, and be computed recursively by $\mathcal{Z}_{[i,i-1]} = 1$ and

$$\mathcal{Z}_{[i,j]} = \mathcal{Z}_{[i+1,j]} + \sum_{l=i+\theta+1}^{j} e^{-\frac{\Delta_{\omega_i,\omega_l}}{RT}} \mathcal{Z}_{[i+1,l-1]} \cdot \mathcal{Z}_{[l+1,j]}. \tag{2}$$

Function GenMuts($i, j, k, \mathbf{w}, \omega$): Returns a sequence/structure couple over interval (i, j) at distance k of ω, drawn with respect to a $\mathbf{w}$-weighted Boltzmann probability.

---

if $i > j$ then return ε (Empty sequence);   // Terminal case
rand $\leftarrow$ Random($\mathcal{Z}_{\left[\substack{i,j \\ k}\right]}$);

for $b \in \mathbb{B}$ do   // Unpaired case
$\quad$ rand $\leftarrow$ rand $- \mathbf{w}^{|b|_{GC}} \cdot \mathcal{Z}_{\left[\substack{i+1,j \\ k-\sigma_{\omega_i,b}}\right]}$;
$\quad$ if rand < 0 then return $\begin{bmatrix} \bullet \\ b \end{bmatrix} \cdot$ GenMuts$(i+1, j, k - \sigma_{\omega_i,b}, \mathbf{w}, \omega)$;

for $b, b' \in \mathbb{B}^2$ do   // Paired case
$\quad$ for $l' \leftarrow i + \theta + 1$ to j do
$\quad\quad$ $\delta \rightarrow l' - (i + \theta + 1)$;   // Boustrophedon search
$\quad\quad$ if δ is even then $l \leftarrow i + \theta + 1 + \left\lfloor \frac{l'}{2} \right\rfloor$ else $l \leftarrow j - \left\lfloor \frac{l'-1}{2} \right\rfloor$;
$\quad\quad$ for $k' \leftarrow 0$ to $k - \sigma_{\omega_i \omega_l, bb'}$ do
$\quad\quad\quad$ rand $\leftarrow$ rand $- \mathbf{w}^{|bb'|_{GC}} \cdot e^{-\frac{\Delta_{b,b'}}{RT}} \cdot \mathcal{Z}_{\left[\substack{i+1,l-1 \\ k'}\right]} \cdot \mathcal{Z}_{\left[\substack{l+1,j \\ k-k'-\sigma_{\omega_i \omega_l, bb'}}\right]}$;
$\quad\quad\quad$ if rand < 0 then
$\quad\quad\quad\quad$ return $\begin{bmatrix} (\\ b \end{bmatrix} \cdot$ GenMuts$(i+1, l-1, k', \mathbf{w}, \omega) \cdot \begin{bmatrix}) \\ b' \end{bmatrix} \cdot$
$\quad\quad\quad\quad$ GenMuts$(l+1, j, k - k' - \sigma_{\omega_i \omega_l, bb'}, \mathbf{w}, \omega)$;

The partition function $\mathcal{Z}_\omega := \mathcal{Z}_{[1,n]}$ can therefore be computed in $\Theta(n^3)/\Theta(n^2)$ time and space. Direct applications of this algorithm include the derivation of base-pairing probabilities [6] and statistical sampling [19].

RNAMutants. For the sake of completeness, let us remind that the RNAMutants algorithm [16] starts from an initial sequence ω and traverses the space of all sequences parameterized by their Hamming distance to ω. A parameterized analogue of the partition function is then obtained by summing over sequences/structures couples that are compatible with a given interval (i, j) and a prescribed number of mutations k.

Let us first remind that the Hamming distance $\sigma : \mathbb{B}^n \times \mathbb{B}^n \to \mathbb{N}$ between two sequences of equal length is defined by $\sigma_{\varepsilon,\varepsilon} = 0$ and by $\sigma_{x.X',y.Y'} = \mathbb{1}_{x \neq y} + \sigma_{X',Y'}$. A partition function over k mutants over a subinterval $[i, j]$ is then defined by

$$\mathcal{Z}_{\left[\substack{i,j \\ k}\right]} = \sum_{\substack{\omega' \text{ s.t.} \\ \sigma_{\omega,\omega'} = k}} \sum_{s \in \mathcal{S}_{\omega'_{i,j},\theta}} \mathcal{B}_{\omega'}(s)$$

and can be recursively computed by $\mathcal{Z}_{\left[\substack{i,i-1 \\ 0}\right]} := 1$, $\mathcal{Z}_{\left[\substack{i,i-1 \\ k}\right]} := 0, \forall k > 0$, and

$$\mathcal{Z}_{\left[\substack{i,j \\ k}\right]} = \sum_{b \in \mathbb{B}} \mathcal{Z}_{\left[\substack{i+1,j \\ k-\sigma_{\omega_i,b}}\right]} + \sum_{b,b' \in \mathbb{B}^2} \sum_{l=i+\theta+1}^{j} \sum_{k'=0}^{k-\sigma_{\omega_i \omega_l,bb'}} e^{-\frac{\Delta_{b,b'}}{RT}} \cdot \mathcal{Z}_{\left[\substack{i+1,l-1 \\ k'}\right]} \cdot \mathcal{Z}_{\left[\substack{l+1,j \\ k-k'-\sigma_{\omega_i \omega_l,bb'}}\right]}.$$
$$(3)$$

A direct computation of the above recursion yields a $\Theta(n^3 \cdot k^2)$ time and $\Theta(n^2 \cdot k)$ space algorithm for computing the sequence/structure partition function, k being the maximal number of mutations.

Improved Statistical Sampling. Statistical sampling was introduced by Ding and Lawrence [19] and implemented within the SFold software. By contrast with previous algorithms which considered only the minimal free energy structure [4] or a deterministic subset of its suboptimals [24], this algorithm performs a stochastic backtrack and generates any suboptimal structure s for a sequence ω with respect to its Boltzmann probability (see Equation 1). Following a general weighted sampling scheme [25], the algorithm starts from an interval $[1, n]$, and chooses at each step one of the possible cases (First base being either unpaired or paired to some l) with probability proportional to the contribution of the case to the local partition function.

A direct adaptation of this principle based on Equation 3 gives Function GenMuts (upon setting $\mathbf{w} := 1$). By contrast with its original implementation [16], this sampling procedure uses a Boustrophedon search [26,27], decreasing the worst-case complexity of the stochastic backtrack from $\Theta(n^2 k)$ [16] to $\Theta(nk \log n)$. Therefore the generation of m structure/sequence couples at Hamming distance k of ω can be performed in $\Theta(n^3 \cdot k^2 + m \cdot nk \log n)$ worst-case complexity.

Reaching Regions of Predefined G+C-content. Now let us address the problem of sampling sequence/structure couples (ω', s') having predefined G+C-content $GC(s) = \frac{\#G(\omega) + \#C(\omega)}{|\omega|}$. The main difficulty here is that the interplay between the Boltzmann distribution and the combinatorial explosion of the number of sequences induces a drift of the expected G+C-content. Furthermore the G+C-content distribution is concentrated around its mean. Therefore a suitable sequence/structure will seldom be obtained by chance if the expected G+C-content does not match the targeted one. Our sampling procedure must also remain unbiased within areas of targeted G+C-content, i.e. generate each sequence/structure (ω', s') such that $\sigma_{\omega',\omega} = k$ and $GC(\omega') = gc^*$ with probability

$$p(\omega', s' \mid k, gc^*) = \frac{\mathcal{B}_{\omega'}(s')}{\sum_{\substack{(\omega'',s'') \text{ s.t.} \\ GC(\omega'')=gc^* \\ \text{and } \sigma_{\omega'',\omega}=k}} \mathcal{B}_{\omega''}(s'')}. \tag{4}$$

Direct Rejection Yields Exponential-Time Sampling. A natural idea for achieving an unbiased sampling consists in sampling from the complete set of structure/sequence and reject sequences of unsuitable G+C-content. Since an unsuitable couple can be generated repeatedly, the worst-case complexity (infinite) of such an algorithm is perhaps not very informative. Therefore we propose an average-case analysis, using methods developed in the *analysis of algorithms* community to determine the asymptotical limit of the G+C-content distribution.

Algorithm 1. Rejection algorithm

Input : RNA ω, targeted G+C-content gc^*, number of samples m, number of
        mutations k and weight $\mathbf{w}$.
Output: Set of m sequence/structure samples

FillMatrices$(\omega,k,\mathbf{w})$;
samples $\leftarrow \varnothing$;
while $|\text{samples}| < m$ **do**
    candidate $\leftarrow$ GenMuts$(1, n, \omega, k, \mathbf{w})$;
    if $GC(\text{candidate}) = gc^*$ **then** samples $\leftarrow$ samples $\cup$ {candidate};

return samples;

Theorem 1. *Assuming an homopolymer model (any base pair can form), a Nussinov-style energy function and an unconstrained number of mutations, the distribution of the number of G +C is asymptotically normal of mean $\mu \cdot n$ and standard deviation $\sigma \sqrt{n}$, for μ and σ positive real constants. The probability of sampling a sequence/structure of G+C-content gc^* is asymptotically equivalent to*

$$p(gc^* \mid n) \sim \frac{1}{\sigma\sqrt{2\pi n}} \cdot e^{-\frac{n(gc^* - \mu)^2}{2\sigma^2}}. \tag{5}$$

Successive attempts of Algorithm 1 are mutually independent, therefore the expected number of calls to GenMuts is the inverse of the probability assigned to a G+C-content of gc^*. It follows that, unless $\mu = gc^*$, the average-case complexity is asymptotically dominated by a term exponential in n, and Algorithm 1 has exponential complexity for some (most) targeted G+C-contents.

A Weighted Sampling Approach. We adapt a general approach recently proposed by Bodini *et al* [28], which uses weights to efficiently bias a random generation process towards areas of interest, while respecting a (renormalized) prior distribution. Namely let $\mathbf{w} \in \mathbb{R}^+$ be a weight associated with each occurrence of G or C, we define the $\mathbf{w}$-weighted partition function as

$$\mathcal{Z}_{[k]}^{[\mathbf{w}]} := \mathcal{Z}_{\begin{bmatrix} 1,n \\ k \end{bmatrix}}^{[\mathbf{w}]} = \sum_{\substack{\omega' \text{ s.t.} \\ \sigma_{\omega,\omega'}=k}} \sum_{s' \in \mathcal{S}_{\omega',\theta}} \mathcal{B}_{\omega'}(s') \cdot \mathbf{w}^{|\omega'|_{GC}}. \tag{6}$$

which can be computed by the following recurrence

$$\mathcal{Z}_{\begin{bmatrix} i,j \\ k \end{bmatrix}}^{[\mathbf{w}]} = \sum_{b \in \mathbb{B}} \mathbf{w}^{|b|_{GC}} \mathcal{Z}_{\begin{bmatrix} i+1,j \\ k-\sigma_{\omega_i,b} \end{bmatrix}}^{[\mathbf{w}]}$$

$$+ \sum_{b,b' \in \mathbb{B}^2} \sum_{l=i+\theta+1}^{j} \sum_{k'=0}^{k-\sigma_{\omega_i\omega_l,bb'}} \mathbf{w}^{|bb'|_{GC}} e^{-\frac{\Delta_{b,b'}}{RT}} \mathcal{Z}_{\begin{bmatrix} i+1,l-1 \\ k' \end{bmatrix}}^{[\mathbf{w}]} \mathcal{Z}_{\begin{bmatrix} l+1,j \\ k-k'-\sigma_{\omega_i\omega_l,bb'} \end{bmatrix}}^{[\mathbf{w}]} \tag{7}$$

where $|x|_{GC} := n \cdot GC(x)$ denotes the number of occurrences of G or C within x. Upon multiplying by a weight $\mathbf{w}$ whenever a Guanine or Cytosine is generated, a **w**-weighted probability distribution is induced on the sequence/structure and any sequence/structure (ω', s') such that $GC(\omega') = gc^*$ and $\sigma_{w,\omega'} = k$ has probability

$$p(\omega', s' \mid \mathbf{w}, k) = \frac{\mathbf{w}^{|x|_{GC}} \cdot \mathcal{B}_{\omega'}(s)}{\mathcal{Z}_{[k]}^{[\mathbf{w}]}}. \tag{8}$$

Function `GenMuts` implements a sampling procedure for the **w**-weighted distribution. Processing its output with Algorithm 1 discards any structure/sequence whose G+C-content differs from gc^*, and the probability of sampling a structure/sequence (ω', s') of G+C-content gc^* is therefore

$$p'(\omega', s' \mid gc^*, \mathbf{w}, k) = \frac{\mathbf{w}^{|\omega'|_{GC}} \mathcal{B}_{\omega'}(s')}{\displaystyle\sum_{\substack{(\omega'', s'') \\ \text{s.t. } GC(\omega'')=gc^* \\ \text{and } \sigma_{\omega'',\omega}=k}} \mathbf{w}^{|\omega''|_{GC}} \mathcal{B}_{\omega''}(s'')} = p(\omega', s' \mid gc^*, k) \tag{9}$$

since $|\omega'|_{GC} = |\omega''|_{GC} = n \cdot gc^*$. Our weighted sampling/rejection pipeline is consequently unbiased within the subset of sequence/structures having targeted G+C-content.

Let us now discuss the algorithmic gain achieved by this approach. Let us assume knowledge of a weight $\mathbf{w}^*$ such that $\mu_{\mathbf{w}^*} = gc^*$. First, let us point out that the proof of Theorem 1 does not rely on any specificity of the energy model/weighted scheme, but rather on intrinsic properties (strong connectedness and aperiodicity) of the context-free grammar used to model the structure/sequence space. It follows that Theorem 1 holds even in the presence of weights, with an additional dependency in **w** for $\mu_{\mathbf{w}}$ the expected G+C-content and $\sigma_{\mathbf{w}}$ its standard deviation. It also follows that the exponential part of the complexity cancels out, and the expected number of calls to `GenMuts` drops to $\Theta(\sqrt{n})$ per sample. Consequently, the generation of m structure/sequence couples at Hamming distance k of ω and G+C-content gc^* can be performed in $\Theta(n^3 \cdot k^2 + m \cdot n\sqrt{n} \cdot k \log n)$ average-case complexity.

Adaptive Weighted Sampling. To conclude, we need to find a weight $\mathbf{w}^*$ such that $gc^* = \mu_{\mathbf{w}^*}$. We claim that $\mathbf{w}^*$ can be computed using a bisection method as illustrated in Figure 1. Namely let $u_{gc^*, k}$ be the cumulated Boltzmann factors over all structures/sequences at distance k of ω, having G+C-content gc^*, then the probability of generating a sequence with G+C-content gc is exactly $p''_{\mathbf{w}, gc, k} := u_{gc, k} \cdot \mathbf{w}^{n \cdot gc} / \mathcal{Z}_{[k]}^{[\mathbf{w}]}$. It follows that

$$\forall k \geq 0, \mu_{\mathbf{w}, k} = \sum_{x=0}^{n} \frac{x}{n} \cdot p''_{\mathbf{w}, gc, k} = \sum_{x=0}^{n} \frac{x}{n} \cdot \frac{u_{x/n, k} \cdot \mathbf{w}^x}{\mathcal{Z}_{[k]}^{[\mathbf{w}]}}$$

$$\Rightarrow \frac{\partial \mu_{\mathbf{w}, k}}{\partial \mathbf{w}} = \sum_{x=0}^{n} \frac{x^2 \cdot u_{x/n, k} \cdot \mathbf{w}^{x-1}}{n \cdot \mathcal{Z}_{[k]}^{[\mathbf{w}]}} > 0.$$

We conclude that $\mu_{\mathbf{w}, k}$ is strictly increasing as a function of **w** and that a bisection search will converge exponentially fast toward the unique solution of $gc^* = \mu_{\mathbf{w}^*}$.

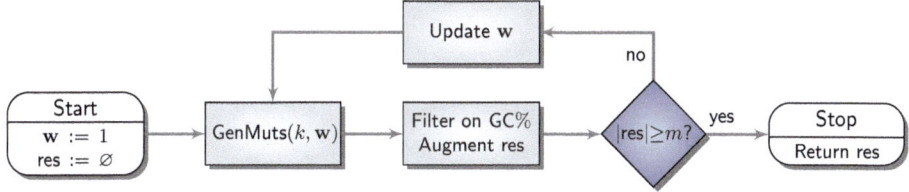

Fig. 1. General flow of the adaptive sampling procedure

Moreover following Equation 9 we know that the emission probability of each structure/sequence couple having suitable G+C-contents is not affected by the weighting scheme. Therefore samples obtained during any iteration of the bisection method can be accumulated into the resulting sample set **res**, and returned as soon as the targeted number of samples m is reached. The unbiasedness of each sampling, in addition with the independence of events, therefore yields an unbiased set of samples.

Implementation Remarks. Our implementation of the adaptive sampling described by Figure 1 uses the sampled sets to estimate expected G+C-contents. Since the G+C-content asymptotically follows a normal law of standard deviation in $\sigma\sqrt{n}$, a sampled set of size $M := K \times m \in \Omega(4n\sigma^2/\varepsilon^2)$, for some $K > 1$, will guarantee a 95% probability of falling within a confidence interval of $[(1 - \varepsilon)\mathbf{w}^*, (1 + \varepsilon)\mathbf{w}^*], \forall \varepsilon > 0$. The generation of such a growing number of samples will however remain negligible compared to the computation of the partition function. An exact expected value of $\mu_\mathbf{w}$ can also be computed exactly in $\Theta(n^3 \cdot k^2)$ using dynamic programming, following ideas underlying Miklos *et al* [29].

The value of $\mathbf{w}^*$ can also be exactly computed. Indeed the partition function can be expressed as $\mathcal{Z}^\mathbf{w}_{[k]} = \sum_{x=0}^{n} u_{x/n,k} \cdot \mathbf{w}^x$, i.e. a polynomial of degree n in $\mathbf{w}$. Therefore it suffices to evaluate $\mathcal{Z}^\mathbf{w}_{[k]}$ for n different values of $\mathbf{w}$ to determine the coefficients $u_{x/n,k}$ using a simple Gaussian elimination. From there, one can use numerical recipes (e.g. Grobner bases [30]) to find the unique root $\mathbf{w}^*$ of the polynom:

$$gc^* = \sum_{x=0}^{n} \frac{x \cdot u_{x/n,k} \cdot \mathbf{w}^{*x}}{n \cdot \mathcal{Z}^{\mathbf{w}^*}_{[k]}} \Leftrightarrow 0 = \sum_{x=0}^{n} (x - n \cdot gc^*) \cdot u_{x/n,k} \cdot \mathbf{w}^{*x}.$$

Also since the weighted partition functions $\mathcal{Z}^\mathbf{w}_{[k]}$ are computed prior to sampling, one can combine these two strategies into an hybrid approach, initially applying the bisection search and then dynamically switching to an exact computation after n computations of $\mathcal{Z}^\mathbf{w}_{[k]}$.

3   Results

Now we illustrate how RNAmutants can be used to explore RNA sequence-structure maps and analyze an evolutionary scenario based on the improvement

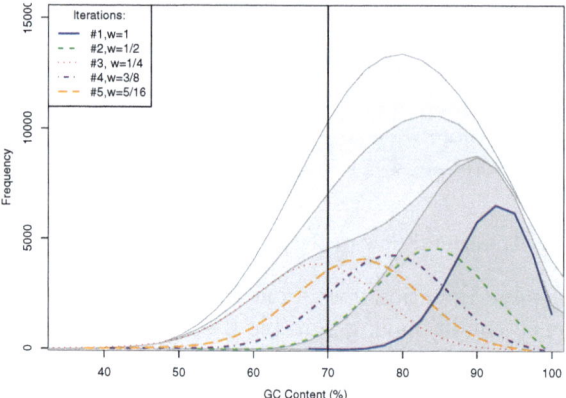

Fig. 2. Typical run of the adaptive sampling algorithm for sampling 10000 sequences having 70% G+C-content at Hamming distance k=16 of an ribosome entry site RNA(PDB: 2HUA_A). The smoothed G+C-content distributions are plotted with solid (unweighted) and dashed thick lines for each iteration of the bisection method. Grey shaded area correspond to their cumulative distribution. Here the adaptive sampling returns a suitable sample set after 5 iterations during which 15.10^4 sequences are generated. Based on observed probabilities an unweighted model would require generating about 3.10^7 sequences.

of the structure stability. This study can be motivated by a recent work of Cowperthwaite *et al.* [12] showing that energetically stable single stem structures correlate with the abundances of RNA sequences in the Rfam database [31].

Benchmark Methodology. In these experiments, we analyzed sequences of size 20, 30 and 40 nucleotides. We also defined five G+C-content regimes at 10%, 30%, 50%, 70% and 90% ($\pm10\%$). For each G+C-content we generated 20 seeds of length 20 and 30, and 10 seeds of length 40. Thus yielding a total of 250 seeds.

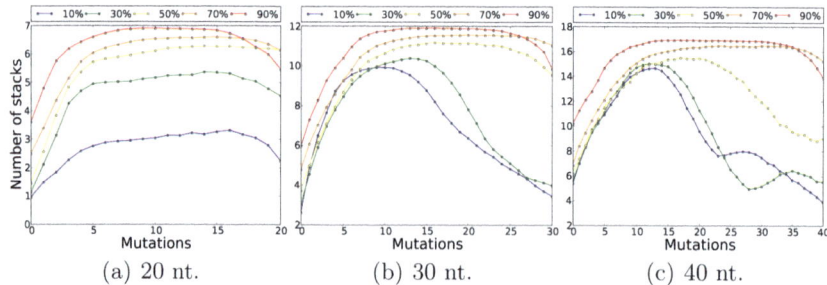

Fig. 3. X-axis: Number of mutations in mutants. Y-axis: Number of stacks in secondary structures. Blue: 10% GC, Green: 30%, Yellow: 50%, Orange: 70%, Red: 90%.

For each seed we ran `RNAmutants` and sampled approximately 1000 secondary structures in each k-neighborhood[1]. Each run explores the complete mutational landscape (i.e. 4^n sequences where n is the length of the sequence) and currently takes less than a minute for a size of 20 nucleotides, about 45 minutes for a 30 nucleotides, and about 5 hours for 40 nucleotides. In each experiment, we report the evolution of four parameters for each value of k (i.e. number of mutations). Namely, the number of stacks in the secondary structures sampled with the mutants (See Fig. 3), the number of bulges and internal loops (See Fig. 4), and the entropy of the sampled sequences (See Fig. 5).

Low G+C-contents Favor Structural Diversity. In these experiments, we seek to characterize how sequences may constrain the variety of structures. RNA secondary structures can be characterized by their number of hairpins, stacks, bulges, internal loops and multi-loops. Here, because our sequences are relatively small, the large majority of the structures have a single stem shape. Thus, they have a single hairpin and no multi-loop, and we choose to report only the number of stacks and loops (bulges and internal loops). In Fig. 3 and 4 we report these statistics in each k-neighborhood of the seed.

Since the number of stacks correlates with single stems structures and thus more stable structures, one could expect that the number of stacks will naturally increase with the number of mutations. This intuition explains the results of simulations performed on sequences of length 20 (See Fig. 3(a)). However, surprisingly, this property does not hold for longer sequences with low G+C-contents. In Fig. 3(b) and 3(c), we observe that the number of stacks increases first, and then drops for large numbers of mutations (approximately $k \geq n/3$). Symmetrically, the number of bulges and internal loops initially drops and then increases.

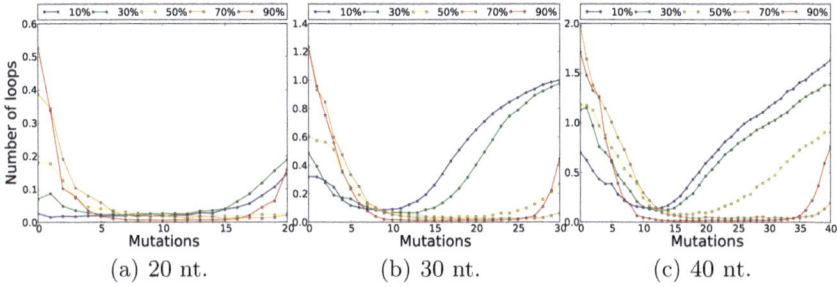

(a) 20 nt.	(b) 30 nt.	(c) 40 nt.

Fig. 4. X-axis: Number of mutations in mutants. Y-axis: Number of bulges and internal loops in secondary structures. Blue: 10% GC, Green: 30%, Yellow: 50%, Orange: 70%, Red: 90%.

---

[1] Our implementation enables us to control the minimal number of sequences to sample at a targeted G+C-content in each k neighborhood.

These experiments enable us estimate the strength of an evolutionary pressure which stems from an improvement of the stability of the folds. Our data indicate that for short period of evolution this "structural" pressure is always dominant. But after longer periods of evolution, low G+C-contents enable more diversity in the structural ensembles. In other words, if we make the assumption that bulges and internal loops represent more sophisticated structures that could be associated to functional shapes. Then, under this scenario, we showed that the structures are first stabilized (i.e. backbone is created) and subsequently refined for functions.

Our results suggest a couple of hypothesis. First the size is an important factor of the structural diversity, and the analysis of sequence-structure maps of sequences of length larger than 20 may result in very different conclusions than those drawn for small sequences [15,12]. Next, sequences with a low G+C-content (below 40%) may allow a broader "choice" of structures. Low G+C-contents seem to favor the apparition of bulges and internal loops, making the apparition of non-canonical interactions and RNA 3D motifs [32,33,34] easier. Such tertiary structure motifs are frequently associated with specific RNA functions, and we conjecture that low G+C-contents favor their synthesis.

High G+C-contents Reduce the Sequence Diversity. Our next analysis aims to reveal how the structural stability (i.e. the folding energy) may influence the diversity of sequences and then the mutational space explored across evolution. We need for that to compute the entropy of the sequences in each k-neighborhood. First, we align all k-mutants and compute the Shannon entropy at position i: $\sigma(i) = \sum_{x=\{A,C,G,U\}} -f_i(x) \cdot \log_4(f_i(x))$, where $f_i(x)$ is the frequency of the nucleotide x in the i-th column of the alignment. Then, we average these measures and compute the average entropy per position $1/N \cdot \sum_{i=i}^{N} \sigma(i)$, where N is the length of the alignment (i.e. also the length the sequences and the target structure since no gaps are allowed). Our results are shown in Fig. 5.

Before discussing these results, we note that the G+C-content biases the entropy values. Indeed, when the distribution of nucleotides is no longer uniform (i.e. when the G+C-content is shifted away from 50%), the maximal entropy value decreases. We report the theoretical limits reachable for G+C-contents of 30% and 70% (approximately 0.94 and indicated with a dotted line in Fig. 5), and 10% and 90% (approximately 0.74 and indicated with a dashed line in Fig. 5). Obviously, the upper bound for a G+C-content of 50% is 1.

Once again, as expected the maximum entropy is reached for sequences with G+C-contents of 50%. Medium G+C-contents offer a larger sequence accessibility. More interestingly, the maximal entropy value reached in these experiments seems to vary between extreme G+C-contents regimes. We observe that sequences at 10% of GC achieve the optimal entropy value, but that sequences at 90% GC significantly fail to explore the complete mutational landscape. This remark suggests that high G+C-contents reduce the evolutionary accessibility and the variety of sequences designed under this scenario. Finally, unlike our previous experiments (cf. section 3), we note that the size of the sequences has no influence on these results.

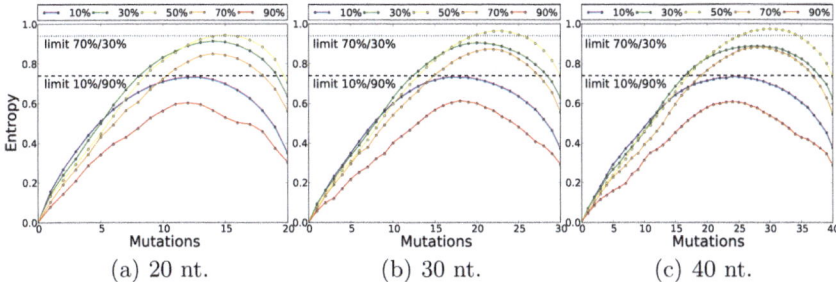

Fig. 5. X-axis: Number of mutations in mutants. Y-axis: Entropy of sampled mutant sequences. Blue: 10% GC, Green: 30%, Yellow: 50%, Orange: 70%, Red: 90%. Dotted line represents the maximal entropy value that can be obtained for GC contents of 30% and 70%. And the dashed line represents the maximal entropy value for GC contents of 10% and 90%.

4   Conclusion

In this paper, we showed how adaptive sampling techniques can be used to explore regions poorly covered by classical sampling algorithms. We applied this methodology to RNAmutants, and showed how regions of the mutational landscape with low G+C-contents could be efficiently sampled and analyzed.

Importantly, the techniques developed in this work can be generalized to many other sequential and structural additive properties, such as the number of mutations, number of base pairs or the free energy. The versatility of these techniques suggests a broad range of novel applications as well as algorithm improvements.

This methodology is well-suited to the exploration of large sequence-structure maps. We expect that their application in various ways will reveal novel properties of the RNA evolutionary landscapes [14,15,1,13]. More practically, as recently reported by Barash and Churkin, our algorithms are also well-suited to predict multiple deleterious mutations in structural RNAs [35]. We expect that our adaptive sampling techniques will help improve our prediction accuracy.

All these algorithms have been implemented in a new version of our RNAmutants software suite available at http://csb.cs.mcgill.ca/RNAmutants. This new distribution includes various new features such as an RNA duplex model for simple hybridizations and weighted substitution events.

Acknowledgement

JW is partially funded by the NSERC discovery program. YP is partially funded by the *ANR 2010 Blanc 0204* MAGNUM grant/project.

References

1. Cowperthwaite, M., Meyers, L.: How mutational networks shape evolution: Lessons from RNA models. Annual Review of Ecology, Evolution, and Systematics, 203–230 (2008)

2. Halvorsen, M., Martin, J.S., Broadaway, S., Laederach, A.: Disease-associated mutations that alter the RNA structural ensemble. PLoS Genet. 6(8) (2010)
3. Westhof, E.: Toward atomic accuracy in RNA design. Nat. Methods 7(4), 272–273 (2010)
4. Zuker, M., Stiegler, P.: Optimal computer folding of large RNA sequences using thermodynamics and auxiliary information. Nucleic Acids Res. 9(1), 133–148 (1981)
5. Parisien, M., Major, F.: The MC-Fold and MC-Sym pipeline infers RNA structure from sequence data. Nature 452(7183), 51–55 (2008)
6. McCaskill, J.S.: The equilibrium partition function and base pair binding probabilities for RNA secondary structure. Biopolymers 29(6-7), 1105–1119 (1990)
7. Turner, D.H., Mathews, D.H.: NNDB: the nearest neighbor parameter database for predicting stability of nucleic acid secondary structure. Nucleic Acids Res. 38(Database issue), D280–D282 (2010)
8. Mathews, D.H.: RNA secondary structure analysis using RNAstructure. Curr. Protoc. Bioinformatics ch. 12, Unit 12.6 (2006)
9. Hofacker, I.L.: RNA secondary structure analysis using the vienna RNA package. Curr. Protoc. Bioinformatics ch. 12, Unit12.2 (June 2009)
10. Markham, N.R., Zuker, M.: UNAFold: software for nucleic acid folding and hybridization. Methods Mol. Biol. 453, 3–31 (2008)
11. Grüner, W., Giegerich, R., Strothmann, D., Reidys, C., Weber, J., Hofacker, I., Stadler, P., Schuster, P.: Analysis of RNA sequence structure maps by exhaustive enumeration i. neutral networks. Monatshefte f. Chemie 127(4), 355–374 (1995)
12. Cowperthwaite, M.C., Economo, E.P., Harcombe, W.R., Miller, E.L., Meyers, L.A.: The ascent of the abundant: how mutational networks constrain evolution. PLoS Comput. Biol. 4(7), e1000110 (2008)
13. Stich, M., Lázaro, E., Manrubia, S.C.: Phenotypic effect of mutations in evolving populations of RNA molecules. BMC Evol. Biol. 10, 46 (2010)
14. Schuster, P., Fontana, W., Stadler, P.F., Hofacker, I.L.: From sequences to shapes and back: a case study in RNA secondary structures. Proc. Biol. Sci. 255(1344), 279–284 (1994)
15. Reidys, C., Stadler, P.F., Schuster, P.: Generic properties of combinatory maps: neutral networks of RNA secondary structures. Bull. Math. Biol. 59(2), 339–397 (1997)
16. Waldispühl, J., Devadas, S., Berger, B., Clote, P.: Efficient algorithms for probing the RNA mutation landscape. PLoS Comput. Biol. 4(8), e1000124 (2008)
17. Waldispühl, J., Behzadi, B., Steyaert, J.M.: An approximate matching algorithm for finding (sub-)optimal sequences in S-attributed grammars. Bioinformatics 18 Suppl 2, S250–S259 (2002)
18. Clote, P., Waldispühl, J., Behzadi, B., Steyaert, J.M.: Energy landscape of k-point mutants of an RNA molecule. Bioinformatics 21(22), 4140–4147 (2005)
19. Ding, Y., Lawrence, C.E.: A bayesian statistical algorithm for RNA secondary structure prediction. Comput. Chem. 23(3-4), 387–400 (1999)
20. Chan, C.Y., Carmack, C.S., Long, D.D., Maliyekkel, A., Shao, Y., Roninson, I.B., Ding, Y.: A structural interpretation of the effect of gc-content on effciency of RNA interference. BMC Bioinformatics 10 Suppl 1, S33 (2009)
21. Wang, H.-c., Hickey, D.A.: Evidence for strong selective constraint acting on the nucleotide composition of 16s ribosomal rna genes. Nucleic Acids Res. 30(11), 2501–2507 (2002)
22. Nussinov, R., Jacobson, A.: Fast algorithm for predicting the secondary structure of single-stranded RNA. Proc. Natl. Acad. Sci. USA 77, 6903–6913 (1980)

23. Waterman, M.S.: Secondary structure of single stranded nucleic acids. Advances in Mathematics Supplementary Studies 1(1), 167–212 (1978)
24. Wuchty, S., Fontana, W., Hofacker, I., Schuster, P.: Complete suboptimal folding of RNA and the stability of secondary structures. Biopolymers 49, 145–164 (1999)
25. Denise, A., Ponty, Y., Termier, M.: Controlled non uniform random generation of decomposable structures. Theoretical Computer Science 411(40-42), 3527–3552 (2010)
26. Flajolet, P., Zimmermann, P., Cutsem, B.V.: A calculus for the random generation of labelled combinatorial structures. Theor. Comput. Sci. 132(2), 1–35 (1994)
27. Ponty, Y.: Efficient sampling of RNAsecondary structures from the boltzmann ensemble of low-energy: The boustrophedon method. Journal of Mathematical Biology 56(1-2), 107–127 (2008)
28. Bodini, O., Ponty, Y.: Multi-dimensional boltzmann sampling of languages. DMTCS Proceedings 0(01) (2010)
29. Moments of the boltzmann distribution for rna secondary structures. Bull. Math. Biol. 67(5), 1031–1047 (September 2005)
30. Faugere, J.C.: A new efficient algorithm for computing Gröbner bases (f4). Journal of Pure and Applied Algebra 139(1-3), 61–88 (1999)
31. Gardner, P.P., Daub, J., Tate, J.G., Nawrocki, E.P., Kolbe, D.L., Lindgreen, S., Wilkinson, A.C., Finn, R.D., Griffiths-Jones, S., Eddy, S.R., Bateman, A.: Rfam: updates to the rna families database. Nucleic Acids Res. 37(Database issue), D136–D140 (2009)
32. Djelloul, M., Denise, A.: Automated motif extraction and classification in RNA tertiary structures. RNA 14(12), 2489–2497 (2008)
33. Lemieux, S., Major, F.: RNA canonical and non-canonical base pairing types: a recognition method and complete repertoire. Nucleic Acids Res. 30(19), 4250–4263 (2002)
34. Leontis, N.B., Lescoute, A., Westhof, E.: The building blocks and motifs of RNA architecture. Curr. Opin. Struct. Biol. 16(3), 279–287 (2006)
35. Barash, D., Churkin, A.: Mutational analysis in RNAs: comparing programs for RNA deleterious mutation prediction. Briefings in Bioinformatics (2010)

Nonparametric Combinatorial Sequence Models

Fabian L. Wauthier[1], Michael I. Jordan[1], and Nebojsa Jojic[2]

[1] University of California, Berkeley
[2] Microsoft Research, Redmond
{flw,jordan}@cs.berkeley.edu, jojic@microsoft.com

Abstract. This work considers biological sequences that exhibit combinatorial structures in their composition: groups of positions of the aligned sequences are "linked" and covary as one unit across sequences. If multiple such groups exist, complex interactions can emerge between them. Sequences of this kind arise frequently in biology but methodologies for analyzing them are still being developed. This paper presents a nonparametric prior on sequences which allows combinatorial structures to emerge and which induces a posterior distribution over factorized sequence representations. We carry out experiments on three sequence datasets which indicate that combinatorial structures are indeed present and that combinatorial sequence models can more succinctly describe them than simpler mixture models. We conclude with an application to MHC binding prediction which highlights the utility of the posterior distribution induced by the prior. By integrating out the posterior our method compares favorably to leading binding predictors.

Keywords: Sequence models, Chinese restaurant process, Chinese restaurant franchise, MHC binding, mixture models.

1 Introduction

Proteins and nucleic acids, polymers whose primary structure can be described by a linear sequence of letters, are found in nature in an astounding diversity. Understanding the diversity of biological sequences has been a major topic in computational biology. Through inheritance, and close functional coupling, the nearby sequence positions in a family of biological sequences are often at a linkage disequilibrium, i.e., the letters at nearby sites tend to covary. However, in their folded form, these molecules also have secondary, tertiary, and quaternary structure, which may reveal geometric proximity, and provide a basis for potential interactions of residues at distant sequence sites and even across different molecules. This creates significant difficulties in modeling diversity of certain families of sequences, where both the nearby and distant sequence positions may exhibit patterns of covariation. This difficulty is exacerbated by the fact that with only a limited number of sequences available for analysis we could arrive at multiple diversity models which are almost equally well supported by data. We model such sequence data starting with a basic componential strategy outlined in Figure 1. We show four aligned subsequences from Influenza HA1

V. Bafna and S.C. Sahinalp (Eds.): RECOMB 2011, LNBI 6577, pp. 516–530, 2011.

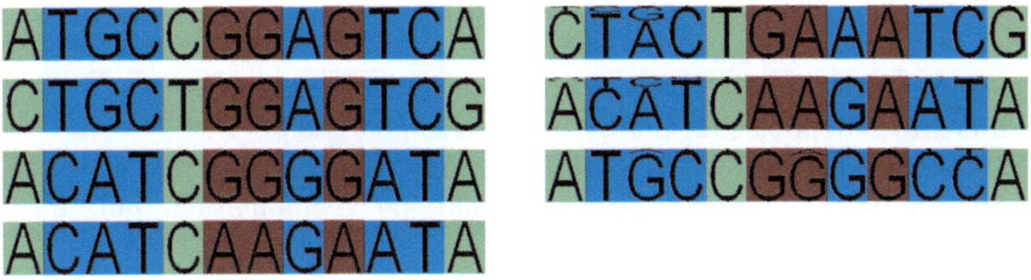

(a) Observations   (b) Profiles

Fig. 1. 1(a) Four aligned short subsections of the sequences exhibiting the combinatorial pattern according to the partition highlighted by color. The blue component, $z_{\text{site}} = 1$, comes in two variants, `TGCATC` and `CATGAT`, while the green component, $z_{\text{site}} = 2$, follows either `ACA` or `CTG`. All four combinations of types of these segments are found in the data. Each of those configurations can be combined with two further variants, `GGG` and `AAA`, in the red component, $z_{\text{site}} = 3$. 1(b) Slight perturbations on the basic types are possible as captured by the profiles inferred whose appropriate sections are shown. The profiles and subsequences correspond to appropriate sections of the Influenza HA1 genes analyzed in Section 4. The sequence sites switch among profiles in groups—the entire component follows one of the three profiles. (In general, some components may be less entropic than others and the sequences may then not be mapped to all three different types.) The four sequences in this example can be represented by the pointers z_{prof} for each of the three components which map the components to the appropriate profiles: 213, 113, 223, and 222. Such compression of the variability can increase statistical power of techniques mapping genotypic and phenotypic variation as we demonstrate for the case of MHC binding prediction in Section 4.

genes whose diversity is well explained by first partitioning the sites into three groups and then representing each partition's induced subsequences by one of several prototypes. The site groupings do not need to follow linear patterns, and distant sites may be grouped together. Assuming that the three types in the three groups can be arbitrarily mixed, the model represents 27 different variants, and could thus also be expressed as a mixture with that many components. However, the use of a traditional mixture model would require considerably more data for training, as having obtained only 50–100 sequences it is likely that we did not see all 27 combinations. On the other hand, it is likely that we observed all three types in all three components multiple times, thus facilitating parameter estimation in a componential model. Furthermore, the componential structure itself may be of importance. If for instance, a phenotype of interest is linked only to one variant of one of the components, then the mixture model would capture this variant in nine components needed to represent the relevant type in combination with three types in each of the other two components. Thus a traditional clustering would lead to nine different statistical tests, lowering statistical power by an order of magnitude. In this sense, the combinatorial structure allows for pooling the traditional mixture components based on the finer-grained patterns of covariation. In this paper we outline a probabilistic model that can be used to discover such structure in several gene and protein families while coping with the dearth of sequence data and the possible additional correlations among the

Table 1. Sequence families exhibiting combinatorial structures

Family	Forces shaping componential diversity
Immunoglobulin/TCR	Clonal V(D)J recombination
Pathogenic proteins	Recombination, mutation
MHC/KIR	Large and small scale recombination, mutation

groups. Such combinatorial diversity is ubiquitous at larger scales such as entire chromosomes. However, some very important biomolecules have relatively short segments that are under significant diversifying selection. In Table 1, we highlight molecules involved in host-pathogen interactions and whose subsequences fit the model discussed above. All these families of molecules have to maintain their biological function, while exhibiting a high degree of variation concentrated in a short subsequence, and the solution to these conflicting requirements has componential structure.

As the first example, we point to the genes encoding immunoglobulin and T cell receptor proteins which are split into multiple gene segments in the germline. These segments are made contiguous by recombination in somatic tissues by the well known V(D)J recombination process [3]. To assemble an antigen receptor gene, one V (variable), one J (joining) and, sometimes, one D (diversity) segment are joined to create an exon that encodes the binding portion of the receptor chain. As there are typically many V, D, and J gene segments, V(D)J recombination creates an immense combinatorial diversity of antibody and TCR binding specificities, responding to the diversity of the immune system's targets.

Pathogen proteins whose subsequences are often targets of immunoglobulin and TCR binding also exhibit combinatorial diversity. For instance, VAR2CSA, a member of the *P. falciparum* erythrocyte membrane 1 protein family and a potential vaccine candidate for pregnancy-associated malaria, contains short segments in which the isolate variation can be well summarized by a small number of very different types. While human-infecting *P. falciparum* isolates exhibit combinatorial diversity resulting from fairly arbitrary mixing of segment types, each type is remarkably conserved across isolates that have them, including isolates of *P. reichenowi* which infects other primates [2]. This indicates a possible role of recombination with other var gene segments in creating combinatorial diversity in the binding domains of these proteins, which have to facilitate adhesion to the placenta while avoiding recognition by the immune system.

The third example we highlight is the major histocompatibility complex (MHC) class I family of molecules which again participate in the interaction between the host immune system and pathogens. In virtually all cells of higher organisms, these molecules present antigenic cellular peptides on the cellular surface for surveillance by cytotoxic T cells. The T cell receptor proteins discussed above may bind to the complex made of the MHC molecule and the antigenic peptide which can lead to the destruction of the infected cell. To properly facilitate the surveillance of the cellular proteome, MHC molecules are again faced with complex requirements: Across different situations, the MHC molecules will encounter a large number of different targets that may need to be carried to the surface, but at any given time, the cellular presentation should be limited to useful targets. Furthermore, as pathogens adapt to immune pressure quickly, a population of hosts is more resilient if it is diverse in its immune surveillance properties. Nature's solution here is somewhat different than in the case of the TCR and immunoglobulin. The immune system needs to learn to tolerate normal self proteins and the variation in binding properties through clonal recombination in one individual would complicate this tolerance. Instead, in humans, three highly diverse loci encode for MHC class I, leading to diversity of MHC binding specificities across individuals, not within one host. The residues forming the peptide binding groove of the MHC molecules have been found to be under a diversifying selection. The statistical study of MHC alleles has yielded evidence of both large-scale recombination events (involving entire exons) and low-scale recombination events (involving apparent exchange of short DNA segments), but convergent evolution in parts of the MHC from different alleles is also supported by the data [5]. Thus, a variety of mutation and recombination events, whose combinations were selected based on the resulting binding properties of the MHC groove lead to the immense diversity at this locus, the most polymorphic in the human genome.

In these three examples, and many more (Figure 1 illustrates diversity in an influenza protein), the functional requirements have created sequence families that exhibit high levels of diversity with combinatorial structure similar to the one illustrated in Figure 1. Models that capture such structure have immediate applications in low-level tasks such as sequencing, haplotype recovery, as well as in higher level tasks involving the matching of the genetic diversity to phenotypic variation. In the case of the immunoglobulin, this structure is essentially encoded in the human genome, and the different V, D, and J variants can be directly read off there. But, when diversity is maintained on a population level, as is the case with most pathogen proteins and RNA molecules, as well as MHC or KIR (receptor on natural killer cells) among human proteins, then we can only recover the structure by analyzing sequences from a number of individuals. This is complicated by two effects: first, the illustration in Figure 1 is a simplification. The groups of sites are only approximately independent of each other. Some residual weak linkage is expected to exist even in the case of the optimal sequence partition. Secondly, due to the high polymorphism in the families of interest the structure in Figure 1 can only be estimated reliably when sufficient data is available. When data is scarce, multiple different solutions are possible that differ little in the data fit.

In this paper we propose a model that differs from existing models in the way it addresses these two issues. In [1,2], the partition is assumed to consist of contiguous segments, a constraint that does not hold for many interesting diversity patterns (cf. Figure 1), and a single optimal segmentation (cf. the pattern library/epitome approach of [7,8]). A combinatorial optimization algorithm for site clustering that does not promote contiguous segments is proposed in [12], but, as the basic generative model creates blocks with limited diversity [6,8], the

result is again a single optimal segmentation which can be sensitive to the size
of the sequence set used to estimate it.

We propose a Bayesian hierarchical site clustering approach with a minimal
number of parameters which not only captures weak linkage among components
at the first level of the clustering hierarchy, but also naturally adjusts to the size
of the dataset. Furthermore, we develop a sampling procedure that produces an
estimate of the posterior over possible sequence partitions. In Section 4 we illustrate for three families of proteins—MHC class I, Influenza HA1 and KIR—that
the componential model discussed here is a better fit than traditional mixture
models, which cluster entire sequences (phylogenetic methods fall in this category). We also show an example where by using the distribution over multiple
partitions we improve on the ability to match the genetic diversity with the phenotype variation. In particular, by representing MHC sequences by the latent
variables in our model we train simple MHC class I–peptide binding estimators.
We show that by integrating over possible MHC sequence representations based
on different partitions we obtain better predictions than when we use the latent
variables for the MAP estimate of the segmentation structure.

2   Model

Most approaches to capturing diversity in sets of aligned sequences treat each sequence as a whole, applying clustering techniques (e.g., neighbor-joining or maximum likelihood approaches) or building a hierarchical clustering of sequences
(e.g., a phylogenetic tree). A special case of such approaches are mixture models
which describe aligned sequences as being sampled from a mixture of a small
number of "latent profiles," also known as "position-specific scoring matrices,"
e.g., [10]. As outlined above, a considerable drawback of a whole sequence mixture model is that each observed sequence corresponds in its entirety to one
latent profile. Our model is a generalized mixture model that relaxes this constraint and allows different sequence positions to correspond to different profiles.
To retain some structure, however, our model introduces a latent partitioning
that groups site positions into linked sites that must be sampled from the same
profile. Each such "site group" thus induces a different mixture model on its
component sites. This allows us to capture combinatorial diversity that is not
captured by a flat mixture model—n site groups with k profiles would need n^k
mixed profiles if the data was to be represented by a flat mixture. Moreover, as
discussed in Section 1, we wish to also couple the mixture models in order to
capture additional weaker links among the site groups. Our model achieves this
by implicitly coupling the mixing proportions of the different mixtures.

When analyzing data with traditional mixture models, one is faced with the
perennial problem of choosing the number of mixture components. Since the
model we are proposing can be thought of as a refined mixture model, it is
not immune to this issue. While information-theoretic techniques do exist for
estimating the structural parameters in mixture models, they are difficult to
justify when the number of components required to represent a large dataset

is large [24]. In a number of biological applications [17,22,23,24] nonparametric methods based on the Chinese restaurant process (CRP), or the closely related Dirichlet process, have been shown to elegantly circumvent such issues by effectively introducing a prior distribution on the number of latent components. A second advantage is that the induced prior automatically accommodates more latent components as the amount of data grows. This allows us to infer conservative representations with few components when little data is available while being flexible enough to represent complex patterns emerging from larger datasets.

Our model relies on a composition of two nonparametric priors—the Chinese restaurant process (CRP) [16] and the related Chinese restaurant franchise (CRF) [21]. By incorporating these two nonparametric priors we circumvent fixing the number of site groups and the number of profile variants a priori, and instead average over these choices under a posterior distribution.

In this section we present our model by means of a sequential, generative description. In this description we use the index s to index sequences and i to index the sites (sequence positions) within a sequence. Let M denote an $S \times I$ matrix of aligned sequences, so that m_s denotes the s-th sequence and $m_{s,i}$ denotes the i-th symbol in the s-th sequence. Our model relies on four sets of latent random variables: $z_{\text{site}}, z_{\text{clust}}, z_{\text{prof}}$ and θ, sampled in top-down fashion according to a CRF that is conditioned on a partition sampled from a CRP.

2.1   Chinese Restaurant Process Linkage Model

The CRP [16] is a nonparametric prior on partitions of a set of items. In its generative form it describes a sequential process that produces a dataset exhibiting clusters. The language of the CRP likens the sequential process to a (potentially endless) stream of customers entering a restaurant one by one. Upon entering, each patron randomly chooses a table to sit at with probability proportional to the number of customers already seated there, or sits at an empty table. Each table is assigned a parameter that is shared by all customers at that table. For clustering, the datapoints are thought of as patrons, and the clusters as tables, which are parameterized by the tables' parameter.

The first step in our model is to sample a partition of the site indices into groups of linked sites. At this level of the model site indices are not yet associated with any data—we only use the CRP seating process to induce a site partitioning. The partition is sampled from a CRP where sites act as customers and site groups as tables. Representing the allocation of sites to groups (tables) by a set of latent variables $z_{\text{site}}(i), i = 1, \ldots, I$, the process operates as follows: Customers (site indices) enter the restaurant one by one and choose to sit either at an existing table or to open a new table. At each step of the sequential process, let the number of existing site tables be denoted by n_{site}, and the number of site indices at table t by $c_{\text{site}}(t), t = 1, \ldots, n_{\text{site}}$. If we parameterize the CRP by α_{site}, then the seating probabilities for site i given the seating assignment for all previous sites $1, \ldots, i-1$ are given as

$$p(z_{\text{site}}(i) = t | z_{\text{site}}(1{:}i-1)) \propto \begin{cases} c_{\text{site}}(t) & \text{if } t \leq n_{\text{site}} \\ \alpha_{\text{site}} & \text{if } t = n_{\text{site}} + 1 \end{cases} . \tag{1}$$

From this definition we see that just as the number of sites visiting the restaurant can in principle be unbounded, so can the number of tables at which they sit. However, as the number of sites grows, it becomes less likely that new tables will be opened; indeed, the growth rate can be shown to be $O(\alpha_{\text{site}} \log i)$. Note the role of the parameter α_{site} in scaling this growth rate in the prior distribution.

In the following, a site group is treated as an inseparable entity which can be grouped further. In the overall process, it is the preliminary site grouping which captures most of the site linkage in the observed data.

2.2   Chinese Restaurant Franchise Observation Model

The second part of our model represents a combinatorial observation model over aligned sequences in the form of a CRF [21] that is conditioned on the initial partitioning z_{site} by the CRP. The CRF is a generalization of the CRP to allow multiple parallel restaurants to share parameters. Specifically, where in the CRP each table is given a parameter which is shared among its occupants, in the CRF these parameters can also be shared across multiple CRPs. It will turn out that the "parameters" that are being shared in our application are pointers to profiles, rather than the profiles themselves. As such, our model can be thought of as an instance of a dependent nonparametric process, discussed by MacEachern [11], where individual parameters are replaced by stochastic processes. In the CRF we interpret each observed sequence as its own restaurant. But instead of thinking of site positions as customers, as in a standard application of the CRF, we now consider the previously induced site groups to be customers. Each restaurant is visited by all site groups, so that the union of the site groups at each restaurant captures the entire set of sequence indices. The CRF is defined as follows. At each sequence m_s the n_{site} site groups indicated in z_{site} are seated at tables a second time according to the rules of a CRP. The seating arrangement of the site groups is represented by latent variables $z_{\text{clust}}(s,t), t = 1, \ldots, n_{\text{site}}$. Denote by $n_{\text{clust}}(s)$ the number of (second-level) tables formed at sequence s at each step of the process, and let $c_{\text{clust}}(s, u), u = 1, \ldots, n_{\text{clust}}(s)$ denote the number of site groups present at the table u. If we parameterize the sequential seating process at each restaurant by α_{clust}, then conditioned on the seating assignment of the site groups $1, \ldots, t - 1$, the seating probabilities for group t are

$$p(z_{\text{clust}}(s,t) = u | z_{\text{clust}}(s, 1{:}t - 1)) \propto \begin{cases} c_{\text{clust}}(s, u) & \text{if } u \leq n_{\text{clust}}(s) \\ \alpha_{\text{clust}} & \text{if } u = n_{\text{clust}}(s) + 1 \end{cases}. \tag{2}$$

In order to produce observed sequences, the CRF model next introduces parameters. Each table u in a sequence restaurant s in the CRF is assigned a latent variable $z_{\text{prof}}(s, u)$, that indicates which of a set of shared parameters θ is used at table u of restaurant s. We will refer to one such shared parameter θ_p as a "sequence profile." As before, at each step of the sequential algorithm, the variable n_{prof} denotes how many distinct profiles the set of z_{prof} variables points to. The function $c_{\text{prof}}(p), p = 1, \ldots, n_{\text{prof}}$ reports how many of the tables in all processed sequence restaurants picked profile p. In the sequential description of the CRF, the choice of profile made by each table is influenced by the number of other tables that have previously chosen that profile. That is, the process can be thought of as another CRP in which distinct profiles can be thought of as tables. If we use parameter α_{prof} to define this CRP, then the probability that table u in restaurant s chooses profile p, given the profile choices of all tables in restaurants $1, \ldots, s-1$ and tables $1, \ldots, u-1$ in restaurant s, is given by

$$p\left(z_{\mathrm{prof}}(s, u){=}p | z_{\mathrm{prof}}(1{:}s{-}1, \cdot), z_{\mathrm{prof}}(s, 1{:}u{-}1)\right) \propto \begin{cases} c_{\mathrm{prof}}(p) \text{ if } p \le n_{\mathrm{prof}} \\ \alpha_{\mathrm{prof}} \quad \text{ if } p = n_{\mathrm{prof}}{+}1. \end{cases} \quad (3)$$

For sequences with an alphabet of size A, each sequence profile $\theta_p, p = 1, \ldots, n_{\mathrm{prof}}$ is comprised of I A-vectors, one for each site index. Each vector $\theta_p(\cdot, i)$ is a probability distribution over the A possible symbols that could be observed at position i. When a new table in one of the restaurants chooses a new profile θ_p which has not yet been chosen before, the profile vectors $\theta_p(\cdot, i), i = 1, \ldots, I$ are sampled from a Dirichlet prior, parameterized by α_{dir}.

Once all latent variables and profiles have been sampled, the observed sequences are generated as follows: given the latent variables $z_{\mathrm{site}}, z_{\mathrm{clust}}, z_{\mathrm{prof}}$ and profiles θ, we generate the symbol at position i in sequence s by sampling from a multinomial with parameter $\theta_p(\cdot, i)$, where $p = z_{\mathrm{prof}}(s, z_{\mathrm{clust}}(s, z_{\mathrm{site}}(i)))$.

The sampling procedure generates data that exhibit the combinatorial structure discussed in Figure 1 and found in a variety of biological sequence families. Of course, our goal is to reverse this process. Starting from the observed sequences we need to reconstruct the latent variables $z_{\mathrm{site}}, z_{\mathrm{clust}}, z_{\mathrm{prof}}$ and the profile sequences, while making explicit our uncertainty over these structures. In the next section we develop an inference algorithm that achieves this by approximating the full posterior over latent structures.

3   Inference

We use a collapsed Gibbs sampler in which the profiles θ are integrated out. The algorithm cycles through resampling the site grouping z_{site}, the secondary grouping of site groups z_{clust} and the assignment of profiles z_{prof}, at each step conditioning on all remaining latent variables. A central property of the CRP and CRF that facilitates this sampling process is *exchangeability*. Exchangeability allows us to treat any customer of a restaurant as if it were the last customer to enter the restaurant. This is consistent with our modeling assumption that sites have unique positions that need to be grouped, but that the ordering of these positions is of little value, since parts may be non-contiguous. The consequence of this exchangeability is that we can now easily sample an updated table seating for any customer in a restaurant. In the following we show the main computations for resampling the site grouping z_{site}. The posteriors for sampling updated variables z_{clust} and z_{prof} can be derived analogously to Teh et al. [21].

3.1   Resampling Site Groupings z_{site}

We denote by $z_{\mathrm{site}}^{-i}, z_{\mathrm{clust}}^{-i}$ and z_{prof}^{-1} the latent variables that remain when site i is removed from the representation. Let n_{site}^{-i} be the number of distinct site tables when site i is removed. Similarly, let $c_{\text{site}}^{-i}(t)$ be the number of site indices seated at table t when site i is removed. Due to the exchangeability of site indices in the top CRP, we may treat site i as if it were the last to enter the restaurant. In order to sample a new site grouping we must compute the probability that a particular site i is seated at a table, given all other relevant information:

$$p\left(z_{\text{site}}(i) = t \mid m_{\cdot,i}, z_{\text{clust}}^{-i}, z_{\text{prof}}^{-i}\right) . \tag{4}$$

Because we treat i as the last customer to enter the restaurant the prior probability of seating site i at table t is given by

$$p\left(z_{\text{site}}(i) = t \mid z_{\text{site}}^{-i}\right) \propto \begin{cases} c_{\text{site}}^{-i}(t) & \text{if } t \leq n_{\text{site}}^{-i} \\ \alpha_{\text{site}} & \text{if } t = n_{\text{site}}^{-i} + 1 . \end{cases} \tag{5}$$

In a collapsed sampler, if t is an existing site table then we compute the likelihood of seating site i at table t by integrating the induced conditional likelihood of sequence symbols at position i against the prior distributions on $\theta_p(\cdot, i), \forall p$. If we define for $z_{\text{site}}(i) = t \leq n_{\text{site}}^{-1}$ (an existing table was chosen) the count that a symbol at position i is of type a and is generated by profile p as

$$c_t(a, p) = \sum_s \mathbf{1}(m_{s,i} = a, z_{\text{prof}}(s, z_{\text{clust}}(s, t)) = p), \tag{6}$$

then for $t \leq n_{\text{site}}^{-1}$ the integrated likelihood of the observed sequence symbols in position i can be computed as

$$p(m_{\cdot,i} \mid z_{\text{site}}(i) = t, z_{\text{clust}}^{-i}, z_{\text{prof}}^{-i}) = \prod_p \frac{\Gamma(\sum_a \alpha_{\text{dir}}(a))}{\prod_a \Gamma(\alpha_{\text{dir}}(a))} \frac{\prod_a \Gamma(\alpha_{\text{dir}}(a) + c_t(a, p))}{\Gamma(\sum_a \alpha_{\text{dir}}(a) + c_t(a, p))}. \tag{7}$$

It is more complicated to compute the likelihood that site index i is seated at a new table $t = n_{\text{site}}^{-1} + 1$ since the creation of a new site index table triggers a cascade of other choices that need to be made for the z_{clust} and z_{prof} variables. In computing the likelihood of a new site table, the parameters $\theta_p(\cdot, i)$, as well as these new choices need to be integrated out. Rather than computing this complicated integral, we adopt a simpler strategy and approximate the likelihood by sampling a set of new assignments for $z_{\text{clust}}(s, n_{\text{site}}^{-1} + 1), s = 1, \ldots, S$ and if necessary also $z_{\text{prof}}(s, z_{\text{clust}}(s, n_{\text{site}}^{-1} + 1)), s = 1, \ldots, S$ by following the sequential generative model outlined before. Once sample allocations have been generated for the proposal that $t = n_{\text{site}}^{-1} + 1$, we can compute the integrated likelihood of the seating proposal by similar means as in equation (7), giving us the last term

$$p(m_{\cdot,i} \mid z_{\text{site}}(i) = n_{\text{site}} + 1, z_{\text{clust}}^{-i}, z_{\text{prof}}^{-i}) . \tag{8}$$

Combining this likelihood with those computed in (7) and the prior in equation (5) allows us to compute the posterior in equation (4) from which we may now sample a new site group allocation for site index i. If an existing site group is chosen, nothing more needs to be done. If a new site group is created we copy the previously sampled allocations into the current state z_{clust} and z_{prof}.

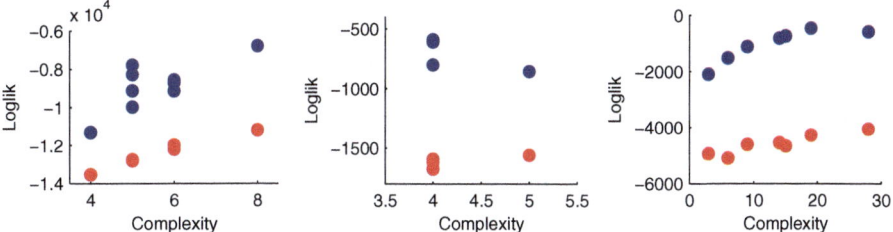

Fig. 2. Comparison of the log likelihood assigned by the combinatorial model (blue scatter) with the log likelihood assigned by a mixture model of comparable complexity (red scatter) as a function of different model complexities. From left to right are shown results for MHC, Flu and KIR sequences. Flu sequences require only relatively few profiles and site groups (cf. Figure 1); thus only two model complexities were explored.

3.2   Resampling z_{clust} and z_{prof}

Once the site partition z_{site} has been resampled, the resampling of z_{clust} and z_{prof} conditioned on z_{site} is performed in similar fashion as in the standard CRF. As before, our implementation integrates out the profile parameters to improve sampling efficiency. The computations can be readily derived from Teh et al. [21].

4   Results

To demonstrate the versatility of our model we applied it to three sequence datasets in which we expect combinatorial patterns to exist. In the following we have focused our analysis on a small number of the most polymorphic sites in each dataset. The first dataset are 526 aligned amino-acid sequences of length 50 for MHC class I proteins from all three alleles A, B, C. The flu dataset comprises aligned 22-long amino-acid sequences for 255 HA1 genes in influenza strains covering the years 1968–2003. The KIR dataset are sequences of unordered (i.e., unphased) pairs of haplotype measurements at 229 SNPs. These SNPs encode variability of a killer cell immunoglobulin-like receptor. If we knew the phase, we could order each pair and turn the data into aligned sequences that could easily be analyzed as outlined before. We have thus extended our model to work with unphased KIR data by introducing extra latent variables z_{phase} that encode the phasing information for each pair. The modified algorithm iterates between sampling phasing variables to turn aligned sequences of pairs into aligned sequences, and then sampling new latent variables $z_{\text{site}}, z_{\text{clust}}$, as well as z_{prof}, as before.

We have carried out experiments to demonstrate that our model successfully isolates combinatorial structures from the data and learns a much more parsimonious sequence model that yields higher log likelihood than a comparable mixture model. For each of the datasets we set up a combinatorial sequence model and computed posterior samples for different settings of the model parameters $\alpha_{\text{site}}, \alpha_{\text{cluster}}, \alpha_{\text{profile}}$, and α_{dir}. Each combination of parameters induces a different nonparametric prior over aligned sequences. We wish to compare the average data likelihoods assigned by the posterior combinatorial models to the likelihoods obtained from flat mixture models. To facilitate this comparison, we ensure that the mixture models we compare against have similar complexity as the nonparametric model. We estimate the complexity of a given model by measuring how many parameters it would take to represent a set of sequences in a typical posterior sample. If for example a sample with n_{site} site tables and n_{prof} profiles of length I with a symbol alphabet of size A was found, we require a total of $I(n_{site} - 1) + n_{prof}I(A - 1)$ parameters as a shared representation across all sequences. The first $I(n_{site} - 1)$ parameters encode which site position is allocated to which site group while the remaining account for the profile parameters. In comparison, a mixture model that links all site positions asserts that $n'_{site} = 1$ and for n'_{prof} profiles requires $n'_{prof}I(A - 1)$ parameters. Assuming that a single set of such parameters is fixed, to encode a set of sequences we would need to also infer for each sequence the posterior distributions over latent variables (mixture components for the mixture model, or profile pointers z_{prof} in our model). Then any remaining uncertainty as to the identity of the letters in individual positions would also have to be collapsed by encoding these individual letters. Information theory prescribes techniques for making the minimum required code length for encoding all this directly dependent on the uncertainties in the data, with less uncertain pieces encoded with shorter messages, so that the total code length in bits reduces to the $\log_2$ likelihood under the model [18]. By adding the cost of encoding the parameters that are shared by the sequences (profiles, partitioning information), we would obtain a description length of the dataset. The cost of encoding parameters would be proportional to the number of the parameters. Similarly, for comparing model fits, statistical literature recommends the use of the Bayesian information criterion (BIC) [19] or the Akaike information criterion (AIC) which combine the log likelihood of the data with a penalty reflecting the number of free parameters. However, rather than comparing the two models by an MDL, BIC or AIC score for only one model complexity, we present a stronger argument here: it turns out that for a wide range of model complexities, the log likelihood of the data is higher under the combinatorial model than under the mixture model.

To show this, for posterior samples of varying complexity under our model, we compute the smallest number of mixture profiles that would exceed it in complexity, i.e., n'_{prof} so that $I(n_{site} - 1) + n_{prof}I(A - 1) \leq n'_{prof}I(A - 1)$. We then fit five mixture models on the data using n'_{prof} profiles and compute the average log likelihood assigned to the data. In Figure 2 we show for the three datasets the average log likelihood of the combinatorial model across samples as a blue scatter and the average log likelihood of the mixture model as a red scatter. For all three datasets, the log likelihood of the combinatorial model exceeds that of the mixture model considerably. Additionally, our model provides a better representation for sequence clustering. The clustering induced by our combinatorial model for Influenza HA1 sequences matches the hemagglutinin inhibition clusters of Smith et al. [20] closer than the clusters obtained by simple mixture modeling, achieving an average adjusted rand index [4] of 0.70 versus

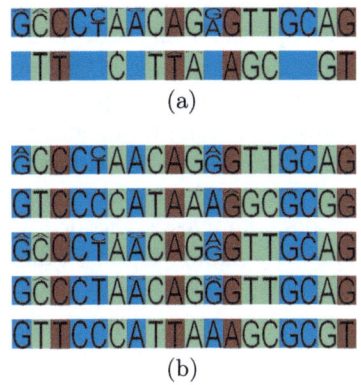

(a)

(b)

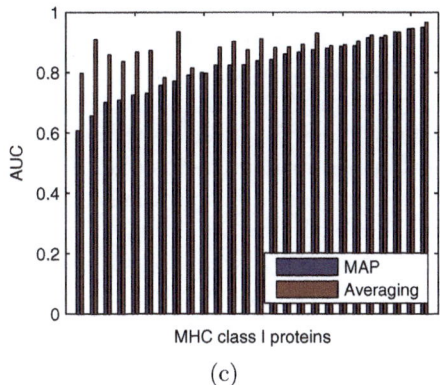

(c)

Fig. 3. (a) Factorized representation for the first 18 SNPs inferred by our model on the KIR data. Empty fields in the profiles denote that no further variants were found for a site group. (b) The 5 profiles for the first 18 SNPs learnt by a mixture model on KIR data. (c) AUC scores for the MHC I binding prediction task across 26 MHC proteins. Averaging regression results across posterior samples significantly improves the AUC score over using only the MAP sample to fit a regression.

0.55^1. In Figure 3(a), we visualize profiles as well as site groups for the first 18 SNPs of the KIR dataset for one posterior sample of the combinatorial model. Figure 3(b) shows relevant parts of the 5 profiles that were inferred by a simpler mixture phasing model. As can be seen, our model factorizes the profiles inferred by the simpler mixture model into a parsimonious form that can still explain the mixture variants. The green group has variants CACGTTA and TCTAGCG, while the red group follows either CAGG or TTAT. Three of the four possible combinations of these patterns occur in the profiles estimated by the mixture model. As a side effect of the compact representation, our model allows for a more careful use of data for profile parameter inference. Mixture models can capture many combinations, but they achieve this by using a substantially greater number of parameters, while still missing many of the combinations outside the region shown. This leads to significantly lower likelihood in comparison with the componential model of similar parametric complexity, as shown in Figure 2.

4.1   MHC Class I Binding Prediction

The latent structure inferred under the model fit to MHC class I sequences above can be used to match these sequences to their binding affinities, and in this way predict epitopes for different MHC molecules. We model the binding affinity (measured in terms of the log IC50 concentration) of an MHC class I protein to an epitope as a linear function that allows sharing across several re-

---

[1] To compute these scores we encoded the sampled latent state of each sequence as a binary vector and clustered these into the same number of clusters as the target clustering. The results were averaged over many samples from the posterior.

lated protein variants. For any particular protein, our sequence model produces a combinatorial representation in terms of site groups and their associated profiles[2]. For a given set of M MHC proteins, we encode this latent structure in binary vectors $b_s, s = 1, \ldots M$. This structure compresses the links produced by co-evolution of the specific sites in the MHC groove. Assuming that some of this co-evolution is driven by selection for particular binding specificity patterns, the latent structure under our model is expected to be useful in binding prediction tasks. For each protein s, a given set of n_s epitopes examples is encoded as binary vectors $e_{sj}, j = 1 \ldots, n_s$. If we denote the corresponding binding affinities as $y_{sj}, j = 1, \ldots, n_s$, then the linear regression we solve in terms of Θ is written as $y_{sj} = e_{sj}^\top \Theta b_s$. The sharing among related proteins is induced by the latent structure b_s. We evaluated two variants of this regression. The first variant uses only the MAP sample from our model posterior to produce a single encoding b_s, while the second fits one regression for each posterior sample (each inducing a different encoding b_s) and then averages the final prediction across samples. The two regression tasks were trained on a total of about 28000 binding affinities over 26 different human MHC molecules. Some MHC molecules were characterized by only a handful of binding measurements, while others were tested against over a thousand different peptides. The results in Figure 3(c) show the AUC score (averaged over five cross-validation runs) obtained from classification into binding and non-binding epitopes. Integration across latent structure significantly boosts the prediction accuracy. Averaged across the 26 MHC variants the averaged predictor yields an AUC score of 0.8846, while the MAP variant achieves a score of only 0.8197. Our result compares favorably with state of the art methods summarized in Peters et al. [14]. The reviewed methods achieve average AUCs of 0.8500 to 0.9146 on a subset of 21 of the 26 proteins for which our averaging method gives a mean of 0.8911. Importantly, the method of Nielsen et al. [13] uses carefully designed nonlinearities and separately known properties of amino-acids to produce improved prediction results. Other leading methods [9,15] use further feature design or exploit the protein structure to boost prediction results. In contrast, even though we use a simple binary representation of epitopes and MHCs, we produce comparable results by virtue of a refined latent sharing structure which is integrated out.

5   Conclusion

This paper presented a nonparametric combinatorial sequence prior that was found to be a good match for a wide range of sequence families. An important feature of the model is that it induces a posterior distribution over latent factorized representations. Our work on MHC binding prediction demonstrates that integrating out this distribution can be an important ingredient in inferences that follow the initial sequence analysis. One way to explain why averaging

---

[2] The parameters used for the combinatorial sequence model were $\alpha_{\text{site}} = 0.1, \alpha_{\text{clust}} = 5, \alpha_{\text{prof}} = 10, \alpha_{\text{dir}} = 0.5$. Posterior samples typically had 3 profiles and 10 site groups over sequences of length 34.

across predictors should be beneficial in the case of MHCs is to consider the potential for suboptimal parsing of the MHC groove. Although many MHC alleles currently present in human populations are known, we cannot directly access the extinct alleles. Thus, our estimate of the site covariation and the resulting optimal sequence partition must suffer from the limited number of sequences used to fit our model. Picking any one segmentation with a high likelihood over MHC sequences may lead to an oversimplification of the sequence representation. A posterior over the partitions, accompanied with latent variables giving sequence types in different parts, reflects more information about a set of amino acids in each MHC sequence than a latent structure based on one optimal segmentation.

Acknowledgements

We would like to thank Daniel Geraghty for providing access to the KIR dataset.

References

1. Bockhorst, J., Jojic, N.: Discovering patterns in biological sequences by optimal segmentation. In: Proceedings of the 23st International Conference on Uncertainty in Artificial Intelligence (UAI) (2007)
2. Bockhorst, J., Lu, F., Janes, J.H., Keebler, J., Gamain, B., Awadalla, P., Su, X.-z., Samurdala, R., Jojic, N., Smith, J.D.: Structural polymorphism and diversifying selection on the pregnancy malaria vaccine candidate VAR2CSA. Molecular and Biochemical Parasitology 155, 103–112 (2007)
3. Fugmann, S.D., Lee, A.I., Shockett, P.E., Villey, I.J., Schatz, D.G.: The RAG proteins and V (D) J recombination: complexes, ends, and transposition. Annual Reviews of Immunology 18, 495–528 (2000)
4. Hubert, L., Arabie, P.: Comparing partitions. Journal of Classification 2(1), 193–218 (1985)
5. Hughes, A.L., Hughes, M.K., Watkins, D.I.: Contrasting roles of interallelic recombination at the HLA-A and HLA-B loci. Genetics 133, 669–680 (1993)
6. Jojic, N., Caspi, Y.: Capturing image structure with probabilistic index maps. In: In Proceedings of the IEEE Conference on Computer Vision and Pattern Recognition, vol. 1, pp. 212–219 (2004)
7. Jojic, N., Jojic, V., Frey, B., Meek, C., Heckerman, D.: Using "epitomes" to model genetic diversity: Rational design of HIV vaccine cocktails. In: Advances in Neural Infomration Processing Systems (NIPS), vol. 18, pp. 587–594 (2006)
8. Jojic, N., Jojic, V., Heckerman, D.: Joint discovery of haplotype blocks and complex trait associations from SNP sequences. In: Proceedings of the 20th International Conference on Uncertainty in Artificial Intelligence (UAI) (2004)
9. Jojic, N., Reyes-Gomez, M., Heckerman, D., Kadie, C., Schueler-Furman, O.: Learning MHC I–peptide binding. Bioinformatics 22(14), e227–e235 (2006)
10. King, O.D., Roth, F.P.: A non-parametric model for transcription factor binding sites. Nucleic Acids Research 31(19), e116 (2003)
11. MacEachern, S.N.: Dependent Nonparametric Processes. In: Proceedings of the Section on Bayesian Statistical Science, pp. 50–55 (1999)
12. Narasimhan, M., Jojic, N., Bilmes, J.: Q-clustering. In: Advances in Neural Infomration Processing Systems (NIPS), vol. 18, pp. 979–986 (2006)

13. Nielsen, M., Lundegaard, C., Worning, P., Lauemoller, S.L., Lamberth, K., Buus, S., Brunak, S., Lund, O.: Reliable prediction of T-cell epitopes using neural networks with novel sequence representations. Protein Science 12, 1007–1017 (2003)
14. Peters, B., Bui, H.-H., Frankild, S., Nielson, M., Lundegaard, C., Kostem, E., Basch, D., Lamberth, K., Harndahl, M., Fleri, W., Wilson, S.S., Sidney, J., Lund, O., Buus, S., Sette, A.: A community resource benchmarking predictions of peptide binding to MHC-I molecules. PLoS Comput. Biol. 2(6), e65 (2006)
15. Peters, B., Tong, W., Sidney, J., Sette, A., Weng, Z.: Examining the independent binding assumption for binding of peptide epitopes to MHC-I molecules. Bioinformatics, 1765–1772 (2003)
16. Pitman, J.: Combinatorial stochastic processes. Springer Lecture Notes in Mathematics. Springer, Heidelberg (2002); Lectures from the 32nd Summer School on Probability Theory held in Saint-Flour (2002)
17. Qin, Z.S.: Clustering microarray gene expression data using weighted Chinese restaurant process. Bioinformatics 22(16), 1988–1997 (2006)
18. Rissanen, J.: Stochastic Complexity in Statistical Inquiry Theory. World Scientific Publishing Co., Inc., River Edge (1989)
19. Schwarz, G.: Estimating the Dimension of a Model. The Annals of Statistics 6(2), 461–464 (1978)
20. Smith, D.J., Lapedes, A., de Jong, J.C., Bestebroer, T.M., Rimmelzwaan, G.F., Osterhause, A.D.M.E., Fouchier, R.A.M.: Mapping the antigenetic and genetic evolution of influenza virus. Science 305, 371–376 (2004)
21. Teh, Y.W., Jordan, M.I., Beal, M.J., Blei, D.M.: Hierarchical Dirichlet processes. Journal of the American Statistical Association 101(476), 1566–1581 (2006)
22. Ting, D., Wang, G., Shapovalov, M., Mitra, R., Jordan, M.I., Dunbrack Jr., R.L.: Neighbor-dependent Ramachandran probability distributions of amino acids developed from a hierarchical Dirichlet process model. PLoS Comput. Biol. 6(4), e1000763 (2010)
23. Xing, E.P., Sharan, R., Jordan, M.I.: Bayesian haplotype inference via the Dirichlet process. In: Proceedings of the 21st International Conference on Machine Learning, pp. 879–886. ACM Press, New York (2004)
24. Xing, E.P., Sohn, K.-A., Jordan, M.I., Teh, Y.W.: Bayesian multi-population haplotype inference via a hierarchical Dirichlet process mixture. In: Proceedings of the 23st International Conference on Machine Learning, pp. 1049–1056. ACM Press, New York (2006)

Algorithms for MDC-Based Multi-locus Phylogeny Inference

Yun Yu[1], Tandy Warnow[2], and Luay Nakhleh[1]

[1] Dept. of Computer Science, Rice University, 6100 Main Street, Houston, TX 77005, USA
{yy9,nakhleh}@cs.rice.edu
[2] Dept. of Computer Sciences, University of Texas at Austin, Austin, TX 78712, USA
tandy@cs.utexas.edu

Abstract. One of the criteria for inferring a species tree from a collection of gene trees, when gene tree incongruence is assumed to be due to incomplete lineage sorting (ILS), is *minimize deep coalescence*, or MDC. Exact algorithms for inferring the species tree from rooted, binary trees under MDC were recently introduced. Nevertheless, in phylogenetic analyses of biological data sets, estimated gene trees may differ from true gene trees, be incompletely resolved, and not necessarily rooted. In this paper, we propose new MDC formulations for the cases where the gene trees are unrooted/binary, rooted/non-binary, and unrooted/non-binary. Further, we prove structural theorems that allow us to extend the algorithms for the rooted/binary gene tree case to these cases in a straightforward manner. Finally, we study the performance of these methods in coalescent-based computer simulations.

1 Introduction

Biologists have long acknowledged that the evolutionary history of a set of species—the *species tree*—and that of a genomic region from those species—the *gene tree*—need not be congruent; e.g., [10]. While many processes can cause gene/species tree incongruence, such as horizontal gene transfer and gene duplication/loss, we focus in this paper on *incomplete lineage sorting*, or ILS, which is best understood under the *coalescent model* [13,20,21], as we illustrate in Fig. 1. The coalescent model views gene lineages moving backward in time, eventually coalescing down to one lineage. In each time interval between species divergences (e.g., t in Fig. 1), lineages entering the interval from a more recent time period may or may not coalesce—an event whose probability is determined largely by the population size and branch lengths.

Thus, a gene tree is viewed as a random variable conditional on a species tree. For the species tree $((AB)C)$, with time t between species divergences, the three possible outcomes for the gene tree topology random variable, along with their probabilities are shown in Fig. 1. With the advent of technologies that make it possible to obtain large amounts of sequence data from multiple species, multi-locus data are becoming widely available, highlighting the issue of gene tree discordance [4,8,14,17,19,25].

Several methods have been introduced for inferring a species tree from a collection of gene trees under ILS-based incongruence. Summary statistics, such as the majority-rule consensus (e.g., [2,8]) and democratic vote (e.g., [1,3,26,27]), are fast to compute

V. Bafna and S.C. Sahinalp (Eds.): RECOMB 2011, LNBI 6577, pp. 531–545, 2011.

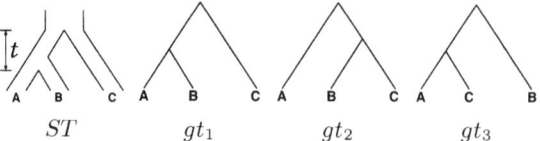

Fig. 1. Gene/species tree incongruence due to ILS. Given species tree ST, with constant population size throughout and time t in coalescent units (number of generations divided by the population size) between the two divergence events, each of the three gene tree topologies gt_1, gt_2, and gt_3 may be observed, with probabilities $1 - (2/3)e^{-t}$, $(1/3)e^{-t}$, and $(1/3)e^{-t}$, respectively.

and provide a good estimate of the species tree in many cases. However, the accuracy of these methods suffer under certain conditions. Further, these methods do not provide explicit reconciliation scenarios; rather, they provide summaries of the gene trees. Recently, methods that explicitly model ILS were introduced, such as Bayesian inference [5,9], maximum likelihood [7], and the maximum parsimony criterion *Minimize Deep Coalescence*, or MDC [10,11,25]. We introduced the first exact algorithms for inferring species trees under the MDC criterion from a collection of rooted, binary gene trees [22,23]. Nevertheless, in phylogenetic analyses of biological data sets, estimated gene trees may differ from the true gene trees, may be incompletely resolved, and may not be rooted. Requiring gene trees to be fully resolved may result in gene trees with wrong branching patterns (e.g., those branches with low bootstrap support) that masquerade as true gene/species tree incongruence, thus resulting in over-, and possibly under-, estimation of deep coalescences.

Here we propose an approach to estimating species trees from estimated gene trees which avoids these problems. Instead of assuming that all gene trees are correct (and hence fully resolved, rooted trees), we consider the case where all gene trees are modified so that they are reasonably likely to be unrooted, edge-contracted versions of the true gene trees. For example, the reliable edges in the gene trees can be identified using statistical techniques, such as bootstrapping, and all low-support edges can be contracted. In this way, the MDC problem becomes one in which the input is a set of gene trees which may not be rooted and may not be fully resolved, and the objective is a rooted, binary species tree and binary rooted refinements of the input gene trees, that optimizes the MDC criterion. We provide exact algorithms and heuristics for inferring species trees for these cases. We have implemented several of these algorithms and heuristics in our PhyloNet software package [24], which is publicly available at http://bioinfo.cs.rice.edu/phylonet, and we evaluate the performance of these algorithms and heuristics on synthetic data.

2   Preliminary Material

Clades and clusters. Throughout this section, unless specified otherwise, all trees are presumed to be rooted binary trees, bijectively leaf-labelled by the elements of $\mathcal{X}$ (that is, each $x \in \mathcal{X}$ labels one leaf in each tree). We denote by $\mathcal{T}_{\mathcal{X}}$ the set of all binary rooted trees on leaf-set $\mathcal{X}$. We denote by $V(T)$, $E(T)$, and $L(T)$ the node-set, edge-set, and leaf-set, respectively, of T. For v a node in T, we define $parent(v)$ to be the parent of

v in T, and $Children(v)$ to be the children of v. A *clade* in a tree T is a rooted subtree of T, which can be identified by the node in T rooting the clade. For a given tree T, we denote the subtree of T rooted at v by $Clade_T(v)$, and when the tree T is understood, by $Clade(v)$. The *clade for node* v is $Clade(v)$, and since nodes can have children, the children of a clade $Clade(v)$ are the clades rooted at the children of v. The set of all clades of a tree T is denoted by *Clades(T)*. The set of leaves in $Clade_T(v)$ is called a *cluster* and denoted by $Cluster_T(v)$ (or more simply by $Cluster(v)$ if the tree T is understood). The clusters that contain either all the taxa or just single leaves are called *trivial*, and the other clusters are called *non-trivial*. The *cluster of node* v is *Cluster(v)*. As with clades, clusters can also have children. If Y is a cluster in a tree T, then the *clade for* Y *within* T, denoted by $Clade_T(Y)$, is the clade of T induced by Y. The set of all clusters of T is denoted by *Clusters(T)*. We say that edge e in gt is *outside* cluster Y if it satisfies $e \notin E(Clade_{gt}(Y))$, and otherwise that it is *inside* Y. Given a set $A \subseteq L(T)$, we define $MRCA_T(A)$ to be the most recent (or least) common ancestor of the taxa in A. Finally, given trees t and T, both on $\mathcal{X}$, we define $H : V(t) \rightarrow V(T)$ by $H_T(v) = MRCA_T(Cluster_t(v))$.

We extend the definitions of *Clades(T)* and *Clusters(T)* to the case where T is unrooted by defining *Clades(T)* to be the set of all clades of all possible rootings of T, and *Clusters(T)* to be the set of all clusters of all possible rootings of T. Thus, the sets *Clades(T)* and *Clusters(T)* depend upon whether T is rooted or not.

Given a cluster $Y \subseteq \mathcal{X}$ of T, the *parent edge of* Y *within* T is the edge incident with the root of the clade for Y, but which does not lie within the clade. When T is understood by context, we will refer to this as the *parent edge of* Y.

A set $\mathcal{C}$ of clusters is said to be *compatible* if there is a rooted tree T on leaf-set S such that $Clusters(T) = \mathcal{C}$. By [18], the set $\mathcal{C}$ is compatible if and only if every pair A and B of clusters in $\mathcal{C}$ are either disjoint or one contains the other.

Valid coalescent histories and extra lineages. Given gene tree gt and species tree ST, a *valid coalescent history* is a function $f : V(gt) \rightarrow V(ST)$ such that the following conditions hold: (1) if w is a leaf in gt, then $f(w)$ is the leaf in ST with the same label; and, (2) if w is a vertex in $Clade_{gt}(v)$, then $f(w)$ is a vertex in $Clade_{ST}(f(v))$. Note that these two conditions together imply that $f(v)$ is a node on the path between the root of ST and the MRCA in ST of $Cluster_{gt}(v)$. Given a gene tree gt and a species tree ST, and given a function f defining a valid coalescent history of gt within ST, the *number of lineages* on each edge in ST can be computed by inspection. An optimal valid coalescent history is one that results in the minimum number of lineages over all valid coalescent histories. We denote the number of *extra lineages* on an edge $e \in E(ST)$ (one less than the number of lineages on e) in an optimal valid coalescent history of gt within ST by $XL(e, gt)$, and we denote by $XL(ST, gt)$ the total number of extra lineages within an optimal valid coalescent history of gt within ST, i.e., $XL(ST, gt) = \sum_{e \in E(ST)} XL(e, gt)$; see Fig. 2. Finally, we denote by $XL(ST, \mathcal{G})$ the total number of extra lineages, or MDC score, over all gene trees in $\mathcal{G}$, so $XL(ST, \mathcal{G}) = \sum_{gt \in \mathcal{G}} XL(ST, gt)$. Given gene tree gt and species tree ST, finding the valid coalescent history that yields the smallest number of extra lineages is achievable in polynomial time, as we now show. Given cluster A in gt and cluster B in ST, we say that A is B-*maximal* if (1) $A \subseteq B$ and (2) for all $A' \in Clusters(gt)$, if

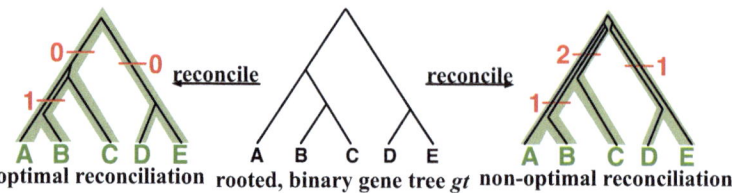

Fig. 2. Illustration of optimal and non-optimal reconciliations of a rooted, binary gene tree *gt* with a rooted, binary species tree *ST*, which yield 1 and 4 extra lineages, respectively

$A \subset A'$ then $A' \not\subseteq B$. We set $k_B(gt)$ to be the number of *B*-maximal clusters within *gt*. Finally, we say that cluster *A* is *ST*-maximal if there is a cluster $B \in Clusters(ST)$ such that $B \neq X$ and *A* is *B*-maximal.

Theorem 1. *(From [22]) Let gt be a gene tree, ST be a species tree, both binary rooted trees on leaf-set X. Let B be a cluster in ST and let e be the parent edge of B in ST. Then $k_B(gt)$ is equal to the number of lineages on e in an optimal valid coalescent history. Therefore, $XL(e, gt) = k_B(gt) - 1$, and $XL(ST, gt) = \sum_B [k_B(gt) - 1]$, where B ranges over the clusters of ST. Furthermore, a valid coalescent history f that achieves this total number of extra lineages can be produced by setting $f(v) = H_{ST}(v)$ (i.e., $f(v) = MRCA_{ST}(Cluster_{gt}(v)))$ for all v.*

In other words, we can score a candidate species tree *ST* with respect to a set $\mathcal{G}$ of rooted binary trees with $XL(ST, \mathcal{G}) = \sum_{gt \in \mathcal{G}} \sum_{B \in Clusters(ST)} [k_B(gt) - 1]$. Finally,

Corollary 1. *Given collection $\mathcal{G}$ of k gene trees and species tree ST, each tree labelled by the species in X, we can compute the optimal coalescent histories relating each gene tree to ST so as to minimize the total number of extra lineages in $O(nk)$ time, and the MDC score of these optimal coalescent histories in $O(n^2 k)$ time, where $|X| = n$.*

The analysis of the running time follows from the following lemma:

Lemma 1. *Given a rooted gene tree gt and a rooted binary species tree ST, we can compute all $H_{ST}(v)$ (letting v range over $V(gt)$) in $O(n)$ time. We can also compute the set of ST-maximal clusters in gt in $O(n^2)$ time.*

2.1   The MDC Problem: Rooted, Binary Gene Trees

The MDC problem is the "minimize deep coalescence" problem; as formulated by Wayne Maddison in [10], this is equivalent to finding a species tree that minimizes the total number of extra lineages over all gene trees in $\mathcal{G}$. Thus, the MDC problem can be stated as follows: given a set $\mathcal{G}$ of rooted, binary gene trees, we seek a species tree *ST* such that $XL(ST, \mathcal{G}) = \sum_{gt \in \mathcal{G}} XL(ST, gt)$ is minimized.

MDC is conjectured to be NP-hard, and no polynomial-time exact algorithm is known for this problem. However, it can be solved exactly using several techniques, as we now show.

Algorithms for MDC. The material in this section is from [22]. The simplest technique to compute the optimal species tree with respect to a set $\mathcal{G}$ of gene trees is to compute a minimum-weight clique of size $n - 2$ (where $|\mathcal{X}| = n$) in a graph which we now describe. Let $\mathcal{G}$ be the set of gene trees in the input to MDC, and let $MDC(\mathcal{G})$ be the graph with one vertex for each non-trivial subset of $\mathcal{X}$ (so $MDC(\mathcal{G})$ does not contain trivial clusters), edges between A and B if the two clusters are compatible (and so $A \cap B = \emptyset$, $A \subset B$, or $B \subset A$). A clique inside this graph therefore defines a set of pairwise compatible clusters, and hence a rooted tree on $\mathcal{X}$. We set the weight of each node A to be $w(A) = \sum_{gt \in \mathcal{G}}[k_A(gt) - 1]$. We seek a clique of size $n - 2$, and among all such cliques we seek one of minimum weight. By construction, the clique will define a rooted, binary tree ST such that $XL(ST, \mathcal{G})$ is minimized.

The graph $MDC(\mathcal{G})$ contains $2^n - n - 1$ vertices, where $n = |\mathcal{X}|$, and is therefore large even for relatively small n. We can constrain this graph size by restricting the allowable clusters to a smaller set, $\mathcal{C}$, of subsets of $\mathcal{X}$. For example, we can set $\mathcal{C} = \cup_{gt \in \mathcal{G}} Clusters(gt)$ (minus the trivial clusters), and we can define $MDC(\mathcal{C})$ to be the subgraph of $MDC(\mathcal{G})$ defined on the vertices corresponding to $\mathcal{C}$. However, the cliques of size $n - 2$ in the graph $MDC(\mathcal{C})$ may not have minimum possible weights; therefore, instead of seeking a minimum weight clique of size $n - 2$ within $MDC(\mathcal{C})$, we will set the weight of node A to be $w'(A) = Q - w(A)$, for some very large Q, and seek a *maximum weight* clique within the graph.

Finally, we can also solve the problem exactly using dynamic programming. For $A \subseteq \mathcal{X}$ and binary rooted tree T on leaf-set A, we define

$$l_T(A, \mathcal{G}) = \sum_{gt \in \mathcal{G}} \sum_{B} [k_B(gt) - 1],$$

where B ranges over all clusters of T. We then set

$$l^*(A, \mathcal{G}) = \min\{l_T(A, \mathcal{G}) : T \in \mathcal{T}_A\}.$$

By Theorem 1, $l^*(\mathcal{X}, \mathcal{G})$ is the minimum number of extra lineages achievable in any species tree on $\mathcal{X}$, and so any tree T such that $l_T(\mathcal{X}, \mathcal{G}) = l^*(\mathcal{X}, \mathcal{G})$ is a solution to the MDC problem on input $\mathcal{G}$. We now show how to compute $l^*(A, \mathcal{G})$ for all $A \subseteq \mathcal{X}$ using dynamic programming. By backtracking, we can then compute the optimal species tree on $\mathcal{X}$ with respect to the set $\mathcal{G}$ of gene trees.

Consider a binary rooted tree T on leaf-set A that gives an optimal score for $l^*(A, \mathcal{G})$, and let the two subtrees off the root of T be T_1 and T_2 with leaf sets A_1 and $A_2 = A - A_1$, respectively. Then, letting B range over the clusters of T, we obtain

$$l_T(A, \mathcal{G}) = \sum_{gt \in \mathcal{G}} \sum_{B} [k_B(gt) - 1] =$$

$$\sum_{gt \in \mathcal{G}} \sum_{B \subseteq A_1} [k_B(gt) - 1] + \sum_{gt \in \mathcal{G}} \sum_{B \subseteq A_2} [k_B(gt) - 1] + \sum_{gt \in \mathcal{G}} [k_A(gt) - 1].$$

If for $i = 1$ or 2, $l_{T_i}(A_i, \mathcal{G}) \neq l^*(A_i, \mathcal{G})$, then we can replace T_i by a different tree on A_i and obtain a tree T' on A such that $l_{T'}(A, \mathcal{G}) < l_T(A, \mathcal{G})$, contradicting the optimality of T. Thus, $l_{T_i}(A_i, \mathcal{G}) = l^*(A_i, \mathcal{G})$ for $i = 1, 2$, and so $l^*(A, \mathcal{G})$ is obtained by taking the minimum over all sets $A_1 \subset A$ of $l^*(A_1, \mathcal{G}) + l^*(A - A_1, \mathcal{G}) + \sum_{gt \in \mathcal{G}}[k_A(gt) - 1]$. In other words, we have proven the following:

Lemma 2. $l^*(A, \mathcal{G}) = min_{A_1 \subset A}\{l^*(A_1, \mathcal{G}) + l^*(A - A_1, \mathcal{G}) + \sum_{gt \in \mathcal{G}}[k_A - 1]\}$.

This lemma suggests the dynamic programming algorithm:

- Order the subsets of $\mathcal{X}$ by cardinality, breaking ties arbitrarily.
- Compute $k_A(gt)$ for all $A \subseteq \mathcal{X}$ and $gt \in \mathcal{G}$.
- For all singleton sets A, set $l^*(A, \mathcal{G}) = 0$.
- For each subset with at least two elements, from smallest to largest, compute $l^*(A, \mathcal{G}) = min_{A_1 \subset A}\{l^*(A_1, \mathcal{G}) + l^*(A - A_1, \mathcal{G}) + \sum_{gt \in \mathcal{G}}[k_A(gt) - 1]\}$.
- Return $l^*(\mathcal{X}, \mathcal{G})$.

There are $2^n - 1$ subproblems to compute (one for each set A) and each takes $O(2^n n)$ time (there are at most 2^n subsets A_1 of A, and each pair A, A_1 involves computing k_A for each $gt \in \mathcal{G}$, which costs $O(n)$ time). Hence, the running time is $O(n2^{2n})$ time. However, Than and Nakhleh showed that using only the clusters of the gene trees would produce almost equally good estimates of the species tree [22,23].

3   MDC on Estimated Gene Trees

Estimating gene trees with high accuracy is a challenging task, particularly in cases where branch lengths are very short (which are also cases under which ILS is very likely to occur). As a result, gene tree estimates are often unrooted, unresolved, or both. To deal with these practical cases, we formulate the problems as estimating species trees and completely resolved, rooted versions of the input trees to optimize the MDC criterion. We show that the clique-based and DP algorithms can still be applied.

3.1   Unrooted, Binary Gene Trees

When reconciling an unrooted, binary gene tree with a rooted, binary species tree under parsimony, it is natural to seek the rooting of the gene tree that results in the minimum number of extra lineages over all possible rootings. In this case, the MDC problem can be formulated as follows: given a set $\mathcal{G} = \{gt_1, gt_2, \ldots, gt_k\}$ of gene trees, each of which is unrooted, binary, with leaf-set $\mathcal{X}$, we seek a species tree ST and set $\mathcal{G}' = \{gt'_1, gt'_2, \ldots, gt'_k\}$, where gt'_i is a rooted version of gt_i, so that $XL(ST, \mathcal{G}')$ is minimum over all such sets $\mathcal{G}'$.

Given a species tree and a set of unrooted gene trees, it is easy to compute the optimal rootings of each gene tree with respect to the given species tree, since there are only $O(n)$ possible locations for the root in an n leaf tree, and for each possible rooting we can compute the score of that solution in $O(n^2)$ time. Thus, it is possible to compute the optimal rooting and its score in $O(n^3)$ time. Here we show how to solve this problem more efficiently – finding the optimal rooting in $O(n)$ time, and the score for the optimal rooting in $O(n^2)$ time, thus saving a factor of n. We accomplish this using a small modification to the techniques used in the case of rooted gene trees.

We begin by extending the definition of B-maximal clusters to the case of unrooted gene trees, for B a cluster in a species tree ST, in the obvious way. Recall that the set $Clusters(gt)$ depends on whether gt is rooted or not, and that $k_B(gt)$ is the number of B-maximal clusters in gt. We continue with the following:

Lemma 3. *Let gt be an unrooted binary gene tree on $\mathcal{X}$ and let ST be a rooted binary species tree on $\mathcal{X}$. Let C^* be the set of ST-maximal clusters in gt. Let e be any edge of gt such that $\forall Y \in C^*, e \notin E(Clade_{gt}(Y))$ (i.e., e is not inside any subtree of gt induced by one of the clusters in C^*). Then the tree gt' produced by rooting gt on edge e satisfies (1) $C^* \subseteq Clusters(gt')$, and (2) $XL(ST, gt') = \sum_{B \in Clusters(ST)} [k_B(gt) - 1]$, which is the best possible. Furthermore, there is at least one such edge e in gt.*

Proof. We begin by showing that there is at least one edge e that is outside Y for all $Y \in C^*$. Pick a cluster $A_1 \in C^*$ that is maximal (i.e., it is not a subset of any other cluster in C^*); we will show that the parent edge of A_1 is outside all clusters in C^*. Suppose e is inside cluster $A_2 \in C^*$. Since A_1 is maximal, it follows that $A_2 \not\subseteq A_1$. However, if the parent edge of A_2 is not inside A_1, then either A_2 is disjoint from A_1 or A_2 contains A_1, neither of which is consistent with the assumptions that A_1 is maximal and the parent edge of A_1 is inside A_2. Therefore, the parent edge of A_2 must be inside A_1. In this case, $A_1 \cap A_2 \neq \emptyset$ and $A_1 \cup A_2 = \mathcal{X}$. Let B_i be the cluster in ST such that A_i is B_i-maximal, $i = 1, 2$. Then $B_1 \cap B_2 \neq \emptyset$, and so without loss of generality $B_1 \subseteq B_2$. But then $A_1 \cup A_2 \subseteq B_1 \cup B_2 = B_2$ and so $B_2 = \mathcal{X}$. But $\mathcal{X}$ is the only $\mathcal{X}$-maximal cluster, contradicting our hypotheses. Hence the parent edge of any maximal cluster in C^* is not inside any cluster in C^*.

We now show that rooting gt on any edge e that is not inside any cluster in C^* satisfies $C^* \subseteq Clusters(gt')$. Let e be any such edge, and let gt' be the result of rooting gt on e. Under this rooting, the two children of the root of gt' define subtrees T_1, with cluster A_1, and T_2, with cluster A_2. Now, suppose $\exists A' \in C^*\text{-}Clusters(gt')$. Since $C^* \subseteq Clusters(gt)$, it follows that A' is the complement of a cluster $B \in Clusters(gt')$. If B is a proper subset of either A_1 or A_2, then the subtree of gt induced by A' contains edge e (since $A' = \mathcal{X} - B$), contradicting how we selected e. Hence, it must be that $B = A_1$ or $B = A_2$. However, in this case, A' is also equal to either A_1 or A_2, and hence $A' \in Clusters(gt')$, contradicting our hypothesis about A'.

We finish the proof by showing that $XL(ST, gt')$ is optimal for all such rooted trees gt', and that all other locations for rooting gt produce a larger number of extra lineages. By Theorem 1, $XL(ST, gt') = \sum_B [k_B(gt') - 1]$, as B ranges over the clusters of ST. By construction, this is exactly $\sum_B [k_B(gt) - 1]$, as B ranges over the clusters of ST. Also note that for *any* rooted version gt^* of gt, $k_B(gt^*) \geq k_B(gt)$, so that this is optimal. Now consider a rooted version gt^* in which the root is on an edge that *is* inside some subtree of gt induced by $A \in C^*$. Let gt^* have subtrees T_1 and T_2 with clusters A_1 and A_2, respectively. Without loss of generality, assume that $A_1 \subset A$, and that $A_2 \cap A \neq \emptyset$. Since $A \in C^*$, there is a cluster $B \in Clusters(ST)$ such that A is B-maximal. But then A_1 is B-maximal. However, since $A - A_1 \neq \emptyset$, there is also at least one B-maximal cluster $Y \subset A$ within T_2. Hence, $k_B(gt^*) > k_B(gt)$. On the other hand, for all other clusters B' of ST, $k_{B'}(gt^*) \geq k_{B'}(gt') = k_{B'}(gt)$. Therefore, $XL(ST, gt^*) > XL(ST, gt')$. In other words, any rooting of gt on an edge that is not within a subtree induced by a cluster in $\mathcal{A}$ is optimal, while any rooting of gt on any other edge produces a strictly larger number of extra lineages.

This theorem allows us to compute the optimal rooting of an unrooted binary gene tree with respect to a rooted binary species tree, and hence gives us a way of computing the score of any candidate species tree with respect to a set of unrooted gene trees:

Corollary 2. *Let ST be a species tree and $\mathcal{G} = \{gt_1, gt_2, \ldots, gt_k\}$ be a set of unrooted binary gene trees. Let $\mathcal{G}' = \{gt'_1, gt'_2, \ldots, gt'_k\}$ be a set of binary gene trees such that gt'_i is a rooted version of gt_i for each $i = 1, 2, \ldots, k$, and which minimizes $XL(ST, \mathcal{G}')$. Then $XL(ST, \mathcal{G}') = \sum_i \sum_{B \in Clusters(ST)} [k_B(gt_i) - 1]$. Furthermore, the optimal $\mathcal{G}'$ can be computed in $O(nk)$ time, and the score of $\mathcal{G}'$ computed in $O(n^2 k)$ time.*

Solving MDC given unrooted, binary gene trees. Let $\mathcal{G} = \{gt_1, gt_2, \ldots, gt_k\}$, as above. We define the MDC-score of a candidate (rooted, binary) species tree ST by $\sum_i \sum_{B \in Clusters(ST)} [k_B(gt_i) - 1]$; by Corollary 2, the tree ST^* that has the minimum score will be an optimal species tree for the MDC problem on input $\mathcal{G}$. As a result, we can use all the techniques used for solving MDC given binary rooted gene trees, since the score function is unchanged.

3.2   Rooted, Non-binary Gene Trees

When reconciling a rooted, non-binary gene tree with a rooted, binary species tree under parsimony, it is natural to seek the refinement of the gene tree that results in the minimum number of extra lineages over all possible refinements; see the illustration in Fig. 3. In this case, the MDC problem can be formulated as follows: given a set $\mathcal{G} = \{gt_1, gt_2, \ldots, gt_k\}$ in which each gt_i may only be partially resolved, we seek a species tree ST and binary refinements gt_i^* of gt_i so that $XL(ST, \mathcal{G}^*)$ is minimized, where $\mathcal{G}^* = \{gt_1^*, gt_2^*, \ldots, gt_k^*\}$. This problem is at least as hard as the MDC problem, which is conjectured to be NP-hard.

A Quadratic Algorithm for Optimal Refinement of Gene Trees Under MDC. We begin with the problem of finding an optimal refinement of a given gene tree gt with respect to a given species tree ST, with both trees rooted.

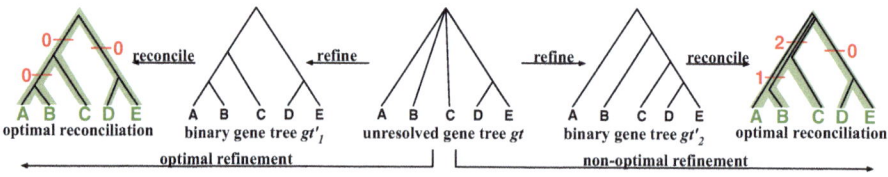

Fig. 3. Illustration of optimal and non-optimal reconciliations of a rooted, non-binary gene tree gt with a rooted, binary species tree ST, which yield 0 and 3 extra lineages, respectively

Definition 1. *(Optimal tree refinement w.r.t. MDC (OTR_{MDC}))*

> **Input:** *Species tree ST and gene tree gt, both rooted and leaf-labelled by set X of taxa.*
> **Output:** *Binary rooted tree gt^* refining gt that minimizes $XL(ST,t)$ over all refinements t of gt. We denote gt^* by $OTR_{MDC}(ST,gt)$.*

We show that $OTR_{MDC}(ST,gt)$ can be solved in $O(n^2)$ time, where n is the number of leaves in either tree. For $B \in Clusters(ST)$ and gene tree gt, we define $F_B(gt)$ to be the number of nodes in gt that have at least one child whose cluster is B-maximal. We will show that for a given rooted gene tree gt and rooted binary species tree ST, the optimal refinement t^* of gt will satisfy $XL(ST,t^*) = \sum_{B \in Clusters(ST)}[F_B(gt) - 1]$. Therefore, to compute the score of the optimal refinement of one gene tree gt, it suffices to compute $F_B(gt)$ for every $B \in Clusters(ST)$.

The algorithm to compute the score of the optimal refinement of gt first computes the set of B-maximal clusters, which takes $O(n)$ time by Lemma 1. It then computes $F_B(gt)$, for each B; this requires an additional $O(n)$ time per B, for a total cost of $O(n^2)$ time.

Algorithm for $OTR_{MDC}(ST,gt)$: To compute the optimal refinement, we have a slightly more complicated algorithm.

Step 1: Preprocessing. We begin by computing $H_{ST}(v)$ for every node $v \in V(gt)$, as described above; this takes $O(n)$ time overall.

Step 2: Refine at every high degree node. We then visit each internal node v of gt that has more than two children, and we modify the tree gt locally at v by replacing the rooted star tree at v by a tree defined by the topology induced in ST by the images under the mapping H_{ST} of v and v's children. The order in which we visit the nodes is irrelevant.

We now make precise how this modification of gt at node v is performed. We denote by $Tree(ST,gt,v)$ the tree formed as follows. First, we compute the subtree of ST induced by the images of v and its children under the H_{ST} mapping. If a child y of v is mapped to an internal node of the induced subtree, we add a leaf l_y and make it a child of $H_{ST}(y)$; in this way, the tree we obtain has all the nodes in $Children(v)$ identified with distinct leaves in $Tree(ST,gt,v)$. (Although ST is assumed to be binary, $Tree(ST,gt,v)$ may not be binary.) After we compute $Tree(ST,gt,v)$, we modify gt by replacing the subtree of gt induced by v and its children with $Tree(ST,gt,v)$. The subtree within the refinement that is isomorphic to $Tree(ST,gt,v)$ is referred to as the *local subtree at v*.

Step 3: Completely refine if necessary. Finally, after the refinement at every node is complete, if the tree is not binary, we complete the refinement with an arbitrary refinement at v.

Theorem 2. *Algorithm $OTR_{MDC}(ST,gt)$ takes $O(n^2)$ time, where ST and gt each have n leaves.*

It is clear that the algorithm is well-defined, so that the order in which we visit the nodes in $V(gt)$ does not impact the output.

Observation 1. *Let gt be an arbitrary rooted gene tree, gt′ a refinement of gt, and ST an arbitrary rooted binary species tree. Then $k_B(gt') \geq F_B(gt)$ for all clusters B of ST.*

Theorem 3. *Let gt be an arbitrary rooted gene tree, ST an arbitrary rooted binary species tree, t the result of the first two steps of $OTR_{MDC}(ST, gt)$, and t^* an arbitrary refinement of t (thus $t^* = OTR_{MDC}(ST, gt)$). Then for all $B \in Clusters(ST)$, $F_B(gt) = F_B(t^*)$ and no node in t or t^* has more than one B-maximal child.*

Proof. Step 2 of $OTR_{MDC}(ST, gt)$ can be seen as a sequence of refinements that begins with gt and ends with t, in which each refinement is obtained by refining around a particular node in gt. The tree $t^* = OTR_{MDC}(ST, gt)$ is then obtained by refining t arbitrarily into a binary tree, if t is not fully resolved. Let the internal nodes of gt with at least three children be $v_1, v_2, \ldots, v_k$. Thus, $gt = gt_0 \to gt_1 \to gt_2 \to \ldots \to gt_k = t \to t^*$, where $gt_i \to gt_{i+1}$ is the act of refining at node v_{i+1}, and $t \to t^*$ is an arbitrary refinement.

We begin by showing that $F_B(gt_i) = F_B(gt_{i+1})$, for $i = 0, 1, 2, \ldots, k-1$. When we refine at node v_i, we modify the tree gt_{i-1} by replacing the subtree immediately below node v_i by $Tree(ST, gt, v_i)$, producing the local subtree below v_i. Fix a cluster $B \in Clusters(ST)$. If the cluster for v_i in gt_{i-1} does not have any B-maximal children, then refining at v_i will not change F_B, and hence $F_B(gt_{i-1}) = F_B(gt_i)$. Otherwise, v_i has at least one B-maximal child in gt_{i-1}. Since v_i is not B-maximal within gt_{i-1}, v_i also has at least one child in gt_{i-1} that is not B-maximal. Hence, the tree gt_i produced by refining gt_{i-1} at v_i (using $Tree(ST, gt_i, v_i)$) contains a node y that is an ancestor of all the B-maximal children of v_i within gt_{i-1} and not the ancestor of any other children of v_i in gt_{i-1}. Therefore, the cluster for y is B-maximal within gt_i, and no other node that is introduced during this refinement is B-maximal within gt_i. Therefore within the local subtree at v_i in gt_i there is exactly one node that defines a B-maximal cluster, and exactly one node that is the parent of at least one B-maximal cluster. As a result, $F_B(gt_{i-1}) = F_B(gt_i)$.

This argument also shows that any node in the local subtree at v_i that is the parent of at least one B-maximal cluster is the parent of exactly one B-maximal cluster. On the other hand, if v_i does not have any B-maximal child in gt_{i-1}, then there is no node in v_i's local subtree that has any B-maximal children. In other words, after refining at node v_i, any node within the local subtree at v_i that has one or more B-maximal children has exactly one such child. As a result, at the end of Step 2 of $OTR_{MDC}(ST, gt)$, every node has at most one B-maximal child, for all $B \in Clusters(ST)$.

The last step of the OTR_{MDC} algorithm produces an arbitrary refinement of $t = gt_k$, if it is not fully resolved. But since no node in gt_k can have more than one B-maximal child, if t^* is a refinement of $t = gt_k$ then $F_B(t) = F_B(t^*)$.

Theorem 4. *Let gt be a rooted gene tree, ST a rooted binary species tree, both on set $\mathcal{X}$, t the result of the first two steps of $OTR_{MDC}(ST, gt)$, and t^* any refinement of t. Then $XL(ST, t^*) = \sum_{B \in Clusters(ST)} [F_B(gt) - 1]$, and t^* is a binary refinement of gt that minimizes $XL(ST, t')$ over all binary refinements t' of gt.*

Proof. Let B be an arbitrary cluster in ST. By Theorem 3, $F_B(t^*) = F_B(gt)$. Also by Theorem 3, no node in t has more than one B-maximal child, and so $k_B(t) =$

$F_B(t)$. Since t^* is an arbitrary refinement of t, it follows that $k_B(t^*) = F_B(t^*)$, and so $k_B(t^*) = F_B(gt)$. By Observation 1, for all refinements t' of gt, $k_B(t') \geq F_B(gt)$. Hence $k_B(t') \geq k_B(t^*)$ for all refinements t' of gt. Since this statement holds for an arbitrary cluster B in ST, it follows that $XL(ST, t') \geq XL(ST, t^*)$ for all refinements t' of gt, establishing the optimality of t^*.

Corollary 3. *Let ST be a species tree and $\mathcal{G} = \{gt_1, gt_2, \ldots, gt_k\}$ be a set of gene trees that may not be resolved. Let $\mathcal{G}^* = \{gt_1^*, gt_2^*, \ldots, gt_k^*\}$ be a set of binary gene trees such that gt_i^* refines gt_i for each $i = 1, 2, \ldots, k$, and which minimizes $XL(ST, \mathcal{G}^*)$. Then $XL(ST, \mathcal{G}^*) = \sum_i \sum_{B \in Clusters(ST)} [F_B(gt_i) - 1]$. Furthermore, the optimal resolution of each gene tree and its score can be computed in $O(n^2 k)$ time.*

Solving MDC given rooted, non-binary gene trees. Corollary 3 allows us to compute the score of any species tree with respect to a set of rooted but unresolved gene trees. We can use this to find optimal species trees from rooted, non-binary gene trees, as we now show. Let $\mathcal{G}$ be a set of rooted gene trees that are not necessarily binary. By Corollary 3, we can formulate the problem as a minimum-weight clique problem. The graph has one vertex for every subset of $\mathcal{X}$, and we set the weight of the vertex corresponding to subset B to be $w(B) = \sum_{gt \in \mathcal{G}} [F_B(gt) - 1]$. We have edges between vertices if the two vertices are compatible (can both be contained in a tree). The solution is therefore a minimum weight clique with $n - 2$ vertices. And, as before, we can describe this as a maximum weight clique problem by having the weight be $w'(B) = Q - w(B)$, for some large enough Q.

However, we can also address this problem using dynamic programming, as before. Let $A \subseteq \mathcal{X}$ and $T \in \mathcal{T}_A$. Let $l_T(A, \mathcal{G}) = \sum_{gt \in \mathcal{G}} \sum_B [F_B(gt) - 1]$, as B ranges over the clusters of T. Let $l^*(A, \mathcal{G}) = \min_{T \in \mathcal{T}_A} \{l_T(A, \mathcal{G})\}$. Then $l^*(\mathcal{X}, \mathcal{G})$ is the solution to the problem of inferring a species tree from rooted, non-binary gene trees.

We set base cases $l^*(\{x\}, \mathcal{G}) = 0$ for all $x \in \mathcal{X}$. We order the subproblems by the size of A, and compute $l^*(A, \mathcal{G})$ only after every $l^*(A', \mathcal{G})$ is computed for $A' \subset A$. The DP formulation is

$$l^*(A, \mathcal{G}) = \min_{A_1 \subset A} \{l^*(A_1, \mathcal{G}) + l^*(A - A_1, \mathcal{G}) + \sum_{gt \in \mathcal{G}} [F_A(gt) - 1]\}.[-5mm]$$

3.3   Unrooted, Non-binary Gene Trees

When reconciling an unrooted and incompletely resolved gene tree with a rooted, binary species tree under parsimony, it is natural to seek the rooting and refinement of the gene tree that results in the minimum number of extra lineages over all possible rootings and refinements; see the illustration in Fig. 4. In this case, the MDC problem can be formulated as follows: given a set $\mathcal{G} = \{gt_1, gt_2, \ldots, gt_k\}$, with each gt_i a tree on $\mathcal{X}$, but not necessarily rooted nor fully resolved, we seek a rooted, binary species tree ST and set $\mathcal{G}' = \{gt_1', gt_2', \ldots, gt_k'\}$ such that each gt_i' is a binary rooted tree that can be obtained by rooting and refining gt_i, so as to minimize $XL(ST, \mathcal{G}')$ over all such $\mathcal{G}'$. As before, the computational complexity of this problem is unknown, but conjectured to be NP-hard.

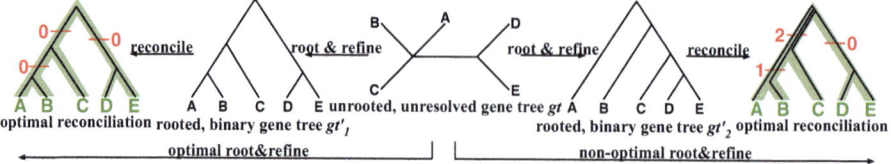

Fig. 4. Illustration of optimal and non-optimal reconciliations of an unrooted, non-binary gene tree gt with a rooted, binary species tree ST, which yield 0 and 3 extra lineages, respectively

Observation 2. *For any gene tree gt and species tree ST, and t^* the optimal refined rooted version of gt that minimizes $XL(ST, t^*)$ can be obtained by first rooting gt at some node, and then refining the resultant rooted tree. Thus, to find t^*, it suffices to find a node $v \in V(gt)$ at which to root the tree t, thus producing a tree t', so as to minimize $\sum_{B \in Clusters(ST)}[F_B(gt') - 1]$.*

From this, the following theorem follows:

Theorem 5. *Let gt be an unrooted, not necessarily binary gene tree on $\mathcal{X}$, and let ST be a rooted species tree on $\mathcal{X}$. Let $A \in Clusters(gt)$ be a largest ST-maximal cluster, and v be the neighbor of the root of the clade for A that is in A. If we root gt at v, then the resultant tree gt' minimizes $\sum_{B \in Clusters(ST)}[F_B(gt') - 1]$ over all rooted versions gt' of t.*

And, therefore,

Theorem 6. *Let $\mathcal{T}$ be a set of gene trees that are unrooted and not necessarily binary. For $B \subset \mathcal{X}$, define t^B to be the rooted version of t formed by rooting t at a node v, as given by Theorem 5. Then, the species tree ST that minimizes $\sum_{t \in \mathcal{T}} \sum_{B \in Clusters(ST)}[F_B(t^B) - 1]$ is an optimal solution to the problem.*

As a result, we can solve the problem using the clique and DP formulations as in the other versions of the MDC problem.

4   Experimental Evaluation

4.1   Methods

Simulated data. We generated species trees using the "Uniform Speciation" (Yule) module in the program Mesquite [12]. Two sets of species trees were generated: one for 8 taxa plus an outgroup, and one for 16 taxa plus an outgroup. Each data set had 500 species trees. All of them have a total branch length of 800,000 generations excluding the outgroup. Within the branch of each species tree, 1, 2, 4, 8, 16, or 32 gene trees were simulated using the "Coalescence Contained Within Current Tree" module in Mesquite with the effective population size N_e equal 100,000. We sampled one allele per species. We used the program Seq-gen [15] to simulate the evolution of DNA sequences of length 2000 under the Jukes-Cantor model [6] down each of the gene trees (these settings are similar to those used in [11]).

Estimated gene trees.   We estimated gene trees from these sequence alignments using default PAUP* heuristic maximum parsimony (MP) methods, returning the strict consensus of all optimal MP trees. We rooted each estimated tree at the outgroup in order to produce rooted estimated trees.

Estimated species trees.   The "heuristic' version of our method uses only the clusters of the input gene trees, and the "exact" version uses all possible clusters on the taxon set. For some analyses using the heuristic MDC algorithms, the estimated species tree is not fully resolved. In this case, we followed this initial analysis with a search through the set of binary resolutions of the initial estimated species tree for a fully resolved tree that optimized the number of extra lineages. This additional step was limited to 5 minutes of analysis. The only cases where this additional search was not applied were when the polytomy (unresolved node) in the species tree was present in all gene trees; in these cases, any resolution is arbitrary and is as good (under the MDC) criterion as any other resolution.

For the 8-taxon data sets, we used both the exact and heuristic versions of all four algorithms. For the 16-taxon data sets, we used only the heuristic versions.

Measurements.   We report the degree of resolution of each estimated gene tree, which is the number of internal branches in t divided by $n - 3$, where t has n leaves. We also report the Robinson-Foulds (RF) error [16] of estimated trees to the true trees, where the RF error is the total number of edges in the two trees that define bipartitions that are not shared by the other tree, divided by $2n - 6$. A value 0 of the RF distance indicates the two trees are identical, and a value of 1 indicates the two trees are completely different (they disagree on every branch).

4.2   Results

The degree of resolution of the reconstructed gene trees was around 0.6 in the case of 8-taxon gene trees, and around 0.5 in the case of 16-taxon gene trees.

With respect to topological accuracy of the estimated gene trees, we found that for 8 taxa, the RF distance is around 0.21. However, 98% of the estimated gene trees have no false positives; thus, all but 2% of the estimated gene trees can be resolved to match the true gene tree. Similarly, the RF distance for the 16-taxon data sets between true gene trees and reconstructed gene trees is around 0.27, but 96% have 0 false positive values.

We now discuss topological accuracy of the species trees estimated using our algorithms for solving the MDC problem. We show results on running the exact and heuristic versions of the algorithms on 8 taxon estimated gene trees in Figure 5. These results show that increasing the number of gene trees improves the accuracy of the estimated species tree, and that very good accuracy is obtainable from a small number of gene trees. We also see that knowing the true root instead of estimating the root is helpful when the number of gene trees is very small, but that otherwise our algorithm is able to produce comparable results even on unrooted gene trees. The results also show that the heuristic version of our algorithm is as accurate as the exact version once there are four or more gene trees (and almost identical in accuracy for two gene trees).

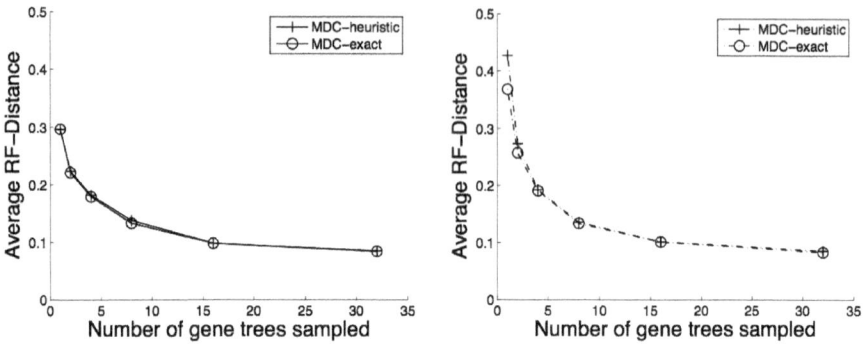

Fig. 5. Performance of MDC methods on estimated gene trees with 8 taxa. Left: MDC on estimated gene trees with correct roots. Right: MDC on unrooted estimated gene trees.

Acknowledgement

This work was supported in part by NSF grant CCF-0622037, grant R01LM009494 from the National Library of Medicine, an Alfred P. Sloan Research Fellowship to LN, a Guggenheim Fellowship to TW, and by Microsoft Research New England support to TW. The contents are solely the responsibility of the authors and do not necessarily represent the official views of the NSF, National Library of Medicine, the National Institutes of Health, the Alfred P. Sloan Foundation, or the Guggenheim Foundation.

References

1. Dawkins, R.: The Ancestor's Tale. Houghton Mifflin, New York (2004)
2. Degnan, J.H., DeGiorgio, M., Bryant, D., Rosenberg, N.A.: Properties of consensus methods for inferring species trees from gene trees. Syst. Biol. 58, 35–54 (2009)
3. Degnan, J.H., Rosenberg, N.A.: Discordance of species trees with their most likely gene trees. PLoS Genet. 2, 762–768 (2006)
4. Degnan, J.H., Rosenberg, N.A.: Gene tree discordance, phylogenetic inference and the multispecies coalescent. Trends Ecol. Evol. 24, 332–340 (2009)
5. Edwards, S.V., Liu, L., Pearl, D.K.: High-resolution species trees without concatenation. PNAS 104, 5936–5941 (2007)
6. Jukes, T.H., Cantor, C.R.: Evolution of protein molecules. In: Munro, H.N. (ed.) Mammalian Protein Metabolism, pp. 21–132. Academic Press, New York (1969)
7. Kubatko, L.S., Carstens, B.C., Knowles, L.L.: STEM: species tree estimation using maximum likelihood for gene trees under coalescence. Bioinformatics 25(7), 971–973 (2009)
8. Kuo, C.-H., Wares, J.P., Kissinger, J.C.: The Apicomplexan whole-genome phylogeny: An analysis of incongruence among gene trees. Mol. Biol. Evol. 25(12), 2689–2698 (2008)
9. Liu, L., Pearl, D.K.: Species trees from gene trees: Reconstructing Bayesian posterior distributions of a species phylogeny using estimated gene tree distributions. Systematic Biology 56(3), 504–514 (2007)
10. Maddison, W.P.: Gene trees in species trees. Syst. Biol. 46, 523–536 (1997)

11. Maddison, W.P., Knowles, L.L.: Inferring phylogeny despite incomplete lineage sorting. Systematic Biology 55(1), 21–30 (2006)
12. Maddison, W.P., Maddison, D.R.: Mesquite: A modular system for evolutionary analysis (2004), version 1.01 http://mesquiteproject.org
13. Nei, M.: Stochastic errors in DNA evolution and molecular phylogeny. In: Gershowitz, H., Rucknagel, D.L., Tashian, R.E. (eds.) Evolutionary Perspectives and the New Genetics, pp. 133–147. Alan R. Liss, New York (1986)
14. Pollard, D.A., Iyer, V.N., Moses, A.M., Eisen, M.B.: Widespread discordance of gene trees with species tree in *Drosophila*: evidence for incomplete lineage sorting. PLoS Genet. 2, 1634–1647 (2006)
15. Rambaut, A., Grassly, N.C.: Seq-gen: An application for the Monte Carlo simulation of DNA sequence evolution along phylogenetic trees. Comp. Appl. Biosci. 13, 235–238 (1997)
16. Robinson, D.R., Foulds, L.R.: Comparison of phylogenetic trees. Math. Biosci. 53, 131–147 (1981)
17. Rokas, A., Williams, B.L., King, N., Carroll, S.B.: Genome-scale approaches to resolving incongruence in molecular phylogenies. Nature 425, 798–804 (2003)
18. Semple, C., Steel, M.: Phylogenetics. Oxford University Press, Oxford (2003)
19. Syring, J., Willyard, A., Cronn, R., Liston, A.: Evolutionary relationships among Pinus (Pinaceae) subsections inferred from multiple low-copy nuclear loci. American Journal of Botany 92, 2086–2100 (2005)
20. Tajima, F.: Evolutionary relationship of DNA sequences in finite populations. Genetics 105, 437–460 (1983)
21. Takahata, N.: Gene genealogy in three related populations: consistency probability between gene and population trees. Genetics 122, 957–966 (1989)
22. Than, C., Nakhleh, L.: Species tree inference by minimizing deep coalescences. PLoS Computational Biology 5(9), e1000501 (2009)
23. Than, C., Nakhleh, L.: Inference of parsimonious species phylogenies from multi-locus data by minimizing deep coalescences. In: Knowles, L.L., Kubatko, L.S. (eds.) Estimating Species Trees: Practical and Theoretical Aspects, pp. 79–98. Wiley-VCH, Chichester (2010)
24. Than, C., Ruths, D., Nakhleh, L.: PhyloNet: a software package for analyzing and reconstructing reticulate evolutionary relationships. BMC Bioinformatics 9, 322 (2008)
25. Than, C., Sugino, R., Innan, H., Nakhleh, L.: Efficient inference of bacterial strain trees from genome-scale multi-locus data. Bioinformatics 24, i123–i131 (2008); Proceedings of the 16th Annual International Conference on Intelligent Systems for Molecular Biology (ISMB 2008)
26. Wu, C.-I.: Inferences of species phylogeny in relation to segregation of ancient polymorphisms. Genetics 127, 429–435 (1991)
27. Wu, C.-I.: Reply to Richard R. Hudson. Genetics 131, 513 (1992)

Rich Parameterization Improves RNA Structure Prediction

Shay Zakov*, Yoav Goldberg*, Michael Elhadad, and Michal Ziv-Ukelson**

Department of Computer Science, Ben-Gurion University of the Negev, Israel
{zakovs,yoavg,elhadad,michaluz}@cs.bgu.ac.il

Abstract. *Motivation.* Current approaches to RNA structure prediction range from physics-based methods, which rely on thousands of experimentally-measured thermodynamic parameters, to machine-learning (ML) techniques. While the methods for parameter estimation are successfully shifting toward ML-based approaches, the model parameterizations so far remained fairly constant and all models to date have relatively few parameters. We propose a move to much richer parameterizations.

Contribution. We study the potential contribution of increasing the amount of information utilized by folding prediction models to the improvement of their prediction quality. This is achieved by proposing novel models, which refine previous ones by examining more types of structural elements, and larger sequential contexts for these elements. We argue that with suitable learning techniques, not being tied to features whose weights could be determined experimentally, and having a large enough set of examples, one could define much richer feature representations than was previously explored, while still allowing efficient inference. Our proposed fine-grained models are made practical thanks to the availability of large training sets, advances in machine-learning, and recent accelerations to RNA folding algorithms.

Results. In order to test our assumption, we conducted a set of experiments that asses the prediction quality of the proposed models. These experiments reproduce the settings that were applied in recent thorough work that compared prediction qualities of several state-of-the-art RNA folding prediction algorithms. We show that the application of more detailed models indeed improves prediction quality, while the corresponding running time of the folding algorithm remains fast. An additional important outcome of this experiment is a new RNA folding prediction model (coupled with a freely available implementation), which results in a significantly higher prediction quality than that of previous models. This final model has about 70,000 free parameters, several orders of magnitude more than previous models. Being trained and tested over the same comprehensive data sets, our model achieves a score of 84% according to the F_1-measure over correctly-predicted base-pairs (i.e. 16% error rate),

---

* These authors contributed equally to the paper.
** Corresponding author.

V. Bafna and S.C. Sahinalp (Eds.): RECOMB 2011, LNBI 6577, pp. 546–562, 2011.
Ⓒ Springer-Verlag Berlin Heidelberg 2011

compared to the previously best reported score of 70% (i.e. 30% error rate). That is, the new model yields an error reduction of about 50%.

Availability. Additional supporting material, trained models, and source code are available through our website at http://www.cs.bgu.ac.il/~negevcb/contextfold

1   Introduction

Within the last few years, non-coding RNAs have been recognized as a highly abundant class of RNAs. These RNA molecules do not code for proteins, but nevertheless are functional in many biological processes, including localization, replication, translation, degradation, regulation and stabilization of biological macromolecules [1,2,3]. It is generally known that much of RNAs functionalities depend on its structural features [3,4,5,6]. Unfortunately, although massive amounts of sequence data are continuously generated, the number of known RNA structures is still limited, since experimental methods such as NMR and Crystallography require expertise and long experimental time. Therefore, computational methods for predicting RNA structures are of significant value [7,8,9]. This work deals with improving the quality of computational RNA structure prediction.

RNA is typically produced as a single stranded molecule, composed as a sequence of *bases* of four types, denoted by the letters A, C, G, and U. Every base can form a hydrogen bond with at most one other base, where bases of type C typically pair with bases of type G, A typically pairs with U, and another weaker pairing can occur between G and U. The set of formed base-pairs is called the *secondary structure*, or the *folding* of the RNA sequence (see Fig. 1), as opposed to the *tertiary structure* which is the actual three dimensional molecule structure. Paired bases almost always occur in a nested fashion in RNA foldings. A folding which sustains this property is called a *pseudoknot-free folding*. In the rest of this work we will consider only pseudoknot-free foldings.

RNA structure prediction (henceforth *RNA folding*) is usually formulated as an optimization problem, where a score is defined for every possible folding of the given RNA sequence, and the predicted folding is one that maximizes this score. While finding a folding which maximizes the score under an arbitrary scoring function is intractable due to the magnitude of the search space, specific classes of scoring functions allow for an efficient solution using dynamic programming [10]. Thus, in the standard scoring approach, the score assigned to a folding is composed as the sum of scores of local structural elements, where the set of local elements are chosen to allow efficient dynamic programming inference.[1]

Several scoring models were introduced over the past three decades, where these models mainly differ in the types of structural elements they examine (the *feature-set*), and the scores they assign to them. A simple example of such a model is the one of Nussinov and Jacobson [10], which defines a single feature

---

[1] Some scoring models also utilize homology information with respect to two or more sequences. Such comparative approaches are beyond the scope of this work.

corresponding to a canonical Watson-Crick base-pair in the structure (i.e. base-pairs of the form *G-C* and *A-U*, and their respective reversed forms). The score of each occurrence of the feature in the structure is 1, and thus the total score of a folding is simply the number of canonical base-pairs it contains. A more complex model, which is commonly referred to as the *Turner99 model*, was defined in [11] and is widely used in many RNA structure prediction systems [7,8,9]. This model distinguishes between several different types of structural elements corresponding to unpaired bases, base-pairs which participate in different types of loops, loop-length elements, etc. In addition, every structural element can be mapped to one of several features, depending on some sequential context (e.g. the type of nucleotides at base-pair endpoints and within their vicinity), or other values (e.g. the specific loop length, internal-loop asymmetry value, etc.).

The parameter values (i.e. scores of each local element) are traditionally obtained from wet-lab experiments [12], reflecting the thermodynamics free energy theory [13,14]. However, the increasing availability of known RNA structures in current RNA databases (e.g. [15]) makes it possible to conduct an improved, fine-tuned parameter estimation based on machine-learning (ML) techniques, resulting in higher prediction accuracies. These methods examine large *training sets*, composed of RNA sequences and their known structures [16,17,18,19].

Do *et al.* [18] proposed to set the parameters by fitting an SCFG-based conditional log-linear model to maximize the conditional log-likelihood of a set of known structures. The approach was extended in [20] to include automatic tuning of regularization hyperparameters. Andronescu et al. [19] and later in [21] used the Turner99 model, and applied Constraint-Generation (CG) and Boltzman-likelihood (BL) methods for the parameter estimation. These

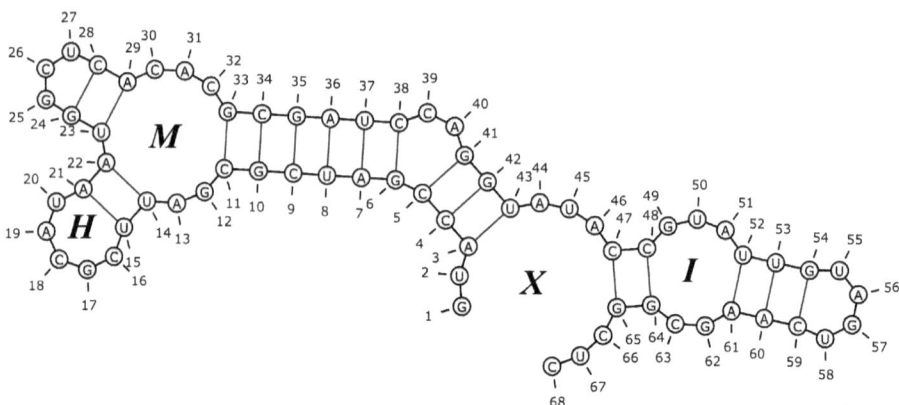

Fig. 1. RNA secondary structure. The figure exemplifies a *secondary structure* of an RNA sequence. Consecutive bases in the sequence are connected with (short) black edges, where base-pairs appear as blue (longer) edges. The labels within the loops stand for loop types, where *H* denotes a *hairpin*, *I* denotes an *internal-loop*, *M* denotes a *multi-loop*, and *X* denotes an *external-loop*. Drawing was made using the VARNA tool [22].

methods start with a set of wet-lab parameter values, and refine them using a training set of RNA sequences and their known structures, and an additional data set containing triplets of sequences, structures and their measured thermodynamic energies. The parameters derived by [21] yield the best published results for RNA folding prediction to date, when tested on a large structural data set.

While the methods for parameter estimation are successfully shifting toward ML-based approaches, the model parameterizations have so far remained fairly constant. Originating from the practice of setting the parameter values using wet-lab measurements, MOST models to date few parameters, where each parameter corresponds to the score of one particular local configuration.

Our Contribution. We propose a move to much richer parameterizations, which is made possible due to the availability of large training sets [23] combined with advances in machine-learning [24,25], and supported in practice by recent accelerations to RNA folding algorithms [26,27]. The scoring models we apply refine previous models by examining more types of structural elements, larger sequential contexts for these elements. Based on this, similar structural elements could get scored differently in different sequential contexts, and different structural elements may get similar scores in similar sequential contexts.

We base our models on the structural elements defined by the Turner99 model in order to facilitate efficient inference. However, in our models, the score assigned to each structural element is itself composed of the sum of scores of many fine-grained local features that take into account portions of larger structural and sequential context. While previous models usually assign a single score to each element (e.g. the base-pair between positions 5 and 41 in Fig. 1), our models score elements as a sum of scores of various features (e.g., the base-pair $(5, 41)$ has the features of being a right-bulge closing base-pair, participating in a stem, having its left endpoint centering a CCG sequence, starting a CGA sequence, and various other contextual factors).

Our final model has about 70,000 free parameters, several orders of magnitude more than previous models. We show that we are still able to effectively set the parameter values based on several thousands of training examples. Our resulting models yield a significant improvement in the prediction accuracy over the previous best results reported by [21], when trained and tested over the same data sets. Our `ContextFold` tool, as well as the various trained models, are freely available on our website and allow for efficient training and inference. In addition to reproducing the results in this work, it also provides flexible means for further experimenting with different forms of richer parameterizations.

2   Preliminaries and Problem Definition

For an RNA sequence x, denote by $\mathcal{Y}_x$ the domain of all possible foldings of x. We represent foldings as sets of index-pairs of the form $(i, j), i < j$, where each pair corresponds to two positions in the sequence such that the bases in these positions are paired. We use the notation (x, y) for a *sequence-folding* pair, where x is an RNA sequence and y is the folding of x. A *scoring model* G is a function that assigns real-values to sequence-folding pairs (x, y). For a given scoring model G, the RNA folding prediction problem is defined as follows[2]: *given an RNA sequence x, find a folding $\hat{y} \in \mathcal{Y}_x$ s.t. $G(x, \hat{y})$ is maximal.* Such a folding $\hat{y}$ will be called an *optimal folding* for x with respect to G. A *folding prediction* (or a *decoding*) algorithm f_G is an algorithm that solves the folding prediction problem, i.e.

$$\hat{y} = f_G(x) = \text{argmax}_{y \in \mathcal{Y}_x} \{G(x, y)\} \qquad (1)$$

Denote by ρ a *cost function* measuring a distance between two foldings, satisfying $\rho(y, y) = 0$ for every y and $\rho(y, y') > 0$ for every $y \neq y'$. This function indicates the cost associated with predicting the structure y' where the real structure is y. For RNA folding, this cost is usually defined in terms of sensitivity, PPV and F-measure (see Sec. 5). Intuitively, a good scoring model G is one such that $\rho(y, f_G(x))$ is small for arbitrary RNA sequences x and their corresponding true foldings y.

In order to allow for efficient computation of f_G, the score $G(x, y)$ is usually computed on the basis of various local features of the pair (x, y). These features correspond to some structural elements induced by y, possibly restricted to appear in some specific sequential context in x. An example of such a feature could be the presence of a stem where the first base-pair in the stem is C-G and it is followed by the base-pair A-U. We denote by Φ the set of different features which are considered by the model, where Φ defines a finite number N of such features. The notation $\Phi(x, y)$ denotes the *feature representation* of (x, y), i.e. the collection of occurrences of features from Φ in (x, y). We assume that every occurrence of a feature is assigned a real-value, which reflects the "strength" of the occurrence. For example, we may define a feature corresponding to the interval of unpaired bases within a hairpin, and define that the value of an occurrence of this feature is the log of the interval length. For binary features such as the stem-feature described above, occurrence values are taken to be 1.

In order to score a pair (x, y), we compute scores for feature occurrences in the pair, and sum up these scores. Each feature in Φ is associated with a corresponding score (or a *weight*), and the score of a specific occurrence of a feature in (x, y) is defined to be the value of the occurrence multiplied by the corresponding feature weight. $\Phi(x, y)$ can be represented as a vector of length N, in which the ith entry ϕ_i corresponds to the ith feature in Φ. Since the same feature may occur several times in (x, y) (e.g., two occurrences of a stem), the value ϕ_i is taken to be the sum of values of the corresponding feature occurrences. Formally, this defines a linear model:

$$G(x, y) = \sum_{\phi_i \in \Phi(x, y)} \phi_i \mathbf{w}_i = \Phi(x, y)^T \cdot \mathbf{w} \qquad (2)$$

---

[2] In models whose scores correspond to free energies, the score optimization is traditionally formulated as a *minimization* problem. This formulation can be easily transformed to the *maximization* formulation that is used here.

where $\mathbf{w}$ is a weight vector in which $\mathbf{w}_i$ is the weight of the ith feature in Φ, and $\cdot$ is the dot-product operator. The vector $\mathbf{w}$ of N feature weights is called the *model parameters*, and Φ is thus referred to as the *model parameterization*. We use the notation $G_{\Phi,\mathbf{w}}$ to indicate a scoring model G with the specific parameterization Φ and parameters $\mathbf{w}$.

The predictive quality of a model of the form $G_{\Phi,\mathbf{w}}$ depends both on the parameterization Φ, defining which features are examined, and on the specific weights in $\mathbf{w}$ which dictate how to score these features. Having fixed a model parameterization Φ, the model parameter values $\mathbf{w}$ can be set based on scientific intuitions and on biological measurements (as done in thermodynamic based models), or based on statistical estimation over observed (x, y) pairs using machine-learning techniques. Aiming to design better models of the form $G_{\Phi,\mathbf{w}}$, there is a need to balance between (a) choosing a rich and informative parameterization Φ so that with optimal weights $\mathbf{w}$ the prediction quality of the model will be as good as possible, (b) allowing for a tractable folding prediction algorithm $f_{G_{\Phi,\mathbf{w}}}$, and (c) being able to estimate optimal (or at least "good") weight parameters $\mathbf{w}$.

3   Feature Representations

We argue that with suitable learning techniques, not being tied to features whose weights could be determined experimentally, and having a large enough set of examples (x, y) such that y is the true folding of x, one could define much richer feature representations than was previously explored, while still allowing efficient inference. These richer representations allow the models to condition on many more fragments of information when scoring the various foldings for a given structure x, and consequently come up with better predictions. This section describes the types of features incorporated in our models.

All examples in this section refer to the folding depicted in Fig. 1, and we assume that the reader is familiar with the standard RNA folding terminology. The considered features broadly resemble those used in the Turner99 model, with some additions and refinements described below, and allow for an efficient Zuker-like dynamic programming folding prediction algorithm [28]. Formal definitions of the terms we use, as well as the exact feature representations we apply in the various models, can be found in the online supplementary material.

We consider two kinds of features: *binary* features, and *real-valued* features.

Binary features. Binary features are features for which occurrence values are always 1, thus the scores of such occurrences are simply the corresponding feature weights. These features occur in a sequence-folding pair whenever some specific *structural element* is present in some specific *sequential context*. The set of structural elements contains base-pairs and unpaired bases, which appear in loops of specific types, for example a multi-loop closing a base-pair, or an unpaired base within a hairpin. A sequential context describes the identities of bases appearing in some given offsets with respect to the location of the structural element in the sequence, e.g. the presence of bases of types C and G at the two endpoints (i,j) of a base-pair. A complete example of such a binary feature is `hairpin_base_0=G_+1=C_-2=U`, indicating the presence of an unpaired-base of type G inside a hairpin at a sequence position i, while positions $i+1$ and $i-2$ contain bases of types C and U respectively. This feature will be generated for the unpaired-bases at positions 17 and 25 in Fig.1.

In contrast to previous models, where each structural element is considered with respect to a *single* sequential context (and producing exactly one scoring term), in our models the score of a structural element is itself a linear combination of different scores of various (possibly overlapping) pieces of information. For example, a model may contain the features `hairpin_base_-1=C_-2=U` and `hairpin_base_0=G_+1=C_-1=C` which will also be generated for the unpaired-base in position 17 (thus differentiating it from the unpaired base at position 25). Note that the appearance of a C-base at relative position -1 appears in both of these features, demonstrating overlapping information regarded by the two features. The decomposition of the sequential context into various overlapping fragments allows us to consider *broader* and *more refined* sequential contexts compared to previous models.

The structural information we allow is also more refined than in previous models: we consider properties of the elements, such as *loop lengths* (e.g. a base-pair which closes a hairpin of length 3 may be scored differently than a base-pair which closes a hairpin of length 4, even if the examined sequential contexts of the two base-pairs are identical), and examine the *two orientations* of each base-pair (e.g. the base-pair $(11, 33)$ may be considered as a *C-G closing* base-pair of the multi-loop marked with an M, and it may also be considered as a *G-C opening* base-pair of the stem that consists of the base-pair $(10, 34)$ in addition to this base-pair). We distinguish between unpaired bases at the "shorter" and "longer" sides of an internal-loop, and distinguish between unpaired bases in external intervals, depending on whether they are at the 5'-end, 3'-end, or neither (i.e. the intervals 1-2, 66-68, and 44-46, respectively). Notably, our refined structural classification allows for the generalization of the concept of "bulges", where, for example, it is possible to define special internal-loop types such that the left length of the loop is exactly k (up to some predefined maximum value for k), and the right length is at least k, and to assign specific features for unpaired bases and base-pairs which participate in such loops.

Real-valued features. Another kind of structural information not covered by the binary unpaired bases and base-pairs features is captured by a set of real-valued *length* features. These features are generated with respect to intervals of unpaired bases, such as the three types of external intervals (as mentioned above), intervals of unpaired bases within hairpins (e.g. the interval 16-20), and intervals of unpaired bases within internal-loops up to some predefined length bound[3] (e.g. the interval 49-51). The value of an occurrence of a length feature

---

[3] In this sense, internal-loop lengths are not restricted here as done in some other models, where arbitrary-length internal-loops are scored with respect to their unpaired bases and terminating base-pairs, and length-depended corrections are added to the scores of relatively "short" loops.

can be any function of the corresponding interval length. In this work, we follow the argumentation of [11] and set the values to be the log of the interval length. As mentioned above, the structural base-pairs and unpaired-bases information is conjoined with various pieces of contextual information. We currently do not consider contextual information for the real-valued length features.

Our features provide varied sources of structural and contextual informations. We rely on a learning algorithm to come up with suitable weights for all these bits and pieces of information.

4   Learning Algorithm

The learning algorithm we use is inspired by the discriminative structured-prediction learning framework proposed by Collins [24] for learning in natural language settings, coupled with a passive-aggressive online learning algorithm [25]. This class of algorithms adapt well to large feature sets, do not require the features to be independent, and were shown to perform well in numerous natural language settings [29,30,31]. Here we demonstrate they provide state of the art results also for RNA folding. The learning algorithm is simple to understand and to implement, and has strong formal guarantees. In addition, it considers one training instance (sequence-folding pair) at a time, making it scale linearly in the number of training examples in terms of computation time, and have a memory requirement which depends on the longest training example.

Recall the goal of the learning algorithm: given a feature representation Φ, a folding algorithm $f_{G_{\Phi,\mathbf{w}}}$, a cost function ρ and a set of training instances S_{train}, find a set of parameter values $\mathbf{w}$ such that the expected cost $\rho(y, f_{G_{\Phi,\mathbf{w}}}(x))$ over unseen sequences x and their true foldings y is minimal.

The algorithm works in rounds. Denote by $\mathbf{w}^0 = 0$ the initial values in the parameter vector maintained by the algorithm. At each iteration i the algorithm is presented with a pair $(x, y) \in S_{train}$. It uses its current parameters $\mathbf{w}^{i-1}$ to obtain $\hat{y} = f_{G_{\Phi,\mathbf{w}^{i-1}}}(x)$, and updates the parameter vector according to:

$$\mathbf{w}^i = \begin{cases} \mathbf{w}^{i-1}, & \rho(y, \hat{y}) = 0, \\ \mathbf{w}^{i-1} + \tau_i \Phi(x, y) - \tau_i \Phi(x, \hat{y}), & \text{otherwise,} \end{cases} \quad (3)$$

where:

$$\tau_i = \min\left(1, \frac{\Phi(x, \hat{y})^T \cdot \mathbf{w}^{i-1} - \Phi(x, y)^T \cdot \mathbf{w}^{i-1} + \sqrt{\rho(y, \hat{y})}}{||\Phi(x, \hat{y}) - \Phi(x, y)||^2}\right).$$

This is the PA-I update for cost sensitive learning with structured outputs described in [25]. Loosely, equation 3 attempts to set $\mathbf{w}^i$ such that the correct structure y would score higher than the predicted structure $\hat{y}$ with *margin* of at least the square-root of the difference between the structures, while trying to minimize the change from $\mathbf{w}^{i-1}$ to $\mathbf{w}^i$. This is achieved by decreasing the weights of features appearing only in the predicted structure, and increasing the weights of features appearing only in the correct structure. Even though one example is considered at a time, the procedure is guaranteed to converge to a good set of parameter values. For the theoretical convergence bounds and proofs see [25]. In practice, due to the finite size of the training data, the algorithm is run for several passes over the training set.

In order to avoid over-fitting of the training data, the final $\mathbf{w}$ is taken to be the average over all $\mathbf{w}^i$ seen in training. That is, $\mathbf{w}^{final} = \frac{1}{K}\sum_{i=1}^{K}\mathbf{w}^i$, where K is the number of processed instances. This widely used practice introduced in [32] improves the prediction results on unseen data.

5   Experiments

In order to test the effect of richer parametrizations on RNA prediction quality, we have conducted 5 learning experiments with increasingly richer model parameterizations, ranging from 226 active features for the simplest model to about 70,000 features for the most complex one.

5.1   Experiment Settings

Feature representations. As described in Section 3, our features combine structural and contextual information. We begin with a baseline model (`Baseline`) which includes a trivial amount of contextual information (the identities of the two bases in a base-pair) and a set of basic structural elements such as *hairpin unpaired base, internal-loop unpaired base, stem closing base-pair, multi-loop closing base-pair, hairpin length*, etc. This baseline model has a potential of inducing up to 1,919 different features, but in practice about 220 features are assigned a non-zero weight after the training process, a number which is comparable to the number of parameters used in previously published models.

We then enrich this basic model with varying amounts of structural (`St`) and contextual (`Co`) information. $\mathtt{St}^{med}$ adds distinction between various kinds of short loops, considers the two orientations of each base-pair, and considers unpaired bases in external intervals, and $\mathtt{St}^{high}$ adds further length-based loop type refinements. Similarly, $\mathtt{Co}^{med}$ considers also the identities of unpaired bases and the base types of the adjacent pair $(i+1, j-1)$ for each base-pair (i, j), while $\mathtt{Co}^{high}$ considers also the neighbors of unpaired bases and more configurations of neighbors surrounding a base-pair.

The models $\mathtt{St}^{med}\mathtt{Co}^{med}$, $\mathtt{St}^{high}\mathtt{Co}^{med}$, $\mathtt{St}^{med}\mathtt{Co}^{high}$ and $\mathtt{St}^{high}\mathtt{Co}^{high}$ can potentially induce about 14k, 30k, 86k and 205k parameters respectively, but in practice much fewer parameters are assigned non-zero values after training, resulting in effective parameter counts of 4k, 7k, 38k and 70k. The exact definition of the different structural elements and sequential contexts considered in each model are provided in the online supplementary material.

Evaluation Measures. We follow the common practice and assess the quality of our predictions based on the *sensitivity, positive predictive value* (PPV), and F_1-*measure* metrics, defined as $\frac{|y \cap \hat{y}|}{|y|}$, $\frac{|y \cap \hat{y}|}{|\hat{y}|}$, and $\frac{2|y \cap \hat{y}|}{|y|+|\hat{y}|}$ respectively, for a known

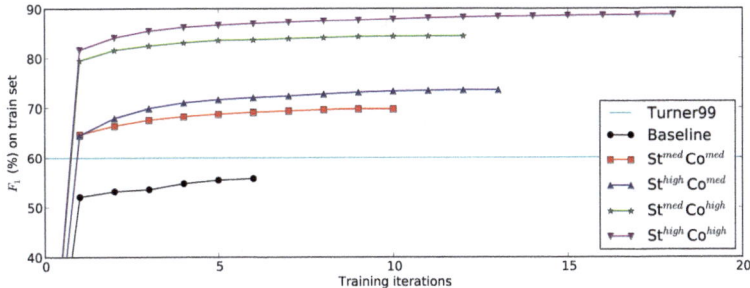

Fig. 2. Performance on S-AlgTrain as a function of the number of training iterations

structure y and a predicted structure $\hat{y}$, where $|y|$ is the number of base-pairs in a structure y, and $|y \cap \hat{y}|$ is the number of base-pairs appearing in both structures. Sensitivity is the proportion of correctly predicted base-pairs among all true base-pairs, PPV is the proportion of correctly predicted base-pairs among all predicted base-pairs, and F_1 is a value which balances sensitivity and PPV. All of the measures range in value from 0 to 1, where a value of 1 indicates that the true and predicted structure are identical, and a value of 0 means that none of the true base-pairs in y are predicted in $\hat{y}$. As in previous works, the reported scores are averaged over the scores of individual predicted structures in the test set.

Folding Prediction Algorithm. We implemented a new folding prediction algorithm, which supports the extended feature representations in our models. This implementation allows for a flexible model design, under which additional models, similarly structured to those presented here, may be defined. In addition, this is the first publicly available implementation to utilize the sparsification techniques, recently reported in [27] for accelerating the running time, over realistic models (weaker sparsification techniques were presented in [26] and applied by [33,34,35]). This yields a significant speedup in folding-time, which enables rapid learning experiments. The code is publicly available on our website.

Learning Setup. The learning algorithm iterates over the training data, halting as soon as the performance over this data does not significantly improve for 3 iterations in a row. The order of the training examples is shuffled prior to each iteration. As the learning algorithm allows for optimization against arbitrary cost functions, we chose the one which is directly related to our evaluation measure, namely $\rho(y, \hat{y}) = 1 - F_1(y, \hat{y})$. The final weight vector is taken to be the average of all computed vectors up to the last iteration. Parameters with absolute value smaller than 0.01 of the maximal absolute parameter value are ignored.

Datasets. Our experiments are based on a large set of known RNA secondary structures. Specifically, we use the exact same data as used in the comprehensive experiments of [21], including the same preprocessing steps, train/test/dev splits and naming conventions. We list some key properties of the data below, and refer the reader to Section 3.1 (for the data and preprocessing steps) and to Section 5.2

Table 1. Performance of final models on the dev set S-AlgTest

Model	# Params	Sens(%)	PPV(%)	F_1(%)
Baseline	226	56.9	55.3	55.8
$St^{med}Co^{med}$	4,054	69.1	66.3	67.4
$St^{high}Co^{med}$	7,075	72.3	70.3	71.0
$St^{med}Co^{high}$	37,846	81.4	80.0	80.5
$St^{high}Co^{high}$	68,606	**83.8**	**83.0**	**83.2**

(for the train/test/dev split) of [21] for the remaining details. The complete data (S-Full) is based on the RNA-Strand dataset [23], and contains known RNA secondary structures for a diverse set of RNA families across various organisms. This data has gone through several preprocessing steps, including the removal of pseudoknots and non-canonical base-pairs. Overall, there are 3245 distinct structures, ranging in length from 10 to 700 nucleotides, with the average length being 269.6. The data is randomly split into S-Train (80%) and S-Test (20%), yielding the train and test sets respectively. S-Train is further split into S-AlgTrain (80% of S-Train) and S-AlgTest (the remaining 20%). We use S-AlgTrain and S-AlgTest (the *dev set*) during the development and for most of the experiments, and reserve S-Train and S-Test for the final evaluation which is presented in Table 3.

5.2 Results

Convergence. Fig. 2 shows the F_1 scores of the various models on the S-AlgTrain training set as a function of the number of iterations. All models converge after less than 20 iterations, where models with more features take more iterations to converge. Training is very fast: complete training of the $St^{med}Co^{med}$ model (about 4k effective features) takes less than half an hour on a single core of one Phenom II CPU, while training the $St^{high}Co^{high}$ model (about 70k effective features) requires about 8.5 hours (in contrast, the CG models described in [19,21] are reported to take between 1.1 and 3.1 days of cpu-time to train, and the BL models take up to 200 days to train). None of the models achieve perfect scores on the training set, indicating that even our richest feature representation does not capture all the relevant information governing the folding process. However, the training set results clearly support our hypothesis: having more features increases the ability of the model to explain the observed data.

Validation accuracy. Train-set performance is not a guarantee of good predictive power. Therefore, the output models of the training procedure were tested on the independent set S-AlgTest. Table 1 shows the accuracies of the various models over this set. The results are expectedly lower than those over the training set, but the overall trends remain: adding more features significantly improves the performance. The contribution of the contextual feature (about 12-13 absolute F_1 points moving from $St^{med}Co^{med}$ to $St^{med}Co^{high}$ and from $St^{high}Co^{med}$ to $St^{high}Co^{high}$) is larger than that of the structural features (about 3 absolute

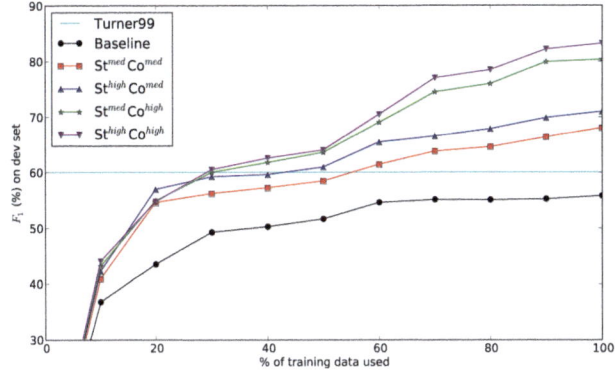

Fig. 3. Effect of training set size on validation-set accuracies

F_1 points moving from $\mathtt{St}^{med}\mathtt{Co}^{med}$ to $\mathtt{St}^{high}\mathtt{Co}^{med}$ and from $\mathtt{St}^{med}\mathtt{Co}^{high}$ to $\mathtt{St}^{high}\mathtt{Co}^{high}$), but the contributions are mostly orthogonal – using richest structural and contextual information ($\mathtt{St}^{high}\mathtt{Co}^{high}$) further increases the results to an F_1 score of 83.2, an absolute F_1 improvement of 27.6 points over the baseline model.

Stability. We performed a 5-fold cross-validation experiment to verify that the results do not depend on a particular train-test split. We randomly shuffled S-TRAIN and performed five test-train splits, each of them reserving a different 20% of the data for testing and training on the rest. The results on the folds are similar to those on the development set with a maximum deviation of about ±1 F_1 points from the numbers in Table 1.

Effect of training-set size. We investigated the effect of the training-set size on the predictive quality of the model by artificially training our models on small subsets of S-ALGTRAIN. Fig. 3 presents the learning curves for these experiments.

Performance clearly increases as more examples are included in the training. The curve for the `Baseline` feature-set flattens out at about 60% of the training data, but the curves of the feature-rich models indicate that further improvement is possible with more training data. 30% of the training data is sufficient for $\mathtt{St}^{med}\mathtt{Co}^{high}$ and $\mathtt{St}^{high}\mathtt{Co}^{high}$ to achieve the performance of the Turner99 model, and all but the `Baseline` feature set surpass the Turner99 performance with 60% of the training data.

Results by RNA family. Table 2 shows the accuracies of the models on the different RNA families appearing in the development set. Interestingly, while the richest $\mathtt{St}^{high}\mathtt{Co}^{high}$ model achieves the highest scores when averaged over the entire dev set, some families (mostly those of shorter RNA sequences) are better predicted by the simpler $\mathtt{St}^{high}\mathtt{Co}^{med}$ and $\mathtt{St}^{med}\mathtt{Co}^{high}$ models. Our machine-learned models significantly outperform the energy-based Turner99 model on all

RNA families, where the effect is especially pronounced on the 5S Ribosomal RNA, Transfer RNA and Transfer Messenger RNA families, for which even the relatively simple $St^{med}Co^{med}$ model already outperform the energy-based model by a very wide margin.

Final Results. Finally, we train our models on the entire training set and evaluate them on the test set. Results on the test set are somewhat higher than on the dev set. In order to put the numbers in context, Table 3 presents the final scores together with the performance of other recent structural prediction systems over the same datasets. The scores of the other systems are taken from [21] and to the best of our knowledge represent the current state-of-the-art for RNA secondary structure prediction.

The `Baseline` model with only 226 parameters achieves scores comparable to those of the Turner-99 model without dangles, despite being very simple, learned completely from data and not containing any physics-based measurements. Our simplest feature-rich model, $St^{med}Co^{med}$, having 4,040 parameters, is already slightly better than all but one of the previously best reported results, where the only better model (BL-FR) being itself a feature-rich model obtained by considering many feature-interactions between the various Truner99 parameters. Adding more features further improves the results, and our richest model, $St^{high}Co^{high}$, achieves a score of 84.1 F_1 on the test set – an absolute improvement of 14.4 F_1-points over the previous best results, amounting to an error reduction of about 50%. Note that the presented numbers reflect the prediction accurcy of the algorithms with respect to a specific dataset. While it is likely that similar accurcy levels would be obtained for new RNAs that belong to RNA families which are covered by the testing data, little can be said about accurcy levels over other RNA families.

Free energy estimates. The free energies associated with RNA structures are also of interest. Unlike the models of [11,19,21], and in particular the DIM-CG model of [21], our models' scores do not represent free energies. However, there is no reason to use the same model for both structure prediction and free energy estimation, which can be considered as two independent tasks. Instead, we

Table 2. F_1 scores (in %) of on the development set, grouped by RNA family. Only families with more than 10 examples in the development set are included. The highest score for each family appears in bold.

Familiy (#instances)	$St^{med}Co^{med}$	$St^{high}Co^{med}$	$St^{med}Co^{high}$	$St^{high}Co^{high}$	Turner99	LAM-CG
Hammerhead Ribozyme(12)	57.9	58.3	69.8	**78.8**	43.9	45.5
Group I Intron(11)	55.2	58.7	**73.5**	70.5	60.4	60.6
Cis-regulatory element(11)	45.9	46.1	81.8	**85.2**	61.1	61.2
Transfer Messenger RNA(70)	55.2	57.6	69.7	**70.8**	37.5	49.5
5S Ribosomal RNA(27)	89.2	90.9	**94.1**	93.9	68.9	79.8
Unknown(48)	93.9	94.1	**95.7**	94.8	91.14	92.2
Ribonuclease P RNA(72)	62.0	70.3	84.7	**87.7**	58.6	61.2
16S Ribosomal RNA(112)	57.9	65.4	81.0	**86.3**	55.2	62.3
Signal Recognition Particle RNA(62)	61.8	62.7	72.6	**76.2**	66.6	64.5
Transfer RNA(80)	91.8	**94.2**	92.2	92.8	60.7	79.5
23S Ribosomal RNA(28)	53.6	54.0	61.2	**68.6**	58.5	60.0
Other RNA(11)	65.9	66.4	71.8	**73.5**	61.1	62.2

Table 3. Final results on the test set. All the models are evaluated on S-TEST. Turner99+Partition is obtained by running Vienna's RNAfold [7] with the -p flag and considering the centroid-structure. Models marked with ‡ are trained on S-TRAIN. Models marked with ⋆ are trained on S-PROCESSED, a larger dataset than S-TRAIN which contains some sequences from S-TEST. In the models marked with †, training is initialized with the Turner99 parameters, and uses additional thermodynamics information regarding the free energies of 1291 known structures.

Model	Desc	# Params	$F_1(\%)$
Turner99+Partition	[11]	363	61.7
Turner99	[11]	363	60.0
Turner99 (no dangles)	[11]	315	56.5
‡ † BL-FR	[21] Ch6	7,726	69.7
‡ † BL*	[21] Ch4.2	363	67.9
‡ † BL (no dangles)	[21] Ch4.2	315	68.0
‡ † LAM-CG (CG*)	[21] Ch4.1	363	67.0
‡ † DIM-CG	[21] Ch4.1	363	65.8
⋆ † CG 1.1	[19]	363	64.0
⋆ CONTRAFold 2.0	[18,20]	714	68.8
‡ $St^{med}Co^{med}$		4040	69.2
‡ $St^{high}Co^{med}$		7150	72.8
‡ $St^{med}Co^{high}$		37866	80.4
‡ $St^{high}Co^{high}$		69,603	**84.1**

can use one model for structure prediction, and then estimate the free energy of the predicted structure using a different model. We predict the structures of the 279 single-molecules appearing in a thermodynamics dataset ([21] Ch 3.2), for which both structure and free energy lab-measurements are available, using the Turner99, CG, DIM-CG and our $St^{high}Co^{high}$ models. The folding accuracy F_1 measures are 89, 92.9, 87 and 97.1, respectively. We then estimate the free energies of the predicted structures using the DIM-CG derived parameters (this model was shown in [21] to provide the best free energy estimates). The RMSE (*Root Mean Squared Error*, lower is better) for the four models are 0.86 (Turner99), 0.90 (CG), 0.87 (DIM-CG), and 0.92 ($St^{high}Co^{high}$). While our model scores slightly worse in terms of RMSE, it is not clear that this difference is significant when considering the standard error and the fact that the other models had access to this test set during their parameter estimation.

6   Discussion

We showed that a move towards richer parameterizations is beneficial to ML-based RNA structure prediction. Indeed, our best model yields an error reduction of 50% over the previously best published results, under the same experimental conditions. Our learning curves relative to the amount of training data indicate that adding more data is likely to increase these already good results. Further improvements are of course possible. First, we considered only four specific richly-parameterized models. It is likely that better parameterizations are possible, and the search for a better richly-parameterized model is a fertile ground for exploration. Second, we considered a single, margin-based, error-driven parameter estimation method. Probabilistic, marginals-based (i.e. partition function based) training and decoding is an appealing alternative.

Our method has some limitations with respect to the physics-based models. In particular, while it is optimized to predict one single best structure, it does not provide estimates of free energies of secondary structures, and cannot compute the partition function, base-pair binding probabilities and centroid structures derived from them. Another shortcoming of our models is that the learned parameter weights are currently not interpretable. We would like to explore methods of analyzing the learned parameters and trying to "make biological sense" of them.

Acknowledgments. The authors are grateful to Mirela Andronescu for her kind help in providing information and pointing us to relevant data. We thank the anonymous referees for their helpful comments. This research was partially supported by ISF grant 478/10 and by the Frankel Center for Computer Science at Ben Gurion University of the Negev.

References

1. Eddy, S.R.: Non–coding RNA genes and the modern RNA world. Nature Reviews Genetics 2, 919–929 (2001)
2. Mandal, M., Breaker, R.R.: Gene regulation by riboswitches. Cell 6, 451–463 (2004)
3. Washietl, S., Hofacker, I.L., Lukasser, M., Huttenhofer, A., Stadler, P.F.: Mapping of conserved RNA secondary structures predicts thousands of functional noncoding RNAs in the human genome. Nature Biotechnology 23, 1383–1390 (2005)
4. Kloc, M., Zearfoss, N.R., Etkin, L.D.: Mechanisms of subcellular mRNA localization. Cell 108, 533–544 (2002)
5. Hofacker, I.L., Stadler, P.F., Stocsits, R.R.: Conserved RNA secondary structures in viral genomes: a survey. Bioinformatics 20, 1495 (2004)
6. Mattick, J.S.: RNA regulation: a new genetics? Pharmacogenomics J. 4, 9–16 (2004)
7. Hofacker, I.L., Fontana, W., Stadler, P.F., Schuster, P.: Vienna RNA package (2002), World Wide Web: http://www.tbi.univie.ac.at/ivo/RNA
8. Zuker, M.: Computer prediction of RNA structure. Methods in Enzymology 180, 262–288 (1989)
9. Zuker, M.: Mfold web server for nucleic acid folding and hybridization prediction. Nucleic Acids Research, 3406–3415 (2003)
10. Nussinov, R., Jacobson, A.B.: Fast algorithm for predicting the secondary structure of single-stranded RNA. PNAS 77, 6309–6313 (1980)
11. Mathews, D.H., Sabina, J., Zuker, M., Turner, D.H.: Expanded sequence dependence of thermodynamic parameters improves prediction of RNA secondary structure. J. Mol. Biol. 288, 911–940 (1999)
12. Mathews, D.H., Burkard, M.E., Freier, S.M., Wyatt, J.R., Turner, D.H.: Predicting oligonucleotide affinity to nucleic acid target. RNA 5, 1458 (1999)

13. Tinoco, I., Uhlenbeck, O.C., Levine, M.D.: Estimation of secondary structure in ribonucleic acids. Nature 230, 362–367 (1971)
14. Tinoco, I., Borer, P.N., Dengler, B., Levine, M.D., Uhlenbeck, O.C., Crothers, D.M., Gralla, J.: Improved estimation of secondary structure in ribonucleic acids. Nature New Biology 246, 40–41 (1973)
15. Griffiths-Jones, S., Moxon, S., Marshall, M., Khanna, A., Eddy, S.R., Bateman, A.: Rfam: annotating non-coding RNAs in complete genomes. Nucleic Acids Research 33, D121 (2005)
16. Eddy, S.R., Durbin, R.: RNA sequence analysis using covariance models. Nucleic Acids Research 22, 2079 (1994)
17. Dowell, R.D., Eddy, S.R.: Evaluation of several lightweight stochastic context-free grammars for RNA secondary structure prediction. BMC Bioinformatics 5, 71 (2004)
18. Do, C.B., Woods, D.A., Batzoglou, S.: CONTRAfold: RNA secondary structure prediction without physics-based models. Bioinformatics 22, e90–e98 (2006)
19. Andronescu, M., Condon, A., Hoos, H.H., Mathews, D.H., Murphy, K.P.: Efficient parameter estimation for RNA secondary structure prediction. Bioinformatics 23, i19 (2007)
20. Do, C.B., Foo, C.S., Ng, A.Y.: Efficient multiple hyperparameter learning for log-linear models. In: Neural Information Processing Systems, vol. 21, Citeseer (2007)
21. Andronescu, M.: Computational approaches for RNA energy parameter estimation. PhD thesis, University of British Columbia, Vancouver, Canada (2008)
22. Darty, K., Denise, A., Ponty, Y.: VARNA: Interactive drawing and editing of the RNA secondary structure. Bioinformatics 25, 1974–1975 (2009)
23. Andronescu, M., Bereg, V., Hoos, H.H., Condon, A.: RNA STRAND: the RNA secondary structure and statistical analysis database. BMC Bioinformatics 9, 340 (2008)
24. Collins, M.: Discriminative training methods for hidden markov models: Theory and experiments with perceptron algorithms. In: Proceedings of the ACL 2002 Conference on Empirical Methods in Natural Language Processing, vol. 10, pp. 1–8. Association for Computational Linguistics (2002)
25. Crammer, K., Dekel, O., Keshet, J., Shalev-Shwartz, S., Singer, Y.: Online passive-aggressive algorithms. The Journal of Machine Learning Research 7, 585 (2006)
26. Wexler, Y., Zilberstein, C., Ziv-Ukelson, M.: A study of accessible motifs and RNA folding complexity. Journal of Computational Biology 14, 856–872 (2007)
27. Backofen, R., Tsur, D., Zakov, S., Ziv-Ukelson, M.: Sparse RNA folding: Time and space efficient algorithms. In: Kucherov, G., Ukkonen, E. (eds.) CPM 2009 Lille. LNCS, vol. 5577, pp. 249–262. Springer, Heidelberg (2009)
28. Zuker, M., Stiegler, P.: Optimal computer folding of large RNA sequences using thermodynamics and auxiliary information. Nucleic Acids Research 9, 133–148 (1981)
29. Chiang, D., Knight, K., Wang, W.: 11,001 new features for statistical machine translation. In: Proceedings of HLT-NAACL 2009, Boulder, Colorado, pp. 218–226. Association for Computational Linguistics (2009)
30. McDonald, R., Crammer, K., Pereira, F.: Online large-margin training of dependency parsers. In: Proceedings of ACL 2009 (2005)
31. Watanabe, Y., Asahara, M., Matsumoto, Y.: A structured model for joint learning of argument roles and predicate senses. In: Proceedings of the ACL 2010 Conference Short Papers, Uppsala, Sweden, pp. 98–102. Association for Computational Linguistics (2010)

32. Freund, Y., Schapire, R.E.: Large margin classification using the perceptron algorithm. Machine Learning 37, 277–296 (1999)
33. Ziv-Ukelson, M., Gat-Viks, I., Wexler, Y., Shamir, R.: A faster algorithm for RNA co-folding. Algorithms in Bioinformatics, 174–185 (2008)
34. Salari, R., Möhl, M., Will, S., Sahinalp, S., Backofen, R.: Time and space efficient RNA-RNA interaction prediction via sparse folding. In: Berger, B. (ed.) RECOMB 2010. LNCS, vol. 6044, pp. 473–490. Springer, Heidelberg (2010)
35. Möhl, M., Salari, R., Will, S., Backofen, R., Sahinalp, S.: Sparsification of RNA Structure Prediction Including Pseudoknots. In: Moulton, V., Singh, M. (eds.) WABI 2010. LNCS, vol. 6293, pp. 40–51. Springer, Heidelberg (2010)

A Bayesian Approach for Determining Protein Side-Chain Rotamer Conformations Using Unassigned NOE Data*

Jianyang Zeng[1], Kyle E. Roberts[3], Pei Zhou[2], and Bruce R. Donald[1,2,3],**

[1] Department of Computer Science, Duke University, Durham, NC 27708, USA
[2] Department of Biochemistry, Duke University Medical Center, Durham, NC 27710, USA
[3] Program in Computational Biology and Bioinformatics, Duke University, Durham NC 27708, USA
Tel.: 919-660-6583; Fax: 919-660-6519
brd+recomb11@cs.duke.edu

Abstract. A major bottleneck in protein structure determination via nuclear magnetic resonance (NMR) is the lengthy and laborious process of assigning resonances and nuclear Overhauser effect (NOE) cross peaks. Recent studies have shown that accurate backbone folds can be determined using sparse NMR data, such as residual dipolar couplings (RDCs) or backbone chemical shifts. This opens a question of whether we can also determine the accurate protein side-chain conformations using sparse or unassigned NMR data. We attack this question by using unassigned nuclear Overhauser effect spectroscopy (NOESY) data, which record the through-space dipolar interactions between protons nearby in 3D space. We propose a Bayesian approach with a Markov random field (MRF) model to integrate the likelihood function derived from observed experimental data, with prior information (i.e., empirical molecular mechanics energies) about the protein structures. We unify the side-chain structure prediction problem with the side-chain structure determination problem using unassigned NMR data, and apply the deterministic *dead-end elimination* (DEE) and A* search algorithms to provably find the global optimum solution that maximizes the posterior probability. We employ a Hausdorff-based measure to derive the likelihood of a rotamer or a pairwise rotamer interaction from unassigned NOESY data. In addition, we apply a systematic and rigorous approach to estimate the experimental noise in NMR data, which also determines the weighting factor of the data term in the scoring function that is derived from the Bayesian framework. We tested our approach on real NMR data of three proteins, including the FF Domain 2 of human transcription elongation factor CA150 (FF2), the B1 domain of Protein G (GB1), and human ubiquitin. The promising results indicate that our approach can be applied in high-resolution protein structure determination. Since our approach does not require any NOE assignment, it can accelerate the NMR structure determination process.

---

* This work is supported by the following grants from National Institutes of Health: R01 GM-65982 and R01 GM-78031 to B.R.D. and R01 GM-079376 to P.Z.
** Corresponding author.

V. Bafna and S.C. Sahinalp (Eds.): RECOMB 2011, LNBI 6577, pp. 563–578, 2011.
© Springer-Verlag Berlin Heidelberg 2011

1 Introduction

Nuclear magnetic resonance (NMR) is an important tool for determining high-resolution protein structures in the solution state. Traditional NMR structure determination approaches [19,21,40,23,35] typically use a dense set of nuclear Overhauser effect (NOE) distance restraints to calculate the 3D coordinates of the protein structure. This process requires nearly complete assignment of both resonances (which serve as IDs of atoms in NMR spectra) and NOE data. Unfortunately, assigning resonances and NOEs is a time-consuming and laborious process, which is a major bottleneck in NMR structure determination. To address this problem, several approaches have been proposed to determine protein structures using sparse experimental data [24,53,54,12,4,7,51,47] or unassigned NMR data [43,46,61,59,62]. These new approaches have shown promising results. In particular, it has been shown that accurate backbone folds can be determined using sparse NMR data, such as residual dipolar couplings (RDCs) [53,54,12,24] or backbone chemical shifts [51,47]. The question remains: After the backbone structure has been solved, can we also determine accurate side-chain conformations using sparse or unassigned NMR data? In this paper, we address this question by using unassigned nuclear Overhauser effect spectroscopy (NOESY) data, which record the through-space dipolar interactions between protons nearby in 3D space. While protein backbones have previously been determined to low resolution [43,44] or even moderate resolution [33,17,18,46] using unassigned NOESY data, it has never been shown, prior to our paper, that high-resolution side-chain conformations can be computed using only unassigned NOESY data. Since our algorithm does not require any NOE assignment, it can shorten the time required in the NMR data analysis, and hence accelerate the NMR structure determination process.

Protein side-chains have been observed to exist in a number of energetically favored conformations, called *rotamers* [42]. Based on this observation, the side-chain structure determination problem can be formulated as a discrete combinatorial optimization problem, in which a set of side-chain conformations are searched over a given rotamer library to optimize a scoring function that represents both empirical molecular mechanics and data restraints. Substantial work has been developed for predicting protein side-chain conformations without using experimental data [52,11,22,31,16,27,3,56,49,30,57,34]. These side-chain structure prediction approaches might be limited by the approximate nature of the employed empirical molecular mechanics energy function, which might not be sufficient to accurately capture the real energetic interactions among atoms in the protein.

Integration of NMR data with the empirical molecular mechanics energy is a challenging problem. Most frameworks for NMR protein structure determination use heuristic models with *ad hoc* parameter settings to incorporate experimental data (which are usually *assigned* NOE data in these approaches) and integrate them with the empirical molecular mechanics energy in a scoring function to compute protein structures. These approaches suffer from the subjective choices in the data treatment, which makes it difficult to objectively calculate high-quality structures. To overcome this drawback, we use a Bayesian approach [48,20,50] and cast the protein side-chain structure determination problem using unassigned NOESY data into a Markov random field (MRF) framework. We treat NMR data as an experimental observation on side-chain rotamer states, and use the MRF to encode prior information about the protein structures, such as empirical molecular mechanics energies. The priors in our framework are in essence parameterized by the random variables representing the side-chain rotamer conformations. The MRF modelling captures atomic interactions among residues both from empirical molecular mechanics energies and geometric restraints from unassigned NOESY data. The derived posterior probability combines prior information and the likelihood model constructed from observed experimental data. Unlike previous *ad hoc* models, our Bayesian framework provides a rational basis to incorporate both experimental data and modelling information, which enables us to develop systematic techniques for computing accurate side-chain conformations.

The side-chain structure determination problem is NP-hard [45,8]. Therefore, a number of algorithms have been developed to address the complexity. Stochastic techniques [52,22,27,49] randomly sample conformation space to generate a set of side-chain rotamer conformations. In contrast, our approach applies deterministic algorithms with provable guarantees [11,41,16,15,9,13] to determine the optimal side-chain rotamer conformations that satisfy both experimental restraints and prior information on the protein structures. We first apply a *dead-end elimination* (DEE) algorithm [11,41, 16] to prune side-chain conformations that are *provably* not part of the optimal solution. After that, an A* search algorithm is employed to find the global optimum solution that best interprets our MRF model.

The guarantee to provably find the global optimum using the DEE/A* algorithms enables us to rigorously and objectively estimate the experimental noise in NMR data and the weighting factor between the empirical molecular mechanics energy and experimental data in the scoring function derived in our Bayesian framework. Specifically, we employ a grid search approach to systematically search over all possible grid point values of the noise parameter, and use the DEE/A* search algorithms to compute the optimal solution that minimizes the scoring function for each grid point. We then compare the best solutions over all grid points and find the globally optimal estimation of the weight parameter. The following contributions are made in this paper:

1. A novel framework to unify the side-chain structure prediction problem with the side-chain structure determination problem using unassigned NOESY data, by applying the provable *dead-end elimination* (DEE) and A* search algorithms to find the global optimum solution;

2. A Bayesian approach with an MRF model to derive the posterior probability of side-chain conformations by combining the likelihood function from observed experimental data with prior information (i.e., empirical molecular mechanics energies) about the protein structures;

3. A systematic and rigorous approach to estimate the experimental noise in NMR data, which determines the weighting factor of the data term in the derived scoring function, by combining grid search and DEE/A* search algorithms;

4. Introduction of a Hausdorff-based measure to derive the likelihood function from unassigned NMR data;

5. Promising test results on real NMR data recorded at Duke University.

1.1 Related Work

In [61, 59], we developed an algorithm, called HANA, that employs a Hausdorff-based pattern matching technique to place the side-chain rotamer conformations on the backbone structures determined mainly using RDC data [53,54,12]. In [62], we proposed an MRF based algorithm, called NASCA, to assign side-chain resonances and compute the side-chain rotamer conformations from unassigned NOESY data without using TOCSY experiments. Neither HANA nor NASCA completely exploits prior information or all the available information from experimental data. For example, HANA only uses the back-computed NOE pattern from side-chain rotamers to backbone to calculate the likelihood of a rotamer. In addition, HANA and NASCA do not take into account the empirical molecular mechanics energy when determining the side-chain rotamer conformations. Thus, the side-chain conformations determined by these two approaches may embrace some bad local geometry such as serious steric clashes. Our current Bayesian approach improves over HANA and NASCA by eliminating all serious steric clashes (Table 3). It is a significant extension of the HANA and NASCA modules, and can be combined with our previously-developed backbone structure techniques [53,54,12,59,62] to determine high-resolution structures, using a protocol similar to [59,62].

Several approaches have been proposed to use backbone chemical shift data [4,7,51, 47] or unassigned NOESY data [33,17,18,43,44,46] in protein structure determination at different resolutions. These frameworks use a generate-and-test strategy or stochastic techniques such as Monte Carlo (MC), simulated annealing (SA), or highly-simplified molecular dynamics (HSMD) to randomly sample conformation space and compute a set of structures that satisfy the data restraints. These approaches suffer from the problem of undersampling conformation space and overfitting to the data. They cannot provide any guarantee on the convergence to the global optimum. In addition, integration of experimental data with the empirical molecular mechanics energy and the parameter settings in these frameworks are usually performed on an *ad hoc* basis.

Unlike a previous Bayesian approach in NMR structure determination [48,20], which requires *assigned* NOE data, our approach works on unassigned NOESY data. Moreover, the Bayesian approach in [48, 20] mainly relies on heuristic techniques, such as Monte Carlo or Gibbs sampling, to randomly sample both conformation space and joint posterior distribution, while our approach employs a systematic and rigorous search method (i.e., a combination of grid search and DEE/A* algorithms) to compute the optimal parameter estimation that is only subject to the resolution used in the grid search.

MRFs offer a mathematically sound framework for describing the dependencies between random variables, and have been widely applied in computer vision [14, 39] and computational structural biology [58,29]. In [29], an MRF was used to estimate the free energy of protein structures, while in [58], a graphical model similar to an MRF was used to predict side-chain conformations. Although both graphical models in [58, 29] provide a reasonable model to describe the protein side-chain rotamer interactions, they do not use any experimental data. In addition, the belief propagation approach used in [58, 29] to search for the low-energy conformations can be trapped into local minima, while our approach computes the global optimum solution.

2   Methods

2.1   Backbone Structure Determination from Residual Dipolar Couplings

In our high-resolution structure determination protocol, we apply our recently-developed algorithms [53, 54, 59, 12] to compute the protein backbone structures using two RDCs per residue (either NH RDCs measured in two media, or NH and CH RDCs measured in a single medium). Details on backbone structure determination from RDCs are available in Supplementary Material (SM) [60] **Section 1** and [53, 54, 12].

2.2   Using Markov Random Fields for Rotamer Assignment

We first use a Markov random field to formulate our side-chain structure determination problem. A Markov random field is a set of random variables defined on an undirected graph, which describes the conditional dependencies among random variables. In our problem, each random variable represents the rotamer state of a residue. Formally, let X_i be a random variable representing the rotamer state at residue i, where $1 \leq i \leq n$, and n is the total number of residues in the protein sequence. Let t_i be the maximum number of rotamer states at residue i. Then each random variable X_i can take a value from set $\{1, \cdots, t_i\}$. We use x_i to represent a specific value taken by random variable X_i. We also call x_i the *rotamer assignment* or *conformation* of residue i. Let $X = \{X_1, \cdots, X_n\}$ be the set of random variables representing the rotamer assignments for all residues $1, \cdots, n$ in the protein sequence. A joint event $\{X_1 = x_1, \cdots, X_n = x_n\}$, abbreviated as $X = x$, is called a *rotamer assignment* or *conformation* for all residues in the protein sequence, where $x = \{x_1, \cdots, x_n\}$.

In our side-chain structure determination problem, we assume that the backbone is rigid. Based on this assumption, it is generally safe to argue that each residue only interacts with other residues within a certain distance threshold or energy cutoff, when considering the pairwise interactions between side-chains. We use a graph $G = (V, E)$ to represent such residue-residue interactions, where each vertex in V represents a residue, and each edge in E represents a possible interaction between two residues (i.e., the minimum distance between atoms from these two residues is within a distance threshold). Such a graph $G = (V, E)$ is called the *residue interaction graph*. Given a residue interaction graph $G = (V, E)$, the *neighborhood* of residue i, denoted by N_i, is defined as $N_i = \{j \mid j \in V, i \neq j, (i, j) \in E\}$. The neighborhood system describes the dependencies between rotamer assignments for all residues in the protein sequence. A Markov random field (MRF), defined based on the neighborhood system of an underlying graph $G = (V, E)$, encodes the following conditional independencies for each variable X_i:

$$\Pr(X_i | X_j, j \neq i) = \Pr(X_i | X_j, j \in N_i). \tag{1}$$

This condition states that each random variable X_i is only dependent on the random variables in its neighborhood.

We use $\Pr(x)$ to represent the *prior* probability for a rotamer assignment $x = \{x_1, \cdots, x_n\}$ of a protein sequence, which is derived from prior information about the protein structures, such as empirical molecular mechanics. Let D be the observation data, which in this case are the unassigned NOESY data. Let σ be the experimental noise in the unassigned NOESY data. The parameter σ is unknown and needs to be estimated. We use $\Pr(D|x, \sigma)$ to represent the *likelihood* function of a rotamer assignment x and a parameter σ given the observation D. We use $\Pr(x, \sigma|D)$ to represent the *a posteriori* probability. Our goal is to find a combination of rotamer assignment x and parameter σ, denoted by (x, σ), that maximizes the *a posteriori* probability (MAP). By Bayes's theorem, the posterior probability can be computed by

$$\Pr(x, \sigma|D) \propto \Pr(D|x, \sigma) \cdot \Pr(x) \cdot \Pr(\sigma). \tag{2}$$

2.3 Deriving the Prior Probability

According to the Hammersley-Clifford theorem [2] on the Markov-Gibbs equivalence, the distribution of an MRF with respect to an underlying graph $G = (V, E)$ can be written in the following Gibbs form:

$$\Pr(x) \propto \exp(-U(x)/\beta), \tag{3}$$

where β is a *global control parameter*, and $U(x)$ is the *prior energy* that encodes prior information about the rotamer interactions in the protein structure. The prior energy can be defined by $U(x) = \sum_{C \in \mathcal{C}} V_C(x)$, where $V_C(\cdot)$ is a *clique potential* and $\mathcal{C}$ is the set of cliques in the neighborhood system of the underlying graph $G = (V, E)$. In our problem, we only focus on one-site and two-site interactions (i.e., with cliques of size 2) in a residue interaction graph $G = (V, E)$. Given an assignment $x = \{x_1, \cdots, x_n\}$ for a residue interaction graph $G = (V, E)$, we use the following empirical molecular mechanics energy function to define the prior energy $U(x)$:

$$U(x) = \sum_{i \in V} E'(x_i) + \sum_{i \in V} \sum_{j \in N_i} E'(x_i, x_j), \tag{4}$$

where $E'(x_i)$ is the *self energy* term for rotamer assignment x_i at residue i, and $E'(x_i, x_j)$ is the *pairwise energy* term for rotamer assignments x_i and x_j at residues i and j respectively. We can use the Boltzmann distribution to further specify the prior probability in Eq. (3) by setting $\beta = k_b T$, where k_b is the Boltzmann constant, and T is the temperature.

2.4 Deriving the Likelihood Function and the Scoring Function

An accurate likelihood function should effectively interpret the observation data, and incorporate experimental uncertainty into the model. In our framework, the likelihood $\Pr(D|x, \sigma)$ is defined as

$$\Pr(D|x, \sigma) = Z(\sigma) \cdot \exp\left(-U(D|x, \sigma)\right), \tag{5}$$

where $Z(\sigma)$ is the *normalizing factor*, and $U(D|x, \sigma)$ is called the *likelihood energy*, which evaluates the likelihood of observed NOESY data given rotamer assignment x and parameter σ.

The likelihood energy $U(D|x, \sigma)$ can be measured by matching the back-computed NOE patterns with experimental cross peaks in unassigned NOESY data D. Given a rotamer assignment x_i at residue i, we can back-compute its NOE pattern between backbone and intra-residue atoms. This NOE pattern is called the *self back-computed NOE pattern*. Similarly, we can back-compute the NOE pattern between a pair of rotamer assignments x_i and x_j at residues i and j respectively. This NOE pattern is called the *pairwise back-computed NOE pattern*. We use a criterion derived from the Hausdorff distance [25, 26], called the *Hausdorff fraction*, to measure the matching score between a back-computed NOE pattern and unassigned NOESY data. Details of deriving the Hausdorff fraction for a back-computed NOE pattern are in SM [60] **Section 2** and [61, 59]. Let $F(x_i)$ and $F(x_i, x_j)$ be the Hausdorff fractions for the self and pairwise back-computed NOE patterns respectively. Then the likelihood energy $U(D|x, \sigma)$ is defined as:

$$U(D|x, \sigma) = \sum_{i \in V} \frac{(1 - F(x_i)/F_0(x_i))^2}{2\sigma^2} + \sum_{i \in V} \sum_{j \in N_i} \frac{(1 - F(x_i, x_j)/F_0(x_i, x_j))^2}{2\sigma^2},$$

(6)

where σ is the experimental noise in unassigned NOESY data, and $F_0(x_i)$ and $F_0(x_i, x_j)$ are the *expected values* of $F(x_i)$ and $F(x_i, x_j)$ respectively. Here we assume that the experimental noise of unassigned NOESY cross peaks follows an independent Gaussian distribution. Thus, σ represents the standard deviation of the Gaussian noise. Such an independent Gaussian distribution provides a good approximation [36, 39] when the accurate noise model to describe the uncertainty in experimental data is not available. In general, it is difficult to obtain the accurate values of the expected Hausdorff fractions $F_0(x_i)$ and $F_0(x_i, x_j)$. In principle, a rotamer conformation should be closer to the native side-chain conformation if its back-computed NOE pattern has a higher Hausdorff fraction (i.e., with higher data satisfaction score). In practice, we use the maximum value of the Hausdorff fraction among the back-computed NOE patterns of all rotamers as the expected value of $F(x_i)$ and $F(x_i, x_j)$.

The function $U(x, \sigma|D) = -\log \Pr(x, \sigma|D)$ is called the *posterior energy* for a rotamer assignment x and parameter σ, given the observed data D. Then maximizing the posterior probability is equivalent to minimizing the posterior energy function. Substituting Eqs. (3), (4) and (6) into Eq. (2), and taking the negative logarithm on both sides of the equation, we have the following form of the posterior energy function:

$$U(x, \sigma|D) \propto \frac{1}{\beta} \left(\sum_{i \in V} E'(x_i) + \sum_{i \in V} \sum_{j \in N_i} E'(x_i, x_j) \right) +$$

$$\left(\sum_{i \in V} \frac{(1 - F(x_i)/F_0(x_i))^2}{2\sigma^2} + \sum_{i \in V} \sum_{j \in N_i} \frac{(1 - F(x_i, x_j)/F_0(x_i, x_j))^2}{2\sigma^2} \right) + \log \frac{Z(\sigma)}{\Pr(\sigma)}.$$

(7)

In Sec. 2.5, we will show how to estimate parameter σ. After σ has been estimated, we have the following form of the posterior energy function:

$$U(x|D) \propto \frac{1}{\beta} \left(\sum_{i \in V} E'(x_i) + \sum_{i \in V} \sum_{j \in N_i} E'(x_i, x_j) \right)$$

$$+ \left(\sum_{i \in V} \frac{(1 - F(x_i)/F_0(x_i))^2}{2\sigma^2} + \sum_{i \in V} \sum_{j \in N_i} \frac{(1 - F(x_i, x_j)/F_0(x_i, x_j))^2}{2\sigma^2} \right). \quad (8)$$

The function $U(x|D)$ is also called the *pseudo energy*. We rewrite the pseudo energy function in Eq. (8). Let $E(x_i) = E'(x_i)/\beta + (1 - F(x_i)/F_0(x_i))^2/2\sigma^2$ and $E(x_i, x_j) = E'(x_i, x_j)/\beta + (1 - F(x_i, x_j)/F_0(x_i, x_j))^2/2\sigma^2$. Then we have

$$U(x|D) = \sum_{i \in V} E(x_i) + \sum_{i \in V} \sum_{j \in N_i} E(x_i, x_j). \quad (9)$$

The pseudo energy function defined in Eq. (9) has the same form as in protein side-chain structure prediction [31,38,49,30,57,34] or protein design [11,41,16,15,9,13]. Thus, we can apply similar algorithms, including the dead-end elimination (DEE) and A* search algorithms, to solve this problem. A brief overview of the DEE/A* algorithms is in SM [60] **Section 3** and [11,41,16,15,9]. Similar to protein side-chain prediction and protein design, the optimal rotamer assignment x^* that minimizes the pseudo energy function in Eq. (9) is called the *global minimum energy conformation (GMEC)*. The DEE/A* algorithms employed in our framework guarantee to find the GMEC with respect to our pseudo energy function. Similar to [15,9,13], we can also extend the original A* search algorithm to compute a gap-free ensemble of conformations such that their energies are all within a user-specified window from the lowest pseudo energy.

2.5   Estimation of Experimental Noise in the NOESY Data

In practice, parameter σ in Eq. (6) is generally unknown, and needs to be estimated for each set of experimental data used in structure calculation. In the likelihood function Eq. (5), the normalizing factor $Z(\sigma)$ is related to the unknown parameter σ. Based on the independent Gaussian distribution assumption on experimental noise in unassigned NOESY data, we have $Z(\sigma) = (2\pi\sigma^2)^{m/2}$, where m is the total number of self and pairwise back-computed NOE patterns. In our problem, m is equal to the size of the residue interaction graph $G = (V, E)$, that is, $m = |V| + |E|$.

Similar to [48,20], we use the Jeffrey prior [28] to represent the prior probability of parameter σ, that is, $\Pr(\sigma) = \sigma^{-1}$. Substituting $Z(\sigma) = (2\pi\sigma^2)^{m/2}$ and $\Pr(\sigma) = \sigma^{-1}$ into Eq. (7), we have

$$U(x, \sigma|D) \propto (m + 1) \log \sigma + \frac{1}{\beta} \left(\sum_{i \in V} E'(x_i) + \sum_{i \in V} \sum_{j \in N_i} E'(x_i, x_j) \right) +$$

$$\left(\sum_{i \in V} \frac{(1 - F(x_i)/F_0(x_i))^2}{2\sigma^2} + \sum_{i \in V} \sum_{j \in N_i} \frac{(1 - F(x_i, x_j)/F_0(x_i, x_j))^2}{2\sigma^2} \right). \quad (10)$$

Now our goal is to find a value of (x, σ) that minimizes the posterior energy in Eq. (10). Here we combine a grid search approach with the DEE/A* search algorithms to compute the optimal estimation of $w = \sigma^{-2}$. Once w is determined, parameter σ can be computed using equation $\sigma = \sqrt{1/w}$. Our parameter estimation approach first incrementally searches the grid points of weighting factor w. For each grid point of w, it uses the DEE and A* search algorithms to find the GMEC that minimizes the pseudo energy function. Finally, it compares all GMEC solutions over all searched grid points, and chooses the optimal value of parameter w that minimizes the posterior energy function in Eq. (10).

In Eq. (10), as the weighting factor w increases (i.e., the data term is weighted more), the first term $(m + 1) \log \sigma$ in Eq. (10) decreases, while the third term representing the data restraints increases. Fig. 1A shows a typical plot of the posterior energy $U(x, \sigma | D)$ vs. the weighting factor w, in which a minimum is usually observed. The performance of our parameter estimation approach is only subject to the resolution used in the grid search. In practice, our approach is sufficient to find the optimal parameter estimation (Fig. 1), as we will show in the Results section.

3   Results

We implemented our Bayesian approach for side-chain structure determination and tested it on NMR data of three proteins: the FF Domain 2 of human transcription elongation factor CA150 (FF2), the B1 domain of Protein G (GB1), and human ubiquitin. The numbers of amino acid residues in these three proteins are 62, 56 and 76 for FF2, GB1 and ubiquitin respectively. The PDB IDs of the NMR reference structures are 2KIQ, 3GB1 and 1D3Z for FF2, GB1, and ubiquitin respectively. The PDB IDs of the X-ray reference structures are 3HFH, 1PGA and 1UBQ for FF2, GB1, and ubiquitin respectively.

Our algorithm uses the following input data: (1) the protein primary sequence; (2) the protein backbone; (3) the 2D or 3D NOESY peak list from both [15]N- and ^{13}C-edited spectra; (4) the resonance assignment list, including both backbone and side-chain resonance assignments; (5) the rotamer library [42]. The empirical molecular mechanics energy function that we used in Eq. (4) consists of the Amber electrostatic, van der Waals (vdW), and dihedral terms, the EEF1 implicit solvation energy term [37], and the rotamer energy term, which represents the frequency of a rotamer that is estimated from high-quality protein structures [42]. All NMR data, except RDCs of GB1 and ubiquitin, were recorded and collected using Varian 600 and 800 MHz spectrometers at Duke University. The NOE cross peaks were picked from 3D ^{15}N- and ^{13}C-edited NOESY-HSQC spectra. Details on the NMR experimental procedures are provided in Supplementary Material [60] **Section 4**. Our tests were performed on a 2.20 GHz Intel core 2 Duo processor with 4 GB memory. The total running time of computing the GMEC solution for a typical medium-size protein, such as GB1, is less than an hour after parameter $w = \sigma^{-2}$ has been estimated.

We used the same rules as in [42] to classify and identify the rotamer conformations, that is, we used a window of $\pm 30°$ to determine most χ angles, except that a few specific values (see Table 1 in [42]) were used in determining the terminal χ angle boundaries for glutamate, glutamine, aspartate, asparagine, leucine, histidine, tryptophan, tyrosine and phenylalanine. Since most rotamer conformations are short, the RMSD is not sufficient to measure the structural dissimilarity between two rotamers. Thus, we did not use the RMSD to compare different rotamers. We used two measurements to evaluate the accuracies of the determined side-chain rotamer conformations. The first one is called the *accuracy of all χ angles*, measuring the percentage of side-chain rotamer conformations in which all χ angles agree with the NMR or X-ray reference structure. The second measurement is called the *accuracy of (χ_1, χ_2) angles*, which measures the percentage of side-chain rotamer conformations whose first two χ angles (i.e., both χ_1 and χ_2) agree with the NMR or X-ray reference structure. We say a determined side-chain conformation is *correct* if all its χ angles agree with the NMR or X-ray reference structure.

3.1   Parameter Estimation

We estimated the weighting factor parameter $w = \sigma^{-2}$ in the posterior energy function using the approach described in Sec. 2.5. Here we used the test on GB1 (Fig. 1) as an example to demonstrate our parameter estimation approach. The parameters for the other two proteins were estimated similarly. For GB1, the optimal weighting factor was 32, where the posterior energy $U(x, \sigma | D)$ met the minimum (Fig. 1A). This optimal weight value corresponded to the best accuracies 77.8% and 87.0% for all χ angles and (χ_1, χ_2) angles respectively (Fig. 1E and Fig. 1F).

Fig. 1C and Fig. 1D show the influence of the weight w on the empirical molecular mechanics energy and the NOE pattern matching score of the GMEC. As expected, as the data restraints were weighted more, the empirical molecular mechanics energy declined while the data satisfaction score was improved for the GMEC solution. At the optimal weight value $w = 32$, the GMEC yielded decent scores for both empirical molecular mechanics energy and NOE pattern matching score. Although the NOE pattern matching score of the GMEC jumped to a higher plateau when $w \geq 110$ (Fig. 1C), the accuracies of all χ angles and (χ_1, χ_2) angles did not increase correspondingly (Fig. 1E and Fig. 1F). Probably this high NOE satisfaction score was caused to some extent by

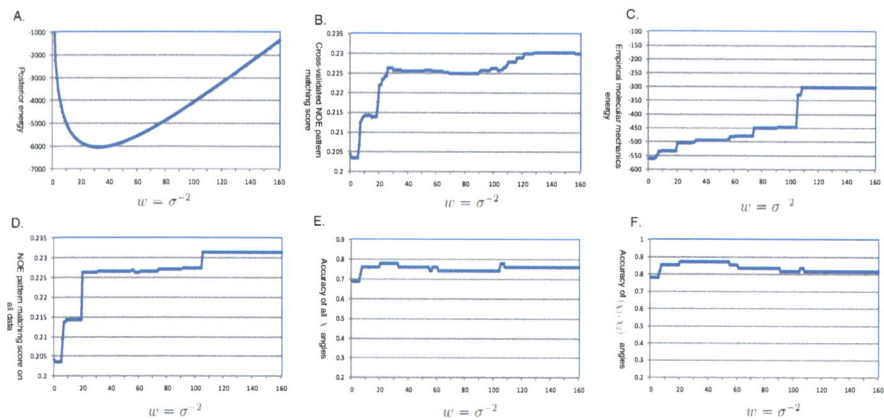

Fig. 1. Estimation of the weighting factor parameter $w = \sigma^{-2}$ for the data term in the posterior energy function for GB1. In plots (B) and (D), the Hausdorff fraction was used to measure the matching score between the back-computed NOE pattern of the GMEC and experimental spectra.

overfitting the side-chain rotamer conformations to experimental data. We also demonstrated that our approach performed better than the cross validation approach [5,6] used in estimating the weighting factor parameter $w = \sigma^{-2}$ (Fig. 1B). Details on the cross validation approach and the comparison results are provided in Supplementary Material [60] **Section 5**.

3.2 Accuracy of Determined Side-Chain Rotamer Conformations

We first tested our side-chain structure determination approach on the backbones from the NMR reference structures (Table 1). To check whether our current side-chain structure determination approach can be combined with our previously-developed backbone structure determination techniques [12,53,54,59] for high-resolution structure determination, we also tested it on the backbones computed mainly using RDC data (Table 2). The RMSD between the input RDC-defined backbone and the NMR reference structure is 0.96 Å, 0.87 Å and 0.97 Å for FF2, GB1 and ubiquitin respectively. In addition to the GMEC, we also computed the top ensemble of 50 conformations with the lowest pseudo energies (Tables 1 and 2), using an extension to the original A* algorithm [15,9,13]. An ensemble of computed structures is important when multiple models may agree with the experimental data [10]. In addition, an ensemble of structures can reflect the conformational difference resulting from different experimental conditions, lack of data, or protein motion in solution [10,1].

In addition to examining the accuracies of the determined side-chain conformations in all residues, we also checked the performance of our approach in *core* residues, which are defined as those residues with solvent accessibility $\leq 10\%$. We used the software MOLMOL [32] with a solvent radius of 2.0 Å to compute solvent accessibility for each residue. Note that in the side-chain structure determination problem using experimental data, we were particularly interested in the accuracies of side-chain conformation determination in core residues because: (1) Biologically the side-chains on the interior and buried regions of the protein play more important roles in studying protein dynamics and determining the accurate structures than other residues on the protein surface; (2) In the X-ray or NMR reference structure, the data for the solvent-exposed side-chains are often missing. Thus, modeling information is often used to compute the side-chain conformations of the residues on the protein surface.

Table 1. Accuracies of the side-chain rotamer conformations determined by our approach on the backbones from the NMR reference structures

Proteins	All residues				Core residues			
	Accuracy of all χ angles (%)		Accuracy of (χ_1, χ_2) angles (%)		Accuracy of all χ angles (%)		Accuracy of (χ_1, χ_2) angles (%)	
	GMEC	Top 50	GMEC	Top 50	GMEC	Top 50	GMEC	Top 50
GB1	77.8	77.8	87.0	87.0	100.0	100.0	100.0	100.0
ubiquitin	75.4	78.3	84.1	85.5	84.0	88.0	88.0	92.0
FF2	71.9	71.9	82.5	86.0	100.0	100.0	100.0	100.0

Table 2. Accuracies of the side-chain rotamer conformations determined by our approach on the RDC-defined backbones computed using the algorithms in [12, 53, 54, 59]

Proteins	All residues				Core residues			
	Accuracy of all χ angles (%)		Accuracy of (χ_1, χ_2) angles (%)		Accuracy of all χ angles (%)		Accuracy of (χ_1, χ_2) angles (%)	
	GMEC	Top 50	GMEC	Top 50	GMEC	Top 50	GMEC	Top 50
GB1	75.9	79.6	81.5	88.9	92.9	100.0	92.9	100.0
ubiquitin	72.5	76.8	79.7	82.6	80.0	84.0	80.0	84.0
FF2	71.9	75.4	80.7	84.2	100.0	100.0	100.0	100.0

Overall, our approach determined more than 70% correct rotamer conformations, and achieved over 80% accuracy for (χ_1, χ_2) angles for all residues (Tables 1 and 2). Our results also show that computing the ensemble of top 50 conformations with the lowest pseudo energies can slightly improve the results (Tables 1 and 2), which indicates that it is necessary to compute an ensemble of conformations rather than a single GMEC solution. In core residues, our approach achieved a high percentage of accurate side-chain conformations. Our approach computed all the correct side-chain conformations in core residues for GB1 and FF2, and had accuracies $\geq$ 84% for ubiquitin, given the backbone structures from the NMR reference structures (Table 1). The tests on the RDC-defined backbones exhibited similar results (Table 2), which indicates that our current Bayesian approach can be combined with our previously-developed backbone structure determination techniques [12, 53, 54, 59] to determine high-resolution protein structures mainly using RDC and unassigned NOESY data.

We also examined the accuracies of the determined side-chain conformations for residues of different lengths (Fig. 2). In general, more short side-chain conformations (i.e., 1-χ and 2-χ side-chains) were determined correctly than the long side-chain conformations (i.e., 3-χ and 4-χ side-chains). On the other hand, although our program assigned a very low percentage of correct 4-χ rotamers (i.e., arginine and lysine), it was able to compute the first two χ angles correctly for most 4-χ side-chains (Fig. 2). In addition to their side-chain flexibility, arginine and lysine are usually exposed to the solvent and undergo many conformational changes. Also, their NOE data are often missing. Therefore, it is generally difficult to compute all the χ angles correctly for these two long side-chains. We further investigated the accuracies of the determined side-chain rotamer conformations for residues with different numbers of available data restraints. We first define the *number of matched NOE peaks* for residue i, denoted by D_i, as follows:

$$D_i = \frac{1}{t_i} \sum_{x_i} \left(f(x_i) + \sum_{j \in N_i} \max_{x_j} f(x_i, x_j) \right), \tag{11}$$

where t_i is the maximum number of rotamer states at residue i, and $f(x_i)$ and $f(x_i, x_j)$ are the numbers of experimental NOE cross peaks that are close to a back-computed NOE peak in the self and pairwise back-computed NOE patterns respectively. Basically D_i measures the degree of available data restraints for residue i averaged over all possible rotamer conformations. We define the value of D_i divided by the number of

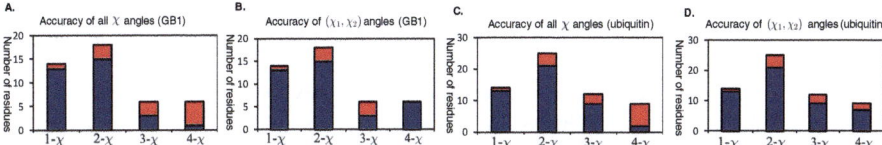

Fig. 2. Accuracies of the determined side-chain rotamer conformations for residues with different lengths (i.e., with different numbers of rotatable χ angles) for GB1 and ubiquitin. The bars represent the number of residues of the indicated type in the protein. The portions marked in blue represent the percentage of rotamers with all χ angles or (χ_1, χ_2) angles that agree with the NMR or X-ray reference structure.

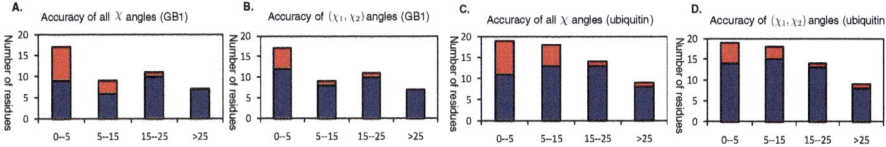

Fig. 3. Accuracies of the determined side-chain rotamer conformations for residue with different numbers of matched NOE peaks per χ angle for GB1 and ubiquitin. Diagrams are shown in the same format as in Fig. 2.

rotatable χ angles in the side-chain as the *number of matched NOE peaks per χ angle* for residue i. As shown in Fig. 3, our approach performed much better on those residues with relatively dense data restraints (i.e., with the number of matched NOE peaks per χ angle ≥ 15) than other residues.

3.3 Improvement on Our Previous Approaches HANA and NASCA

In our previous approaches, HANA [61, 59] and NASCA [62], only experimental data were used in determining side-chain conformations. Thus, they did not consider the empirical molecular mechanics energy when packing side-chain conformations. Thus, the side-chain structures computed by HANA and NASCA can contain steric clashes. Our new approach solves this problem by taking into account a molecular mechanics potential, which sharply penalizes physically unrealistic conformations. As shown in Table 3, our new approach eliminated all the serious steric clash overlaps (> 0.9 Å), which appeared previously in the side-chain conformations computed by HANA and NASCA.

3.4 Comparisons with SCWRL4

SCWRL4 [34] is one of the most popular programs for predicting side-chain rotamer conformations given a backbone structure. Note that our approach uses unassigned NOESY data, while SCWRL4 does not use any experimental data. We compared the performance of our approach with that of SCWRL4 on GB1 using different input backbone structures (Table 4). The comparison showed that our approach outperformed SCWRL4

Table 3. Comparison between our current Bayesian approach and HANA and NASCA on the number of serious steric clash overlaps (> 0.9 Å) in the determined side-chain conformations

Proteins	Current Bayesian approach	HANA	NASCA
GB1	0	10	14
ubiquitin	0	16	21
FF2	0	2	14

Table 4. Comparison with the side-chain structure prediction program SCWRL4 on GB1 using different input backbone structures. The backbone RMSD from 2GB1, 1GB1 , 1P7E, 1PGA and 1PGB to 3GB1 is 1.01 Å, 1.00 Å, 0.44 Å, 0.54 Å and 0.56 Å respectively. The program REDUCE [55] was used to add hydrogens to the X-ray backbone structures 1PGA and 1PGB. In our approach, the GMEC was computed for this comparison.

Backbones	All residues				Core residues			
	Accuracy of all χ angles (%)		Accuracy of (χ_1, χ_2) angles (%)		Accuracy of all χ angles (%)		Accuracy of (χ_1, χ_2) angles (%)	
	Our approach	SCWRL4	Our approach	SCWRL4	Our approach	SCWRL4	Our approach	SCWRL4
3GB1	77.8	72.2	85.2	79.4	100.0	85.7	100.0	85.7
2GB1	72.1	68.5	81.3	74.5	92.9	78.6	92.9	78.6
1GB1	74.1	70.4	83.3	77.8	92.9	78.6	92.9	78.6
1P7E	74.1	70.4	83.3	75.9	92.9	78.6	92.9	78.6
1PGA	70.4	64.8	79.6	70.4	92.9	71.4	92.9	71.4
1PGB	75.9	74.1	83.3	77.8	100.0	85.7	100.0	85.7

for all input backbone structures, especially on the core regions (Table 4). For core residues, our approach achieved accuracies between 92.9-100.0%, while SCWRL4 only achieved accuracies up to 85.7%. As we discussed previously, the correctness of the side-chain conformations on the core regions is crucial for determining the accurate global fold of a protein. Thus, in order to meet the requirement of high-resolution structure determination, the data restraints must be incorporated for packing the side-chain conformations in core residues.

4   Conclusions

In this paper, we unified the side-chain structure prediction problem with the side-chain structure determination problem using unassigned NOESY data. We proposed a Bayesian approach to integrate experimental data with modeling information, and used the provable algorithms to find the optimal solution. Tests on real NMR data demonstrated that our approach can determine a high percentage of accurate side-chain conformations. Since our approach does not require any NOE assignment, it can accelerate NMR structure determination.

Availability

The source code of our program is available by contacting the authors, and is distributed open-source under the GNU Lesser General Public License (Gnu, 2002). The source code can be freely downloaded after publication of this paper.

Acknowledgements

We thank Mr. Pablo Gainza and Ms. Swati Jain for helping us set up the DEE/A* code. We thank all members of the Donald and Zhou Labs for helpful discussions and comments.

References

1. Andrec, M., et al.: Proteins 69(3), 449–465 (2007)
2. Besag, J.: J. Royal Stat. Soc. B 36 (1974)
3. Bower, M.J., et al.: J. Mol. Biol. 267(5), 1268–1282 (1997)
4. Bowers, P.M., et al.: J. Biomol. NMR 18(4), 311–318 (2000)
5. Brünger, A.T.: Nature 355(6359), 472–475 (1992)
6. Brünger, A.T., et al.: Science 261(5119), 328–331 (1993)
7. Cavalli, A., et al.: Proc. Natl. Acad. Sci. USA 104(23), 9615–9620 (2007)
8. Chazelle, B., et al.: INFORMS J. on Computing 16(4), 380–392 (2004)
9. Chen, C.Y., et al.: Proc. Natl. Acad. Sci. USA 106, 3764–3769 (2009)
10. De Pristo, M.A., et al.: Structure 12(5), 831–838 (2004)
11. Desmet, J., et al.: Nature 356, 539–542 (1992)
12. Donald, B.R., Martin, J.: Progress in NMR Spectroscopy 55, 101–127 (2009)
13. Frey, K.M., et al.: Proc. Natl. Acad. Sci. USA 107(31), 13707–13712 (2010)
14. Geman, S., Geman, D.: IEEE Trans. Pattern Anal. Mach. Intell., 721–741 (1984)
15. Georgiev, I., et al.: Journal of Computational Chemistry 29, 1527–1542 (2008)
16. Goldstein, R.F.: Biophysical Journal 66, 1335–1340 (1994)
17. Grishaev, A., Llinás, M.: Proc. Natl. Acad. Sci. USA 99, 6707–6712 (2002)
18. Grishaev, A., Llinás, M.: Proc. Natl. Acad. Sci. USA 99, 6713–6718 (2002)
19. Güntert, P.: Progress in Nuclear Magnetic Resonance Spectroscopy 43, 105–125 (2003)
20. Habeck, M., et al.: Proc. Natl. Acad. Sci. USA 103(6), 1756–1761 (2006)
21. Herrmann, T., et al.: Journal of Molecular Biology 319(1), 209–227 (2002)
22. Holm, L., Sander, C.: Proteins 14(2), 213–223 (1992)
23. Huang, Y.J., et al.: Proteins 62(3), 587–603 (2006)
24. Hus, J.C., et al.: J. Am. Chem. Soc. 123(7), 1541–1542 (2001)
25. Huttenlocher, D.P., Kedem, K.: Distance Metrics for Comparing Shapes in the Plane. In: Donald, B.R., et al. (eds.) Symbolic and Numerical Computation for Artificial Intelligence, pp. 201–219. Academic press, London (1992)
26. Huttenlocher, D.P., et al.: IEEE Trans. Pattern Anal. Mach. Intell. 15(9), 850–863 (1993)
27. Hwang, J.K., Liao, W.F.: Protein Eng. 8(4), 363–370 (1995)
28. Jeffreys, H.: Proceedings of the Royal Society of London (Series A) 186, 453–461 (1946)
29. Kamisetty, H., et al.: Journal of Computational Biology 15, 755–766 (2008)
30. Kingsford, C.L., et al.: Bioinformatics 21(7), 1028–1036 (2005)
31. Koehl, P., Delarue, M.: J. Mol. Biol. 239(2), 249–275 (1994)
32. Koradi, R., et al.: J. Mol. Graph. 14(1) (1996)
33. Kraulis, P.J.: J. Mol. Biol. 243(4), 696–718 (1994)
34. Krivov, G.G., et al.: Proteins 77(4), 778–795 (2009)
35. Kuszewski, J., et al.: J. Am. Chem. Soc. 126(20), 6258–6273 (2004)
36. Langmead, C.J., Donald, B.R.: J. Biomol. NMR 29(2), 111–138 (2004)
37. Lazaridis, T., Karplus, M.: Proteins 35(2), 133–152 (1999)
38. Leach, A.R., Lemon, A.P.: Proteins 33(2), 227–239 (1998)
39. Li, S.Z.: Markov random field modeling in computer vision. Springer, London (1995)

40. Linge, J.P., et al.: Bioinformatics 19(2), 315–316 (2003)
41. Looger, L.L., Hellinga, H.W.: J. Mol. Biol. 3007(1), 429–445 (2001)
42. Lovell, S.C., et al.: Proteins: Structure Function and Genetics 40, 389–408 (2000)
43. Meiler, J., Baker, D.: Proc. Natl. Acad. Sci. USA 100(26), 15404–15409 (2003)
44. Meiler, J., Baker, D.: J. Magn. Reson. 173(2), 310–316 (2005)
45. Pierce, N.A., Winfree, E.: Protein Eng. 15(10), 779–782 (2002)
46. Raman, S., et al.: J. Am. Chem. Soc. 132(1), 202–207 (2010)
47. Raman, S., et al.: Science 327(5968), 1014–1018 (2010)
48. Rieping, W., et al.: Science 309, 303–306 (2005)
49. Rohl, C.A., et al.: Proteins 55(3), 656–677 (2004)
50. Russell, S., Norvig, P.: Artificial Intelligence: A Modern Approach. Prentice Hall, Engle-
    wood Cliffs (2002)
51. Shen, Y., et al.: Proc. Natl. Acad. Sci. USA 105(12), 4685–4690 (2008)
52. Tuffery, P., et al.: J. Biomol. Struct. Dyn. 8(6), 1267–1289 (1991)
53. Wang, L., Donald, B.R.: Jour. Biomolecular NMR 29(3), 223–242 (2004)
54. Wang, L., et al.: Journal of Computational Biology 13(7), 1276–1288 (2006)
55. Word, J.M., et al.: J. Mol. Biol. 285(4), 1735–1747 (1999)
56. Xiang, Z., Honig, B.: J. Mol. Biol. 311(2), 421–430 (2001)
57. Xu, J., Berger, B.: Journal of the ACM 53(4), 533–557 (2006)
58. Yanover, C., Weiss, Y.: In: NIPS (2002)
59. Zeng, J., et al.: Journal of Biomolecular NMR 45(3), 265–281 (2009)
60. Zeng, J., et al. A Bayesian Approach for Determining Protein Side-Chain Rotamer Confor-
    mations Using Unassigned NOE Data–Supplementary Material (2011), http://www.cs.
    duke.edu/donaldlab/Supplementary/recomb11/bayesian/
61. Zeng, J., et al. In: Proceedings of CSB 2008, Stanford CA (2008) PMID: 19122773
62. Zeng, J., et al.: A markov random field framework for protein side-chain resonance assign-
    ment. In: Berger, B. (ed.) RECOMB 2010. LNCS, vol. 6044, pp. 550–570. Springer, Heidel-
    berg (2010)

Author Index